Fundamentals of Photon Physics

The photon, an abstract concept belonging to a global vacuum, only manifests itself during interaction with matter. Fundamentals of Photon Physics describes the richly faceted, basic theory of photon-matter interaction, selecting a wide number of topics. Together with the author's book Light -- The Physics of the Photon (CRC, 2014), both written on a scholarly level, the reader is given a comprehensive exposition of photon wave mechanics, quantum optics and quantum electrodynamics (QED).

Divided into 10 parts, the book begins by exploring the relation between photon wave mechanics and quantum field theory. It then describes the theories of zero- and one-photon states and that of bi-photons. After discussing conservation laws, Lagrangian formulations, geometric phase and topology, the author turns towards the theory of photon scattering, emphasizing a density matrix operator approach and the role of microscopic extinction theorems. The book next focuses on mesoscopic QED, devoting particular attention to collective jellium excitations and photon-spin interactions. Special attention is given to the basics of the photon-magnon interaction and nonlinear superconductor electrodynamics, including the nonlinear Meissner rectification phenomenon, before studying the theory of transverse photons tied to (dressing) massive particles.

The last three parts take the reader on a journey to topics usually not treated in books on photon-matter interaction. Beginning with photons in curved space-time structures and in spatially curved media, e.g. Möbius bands, the author discusses the extension of QED to the electro-weak interaction at an introductory level. Fundamentals of Photon Physics ends with the establishment of the set of isovector Maxwell equations in non-Abelian SO(3) gauge theory, leading to the celebrated hedgehog monopole model.

Series in Optics and Optoelectronics

Detection of Optical Signals
Antoni Rogalski & Zbigniew Bielecki

Handbook of Optoelectronics, Second Edition
Applied Optical Electronics (Volume Three)
John P. Dakin, Robert G. W. Brown

Handbook of Optoelectronic Device Modeling and Simulation
Lasers, Modulators, Photodetectors, Solar Cells, and Numerical Methods, Vol. 2
Joachim Piprek

Handbook of Optoelectronics, Second Edition
Concepts, Devices, and Techniques (Volume One)
John P. Dakin, Robert Brown

Handbook of GaN Semiconductor Materials and Devices
Wengang (Wayne) Bi, Haochung (Henry) Kuo, Peicheng Ku, Bo Shen

Handbook of Optoelectronic Device Modeling and Simulation (Two-Volume Set)
Joachim Piprek

Handbook of Optoelectronics, Second Edition (Three-Volume Set)
John P. Dakin, Robert G. W. Brown

Optical MEMS, Nanophotonics, and Their Applications
Guangya Zhou, Chengkuo Lee

Thin-Film Optical Filters, Fifth Edition
H. Angus Macleod

Laser Spectroscopy and Laser Imaging
An Introduction
Helmut H. Telle, Ángel González Ureña

Fourier Optics in Image Processing
Neil Collings

Holography
Principles and Applications
Raymond K. Kostuk

An Introduction to Quantum Optics, Second Edition
Photon and Biphoton Physics
Yanhua Shih

Polarized Light and the Mueller Matrix Approach, Second Edition
José J. Gil, Razvigor Ossikovski

Introduction to Holography, Second Edition
Vincent Toal

Optical Coherence Tomography in Dentistry
Scientific Developments to Clinical Applications
Anderson S. L. Gomes, Denise M. Zezell, Cláudia C. B. O. Mota, John M. Girkin

Fundamentals of Photon Physics
Ole Keller

For more information about this series, please visit:
https://www.crcpress.com/Series-in-Optics-and-Optoelectronics/book-series/TFOPTICSOPT

Fundamentals of Photon Physics

Ole Keller

CRC Press
Taylor & Francis Group
Boca Raton London New York

CRC Press is an imprint of the
Taylor & Francis Group, an **informa** business

Designed cover image: Ole Keller

First edition published 2025
by CRC Press
2385 NW Executive Center Drive, Suite 320, Boca Raton FL 33431

and by CRC Press
4 Park Square, Milton Park, Abingdon, Oxon, OX14 4RN

CRC Press is an imprint of Taylor & Francis Group, LLC

ISBN: 978-0-367-45726-6 (hbk)
ISBN: 978-1-032-82498-7 (pbk)
ISBN: 978-1-003-02945-8 (ebk)

DOI: 10.1201/9781003029458

Typeset in Latin Modern font
by KnowledgeWorks Global Ltd.

Publisher's note: This book has been prepared from camera-ready copy provided by the authors.

*In memory of my grand parents, Anne Marie
and Andreas Kristian Keller.*

Contents

Foreword

A decade ago my Taylor & Francis book entitled "LIGHT. The Physics of the Photon" was published. Space and time did not allow me to treat in that book a number of fundamental subjects in classical and quantum electrodynamics (optics). I hope that the reader will find that some of these are covered in the present book "Fundamentals of Photon Physics". The inevitably present photon-matter interaction makes even a theoretical exposition of just fundamental aspects of photon physics a diverse subject. I have tried to balance my description in such a manner that both the deeper structure of photon physics, based on just a few "rules", and the subjects diversity appear. Parts of the book can be studied by readers with a general background in classical and quantum electrodynamics. Topics such as photons in curved structures, nonlinear electromagnetic rectification in superconductors and magnetic monopole physics, which may appear more demanding, are treated on an introductory level.

New research results in fundamental photon physics have appeared in recent years, and in a small part of these I have had the fortune to participate. A glimpse of our new theoretical understanding e.g. of (i) the photon in what I have called Dynamic Structural Vacuum, (ii) photon-spin electrodynamics in mesoscopic media, (iii) the photon's role in jellium electrodynamics and (iv) photon propagation in curved structures, e.g., along a mesoscopic Möbius band, is presented.

In some of these theoretical studies and numerical studies my friend and collaborator, Dr. Jesper Jung has taken a ride together with me over the last ten years. Furthermore, Jesper has helped me write the LaTex version of my handwritten manuscript. I want to acknowledge here Jesper's comprehensive work during the four years I have worked on the book manuscript.

Author Biography

Ole Keller is professor emeritus of theoretical physics at Aalborg University, Denmark. He earned his Licentiate ($\sim$ PhD) degree in semiconductor physics from the Danish Technical University in Copenhagen in 1972, and the Doctor of Science degree from the University of Aarhus (1990). In 1989 he was appointed as the first professor in physics at Aalborg University by Margrethe Den Anden, queen of Denmark. The same year he was admitted to Kraks Blaa Bog, a prestigious Danish biographical dictionary which (citatum) "Includes men and women, whose life story could have an interest for a wider public". He is a fellow of the Optical Society of America.

He has written the books entitled Quantum Theory of Near-Field Electrodynamics (Springer, 2011) and LIGHT - The Physics of the Photon (CRC, 2014), as well as the monographs Local Fields in the Electrodynamics of Mesoscopic Media (Physics Reports, 1996) and On the Theory of Spatial Localization of Photons (Physics Reports, 2005). He is the editor of the books Nonlinear Optics in Solids (Springer, 1990), Studies in Classical and Quantum Nonlinear Optics (Nova Science, 1995) and Notions and Perspectives of Nonlinear Optics (World Scientific, 1996).

In recent years he has carried out theoretical research in fundamental photon physics, microscopic few-photon diffraction, mesoscopic and Möbius band electrodynamics, and studied magnetic monopole theory based on QED and the isovector Maxwell equations in non-Abelian gauge symmetry.

Preface

Microscopic electrodynamics is that branch of physics which deals with the field-mediated interaction between electrically charged microscopic massive particles, such as electrons and nuclei. In a certain domain of microscopic electrodynamics the quantum features of light matter, and this domain is characterized by the name Quantum Electrodynamics, abbreviated QED. In the nonrelativistic domain instead of the label QED one most often uses the name Quantum Optics in appreciation of the circumstance that the field quantization here often appears as a branch of optics.

In this book, entitled "Fundamentals of Photon Physics", I emphasize the fingerprint the particle-field coupling leaves on the field variables. With this in mind I use the name Photon Physics. Although the book begins with the word "Fundamentals", it is to a certain extent a personal point of view what reasonably may be characterized as the fundamentals of QED. In this book I treat the physical theory of a number of quite distinct topics all of which I consider as belonging to fundamental photon physics.

It is not possible to cover the framework of theoretical photon physics in a single volume. In my previously published CRC book entitled "LIGHT. The Physics of the Photon" other aspects of basic photon physics are treated. My two books on the theory of photon physics supplement each other with only minor overlap of subjects. I have aimed at using to some extent the same notational frames in the two books in order to make it easier for reader to switch her/his reading between the books.

The present book is divided into ten main sections (I-X) devoted to the wave mechanical and quantum field theories of photons (I), zero- and one-photon states and bi-photons (II), conservation laws, Lagrangians, geometric phase and topology (III), photon scattering theory and the density operator formalism (IV). Each of the Section I-IV is subdivided into four chapters. Section V, which deals with the basics of mesoscopic electrodynamics, has six chapters centered on collective electron-photon oscillations in jellium systems, and photon-spin interaction. The subsequent two sections (VI, VII) are each divided into just three chapters. Section VI deals with the basics of magnon electrodynamics and the present authors theory for nonlinear electromagnetic rectification in BCS superconductors. In the two aforementioned fields the photon-spin interaction plays a central role. Photons can be tied to massive particles, and in Section VIII we discuss examples of how this phenomenon can manifest itself in single-particle (electron, atom) electrodynamics, and in the form of a photon cloud attached to a plasmariton quasiparticle. In Section VIII we turn our attention towards the theory of photons in curved structures. Starting from the electrodynamics in general relativity and non-inertial frames, we specialize in structures possessing only spatial curvature. With the silent assumption that the general reader of this book has a theoretical background mainly in electrodynamics, I present for such readers a brief description of the manner in which quantum electrodynamics has been shown to appear as a sector of the electro-weak theory (Section IX). The final section (X) of this book gives the reader an introduction to the theory of the magnetic monopole, a particle predicted by Dirac in 1931 and searched for experimentally ever since, so far with no success. The photon plays a crucial role in the electrodynamic monopole theory, and in the non-Abelian gauge theory is a set of dynamical equations among the field isovectors $(\mathbf{E}_{\mathrm{iso}}, \mathbf{B}_{\mathrm{iso}})$ turns out to be form-identical to those governing the Maxwell-Dirac theory of electrodynamics. In the non-Abelian gauge theory the magnetic monopole may appear in the form of a static "hedgehog" solution to the set of isovector relations. Sections VIII-X each is subdivided into four chapters.

In classical microscopic electrodynamics the electric ($\mathbf{E}$) and magnetic ($\mathbf{B}$) fields usually are considered to be the primary field quantities, and the vector ($\mathbf{A}$) and scalar (ϕ) potentials are taken as auxiliary fields, yet of great importance in many practical calculations. In textbooks this point of view often is ascribed to the circumstance that the potentials are not gauge invariant, and strengthened by the fact that the dynamics of a point-like charged particle is driven by the Lorentz force $[q(\mathbf{E}+\mathbf{v}\times\mathbf{B})]$ in the microscopic Newton-Lorentz equation. Part of the $(\mathbf{E}, \mathbf{B})$-field stems from (all) other point particles in uniform and non-uniform motion, as described by the Liénard-Wiechert expression for $\mathbf{E}$- and $\mathbf{B}$-fields. [I here remind the reader that these fields adequately are obtained via the $(\mathbf{A}, \phi)$-potential in the Lorenz gauge]. Part of the electromagnetic interaction originates in the classical radiation reaction associated with the fact that part of the radiation acts back on the particle, and modifies its motion. This classical radiation reaction is described in the nonrelativistic region by the Abraham-Lorentz equation of motion, and in the relativistic region by the Lorentz-Dirac equation. An unaccelerated charged point-like particle carries in its rest frame a Coulomb field which gives rise to an electromagnetic contribution, m_{em}, to the particle ($\sim$ electron's) bare mass, m_{bare}. The distortion of the self-four-momentum can be studied via the Lorentz-Dirac equation. However, it must be emphasized that a rigorous description of radiation reaction and self-field effects requires a quantum mechanical approach.

In semiclassical microscopic electrodynamics the Newton-Lorentz equation is replaced by the field coupled Schrödinger [or Pauli or Dirac] equation. Now, the potentials take the primary role, cf., the Minimal Coupling Substitution Principle, $\hat{p}_\mu \Rightarrow \hat{p}_\mu - e\hat{A}_\mu$, for the four-momentum of a particle with electric charge e. At this point a dilemma seems to appear: If one tries to eliminate the potentials in favour of the field one runs into great troubles, and yet the four-potential is gauge dependent. The solution to this problem dissolves itself by the fact that the divergence-free [named transverse (T)] part ($\mathbf{A}_T$) of the vector potential ($\mathbf{A}$) is gauge invariant. In the Coulomb gauge $\mathbf{A} = \mathbf{A}_T$, and the scalar potential is eliminated as a dynamical variable, amounting to the fact that the potential energy appears as the Coulomb self-energy among the individual particles in the Hamiltonian operator of the many-body Schrödinger equation.

In the present book I advocate the following point of view: *In photon physics the transverse vector potential ($\mathbf{A}_T$) is the primary physical quantity, and the associated* *electromagnetic fields,* $\mathbf{E}_T = -\partial \mathbf{A}_T/\partial t$ *and* $\mathbf{B} = \boldsymbol{\nabla} \times \mathbf{A}_T,$ *are nothing more than auxiliary quantities.* Although transversality is not a frame-independent quality the point of view above is still advocated since a given observation always must be carried out in a particular reference frame.

The pivotal role of the transverse vector potential appears on the horizon in the dynamic description of a free photon. Thus, we show in Chapter 2 that a two-spinor photon wave function, defined by

$$\boldsymbol{\mathcal{A}}_T^{(+)}(\mathbf{r}, t) = \begin{pmatrix} \mathbf{A}_{T,+}^{(+)}(\mathbf{r}, t) \\ -\mathbf{A}_{T,-}^{(+)}(\mathbf{r}, t) \end{pmatrix} \tag{1}$$

satisfies a dynamical evolution equation of the following form

$$i\hbar \frac{\partial}{\partial t} \boldsymbol{\mathcal{A}}_T^{(+)}(\mathbf{r}, t) = (c\hat{\mathbf{p}} \cdot \boldsymbol{\sigma}) \boldsymbol{\mathcal{A}}_T^{(+)}(\mathbf{r}, t). \tag{2}$$

To the best of my knowledge this equation has not appeared in the scientific literature before, and the extension of this form of dynamical evolution theory for theoretical studies of photon-matter interaction has turned into an active field of research recently [130]. The evolution equation is of first order in the time derivative which means that it is sufficient to know $\boldsymbol{\mathcal{A}}_T^{(+)}(\mathbf{r}, t)$ at a particular time to determine it at a later (prior) time. The dynamical equation in Eq. (2) is a Schrödinger-like representation with Hamilton operator $\hat{H} = c\hat{\mathbf{p}} \cdot \boldsymbol{\sigma}$, where $\boldsymbol{\sigma}$ is the Cartesian and dimensionless spin-one vector (and operator $\boldsymbol{\sigma} = \hat{\boldsymbol{\sigma}}$). The superscript $(+)$ on $\boldsymbol{\mathcal{A}}_T^{(+)}(\mathbf{r}, t)$ [and its two spinor components] means that only the analytic (positive frequency) part of $\boldsymbol{\mathcal{A}}_T(\mathbf{r}, t)$ enters the formalism. The negative frequency part of the spinor, $\boldsymbol{\mathcal{A}}_T^{(-)}(\mathbf{r}, t)$, is associated with the antiphoton. Since the photon is its own antiparticle, the negative frequency part of $\boldsymbol{\mathcal{A}}_T(\mathbf{r}, t)$ is redundant.

In the wake of the studies carried out in 1990s in particular by Bialynicki-Birula, the interest in a photon wave mechanical description based on an energy wave function formalism, initiated already in 1931 by Oppenheimer, increased significantly. The dynamical equations for the energy wave functions positive (subscript: $+$) and negative (subscript: $-$) helicity components can be obtained from Eq. (2). Hence, by means of the auxiliary fields $\boldsymbol{\mathcal{E}}_{T,\pm}^{(+)}(\mathbf{r}, t) = -\partial \mathbf{A}_{T,\pm}(\mathbf{r}, t)$ and $\boldsymbol{\mathcal{B}}_{\pm}^{(+)}(\mathbf{r}, t) = \boldsymbol{\nabla} \times \mathbf{A}_{T,\pm}^{(+)}(\mathbf{r}, t)$, it appears that

$$\boldsymbol{\mathcal{E}}_{T,\pm}^{(+)}(\mathbf{r}, t) \mp ic\boldsymbol{\mathcal{B}}_{\pm}^{(+)}(\mathbf{r}, t) = \mathbf{0}, \tag{3}$$

and from Eq. (3) one can arrive at the Oppenheimer-Bialynicki dynamical equations for the analytical parts of the two helicities of the Riemann-Silberstein field using a paradigmatic shift $(\boldsymbol{\mathcal{E}}_{T,\pm}^{(+)}, \boldsymbol{\mathcal{B}}_{\pm}^{(+)}) \Rightarrow (\mathbf{E}_T^{(+)}, \mathbf{B}^{(+)})$.

It is not possible to localize a photon completely in space at a given time, as I have discussed in a Physics Reports review almost twenty years ago [153]. In a sense, this circumstance is related to the fact that the photon is a relativistic object. In a non-relativistic setting it seems possible to localize an electron (or other massive charged particles) completely in space. However, taking into account relativistic effects neither the electron can be localized perfectly. Nevertheless there is a quantitative difference between the electron and photon localizability: Electrons are exponentially localizable, whereas photons are only algebraically localizable. In my CRC book "LIGHT" I advocated for the point of view that a suitable "mean" position operator, $\hat{\mathcal{R}}$, can be introduced for the photon. The eigenstates, defined by

$$|\mathcal{R}\rangle(\mathbf{r}, t) = \left(\frac{2\varepsilon_0 c}{\hbar}\right)^{1/2} \hat{\mathbf{A}}_T^{(-)}(\mathbf{r}, t)|0\rangle, \tag{4}$$

in Hilbert space are non-orthogonal, but obey the weak localization property $\langle\mathcal{R}(\mathbf{r}, t)|\mathcal{R}(\mathbf{r}, t)\rangle = [\pi|\mathbf{r} - \mathbf{r}'|]^{-1}$, noting that the dimension of the position operator is $(\text{length})^{-1}$. [If wished, the inner product can be made dimensionless by multiplication by L^2, using quantization in a cubic cavity of volume L^3]. The development of a photon concept based on the two-spinor $\boldsymbol{\mathcal{A}}_T^{(+)}(\mathbf{r}, t)$, [Eq. (1)] in the last decade has strengthened the operational usefulness of considering $|\mathcal{R}\rangle(\mathbf{r}, t)$ as the best position state for the photon.

Polychromatic one-photon states are discussed in Section II, and the transition from single-photon states

$$\left|A_{T,\pm}^{(+)}\right\rangle = L^{-3/2} \sum_{\mathbf{q}} \phi_{\mathbf{q}\pm} \hat{a}_{\mathbf{q}\pm}^{\dagger}|0\rangle, \tag{5}$$

to photon wave mechanics is obtained via

$$\mathcal{N}\mathbf{A}_{T,\pm}^{(+)}(\mathbf{r}, t) = \left\langle\mathcal{R}\left|A_{T,\pm}^{(+)}\right\rangle(\mathbf{r}, t), \tag{6}$$

$\mathcal{N}$ is a global normalization constant. Normalization in the relativistic sense turns out to be given by

$$\mathcal{N}^2 \int_{-\infty}^{\infty} \left[\boldsymbol{\mathcal{A}}_T^{(+)}(\mathbf{r}, t)\right]^{\dagger} \cdot \left[\boldsymbol{\mathcal{A}}_T^{(+)}(\mathbf{r}, t)\right] d^3r = \frac{1}{(2\pi)^3} \int_{-\infty}^{\infty} \left[|\Phi_+(\mathbf{q})|^2 + |\Phi_-(\mathbf{q})|^2\right] \frac{d^3q}{q} = 1, \tag{7}$$

where the quantity $\Phi_{\mathbf{q}\pm} = q^{1/2}\phi_{\mathbf{q}\pm}$ is the scalar part of the wave function $[\Phi_{\mathbf{q}\pm}\mathbf{e}_\pm(\boldsymbol{\kappa})]$ belonging to the plane-wave mode (wave vector $\mathbf{q} = q\boldsymbol{\kappa}$) with helicity vector $\mathbf{e}_\pm(\boldsymbol{\kappa})$.

The correlation between mean position states associated with the space-time points $x = (\mathbf{r}, ct)$ and $x' = (\mathbf{r}', ct')$ may be characterized by the overlap matrix

$$\frac{\hbar}{2\varepsilon_0 c} \langle\mathcal{R}(x)|\mathcal{R}(x')\rangle = \langle 0|\hat{\mathbf{A}}_T^{(+)}(x)\hat{\mathbf{A}}_T^{(-)}(x')|0\rangle, \tag{8}$$

and it turns out that the time-ordered position correlation $[T_>:$ the time-ordering symbol] for $t > t'$, is given by

$$T_>(x, x') = \Theta(t - t')\langle 0|T\{\hat{\mathbf{A}}_T(x)\hat{\mathbf{A}}_T(x')\}|0\rangle = \frac{\hbar}{i\varepsilon_0 c^2}\mathbf{G}_T(\mathbf{R}, \tau). \tag{9}$$

Here [with $\mathbf{R} = \mathbf{r} - \mathbf{r}'$, $\tau = t - t'$]

$$\mathbf{G}_T(\mathbf{R}, \tau) = \frac{1}{4\pi R} \delta\left(\frac{R}{c} - \tau\right)(\mathbf{U} - \mathbf{e_R}\mathbf{e_R}) - \frac{c^2\tau}{4\pi R^3}\Theta(\tau)\Theta\left(\frac{R}{c} - \tau\right)(\mathbf{U} - 3\mathbf{e_R}\mathbf{e_R})$$
(10)

is the transverse photon propagator, describing the photon-field propagation from x' to x. For a definition of the various quantities used in Eq. (10) the reader is referred to the book's main text [Chapter 3].

The transverse vector potential also plays a pivotal role in quantum cavity electrodynamics. For space points ($\mathbf{r}$) inside a closed cavity with the enclosing surface denoted by Σ, an Ewald-Oseen-like extinction theorem extended to the quantum operator level results in a self-consistency requirement for $\hat{\mathbf{A}}_T(\mathbf{r}; \omega)$ in the space-frequency domain, viz.,

$$\hat{\mathbf{A}}_T(\mathbf{r}; \omega) = \int^{\Sigma}\left[g(R; \omega)\frac{\partial\hat{\mathbf{A}}_T(\mathbf{r}; \omega)}{\partial n'} - \hat{\mathbf{A}}_T(\mathbf{r}; \omega)\frac{\partial g(R; \omega)}{\partial n'}\right] dS',$$
(11)

where $g(|\mathbf{r} - \mathbf{r}'|; \omega)$ is the Huygens scalar propagator, and $\partial(...)/\partial n'$ denotes a derivative in the normal direction to the cavity surface. We shall discuss certain aspects of cavity electrodynamics in photon-empty cavities in Chapter 6.

In a group-theoretical setting we shall introduce the photon little group using a gauge transformation within the Lorenz gauge to eliminate the longitudinal and time-like components of the four-potential vector. The transverse components of the four-potential are left invariant. Hence, without loss of generality it appears that the photon four-potential may satisfy the photon little group criterion.

In theoretical studies of the Aharanov-Bohm effect and magnetic monopoles the transverse vector potential plays an utmost important role as it enters the Dirac phase factor,

$$\Phi(C) = \exp\left[i\frac{e}{\hbar}\oint_C \mathbf{A}_T(\mathbf{r}, t)\cdot d\mathbf{r}\right],$$
(12)

via a line integral along a closed curve (C). One is led to this expression for $\Phi(C)$ from the Principle of Local Phase Invariance which [above for a fixed time loop], is a principle which ultimately results in identifying the photon as a local gauge particle related to $\mathbf{A}_T$. The Aharanov-Bohm theory tells us that there will be an observable effect on a charged particle ($\sim$ electron) moving in a region of space where both the electric ($\mathbf{E}$) and magnetic ($\mathbf{B}$) fields vanish. In its purest form the Aharanov-Bohm phenomenon is a magnetostatic effect, where the low-frequency (ω) limit

$$c\mathbf{B}(\mathbf{r}; \omega \to 0) = \mathbf{E}(\mathbf{r}; \omega \to 0) = \mathbf{0},$$
(13)

but $\mathbf{A}_T(\mathbf{r}; \omega \to 0) \neq \mathbf{0}$. If the conditions in Eq. (13) are satisfied in a certain region (V) of space, I call V a domain of Static Structural Vacuum. Since the static ($\omega \to 0$) limit from a rigorous physical point of view is an abstraction, one should be able to understand the Aharanov-Bohm effect in a broader perspective. My advocation for considering $\mathbf{A}_T(\mathbf{r},t)$ as the primary physical quantity, and $[\mathbf{E}(\mathbf{r},t),\mathbf{B}(\mathbf{r},t)]$ as auxiliary quantities, implies that the V-domain in general, i.e., for all frequencies, is considered as a Dynamical Structural Vacuum. This structural vacuum is a reservoir for $\mathbf{A}_T$-gauge photons, and in relation to the Aharonow-Bohm effect these photons may be launched, in the perhaps simplest case, from a rectilinear magnetic dipole string in the form of polychromatic single-photon wave packets. When the center frequency of the wave packet becomes sufficiently low ($\omega \approx 0$), one obtains the Aharonov-Bohm effect. The gauge photon $[\boldsymbol{\mathcal{A}}_T^{(+)}(\mathbf{r},t)]$ point of view for the Aharanov-Bohm effect constitutes a field of active research [130].

In studies of persistent currents and angular momentum photon drag in mesoscopic rings and cylinders centered on the magnetic dipole string, the Dirac phase factor

$$\Phi(C) = \exp\left[-2\pi i \frac{\Phi}{\Phi_0}\right] \tag{14}$$

becomes a periodic function of the magnetic flux (Φ) in the string, the period being the fundamental (elementary flux quantum) $\Phi_0 = h/|e|$.

In the Dirac string theory of magnetic monopoles, the presence of a magnetic monopole implies that in the phase shift ($\Delta\alpha$) calculation

$$\Delta\alpha = \frac{e}{\hbar}\oint \mathbf{A}_T(\mathbf{r},t)\cdot d\mathbf{l} \neq \frac{e}{\hbar}\int \boldsymbol{\nabla}\times\mathbf{A}_T(\mathbf{r},t)\cdot d\mathbf{S} \tag{15}$$

one cannot uphold the usual relation between the line and surface integrals [indicated by $\neq$ above]. This forced Dirac to conclude that the vector potential, $\mathbf{A}_T$, must be singular along a certain line. This, in general curved line, is known as the Dirac String and it stretches from infinity to the position of the magnetic monopole. The Minimal Coupling Principle, $\mathbf{p} \to \mathbf{p} - e\mathbf{A}_T$, thus cannot be upheld in a quantum theory with a Dirac string. However, the unphysical Dirac string can be avoided using a double potential formalism with transverse electric ($\mathbf{A}_T^e$) and magnetic ($\mathbf{A}_T^m$) vector potentials. Recently, it has been shown that even the Minimal Coupling Principle can be upheld provided a certain nonlocal transformation is carried out to obtain new transverse vector potentials [$\boldsymbol{\mathcal{A}}_T^\alpha$, $\alpha = e, m$], curiously enough related to the old ones by an equation

$$-\frac{1}{c}\frac{\partial}{\partial t}\mathbf{A}_T^\alpha(\mathbf{r},t) = \boldsymbol{\nabla}\times\boldsymbol{\mathcal{A}}_T^\alpha(\mathbf{r},t). \tag{16}$$

Curiously, because Eq. (16) shows a certain resemblance to the dynamic evolution equations for the two helicity species (subscript $+,-$) of the analytical (superscript $(+)$) transverse vector potentials [$\mathbf{A}_+^{(+)}, \mathbf{A}_-^{(+)}$], viz.,

$$\frac{i}{c}\frac{\partial}{\partial t}\mathbf{A}_{T,\pm}^{(+)}(\mathbf{r},t) = \pm\boldsymbol{\nabla}\times\mathbf{A}_{T,\pm}^{(+)}(\mathbf{r},t). \tag{17}$$

For the e-particle the Minimal Coupling Principle thus finally takes the form

$$\mathbf{p}^e \to \mathbf{p}^e - e[\mathbf{A}_T^e + \boldsymbol{\mathcal{A}}_T^m]. \tag{18}$$

A similar one holds for the m-particle, namely

$$\mathbf{p}^m \to \mathbf{p}^m - g[\mathbf{A}_T^m - \boldsymbol{\mathcal{A}}_T^e]. \tag{19}$$

The transverse vector potential is the key quantity for understanding the particle dressing by a cloud of transverse photons. As we shall describe in this book the transverse photon dressing is of importance for understanding quite diverse physical phenomena such as (i) the renormalization of the bare electron mass, (ii) the quantized near-field light emission from a single-electron atom and (iii) the localized plasmariton quasiparticle appearing in jellium electrodynamics.

In order to describe the electrodynamics of photons in curved structures [3D- and 4D-curvature], the formalism based on the transverse vector potential is of central importance, and recent research of mine aims at an understanding of what one would expect for photons tied to a Möbius band with a view to the Dirac phase factor $\Phi(C)$ and the knowledge of the gauge photon field associated with plasmariton quasiparticle propagation along the Möbius band.

In non-Abelian gauge symmetry of the SO(3) type, the magnetic monopole may appear as a static hedgehog solution, where the local Higgs field in isospace everywhere points in the radial direction. The magnetic near field of the magnetic monopole is carried alone by the Higgs field. In an electrodynamic theory extended to SO(3) the Maxwell fields become isovectors (iso) and the symmetrized set of Maxwell equations take the following form:

$$\mathbf{D} \cdot \mathbf{E}_{\mathrm{iso}} = e\mathbf{J}^{e,0}, \tag{20}$$

$$\mathbf{D} \cdot \mathbf{B}_{\mathrm{iso}} = g\mathbf{J}^{m,0}, \tag{21}$$

$$\mathbf{D} \times \mathbf{B}_{\mathrm{iso}} = D_0\mathbf{E}_{\mathrm{iso}} + e\mathbf{J}^e_{\mathrm{iso}}, \tag{22}$$

$$-\mathbf{D} \times \mathbf{E}_{\mathrm{iso}} = D_0\mathbf{B}_{\mathrm{iso}} + g\mathbf{J}^m_{\mathrm{iso}}. \tag{23}$$

The reader may note that this set of isovector equations is form-identical to the perhaps well-known symmetrized set of U(1) Maxwell equations. A Dynamic Isovector Structural Vacuum can be introduced on the basis of Eq. (20)–(23). In this vacuum "isovector photons" may propagate in a vacuum filled with Higgs modes, which can be excited as determined by the isovector photon-source entanglement as recent preliminary research seems to indicate. It is worthwhile to mention here that the non-Abelian SO(3) theory is incompatible with the SU(2)×U(1) theory describing electro-weak interactions.

Throughout "Fundamentals of Photon Physics" I have wherever possible made use of the transverse gauge photon concept, and underlined its pivotal role in quantum electrodynamics and photon wave mechanics.

I

Photons: Wave Mechanics and Quantum Field Theory

CHAPTER 1

Bird's-Eye View of Electrodynamics as a Gauge Theory

1.1 KINDERGARTEN STORY OF GAUGE TRANSFORMATIONS

In classical electromagnetism it is often convenient to introduce a three-vector potential, $\mathbf{A}$, and a scalar potential, ϕ, in place of the electric and magnetic fields, $\mathbf{E}$ and $\mathbf{B}$, as follows:

$$\mathbf{B} = \boldsymbol{\nabla} \times \mathbf{A}, \quad \mathbf{E} = -\frac{\partial \mathbf{A}}{\partial t} - \boldsymbol{\nabla}\phi. \tag{1.1}$$

With these definitions, the homogeneous Maxwell equations

$$\boldsymbol{\nabla} \cdot \mathbf{B} = 0, \quad \boldsymbol{\nabla} \times \mathbf{E} = -\frac{\partial \mathbf{B}}{\partial t} \tag{1.2}$$

are automatically satisfied. Clearly $\mathbf{A}$ and ϕ can be changed by a *gauge transformation*

$$\mathbf{A} \rightarrow \mathbf{A}' = \mathbf{A} + \boldsymbol{\nabla}\chi, \tag{1.3}$$

$$\phi \rightarrow \phi' = \phi - \frac{\partial \chi}{\partial t}, \tag{1.4}$$

with an arbitrary *gauge function* χ, yet preserving the fields; $\mathbf{E}' = \mathbf{E}$, $\mathbf{B}' = \mathbf{B}$. This arbitrariness in the potentials inclines one to believe that these are unphysical quantities, $\mathbf{E}$ and $\mathbf{B}$ being the *real* physical quantities.

However, in order to qualify as real quantities in physics, one must be able to *measure* $\mathbf{E}$ and $\mathbf{B}$, as these appear in a given reference (inertial) frame. Now, every electromagnetic field measurement requires interaction with a least one charged particle, and it is via the induced motion of a charge (or charges) that the electromagnetic field manifests itself. In *classical electrodynamics* the motion of a given particle with

DOI: 10.1201/9781003029458-1

a mass m and an electric charge e is determined by Newton's second law, driven by the Lorentz force

$$\mathbf{F} = e(\mathbf{E} + \mathbf{v} \times \mathbf{B}), \tag{1.5}$$

$\mathbf{v}$ being the instantaneous particle velocity. The dynamics of charged particles enters the two inhomogeneous Maxwell-Lorentz equations via related charge and current densities. The form of the Lorentz force strengthens one's belief that the potentials are unphysical quantities, and as such, together with the associated gauge transformations, at best convenient as auxiliary tools in "practical" calculations.

1.2   GAUGE INVARIANCE IN QUANTUM MECHANICS

In quantum mechanics a gauge transformation of the electromagnetic potentials must be accompanied by a unitary transformation of the scalar wave function, viz.

$$\psi(x) \to \psi'(x) = \exp\left(i\frac{e}{\hbar}\chi(x)\right)\psi(x), \tag{1.6}$$

where in covariant notation $x = \{x^\mu\}$, $\mu = 0 - 3$. Together, the transformations in Eqs. (1.3), (1.4) and (1.6) result in a form invariance of the field-coupled Schrödinger equation. The transformation in Eq. (1.6) is called a gauge transformation of the second kind. As we shall realize in Section 2.1, the physically important quantity relates to the round-trip phase shift against the clock, namely

$$\Delta = \frac{e}{\hbar} \oint A_\mu(x)dx^\mu, \tag{1.7}$$

where in covariant notation, $\{A_\mu\} = (-\phi/c, \mathbf{A})$. The phase shift Δ is zero if the generalized curl of the gauge potential, $\{F_{\mu\nu}\}$, given by

$$F_{\mu\nu}(x) = \partial_\mu A_\nu(x) - \partial_\nu A_\mu(x), \tag{1.8}$$

is zero over any surface (Σ) bounded by the closed curve in Eq. (1.7) $[\Delta = 0$ for $F_{\mu\nu}(x) = 0$ on $\Sigma]$. It turns out that the *physical observable* is not the phase shift in Eq. (1.7) but the *Dirac phase factor*

$$\Phi(C) = \exp\left[i\frac{e}{\hbar} \oint_C A_\mu(x)dx^\mu\right] \tag{1.9}$$

taking over the closed curve C. We shall highlight this result when we discuss Aharonov-Bohm effect in Chapter 11.

The gauge potential formalism can be said to represent a *structural theory for a particle vacuum*. The dynamics of the structural vacuum satisfies the wave equation

$$\partial^\mu\partial_\mu A_\nu(x) - \partial_\nu(\partial^\mu A_\mu(x)) = 0, \tag{1.10}$$

with $\{\partial^\mu\} = (-c^{-1}\partial/\partial t, \boldsymbol{\nabla})$.

1.3 PHOTON WAVE MECHANICS BASED ON POTENTIALS

In a given inertial frame the transverse (T) part of the gauge potential $(\mathbf{A}_T(x))$, which is a gauge invariant quantity $[\mathbf{A}'_T(x) = \mathbf{A}_T(x)]$, satisfies the wave equation

$$\partial^\mu \partial_\mu \mathbf{A}_T(x) = \mathbf{0}. \tag{1.11}$$

In order to unify the quantum electrodynamic theory of photons and other elementary particles (e.g. electrons), we shall see in Chapter 2 that it is useful to focus the attention on the analytic (positive frequency) part of $\mathbf{A}_T$, denoted by $\mathbf{A}_T^{(+)}$, and further divide the analytical signal into its positive (subscript $+$) and negative (subscript $-$) helicity species. Thus $\mathbf{A}_T^{(+)} = \mathbf{A}_{T,+}^{(+)} + \mathbf{A}_{T,-}^{(+)}$. The two helicity species satisfy the dynamical evolution equations

$$i\frac{\partial}{\partial t}\mathbf{A}_{T,\pm}^{(+)} = \pm c\boldsymbol{\nabla} \times \mathbf{A}_{T,\pm}^{(+)}. \tag{1.12}$$

The two-spinor wave function, $\boldsymbol{\mathcal{A}}_T^{(+)} \equiv (\mathbf{A}_{T,+}^{(+)}, -\mathbf{A}_{T,-}^{(+)})$, satisfies the Hamiltonian (Schrödinger-like) wave equation

$$i\hbar\frac{\partial}{\partial t}\boldsymbol{\mathcal{A}}_T^{(+)}(\mathbf{r}, t) = (c\hat{\mathbf{p}} \cdot \boldsymbol{\sigma})\boldsymbol{\mathcal{A}}_T^{(+)}(\mathbf{r}, t), \tag{1.13}$$

$\hat{\mathbf{p}} = (\hbar/i)\boldsymbol{\nabla}$ and $\boldsymbol{\sigma}$ being the momentum operator and Cartesian (dimensionless) spin-one vector, respectively. The dynamical evolution equation above [Eq. (1.13)] may be considered as the basic equation for a photon wave mechanical potential description of unquantized fields in structural vacuum.

1.4 EXTENSION TO QUANTUM FIELD THEORY

The electrodynamic gauge theory based on potentials can be extended to a field-quantized formulation by replacing the gauge invariant transverse vector potential by a corresponding quantum field operator, i.e.,

$$\mathbf{A}_T(x) \Rightarrow \hat{\mathbf{A}}_T(x). \tag{1.14}$$

The Hamiltonian wave equations for the two helicity species now are replaced by the operator relations

$$i\frac{\partial}{\partial t}\hat{\mathbf{A}}_{T,\pm}^{(+)}(\mathbf{r}, t) = \pm c\boldsymbol{\nabla} \times \hat{\mathbf{A}}_{T,\pm}^{(+)}(\mathbf{r}, t), \tag{1.15}$$

as we shall realize in Chapter 3.

1.5 FIELD-MATTER INTERACTION

Once the interaction between charged massive particles and photons is turned on, Eq. (1.15) must be replaced by the following inhomogeneous dynamical equations:

$$i\frac{\partial}{\partial t}\hat{\mathbf{A}}_{T,\pm}^{(+)}(\mathbf{r}, t) \mp c\boldsymbol{\nabla} \times \hat{\mathbf{A}}_{T,\pm}^{(+)}(\mathbf{r}, t)$$

$$= -\frac{1}{2\varepsilon_0 c}\int_{-\infty}^{\infty} \mathbf{e}_\pm(\boldsymbol{\kappa})\mathbf{e}_\pm^*(\boldsymbol{\kappa}) \cdot \hat{\mathbf{J}}_T(\mathbf{q}; t)e^{i\mathbf{q}\cdot\mathbf{r}}\frac{d^3q}{(2\pi)^3 q}. \tag{1.16}$$

The time evolution of the analytical quantum potentials associated with the two helicity species is driven (in a self-consistent manner) by the transverse particle current density operator $\hat{\mathbf{J}}_T(\mathbf{r}, t)$, and it is the spatial Fourier transform of this $[\hat{\mathbf{J}}_T(\mathbf{q}; t)]$ which enters Eq. (1.16). The quantities $\mathbf{e}_\pm(\boldsymbol{\kappa})$ are helicity unit vectors belonging to a plane-wave mode in the direction $\boldsymbol{\kappa} = \mathbf{q}/q$, and in Eq. (1.16) these appear in the dyadic product $\mathbf{e}_\pm(\boldsymbol{\kappa})\mathbf{e}_\pm^*(\boldsymbol{\kappa})$.

CHAPTER 2

Dynamical Evolution Equations of Gauge Photons

2.1 PHOTON: A LOCAL GAUGE PARTICLE

A photon in matter-free space is a ghost, impossible to pin down unless massive charged particles are introduced in regions of space-time. The study of this pinning down process is PHOTON PHYSICS. In order to clarify the basic concepts, we begin from wave mechanics first for a charged scalar particle, then for a transverse photon. Later on, second quantization of the matter and photon fields extends the FUNDAMENTALS OF PHOTON PHYSICS to a fledged bird.

The first and simplest local gauge theory known to physicists is electromagnetism. This theory offered physicists a new perspective on the century-old field dealing with the interaction of charged quantum particles and the electromagnetic field. To understand the idea of photon gauge theory, it is rewarding to turn to a simple example for reference.

Hence, using the covariant notation let us consider a scalar wave function of a massive charged particle, $\psi(x)$. From elementary quantum mechanics we know that a global space-time independent phase transformation of $\psi(x)$, called a gauge transformation of the first kind, does not change any physical observations. What happens if one allows the phase factor to depend on x? A local gauge transformation is carried out by rotating the phase of the wave function at the space-time point x by an amount $(e/\hbar)\chi(x)$ depending on x, namely:

$$\psi(x) \to \psi'(x) = \exp\left(i\frac{e}{\hbar}\chi(x)\right)\psi(x). \tag{2.1}$$

The transformation in Eq. (2.1) is called a gauge transformation of the second kind. The reason for pulling out the factor $e/\hbar$ [= particle charge/(Planck's constant/(2π))] appears later.

If we allowed independent phase transformations in the various space-time points, and additionally required that such arbitrary phase rotations have no physical significance, we would end up in a hopeless situation. A fruitful generalization of the first kind of gauge transformations appears if one assumes that the phase $(e/\hbar)\chi(x)$

DOI: 10.1201/9781003029458-2

needs not to have a definite value at a particular point, but only *a definite difference for neighbouring points x and $x + dx$*. This idea played a central role in Dirac's monumental article on magnetic monopoles [71]. Aspects of Dirac's theory are described in Chapter 37.

Observations in quantum mechanics relate to certain Hermitian operators (observables) $\hat{O} = \hat{O}^\dagger$, where † stands for Hermitian conjugation. In our simple example it is the absolute square of the transition matrix element between different single-particle quantum states which carries the physics, viz.,

$$|\langle\phi|\hat{O}|\psi\rangle|^2 = |\int_{-\infty}^{\infty} \phi^*(x)\hat{O}(x)\psi(x)\mathrm{d}^4x|^2 \tag{2.2}$$

for Hilbert states $|\psi\rangle$ and $|\phi\rangle$ in the Dirac notation. The local gauge prescription, with correlation of neighbouring phases, obviously only makes sense if the phase correlation is the same for all wave functions. An excellent discussion of observables is given in [62]. The reader may find a thoughtful analysis of the epistemology of quantum mechanics in the collected works of E. P. Wigner [274].

In order to compare the phases at two neighbouring points one introduces a set ($\mu = 0, 1, 2, 3$) of so-called gauge functions (potentials) $A_\mu(x)$, and stipulates that the local phases at x and $x + \mathrm{d}x$ differ by an amount $(e/\hbar)A_\mu(x)\mathrm{d}x^\mu$. Using the $(e/\hbar)A_\mu(x)$-prescription, we say that a parallel transport of the phase from x to $x + \mathrm{d}x$ has been performed. The term *parallel transport* is boroughed from general relativity, where parallel transport of vectors and higher rank tensors plays a central role; see e.g. [258]. It appears that the phase in a neighbouring point $x+\mathrm{d}x$ will change by an amount $(e/\hbar)[\chi(x) + \partial_\mu\chi(x)\mathrm{d}x^\mu]$ due to the transformation in Eq. (2.1), which gives

$$\psi'(x + \mathrm{d}x) = \exp\left[i\frac{e}{\hbar}(\chi(x) + \partial_\mu\chi(x)\mathrm{d}x^\mu)\right]\psi(x + \mathrm{d}x). \tag{2.3}$$

In order that the principle of parallel transport has the same meaning for $\psi(x)$ and $\psi'(x)$, the gauge functions belonging to the various μ's must transform as

$$A'_\mu(x) = A_\mu(x) + \partial_\mu\chi(x). \tag{2.4}$$

Conclusion: For a gauge function $A_\mu(x)$, the phase of the particle wave function thus changes by an amount $\delta\Delta(x) = (e/\hbar)A_\mu\mathrm{d}x^\mu$ by parallel transport from x to $x + \mathrm{d}x$.

By applying the infinitesimal parallel transport repeatedly, one can obtain the change in the phase of the wave function over a finite distance from a point P to a point Q along any path. Along a path Γ, the wave function thus acquires a change (Δ_Γ) from the local value chosen at P given by the line integral

$$\Delta_\Gamma = \frac{e}{\hbar}\int_{\Gamma P}^{Q} A_\mu(x)\mathrm{d}x^\mu. \tag{2.5}$$

The difference between the phases of the wave functions obtained by parallel transport along two distinct paths Γ_1 and Γ_2 becomes

$$\Delta_{\Gamma_2} - \Delta_{\Gamma_1} = \frac{e}{\hbar}\left[\int_{\Gamma_2 P}^{Q} A_\mu(x)\mathrm{d}x^\mu - \int_{\Gamma_1 P}^{Q} A_\mu(x)\mathrm{d}x^\mu\right]$$

$$= \frac{e}{\hbar}\oint_{\Gamma_2 - \Gamma_1} A_\mu(x)\mathrm{d}x^\mu. \tag{2.6}$$

The last expression is the phase shift $\Delta = \Delta_{\Gamma_2} + \Delta_{-\Gamma_1}$ obtained by the round trip $PQ(\Gamma_2) + QP(\Gamma_1)$ (against the clock), see Fig. 2.1. The resulting phase shift of the wave function around a closed path is in general non-zero.

By assuming that a generalized version of Stokes' theorem can be applied, the line closed integral can be rewritten as a surface integral over any surface Σ bounded by the closed curve $\Gamma_2 - \Gamma_1$. Thus,

$$\oint_{\Gamma_2 - \Gamma_1} A_\mu(x)\mathrm{d}x^\mu = \int_\Sigma F_{\mu\nu}(x)\mathrm{d}\sigma^{\mu\nu}, \tag{2.7}$$

where

$$F_{\mu\nu}(x) = \partial_\mu A_\nu(x) - \partial_\nu A_\mu(x) \tag{2.8}$$

is the $\mu\nu$'th component of the generalized curl of the gauge potential (I urge the reader to prove Eq. (2.7) herself). Thus, it appears that the phase shift obtained at Q by parallel transport from P is independent of the path if the gauge potential is curl-free everywhere, $F_{\mu\nu} = 0$.

The quantity $\{F_{\mu\nu}\}$ is known as the electromagnetic field tensor and it is a gauge invariant quantity:

$$F'_{\mu\nu}(x) = F_{\mu\nu}(x) + \partial_\mu(\partial_\nu\chi(x)) - \partial_\nu(\partial_\mu\chi(x)) = F_{\mu\nu}(x) \tag{2.9}$$

At this stage is is interesting to ask the conceptual question: What are the actual physical quantities described by the gauge theory? Wave function interference is a cornerstone in quantum mechanics, and the fundamental physical experiment which is basic for checking the gauge structure of electrodynamics is the Aharonov-Bohm experiment. The theory behind this experiment is discussed in Sections 11.2 and 11.3, and the Aharanov-Bohm theory's relation to magnetic monopole electrodynamics is treated in Chapt's 37 and 38. In this section it is sufficient to note that the amplitude interference between the parts of a charged particle wave function arriving at the (observation) point Q and coming from the (source) point P along parts Γ_1 and Γ_2 relates directly to the phase difference in Eq. (2.6). The diffraction pattern observed when displacing the point of observation in space-time which is a physical phenomenon in the Bohr understanding tells us that the gauge potential $A_\mu(x)$ has the disadvantage of over-describing the physics in the sense that different values of $A_\mu(x)$ can correspond to the same diffraction pattern. Indeed, the connection in Eq. (2.4) shows that $A\mu(x)$ and $A'_\mu(x) = A_\mu(x) + \partial_\mu\chi(x)$ $(\mu = 0, 1, 2, 3)$ for any function $\chi(x)$ of the space-time point x result in the same round trip phase shift since

$$\oint \partial_\mu\chi(x)\mathrm{d}x^\mu = 0. \tag{2.10}$$

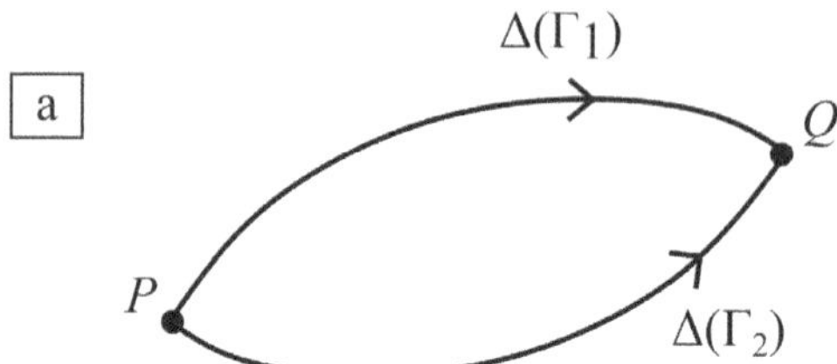

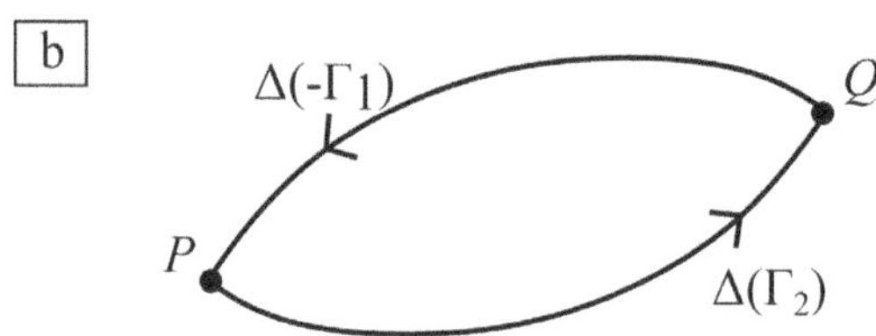

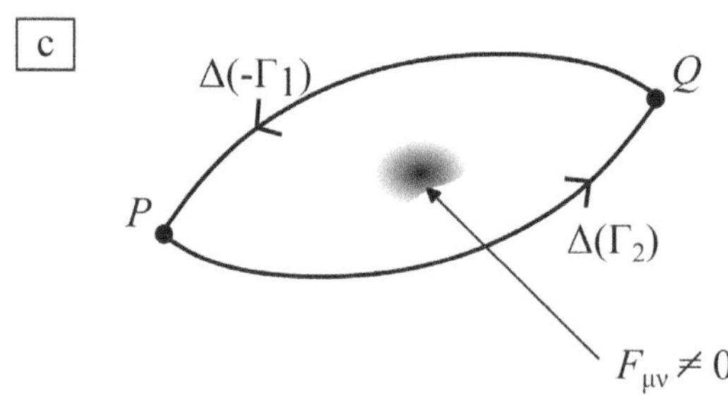

Figure 2.1 The change $\delta\Delta(x) = (e/\hbar)A_\mu(x)dx^\mu$ of the phase $\Delta(x)$ between neighbouring space-time points x and $x + dx$ must be the same for all particle wave functions, and is expressed in terms of the gauge function $A_\mu(x)$. Fig. a: The changes $\Delta(\Gamma_1)$ and $\Delta(\Gamma_2)$ of the phase from point P to point Q along the two paths Γ_1 and Γ_2. Fig. b: Round trip phase shift (against the clock) $\Delta = \Delta(\Gamma_2) + \Delta(-\Gamma_1)$. If the field tensor $F_{\mu\nu}(x) = \partial_\mu A_\nu(x) - \partial_\nu A_\mu(x) = 0$ everywhere, $\Delta = 0$ [or $\Delta = 2\pi n$, n being an integer]. Fig. c: $\Delta \neq 0$ [or $2\pi n$] if $F_{\mu\nu}(x) \neq 0$ somewhere on a surface Σ bounded by the closed curve $\Gamma_2 + (-\Gamma_1)$. The physical observable is the Dirac phase factor $\Phi(C) = \exp\{i\Delta\}$ over a closed curve, here $\Gamma_2 + (-\Gamma_1)$.

Even the phase difference in Eq. (2.6) is not an observable because phase changes differing by an integral multiple of 2π will result in the same diffraction pattern. The physical observable in the Aharanov-Bohm effect is the *Dirac phase factor*

$$\Phi(C) = \exp\left[i\frac{e}{\hbar}\oint_C A_\mu(x)dx^\mu\right], \tag{2.11}$$

taken over a closed curve C (such as $\Gamma_2 - \Gamma_1$ in Fig. 2.1). The gauge invariant $\Phi(C)$ plays a central role in magnetic monopole electrodynamics (Part X). As we shall realize in Section 11.2, $\{F_{\mu\nu}\}$ is unable to explain the Aharonov-Bohm phenomenon, since the interfering wave function parts move along paths (Γ_1, Γ_2) where $F_{\mu\nu}(x)$ is zero everywhere.

From the definition of the field tensor in Eq. (2.8) follows the identity ($\alpha \neq \beta \neq \gamma \neq \alpha$)

$$\partial_\alpha F_{\beta\gamma} + \partial_\beta F_{\gamma\alpha} + \partial_\gamma F_{\alpha\beta} = 0. \tag{2.12}$$

The important result in Eq. (2.12) is easy to remember since the three terms are related by cyclic permutation ($\alpha\beta\gamma \to \beta\gamma\alpha \to \gamma\alpha\beta$).

Apart from a numerical factor, $e/\hbar$, proportional to the charge of the massive particle, our parallel transport acts on all wave functions in the same manner. Therefore, the gauge potential formalism can be said to represent a structural theory for a particle vacuum, just called vacuum below.

In order to introduce the photon in the local gauge theory, a dynamical equation is needed for propagation effects in our structural vacuum. We postulate (in hindsight) that the dynamics of the structural vacuum satisfies

$$\partial^\mu F_{\mu\nu}(x) = 0, \tag{2.13}$$

or equivalently

$$\partial^\mu \partial_\mu A_\nu(x) - \partial_\nu(\partial^\mu A_\mu(x)) = 0. \tag{2.14}$$

Dynamical states described by Eq. (2.13) satisfy the identity in Eq. (2.12). Eq. (2.14) is the dynamical basis for the gauge field evolution in the structural vacuum. To the extent that it is possible to consider the electric ($\mathbf{E}$) and magnetic ($\mathbf{B}$) fields as primary field quantities the set in Eqs. (2.12) and (2.13) is identical to the free-space Maxwell equations. As the reader may verify, Eq. (2.12): $\nabla \times \mathbf{E} = -\partial \mathbf{B}/\partial t$, $\nabla \cdot \mathbf{B} = 0$; Eq. (2.13): $\nabla \cdot \mathbf{E} = 0$, $c^2 \nabla \times \mathbf{B} = \partial \mathbf{E}/\partial t$.

The four-divergence of Eq. (2.4),

$$\partial^\mu A'_\mu(x) = \partial^\mu A_\mu(x) + \partial^\mu \partial_\mu \chi(x), \tag{2.15}$$

shows that the gauge potential can be chosen to satisfy the Lorenz condition

$$\partial^\mu A_\mu(x) = 0, \tag{2.16}$$

if one takes $\chi(x)$ as a solution to the inhomogenous wave equation $\partial^\mu \partial_\mu \chi(x) = \partial^\mu A'_\mu(x)$. In the Lorenz gauge Eq. (2.14) is simplified to (with index interchange $\mu \leftrightarrow \nu$)

$$\partial^\nu \partial_\nu A_\mu(x) = 0, \ \mu = 0 - 3. \tag{2.17}$$

Together, Eqs. (2.16) and (2.17) describe the gauge field dynamics in a manifest covariant manner.

An important classification of the solution $\{A_\mu(x)\}$ is obtained starting from the Fourier integral decomposition

$$A_\mu(x) = (2\pi)^{-4} \int_{-\infty}^{\infty} A_\mu(q) e^{iq^\mu x_\mu} \mathrm{d}^4 q. \tag{2.18}$$

By inserting Eq. (2.18) in Eq. (2.17), it appears that the wave four-vector $\{q_\mu\} = (-\omega/c, \mathbf{q})$ satisfies the vacuum dispersion relation

$$q_\mu q^\mu \left[= \mathbf{q} \cdot \mathbf{q} - \left(\frac{\omega}{c}\right)^2 \right] = 0. \tag{2.19}$$

If one takes the 3-vector divergence of the spatial component $\mathbf{A}(x)$ of $\{A_\mu(x)\} = (A_0, \mathbf{A})$, viz. (with $i = 1 - 3$),

$$\partial^i A_i(x) = i(2\pi)^{-4} \int_{-\infty}^{\infty} q^i A_i(q) e^{iq^\mu x_\mu} \mathrm{d}^4 q, \tag{2.20}$$

and then divides $\mathbf{A}(q) = \{A_i(q)\}$ into components parallel (L) and transverse (T) to the $\mathbf{q}$-direction,

$$\mathbf{A}(q) = \mathbf{A}_\mathrm{L}(q) + \mathbf{A}_\mathrm{T}(q), \tag{2.21}$$

where

$$\mathbf{q} \times \mathbf{A}_\mathrm{L}(q) = 0, \;\; \mathbf{q} \cdot \mathbf{A}_\mathrm{T}(q) = 0, \tag{2.22}$$

it appears that the covariant wave equation in Eq. (2.17) can be divided into three. A scalar wave equation for A_0

$$\partial^\mu \partial_\mu A_0(x) = 0. \tag{2.23}$$

A wave equation for the longitudinal gauge dynamics:

$$\partial^\mu \partial_\mu \mathbf{A}_\mathrm{L}(x) = \mathbf{0}, \tag{2.24}$$

and a wave equation for the transverse part of the gauge potential:

$$\partial^\mu \partial_\mu \mathbf{A}_\mathrm{T}(x) = \mathbf{0}. \tag{2.25}$$

As we shall realize below, and in combination with the analysis in Section 2.4, it is the wave equation for the transverse part of the 3-vector potential which makes the bridge to the photon concept in structural vacuum. In this context it is important to notice that $\mathbf{A}_\mathrm{T}(x)$ is gauge invariant, $\mathbf{A}'_\mathrm{T}(x) = \mathbf{A}_\mathrm{T}(x)$ since $\boldsymbol{\nabla} \times \boldsymbol{\nabla}\chi(x) = 0$; c.f. Eq. (2.4).

In order to unify the quantum electrodynamic (QED) photons and other elementary particles (e.g. electrons), it is convenient to compose photon states of positive frequency ($\omega > 0$) components, and antiphoton states of negative frequency ($\omega < 0$) components, although photons and antiphotons are physically identical [149]. The electron and its antiparticle, the positron, are physically distinct. In standard notation [wave vector($\mathbf{q}$)-frequency(ω), and space($\mathbf{r}$)-time(t)] the analysis proceeds in the following manner. First, we introduce the analytic (positive frequency) transverse potential

$$\mathbf{A}_\mathrm{T}^{(+)}(\mathbf{q}, \omega) = \mathbf{A}_\mathrm{T}(\mathbf{q}, \omega)\Theta(\omega) \tag{2.26}$$

where the superscript (+) here, and in what follows, stands for analytic and $\Theta(\omega)$ is the Heaviside unit step function. Next, we divide $\mathbf{A}_T(\mathbf{q}, \omega)$ into helicity components with complex unit vectors $\mathbf{e}_+(\boldsymbol{\kappa})$ [positive helicity] and $\mathbf{e}_-(\boldsymbol{\kappa})$ [negative helicity]. The orthonormalized triplet of unit vectors, $\boldsymbol{\kappa} \equiv \mathbf{q}/q$, $\mathbf{e}_+(\boldsymbol{\kappa})$, and $\mathbf{e}_-(\boldsymbol{\kappa})$ satisfy the relations [158]

$$\pm i\boldsymbol{\kappa} \times \mathbf{e}_\pm(\boldsymbol{\kappa}) = \mathbf{e}_\pm(\boldsymbol{\kappa}), \tag{2.27}$$

and

$$\mathbf{e}_\pm(\boldsymbol{\kappa}) \cdot \mathbf{e}_\pm^*(\boldsymbol{\kappa}) = 1; \ \ \mathbf{e}_\pm(\boldsymbol{\kappa}) \cdot \mathbf{e}_\pm(\boldsymbol{\kappa}) = \mathbf{e}_\pm(\boldsymbol{\kappa}) \cdot \mathbf{e}_\mp^*(\boldsymbol{\kappa}) = 0. \tag{2.28}$$

The analytic potential in Eq. (2.26) finally can be decomposed as follows:

$$\begin{aligned}
\mathbf{A}_T^{(+)}(\mathbf{q}, \omega) &= \Theta(\omega) \left[A_{T,+}(\mathbf{q}, \omega)\mathbf{e}_+(\boldsymbol{\kappa}) + A_{T,-}(\mathbf{q}, \omega)\mathbf{e}_-(\boldsymbol{\kappa}) \right] \\
&= A_{T,+}^{(+)}(\mathbf{q}, \omega)\mathbf{e}_+(\boldsymbol{\kappa}) + A_{T,-}^{(+)}(\mathbf{q}, \omega)\mathbf{e}_-(\boldsymbol{\kappa}) \\
&= \mathbf{A}_{T,+}^{(+)}(\mathbf{q}, \omega) + \mathbf{A}_{T,-}^{(+)}(\mathbf{q}, \omega).
\end{aligned} \tag{2.29}$$

The space-time analytic transverse gauge potential,

$$\mathbf{A}_T^{(+)}(\mathbf{r}, t) = \mathbf{A}_{T,+}^{(+)}(\mathbf{r}, t) + \mathbf{A}_{T,-}^{(+)}(\mathbf{r}, t), \tag{2.30}$$

we hence has divided into two parts,

$$\mathbf{A}_{T,\pm}^{(+)}(\mathbf{r}, t) = (2\pi)^{-4} \int_{-\infty}^{\infty} \mathbf{A}_{T,\pm}^{(+)}(\mathbf{q}, \omega) e^{i(\mathbf{q}\cdot\mathbf{r}-\omega t)} d^3 q d\omega, \tag{2.31}$$

composed of positive (subscript +) and negative (subscript -) helicity species, respectively.

Since each Fourier mode, $\sim \exp[i(\mathbf{q} \cdot \mathbf{r} - \omega t)]$, satisfies Eq. (2.25), each of the analytic potentials in Eq. (2.31) will satisfy the same type of wave equation, viz.,

$$\Box \mathbf{A}_{T,\pm}^{(+)}(\mathbf{r}, t) = \mathbf{0}, \tag{2.32}$$

where

$$\Box = \nabla^2 - \frac{1}{c^2}\frac{\partial^2}{\partial t^2} \tag{2.33}$$

is the d' Lambertian operator

As the reader may show to herself (remembering that $\nabla \cdot \mathbf{A}_{T,\pm}^{(x)} = \mathbf{0}$), the wave equation in Eq. (2.32) can be written in the symbolic form

$$\left(\frac{i}{c}\frac{\partial}{\partial t} + \nabla\times\right)\left(\frac{i}{c}\frac{\partial}{\partial t} - \nabla\times\right)\mathbf{A}_{T,\pm}^{(+)}(\mathbf{r}, t) = \mathbf{0}. \tag{2.34}$$

The operator product form is the key in establishing a photon gauge theory which *also* is valid in structural vacuum. Since the gauge potential is unquantized at this stage, the formalism below must be classified as a PHOTON WAVE MECHANICAL

THEORY. Notwithstanding the fact that is has been advocated previously [158] that $\mathbf{A}_{T,\pm}^{(+)}(\mathbf{r},t)$ should be considered the primary candidate for a transverse photon wave function (with the true (+,-) spinorial helicity species) the line of reasoning below is new and rigorous.

In a sense, it becomes unavoidable not to consider the Hamiltonian wave equation established below as *the correct* dynamical equation for the transverse gauge photon. Furthermore, the theory extends the often preferred Riemann-Silberstein-Oppenheimer-Bialynicki photon energy wave function [25, 26, 156, 158, 198] to structural vacuum in an appealing way, as we shall realize in Section 2.2.

The wave equation in Eq. (2.32) can also be written in a symbolic form with the differential operators written in opposite order, i.e. as

$$\left(\frac{i}{c}\frac{\partial}{\partial t} - \boldsymbol{\nabla}\times\right)\left(\frac{i}{c}\frac{\partial}{\partial t} + \boldsymbol{\nabla}\times\right)\mathbf{A}_{T,\pm}^{(+)}(\mathbf{r},t) = \mathbf{0}. \tag{2.35}$$

Perhaps, it may come as a surprise to the reader that the operator order matters in the Hamiltonian description of photon wave mechanics.

To understand this, let us make a Fourier integral decomposition of $\mathbf{A}_{T,\pm}^{(+)}(\mathbf{r},t)$ in plane waves in space-time, namely

$$\mathbf{A}_{T,\pm}^{(+)}(\mathbf{r},t) = (2\pi)^{-4}\int_{-\infty}^{\infty}\mathbf{e}_{\pm}(\boldsymbol{\kappa})A_{T,\pm}^{(+)}(\mathbf{q},\omega)e^{i(\mathbf{q}\cdot\mathbf{r}-\omega t)}\mathrm{d}^3 q\mathrm{d}\omega. \tag{2.36}$$

cf. Eq (2.29). Now it follows, with the help of Eq. (2.32) that

$$\boldsymbol{\nabla}\times\mathbf{A}_{T,\pm}^{(+)}(\mathbf{r},t) = \pm(2\pi)^{-4}\int_{-\infty}^{\infty}q A_{T,\pm}^{(+)}(\mathbf{q},\omega)e^{i(\mathbf{q}\cdot\mathbf{r}-\omega t)}\mathrm{d}^3 q\mathrm{d}\omega. \tag{2.37}$$

It appears that every solution to the first-order equation

$$\left(\frac{i}{c}\frac{\partial}{\partial t} - \boldsymbol{\nabla}\times\right)\mathbf{A}_{T,\pm}^{(+)}(\mathbf{r},t) = \mathbf{0}. \tag{2.38}$$

also is a solution to Eq. (2.34) [equivalent to the wave equation in Eq. (2.32)]. By means of Eq. (2.37) and the associated $\partial/\partial t \to -i\omega$, Eq. (2.38) is equivalent to

$$\int_{-\infty}^{\infty}\left(\frac{\omega}{c}\mp q\right)\mathbf{A}_{T,\pm}^{(+)}(\mathbf{q},\omega)e^{i(\mathbf{q}\cdot\mathbf{r}-\omega t)}\mathrm{d}^3 q\mathrm{d}\omega = \mathbf{0}. \tag{2.39}$$

Because $cq = \omega(>0)$, the vacuum dispersion relation for the analytical signal, one has $\omega + cq \neq 0$. In consequence, all negative helicity Fourier amplitudes, $\mathbf{A}_{T,-}^{(+)}(\mathbf{q},\omega)$, must be zero to satisfy Eq. (2.39). On the other hand, $\omega = cq$ tells us that Eq. (2.39) is satisfied for nonvanishing positive helicity, $\mathbf{A}_{T,+}^{(+)}(\mathbf{q},\omega)$, modes.

The considerations above lead to the following conclusion: If one considers $\mathbf{A}_{T,+}^{(+)}(\mathbf{r},t)$ as the genuine transverse photon wave function for positive helicity species in space-time, then

$$i\frac{\partial}{\partial t}\mathbf{A}_{T,+}^{(+)}(\mathbf{r},t) = c\boldsymbol{\nabla}\times\mathbf{A}_{T,+}^{(+)}(\mathbf{r},t) \tag{2.40}$$

is the Hamiltonian (Schrödinger-like) wave equation governing its dynamic evolution.

With the linear differential operators in opposite order [Eq. (2.35)] a line of reasoning analogous to the one given above leads to the following conclusion: The transverse potential, $\mathbf{A}_{T,-}^{(+)}(\mathbf{q},\omega)$, describes negative helicity photon species, satisfying the Hamiltonian wave equation

$$i\frac{\partial}{\partial t}\mathbf{A}_{T,-}^{(+)}(\mathbf{r},t) = -c\boldsymbol{\nabla}\times\mathbf{A}_{T,-}^{(+)}(\mathbf{r},t) \tag{2.41}$$

in space-time.

In the interaction with matter the two gauge photon helicity species often are coupled. This fact makes it useful to introduce a two-spinor wave function (which components are three-vectors) $\mathcal{A}_T^{(+)}(\mathbf{r},t)$, defined by

$$\mathcal{A}_T^{(+)}(\mathbf{r},t) \equiv \begin{pmatrix} \mathbf{A}_{T,+}^{(+)}(\mathbf{r},t) \\ -\mathbf{A}_{T,-}^{(+)}(\mathbf{r},t) \end{pmatrix}. \tag{2.42}$$

It an easily understood compact symbolic notation, Eqs. (2.40) and (2.41) can be written as one equation, namely,

$$i\frac{\partial}{\partial t}\mathcal{A}_T^{(+)}(\mathbf{r},t) = c\boldsymbol{\nabla}\times\mathcal{A}_T^{(+)}(\mathbf{r},t). \tag{2.43}$$

With the help of the Cartesian and dimensionless spin-one vector $\boldsymbol{\sigma} = (\sigma_1,\sigma_2,\sigma_3)$, which components σ_i $(i = 1 - 3)$ are 3x3 matrices, given by

$$(\sigma_i)_{jk} = i^{-1}\epsilon_{ijk}, \tag{2.44}$$

the curl of $\mathbf{A}_{T,\pm}^{(+)}(\mathbf{r},t)$ can be written in the form

$$\boldsymbol{\nabla}\times\mathbf{A}_{T,\pm}^{(+)}(\mathbf{r},t) = \frac{1}{i}(\boldsymbol{\sigma}\cdot\boldsymbol{\nabla})\mathbf{A}_{T,\pm}^{(+)}(\mathbf{r},t). \tag{2.45}$$

If one introduces also the particle momentum operator $\hat{\mathbf{p}} \equiv (\hbar/i)\boldsymbol{\nabla}$ (in $\mathbf{r}$-representation) Eq. (2.43) can be given in the following form:

$$i\hbar\frac{\partial}{\partial t}\mathcal{A}_T^{(+)}(\mathbf{r},t) = (c\hat{\mathbf{p}}\cdot\boldsymbol{\sigma})\mathcal{A}_T^{(+)}(\mathbf{r},t). \tag{2.46}$$

The reader may notice that $\hbar$ (Planck's constant divided by 2π) has entered the formalism. At the present photon wave mechanical (field-unquantized) level, $\hbar$ appears as a "decorative" factor only, $\hat{\mathbf{p}}/\hbar = i^{-1}\boldsymbol{\nabla}$.

In Chapt's 7 and 8, where we use a well-known group theoretical approach to classify single-particle states, we shall have more to say about Eq. (2.46).

2.2 FROM STRUCTURAL VACUUM TO PARTICLE VACUUM

We have realized in the previous section that Eq. (2.34), with the linear space-time operators in the indicated order, describes the dynamics of the positive helicity transverse gauge photon via a first order-partial differential equation [Eq. (2.40)]. Let us now introduce the following two *definitions*:

$$\boldsymbol{\mathcal{E}}_{T,\pm}^{(+)}(\mathbf{r},t) \equiv -\frac{\partial}{\partial t}\mathbf{A}_{T,\pm}^{(+)}(\mathbf{r},t), \tag{2.47}$$

$$\boldsymbol{\mathcal{B}}_{\pm}^{(+)} \equiv \boldsymbol{\nabla} \times \mathbf{A}_{T,\pm}^{(+)}(\mathbf{r},t). \tag{2.48}$$

By means of these, the dynamical equation for $\mathbf{A}_{T,+}^{(+)}(\mathbf{r},t)$ [Eq. (2.40)] can be written as

$$\boldsymbol{\mathcal{E}}_{T,+}^{(+)}(\mathbf{r},t) - ic\boldsymbol{\mathcal{B}}_{+}^{(+)}(\mathbf{r},t) = \mathbf{0}. \tag{2.49}$$

With the differential operators in the reverse order [Eq. (2.35)] the dynamical equation for the negative helicity transverse gauge function [Eq. (2.41)] can [with the definitions in Eqs. (2.47) and (2.48)] be given the form

$$\boldsymbol{\mathcal{E}}_{T,-}^{(+)}(\mathbf{r},t) + ic\boldsymbol{\mathcal{B}}_{-}^{(+)}(\mathbf{r},t) = \mathbf{0}. \tag{2.50}$$

It is important to emphasize that the $\boldsymbol{\mathcal{E}}_{T,\pm}^{(+)}(\mathbf{r},t)$ and $\boldsymbol{\mathcal{B}}_{\pm}^{(+)}(\mathbf{r},t)$ auxiliary fields *are not* our familiar (anaytical and transverse) electric ($\mathbf{E}_{T}^{(+)}$) and magnetic ($\mathbf{B}^{(+)}$) fields of the Maxwell equations.

Why not? Simply because we are analyzing the gauge structure of the structural vacuum in $\mathbf{E} = c\mathbf{B} = \mathbf{0}$ $[\Rightarrow \mathbf{E}_{T} = c\mathbf{B} = \mathbf{0}]$.

Let us at this stage make an intellectual PARADIGMATIC SHIFT from structural vacuum to a standard particle vacuum in which $\mathbf{E}$ and $\mathbf{B}$ fields are present but no massive particles. In schematic form

$$\left(\boldsymbol{\mathcal{E}}_{T,\pm}^{(+)}, \boldsymbol{\mathcal{B}}_{\pm}^{(+)}\right) \Rightarrow \left(\mathbf{E}_{T}^{(+)}, \mathbf{B}^{(+)}\right). \tag{2.51}$$

$$\text{stuctural vacuum} \Rightarrow \text{particle vacuum}$$

The reader may note that the two helicity components of $(\boldsymbol{\mathcal{E}}_{T,\pm}^{(+)}, \boldsymbol{\mathcal{B}}_{\pm}^{(+)})$ have been associated with the same $(\mathbf{E}_{T}^{(+)}, \mathbf{B}^{(+)})$. The reason for this follows from Eqs. (2.52) and (2.53) below: The combinations $\mathbf{E}_{T}^{(+)} \pm ic\mathbf{B}^{(+)}$ are sufficient for dealing with the two helicity states in particle vacuum.

When I say that $\mathbf{E}_{T}$ and $\mathbf{B}$ fields are "present" in particle vacuum, my language is not sufficiently precis. Thus, the point of view I advocate (and show by rigorous argumentation throughout this book) is that electric and magnetic fields *always* can be considered as auxiliary fields for the gauge photon field. However, one must not forget that these fields are helpful in dealing with (simplify) a vast number of theoretical analysis in low-energy QED, quantum optics and classical electrodynamics.

Let us again take a look of Eqs. (2.34) and (2.35), and make the paradigmatic shift [Eq. (2.51)] from structural vacuum to particle vacuum. If one introduces the two analytic and complex Riemann-Silberstein (RS) vectors $\mathbf{F}_{\pm}^{(+)}(\mathbf{r},t)$ belonging to the two helicity species by the definitions [here the factor of $(\varepsilon_0/2)^{1/2}$ comes from our choice of SI-units]

$$\mathbf{F}_{\pm}^{(+)}(\mathbf{r},t) \equiv \sqrt{\frac{\varepsilon_0}{2}}[\mathbf{E}_{T}^{(+)}(\mathbf{r},t) \pm ic\mathbf{B}^{(+)}(\mathbf{r},t)], \tag{2.52}$$

Eqs. (2.34) and(2.35) are transformed to

$$i\frac{\partial}{\partial t}\mathbf{F}_{\pm}^{(+)}(\mathbf{r},t) = \pm c\boldsymbol{\nabla} \times \mathbf{F}_{\pm}^{(+)}(\mathbf{r},t). \tag{2.53}$$

The two equations in (2.53) are the well-known dynamical equations for the positive and negative helicity RS vectors [25,26]. In the wake of Oppenheimer's photon particle picture [156,198], it has with success been advocated by Bialyniciki and others that the $\mathbf{F}_{\pm}^{(+)}(\mathbf{r},t)$'s may be considered as so-called ENERGY WAVE FUNCTIONS (for the two helicity species) in PHOTON WAVE MECHANICS [26]. The RS vector $\mathbf{F}(\mathbf{r},t) \equiv (\varepsilon_0/2)^{1/2}[\mathbf{E}(\mathbf{r},t) + ic\mathbf{B}(\mathbf{r},t)]$ was originally introduced in classical physics to rewrite the Maxwell equations in a complex form [239–241,256].

The Riemann-Silberstein-Oppenheimer-Bialynicki (RSOB) formalism can be extended to the QED level [Chapter 3], and several useful applications of the description have appeared in the last two-three decades.

In Sections 11.2 and 11.3, I shall analyze the Aharonov-Bohm effect (its existence first postulated in 1959 [3,39,207]) including the dynamic effects of transverse (T),Longitudinal (L) and scalar (O) gauge photons. The Aharonov-Bohm effect is a typical structural vacuum phenomenon. At first sight the reader may be tempted to believe that the magnetostatic case, involving the flux through an open surface of the particle vacuum magnetic field $\mathbf{B}(\mathbf{r}) = \boldsymbol{\nabla} \times \mathbf{A}_T(\mathbf{r})$, necessitates the presence of a "real" magnetic field. In the magnetostatic case, as well as in all dynamic cases, this is not correct. Because both the divergence and curl of the magnetic field are known $[\boldsymbol{\nabla} \cdot \mathbf{B} = 0, \ \boldsymbol{\nabla} \times \mathbf{B} = \mu_0\mathbf{J}]$ the magnetic field itself is known (up to an unphysical constant). By means of Helmholtz' theorem one immediately obtains the (textbook) result

$$\mathbf{B}(\mathbf{r}) = \frac{\mu_0}{4\pi}\boldsymbol{\nabla} \times \int_{-\infty}^{\infty} \frac{\mathbf{J}(\mathbf{r}')\mathrm{d}^3 r'}{|\mathbf{r} - \mathbf{r}'|}. \tag{2.54}$$

Together, Eqs. (2.6) [for the magnetostatic case] and (2.54) show that the structural vacuum static gauge field is generated by the $\mathbf{J}(\mathbf{r})$-distribution. No reel need for a magnetic field, indeed. I urge the reader herself to analyze the classical case where the surface current of a infinitely long solenoid gives rise to the charged particle ($\sim$ electron) diffraction pattern, originally observed in a magnetized iron whisker experiment carried out by Chambers in 1960 [59,207].

2.3 MINIMAL COUPLING. MOMENTUM OF TRANSVERSE GAUGE PHO-TON

The transverse gauge photon is generated from the prevailing massive charged-particle current density. This is, as we shall realize, related to a bi-linear product of the (complex adjoint) particle wave function and its gradient [charged Klein-Gordon equation, Dirac equation, Pauli and Schrödinger equations]. In order to make the gauge theory consistent it is necessary that a new, so-called gauge (superscript G) covariant gradient $\{\nabla_{\mu}^{G}\}$, enters the formalism. The condition that the gradient (new) of the (here scalar) particle wave function (ψ) transforms in the same manner as ψ itself [Eq. (2.1)] means that

$$\nabla_\mu^G \psi(x) \to [\nabla_\mu^G \psi(x)]' = \exp\left(i\frac{e}{\hbar}\chi(x)\right)\nabla_\mu^G \psi(x). \tag{2.55}$$

As the reader may convince herself of, one finds

$$\nabla_\mu^G = \partial_\mu - i\frac{e}{\hbar}A_\mu. \tag{2.56}$$

Invariance of the theory of electrodynamics under general gauge transformations hence is expressed by Eqs. (2.1), (2.4) and (2.55). Multiplication of Eq. (2.56) by a factor $\hbar/i$ gives

$$\pi_\mu \equiv \frac{\hbar}{i}\nabla_\mu^G = p_\mu - eA_\mu, \tag{2.57}$$

where $\{\pi_\mu\}$ and $\{p_\mu\} = \{(\hbar/i)\nabla_\mu\}$ are the kinematical kinetic (gauge invariant) and canonical momenta, respectively.

Eq. (2.57) expresses the renowned MINIMAL COUPLING PRINCIPLE in electrodynamics. The transverse part of the four-vector potential multiplied by e, namely

$$\{p_\mu^{\text{FIELD}}(x)\} = e\{A_{T,\mu}(x)\} \tag{2.58}$$

can be identified as the electromagnetic part of the particle's total momentum in the chosen inertial system at a specific space-time point. The particle's total canonical momentum, $\{p_\mu\}$, is the sum of the particle's mechanical momentum $\{\pi_\mu\}$, and the electromagnetic momentum, $e\{A_{T,\mu}\}$.

2.4 LONGITUDINAL AND SCALAR GAUGE PHOTONS

To a certain extent, it is possible to establish wave equations for longitudinal and scalar gauge photons in structural vacuum following the line reasoning employed for the transverse gauge photon.

We start with the introduction of the operator $\sqrt{-\nabla^2}$. The manner in which this operator acts on a (vector) function, $\mathbf{F}(\mathbf{r})$, given in direct ($\mathbf{r}$) space, is defined indirectly by the demand that its transform in wave-vector ($\mathbf{q}$) space is $\mathbf{qF}(\mathbf{q})$, $\mathbf{F}(\mathbf{q})$ being the Fourier transform $\mathbf{F}(\mathbf{r})$. With the help of the $\sqrt{-\nabla^2}$ operator the wave equation for the longitudinal gauge field [Eq. (2.24)] can be written in the symbolic form

$$\left(\frac{i}{c}\frac{\partial}{\partial t} + \sqrt{-\nabla^2}\right)\left(\frac{i}{c}\frac{\partial}{\partial t} - \sqrt{-\nabla^2}\right)\mathbf{A}_L^{(+)}(\mathbf{r},t) = \mathbf{0}. \tag{2.59}$$

Hereafter, $\mathbf{A}_L^{(+)}(\mathbf{r},t)$ is considered as a longitudinal photon wave function, satisfying in structural vacuum a first-order wave equation

$$i\frac{\partial}{\partial t}\mathbf{A}_L^{(+)}(\mathbf{r},t) = c\sqrt{-\nabla^2}\mathbf{A}_L^{(+)}(\mathbf{r},t). \tag{2.60}$$

All solutions to Eq. (2.60) are solutions also to the L-wave equation [Eqs. (2.24) or (2.59)].

It is obvious that Eq. (2.23) leads to the following first-order wave equation for $A_0^{(+)}(\mathbf{r}, t)$, to be considered as the scalar photon wave function in structural vacuum:

$$i\frac{\partial}{\partial t}A_0^{(+)}(\mathbf{r}, t) = c\sqrt{-\nabla^2}A_0^{(+)}(\mathbf{r}, t). \tag{2.61}$$

It appears from Eqs. (2.40), (2.41) [both with the form in Eq. (2.42) inserted], (2.60) and (2.61) that we have obtained four Hamiltonian wave equations for the $T+$, $T-$, L and S photons in structural vacuum. The Hamilton operators associated with these photon types are $\pm i^{-1}c\hbar\boldsymbol{\sigma}\cdot\boldsymbol{\nabla}$ $[T+, T-]$ and $c\hbar\sqrt{-\nabla^2}$ $[L, S]$.

In the wave-vector($\mathbf{q}$)-time(t) domain Eqs. (2.60) and (2.61) have the following forms:

$$\left(i\frac{\partial}{\partial t} - cq\right) A_L^{(+)}(\mathbf{q}; t) = 0, \tag{2.62}$$

[remembering that $\mathbf{A}_L^{(+)}(\mathbf{q}; t) = \boldsymbol{\kappa}A_L^{(+)}(\mathbf{q}; t)$, since $\boldsymbol{\nabla}\times\mathbf{A}_L^{(+)}(\mathbf{r}, t) = \mathbf{0}$], and

$$\left(i\frac{\partial}{\partial t} - cq\right) A_0^{(+)}(\mathbf{q}; t) = 0. \tag{2.63}$$

Since the common operator, $i\partial/\partial t - cq$, of Eqs. (2.62) and (2.63) transformed to the frequency domain $(\partial/\partial t \to -i\omega)$ is $\omega - cq = 0$. [vacuum dispersion relation for photons $(\omega > 0)$], $A_L^{(+)}(\mathbf{q}; t)$ and $A_0^{(+)}(\mathbf{q}; t)$ do not vanish identically, *in general.* We emphasize the words *in general,* because a certain gauge transformation within the Lorenz gauge can eliminate either the $L-$ or the $S-$photon.

In the $(\mathbf{q}; t)$-domain the Lorenz condition [Eq. (2.16)] is given by

$$iqA_L^{(+)}(\mathbf{q}; t) + \frac{1}{c}\frac{\partial}{\partial t}A_0^{(+)}(\mathbf{q}; t) = 0. \tag{2.64}$$

A gauge transformation to a new (superscript NEW) set of potentials, given by

$$A_L^{(+)\mathrm{NEW}}(\mathbf{q}; t) = A_L^{(+)}(\mathbf{q}; t) + iq\chi^{(+)}(\mathbf{q}; t), \tag{2.65}$$

$$A_0^{(+)\mathrm{NEW}}(\mathbf{q}; t) = A_0^{(+)}(\mathbf{q}; t) + iq\chi^{(+)}(\mathbf{q}; t), \tag{2.66}$$

allows one to satisfy the Lorenz condition also for the new set of potentials, i.e. $icqA_L^{(+)\mathrm{NEW}} + \partial A_0^{(+)\mathrm{NEW}}/\partial t = 0$, provided the analytic gauge function, $\chi^{(+)}(\mathbf{q}, t)$, obeys the wave equation

$$\left(q^2 + \frac{1}{c^2}\frac{\partial^2}{\partial t^2}\right)\chi^{(+)}(\mathbf{q}; t) = 0, \tag{2.67}$$

as the reader may prove. From the equivalent form

$$\left(\frac{i}{c}\frac{\partial}{\partial t} + q\right)\left(\frac{i}{c}\frac{\partial}{\partial t} - q\right)\chi^{(+)}(\mathbf{q}; t) = 0, \tag{2.68}$$

it appears that all nonvanishing $\chi^{(+)}(\mathbf{q};t)$'s must satisfy the first-order differential equation

$$\left(\frac{i}{c}\frac{\partial}{\partial t} - q\right)\chi^{(+)}(\mathbf{q};t) = 0, \tag{2.69}$$

since a transformation to the ω-domain $[i\partial/\partial t - cq \rightarrow \omega - cq(=0)]$. In conclusion: Provided the analytic gauge function $\chi^{(+)}(\mathbf{q};t)$ satisfies Eq. (2.69), all gauge transformed sets of analytic potentials, $(A_L^{(+)}, A_0^{(+)})$, satisfy the Lorenz condition. Eq. (2.69) also tells one that the transformations in Eqs. (2.65) and (2.66) can be written in the compact form with a common, $iq\chi^{(+)}$, term:

$$A_{L,0}^{(+)\text{NEW}}(\mathbf{q};t) = A_{L,0}^{(+)}(\mathbf{q};t) + iq\chi^{(+)}(\mathbf{q};t). \tag{2.70}$$

The $\chi^{(+)}$ choices $A_L^{(+)} = -iq\chi^{(+)}$ and $A_0^{(+)} = iq\chi^{(+)}$ makes $A_L^{(+)\text{NEW}} = 0$ and $A_0^{(+)\text{NEW}} = 0$, respectively.

2.5   NEAR-FIELD PHOTON: A VIRTUAL LOCAL GAUGE PARTICLE

One cannot consider the longitudinal and/or the scalar photon wave functions as *genuine* photon types in structural vacuum since they are gauge dependent. There does exist in addition to the transverse photons $[T+, T-]$ another gauge invariant photon type in structural vacuum, namely the so-called NEAR-FIELD PHOTON [NF] as we shall realize in what follows. Originally, the near-field photon concept was introduced in a study of photon wave mechanics in the Lorenz gauge [155]. It is the NF-photon which is responsible for the Aharonov-Bohm effect in a static magnetic coil field; see Chapter 11.

It is common in all branches of physics to use the phrases *longitudinal photon* and *scalar photon* for the esoteric particles, which quantum mechanical wavefunctions $\mathbf{A}_L^{(+)}$ and $A_0^{(+)}$ satisfy the positive-frequency parts of the wave equations in Eqs. (2.24) and (2.23), and, on a bit deeper level, the Hamiltonian first-order equations of motions in Eqs. (2.60) and (2.61). However, I consider the label *photon* for the L- and S-quanta a misnomer, and later in the book, when the physics of these fields in structural vacuum become transparent, I shall suggest use of another (better) label. But for the time being we stick to the conventional term.

Let us now address the following two urgent problems facing one in relation to Eqs. (2.60) and (2.61): (i) The meaning of the strange (unusual) operator $\sqrt{-\nabla^2}$, and (ii) the fact that the balance between $\mathbf{A}_L^{(+)}(\mathbf{r},t)$ and $A_0^{(+)}(\mathbf{r},t)$ can be changed by a gauge transformation within the Lorenz gauge.

First we discuss point (i). The definition

$$\sqrt{-\nabla^2}\mathbf{F}(\mathbf{r}) \equiv \int_{-\infty}^{\infty} q\mathbf{F}(\mathbf{q})e^{i\mathbf{q}\cdot\mathbf{r}}\frac{d^3q}{(2\pi)^3} \tag{2.71}$$

allows one to obtain a satisfactory description of the wave mechanics in the momentum representation $[(\mathbf{q};t)\text{-domain}]$ cf. Eqs (2.62) and (2.63). The Hamilton operator (with $\hbar$) $\hbar cq$, in the particle language, gives on the correct deBroglie energy-frequency relation $E = \hbar\omega[= \hbar cq]$ for a massless particle, and with a $\mathbf{q}$-representation of the momentum operator $\hat{\mathbf{p}} = \hbar\mathbf{q}$, also the de Broglie momentum-wave vector relation $\mathbf{p} = \hbar\mathbf{q}$. Despite of this, the formalism based on Eqs. (2.60) and (2.61), with the associated Hamiltonian $(\mathbf{q};t)$ equations in Eqs. (2.62) and (2.63), is unsatisfactory. In quantum physics the superposition principle is an indispensable cornerstone. Superposition of monochromatic ($\omega = cq$ in structural vacuum) plane-wave states *must* lead to a satisfactory wave mechanical description in direct (real) space. However, the Hamilton operator, given as $\hat{H} = c\hbar\sqrt{-\nabla^2}$ in real space is unacceptable.

In order to understand this we use the fact that the Fourier integral transform of the product form $qA_L^{(+)}(\mathbf{q};t)$ $[qA_0^{(+)}(\mathbf{q};t)]$ results in a folding integral in direct space. Hence, one obtains a dynamical time evolution of the L-wave function governed by

$$i\hbar\frac{\partial}{\partial t}A_L^{(+)}(\mathbf{r},t) = \int_{-\infty}^{\infty} \hat{H}(|\mathbf{r} - \mathbf{r}'|)A_L^{(+)}(\mathbf{r}',t)\mathrm{d}^3r', \tag{2.72}$$

where [with $\mathbf{R} = \mathbf{r} - \mathbf{r}'$]

$$\hat{H}(|\mathbf{R}|) = \int_{-\infty}^{\infty} hcqe^{i\mathbf{q}\cdot\mathbf{R}}\frac{\mathrm{d}^3q}{(2\pi)^3}. \tag{2.73}$$

Although an integro-differential equation for the wave function dynamics *perhaps* is acceptable, the form of the Hamilton operator is not, for certainty. Using spherical coordinates in $\mathbf{q}$-space, and with the polar axis along $\mathbf{R}$, one obtains $[R = |\mathbf{R}|]$

$$\begin{aligned}
\hat{H}(R) &= \frac{ic\hbar}{2\pi}\frac{1}{R}\frac{d^2}{dR^2}\left[\delta^{(+)}(R) - (\delta^{(+)}(R))^*\right] \\
&= -\frac{c\hbar}{2\pi}\frac{1}{R}\frac{d^2}{dR^2}\left(\mathcal{P}\frac{1}{R}\right)
\end{aligned} \tag{2.74}$$

where

$$\delta^{(+)}(R) = (2\pi)^{-1}\int_0^{\infty} e^{iqR}\mathrm{d}q = \frac{1}{2}\left[\delta(R) + \frac{i}{\pi}\mathcal{P}\frac{1}{R}\right] \tag{2.75}$$

is the positive wave-number part of the Dirac delta function, $\delta(R)$. Mathematically, Eq. (2.75) is analogous to the positive frequency part of the Dirac delta function discussed in relation to a study of complex analytical signals in Section 3.1 of [185]. In the two equations above $\mathcal{P}\frac{1}{R}$ stands for the principal part of $\frac{1}{R}$. The singular nature of the Hamilton operator $\hat{H}(R)$ makes the whole wave-function formalism for $A_L^{(+)}(\mathbf{r},t)$ [and of course also for $A_0^{(+)}(\mathbf{r},t)$, which satisfy a form-identical dynamical equation] untenable. The Hamilton operator in Eq. (2.74) is the one obtained in photon wave mechanics based on complex field vectors accentuating the photon helicity, a formalism studied in detail in [153]. In the Riemann-Silberstein-Oppenheimer-Bialynicki approach one comes to a certain degree formally around the problem by relating the Hamilton operators in $\mathbf{q}$- and $\mathbf{r}$-space by a *scaled* Fourier integral transformation [149,242]. Despite this, we shall see that the basic physical problem still exists.

Let us now turn the attention towards the second point (ii) [mentioned above Eq. (2.71)]. Below, we shall realize that the scalar and longitudinal photon variables can be replaced by two others, the so-called near-field and gauge photon variables. The near-field photon is gauge invariant (not only against gauge transformations within the Lorenz gauge), and the gauge photon can be completely eliminated from the description. Though a satisfactory solution can be obtained for the second point (ii), the problem faced in point (i) still exists for the near-field photon. In a pre-conclusion: An entirely different interpretation of the physical role of the "near-field photon" has to be sought. The NF-photon is virtual in the sense that it is always coupled to matter. This coupling is studied in detail in Chapt's 19-21 in relation to a description of collective-mode electrodynamics in jellium.

A unitary transformation $(\mathbf{U}^{-1} = \mathbf{U}^{\dagger})$ given by

$$\mathbf{U} = \frac{1}{\sqrt{2}} \begin{pmatrix} i & -i \\ 1 & 1 \end{pmatrix} \tag{2.76}$$

is used to replace the L and S photon variables in the q-representation, $A_L^{(+)}(\mathbf{q};t)$ and $A_0^{(+)}(\mathbf{q};t)$, by two new ones, $A_{NF}^{(+)}(\mathbf{q};t)$ and $A_G^{(+)}(\mathbf{q};t)$, as follows:

$$\begin{pmatrix} A_{NF}^{(+)}(\mathbf{q};t) \\ A_G^{(+)}(\mathbf{q};t) \end{pmatrix} = \mathbf{U} \begin{pmatrix} A_L^{(+)}(\mathbf{q};t) \\ A_0^{(+)}(\mathbf{q};t) \end{pmatrix} \tag{2.77}$$

For reasons to be given below, the new gauge photon variables are named the near-field (NF) and gauge (G) photon potentials. In a first step, it appears from Eqs. (2.62) and (2.63) that the NF and the G photon potentials must satisfy the dynamical equations

$$\left(i\frac{\partial}{\partial t} - cq \right) A_{NF}^{(+)}(\mathbf{q};t) = 0, \tag{2.78}$$

$$\left(i\frac{\partial}{\partial t} - cq \right) A_G^{(+)}(\mathbf{q};t) = 0. \tag{2.79}$$

The Lorenz gauge condition in Eq. (2.64) can be expressed as a coupling condition between the NF and G variables. By means of the inverse transformation to Eq. (2.77), the Lorenz condition takes the form

$$\left(cq - i\frac{\partial}{\partial t} \right) A_G^{(+)}(\mathbf{q};t) = i \left(cq + i\frac{\partial}{\partial t} \right) A_{NF}^{(+)}(\mathbf{q};t). \tag{2.80}$$

By inserting Eq. (2.79) into Eq. (2.80) one obtains

$$\left(i\frac{\partial}{\partial t} + cq \right) A_{NF}^{(+)}(\mathbf{q};t) = 0. \tag{2.81}$$

In order to satisfy both Eqs. (2.78) and (2.81) it follows that

$$A_{NF}^{(+)}(\mathbf{q};t) = 0. \tag{2.82}$$

The same conclusion is reached by transforming Eq. (2.81) to the $(\mathbf{q}; \omega)$-domain. Here $(\omega + cq)A_{NF}^{(+)}(\mathbf{q}, \omega) = 0$, and since $\omega + cq > 0$, each Fourier component $A_{NF}^{(+)}(\mathbf{q}, \omega)$ must vanish. In consequence, Eq. (2.82) is obtained. Although the near-field potential is identically zero in a structural vacuum, $A_{NF}^{(+)}(\mathbf{q}; t)$ will not vanish in photon-matter interaction processes. For this reason one may call this photon a *virtual photon.*

The result in Eq. (2.82) only makes sense if $A_{NF}^{(+)}(\mathbf{q}; t)$ is gauge invariant. By means of the gauge transformation of the L and S potentials [Eqs. (2.65) and (2.66)] one obtains

$$A_{NF}^{(+)\mathrm{NEW}}(\mathbf{q}; t) = A_{NF}^{(+)}(\mathbf{q}; t) + \frac{i}{\sqrt{2}}\left(q - \frac{i}{c}\frac{\partial}{\partial t}\right)\chi^{(+)}(\mathbf{q}, t), \qquad (2.83)$$

and remembering that the analytic gauge function satisfies Eq. (2.69), one reaches the required result

$$A_{NF}^{(+)\mathrm{NEW}}(\mathbf{q}; t) = A_{NF}^{(+)}(\mathbf{q}; t). \qquad (2.84)$$

The virtual near-field photon is a gauge-invariant object, indeed.

The effect on $A_G^{(+)}(\mathbf{q}; t)$ of the gauge transformations of the L and S potentials, conveniently written in the compact form given in Eq. (2.70), immediately leads to

$$A_G^{(+)\mathrm{NEW}}(\mathbf{q}; t) = A_G^{(+)}(\mathbf{q}; t) + i\sqrt{2}q\chi^{(+)}(\mathbf{q}; t). \qquad (2.85)$$

The name gauge (G) photon wave function for $A_G^{(+)}(\mathbf{q}; t)$ refers to the fact that this analytic potential can be completely eliminated from the formalism, i.e.,

$$A_G^{(+)\mathrm{NEW}}(\mathbf{q}; t) = 0. \qquad (2.86)$$

Starting from a given nonvanishing $A_G^{(+)}(\mathbf{q}; t)$, the analytic gauge function choice

$$\chi^{(+)}(\mathbf{q}; t) = \frac{i}{\sqrt{2}q}A_G^{(+)}(\mathbf{q}; t) \qquad (2.87)$$

gives one Eq. (2.86).

The analysis of Section 2.5 has led to the conclusion that the *two* gauge dependent longitudinal and scalar photon wave functions can be replaced by a *single* so-called near-field photon. This photon is *virtual* in the sense that it exists only in the presence of field-matter interaction. It is known [157, 158] that the range of the near-field photon field does not extend beyond the rim ($\sim$near-field) zone of the source (or detector) particle current density distribution.

2.6   STRUCTURAL VACUUM IN THE RIM ZONE OF MATTER

In partial analogy with the definitions in Eqs. (2.47) and (2.48), we now introduce the following *definition* in space-time:

$$\mathcal{E}_L^{(+)}(\mathbf{r}, t) \equiv -\frac{\partial}{\partial t}\mathbf{A}_L^{(+)}(\mathbf{r}, t) - c\boldsymbol{\nabla}A_0^{(+)}(\mathbf{r}, t). \qquad (2.88)$$

In the wave vector-time domain, where $\boldsymbol{\mathcal{E}}_L(\mathbf{q}, t) = \mathcal{E}_L(\mathbf{q}, t)\boldsymbol{\kappa}$, Eq. (2.88) reduces to the scalar form

$$\mathcal{E}_L^{(+)}(\mathbf{q}; t) = -\frac{\partial}{\partial t} A_L^{(+)}(\mathbf{q}; t) - icq A_0^{(+)}(\mathbf{q}; t). \tag{2.89}$$

Since $A_{NF}^{(+)}(\mathbf{q}; t)$ is gauge invariant [see Eq. (2.84)] we can put

$$A_L^{(+)}(\mathbf{q}; t) = -A_0^{(+)}(\mathbf{q}; t) = \frac{1}{i\sqrt{2}} A_{NF}^{(+)}(\mathbf{q}; t) \tag{2.90}$$

working in the gauge where $A_G^{(+)}(\mathbf{q}; t) = 0$. By combining Eqs. (2.90) and (2.89) one obtains the following relation between $\mathcal{E}_L^{(+)}(\mathbf{q}; t)$ and $A_{NF}^{(+)}(\mathbf{q}; t)$:

$$\mathcal{E}_L^{(+)}(\mathbf{q}; t) = \frac{c}{\sqrt{2}} \left(\frac{i}{c}\frac{\partial}{\partial t} + q \right) A_{NF}^{(+)}(\mathbf{q}; t) \tag{2.91}$$

For $A_{NF}^{(+)}(\mathbf{q}; t) = 0$, we also have $\mathcal{E}_L^{(+)}(\mathbf{q}; t) = 0$, of course, but if a localized charged-particle current density distribution is embedded in a region of space-time, the structural matter field $\mathcal{E}_L^{(+)}(\mathbf{r}, t)$ does not vanish in the rim zone.

The PARADIGMATIC SHIFT

$$\mathcal{E}_L^{(+)} \Rightarrow \mathbf{E}_L^{(+)} \tag{2.92}$$

relates the rime zone of structural vacuum and standard particle vacuum. The paradigmatic shift leads to the identification

$$\mathbf{E}_L^{(+)}(\mathbf{r}, t) = -\frac{\partial}{\partial t} \mathbf{A}_L^{(+)}(\mathbf{r}, t) - c\boldsymbol{\nabla} A_0^{(+)}(\mathbf{r}, t). \tag{2.93}$$

This relation together with $\mathbf{B}^{(+)}(\mathbf{r}, t) = \boldsymbol{\nabla} \times \mathbf{A}_T^{(+)}(\mathbf{r}, t)$ constitute (for the analytical fields) the well-known connection between fields $(\mathbf{E}_L, \mathbf{B})$ and potentials $(\mathbf{A}_L, \Phi = cA_0)$ in the Maxwell theory.

CHAPTER 3

Quantum Theory of Gauge Photons in Structural Vacuum

3.1 FROM PHOTON WAVE MECHANICS TO QUANTUM FIELD THEORY

The second quantization of the photon wave mechanical theory of the transverse gauge photons in structural vacuum follows the usual (well known) strategy [185, 230]. Thus, we replace the gauge invariant transverse vector potential, $\mathbf{A}_T(x)$, by a corresponding quantum field operator, $\hat{\mathbf{A}}_T(x)$, which depends on the continuous space-time index, x:

$$\mathbf{A}_T(x) \Rightarrow \hat{\mathbf{A}}_T(x). \tag{3.1}$$

In order to connect the field operator formalism to the kinetic Einstein-de Broglie particle-wave relations, a monochromatic plane wave expansion of $\hat{\mathbf{A}}_T(x)$ is made. Hence,

$$\hat{\mathbf{A}}_T(x) = \hat{\mathbf{A}}_{T,+}^{(+)}(x) + \hat{\mathbf{A}}_{T,-}^{(+)}(x) + \hat{\mathbf{A}}_{T,+}^{(-)}(x) + \hat{\mathbf{A}}_{T,-}^{(-)}(x) \tag{3.2}$$

where

$$
\begin{aligned}
\hat{\mathbf{A}}_{T,\pm}^{(+)}(x) &= \left(\frac{\hbar}{2\varepsilon_0 L^3}\right)^{1/2} \sum_{\mathbf{q}} \frac{1}{\omega^{1/2}} \mathbf{e}_{\boldsymbol{\kappa}\pm}\hat{a}_{\mathbf{q}\pm}e^{iq_\mu x^\mu} \\
&\to \left(\frac{\hbar}{2\varepsilon_0}\right)^{1/2} \int_{-\infty}^{\infty} \frac{1}{\omega^{1/2}} \mathbf{e}_\pm(\boldsymbol{\kappa})\hat{a}_\pm(\mathbf{q})e^{iq_\mu x^\mu} \frac{\mathrm{d}^3 q}{(2\pi)^{3/2}},
\end{aligned}
\tag{3.3}
$$

and

$$
\begin{aligned}
\hat{\mathbf{A}}_{T,\pm}^{(-)}(x) &= \left(\frac{\hbar}{2\varepsilon_0 L^3}\right)^{1/2} \sum_{\mathbf{q}} \frac{1}{\omega^{1/2}} \mathbf{e}_{\boldsymbol{\kappa}\pm}^*\hat{a}_{\mathbf{q}\pm}^\dagger e^{-iq_\mu x^\mu} \\
&\to \left(\frac{\hbar}{2\varepsilon_0}\right)^{1/2} \int_{-\infty}^{\infty} \mathbf{e}_\pm^*(\boldsymbol{\kappa})\hat{a}_\pm^\dagger(\mathbf{q})s e^{-iq_\mu x^\mu} \frac{\mathrm{d}^3 q}{(2\pi)^{3/2}}.
\end{aligned}
\tag{3.4}
$$

DOI: 10.1201/9781003029458-3

The symbol $\dagger$ in Eq. (3.4) stands for Hermitian conjugation, and $\omega = cq$ in Eqs. (3.3) and (3.4). The positive- and negative-frequency parts of the vector potential operator are each other's Hermitian conjugate,

$$\left[\hat{\mathbf{A}}_{T\pm}^{(-)}(x)\right]^{\dagger} = \hat{\mathbf{A}}_{T\pm}^{(+)}(x). \tag{3.5}$$

The operator expressions in Eqs. (3.3) and (3.4) are given both in discrete mode (box) quantization [cubic box with volume L^3], and continuous mode (infinite space) quantization. In order to emphasize the difference between the two quantization schemes we have written $\mathbf{e}_{\kappa\pm} \to \mathbf{e}_{\pm}(\boldsymbol{\kappa})$ for the polarization unit vectors of the two helicities, and $\hat{a}_{\mathbf{q}\pm}$ $[\hat{a}_{\mathbf{q}\pm}^{\dagger}] \to \hat{a}_{\pm}(\mathbf{q})$ $[\hat{a}_{\pm}^{\dagger}(\mathbf{q})]$ for the mode annihilation [creation] operators. All these operators commute mutually, unless the two operators are annihilation and creation operators belonging to the same $\mathbf{q}$-mode and helicity. With the discrete helicity indices denoted $s = +$ or $-$, and $s' = +$ or $-$, the essential commutators are given by

$$[\hat{a}_{\mathbf{q}s}, \hat{a}_{\mathbf{q}'s'}^{\dagger}] = \delta_{\mathbf{q}\mathbf{q}'}\delta_{ss'} \tag{3.6}$$

in the discrete case [$\delta_{\mathbf{q}\mathbf{q}'}$ and $\delta_{ss'}$ being 3D and 1D Krönicker delta's], and

$$[\hat{a}_{s}(\mathbf{q}), \hat{a}_{s'}^{\dagger}(\mathbf{q}')] = \delta(\mathbf{q} - \mathbf{q}')\delta_{ss'}, \tag{3.7}$$

where $\delta(\mathbf{q} - \mathbf{q}')$ is the 3D Dirac delta function. In structural vacuum

$$e^{iq_{\mu}x^{\mu}} = \exp[i(\mathbf{q} \cdot \mathbf{r} - cqt)], \tag{3.8}$$

since $\omega = cq$.

In free-space relativistic quantum theory as such, the solution types

$$\exp[i(\mathbf{q} \cdot \mathbf{r} - \omega t)] \text{ and } \exp[i(\omega t - \mathbf{q} \cdot \mathbf{r})], \tag{3.9}$$

describe different physical entities, viz., particle $[\sim e^{-i\omega t}]$ and antiparticle $[\sim e^{i\omega t}]$. Photons and antiphotons are, as far as we know, identical, and therefore, the information carried by negative [superscript (-)]- and positive [superscript (+)]-frequency solutions is the same.

The Hamiltonian photon wave equations for the two helicity species [Eqs. (2.40) and (2.41)] are replaced by the operator relations

$$i\frac{\partial}{\partial t}\hat{\mathbf{A}}_{T,\pm}^{(+)}(\mathbf{r}, t) = \pm c\boldsymbol{\nabla} \times \hat{\mathbf{A}}_{T,\pm}^{(+)}(\mathbf{r}, t). \tag{3.10}$$

I urge the reader to verify, using Eqs. (3.3) and (2.27), that Eq.(3.10) is correct.

3.2 COMMUTATOR RELATION BETWEEN FIELD COORDINATE AND CONJUGATE MOMENTUM

In the Lagrangian approach to the transverse gauge photon dynamics, the classical field (subscript F) Lagrangian density is given by [see Chapter 12, in which a brief account of the Lagrangian approach to electrodynamics is presented]

$$\mathcal{L}_F = \frac{\varepsilon_0}{2}\left[(\dot{\mathbf{A}}_T(\mathbf{r}, t))^2 - c^2(\boldsymbol{\nabla} \times \mathbf{A}_T(\mathbf{r}, t))^2\right], \tag{3.11}$$

where $\mathbf{A}_T(\mathbf{r}, t)$ is the generalized (vectorial) "coordinate", which depends on the continuous space ($\mathbf{r}$) and time (t) indices, and $\dot{\mathbf{A}}_T(\mathbf{r}, t) \equiv \partial \mathbf{A}_T(\mathbf{r}, t) \partial t$ is the associated "velocity". The conjugate transverse field momentum, $\mathbf{\Pi}_T(\mathbf{r}, t)$, defined by

$$\mathbf{\Pi}_T(\mathbf{r}, t) = \frac{\partial \mathcal{L}}{\partial \dot{\mathbf{A}}_T}, \tag{3.12}$$

thus becomes

$$\mathbf{\Pi}_T(\mathbf{r}, t) = \varepsilon_0 \dot{\mathbf{A}}_T(\mathbf{r}, t). \tag{3.13}$$

The extension to the second-quantized operator level is done by the replacements

$$\mathbf{A}_T(\mathbf{r}, t) \Rightarrow \hat{\mathbf{A}}_T(\mathbf{r}, t), \tag{3.14}$$

$$\mathbf{\Pi}_T(\mathbf{r}, t) \Rightarrow \hat{\mathbf{\Pi}}_T(\mathbf{r}, t) = \varepsilon_0 \frac{\partial \hat{\mathbf{A}}_T(\mathbf{r}, t)}{\partial t}. \tag{3.15}$$

From a fundamental point of view the free-field correlation effects associated with the coordinate-momentum commutator $[\hat{\mathbf{A}}_T(\mathbf{r}, t), \hat{\mathbf{\Pi}}_T(\mathbf{r}', t')]$ are of particular physical importance, because this commutator has a space-like part (nonvanishing in the front of the light cone). As we shall realize below this part stems from the inability to localize a transverse photon completely in space-time. The reader may find a detailed account of the theory of spatial photon localization in [153].

On the basis of the monochromatic plane-wave expansion for $\hat{\mathbf{A}}_T(\mathbf{r}, t)$ [e.g. in the helicity basis; Eqs. (3.2)–(3.4)] a somewhat lengthy calculation [63, 157, 158] gives the following result for the commutator associated with the two different space-time points $[(\mathbf{r}, t); (\mathbf{r}', t')]$:

$$[\hat{\mathbf{A}}_T(\mathbf{r}, t), \hat{\mathbf{\Pi}}_T(\mathbf{r}', t')] = \frac{\hbar}{4\pi i} \left((\mathbf{U} - \mathbf{e}_\mathbf{R}\mathbf{e}_\mathbf{R}) \frac{1}{R} [\delta'(R - c\tau) + \delta'(R + c\tau)] \right.$$

$$+ (3\mathbf{e}_\mathbf{R}\mathbf{e}_\mathbf{R} - \mathbf{U}) \left\{ \frac{1}{R^2} [\delta(R - c\tau) + \delta(R + c\tau)] \right.$$

$$\left. \left. \frac{1}{R^3} [\theta(c\tau - R) - \theta(c\tau + R)] \right\} \right), \tag{3.16}$$

where θ, δ and δ' are the Heaviside unit step function, the Dirac delta function and its first derivative [with respect to its argument], respectively. The translational invariance in space and time of the structural vacuum is the reason that the commutator only depends on the differences $\mathbf{R} = \mathbf{r} - \mathbf{r}'$ [through the magnitude $R = |\mathbf{R}|$ and the unit vector $\mathbf{e}_\mathbf{R} = \mathbf{R}/R$] and $\tau = t - t'$. Readers interested in the details of the derivation may consult e.g. Chapter 24 of [157] where a closely related commutator relation $[\hat{\mathbf{A}}_T(\mathbf{r}, t), \hat{\mathbf{E}}_T(\mathbf{r}', t')]$, with $\hat{\mathbf{E}}_T(\mathbf{r}', t') = -\partial \hat{\mathbf{A}}_T(\mathbf{r}', t')/\partial t'$, is derived.

It appears from Eq. (3.16) that for far ($\sim R^{-1}$)- and mid ($\sim R^{-2}$)- field separations between the space points, $\hat{\mathbf{A}}_T(\mathbf{r}.t)$ and $\hat{\mathbf{\Pi}}_T(\mathbf{r}', t')$ are correlated only on the forward ($R = c\tau$) and backward ($R = -c\tau$) light cones, as one would expect. However, it appears that the commutator for near ($\sim R^{-3}$)-field separations has a spacelike term given as

$$[\hat{\mathbf{A}}_T(\mathbf{r}, t), \hat{\mathbf{\Pi}}_T(\mathbf{r}', t')]_{\text{SPACELIKE}} = \frac{\hbar}{i} \frac{3\mathbf{e}_\mathbf{R}\mathbf{e}_\mathbf{R} - \mathbf{U}}{4\pi R^3} [\theta(c\tau - R) - \theta(c\tau + R)]. \tag{3.17}$$

The commutator in Eq. (3.17) is spacelike indeed, and furthermore independent of τ, since

$$\theta(c\tau - R) - \theta(c\tau + R) = \begin{cases} 0 & \text{for } c|\tau| > R \\ -1 & \text{for } c|\tau| < R \end{cases}. \tag{3.18}$$

By introduction of the explicit expression for the transverse delta function, $\boldsymbol{\delta}_T(\mathbf{R})$, in spherical contraction, viz.,

$$\boldsymbol{\delta}_T(\mathbf{R}) = \frac{2}{3}\mathbf{U}\delta(\mathbf{R}) + \frac{3\mathbf{e_R e_R} - \mathbf{U}}{4\pi R^3}, \tag{3.19}$$

the space-like part of the coordinate-momentum commutator can be written in the compact form

$$[\hat{\mathbf{A}}_T(\mathbf{r}, t), \hat{\mathbf{\Pi}}_T(\mathbf{r}', t')]_{\text{SPACELIKE}} = i\hbar\boldsymbol{\delta}_T(\mathbf{R})\theta\left(\frac{R}{c} - |\tau|\right), \mathbf{R} \neq \mathbf{0}. \tag{3.20}$$

The result in Eq. (3.20) reduces to the equal time ($\tau = 0$) commutator $[\hat{\mathbf{A}}_T(\mathbf{r}, t), \hat{\mathbf{\Pi}}_T(\mathbf{r}', t)]$, examined in some detail in [183]. Of course, it is also the result one would obtain in the Schrödinger Picture, where the operators are time independent, $[\hat{\mathbf{A}}_T(\mathbf{r}), \hat{\mathbf{\Pi}}_T(\mathbf{r}')]$. The fact that the spacelike commutator essentially is independent of τ, but depends on $\mathbf{R}$, indicates that the correlation in Eq. (3.20), existing only for near-field separations of $\hat{\mathbf{A}}_T$ and $\hat{\mathbf{\Pi}}_T$, cannot originate in free-space propagation effects. Such effects only link certain field components on the light cone, cf. Eq. (3.16).

Ultimatively, all free-field correlation phenomena originate in the correlations existing in the massive particle system (charged particle source) emitting the transverse gauge photons. Readers familiar with classical and/or quantum-field diffraction theory will know that light diffraction studies (say from two holes in a screen) can be used to obtain information on electronic correlations in the light source [124, 125, 182, 185, 235].

In Section 3.4 we shall see how a particular (and quite extraordinary) form of electronic source correlation can account for the result in Eq. (3.17).

3.3 TRANSVERSE PHOTON CORRELATIONS. FEYNMAN T-PHOTON GREEN TENSOR

Let us now take a look at the light-cone and spacelike correlations, appearing in the coordinate ($\hat{\mathbf{A}}_T(x)$)- momentum ($\mathbf{\Pi}_T(x')$) commutation relation [Eqs. (3.16) and (3.17), from a somewhat different perspective.

In my book *Light - The Physics of the Photon* [158] the space-time localization of photons was studied in some detail, and it was realized that it is meaningful to consider the quantity

$$|\mathcal{R}(\mathbf{r}, t)\rangle = \left(\frac{2\varepsilon_0 c}{\hbar}\right)^{1/2} \hat{\mathbf{A}}_T^{(-)}(\mathbf{r}, t)|0\rangle \tag{3.21}$$

as a kind of mean position state for transverse gauge photons in Hilbert space. The state $|0\rangle$ denotes the global vacuum state of T-photons in structural vacuum.

The correlation between mean position states associated with the space-time points $x = (\mathbf{r}, ct)$ and $x' = (\mathbf{r}', ct')$ may be characterized by the overlap matrix

$$\frac{\hbar}{2\varepsilon_0 c} \langle \mathcal{R}(x)|\mathcal{R}(x')\rangle = \langle 0|\, \hat{\mathbf{A}}_T^{(+)}(x)\hat{\mathbf{A}}_T^{(-)}(x')\, |0\rangle. \tag{3.22}$$

Each of the two operators $\hat{\mathbf{A}}_T^{(+)}$ and $\hat{\mathbf{A}}_T^{(-)}$ involves a summation over the helicities; see Eq. (3.2). Since part of the position correlation relates to light-cone propagation, it is useful to study the properties of the time-ordered position correlation. For this purpose, one introduces the correlation matrix

$$\mathbf{T}(x, x') = \langle 0|\, T\{\hat{\mathbf{A}}_T(x)\hat{\mathbf{A}}_T(x')\}\, |0\rangle, \tag{3.23}$$

where T is the time-ordering symbol, i.e.,

$$T\{\hat{\mathbf{A}}_T(x)\hat{\mathbf{A}}_T(x')\} = \Theta(t - t')\hat{\mathbf{A}}_T(x)\hat{\mathbf{A}}_T(x') + \Theta(t' - t)\hat{\mathbf{A}}_T(x')\hat{\mathbf{A}}_T(x). \tag{3.24}$$

Since

$$\hat{\mathbf{A}}_T(x)\, |0\rangle = (\hat{\mathbf{A}}_T^{(+)}(x) + \hat{\mathbf{A}}_T^{(-)}(x))\, |0\rangle = \hat{\mathbf{A}}_T^{(-)}(x)\, |0\rangle, \tag{3.25}$$

one obtains

$$\mathbf{T}(x, x') = \Theta(t - t')\mathbf{T}_>(x, x') + \Theta(t' - t)\mathbf{T}_<(x, x') \tag{3.26}$$

where

$$\mathbf{T}_>(x, x') = \langle 0|\, \hat{\mathbf{A}}_T^{(+)}(x)\hat{\mathbf{A}}_T^{(-)}(x')\, |0\rangle, \tag{3.27}$$

and

$$\mathbf{T}_<(x, x') = \langle 0|\, \hat{\mathbf{A}}_T^{(+)}(x')\hat{\mathbf{A}}_T^{(-)}(x)\, |0\rangle. \tag{3.28}$$

Because $\hat{\mathbf{A}}_T^{(+)}(x)\, |0\rangle = 0$, it is obvious that $\mathbf{T}_>(x, x')$ is the mean value of the commutator $[\hat{\mathbf{A}}_T^{(+)}(x), \hat{\mathbf{A}}_T^{(-)}(x')]$ in the vacuum state:

$$\mathbf{T}_>(x, x') = \langle 0|\, [\hat{\mathbf{A}}_T^{(+)}(x), \hat{\mathbf{A}}_T^{(-)}(x')]\, |0\rangle. \tag{3.29}$$

It can be shown [158] that the commutator in Eq. (3.29), is given by

$$[\hat{\mathbf{A}}_T^{(+)}(x), \hat{\mathbf{A}}_T^{(-)}(x')] = \frac{\hbar}{i\varepsilon_0 c^2}\mathbf{G}_T(\mathbf{R}, \tau), \tag{3.30}$$

where

$$\mathbf{G}_T(\mathbf{R}, \tau) = \frac{1}{4\pi R}\delta\left(\frac{R}{c} - \tau\right)(\mathbf{U} - \mathbf{e_R}\mathbf{e_R}) - \frac{c^2\tau}{4\pi R^3}\Theta(\tau)\Theta\left(\frac{R}{c} - \tau\right)(\mathbf{U} - 3\mathbf{e_R}\mathbf{e_R}) \tag{3.31}$$

is the transverse photon "propagator". The quotation mark refers to the circumstance that $\mathbf{G}_T(\mathbf{R}, \tau)$ besides a far-field ($\approx R^{-1}$) term, non-vanishing only on the light cone, has a space-like term different from zero for the near-field ($\approx R^{-3}$) separations of $\mathbf{r}$ and $\mathbf{r}'$, yet only in front of the light cone. The commutator in Eq. (3.30) is a c-number, $\langle 0| [..,..] |0\rangle = [..,..] \langle 0|0\rangle = [..,..]$. Hence,

$$\mathbf{T}_>(x, x') = \frac{\hbar}{i\varepsilon_0 c^2} \mathbf{G}_T(\mathbf{R}, \tau). \tag{3.32}$$

The transverse part of the Feynman photon propagator is $\mathbf{G}_T(\mathbf{R}, \tau)/c$; see [158]. In front of the forward light cone ($R > c\tau$), the spatial part of $\mathbf{G}_T(\mathbf{R}, \tau)$ is given by

$$[\mathbf{G}_T(\mathbf{R}, \tau)]_{\text{SPACELIKE}} = c^2 \tau \boldsymbol{\delta}_T(\mathbf{R}), \tag{3.33}$$

and hence

$$[\mathbf{T}_>(x, x')]_{\text{SPACELIKE}} = \frac{\hbar \tau}{i\varepsilon_0} \boldsymbol{\delta}_T(\mathbf{R}). \tag{3.34}$$

A comparison of Eqs. (3.20) and (3.34) shows that

$$[\hat{\mathbf{A}}_T(\mathbf{r}, t), \hat{\boldsymbol{\Pi}}_T(\mathbf{r}', t')] = -\varepsilon_0 \frac{\partial}{\partial t} [\mathbf{T}_>(x, x')]_{\text{SPACELIKE}}. \tag{3.35}$$

Since the transverse propagator essentially describes the T-photon emission from a charged-particle source current density, the spacelike correlation in Eq. (3.35) must have their origin in some kind of electronic correlation in the source. As we shall realize in Section 3.4, this correlation is of fundamental nature.

A comparison of Eqs. (3.27) and (3.28) makes it clear that the expression for $\mathbf{T}_<(x, x')$ can be obtained from Eq. (3.32) using replacements $\mathbf{R} \to -\mathbf{R}$ and $\tau \to -\tau$. Hence,

$$\mathbf{T}_<(x, x') = \frac{\hbar}{i\varepsilon_0 c^2} \mathbf{G}_T(-\mathbf{R}, -\tau). \tag{3.36}$$

Altogether, it appears that the time-ordered correlation matrix is given by

$$\langle 0| T\{\hat{\mathbf{A}}_T(x)\hat{\mathbf{A}}_T(x')\} |0\rangle = \frac{\hbar}{i\varepsilon_0 c^2} [\Theta(t - t')\mathbf{G}_T(\mathbf{R}, \tau) + \Theta(t' - t)\mathbf{G}_T(-\mathbf{R}, -\tau)]. \tag{3.37}$$

The position correlation in the x and x' space-time points hence relates directly to the transverse photon propagator. For far-field separation of x and x' the correlation is limited to the light cone, but for near-field separation an off-light-cone space-like correlation is present. In compact form, this correlation is given by

$$\langle 0| T\{\hat{\mathbf{A}}_T(x)\hat{\mathbf{A}}_T(x')\} |0\rangle_{\text{SPACELIKE}} = \frac{i\hbar|\tau|}{4\pi\varepsilon_0 R^3} \theta\left(\frac{R}{c} - |\tau|\right) (\mathbf{U} - 3\mathbf{e_R}\mathbf{e_R}), \tag{3.38}$$

utilizing that $\mathbf{e}_{-\mathbf{R}}\mathbf{e}_{-\mathbf{R}} = \mathbf{e}_{\mathbf{R}}\mathbf{e}_{\mathbf{R}}$. In order to underline the fact that the off-light-cone correlation has no time-like part a factor $\Theta\left(\frac{R}{c} - |\tau|\right)$ appears in Eq. (3.38). In terms of the transverse delta function

$$\langle 0| \, T\{\hat{\mathbf{A}}_T(x)\hat{\mathbf{A}}_T(x')\} \, |0\rangle_{\text{SPACELIKE}} = \frac{i\hbar|\tau|}{i\varepsilon_0}\boldsymbol{\delta}_T(\mathbf{R})\Theta\left(\frac{R}{c} - |\tau|\right), \qquad (3.39)$$

a form which shows the close relation between the space-like structure of the coordinate-momentum commutator [Eq. (3.20)] and the time-ordered mean position correlation matrix [Eq. (3.23)].

3.4 SOURCE CORRELATIONS ASSOCIATED WITH TRANSVERSE GAUGE PHOTON EMISSION

In order to launch transverse photons in a structural vacuum one must introduce a source of charged massive particles in a region of space-time. We describe the source by a current density distribution $\mathbf{J}(\mathbf{r}, t)$. For what follows it is not necessary to present any explicit expression for $\mathbf{J}(\mathbf{r}, t)$. However, it is important to understand that the T-photon emission from the source effectively may be considered as driven by the transverse part of $\mathbf{J}(\mathbf{r}, t)$, i.e.,

$$\mathbf{J}_T(\mathbf{r}, t) = \int_{-\infty}^{\infty} \boldsymbol{\delta}_T(\mathbf{r} - \mathbf{r}') \cdot \mathbf{J}(\mathbf{r}, t)\mathrm{d}^3r', \qquad (3.40)$$

$\boldsymbol{\delta}_T(\mathbf{r} - \mathbf{r}')$ being the transverse delta function, given in spherical contraction in Eq. (3.19). Although, Eq. (3.40) relates $\mathbf{J}$ and $\mathbf{J}_T$ in a spatially nonlocal manner, there is no propagation effects associated with the relation; it is the same time (t) entering the two current densities.

In classical electrodynamics the wave equation for the transverse part of the vector potential has the form

$$\Box\mathbf{A}_T(\mathbf{r}, t) = -\mu_0\mathbf{J}_T(\mathbf{r}, t). \qquad (3.41)$$

The inhomogeneous solution to Eq. (3.41), which gives the source field emission, can be presented in a physically appealing manner by means of the Huygens scalar propagator

$$g(R, \tau) = \frac{1}{4\pi R}\delta\left(\frac{R}{c} - \tau\right). \qquad (3.42)$$

Thus,

$$\mathbf{A}_T(\mathbf{r}, t) = \mu_0 \int_{-\infty}^{\infty} g(R, \tau)\mathbf{J}_T(\mathbf{r}', t')\mathrm{d}^3r'\mathrm{d}t', \qquad (3.43)$$

with $R = |\mathbf{r} - \mathbf{r}'|$ and $\tau = t - t'$. The presence of the Dirac delta function $\delta(R/c - \tau)$ shows that the transverse vector potential in the space-time "observation" point, $(\mathbf{r}, t)$, is synthesized by Einsteinian retarded contributions from the various source points, $(\mathbf{r}', t)$, in the given $\mathbf{J}_T(\mathbf{r}', t')$-distribution. The manner in which the self-consistent current density is created in the first place will be discussed later. In the description offered by Eq. (3.43) only light-cone couplings relate $\mathbf{J}_T(\mathbf{r}', t')$ and $\mathbf{A}_T(\mathbf{r}, t)$.

One can relate the transverse vector potential to the current density itself by inserting Eq. (3.40) into Eq. (3.43). A perhaps amazing result then follows, viz.,

$$\mathbf{A}_T(\mathbf{r}, t) = \mu_0 \int_{-\infty}^{\infty} \mathbf{G}_T(\mathbf{R}, \tau) \cdot \mathbf{J}(\mathbf{r}', t') \mathrm{d}^3 r' \mathrm{d}t', \tag{3.44}$$

where

$$\mathbf{G}_T(\mathbf{R}, \tau) = \int_{-\infty}^{\infty} g(|\mathbf{r} - \mathbf{r}''|, \tau) \boldsymbol{\delta}_T(\mathbf{r}'' - \mathbf{r}') \mathrm{d}^3 r''. \tag{3.45}$$

As indicated by the notation in Eq. (3.45) $\mathbf{G}_T(\mathbf{R}, \tau)$, is just the transverse part of the Feynman photon propagator (multiplied by c), given in Eq. (3.31). By a change of the integration variable in Eq. (3.45) from $\mathbf{r}''$ to $\mathbf{R}_0 = \mathbf{r}'' - \mathbf{r}'$, one obtains

$$\mathbf{G}_T(\mathbf{R}, \tau) = \int_{-\infty}^{\infty} g(|\mathbf{R} - \mathbf{R}_0|, \tau) \boldsymbol{\delta}_T(\mathbf{R}_0) \mathrm{d}^3 R_0, \tag{3.46}$$

where $\mathbf{R} = \mathbf{r} - \mathbf{r}'$. Written in this form, it is manifest that $\mathbf{G}_T(\mathbf{R}, \tau)$ is a function of the space-time separation of $\mathbf{J}(\mathbf{r}, t)$ and $\mathbf{A}_T(\mathbf{r}, t)$, only. A derivation of Eq. (3.45) is given in [147].

The results in Eqs. (3.43) and (3.44) are from a physical point of view identical, although the perspectives appear different: The truncation of the source domain from $\mathbf{J}_T(\mathbf{r}, t)$ to $\mathbf{J}(\mathbf{r}, t)$, by means of the only spatially nonlocal transverse delta function, necessarily implies that an off-light cone near-field term must appear in the $\mathbf{G}_T$-propagator. It is known that spacelike couplings arise from our inability to localize a T-photon completely in space at any fixed time, as mentioned before.

3.5   LAGRANGIAN AND HAMILTONIAN APPROACHES

The dynamics of the transverse gauge photons can be described starting from the Lagrangian formalism summarized in the following, Since we are not interested in the dynamics of the massive particles in this subsection, we can omit the particle part of the Lagrangian from the analysis. Hence, the starting point is the classical Lagrangian

$$L = \int_{-\infty}^{\infty} \mathcal{L}_{F+I} \mathrm{d}^3 r, \tag{3.47}$$

with density

$$\mathcal{L}_{F+I} = \mathcal{L}_F + \mathcal{L}_I = \frac{\varepsilon_0}{2} [(\dot{\mathbf{A}}_T(\mathbf{r}, t))^2 - c^2 (\boldsymbol{\nabla} \times \mathbf{A}_T(\mathbf{r}, t))^2] + \mathbf{J}_T(\mathbf{r}, t) \cdot \mathbf{A}_T(\mathbf{r}, t). \tag{3.48}$$

In the interaction Lagrangian density

$$\mathcal{L}_I = \mathbf{J}_T(\mathbf{r}, t) \cdot \mathbf{A}_T(\mathbf{r}, t), \tag{3.49}$$

if wished, one can replace $\mathbf{J}_T(\mathbf{r}, t)$ with $\mathbf{J}(\mathbf{r}, t) = \mathbf{J}_T(\mathbf{r}, t) + \mathbf{J}_L(\mathbf{r}, t)$ since the integral

$$\int_{-\infty}^{\infty} \mathbf{J}_L(\mathbf{r}, t) \cdot \mathbf{A}_T(\mathbf{r}, t) \mathrm{d}^3 r = 0 \tag{3.50}$$

due to the orthogonality of longitudinal (L) and transverse (T) vector fields. The reader may prove Eq. (3.50) transferring the integral to an integral over reciprocal space, where $\mathbf{J}_L$ and $\mathbf{A}_T$ are orthogonal in the geometrical sense.

The Lagrange equation for the three Cartesian components ($i = x, y, z$) of the transverse vector potential are [63]

$$\frac{d}{dt}\left(\frac{\partial \mathcal{L}_{F+I}}{\partial \dot{A}_{T,i}}\right) = \frac{\partial \mathcal{L}_{F+I}}{\partial A_{T,i}} - \sum_{j=x,y,z} \partial_j \left[\frac{\partial \mathcal{L}_{F+I}}{\partial(\partial_j A_{T,i})}\right]. \tag{3.51}$$

For the x-component one obtains

$$\varepsilon_0 \ddot{A}_{T,x} = J_{T,x} + \varepsilon_0 c^2 \nabla^2 A_{T,x}, \tag{3.52}$$

as the reader may show, remembering the transversality of the vector potential, $\sum_j \partial_j A_{T,j} = 0$. Form-identical equations hold for the y- and z-components of $\mathbf{A}_T$ of course. Altogether, the three Lagrange equations lead to Eq. (3.41) $[\mu_0 = (\varepsilon_0 c^2)^{-1}]$.

Extended to the quantum field operator level the Lagrange formalism for the dynamical variable $\hat{\mathbf{A}}_T$ thus gives the dynamical equation

$$\Box \hat{\mathbf{A}}_T(\mathbf{r}, t) = -\mu_0 \hat{\mathbf{J}}_T(\mathbf{r}, t), \tag{3.53}$$

with

$$\boldsymbol{\nabla} \cdot \hat{\mathbf{A}}_T(\mathbf{r}, t) = 0, \tag{3.54}$$

and $\boldsymbol{\nabla} \cdot \hat{\mathbf{J}}_T(\mathbf{r}, t) = 0$.

The classical Hamiltonian

$$H = \int_{\infty}^{\infty} \mathcal{H}_{F+I} \mathrm{d}^3 r \tag{3.55}$$

has density

$$\mathcal{H}_{F+I} = \boldsymbol{\Pi}_T(\mathbf{r}, t) \cdot \dot{\mathbf{A}}_T(\mathbf{r}, t) - \mathcal{L}_{F+I}, \tag{3.56}$$

with the conjugate transverse field momentum given by Eq. (3.13). The introduction of the Hamiltonian density

$$\mathcal{H}_{F+I} = \frac{\varepsilon_0}{2}\left[\left(\frac{\boldsymbol{\Pi}_T(\mathbf{r}, t)}{\varepsilon_0}\right)^2 + c^2(\boldsymbol{\nabla} \times \mathbf{A}_T(\mathbf{r}, t))^2\right] - \mathbf{J}_T(\mathbf{r}, t) \cdot \mathbf{A}_T(\mathbf{r}, t), \tag{3.57}$$

with the transverse variables $\mathbf{A}_T(\mathbf{r}, t)$ and $\boldsymbol{\Pi}_T(\mathbf{r}, t)$, gives (using the standard method) the following set of classical equations for the Cartesian components ($i = x, y, x$):

$$\dot{A}_{T,i} = \frac{\partial \mathcal{H}_{F+I}}{\partial \Pi_{T,i}} \tag{3.58}$$

$$\dot{\Pi}_{T,i} = -\frac{\partial \mathcal{H}_{F+I}}{\partial A_{T,i}} + \sum_{j=x,y,z} \partial_j \left[\frac{\partial \mathcal{H}_{F+I}}{\partial(\partial_j A_{T,i})} \right]. \tag{3.59}$$

Nothing new is obtained from the Hamiltonian equations of motion. Thus, the three in each of Eqs. (3.58) and (3.59) lead to Eqs. (3.13) and (3.41), respectively.

Without specifying any explicit formula for the transverse current density, it appears that the Hamiltonian operator associated with the transverse gauge photon field ($\hat{H}_F$) plus this field's interaction with the massive particle system ($\hat{H}_I$) is given by the following expression in the Heisenberg picture:

$$\hat{H} = \hat{H}_F + \hat{H}_I = \frac{\varepsilon_0}{2} \int_{-\infty}^{\infty} \left[\left(\frac{\hat{\mathbf{\Pi}}_T(\mathbf{r},t)}{\varepsilon_0} \right)^2 + c^2 (\nabla \times \hat{\mathbf{A}}_T(\mathbf{r},t))^2 \right] \mathrm{d}^3 r$$

$$- \int_{-\infty}^{\infty} \hat{\mathbf{J}}_T(\mathbf{r},t) \cdot \hat{\mathbf{A}}_T(\mathbf{r},t) \mathrm{d}^3 r. \tag{3.60}$$

One of the advantages of the Heisenberg point of view is that this allows one to discuss easily the analogous and differences between the classical (CED) and quantum (QED) electrodynamic theories.

3.6 HEISENBERG OPERATOR EQUATIONS FOR TRANSVERSE GAUGE PHOTON DYNAMICS

In the Heisenberg Picture the state vector of the global system under study remains fixed in time, whereas the operator (not only observables) $\hat{O}(t)$ of the system evolve according to the Heisenberg equation [63, 186]

$$i\hbar\frac{d\hat{O}(t)}{dt} = [\hat{O}(t), \hat{H}(t)], \tag{3.61}$$

where $\hat{H}(t)$ is the relevant Hamilton operator.

In relation to our investigation of the dynamics of transverse gauge photons $\hat{H}(t)$ is given by Eq. (3.60). In the absence of the particle part of the Hamiltonian, $\hat{H}(t)$ is a global quantity, independent of $\mathbf{r}$, whereas the transverse position ($\hat{\mathbf{A}}_T(\mathbf{r},t)$)- and momentum ($\boldsymbol{\pi}_T(\mathbf{r},t)$)-observables are examples of local ($\mathbf{r}$-dependent) quantities. With the omission of the particle Hamiltonian, Heisenberg equations for particle operators cannot be set up.

In the presence of field-matter interaction the Hamiltonian operator relation in Eq. (3.10) must be generalized since the number of transverse photons is no longer a conserved quantity. In order to determine this generalization, it is convenient to start from the Heisenberg equations for the annihilation operator $\hat{a}_i(t)$. With reference to discrete mode quantization, the index i designates the set $(\mathbf{q}_i, \boldsymbol{\varepsilon}_i)$, $\boldsymbol{\varepsilon}_i$ being the relevant polarization unit vector belonging to the mode wave vector $\mathbf{q}_i$. The equation of motion for $\hat{a}_i(t)$, viz.,

$$i\hbar\frac{d\hat{a}_i(t)}{dt} = [\hat{a}_i(t), \hat{H}_F(t) + \hat{H}_I(t)] \tag{3.62}$$

is useful in the derivation of the Maxwell equations between the field operators; see e.g. [63] and the derivation in Section 4.2.

The operator associated with the transverse vector potential at each point $\mathbf{r}$ of space thus is given by

$$\hat{\mathbf{A}}_T(\mathbf{r}, t) = \left(\frac{\hbar}{2\varepsilon_0 L^3}\right)^{1/2} \sum_i \frac{1}{\omega_i^{1/2}} [\hat{a}_i(t)\boldsymbol{\varepsilon}_i e^{i\mathbf{q}\cdot\mathbf{r}} + \hat{a}_i^\dagger(t)\boldsymbol{\varepsilon}_i^* e^{-i\mathbf{q}\cdot\mathbf{r}}], \qquad (3.63)$$

where $\omega_i = cq_i$. The destruction and creation operators $\hat{a}_i(t)$ and $\hat{a}_i^\dagger(t)$ satisfy the equal-time commutation relations

$$[\hat{a}_i(t), \hat{a}_j(t)] = [\hat{a}_i^\dagger(t), \hat{a}_j^\dagger(t)] = 0, \qquad (3.64)$$

$$[\hat{a}_i(t), \hat{a}_j^\dagger(t)] = \delta_{ij}, \qquad (3.65)$$

cf. Eq. (3.6).

By means of Eqs. (3.63)–(3.65), the field part of the Hamilton operator in Eq. (3.61) can be written as a sum of contributions from a set of fictitious harmonic oscillators, i.e.,

$$\begin{aligned}
\hat{H}_F(t) &= \frac{\varepsilon_0}{2} \int_{L^3} \left[\left(\frac{\hat{\boldsymbol{\Pi}}_T(\mathbf{r}, t)}{\varepsilon_0}\right)^2 + c^2 (\boldsymbol{\nabla} \times \hat{\mathbf{A}}_T(\mathbf{r}, t))^2 \right] \mathrm{d}^3 r \\
&= \sum_i \hbar\omega_i \left(\hat{a}_i^\dagger(t)\hat{a}_i(t) + \frac{1}{2} \right),
\end{aligned} \qquad (3.66)$$

a well-known result in elementary quantum optics [182, 183, 185, 230]. In turn, one obtains

$$[\hat{a}_i(t), \hat{H}_F(t)] = \hbar\omega_i \hat{a}_i(t). \qquad (3.67)$$

Since the particle and field dynamical variables commute mutually, one gets

$$-[\hat{a}_i(t), \hat{H}_I(t)] = \int_{L^3} \hat{\mathbf{J}}_T(\mathbf{r}, t) \cdot [\hat{a}_i(t), \hat{\mathbf{A}}_T(\mathbf{r}, t)] \mathrm{d}^3 r. \qquad (3.68)$$

In later chapters we shall study the Maxwell field's interaction with Dirac (half-integer spin) and Klein-Gordon (integer spin) fields. By means of the mode expansion for $\hat{\mathbf{A}}_T(\mathbf{r}, t)$ [Eq. 3.63] with running index j] we obtain

$$[\hat{a}_i(t), \hat{\mathbf{A}}_T(\mathbf{r}, t)] = \left(\frac{\hbar}{2\varepsilon_0 L^3}\right)^{1/2} \frac{1}{\omega_i^{1/2}} \boldsymbol{\varepsilon}_i^* e^{-i\mathbf{q}_i\cdot\mathbf{r}}, \qquad (3.69)$$

and then

$$-[\hat{a}_i(t), \hat{H}_I(t)] = \left(\frac{\hbar}{2\varepsilon_0 L^3}\right) \frac{1}{\omega_i^{1/2}} \boldsymbol{\varepsilon}_i^* \cdot \left[\int_{L^3} \hat{\mathbf{J}}_T(\mathbf{r}, t) e^{-i\mathbf{q}_i\cdot\mathbf{r}} \mathrm{d}^3 r \right] \qquad (3.70)$$

Insertion of Eqs. (3.67) and (3.70) into Eq. (3.62) gives for $\hat{a}_i(t)$ the following Heisenberg equations

$$\frac{d\hat{a}_i(t)}{dt} + i\omega_i\hat{a}_i(t) = \frac{i}{(2\varepsilon_0\hbar\omega_i L^3)^{1/2}}\boldsymbol{\varepsilon}_i^* \cdot \left[\int_{L^3}\hat{\mathbf{J}}_T(\mathbf{r},t)e^{-i\mathbf{q}_i\cdot\mathbf{r}}\mathrm{d}^3r\right]. \tag{3.71}$$

In order to obtain the generalization of the operator relation in Eq. (3.10), one applies Eq. (3.71) to the two helicity species [$\boldsymbol{\varepsilon}_i \to \mathbf{e}_{\kappa\pm}, \hat{a}_i \to \hat{a}_{\mathbf{q}\pm}$, and $\omega_i \equiv \omega = cq$ for brevity]. Thus,

$$\frac{d}{dt}\hat{a}_{\mathbf{q}\pm}(t) + i\omega\hat{a}_{\mathbf{q}\pm}(t) = \frac{i}{(2\varepsilon_0\hbar\omega L^3)^{1/2}}\mathbf{e}_{\kappa\pm}^* \cdot \hat{\mathbf{J}}_T(\mathbf{q};t), \tag{3.72}$$

where

$$\hat{\mathbf{J}}_T(\mathbf{q};t) = \int_{L^3}\hat{\mathbf{J}}_T(\mathbf{r},t)e^{-i\mathbf{q}\cdot\mathbf{r}}\mathrm{d}^3r, \tag{3.73}$$

is the spatial Fourier transform of the transverse current density operator. Since

$$\hat{\mathbf{A}}_{T,\pm}^{(+)}(\mathbf{r},t) = \left(\frac{\hbar}{2\varepsilon_0 L^3}\right)^{1/2}\sum_{\mathbf{q}}\frac{1}{\omega^{1/2}}\mathbf{e}_{\kappa,\pm}\hat{a}_{\mathbf{q}\pm}(t)e^{i\mathbf{q}\cdot\mathbf{r}}, \tag{3.74}$$

the Heisenberg equation for $\hat{\mathbf{A}}_{T,\pm}^{(+)}(\mathbf{r},t)$ is obtained by multiplying Eq. (3.72) by the vectorial factor $[\hbar/(2\varepsilon_0 L^3)]^{1/2}\omega^{-1/2}\mathbf{e}_{\kappa\pm}\exp\{i\mathbf{q}\cdot\mathbf{r}\}$ and afterwards carrying out a summation over the $\mathbf{q}$-modes. From the first term on the left-hand side of Eq. (3.72) one gets

$$\left(\frac{\hbar}{2\varepsilon_0 L^3}\right)^{1/2}\sum_{\mathbf{q}}\frac{1}{\omega^{1/2}}\mathbf{e}_{\kappa\pm}\left(\frac{d}{dt}\hat{a}_{\mathbf{q}\pm}(t)\right)e^{i\mathbf{q}\cdot\mathbf{r}} = \frac{\partial}{\partial t}\mathbf{A}_{T,\pm}^{(+)}(\mathbf{r},t). \tag{3.75}$$

By utilizing that [Eq. (2.27)]

$$\mathbf{e}_{\kappa\pm} = \pm i\boldsymbol{\kappa} \times \mathbf{e}_{\kappa\pm}, \tag{3.76}$$

the second term gives

$$\left(\frac{\hbar}{2\varepsilon_0 L^3}\right)^{1/2}\sum_{\mathbf{q}}\frac{icq}{\omega^{1/2}}\hat{a}_{\mathbf{q}\pm}(t)\mathbf{e}_{\kappa\pm}e^{i\mathbf{q}\cdot\mathbf{r}} = \pm ic\left(\frac{\hbar}{2\varepsilon_0 L^3}\right)^{1/2}\sum_{\mathbf{q}}\frac{i\mathbf{q} \times \mathbf{e}_{\kappa\pm}}{\omega^{1/2}}\hat{a}_{\mathbf{q}\pm}(t)e^{i\mathbf{q}\cdot\mathbf{r}}$$

$$= \pm ic\boldsymbol{\nabla} \times \hat{\mathbf{A}}_{T,\pm}^{(+)}(\mathbf{r},t). \tag{3.77}$$

Finally, one obtains from the term on the right side of Eq. (3.72) the contribution

$$\left(\frac{\hbar}{2\varepsilon_0 L^3}\right)^{1/2}\sum_{\mathbf{q}}\frac{1}{\omega^{1/2}}\frac{i}{(2\varepsilon_0\hbar\omega L^3)^{1/2}}\mathbf{e}_{\kappa\pm}\mathbf{e}_{\kappa\pm}^* \cdot \hat{\mathbf{J}}_T(\mathbf{q};t)e^{i\mathbf{q}\cdot\mathbf{r}}$$

$$= \frac{i}{2\varepsilon_0 c}\frac{1}{L^3}\sum_{\mathbf{q}}\frac{1}{q}\mathbf{e}_{\kappa\pm}\mathbf{e}_{\kappa\pm}^* \cdot \hat{\mathbf{J}}_T(\mathbf{q};t)e^{i\mathbf{q}\cdot\mathbf{r}}$$

$$\to \frac{i}{2\varepsilon_0 c}\int_{-\infty}^{\infty}\mathbf{e}_{\pm}(\boldsymbol{\kappa})\mathbf{e}_{\pm}^*(\boldsymbol{\kappa}) \cdot \hat{\mathbf{J}}_T(\mathbf{q};t)e^{i\mathbf{q}\cdot\mathbf{r}}\frac{\mathrm{d}^3q}{(2\pi)^3 q}. \tag{3.78}$$

The last expression in Eq. (3.78) is the result of continuous mode quantization, and $\mathrm{d}^3q/[(2\pi)^3 q]$ is the relativistically invariant volume element [158].

By gathering the information in Eqs. (3.75), (3.77) and (3.78), it appears that the Hamiltonian T-photon wave equations satisfied by the operators $\hat{\mathbf{A}}_{T,\pm}^{(+)}(\mathbf{r}, t)$ in structural vacuum [Eq. (3.10)] must be replaced by the dynamical equations

$$i\frac{\partial}{\partial t}\hat{\mathbf{A}}_{T,\pm}^{(+)}(\mathbf{r}, t) \mp c\boldsymbol{\nabla} \times \hat{\mathbf{A}}_{T,\pm}^{(+)}(\mathbf{r}, t) = -\frac{1}{2\varepsilon_0 c}\int_{-\infty}^{\infty} \mathbf{e}_\pm(\boldsymbol{\kappa})\mathbf{e}_\pm^*(\boldsymbol{\kappa}) \cdot \hat{\mathbf{J}}_T(\mathbf{q}; t)e^{i\mathbf{q}\cdot\mathbf{r}}\frac{\mathrm{d}^3 q}{(2\pi)^3 q}$$

$$(3.79)$$

in the presence of field-matter interaction.

CHAPTER 4

Maxwell Operator Equations. Einstein Causality

4.1 FROM STRUCTURAL VACUUM TO PARTICLE VACUUM

Let us return to the operator associated with the transverse vector potential [Eq. (3.63)]. In the continuum description the plane-mode quantization of $\hat{\mathbf{A}}_T(\mathbf{r}, t)$ has the form

$$\hat{\mathbf{A}}_T(\mathbf{r}, t) = \left(\frac{\hbar}{2\varepsilon_0} \right)^{1/2} \sum_i \int_{-\infty}^{\infty} \frac{1}{\omega^{1/2}} [\hat{a}_i(\mathbf{q}; t)\boldsymbol{\varepsilon}_i(\boldsymbol{\kappa})e^{i\mathbf{q}\cdot\mathbf{r}} + h.c.] \frac{\mathrm{d}^3 q}{(2\pi)^{3/2}}, \qquad (4.1)$$

where $\omega = cq$, and $h.c.$ stands for Hermitian conjugate. The paradigmatic shift from structural vacuum to particle vacuum, and then to the transverse electromagnetic electric ($\hat{\mathbf{E}}_T(\mathbf{r}, t)$) and magnetic ($\mathbf{B}(\mathbf{r}, t)$) field operators, is based on the following prescriptions:

$$\hat{\mathbf{E}}_T(\mathbf{r}, t) = \left(\frac{\hbar}{2\varepsilon_0} \right)^{1/2} \sum_i \int_{-\infty}^{\infty} \omega^{1/2} [i\hat{a}_i(\mathbf{q}; t)\boldsymbol{\varepsilon}_i(\boldsymbol{\kappa})e^{i\mathbf{q}\cdot\mathbf{r}} + h.c.] \frac{\mathrm{d}^3 q}{(2\pi)^{3/2}}, \qquad (4.2)$$

and

$$c\hat{\mathbf{B}}_T(\mathbf{r}, t) = \left(\frac{\hbar}{2\varepsilon_0} \right)^{1/2} \sum_i \int_{-\infty}^{\infty} \omega^{1/2} [i\hat{a}_i(\mathbf{q}; t)\boldsymbol{\kappa} \times \boldsymbol{\varepsilon}_i(\boldsymbol{\kappa})e^{i\mathbf{q}\cdot\mathbf{r}} + h.c.] \frac{\mathrm{d}^3 q}{(2\pi)^{3/2}}, \qquad (4.3)$$

The prescription above relates to the Heisenberg Picture, and in the presence of field matter interaction, the equation of motion for the annihilation operator $[\hat{a}_i(\mathbf{q}; t)]$ is given by [63]

$$\frac{d}{dt}\hat{a}_i(\mathbf{q}; t) + i\omega\hat{a}_i(\mathbf{q}; t) = \frac{i}{(2\varepsilon_0\hbar\omega)^{1/2}} \frac{1}{(2\pi)^{3/2}} \boldsymbol{\varepsilon}_i^*(\boldsymbol{\kappa}) \cdot \hat{\mathbf{J}}_T(\mathbf{q}; t), \qquad (4.4)$$

where

$$\hat{\mathbf{J}}_T(\mathbf{q}; t) = \int_{-\infty}^{\infty} \hat{\mathbf{J}}_T(\mathbf{r}, t)e^{-i\mathbf{q}\cdot\mathbf{r}}\mathrm{d}^3 r \qquad (4.5)$$

DOI: 10.1201/9781003029458-4

is the spatial Fourier-integral transform of the transverse current density operator, $\hat{\mathbf{J}}_T(\mathbf{r}, t)$. As we have realized in Section 3.6, the equation of motion for the annihilation operator was derived from the Heisenberg equation given in Eq. (3.62). In passing, we note that the Heisenberg equation for the creation operator, $\hat{a}_i^\dagger(t)$, has the form

$$i\hbar \frac{d\hat{a}_i^\dagger(t)}{dt} = [\hat{a}_i^\dagger(t), \hat{H}_F(t) + \hat{H}_I(t)], \tag{4.6}$$

since the Hamilton operator is Hermitian. In Eq. (4.6) $\hat{a}_i^\dagger(t) \equiv \hat{a}_{\mathbf{q}s}^\dagger(t) \to \hat{a}_s^\dagger(\mathbf{q}; t)$ [and then the notional change $s \to i$].

In particle vacuum, where $d\hat{a}_i(\mathbf{q}; t)/dt = -i\omega\hat{a}_i(\mathbf{q}; t)$, the prescriptions in Eqs. (4.2) and (4.3) are in a sense *not* a prescription since these equations can be *derived* from Eq. (4.1) using the relations

$$\hat{\mathbf{E}}_T(\mathbf{r}, t) = -\frac{\partial}{\partial t}\hat{\mathbf{A}}_T(\mathbf{r}, t), \tag{4.7}$$

$$\hat{\mathbf{B}}(\mathbf{r}, t) = \nabla \times \hat{\mathbf{A}}_T(\mathbf{r}, t), \tag{4.8}$$

as the reader may prove to herself. Perhaps, this fact is not surprising due to the circumstance that the electromagnetic field $[\hat{\mathbf{E}}_T(\mathbf{r}, t), \hat{\mathbf{B}}(\mathbf{r}, t)]$ has its own autonomy in particle vacuum in quantum theory. In the presence of field-matter interaction, where the annihilation operator $\hat{a}_i(\mathbf{q}; t)$ satisfies Eq. (4.4), the expressions for the transverse electric [Eq. (4.2)] and magnetic [Eq. (4.3)] field operators are *genuine prescriptions* originating in the replacement of classical transverse field variables by quantum operators [[63]]. In the transverse Maxwellian operator theory, to be established and discussed below, *only* the operators $\hat{\mathbf{E}}_T(\mathbf{r}, t)$, $\hat{\mathbf{B}}(\mathbf{r}, t)$ and $\hat{\mathbf{J}}_T(\mathbf{r}, t)$ enter. Conclusion: In the potential description one uses the operator set $[\hat{\mathbf{A}}_T(\mathbf{r}, t), \hat{\mathbf{J}}_T(\mathbf{r}, t)]$, and in the Maxwellian formulation the set $[\hat{\mathbf{E}}_T(\mathbf{r}, t), \hat{\mathbf{B}}(\mathbf{r}, t), \hat{\mathbf{J}}_T(\mathbf{r}, t)]$ is employed. Transition between the two descriptions is done using first the classical field version of Eqs. (4.7) and (4.8) and then the quantization scheme.

4.2 HEISENBERG EQUATIONS FOR $\hat{\mathbf{E}}_T(\mathbf{R}, T)$, $\hat{\mathbf{B}}(\mathbf{R}, T)$

The set operators relations in Eqs. (4.7) and (4.8) has its roots in the standard potential formulation of classical electrodynamics [118], where

$$\mathbf{E}(\mathbf{r}, t) = -\frac{\partial}{\partial t}\mathbf{A}(\mathbf{r}, t) - \nabla\Phi(\mathbf{r}, t), \tag{4.9}$$

$$\mathbf{B}(\mathbf{r}, t) = \nabla \times \mathbf{A}(\mathbf{r}, t), \tag{4.10}$$

$\Phi(\mathbf{r}, t)$ being the scalar potential. By a division of the vector potential into its divergence-free (transverse), $\mathbf{A}_T(\mathbf{r}, t)$, and rotational-free (longitudinal), $\mathbf{A}_L(\mathbf{r}, t)$, parts, i.e.,

$$\mathbf{A}(\mathbf{r}, t) = \mathbf{A}_T(\mathbf{r}, t) + \mathbf{A}_L(\mathbf{r}, t), \tag{4.11}$$

with $\nabla \cdot \mathbf{A}_T(\mathbf{r}, t) = 0$, and $\nabla \times \mathbf{A}_L(\mathbf{r}, t) = \mathbf{0}$, the relations in Eqs. (4.9) and (4.10) are split into two sets. Thus, the set

$$\mathbf{E}_T(\mathbf{r}, t) = -\frac{\partial}{\partial t}\mathbf{A}_T(\mathbf{r}, t), \tag{4.12}$$

$$\mathbf{B}(\mathbf{r}, t) = \nabla \times \mathbf{A}_T(\mathbf{r}, t), \tag{4.13}$$

is the one entering transverse quantum electrodynamics upon an extension to the operator level, Eqs. (4.7) and (4.8). The longitudinal part of Eq. (4.9), viz.,

$$\mathbf{E}_L(\mathbf{r}, t) = -\frac{\partial}{\partial t}\mathbf{A}_L(\mathbf{r}, t) - \nabla\Phi(\mathbf{r}, t), \tag{4.14}$$

relates to the dynamics of the charged particles, and enters the field-coupled Dirac, Klein-Gordon, Pauli and Schrödinger equations in various equivalent forms. The quantities $\mathbf{A}_L(\mathbf{r}, t)$ and $\Phi(\mathbf{r}, t)$ are gauge dependent, c.f. the analysis in Section 2.4, and a definite gauge choice is made in accordance with what will be convenient in a specific application. Two extreme cases are often used: (i) The minimal coupling gauge, in which $\nabla\Phi(\mathbf{r}, t) = \mathbf{0}$ [briefly discussed in Section 2.3], and the Coulomb gauge, with $\partial\mathbf{A}_L)(\mathbf{r}, t)/\partial t = 0$. Outside the rim zone of matter the longitudinal part of the electric field vanishes, but Eq. (4.14) *always* is important in physics because it enters the source and detector problem for transverse photons [147, 153, 157, 158].

A combination of Eqs. (4.7) and (4.8) immediately gives the operator relation

$$\nabla \times \hat{\mathbf{E}}_T(\mathbf{r}, t) = -\frac{\partial}{\partial t}\hat{\mathbf{B}}(\mathbf{r}, t). \tag{4.15}$$

In a slightly different perspective, Eq. (4.15) appears as the Heisenberg equation for the magnetic field operator, i.e.,

$$i\hbar\frac{d}{dt}\hat{\mathbf{B}}(\mathbf{r}, t) = [\hat{\mathbf{B}}(\mathbf{r}, t), \hat{H}_F(t) + \hat{H}_I(t)] = [\hat{\mathbf{B}}(\mathbf{r}, t), \hat{H}_F(t)], \tag{4.16}$$

since $\hat{\mathbf{A}}_T(\mathbf{r}, t)$ and $\nabla \times \hat{\mathbf{A}}_T(\mathbf{r}, t)$ commute. The commutator $[\hat{\mathbf{B}}(\mathbf{r}, t), \hat{H}_F(t)]$ can be calculated by combining Eqs (3.67) and (4.3). The result is identical to the one obtained by a calculation of $(k/i)\nabla\times\hat{\mathbf{E}}_T(\mathbf{r}, t)$ from Eq. (4.2). The Heisenberg equation for $\hat{\mathbf{B}}(\mathbf{r}, t)$ hence leads to Eq. (4.15); a satisfactory finding.

Remembering that $\boldsymbol{\kappa} \times (\boldsymbol{\kappa} \times \boldsymbol{\varepsilon}_i = -\boldsymbol{\varepsilon}_i)$, it easily follows from Eqs. (4.2) and (4.3) that

$$\nabla \times \hat{\mathbf{B}}(\mathbf{r}, t) - \frac{1}{c^2}\frac{\partial\hat{\mathbf{E}}_T(\mathbf{r}, t)}{\partial t}$$
$$= \left(\frac{\hbar}{2\varepsilon_0}\right)^{1/2}\sum_i\int_{-\infty}^{\infty}\frac{\omega^{1/2}}{c^2}\left\{\frac{1}{i}\left[\frac{d}{dt}\hat{a}_i(\mathbf{q}; t) + i\omega\hat{a}_i(\mathbf{q}; t)\right]\boldsymbol{\varepsilon}_i(\boldsymbol{\kappa})e^{i\mathbf{q}\cdot\mathbf{r}} + h.c.\right\}\frac{d^3q}{(2\pi)^{3/2}}. \tag{4.17}$$

The quantity inside the square bracket of Eq. (4.17) is just the left-hand side of Eq. (4.4). By combining Eqs. (4.4) and (4.17), one obtains after a few algebraic calculations

$$\boldsymbol{\nabla} \times \hat{\mathbf{B}}(\mathbf{r}, t) - \frac{1}{c^2} \frac{\partial \hat{\mathbf{E}}_T(\mathbf{r}, t)}{\partial t}$$
$$= \frac{1}{2\varepsilon_0 c^2} \sum_i \int_{-\infty}^{\infty} [\boldsymbol{\varepsilon}_i^*(\boldsymbol{\kappa})\boldsymbol{\varepsilon}_i(\boldsymbol{\kappa}) \cdot \hat{\mathbf{J}}_T(\mathbf{q}; t)e^{i\mathbf{q}\cdot\mathbf{r}} + h.c.] \frac{\mathrm{d}^3 q}{(2\pi)^3} \tag{4.18}$$

The result can be simplified by utilizing the following dyadic expansion of the unit tensor

$$\mathbf{U} = \boldsymbol{\kappa}\boldsymbol{\kappa} + \sum_i \boldsymbol{\varepsilon}_i^*(\boldsymbol{\kappa})\boldsymbol{\varepsilon}_i(\boldsymbol{\kappa}). \tag{4.19}$$

Thus, since $\varepsilon_0 c^2 = \mu_0^{-1}$

$$\boldsymbol{\nabla} \times \hat{\mathbf{B}}(\mathbf{r}, t) - \frac{1}{c^2} \frac{\partial \hat{\mathbf{E}}_T(\mathbf{r}, t)}{\partial t}$$
$$= \frac{\mu_0}{2} \int_{-\infty}^{\infty} [(\mathbf{U} - \boldsymbol{\kappa}\boldsymbol{\kappa}) \cdot \hat{\mathbf{J}}_T(\mathbf{q}; t)e^{i\mathbf{q}\cdot\mathbf{r}} + h.c.] \frac{\mathrm{d}^3 q}{(2\pi)^3}$$
$$= \mu_0 \int_{-\infty}^{\infty} \frac{1}{2} [\hat{\mathbf{J}}_T(\mathbf{q}; t)e^{i\mathbf{q}\cdot\mathbf{r}} + \hat{\mathbf{J}}_T^{\dagger}(\mathbf{q}; t)e^{-i\mathbf{q}\cdot\mathbf{r}}] \frac{\mathrm{d}^3 q}{(2\pi)^3} \tag{4.20}$$

If the integration variable in the last term on the right side of Eq. (4.20) is changed from $\mathbf{q}$ to $-\mathbf{q}$, one finally gets

$$\boldsymbol{\nabla} \times \hat{\mathbf{B}}(\mathbf{r}, t) - \frac{1}{c^2} \frac{\partial \hat{\mathbf{E}}_T(\mathbf{r}, t)}{\partial t} = \mu_0 \hat{\mathbf{J}}_T^{\mathrm{sym}}(\mathbf{r}, t), \tag{4.21}$$

where

$$\hat{\mathbf{J}}_T^{\mathrm{sym}}(\mathbf{r}, t) = \frac{1}{2} \int_{-\infty}^{\infty} [\hat{\mathbf{J}}_T(\mathbf{q}; t) + \hat{\mathbf{J}}_T^{\dagger}(-\mathbf{q}; t)]e^{i\mathbf{q}\cdot\mathbf{r}} \frac{\mathrm{d}^3 q}{(2\pi)^3} \tag{4.22}$$

is the symmetrized (sym) transverse current density operator in the space-time domain. This operator is Hermitian, i.e.,

$$[\hat{\mathbf{J}}_T^{\mathrm{sym}}(\mathbf{r}, t)]^{\dagger} = \hat{\mathbf{J}}_T^{\mathrm{sym}}(\mathbf{r}, t), \tag{4.23}$$

and appears in Eq. (4.22) as a spatial Fourier integral of the operator $\frac{1}{2}[\hat{\mathbf{J}}_T(\mathbf{q}; t) + \hat{\mathbf{J}}_T^{\dagger}(-\mathbf{q}; t)]$. Since $\hat{\mathbf{B}}(\mathbf{r}, t)$ and $\hat{\mathbf{E}}_T(\mathbf{r}, t)$ *are* Hermitian operators as it appears from the prescription in Eqs. (4.2) and (4.3), it was clear from the outset that one had to end up with a Hermitian quantity on the right side of Eq. (4.21).

The result in Eq. (4.21) can also be obtained from the Heisenberg equation for the transverse electric field operator [147], viz.,

$$i\hbar \frac{d}{dt} \hat{\mathbf{E}}_T(\mathbf{r}, t) = [\hat{\mathbf{E}}_T(\mathbf{r}, t), \hat{H}_F(t) + \hat{H}_I(t)], \tag{4.24}$$

as the reader may prove, starting from the plane-mode expansion of $\hat{\mathbf{E}}_T(\mathbf{r}, t)$, and using thereafter the commutator relations in Eqs. (3.67) and (3.70) [in their continuum form].

Let us summarize the results obtained in this section as follows: The four operator equations

$$\boldsymbol{\nabla} \cdot \hat{\mathbf{B}}(\mathbf{r}, t) = 0, \tag{4.25}$$

$$\boldsymbol{\nabla} \cdot \hat{\mathbf{E}}_T(\mathbf{r}, t) = 0, \tag{4.26}$$

$$\boldsymbol{\nabla} \times \hat{\mathbf{E}}_T(\mathbf{r}, t) = \frac{\partial}{\partial t} \hat{\mathbf{B}}(\mathbf{r}, t), \tag{4.27}$$

$$\boldsymbol{\nabla} \times \hat{\mathbf{B}}(\mathbf{r}, t) = \mu_0 \hat{\mathbf{J}}_T^{\mathrm{sym}}(\mathbf{r}, t) + \frac{1}{c^2} \frac{\partial}{\partial t} \hat{\mathbf{E}}_T(\mathbf{r}, t), \tag{4.28}$$

constitute the so-called transverse set of Maxwell equations on operator form. The set is form-identical to the corresponding set of classical Maxwell equations, provided one remembers that quantum electrodynamics necessarily requires that one works with a symmetrized transverse current density operator. Eq. (4.25) originates in the transversality of the magnetic field in the absence of magnetic monopoles, Eq. (4.26) is the definition of the transversality concept, and Eqs. (4.27) and (4.28) are the equations of motion for $\hat{\mathbf{B}}(\mathbf{r}, t)$ and $\hat{\mathbf{E}}_T(\mathbf{r}, t)$ in the Heisenberg Picture.

4.3 PHOTON ENERGY WAVE FUNCTION FORMALISM

Let us now introduce the following quantum field operators

$$\hat{\boldsymbol{\mathcal{F}}}_{\pm}(\mathbf{r}, t) = \sqrt{\frac{\varepsilon_0}{2}} [\hat{\mathbf{E}}_T(\mathbf{r}, t) \pm ic\hat{\mathbf{B}}(\mathbf{r}, t)], \tag{4.29}$$

here written in the Heisenberg Picture. These operators are transverse, i.e.,

$$\boldsymbol{\nabla} \cdot \hat{\boldsymbol{\mathcal{F}}}_{\pm}(\mathbf{r}, t) = 0, \tag{4.30}$$

c.f. Eqs. (4.25) and (4.26). By combining Eqs. (4.27) and (4.28), two operator equations can be obtained, viz.,

$$i\frac{\partial}{\partial t} \hat{\boldsymbol{\mathcal{F}}}_{\pm}(\mathbf{r}, t) \mp c\boldsymbol{\nabla} \times \hat{\boldsymbol{\mathcal{F}}}_{\pm}(\mathbf{r}, t) = -\frac{i}{(2\varepsilon_0)^{1/2}} \hat{\mathbf{J}}_T^{\mathrm{sym}}(\mathbf{r}, t). \tag{4.31}$$

Eqs. (4.30) and (4.31) [in fact only either the + or - field operator is necessary] represent a compact (and complex) form of the set of transverse Maxwell equations in operator form.

In order to form the bridge to the formalism of quantum field radiation by classical sources, let us assume that the current density operator is a *prescribed* quantity. Strictly speaking $\hat{\mathbf{J}}_T^{\mathrm{sym}}(\mathbf{r}, t)$ is a function of the particle variables forming the source. These themselves evolve under the influence of the forces exerted on them by the prevailing fields; c.f. the minimum coupling principle discussed in Section 2.3. In the Heisenberg equation of motion for $\hat{a}_i(\mathbf{q}; t)$ [Eq. (4.4)] the current density operator

$\hat{\mathbf{J}}_T(\mathbf{q};t)$ hence also couples the evolution equation for $\hat{a}_i(\mathbf{q};t)$ to those of all other modes $(\mathbf{q};j)$. In Section 4.4, we shall see how response theory and causality considerations can enable one to obtain progress in the direction of replacing the unknown dynamical operator $\hat{\mathbf{J}}_T$ by a prescribed quasi-classical (qc) c-number current density, $\hat{\mathbf{J}}_T^{qc}(\mathbf{r},t)$. In Part V, we shall study the problem further, paying particular attention to mesoscopic electrodynamics [Recent studies [119–122,124,125] show the formalism at work in the case of elastic electrodynamic scattering from a mesoscopic hole in a flat quantum-well screen].

In [12,97,98,153,185] the globally coherent (COH) field quantum state $|\text{COH}\rangle$ is discussed, and it is shown that

$$\mathbf{J}_T^{qc}(\mathbf{r},t) = \langle\text{COH}|\,\hat{\mathbf{J}}_T^{\,\text{sym}}(\mathbf{r},t)\,|\text{COH}\rangle. \tag{4.32}$$

When the sources are quasi-classical, the coherent radiation approximates the classical radiation. The only deviation is due to quantum fluctuations, which in the coherent state are minimal, and originate in the vacuum state of the electromagnetic field. With

$$\mathbf{F}_\pm(\mathbf{r},t) = \langle\text{COH}|\,\hat{\boldsymbol{\mathcal{F}}}_\pm(\mathbf{r},t)\,|\text{COH}\rangle, \tag{4.33}$$

one obtains from Eqs. (4.31)–(4.33) the following quasi-classical result

$$i\frac{\partial}{\partial t}\boldsymbol{F}_\pm(\mathbf{r},t) \mp c\boldsymbol{\nabla}\times\boldsymbol{F}_\pm(\mathbf{r},t) = -\frac{i}{(2\varepsilon_0)^{1/2}}\mathbf{J}_T^{qc}(\mathbf{r},t). \tag{4.34}$$

If one neglects the vacuum fluctuations, the analytical part of Eq. (4.34) equals the evolution equation for a single transverse photon in the photon energy wave function formalism [25,26,149]. In free space one regains the dynamical RS-equations for the positive and negative helicity species, given in Eq. (2.53).

Although the field vacuum fluctuations are neglected, the particle current density may still fluctuate due to electronic effects. This kind of fluctuation can be quite significant particularly in mesoscopic systems.

4.4 SELFCONSISTENCY, RESPONSE THEORY AND CAUSALITY

A certain combination of Eqs. (4.26)–(4.28) shows that the transverse electric field operator satisfies the inhomogenous wave equation

$$\Box\hat{\mathbf{E}}_T(\mathbf{r},t) = \mu_0\frac{\partial}{\partial t}\hat{\mathbf{J}}_T^{\,\text{sym}}(\mathbf{r},t). \tag{4.35}$$

Physical insight into the structure of the field emission (absorption) associated with the transverse current density distribution can be obtained by employing a scalar (Huygens) propagator (g) formalism in space-time. The propagator satisfies the singular inhomogenous differential equation

$$\Box g(\mathbf{r},t;\mathbf{r}',t') = -\delta(\mathbf{r}-\mathbf{r}')\delta(t-t'), \tag{4.36}$$

and among the solutions to Eq. (4.36) the well-known retarded (superscript: R) and advanced (superscript: A) solutions, viz.,

$$g^R(R,\tau)[\equiv g(R,\tau)] = \frac{\delta\left(\frac{R}{c} - \tau\right)}{4\pi R}, \tag{4.37}$$

and

$$g^A(R,\tau) = \frac{\delta\left(\frac{R}{c} + \tau\right)}{4\pi R}, \tag{4.38}$$

with $R = |\mathbf{r} - \mathbf{r}'|$ and $\tau = t - t'$, are of most importance from a fundamental point of view. The retarded Huygens propagator has already been introduced in Section 3.4 [Eq. (3.42)] in relation to the discussion of the transverse part of the Feynman photon propagator.

With the help of the $g^R(R,\tau)$ and $g^A(R,\tau)$ scalar propagators, the general solution to Eq. (4.35) can be written as follows:

$$\hat{\mathbf{E}}_T(\mathbf{r},t) = \hat{\mathbf{E}}_T^{\mathrm{vac}}(\mathbf{r},t) - \mu_0 \int_{-\infty}^{\infty} [c^R g^R(R,\tau) + c^A g^A(R,\tau)]\frac{\partial}{\partial t'}\hat{\mathbf{J}}_T^{\mathrm{sym}}(\mathbf{r}',t')\mathrm{d}^3 r'\mathrm{d}t'. \tag{4.39}$$

The term $\hat{\mathbf{E}}_T^{\mathrm{vac}}$ represents the general solution to the homogeneous part of Eq. (4.35). In quantum electrodynamics T-photons originate in the time-varying transverse current flow of charges, and rigorously speaking the current density operator in Eq.(4.35) therefore comprises the transverse flow of *all* charges (even on a cosmological scale, in principle). In the photon-free ground state of the quantum field the variances of the transverse electric and magnetic fields are nonvanishing, and these so-called vacuum (superscript: vac) fluctuations enter QED via the operator $\hat{\mathbf{E}}_T^{\mathrm{vac}}(\mathbf{r},t)$. One may argue that $\hat{\mathbf{E}}_T^{\mathrm{vac}}(\mathbf{r},t)$ in an indirect manner depends on matter. Thus, the collection of the spatial (classical) eigenstates for the free field determines the spectrum of the vacuum fluctuations, see Chapter 6. In microcavities parts of the vacuum fluctuations can be suppressed.

The integral term in Eq. (4.39) represents a particular solution to the inhomogeneous wave equation, given in Eq. (4.35). Both retarded and advanced field contributions are incorporated, and the balance between them is parameterized via the constants c^R and c^A. If Eq. (4.39) is inserted into Eq. (4.35), and use is made of Eq. (4.36), it appears that the constants are constrained by the relation

$$c^R + c^A = 1. \tag{4.40}$$

To be meaningful in a given theoretical analysis, i.e., make sense for a study of a physical phenomenon (in the Bohr sense of the word [158]) only a certain part of the total current density distribution can enter the theoretical considerations. Hence, let us briefly discuss the case where $\hat{\mathbf{J}}_T^{\mathrm{sym}}(\mathbf{r},t)$ to a good approximation can be replaced by the sum of two *spatially separated* flows $\hat{\mathbf{J}}_{T,1}^{\mathrm{sym}}(\mathbf{r},t)$ and $\hat{\mathbf{J}}_{T,2}^{\mathrm{sym}}(\mathbf{r},t)$. By spatially separated is meant that the effective volumes $V_{T,1}$ and $V_{T,2}$, occupied by the $\hat{\mathbf{J}}_{T,1}^{\mathrm{sym}}$

and $\hat{\mathbf{J}}_{T,2}^{\text{sym}}$ distributions never overlap in space. In general one must imagine that the two flows have to be determined in a self-consistent manner, since the radiation field provides an electromagnetic coupling between them. In may happen (or we may chose the flow so) that the coupling is essentially a one-way coupling. To investigate the electromagnetics of a system experimentally one must always excite the system from some sort of external source, the electromagnetics of which is unaffected by the system under study. This one-way coupling from the source to the system being probed is the first of two fundamental assumptions in so-called *response theory* (linear or nonlinear).The second assumption is based on the (apparently) never-violated experimental observation that the response of the system always is delayed (retarded with the speed c) in time with respect to the onset of the (nonuniform) charge motion in the source (driving system). This so-called Einsteinian retardation, implies that one must choose $c^A = 0$ in Eq. (4.39). Setting $\hat{\mathbf{J}}_{T,1}^{\text{sym}}(\mathbf{r},t) = \hat{\mathbf{J}}_{T,\text{source}}^{\text{sym}}(\mathbf{r},t)$ and $\hat{\mathbf{J}}_{T,2}^{\text{sym}}(\mathbf{r},t) = \hat{\mathbf{J}}_{T}^{\text{sym}}(\mathbf{r},t)$ [system transverse current density] Eq. (4.39) is reduced to the form

$$\hat{\mathbf{E}}_T(\mathbf{r},t) = \hat{\mathbf{E}}_T^{\text{ext}}(\mathbf{r},t) - \mu_0 \int_{V_T} g^R(R,\tau)\frac{\partial}{\partial t'}\hat{\mathbf{J}}_T^{\text{sym}}(\mathbf{r},t)\mathrm{d}^3r'\mathrm{d}t', \qquad (4.41)$$

where

$$\hat{\mathbf{E}}_T^{\text{ext}}(\mathbf{r},t) = \hat{\mathbf{E}}_T^{\text{vac}}(\mathbf{r},t) - \mu_0 \int_{V_T^{\text{source}}} g^R(R,\tau)\frac{\partial}{\partial t'}\hat{\mathbf{J}}_{T,\text{source}}^{\text{sym}}(\mathbf{r}',t')\mathrm{d}^3r'\mathrm{d}t', \qquad (4.42)$$

is the so-called external (superscript: ext) transverse electric field operator driving the system under investigation.

The current density in Eq. (4.42) evolves in time under the influence of the self-consistent field acting on the charges. This field is the sum of the external driving field ($\hat{\mathbf{E}}_T^{\text{ext}}$) and the field induced (ind) ($\hat{\mathbf{E}}_T^{\text{ind}}$) by the moving charges in the system. In the minimal coupling one may write in *symbolic* notation

$$\hat{\mathbf{J}}_T^{\text{sym}} = \hat{\mathbf{J}}_T^{\text{sym}}(\{p_\mu - e(A_\mu^{\text{ext}} + A_\mu^{\text{ind}})\}). \qquad (4.43)$$

The potential field $\{A_\mu^{\text{ext}} + A_\mu^{\text{ind}}\}$ is the so-called local field in the system. One cannot conclude that $\hat{\mathbf{J}}_T^{\text{sym}}$ is delayed in time with respect to the driving field $\{A_\mu^{\text{ext}} + A_\mu^{\text{ind}}\}$. However, if one imagines that the local field can be eliminated in favour of the external driving field $\{A_T^{\text{ext}}\}$, and thus obtain a new so-called *causal* transverse system current density

$$\hat{\mathbf{J}}_T^{\text{sym}}(\{p_\mu - e(A_\mu^{\text{ext}} + A_\mu^{\text{ind}})\}) \Rightarrow \hat{\mathbf{J}}_{T,\text{cau}}^{\text{sym}}(\{p_\mu - eA_\mu^{\text{ext}}\}), \qquad (4.44)$$

$\hat{\mathbf{J}}_{T,\text{cau}}^{\text{sym}}$ would be related in a causal manner to $\{A_\mu^{\text{ext}}\}$. The quantum physical calculation needed to achieve the transition in Eq. (4.44) implies that the word causal refers both to Einsteinian causality and the microcausality of the particle quantum physics. It is rather obvious that the calculation needed to achieve the transition in Eq. (4.44) can be carried out only in particular simple cases. In Part IV we shall show how the causal transverse current density can be determined (approximately) in certain mesoscopic systems.

TABLE 4.1 Transverse photon, electron and mixed photon-electron propagators. 1. The generalized scalar (Huygens) photon propagator, $c^R g^R + (1 - c^R) g^A$, is a linear superposition of the retarded (g^R) and advanced (g^A) scalar photon propagators. 2. Although there are applications in which an advanced propagator is useful, e.g., in scattering processes where the out-state is specified, and one wants to determine the possible in-states. However, it is believed that the genuine scalar propagator obeys Einsteinian causality (retardation with the vacuum speed of light). Thus with $c^R = 1$, we are left with g^R. 3. For g^R the transverse part of the current density, $\mathbf{J}_T$, is the source domain. In this case, the transverse part of the dyadic Feynman photon propagator, $\mathbf{G}_T$, must be used. From a physical point of view the description with g^R and $\mathbf{G}_T$ are equivalent. I have called the two approaches, respectively, the photon-eye view (with g^R) and the electron-eye view (with $\mathbf{G}_T$). In linear, nonlocal response theory the dyadic electron propagator we call $\mathbf{S}^{-1}$, and this propagator is studied in some details in Chapter 14. Kramers-Krönig relations do not relate the real and imaginary part of $\mathbf{S}^{-1}$ (nor $\mathbf{S}$), since the $\mathbf{S}^{-1}$-response does not obey microcausality. 5. By means of the dyadic local-field factor $\mathbf{\Gamma}$, a microcausal dyadic electron propagator $[\mathbf{S}^{\text{CAU}}]^{-1}$ may be constructed, at least formally. Thus, $[\mathbf{S}^{\text{CAU}}]^{-1} \equiv \int \mathbf{G}_T \cdot \mathbf{S}^{\text{CAU}}$ in symbolic form, where the integration is over all common points for $\mathbf{G}_T$ and $\mathbf{S}^{\text{CAU}}$. 6. The mixed and causal + microcausal dyadic transverse photon-electron propagator $\mathbf{R}^{\text{CAU}} \equiv \int \mathbf{G}_T \cdot \mathbf{S}^{\text{CAU}}$ (in symbolic form) is a field-field propagator describing the retarded propagation giving the self-consistent transverse vector potential, $\mathbf{A}_T(\mathbf{r}; \omega)$, at the space point $\mathbf{r}$ from a knowledge of the externally impressed (external) transverse vector potential, $\mathbf{A}_T^{\text{ext}}(\mathbf{r}', \omega)$ at the space-point $\mathbf{r}'$, both fields taken in the frequency (ω) domain [see Chapter 14 for details on $\mathbf{R}^{\text{CAU}}$].

1.	$c^R g^R + (1 - c^R) g^A$	Gen. scalar prop.
2.	g^R	Einstein causal scalar prop.
3.	$\mathbf{G}_T$	Transverse Feynman photon prop.
4.	$\mathbf{S}^{-1}$	Electron prop. (linear response theory)
5.	$[\mathbf{S}^{\text{CAU}}]^{-1} \equiv [\int \mathbf{S} \cdot \mathbf{\Gamma}]^{-1}$	Causal electron prop.
6.	$\mathbf{R}^{\text{CAU}} \equiv \int \mathbf{G}_T \cdot \mathbf{S}^{\text{CAU}}$	Field-field prop.

Although there seems to be no room for the advanced propagator (and the associated charge-charge coupling) in fundamental physics, this does not mean that the (mathematical) scalar propagator, given in Eq. (4.38), is not useful in physics as such. We shall realize this in connection to our studies of asymptotic in- and out-states in photon scattering theory (Part IV).

In Table 4.1 the fundamentals of selfconsistency, response theory and causality are illustrated in schematic form.

II

Zero - and One-Photon States. Bi-photons

CHAPTER 5

From Gauge-Photon Vacuum to Bi-Photons

5.1 CAVITY ELECTRODYNAMICS

Notwithstanding the fact that methods in theoretical quantum optics, which examine the photon field in free space, have contributed enormously to our understanding of basic properties of the electromagnetic field it should be remembered that the free-photon concept is an abstraction of our mind. In the absence of charges (and associated currents) we are left with the problem of solving the free-wave equation for the transverse vector potential $\mathbf{A}_T(\mathbf{r}, t)$. This can be done either in the form of a Fourier integral or a Fourier series, leading to a discrete or a continuous decomposition of $\mathbf{A}_T(\mathbf{r}, t)$, respectively. Subsequently, a canonical quantization of the field gives one the basis for the introduction of the photon concept. If one imagines that the electromagnetic field is contained in a large rectangular box, and we impose periodic boundary conditions on the field, the canonical quantization in plane-wave modes results in the following most celebrated expression for the Hamiltonian $(\hat{H}_T)$ of the quantized radiation field:

$$\hat{H}_T = \sum_{\mathbf{q},s} \hbar\omega_{qs} \left[\hat{a}^\dagger_{\mathbf{q}s}\hat{a}_{\mathbf{q}s} + \frac{1}{2} \right]. \tag{5.1}$$

In the oscillator mode with wave vector $\mathbf{q}$ and polarization s the photon energy is $\hbar\omega_{qs} = \hbar c|\mathbf{q}|$, independent of s. The Hermitian operator $\hat{a}^\dagger_{\mathbf{q}s}\hat{a}_{\mathbf{q}s}$ is a particularly important one since its spectrum is the set of integers, 0,1,2,.., and hence it counts the number of photons in the mode $\mathbf{q}s$. The global vacuum is an eigenstate for $\hat{H}_T$ with eigenenergy

$$H^0_T = \frac{\hbar}{2} \sum_{\mathbf{q}s} \omega_{qs} = \hbar c \sum_{\mathbf{q}} |\mathbf{q}|. \tag{5.2}$$

In cavity electrodynamics the fictitious box is "replaced" by a physical box most often made of a solid material. To study the atom-field interaction inside the cavity it is often assumed that the cavity wall is a perfect reflector. In essence this amounts to a complete decoupling of the field from the particle in the wall. Perhaps, such an

DOI: 10.1201/9781003029458-5

approximation works best for a superconducting cavity and a spectrum of photons with frequencies below the superconducting gap frequency.

To investigate the electromagnetic field in an empty cavity, yet, coupled in a dynamic manner to the charged massive particles of the cavity walls, we establish an operator extinction theorem. For space points ($\mathbf{r}$) inside a closed cavity the extinction theorem results in a selfconsistency requirement for the transverse vector potential operator, viz.,

$$\hat{\mathbf{A}}_T(\mathbf{r};\omega) = \int^{\Sigma} \left[g(R;\omega)\frac{\partial \hat{\mathbf{A}}_T(\mathbf{r};\omega)}{\partial n'} - \hat{\mathbf{A}}_T(\mathbf{r}';\omega)\frac{\partial g(R;\omega)}{\partial n'} \right] dS'. \tag{5.3}$$

Since the integral $\int^{\Sigma}(\ldots)dS'$ is over the closed wall surface one expects that Eq. (5.3) has nonvanishing solutions for discrete ω-values. The scalar $g(R;\omega) = g(|\mathbf{r}-\mathbf{r}'|;\omega)$ is the Huygens propagator and $\partial(\ldots)/\partial n'$ denotes a derivative in the normal direction to the cavity surface.

As originally emphasized and discussed by Landau and Peierls [177], and by Bohr and Rosenfeld [42, 43], it is not possible to measure the state of the photon field at a specific point $\mathbf{r}$. Alone, the lack of our ability to obtain a perfect spatial photon localization prevents this [153]. The operator extinction theorem [Eq. (5.3)] can be used to study the mean-field operator

$$\underline{\hat{\mathbf{A}}}_T(\mathbf{r},t) = \int_V \mathbf{f}(\boldsymbol{\rho}) \cdot \hat{\mathbf{A}}_T(\mathbf{r}-\boldsymbol{\rho},t)d^3\rho \tag{5.4}$$

with a suitable chosen time independent *tensorial weight factor*, $\mathbf{f}(\boldsymbol{\rho})$.

In a description where the electromagnetic field is treated as a classical quantity, the field energy can be written in the compact form

$$\mathcal{H}_T = \frac{1}{2}\sum_m [\dot{q}_m^2(t) + \omega_m^2 q_m^2(t)] + \frac{1}{2\mu_0}\int^{\Sigma} \mathbf{A}_T(\mathbf{r},t) \times [\boldsymbol{\nabla} \times \mathbf{A}_T(\mathbf{r},t)] \cdot d\mathbf{S}, \tag{5.5}$$

where the first part represents the radiative field *inside* the cavity, here written as a sum of the energies of harmonic field oscillators with a generalized coordinate $q_m(t)$ in mode number m. The second part is the surface energy contribution. This part is related to the *inevitably* present field-matter coupling. In a metallic cavity this part has a large (dominant) contribution from the *surface eigenmodes* in the wall.

A photon empty cavity, with energy given by H_T^0 [Eq. 5.2], *thus does not exist in a rigorous sense.*

5.2 SINGLE-PHOTON WAVE PACKETS

To some extent it appears meaningful to localize a single photon in space-time. As for electrons and other massive particles, a formalism for photon localization starts from the introduction of a relevant position operator ($|\mathcal{R}\rangle\,(\mathbf{r},t)$) in Hilbert space. For a transverse gauge photon an important kind of mean position state is defined by

$$|\mathcal{R}(\mathbf{r},t)\rangle = \left(\frac{2\varepsilon_0 c}{\hbar}\right)^{1/2} \hat{\mathbf{A}}_T^{(-)}(\mathbf{r},t)\,|0\rangle, \tag{5.6}$$

where $\hat{\mathbf{A}}_T^{(-)}(\mathbf{r},t)$ is the negative frequency part of the transverse vector potential operator, and $|0\rangle$ is the global vacuum state for transverse gauge photons. For a polychromatic Hilbert photon state

$$|\Phi\rangle(t) = \sum_i c_i(t)\hat{a}_i^\dagger |0\rangle, \tag{5.7}$$

synthesized from plane-wave states with amplitude coefficients $c_i(t)$ satisfying $\sum_i |c_i(t)|^2 = 1$, the related photon wave function in direct space is given by

$$\boldsymbol{\Phi}(\mathbf{r},t) = \langle \mathcal{R}|\Phi\rangle(\mathbf{r},t). \tag{5.8}$$

The free photon wave function in direct space satisfies a dynamical equation of the Hamiltonian-like form, viz.,

$$i\hbar\frac{\partial}{\partial t}\langle\mathcal{R}|\Phi\rangle(\mathbf{r},t) = \langle\mathcal{R}|\hat{\mathcal{H}}_{\mathrm{ph}}|\Phi\rangle(\mathbf{r},t), \tag{5.9}$$

where

$$\hat{\mathcal{H}}_{\mathrm{ph}} = \hat{\mathcal{H}} - \hat{\mathcal{H}}_0 = \sum_i \hbar\omega_i \hat{a}_i^\dagger \hat{a}_i \tag{5.10}$$

is the photon (ph) Hamiltonian, obtained by subtraction of the zero-point (vacuum) part $\mathcal{H}_0$ from the total Hamiltonian of the quantized radiation field. The normalization of the wave function is given by the relativistic condition

$$\int_{-\infty}^{\infty} \boldsymbol{\Phi}^\dagger(\mathbf{r},t)\cdot\left[\frac{i}{c}\frac{\partial}{\partial t}\boldsymbol{\Phi}(\mathbf{r},t)\right]d^3r = 1. \tag{5.11}$$

Finally, it should be noted that it is possible to introduce a wave-packet basis for gauge photons in direct space.

5.3 BI-PHOTONS: ENTANGLED TWO-PHOTON STATES

Once our theoretical study is extended from single-photon wave packets to two-photon wave packets entangled states of two gauge photons can occur.

Using wave-packet bases $[\{|W_i(1)\rangle\}, \{|W_j(2)\rangle\}]$, the most general Hilbert states of the two photons (denoted 1 and 2) are

$$|\Phi(1)\rangle = \sum_i a_i |W_i(1)\rangle, \tag{5.12}$$

$$|\Phi(2)\rangle = \sum_j b_j |W_j(2)\rangle, \tag{5.13}$$

In the union $(\mathcal{H}_1 \otimes \mathcal{H}_2)$ of the two single-particle Hilbert spaces ($\mathcal{H}_1$ and $\mathcal{H}_2$) the basis vectors are

$$|W_{ij}\rangle = \frac{1}{\sqrt{2!}}|W_i(1)\rangle \otimes |W_j(2)\rangle. \tag{5.14}$$

Now in the combined Hilbert space every state can be expanded as follows:

$$|\Psi(1,2)\rangle = \sum_{ij} c_{ij} \, |W_i(1)\rangle \otimes |W_j(2)\rangle . \tag{5.15}$$

Among the two-photon states in Eq. (5.15) one has factored (FAC) states

$$\left|\Psi^{\mathrm{FAC}}(1,2)\right\rangle = |\Phi(1)\rangle \otimes |\Phi(2)\rangle = \sum_{ij} a_i b_j \, |W_i(1)\rangle \otimes |W_j(2)\rangle . \tag{5.16}$$

These wave-packet states are combined states of *two individual gauge photons*. However, the elements of the arbitrary matrix $\{c_{ij}\}$ in Eq. (5.15) cannot be expressed as the elements $\{a_i b_j\}$ of the dyadic product of $\{a_i\}$ and $\{b_j\}$ elements. In consequence, there exists so-called *entangled gauge photon states* among the $|\Psi(1,2)\rangle$ states. An entangled two-particle state I have called a *bi-photon*.

The wave-packet two-photon formalism described above can be transformed to direct space introducing a two-photon "mean" position state given as

$$|\mathcal{R}_2\rangle\,(\mathbf{r},\mathbf{r}',t) = \frac{2\varepsilon_0 c}{\hbar}\,\hat{\mathbf{A}}_T^{(-)}(\mathbf{r},t)\hat{\mathbf{A}}_T^{(-)}(\mathbf{r}',t)\,|0\rangle . \tag{5.17}$$

For a general two-photon state, $|\Psi(1,2)\rangle \equiv |\Phi_2\rangle$, the associated two-point wave function is given by

$$\Phi_2(\mathbf{r},\mathbf{r}',t) = \langle\mathcal{R}_2|\Phi_2\rangle\,(\mathbf{r},\mathbf{r}',t), \tag{5.18}$$

in analogy with Eq. (5.8) for a single photon. The two-photon wave function satisfies the dynamical equation

$$i\hbar\frac{\partial}{\partial t}\,\langle\mathcal{R}_2|\Phi_2\rangle\,(\mathbf{r},\mathbf{r}',t) = \langle\mathcal{R}_2|\,\hat{\mathcal{H}}_{\mathrm{ph}}\,|\Phi_2\rangle\,(\mathbf{r},\mathbf{r}',t) \tag{5.19}$$

in the two-coordinate $[\mathbf{r},\mathbf{r}']$ configuration space.

CHAPTER 6

Operator Extinction Theorem: Mean Fields, Correlations and Fluctuations in Free Space

6.1 FIELD EXTINCTION IN PARTICLE VACUUM

In particle vacuum the transverse electric and magnetic field operators satisfy the set of Maxwell equations in Eqs. (4.25)–(4.28) with $\hat{\mathbf{J}}_T^{\mathrm{sym}}(\mathbf{r}, t) = \mathbf{0}$. From this set and Eqs. (4.7) and (4.8) it follows that the transverse vector potential operator satisfies the following homogeneous wave equation in the space-frequency domain:

$$(\nabla^2 + q_0^2)\hat{\mathbf{A}}_T(\mathbf{r}; \omega) = \mathbf{0}, \tag{6.1}$$

where $q_0 = \omega/c$ is the vacuum wave vector. The scalar propagator $g(\mathbf{r} - \mathbf{r}'; \omega)$ satisfies the inhomogeneous (singular) differential equation

$$(\nabla^2 + q_0^2)g(\mathbf{r} - \mathbf{r}'; \omega) = -\delta(\mathbf{r} - \mathbf{r}'). \tag{6.2}$$

In free space $g(\mathbf{r} - \mathbf{r}'; \omega) = g(|\mathbf{r} - \mathbf{r}'|; \omega) = g(R; \omega)$, where $R = \mathbf{r} - \mathbf{r}'$.

Let the ith Cartesian component of $\hat{\mathbf{A}}_T(\mathbf{r}; \omega)$ be denoted $\hat{A}_{T,i}(\mathbf{r}; \omega)$. With $\boldsymbol{\nabla}' \cdot \boldsymbol{\nabla}' \equiv (\nabla')^2$ the identity

$$\boldsymbol{\nabla}' \cdot \left[\hat{A}_{T,i}(\mathbf{r}'; \omega)\boldsymbol{\nabla}'g(R; \omega) - g(R; \omega)\boldsymbol{\nabla}'\hat{A}_{T,i}(\mathbf{r}'; \omega) \right] =$$
$$\hat{A}_{T,i}(\mathbf{r}'; \omega)(\nabla')^2 g(R; \omega) - g(R; \omega)(\nabla')^2 \hat{A}_{T,i}(\mathbf{r}'; \omega), \tag{6.3}$$

in combination with Eqs. (6.1) and (6.2), give for $\mathbf{r} \neq \mathbf{r}'$

$$\boldsymbol{\nabla}' \cdot \left[\hat{A}_{T,i}(\mathbf{r}'; \omega)\boldsymbol{\nabla}'g(R; \omega) - g(R; \omega)\boldsymbol{\nabla}'\hat{A}_{T,i}(\mathbf{r}'; \omega) \right] = 0, \ \mathbf{r} \neq \mathbf{r}'. \tag{6.4}$$

DOI: 10.1201/9781003029458-6

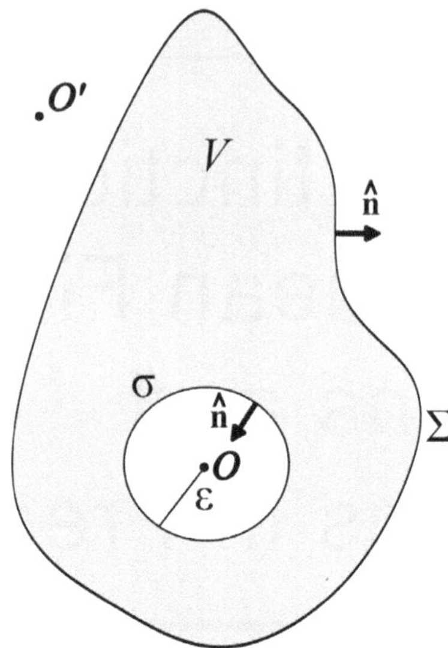

Figure 6.1 Illustration of the geometry used to obtain the field-operator extinction theorem in a particle vacuum. The (arbitrary) integration volume V (grey-toned) is bounded by the closed surface Σ. For field points $\mathcal{O}$ inside V, a spherical exclusion volume of radius ε [$\to 0$, at the end of the calculation] and centered on $\mathcal{O}$ is cut off from V. The sphere surface is denoted by σ. For field points $\mathcal{O}'$ outside V the exclusion volume is absent. The normal unit vectors, $\hat{\mathbf{n}}$, on Σ and σ are directed outwards from the grey-toned volume.

With reference to Fig. 6.1, Gauss' theorem applied to Eq. (6.4), and with integration over the singularity-free grey-toned volume (V), leads to the result

$$\int^{\Sigma+\sigma} \left\{ \hat{A}_{T,i}(\mathbf{r}';\omega)[\mathbf{n}(\mathbf{r}') \cdot \boldsymbol{\nabla}']g(R:\omega) - g(R;\omega)[\mathbf{n}(\mathbf{r}') \cdot \boldsymbol{\nabla}']\hat{A}_{T,i}(\mathbf{r}';\omega) \right\} \mathrm{d}S' = 0.$$

$$(6.5)$$

The surface integral is over the outer surface (Σ) plus the spherical surface (σ) centred on the singular point $\mathbf{r} = \mathbf{r}'$ in Eq. (6.2). The unit normal vector $\mathbf{n}(\mathbf{r}')$ is directed outward from the gray-toned volume everywhere on the surfaces Σ and σ. The radius of the (small) spherical volume is denoted by ε. By multiplication of Eq. (6.5) with a unit vector ($\mathbf{i}$) in the i-direction followed by summation over i, one obtains for $\hat{\mathbf{A}}_T(\mathbf{r}';\omega) = \sum_i \hat{A}_{T,i}(\mathbf{r}';\omega)\mathbf{i}$ the temporary result

$$\int^{\Sigma+\sigma} \left[\hat{\mathbf{A}}_T(\mathbf{r}';\omega)\frac{\partial g(R;\omega)}{\partial n'} - g(R;\omega)\frac{\partial \hat{\mathbf{A}}_T(\mathbf{r}';\omega)}{\partial n'} \right] \mathrm{d}S' = 0, \qquad (6.6)$$

where

$$\frac{\partial}{\partial n'} \equiv \mathbf{n}(\mathbf{r}') \cdot \boldsymbol{\nabla}'. \qquad (6.7)$$

As a final step we calculate the surface integral over σ in the limit $\varepsilon \to 0$. For a sufficiently small ε, the spatial variation of the nonsingular vector potential on σ becomes negligible $[\hat{\mathbf{A}}_T(\mathbf{r}';\omega) = \hat{\mathbf{A}}_T(\mathbf{r};\omega), \varepsilon \text{ small}]$. The remaining relevant surface integrals are given by

$$\lim_{\varepsilon \to 0} \left[\int^{\sigma} \mathbf{n}(\mathbf{r}') \cdot \boldsymbol{\nabla}' g(R;\omega) \mathrm{d}S' \right] = 1 \tag{6.8}$$

and

$$\lim_{\varepsilon \to 0} \left[\int^{\sigma} g(R;\omega) \mathbf{n}(\mathbf{r}') \mathrm{d}S' \right] = 0 \tag{6.9}$$

as the reader may show (using polar coordinates, e.g.). When Eqs (6.8) and (6.9) are used in Eq. (6.5), it is obvious that the surface integral over σ, in the limit $\varepsilon \to 0$, gives $\hat{\mathbf{A}}_T(\mathbf{r},\omega)$. Hence, one can conclude that

$$\hat{\mathbf{A}}_T(\mathbf{r},\omega) = \hat{\boldsymbol{\Sigma}}_T(\mathbf{r};\omega), \ \mathbf{r} \in V, \ \mathbf{r}' \in V, \tag{6.10}$$

where

$$\hat{\boldsymbol{\Sigma}}_T(\mathbf{r};\omega) = \int^{\Sigma} \left[g(R;\omega) \frac{\partial \hat{\mathbf{A}}_T(\mathbf{r}';\omega)}{\partial n'} - \hat{\mathbf{A}}_T(\mathbf{r}';\omega) \frac{\partial g(R;\omega)}{\partial n'} \right] \mathrm{d}S'. \tag{6.11}$$

If $\mathbf{r}$ is outside V ($\mathbf{r} \notin V$), and one still integrates over V ($\mathbf{r}' \in V$), $\delta(\mathbf{r} - \mathbf{r}') = 0$, which means that there is no need to cut out a small ($\varepsilon \to 0$) spherical volume before applying Gauss' theorem. Thus, one obviously obtains

$$0 = \hat{\boldsymbol{\Sigma}}_T(\mathbf{r};\omega), \ \mathbf{r} \notin V, \ \mathbf{r}' \in V. \tag{6.12}$$

In Chapter 16 the calculation will be extended in two ways: (i) we shall divide space into three nonoverlapping domains V, $\tilde{V}$ and V^0 and (ii) assume that particle current densities are present inside V^0 (source domain) and V (medium with quantum electrodynamic properties being investigated). If the closed surface separating $\tilde{V}$ and V^0 is denoted Σ^0, and the related integral $\hat{\boldsymbol{\Sigma}}_T^0(\mathbf{r};\omega)$ [unit normal vector pointing into $\tilde{V}$] one obtains for integration over $\tilde{V}$

$$\hat{\mathbf{A}}_T(\mathbf{r};\omega) = - \hat{\boldsymbol{\Sigma}}_T(\mathbf{r};\omega) - \hat{\boldsymbol{\Sigma}}_T^0(\mathbf{r};\omega), \ \mathbf{r} \in \tilde{V}, \ \mathbf{r}' \in \tilde{V}, \tag{6.13}$$

$$0 = - \hat{\boldsymbol{\Sigma}}_T(\mathbf{r};\omega) - \hat{\boldsymbol{\Sigma}}_T^0(\mathbf{r};\omega), \ \mathbf{r} \notin \tilde{V}, \ \mathbf{r}' \in \tilde{V}, \tag{6.14}$$

and for integration over V^0

$$\hat{\mathbf{A}}_T(\mathbf{r};\omega) = \hat{\boldsymbol{\Sigma}}_T^0(\mathbf{r};\omega), \ \mathbf{r} \in V^0, \ \mathbf{r}' \in V^0, \tag{6.15}$$

$$0 = \hat{\boldsymbol{\Sigma}}_T^0(\mathbf{r};\omega), \ \mathbf{r} \notin V^0, \ \mathbf{r}' \in V^0. \tag{6.16}$$

When particle current density distributions are present in V and V^0, Eqs (6.10), (6.12), (6.15) and (6.16) are modified. Since there are no charged particles in $\tilde{V}$, Eqs.

(6.13) and (6.14) keep the same *forms*, although the *explicit* $\mathbf{r}$-dependence of $\hat{\boldsymbol{\Sigma}}_T(\mathbf{r};\omega)$ and $\hat{\boldsymbol{\Sigma}}_T^0(\mathbf{r};\omega)$ will change.

The set in Eqs. (6.10), (6.12) and (6.13)–(6.16) constitutes a field operator version of a corresponding set in semiclassical electrodynamics. When extended to include current density distributions (here in V and V^0) certain combinations of set members are called Ewald-Oseen extinction theorems [46, 79, 80, 193, 199, 276]. In the title of this chapter the phrase "OPERATOR EXTINCTION THEOREM" refers for brevity to the set as such.

6.2 MEAN FIELD OPERATOR

The various field operators $[\hat{\mathbf{A}}(\mathbf{r},t),\ \hat{\mathbf{E}}_T(\mathbf{r},t),\ \hat{\mathbf{B}}(\mathbf{r},t),\ \hat{\boldsymbol{F}}_\pm(\mathbf{r},t)]$ relate to the fields at a specific point $\mathbf{r}$. However, it is not possible to measure the state of the photon field at a given point, nor is it possible in a photon emission process to claim that a given photon starts its journey in space-time from a given $\mathbf{r}$-point. The spatial photon localization problem one is facing in source emission and detector absorption processes is treated in Chapter 27.

Independent of the physical details in a given situation, rather than considering the field operators of a point $\mathbf{r}$ we average the given field operator over an essentially finite volume (V) surrounding the point $\mathbf{r}$. In order to illustrate the main principle let us consider the transverse vector potential operator and introduce the mean field operator

$$\hat{\underline{\mathbf{A}}}_T(\mathbf{r},t) = \int_V \mathbf{f}(\boldsymbol{\rho}) \cdot \hat{\mathbf{A}}_T(\mathbf{r} - \boldsymbol{\rho}, t)\mathrm{d}^3\rho, \tag{6.17}$$

where $\mathbf{f}(\boldsymbol{\rho})$ is a time-independent tensorial weight factor. Although the mean field in Eq. (6.17) relates to a given time (t), all physical photon emission and absorption processes require some kind of averaging over a finite time interval. In general, it is useful to work with normalized weight functions such that

$$\int_{-\infty}^{\infty} \mathbf{f}(\boldsymbol{\rho})\mathrm{d}^3\rho = \mathbf{U}. \tag{6.18}$$

In the discrete plane-mode quantization of the transverse vector potential [Eq. (3.63)] the mean field operator is given by

$$\hat{\underline{\mathbf{A}}}_T(\mathbf{r},t) = \left(\frac{\hbar}{2\varepsilon_0 L^3}\right)^{1/2} \sum_i \frac{1}{\omega_i^{1/2}}[\hat{a}_i(t)\mathbf{f}(\mathbf{q}_i) \cdot \boldsymbol{\varepsilon}_i e^{i\mathbf{q}_i \cdot \mathbf{r}} + h.c.], \tag{6.19}$$

where

$$\mathbf{f}(\mathbf{q}_i) = \int_{-\infty}^{\infty} \mathbf{f}(\boldsymbol{\rho})e^{-i\mathbf{q}_i \cdot \boldsymbol{\rho}}\mathrm{d}^3\rho \tag{6.20}$$

is the Fourier integral transform belonging to $\mathbf{q}_i$.

As a heuristic example, let us consider the case where the weight function is isotropic, and constant inside a sphere of radius r_0 and zero outside, i.e.,

$$\mathbf{f}(\boldsymbol{\rho}) = \frac{3}{4\pi r_0^3}\Theta(r_0 - \rho)\mathbf{U}, \tag{6.21}$$

on normalized form. From a strict physical point of view, a weight function with a sharp cutoff $[\Theta(r_0 - \rho)]$ never appears. As the reader may show, the Fourier transform of $\mathbf{f}(\boldsymbol{\rho})$ is

$$\mathbf{f}(\mathbf{q}) = \frac{3\mathbf{U}}{(qr_0)^3}[\sin(qr_0) - qr_0\cos(qr_0)], \tag{6.22}$$

with a range $\sim (qr_0)^{-2}$, and $\mathbf{f}(|\mathbf{q}| \to 0) = \mathbf{U}$.

A particular choice of weight function stands out, namely

$$\mathbf{f}(\boldsymbol{\rho}) = \frac{3}{2}\boldsymbol{\delta}_T(\boldsymbol{\rho}). \tag{6.23}$$

This is so because the transverse delta function relates directly to the best possible spatial photon localization in an emission process from a source, c.f. the analysis in Sections 3.2–3.4. The factor $3/2$ in Eq. (6.23) is needed for normalization of $\mathbf{f}(\boldsymbol{\rho})$ [Eq. (6.18)], as the reader may show using Eq. (3.19).

In a particular vacuum domain V, we know that

$$\hat{\mathbf{A}}_T(\mathbf{r};\omega) = \hat{\boldsymbol{\Sigma}}_T(\mathbf{r};\omega),\ \mathbf{r} \in V, \tag{6.24}$$

a result already given in Eq. (6.10) and obtained by $\mathbf{r}'$-integration over V [see Section 6.1]. If one lets the volume of average (v) coincide with V, the mean field operator related to $\hat{\mathbf{A}}_T$ is given by

$$\underline{\hat{\mathbf{A}}}_T(\mathbf{r};\omega) = \frac{3}{2}\int_V \boldsymbol{\delta}_T(\mathbf{r} - \boldsymbol{\rho}) \cdot \hat{\mathbf{A}}_T(\boldsymbol{\rho};\omega)\mathrm{d}^3\rho = \frac{3}{2}\int_V \boldsymbol{\delta}_T(\mathbf{r} - \boldsymbol{\rho}) \cdot \hat{\boldsymbol{\Sigma}}_T(\boldsymbol{\rho};\omega)\mathrm{d}^3\rho \tag{6.25}$$

in the space-frequency domain. By inserting the explicit expression given in Eq. (6.11) into Eq.(6.25) one obtains

$$\underline{\hat{\mathbf{A}}}_T(\mathbf{r};\omega) = \frac{3}{2}\int^\Sigma \left\{ \boldsymbol{\mathcal{G}}_T^V(\mathbf{r} - \mathbf{r}';\omega)\frac{\partial}{\partial n'}\hat{\mathbf{A}}_T(\mathbf{r}';\omega) - \left[\frac{\partial}{\partial n'}\boldsymbol{\mathcal{G}}_T^V(\mathbf{r} - \mathbf{r}';\omega)\right] \cdot \hat{\mathbf{A}}_T(\mathbf{r}';\omega) \right\}\mathrm{d}S', \tag{6.26}$$

where

$$\boldsymbol{\mathcal{G}}_T^V(\mathbf{r} - \mathbf{r}';\omega) = \int_V \boldsymbol{\delta}_T(\mathbf{r} - \boldsymbol{\rho})g(|\boldsymbol{\rho} - \mathbf{r}'|;\omega)\mathrm{d}^3\rho. \tag{6.27}$$

If the volume V is taken as the *effective* range of $\boldsymbol{\delta}_T$ [the rim zone in the case of the presence of a particle current density $\mathbf{J}(\mathbf{r};\omega)$] one may replace $\boldsymbol{\mathcal{G}}_T^V(\mathbf{r} - \mathbf{r}';\omega)$ in Eq. (6.26) by the transverse part of the Feynman photon propagator

$$\mathbf{G}_T(\mathbf{r} - \mathbf{r}';\omega) = \lim_{V \to \infty} \boldsymbol{\mathcal{G}}_T^V(\mathbf{r} - \mathbf{r}';\omega). \tag{6.28}$$

In a specific application the extension of the rim zone V must be chosen in such a manner that the contribution outside V is negligible ["$V \to \infty$" means $V \to V_{\text{eff}}$, the effective rim zone volume]. Our final expression for the mean field operator with weight factor $3\rho_T(\rho)/2$ hence takes the form

$$\underline{\hat{\mathbf{A}}}_T(\mathbf{r};\omega) = \frac{3}{2} \int^{\Sigma} \left\{ \mathbf{G}_T(\mathbf{r}-\mathbf{r}';\omega)\frac{\partial}{\partial n'}\hat{\mathbf{A}}_T(\mathbf{r}';\omega) - \left[\frac{\partial}{\partial n'}\mathbf{G}_T(\mathbf{r}-\mathbf{r}';\omega)\right]\cdot\hat{\mathbf{A}}_T(\mathbf{r}';\omega)\right\} \mathrm{d}S'. \tag{6.29}$$

The mean field of the transverse electric field operator, $\underline{\hat{\mathbf{E}}}_T(\mathbf{r};\omega) = i\omega\underline{\hat{\mathbf{A}}}_T(\mathbf{r};\omega)$, follows immediately from Eq. (6.29), and for the magnetic field $\hat{\mathbf{B}}(\mathbf{r};\omega) = \mathbf{\nabla} \times \underline{\hat{\mathbf{A}}}_T(\mathbf{r};\omega)$ an expression form-identical to the right-hand side of Eq. (6.29) is obtained with the replacement $\mathbf{\nabla} \times \mathbf{G}_T(\mathbf{r}-\mathbf{r}';\omega) \to i(\omega/c)\mathbf{G}_M(\mathbf{r}-\mathbf{r}';\omega)$, where $\mathbf{G}_M$ is the magnetic field propagator see [153, 157].

6.3 CAVITY FIELDS. ZERO-PHOTON STATES

In a multimode electromagnetic field the globally coherent state

$$|\mathrm{COH}\rangle = \Pi_{\mathbf{q},s}\,|\alpha_{\mathbf{q},s}\rangle\,, \tag{6.30}$$

which is the direct product of $|\alpha_{\mathbf{q},s}\rangle$ state vectors of all $i = (\mathbf{q}, s)$ modes, has a mean value

$$< \hat{\mathbf{A}}_T(\mathbf{r}, t) > \equiv \langle\mathrm{COH}|\,\hat{\mathbf{A}}_T(\mathbf{r}, t)\,|\mathrm{COH}\rangle = \mathbf{A}_T(\mathbf{r}, t) \tag{6.31}$$

identical to the expression for a classical transverse vector potential. The fluctuation in the potential characterized by

$$\sigma_{\mathbf{A}_T} \equiv [\langle\mathrm{COH}|\,\hat{\mathbf{A}}_T\cdot\hat{\mathbf{A}}_T\,|\mathrm{COH}\rangle - (\langle\mathrm{COH}|\,\hat{\mathbf{A}}_T\,|\mathrm{COH}\rangle)^2]^{1/2} \tag{6.32}$$

is independent of the amplitude of $\mathbf{A}_T(\mathbf{r}, t)$. The fluctuations in the global vacuum $(|0\rangle)$ and in the coherent state are identical [182]. When the average number of $(\mathbf{q}, s)$-photons in the individual modes becomes sufficiently large, the coherent state mean value $< \hat{\mathbf{A}}_T(\mathbf{r}, t) >$ approaches $\mathbf{A}(\mathbf{r}, t)$ of semiclassical physics.

In semiclassical electromagnetics (particle vacuum) the operator relation in Eq. (6.10) [with Eq. (6.11)] is replaced by

$$\mathbf{A}_T(\mathbf{r};\omega) = \int^{\Sigma} \left[g(R;\omega)\frac{\partial\mathbf{A}_T(\mathbf{r}',\omega)}{\partial n'} - \mathbf{A}_T(\mathbf{r}';\omega)\frac{\partial g(R;\omega)}{\partial n'}\right] \mathrm{d}S'. \tag{6.33}$$

In (microscopic) electrodynamics all electromagnetic fields are differentiable, basically, and Eq. (6.33) represents a selfconsistency requirement for the transverse vector potential. Since the integral $\int^{\Sigma}(...)\mathrm{d}S'$ is over a closed surface it is expected that Eq. (6.33) has nonvanishing solutions for discrete ω-values. In the modern interpretation of the (microscopic) Ewald-Ossen extinction theorem the integral term in Eq. (6.33) is considered as a spatially nonlocal self-consistent boundary condition for the field in V.

In eletromagnetic diffraction problems, and cavity electrodynamics, the "Gordian knot" in Eq. (6.33) is cut by making a clever choice for $\mathbf{A}_T(\mathbf{r}';\omega)$ and $\partial\mathbf{A}_T(\mathbf{r}';\omega)/\partial n'$

on Σ. The choice, which in general breaks the selfconsistency leads to an approximate solution for $\mathbf{A}_T(\mathbf{r};\omega)$ in the cavity V. The choice constitutes what commonly is called the boundary conditions for the field in the vacuum domain inside Σ.

Since

$$(\nabla^2 + q_0^2)\boldsymbol{\Sigma}_T(\mathbf{r};\omega) = \int^{\Sigma}\left[\mathbf{A}_T(\mathbf{r}'\omega)\frac{\partial\delta(\mathbf{r}-\mathbf{r}')}{\partial n'} - \delta(\mathbf{r}-\mathbf{r}')\frac{\partial\mathbf{A}_T(\mathbf{r}';\omega)}{\partial n'}\right]dS' = 0,$$

(6.34)

a result which follows by use of the singular differential equation for the Huygens propagator [Eq. (6.2)], and the fact that $\delta(\mathbf{r}-\mathbf{r}') = 0$ for points in the domain inside Σ. The extinction theorem $\mathbf{A}_T(\mathbf{r};\omega) = \boldsymbol{\Sigma}_T(\mathbf{r};\omega)$ in turn gives

$$(\nabla^2 + q_0^2)\mathbf{A}_T(\mathbf{r};\omega) = \mathbf{0}.$$

(6.35)

The Helmholtz equation for $\mathbf{A}_T(\mathbf{r};\omega)$ just confirms that Eq. (6.33) leads back to the semiclassical form of Eq. (6.1); $\hat{\mathbf{A}}_T(\mathbf{r};\omega) \to \mathbf{A}_T(\mathbf{r};\omega)$.

In order to investigate the field energy in a matter-empty cavity we start form the classical wave equation for the transverse vector potential in the space-time domain [Eq. (2.25)]. The general solution to this equation, including appropriate field conditions on the wall of the cavity, we resolve into a sum of standing modes in the usual manner. Thus,

$$\mathbf{A}_T(\mathbf{r}, t) = \sum_m \frac{1}{\sqrt{\varepsilon_0 V_m}}q_m(t)\mathbf{u}_m(\mathbf{r}),$$

(6.36)

where V_m is a kind of effective mode volume [230]. In the special case where the cavity is a box of volume V, V_m is independent of the mode index m and given by $V_m = V/8$ see [230]. The real spatial mode functions are dimensionless, and satisfy an orthonomality condition

$$\frac{1}{\sqrt{V_m V_n}}\int_{\text{cav}}\mathbf{u}_m(\mathbf{r})\cdot\mathbf{u}_n(\mathbf{r})\mathrm{d}^3r = \delta_{mn}.$$

(6.37)

The factor $\varepsilon_0^{-1/2}$ in Eq. (6.36) is introduced for later convenience. From Eq. (6.36) one obtains the following expressions for the transverse electric field and the magnetic field:

$$\mathbf{E}_T(\mathbf{r}, t) = -\frac{\partial\mathbf{A}_T(\mathbf{r}, t)}{\partial t} = -\sum_m \frac{1}{\sqrt{\varepsilon_0 V_m}}\dot{q}_m(t)\mathbf{u}_m(\mathbf{r}),$$

(6.38)

$$\mathbf{B}(\mathbf{r}, t) = \nabla\times\mathbf{A}_T(\mathbf{r}, t) = \sum_m \frac{1}{\sqrt{\varepsilon_0 V_m}}q_m(t)\nabla\times\mathbf{u}_m(\mathbf{r}).$$

(6.39)

The classical field energy in the cavity plus wall ($\mathcal{H}_T$) hence becomes

$$\mathcal{H}_T = \frac{\varepsilon_0}{2}\int_{\text{cav}}[\mathbf{E}_T^2(\mathbf{r}, t) + c^2\mathbf{B}^2(\mathbf{r}, t)]\mathrm{d}^3r = \frac{1}{2}\sum_{m,n}\dot{q}_m(t)\dot{q}_n(t)\frac{1}{\sqrt{V_m V_n}}\int_{\text{cav}}\mathbf{u}_m(\mathbf{r})\cdot\mathbf{u}_n(\mathbf{r})\mathrm{d}^3r$$

$$+ \frac{c^2}{2}\sum_{m,n}q_m(t)q_n(t)\frac{1}{\sqrt{V_m V_n}}\int_{\text{cav}}[\nabla\times\mathbf{u}_m(\mathbf{r})]\cdot[\nabla\times\mathbf{u}_n(\mathbf{r})]\mathrm{d}^3r.$$

(6.40)

The part of $\mathcal{H}_T$ which contains the magnetic energy is rewritten using the relation

$$
\begin{aligned}
(\nabla \times \mathbf{u}_m)(\nabla \times \mathbf{u}_n) &= \mathbf{u}_n \cdot \nabla \times (\nabla \times \mathbf{u}_m) + \nabla \cdot [\mathbf{u}_n \times (\nabla \times \mathbf{u}_m)] \\
&= -\mathbf{u}_n \cdot \nabla^2 \mathbf{u}_m + \nabla \cdot [\mathbf{u}_n \times (\nabla \times \mathbf{u}_m)].
\end{aligned}
\tag{6.41}
$$

The last member of Eq. (6.41) follows from the fact that $\mathbf{u}_m(\mathbf{r})$ is divergence-free

$$
\nabla \cdot \mathbf{u}_m(\mathbf{r}) = 0, \quad \forall m,
\tag{6.42}
$$

cf. Eq. (6.36). The contribution to $\mathcal{H}_T$ from the electric energy simplifies immediately due to the mode orthonormality [Eq. (6.37)]. Altogether, one obtains

$$
\begin{aligned}
\mathcal{H}_T =& \frac{1}{2} \sum_m \dot{q}_m^2 - \frac{c^2}{2} \sum_{m,n} q_m q_n \frac{1}{\sqrt{V_m V_n}} \int_{\text{cav}} \mathbf{u}_n \cdot \nabla^2 \mathbf{u}_m \mathrm{d}^3 r \\
&+ \frac{c^2}{2} \sum_{m,n} q_m q_n \frac{1}{\sqrt{V_m V_n}} \int_{\text{cav}} \nabla \cdot [\mathbf{u}_n \times (\nabla \times \mathbf{u}_m)] \mathrm{d}^3 r.
\end{aligned}
\tag{6.43}
$$

The individual cavity modes satisfy the partial differential equation

$$
q_m(t) \nabla^2 \mathbf{u}_m(\mathbf{r}) + \frac{1}{c^2} \ddot{q}_m(t) \mathbf{u}_m(\mathbf{r}) = \mathbf{0},
\tag{6.44}
$$

and hence

$$
\left[\nabla^2 + \left(\frac{\omega_m}{c} \right)^2 \right] \mathbf{u}_m(\mathbf{r}) = \mathbf{0},
\tag{6.45}
$$

and

$$
\frac{d^2 q_m(t)}{dt^2} + \omega_m^2 q_m(t) = 0,
\tag{6.46}
$$

where ω_m is the mode eigenfrequency.

Utilizing Eq. (6.45), the orthonormality [Eq. (6.37)], the expression for $\mathcal{H}_T$ [Eq. (6.43)] is reduced to

$$
\mathcal{H}_T = \frac{1}{2} \sum_m (\dot{q}_m^2 + \omega_m^2 q_m^2) + \frac{1}{2\mu_0} \int^\Sigma \left[\sum_n \frac{1}{\sqrt{\varepsilon_0 V_n}} q_n \mathbf{u}_n \right] \times \left[\sum_m \frac{1}{\sqrt{\varepsilon_0 V_m}} q_m \nabla \times \mathbf{u}_m \right] \cdot \mathrm{d}\mathbf{S}.
\tag{6.47}
$$

Use of Eq. (6.36) allows one to write the field energy in compact form

$$
\mathcal{H}_T = \frac{1}{2} \sum_m [\dot{q}_m^2(t) + \omega_m^2 q_m^2(t)] + \frac{1}{2\mu_0} \int^\Sigma \mathbf{A}_T(\mathbf{r}, t) \times [\nabla \times \mathbf{A}_T(\mathbf{r}, t)] \cdot \mathrm{d}\mathbf{S}.
\tag{6.48}
$$

Let us now reflect on the physics of the result for $\mathcal{H}_T$. The energy of the radiation field in the cavity is the sum of the energies

$$
\mathcal{H}_m = \frac{1}{2} [\dot{q}_m^2(t) + \omega_m^2 q_m^2(t)]
\tag{6.49}
$$

of harmonic field oscillations with a generalized coordinate $q_m(t)$, plus a surface (Σ) energy part. In a sense the surface energy contribution is the most complicated to handle for at least the following reasons: (i) The interchange of energy between the surroundings (charged particles) and the cavity wall is described via the surface term, and *in general* one would therefore expect a time dependence of the cavity plus wal (resonator) energy, $\mathcal{H}_T = \mathcal{H}_T(t)$. (ii) A sharp cavity boundary is a physical idealization: A transition region where the electronic (particle) density changes always is present, and the electromagnetics in the transition layer is extremely difficult to describe; see [136] and references therein. (iii) The surface integral often is considered as a boundary condition on the electromagnetic field. If so, how do we "chose" the correct boundary conditions for the $\mathbf{E}_T$ and $\mathbf{B}$ fields? Simple textbook choices often are incorrect. In our description of the basics of mesoscopic electrodynamics [Chapt's 19-21] we shall come back to this question.

In many cases the cavity wall is metallic, and its bulk is treated as a "perfect" conductor. Perfect means that the bulk conductivity $[\sigma(\omega)]$ is assumed to be infinite ($\sigma(\omega) \to \infty$). Although this approximation may be good at long ($\sim$ infrared - microwave) wavelengths it certainly fails at sufficiently short ($\sim$ ultraviolet) wavelengths. In the perfect conductor limit the electromagnetic field vanishes in the bulk. However, a surface (superscript S) current density

$$\mathbf{J}^S(\mathbf{r}_s; \omega) = \mathbf{J}_{\|}(\mathbf{r}_s; \omega) + J_{\perp}(\mathbf{r}_s; \omega)\mathbf{n}(\mathbf{r}_s) \tag{6.50}$$

having components both parallel ($\mathbf{J}_{\|}$) and perpendicular ($J_{\perp}\mathbf{n}$) to the cavity surface at the (arbitrary) point $\mathbf{r}_s$ is present in general. In the textbook literature $J_{\perp}\mathbf{n}$ is zero, but the argumentation used to reach this conclusion only holds in the static limit, $J_{\perp}(\mathbf{r}_s; \omega \to 0)\mathbf{n}(\mathbf{r}_s) = \mathbf{0}$. It can be shown [119, 141] that the local boundary conditions for the electric and magnetic fields

$$\mathbf{E}(\mathbf{r}_s; \omega) = \mathbf{E}_{\|}(\mathbf{r}_s; \omega) + \mathbf{n}(\mathbf{r}_s)E_{\perp}(\mathbf{r}_s; \omega), \tag{6.51}$$

$$\mathbf{B}(\mathbf{r}_s; \omega) = \mathbf{B}_{\|}(\mathbf{r}_s; \omega) + \mathbf{n}(\mathbf{r}_s)B_{\perp}(\mathbf{r}_s; \omega), \tag{6.52}$$

take the forms

$$\mathbf{E}_{\|}(\mathbf{r}_s; \omega) = \frac{1}{i\varepsilon_0\omega}\boldsymbol{\nabla}_{\|}(\mathbf{r}_s)J_{\perp}^S(\mathbf{r}_s; \omega), \tag{6.53}$$

$$E_{\perp}(\mathbf{r}_s; \omega) = \frac{1}{i\varepsilon_0\omega}\boldsymbol{\nabla}_{\|}(\mathbf{r}_s) \cdot \mathbf{J}_{\|}^S(\mathbf{r}_s; \omega), \tag{6.54}$$

$$\mathbf{B}_{\|}(\mathbf{r}_s; \omega) = \mu_0\mathbf{J}_{\|}^S(\mathbf{r}_s; \omega) \times \mathbf{n}(\mathbf{r}_s), \tag{6.55}$$

$$B_{\perp}(\mathbf{r}_s; \omega) = 0, \tag{6.56}$$

where $\boldsymbol{\nabla}(\mathbf{r}_s) = \boldsymbol{\nabla}_{\|}(\mathbf{r}_s) + \mathbf{n}(\mathbf{r}_s)\nabla_{\perp}(\mathbf{r}_s)$. In Part V, we shall study the electrodynamics in mesoscopic media in which surface currents are present at interfaces.

Let us return to $\mathcal{H}_T$ [Eq. (6.48)] and rewrite the surface term as follows:

$$\frac{1}{2\mu_0} \int^{\Sigma} \mathbf{A}_T(\mathbf{r}_s, t) \times [\boldsymbol{\nabla} \times \mathbf{A}_T(\mathbf{r}_s, t)] \cdot \mathbf{n}(\mathbf{r}_s) \mathrm{d}S$$

$$= \frac{1}{2\mu_0} \int^{\Sigma} \mathbf{A}_{T,\|}(\mathbf{r}_s, t) \times \mathbf{B}_{\|}(\mathbf{r}_s, t) \cdot \mathbf{n}(\mathbf{r}_s) \mathrm{d}S$$

$$= -\frac{1}{2} \int^{\Sigma} \mathbf{J}_{\|}^S(\mathbf{r}_s, t) \cdot \mathbf{A}_{T,\|}(\mathbf{r}_s, t) \mathrm{d}S. \tag{6.57}$$

The last expression is obtained using Eq. (6.55) transformed to the space-time domain. Schematic diagrams showing important aspects of cavity electrodynamics is shown in Fig. 6.2. If one neglects the normal part of the surface current density $[J_\perp^S(\mathbf{r}_s; \omega) = 0]$ one gets $\mathbf{E}_{\|}(\mathbf{r}_s; \omega) = \mathbf{0}$. In this approximation there will be no radiative energy transport into (or from) the interior of the cavity. Since, in turn $\mathbf{A}_{\|}(\mathbf{r}_s; \omega) = \mathbf{E}_{\|}(\mathbf{r}_s; \omega)/(i\omega) = 0$, the surface contribution to $\mathcal{H}_T$ vanishes. Physically the approximation $\mathbf{E}_{\|}(\mathbf{r}_s, t) = \mathbf{0}$ [and always $B_\perp(\mathbf{r}_s; \omega) = 0$] means that one assumes a complete decoupling of the radiative cavity field from the particle system. In case, the transverse field energy in the cavity, $\mathcal{H}_T = \sum_m \mathcal{H}_m$, is just the sum of the energies in the various harmonic oscillator modes. The energy in each mode is of course time independent

$$\frac{d\mathcal{H}_m}{dt} = 0. \tag{6.58}$$

If one to a good approximation can adopt a complete particle-transverse field decoupling, the radiative energy in the cavity is given by the sum of the harmonic oscillator energies [Eq. (6.49)]. It is shown in most textbooks on quantum mechanics that q_m and $p_m \equiv \dot{q}_m$ are conjugate position and momentum variables. It appears from Eq. (6.36) that the magnetic field, $\boldsymbol{\nabla} \times \mathbf{A}_T$, is proportional to q_m and the electric field, $-\partial \mathbf{A}_T/\partial t$, is proportional to $\dot{q}_m$. The two fields hence behave as a sort of conjugate position and momentum variables. The field oscillators are quantized in the same manner as mechanical oscillators, as briefly summarized below.

We start by a linear transformation in which the real quantities (q_m, p_m) are replaced by a combination of complex amplitude quantities (a_m, a_m^*):

$$q_m = \sqrt{\frac{\hbar}{2\omega_m}}(a_m + a_m^*), \tag{6.59}$$

$$p_m = \frac{1}{i}\sqrt{\frac{\hbar\omega_m}{2}}(a_m - a_m^*). \tag{6.60}$$

By combining Eqs. (6.49), (6.59) and (6.60) one finds

$$\mathcal{H}_m = \frac{1}{2}\hbar\omega_m(a_m^* a_m + a_m a_m^*), \tag{6.61}$$

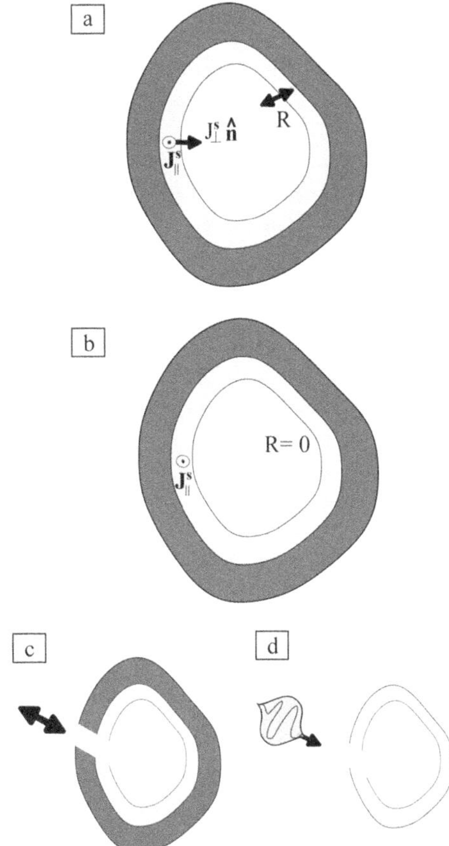

Figure 6.2 Schematic diagrams showing the extinction theorem's application to cavity electrodynamics. The cavity wall is assumed to be an ideal metal (dark grey-toned domain) carrying a surface current density distribution in the metal-vacuum (cavity) density profile (light grey-toned domain). a: Active cavity. The presence of a time-dependent normal component of the surface current density $[J_\perp^S(\mathbf{r}_s)\hat{\mathbf{n}}(\mathbf{r}_s)$, with local (at $\mathbf{r}_s$) normal unit vector $\hat{\mathbf{n}}(\mathbf{r}_s)$, results in a *radiative coupling* (R) between the particle-empty interior cavity and the wall surface. b: Passive cavity. If the surface carries only a current density distribution perpendicular to $\hat{\mathbf{n}}$ $[\mathbf{J}_S^\|(\mathbf{r}_s)]$ a complete particle-transverse field decoupling results $[R = 0]$. Decoupling can from a physical point of view only occur approximately. c: A mesoscopic hole in the cavity wall allows one to excite the cavity from outside, or couple a radiative field out of the cavity. d: In a simplified approach the in- and out-coupling of single-photon wave packets through a mesoscopic hole can be studied quantitatively neglecting the ideal bulk metal part of the cavity wall. The single-photon diffraction from a locally flat wall has been studied theoretically in [124]. In this article the importance of $\mathbf{J}_\|^s$ and $J_\perp^s\hat{\mathbf{n}}$ is calculated assuming that the dynamics of the mesoscopic hole plus its neighbourhood can be treated quite accurately in a perturbative manner. Thus, only electric-dipole [ED], magnetic-dipole [MD] and electric-quadropole [EQ] are kept in general diffraction analysis [122]. The photon tunneling process also may play an important role in the near-field zone of the mesoscopic hole [121].

keeping track of the order in which a_m and a_m^* enter the Hamiltonian. This is important when the amplitudes are replaced by operators in the usual manner

$$a_m \Rightarrow \hat{a}_m, \quad a_m^* \Rightarrow \hat{a}_m^\dagger, \tag{6.62}$$

cf. Eq. (3.1). The destruction ($\hat{a}_m$) and creation ($\hat{a}_m^\dagger$) operators satisfy the commutator relations

$$[\hat{a}_m, \hat{a}_n^\dagger] = \delta_{mn} \tag{6.63}$$

$$[\hat{a}_m, \hat{a}_n] = [\hat{a}_m^\dagger, \hat{a}_n^\dagger] = 0 \tag{6.64}$$

provided the set of operators $\{\hat{q}_m\}$ and $\{\hat{p}_m\}$ of the generalized coordinates and momenta are postulated to satisfy the usual commutation relations for the analogue mechanical oscillators. The Hamilton operator of the transverse field hence is given by

$$\hat{\mathcal{H}}_T = \sum_m \frac{1}{2}\hbar\omega_m(\hat{a}_m^\dagger\hat{a}_m + \hat{a}_m\hat{a}_m^\dagger) = \sum_m \hbar\omega_m(\hat{a}_m^\dagger\hat{a}_m + \frac{1}{2}). \tag{6.65}$$

For a cavity field in the global vacuum state ($|0\rangle$) defined by

$$\hat{a}_m|0\rangle = 0, \quad \forall m, \tag{6.66}$$

one obtains

$$\hat{\mathcal{H}}_T|0\rangle = \left(\frac{\hbar}{2}\sum_m \omega_m\right)|0\rangle. \tag{6.67}$$

The global vacuum state thus is an eigenstate for $\hat{\mathcal{H}}_T$ with eigenenergy

$$H_T^0 = \frac{\hbar}{2}\sum_m \omega_m. \tag{6.68}$$

Resolved into monochromatic plane-wave modes

$$H_T^0 = \frac{\hbar}{2}\sum_{\mathbf{q},s} \omega_q = \hbar c\sum_{\mathbf{q}} |\mathbf{q}|, \tag{6.69}$$

since the summation over the polarizations gives a factor of two.

6.4 PHOTON-EMPTY CAVITY

An extension of the classical standing mode expression given in Eqs. (6.38) and (6.39) to the field quantized level readily leads to the following results for the transverse electric and magnetic field operators:

$$\hat{\mathbf{E}}_T(\mathbf{r}, t) = -\sum_m \frac{1}{\sqrt{\varepsilon_0 V_m}}\hat{p}_m(t)\mathbf{u}_m(\mathbf{r})$$

$$= i\sum_m \frac{1}{\sqrt{\varepsilon_0 V_m}}\sqrt{\frac{\hbar\omega_m}{2}}(\hat{a}_m(t) - \hat{a}_m^\dagger(t))\mathbf{u}_m(\mathbf{r}), \tag{6.70}$$

and

$$\hat{\mathbf{B}}_T(\mathbf{r},t) = \sum_m \frac{1}{\sqrt{\varepsilon_0 V_m}} \hat{q}_m(t)\boldsymbol{\nabla} \times \mathbf{u}_m(\mathbf{r})$$

$$= \sum_m \frac{1}{\sqrt{\varepsilon_0 V_m}} \sqrt{\frac{\hbar}{2\omega_m}}(\hat{a}_m(t) + \hat{a}_m^\dagger(t))\boldsymbol{\nabla} \times \mathbf{u}_m(\mathbf{r}). \qquad (6.71)$$

In the vacuum state the mean values of the electric and magnetic fields both vanish at every space-time point inside the cavity,i.e.,

$$\langle 0 | \hat{\mathbf{E}}_T(\mathbf{r},t) | 0 \rangle = \langle 0 | \hat{\mathbf{B}}(\mathbf{r},t) | 0 \rangle = 0. \qquad (6.72)$$

However, the variances of these fields are not zero. By utilizing that

$$\langle 0 | (\hat{a}_m \mp \hat{a}_m^\dagger)(\hat{a}_n \mp \hat{a}_n^\dagger) | 0 \rangle = \mp\delta_{mn} \qquad (6.73)$$

one obtains for the transverse electric field

$$\langle 0 | \hat{\mathbf{E}}_T^2(\mathbf{r},t) | 0 \rangle = \sum_m \frac{\hbar\omega_m}{2\varepsilon_0 V_m} \mathbf{u}_m(\mathbf{r}) \cdot \mathbf{u}_m(\mathbf{r}). \qquad (6.74)$$

Thus, the variance of the electric field is time independent but different at the various field points in the cavity. For the magnetic field (multiplied by c), use of Eq. (6.73) gives

$$\langle 0 | c^2\hat{\mathbf{B}}^2(\mathbf{r},t) | 0 \rangle = \sum_m \frac{\hbar c^2}{2\varepsilon_0 V_m} \frac{1}{\omega_m}[\boldsymbol{\nabla} \times \mathbf{u}_m(\mathbf{r})] \cdot [\boldsymbol{\nabla} \times \mathbf{u}_m(\mathbf{r})], \qquad (6.75)$$

also a space-dependent variance which is constant in time Since the global vacuum possesses time-translational invariance, it is without carrying out any calculations clear that the above variances cannot depend on time. For a cavity with walls of infinite conductivity, one may make the replacement

$$[\boldsymbol{\nabla} \times \mathbf{u}_m(\mathbf{r})] \cdot [\boldsymbol{\nabla} \times \mathbf{u}_m(\mathbf{r})] \Rightarrow -\mathbf{u}_m(\mathbf{r}) \cdot \nabla^2\mathbf{u}_m(\mathbf{r}) = \left(\frac{\omega_m}{c}\right)^2 \mathbf{u}_m(\mathbf{r}) \cdot \mathbf{u}_m(\mathbf{r}), \quad (6.76)$$

as shown in Section 6.3. By inserting the replacement in Eq.(6.76) into Eq. (6.75) one obtains a magnetic field variance

$$\langle 0 | c^2\hat{\mathbf{B}}^2(\mathbf{r},t) | 0 \rangle = \sum_m \frac{\hbar\omega_m}{2\varepsilon_0 V_m} \mathbf{u}_m(\mathbf{r}) \cdot \mathbf{u}_m(\mathbf{r}). \qquad (6.77)$$

The variances of $\mathbf{E}_T$ and $c\mathbf{B}$ hence are equal in global vacuum if the field-particle coupling is neglected.

The vacuum energy inside the cavity is given by

$$H_T^0 = \langle 0 | \frac{\varepsilon_0}{2} \int_{\text{cav}} [\hat{\mathbf{E}}_T^2(\mathbf{r},t) + c^2\hat{\mathbf{B}}^2(\mathbf{r},t)]\mathrm{d}^3 r | 0 \rangle$$

$$= \frac{\hbar}{2} \sum_m \omega_m \frac{1}{V_m} \int_{\text{cav}} \mathbf{u}_m(\mathbf{r}) \cdot \mathbf{u}_m(\mathbf{r})\mathrm{d}^3 r, = \frac{\hbar}{2} \sum_m \omega_m \qquad (6.78)$$

in agreement with Eq. (6.69). The last result in Eq. (6.78) is obtained using the normalization of the spatial modes [see Eq. (6.37)].

The infinite energy of the vacuum field in the cavity must be taken with a grain of salt. Although the perfect-conductor model may be a fair approximation at long wavelengths it certainly fails at higher frequencies. Thus, at frequencies above the plasma frequency ω_p the cavity walls are essentially transparent. So only modes $\omega_m \leq \omega_p$ can at best be confined in the cavity. From an experimental point view real effects are related to externally induced *changes* in the vacuum energy, and these changes are always finite. In the Casimir effect [57, 58, 216] a force resulting from a change in cavity geometry follows from the subtraction of two infinite cavity vacuum energies [230]. The difference is finite, and even with the infinite-conductivity wall assumption rather good agreement between theory and experiment is obtained. For fundamental reasons, the electric and magnetic field variances (and thus the vacuum field energy) must be calculated from the mean-field operators; see Section 6.2. The variance of the transverse electric mean field, $\hat{\underline{\mathbf{E}}}_T(\mathbf{r}, t)$, thus is given by

$$\langle 0| \hat{\underline{\mathbf{E}}}_T^2(\mathbf{r}, t) |0\rangle = \sum_m \frac{\hbar \omega_m}{2\varepsilon_0 V_m} \underline{\mathbf{u}}_m(\mathbf{r}) \cdot \underline{\mathbf{u}}_m(\mathbf{r}), \tag{6.79}$$

where

$$\underline{\mathbf{u}}_m(\mathbf{r}) = \int_{\text{cav}} \mathbf{f}(\boldsymbol{\rho}) \cdot \mathbf{u}_m(\mathbf{r} - \boldsymbol{\rho}) \mathrm{d}^3\rho. \tag{6.80}$$

Since the mean value $\underline{\mathbf{u}}_m(\mathbf{r})$ tends rapidly towards zero for increasing mode index number (m), the high wave number (ω_m/c) terms in the expression for the variance in Eq.(6.79) are suppressed.

CHAPTER 7

Group Theory of One-Photon States

7.1 SYMMETRIES IN A QUANTUM SETTING

A symmetry transformation is a change in our point of view that does not change the results of possible experiments carried out on a given system (event). Physical states are represented by rays in Hilbert space: Normalized state vectors $|\psi\rangle$ and $|\psi'\rangle$ belong to the same ray if $|\psi'\rangle = \alpha\,|\psi\rangle$, where α is an arbitrary complex number with $|\alpha| = 1$ Thus $\langle\psi'|\psi'\rangle = |\alpha|^2\,\langle\psi|\psi\rangle = \langle\psi|\psi\rangle = 1$. If an observer $\mathcal{O}$ finds (measures) the probability $P(\mathcal{R}_m \to \mathcal{R}_n)$ for transition from ray $\mathcal{R}_m$ to ray $\mathcal{R}_n$, then an *equivalent* observer $\mathcal{O}'$ who looks at the *same* system will observe it in two different states represented by rays $\mathcal{R}'_m$ and $\mathcal{R}'_n$, but the two observers necessarily *must* find the same transition probabilities:

$$P(\mathcal{R}_m \to \mathcal{R}_n) = P(\mathcal{R}'_m \to \mathcal{R}'_n). \tag{7.1}$$

It is known from a theorem of Wigner [270,272] that for any such transformation $\mathcal{R} \to \mathcal{R}'$ of rays one may introduce a unitary and linear operator U on Hilbert space such that if $|\psi\rangle$ is in ray $\mathcal{R}$ then $|\psi'\rangle = U\,|\psi\rangle$ is in the ray $\mathcal{R}'$. The theorem of Wigner is somewhat more general than indicated above. Thus, the operator U may instead be antiunitary and antilinear. A proof of the Wigner theorem can be found in [261], e.g. The trivial symmetry transformation $\mathcal{R} \to \mathcal{R}$ is represented by the identity operator $U = 1$. This operator is, of course, unitary $U^\dagger = U^{-1}$ and linear. Symmetry transformations that connect to $U = 1$ by a continuous change of some parameters (like angle, distances or velocity) thus must be represented by a linear unitary operator. Only continuous symmetries (like a rotation, translation or Lorentz transformation) are of interest in the following. Symmetry that is infinitesimal close to $U = 1$ can be represented by

$$U = 1 + i\varepsilon t, \tag{7.2}$$

where ε is a real infinitesimal. Unitarity of U ($U^\dagger U = 1$),

$$1 = (1 - i\varepsilon t^\dagger)(1 + i\varepsilon t) = 1 + i\varepsilon(t - t^\dagger), \tag{7.3}$$

DOI: 10.1201/9781003029458-7

requires that the operator t (called the generator) is linear and Hermitian,

$$t = t^\dagger. \tag{7.4}$$

Since observables in quantum physics are Hermitian (with eigenvectors forming a complete set) [62,274] t is a candidate for an observable.

A given set of symmetry transformations, T, will have properties that define it as a *group*: (*i*) If T_1 takes $\mathcal{R}_i$ into $\mathcal{R}'_i$ [$T_1 : \mathcal{R}_i \to \mathcal{R}'_i$] and $T_2 : \mathcal{R}'_i \to \mathcal{R}''_i$ then the combined transformation (multiplication) $T_2 T_1$ is another symmetry transformation which takes $\mathcal{R}_i$ into $\mathcal{R}''_i$ [$T_1 T_2 : \mathcal{R}_i \to \mathcal{R}''_i$]. (*ii*) The multiplication is associative $(T_3 T_2)T_1 = T_3(T_2 T_1)$. (*iii*) A symmetry transformation $T : \mathcal{R}_n \to \mathcal{R}'_n$ has an inverse $T^{-1} : \mathcal{R}'_n \to \mathcal{R}_n$. (*iv*) There is an identity transformation (I) such that $IT = TI = T$ for all elements T in the group. The unitary operators $U(T)$ mirror this group structure up to a phase factor. Thus, if $T_1 : \mathcal{R}_i \to \mathcal{R}'_i$ then $U(T_1) : |\psi_1\rangle \to |\psi'_i\rangle = U(T_1)|\psi_i\rangle$ yields a Hilbert vector in the ray $\mathcal{R}'_i$. Next, if $T_2 : \mathcal{R}'_i \to \mathcal{R}''_i$ then $U(T_2)|\psi'_i\rangle = U(T_2)U(T_1)|\psi_i\rangle$ must yield a Hilbert vector in $\mathcal{R}''_i$. Now, $U(T_2 T_1)|\psi_i\rangle$ is also in this ray, so the two Hilbert vectors can differ only by a phase $\phi_i(T_2, T_1)$:

$$U(T_2)U(T_1)|\psi_i\rangle = e^{i\phi_i(T_2,T_1)} U(T_2 T_1)|\psi_i\rangle. \tag{7.5}$$

By applying Eq. (7.5) to a sum $|\psi_{i+j}\rangle$ of two linearly independent state, $|\psi_i\rangle$ and $|\psi_j\rangle$ [$\langle\psi_i|\psi_j\rangle = \delta_{ij}$], namely, $|\psi_{i+j}\rangle \equiv \frac{1}{\sqrt{2}}(|\psi_i\rangle + |\psi_j\rangle)$ it is easy to show that the phase factor is independent of the state vector (subscript: $i, j...$). Hence Eq. (7.5) can be written as (replaced by) an operator relation

$$U(T_2)U(T_1) = e^{i\phi_i(T_2,T_1)} U(T_2 T_1). \tag{7.6}$$

For $\phi = 0$, $U(T)$ furnishes a representation of the given group of symmetry transformations. For general phases $\phi(T_2, T_1)$ the representation is called a projective representation (or a representation up to a phase). A discussion of projective representations has been given by Wigner [270] and Weinberg [261], e.g.

7.2 CONNECTED LIE GROUP. LIE ALGEBRA

In relation to our study of quantum Lorentz transformations, the *connected Lie group* is of central importance. The group of transformations $T(\Gamma)$ we shall be interested in is described by a finite set of real continuous parameters, Γ^a, and each element of the group is connected to the identity by a path $[f(\bar{\Gamma}, \Gamma)]$ within the group. The group multiplication law then takes the (formal) form

$$T(\bar{\Gamma})T(\Gamma) = T(f(\bar{\Gamma}, \Gamma)). \tag{7.7}$$

At least in a finite neighbourhood of the identity transformation [$\Gamma^a = 0$, $\forall a$] the unitary operator $U(T(\Gamma))$ representing the Lie group in Eq. (7.7) on the physical Hilbert space can be represented by a power series (summation over repeated upper and lower indices implicit in the notation)

$$U(T(\Gamma)) = 1 + i\Gamma^a t_a + \frac{1}{2!}\Gamma^b \Gamma^c t_{bc} + ..., \tag{7.8}$$

where $t_{bc} = t_{cb}$, and t_a (as we already know) are Hermitian operators independent of the Γ^a's. Unitarity, $U^\dagger U = 1$, gives in lowest order when the power series in Eq. (7.8) is inserted for the imaginary part $t_a = t_a^\dagger$, and for the real part $t_{bc} = -t_a^\dagger t_b$, and hence

$$t_{bc} = t_{bc}^\dagger. \tag{7.9}$$

Suppose that the $U(T(\Gamma))$'s form an ordinary representation [cf. $\phi(T_2, T_1) = 0$ in Eq. 7.6], i.e.,

$$U(T(\bar{\Gamma}))U(T(\Gamma)) = U(T(f(\bar{\Gamma}, \Gamma))). \tag{7.10}$$

The expansion of $f^a(\bar{\Gamma}, \Gamma)$ must take the form

$$f^a(\bar{\Gamma}, \Gamma) = \Gamma^a + \bar{\Gamma}^a + f_{bc}^a \bar{\Gamma}^b \Gamma^c + \dots \tag{7.11}$$

since

$$f^a(\bar{\Gamma}, 0) = \bar{\Gamma}^a, \quad f^a(0, \Gamma^a) = \Gamma^a. \tag{7.12}$$

The results in Eq. (7.12) follow immediately from Eq. (7.7). There are no two terms of order $(\Gamma^a)^2$ and $(\bar{\Gamma}^a)^2$ in Eq. (7.11) since such terms would violate Eqs. (7.12). When the expansions in Eqs. (7.8) and (7.12) are inserted in Eq. (7.10) terms of order 1, Γ, $\bar{\Gamma}$, Γ^2 and $\bar{\Gamma}^2$ automatically match. The $\Gamma\bar{\Gamma}$ terms only match if the condition

$$t_{bc} = -t_b t_c - i f_{bc}^a t_a \tag{7.13}$$

is satisfied. Thus, it follows that a knowledge of the quadratic coefficients f_{bc}^a in the power expansion of the group structure function allows one to calculate the t_{bc}'s from the generators t_a. Since $t_{cb} = t_{bc}$, Eq. (7.13), requires that the commutator of t_b and t_c is given by

$$[t_b, t_c] = i C_{bc}^a t_a, \tag{7.14}$$

where

$$C_{bc}^a = f_{cb}^a - f_{bc}^a \tag{7.15}$$

is a set of real constants known as the *structure constants*. The set of commutation relations in Eq. (7.14) is known as the *Lie Algebra* for the structure In Section 7.4, we shall derive the Lie algebra for the Poincare group (the inhomogenous Lorentz group).

In the special case where

$$f^a(\bar{\Gamma}, \Gamma) = \Gamma^a + \bar{\Gamma}^a \tag{7.16}$$

the structure coefficients f_{bc}^a in Eq. (7.11) vanish. The generators then all commute

$$[t_b, t_c] = 0, \tag{7.17}$$

and the group is called *Abelian*. An Abelian group besides the general properties needed to define a set of symmetry transformations as a group [given by points (i)-(iv); see the text in Section 7.1 just below Eq. (7.4).] satisfies $T_2 T_1 = T_1 T_2$. Translations in space-time and rotations about any one fixed axis satisfy Eq. (7.16). Together, such a set of symmetry transformations is not Abelian, as is well known.

It follows from Eq. (7.10) and Eq. (7.16) that an Abelian one-parameter group satisfy

$$U(T(\bar{\Gamma}))U(T(\Gamma)) = U(T(\Gamma + \bar{\Gamma})), \tag{7.18}$$

and hence

$$U(T(\Gamma)) = \exp(i\Gamma t). \tag{7.19}$$

Generalized to a finite set of parameters

$$U(T(\Gamma)) = \exp(i\Gamma^a t_a). \tag{7.20}$$

7.3 QUANTUM LORENTZ TRANSFORMATIONS

In Special Relativity, with the diagonal metric tensor $\{\eta_{\mu\nu}\}$ having the elements

$$\eta_{ii} = +1(i = 1 - 3), \ \eta_{00} = -1, \tag{7.21}$$

the invariance of the proper time implies that a coordinate transformation $x^\mu \to x'^\mu$ [with x^μ ($\mu = 1 - 3$) being Cartesian space coordinates and $x^0 = t$ (speed of light $c = 1$) a time coordinate] between inertial systems must satisfy

$$\eta_{\mu\nu}dx'^\mu dx'^\nu = \eta_{\mu\nu}dx^\mu dx^\nu, \tag{7.22}$$

in a notation where the summation over repeated upper and lower indices is in force.

Any coordinate transformation that satisfies Eq. (7.22) is linear, i.e.,

$$x'^\mu = \Lambda^\mu_\nu x^\nu + a^\mu, \tag{7.23}$$

where $\{a^\mu\}$ are arbitrary constants, and $\{\Lambda^\mu_\nu\}$ a constant matrix satisfying the Lorentz transformation condition

$$\eta_{\mu\nu}\Lambda^\mu_\rho \Lambda^\nu_\sigma = \eta_{\rho\sigma}. \tag{7.24}$$

The inhomogeneous Lorentz transformations in Eq. (7.23) form a group, called the Poincare group. The two Lorentz transformations ($x^\mu \to x'^\mu \to x''^\mu$), giving

$$x''^\mu = \overline{\Lambda}^\mu_\rho x'^\rho + \overline{a}^\mu = \overline{\Lambda}^\mu_\rho(\Lambda^\rho_\nu x^\nu + a^\rho) + \overline{a}^\mu, \tag{7.25}$$

is the same as a single Lorentz transformation ($x^\mu \to x''^\mu$)

$$x''^\mu = (\overline{\Lambda}^\mu_\rho \Lambda^\rho_\nu)x^\nu + (\overline{\Lambda}^\mu_\rho a^\rho + \overline{a}^\mu). \tag{7.26}$$

Note that $\overline{\Lambda}^\mu_\rho \Lambda^\rho_\nu$ satisfies Eq. (7.24) since Λ^μ_ν and $\overline{\Lambda}^\mu_\nu$ both do.

The transformations $T(\Lambda, a)$ therefore satisfy the group composition rule (i)

$$T(\overline{\Lambda}, \overline{a})T(\Lambda, a) = T(\overline{\Lambda}\Lambda, \overline{\Lambda}a + \overline{a}), \qquad (7.27)$$

and the associate rule of multiplication (ii). Since the determinant squared is

$$(\text{Det}\Lambda)^2 = 1, \qquad (7.28)$$

as one may show from Eq. (7.25), Λ has an inverse Λ^{-1} (iii), and, of course an identity transformation $T(1,0)$, (iv). The inverse of the transformation $T(\Lambda, a)$ can be obtained from Eq. (7.27). Thus with $\overline{\Lambda} = \Lambda^{-1}$ one gets

$$T(1,0) = T(\Lambda^{-1}, \overline{a})T(\Lambda, a) = T(\Lambda^{-1}\Lambda, \Lambda^{-1}a + \overline{a}) \qquad (7.29)$$

and hence

$$\overline{a} = -\Lambda^{-1}a. \qquad (7.30)$$

The inverse of the transformation $T(\Lambda, a)$ therefore becomes $T(\Lambda^{-1}, -\Lambda^{-1}a)$.

The unitary operators relating inhomogeneous Lorentz transformations to the linear transformation of state vectors in Hilbert space in consequence satisfy the group multiplication rule

$$U(\overline{\Lambda}, \overline{a})U(\Lambda, a) = U(\overline{\Lambda}\Lambda, \overline{\Lambda}a + \overline{a}), \qquad (7.31)$$

assuming as before that the group representation is non-projective $[\phi = 0$, in Eq. (7.6)]. The group of transformations $T(\Lambda, a)$ is known as the *inhomogeneous Lorentz group* or *Poincare group*. The set of homogeneous transformations $\{a^\mu\} = 0$ obviously forms a subgroup with multiplication rule

$$T(\overline{\Lambda}, 0)T(\Lambda, 0) = T(\overline{\Lambda}\Lambda, 0). \qquad (7.32)$$

The group is called the *homogeneous Lorentz group*. Furthermore, Eq. (7.28) tells us that $\text{Det}(\Lambda) = +1$ or $\text{Det}(\Lambda) = -1$. The transformations with $\text{Det}(\Lambda) = +1$ form a subgroup for both the homogeneous and the inhomogeneous Lorentz groups. Transformations with $\text{Det}(\Lambda) = +1$ are named *proper transformations*, and those with $\text{Det}(\Lambda) = -1$ are *improper transformations*. From the $\rho = \sigma = 0$ component of Eq. (7.24), one has

$$(\Lambda_0^0)^2 = 1 + \Lambda_o^i\Lambda_0^i, \qquad (7.33)$$

with i summed over 1,2 and 3. Hence $\Lambda_0^0 \geq +1$ (orthochronous) or $\Lambda_0^0 \leq -1$ (non-orthochronous). Together with the condition on the determinant, there are therefore *four classes* of homogeneous Lorentz transformations that are disconnected in the sense that one cannot go from one of the classes to another by a continuous change of parameters. Lorentz transformations in each class can be continuously deformed into any other Lorentz transformation in that class.

Let us label the four classes as follows: $L_+^\uparrow$ ($\Lambda_0^0 \geq 1, \mathrm{Det}(\Lambda) = +1$), $L_+^\downarrow$ ($\Lambda_0^0 \leq -1, \mathrm{Det}(\Lambda) = +1$), $L_-^\uparrow$ ($\Lambda_0^0 \geq 1, \mathrm{Det}(\Lambda) = -1$), $L_-^\downarrow$ ($\Lambda_0^0 \leq -1, \mathrm{Det}(\Lambda) = -1$). The class $L_+^\uparrow$ is a subgroup, but none of the others are groups because they have no identity element: It is the $L_+^\uparrow$ class which holds the identity element. The four classes can be connected by discrete transformations related to time inversion ($\mathcal{T}$) and space inversion ($\mathcal{P}$) [parity]. The nonvanishing elements of these transformations are [with $i = 1 - 3$]

$$\mathcal{T}_0^0 = -1, \ \mathcal{T}_i^i = 1, \tag{7.34}$$

and

$$\mathcal{P}_0^0 = 1, \ \mathcal{P}_i^i = -1. \tag{7.35}$$

It can be shown that the mapping from the class $L_+^\uparrow$ to the three others are

$$\mathcal{T}L_+^\uparrow = L_-^\downarrow, \tag{7.36}$$
$$\mathcal{P}L_+^\uparrow = L_-^\uparrow, \tag{7.37}$$
$$\mathcal{T}\mathcal{P}L_+^\uparrow = L_+^\downarrow, \tag{7.38}$$

and there it is sufficient to study the group of continuous transformations, $L_+^\uparrow$, and the two discrete transformations, $\mathcal{T}$ and $\mathcal{P}$. The time and space inversions commute, i.e.,

$$[\mathcal{T}, \mathcal{P}] = 0, \tag{7.39}$$

and $\mathcal{P}^2 = \mathcal{T}^2 = U$, of course.

7.4 THE POINCARE GROUP ALGEBRA

As we realized in Section 7.2, much of the information about the connected Lie group can be obtained from studies of the group elements infinitesimally close to the identity element. For the Poincare group, those transformations have the form

$$\Lambda_\nu^\mu = \delta_\nu^\mu + \omega_\nu^\mu, \ a^\mu = \epsilon^\mu, \tag{7.40}$$

where ω_ν^μ and ϵ^μ are our infinitesimal quantities. By inserting the expression for Λ_ν^μ in Eq. (7.40) into the Lorentz transformation condition [Eq. (7.24)] one obtains to first order in $\{\omega_\nu^\mu\}$

$$0 = \eta_{\mu\nu}(\delta_\rho^\mu \omega_\sigma^\nu + \omega_\rho^\mu \delta_\sigma^\nu) = \eta_{\rho\nu}\omega_\sigma^\nu + \eta_{\mu\sigma}\omega_\rho^\mu = \omega_{\rho\sigma} + \omega_{\sigma\nu}. \tag{7.41}$$

In the last step of Eq. (7.41), we have used that indices may be lowered (or raised) by contraction with the metric tensor. To first order in $\{\omega_{\mu\nu}\}$, the Lorentz condition [Eq.(7.24)] thus reduces to antisymmetry of $\omega_{\mu\nu}$

$$\omega_{\nu\mu} = -\omega_{\mu\nu}. \tag{7.42}$$

The antisymmetry implies that the homogeneous Lorentz transformation has $4 \times 3/2 = 6$ independent parameters [spatial rotations: 3, boosts 3]. Including the four space-time translations of $\{\epsilon^\mu\}$, it follows that the Poincare group is described by $6+4 = 10$ parameters.

Since $U(1,0)$ carries any ray into itself, it must be proportional to the unit operator, and can by a choice of phase be made equal to it [use the same reasoning leading from Eq. (7.5) to Eq. (7.6), and neglect the phase factor needed in projective representations]. Hence, we take

$$U(1+\omega, \epsilon) = 1 + \frac{i}{2}\omega_{\rho\sigma}J^{\rho\sigma} - i\epsilon_\rho P^\rho + \ldots \tag{7.43}$$

Unitarity (to first order) then gives

$$(J^{\rho\sigma})^\dagger = J^{\rho\sigma}, \ (P^\rho)^\dagger = P^\rho. \tag{7.44}$$

By dividing $J^{\rho\sigma}$ into symmetric and antisymmetric parts, it follows from the antisymmetry of $\omega_{\rho\sigma}$ that one can let $\{J^{\rho\sigma}\}$ be antisymmetric also

$$J^{\rho\sigma} = -J^{\sigma\rho}. \tag{7.45}$$

Now, let $U(\Lambda, a)$ be the unitary operator representing a *general* Lorentz transformation. Utilizing that $U(\Lambda^{-1}, -\Lambda^{-1}a)$ is the inverse of $U(\Lambda, a)$ [cf. the analysis in Section 7.3] it follows from Eq. (7.31) that

$$U(\Lambda, a)U(1+\omega, \epsilon)U^{-1}(\Lambda, a) = U(\Lambda(1+\omega)\Lambda^{-1}, \Lambda\epsilon - \Lambda\omega\Lambda^{-1}a). \tag{7.46}$$

By inserting the first-order form of $U(1+\omega, \epsilon)$ [given in Eq. (7.43)] in the equation above, and equating thereafter the coefficients of $\omega_{\rho\sigma}$ and ϵ_ρ, one obtains

$$U(\Lambda, a)J^{\rho\sigma}U^{-1}(\Lambda, a) = (\Lambda^{-1})^\rho_\mu(\Lambda^{-1})^\sigma_\nu(J^{\mu\nu} - a^\mu P^\nu + a^\nu P^\mu), \tag{7.47}$$

and

$$U(\Lambda, a)P^\rho U^{-1}(\Lambda, a) = (\Lambda^{-1})^\rho_\mu P^\mu. \tag{7.48}$$

From Eq. (7.24) it follows that the elements of the reciprocal transformation to Λ is given by $(\Lambda^{-1})^\rho_\mu = \eta_{\mu\nu}\eta^{\rho\sigma}\Lambda^\nu_\sigma$. If we now apply Eqs. (7.47) and (7.48) to a transformation $U(\Lambda, a)$ that is itself infinitesimal [Eq. (7.43)], and again equate coefficients of $\omega_{\mu\nu}$ and ϵ_μ, we obtain the following commutation relations

$$i[J^{\mu\nu}, J^{\rho\sigma}] = \eta^{\nu\rho}J^{\mu\sigma} - \eta^{\mu\rho}J^{\nu\sigma} - \eta^{\sigma\mu}J^{\rho\nu} + \eta^{\sigma\nu}J^{\rho\mu}, \tag{7.49}$$

$$i[P^\mu, J^{\rho\sigma}] = \eta^{\mu\rho}P^\sigma - \eta^{\mu\sigma}P^\rho, \tag{7.50}$$

$$[P^\mu, P^\rho] = 0. \tag{7.51}$$

Eqs. (7.49)–(7.51) is the Lie algebra of the Poincare group in four-dimensional notation.

By introduction of the momentum three-vector

$$\mathbf{P} = (P^1, P^2, P^3), \tag{7.52}$$

the energy $H = \hbar c P^0$, the angular-momentum three-vector

$$\mathbf{J} = (J^{23}, J^{31}, J^{12}), \tag{7.53}$$

and the so-called boost three-vector

$$\mathbf{K} = (J^{10}, J^{20}, J^{30}), \tag{7.54}$$

the commutation rules in Eqs. (7.49)–(7.51) may be written as follows:

$$[J_i, J_j] = i\varepsilon_{ijk} J_k, \tag{7.55}$$
$$[J_i, K_j] = i\varepsilon_{ijk} K_k, \tag{7.56}$$
$$[K_i, K_j] = -i\varepsilon_{ijk} J_k, \tag{7.57}$$
$$[J_i, P_j] = i\varepsilon_{ijk} P_k, \tag{7.58}$$
$$[K_i, P_j] = i\delta_{ij} H, \tag{7.59}$$
$$[J_i, H] = [P_i, H] = [H, H] = 0, \tag{7.60}$$
$$[K_i, H] = iP_i. \tag{7.61}$$

In Eqs. (7.55)–(7.58), ε_{ijk} is the totally antisymmetric Levi-Civita symbol $[\varepsilon_{123} = +1]$. The commutation relation in Eq. (7.55) will be recognized as that of the angular-momentum operator.

Pure space-time translations and three-dimensional spatial rotations form two subgroups of the Poincare group with group multiplication rule given by Eq. (7.18). The solution in Eq. (7.20) therefore shows that finite translations and spatial rotations are represented by

$$U(1, a) = \exp(-iP^\mu a_\mu) = \exp[i(\mathbf{q} \cdot \mathbf{r} - \omega t)], \tag{7.62}$$

and

$$U(R_{\boldsymbol{\theta}}, 0) = \exp(i\mathbf{J} \cdot \boldsymbol{\theta}), \tag{7.63}$$

respectively. The $R_{\boldsymbol{\theta}}$ is around the direction $\hat{\boldsymbol{\theta}} = \boldsymbol{\theta}/\theta$, by an angle θ. It follows from the Heisenberg equation of motions for the operators $\mathbf{J}, \mathbf{P}$ and H and the commutator relations in Eq. (7.60) that $\mathbf{J}, \mathbf{P}$ and H are conserved quantities.

7.5 PHOTON LITTLE GROUP: SE(2)

Since the components of the energy-momentum four vector all commute [Eq. (7.51)] it is natural to express the (here elementary) one-particle state vectors $(|\psi_{p,\sigma}\rangle)$ as eigenvectors of the four-momentum, i.e.

$$\hat{P}^\mu |\psi_{p,\sigma}\rangle = p^\mu |\psi_{p,\sigma}\rangle \tag{7.64}$$

in a notation where we have reinserted the usual operator symbol $(\hat{\ })$ on the four-momentum operator. The eigenvalue of $\hat{P}^{\mu}$ is p^{μ}. The label σ, which we shall assume is discrete, characterizes the internal state of the one-particle state. The only quantity that is left invariant by all proper $(p^0 \geq 0)$ orthochromous Lorentz transformation is $p^2 = \eta_{\mu\nu}p^{\mu}p^{\mu}$, [the scalar product $\eta_{\mu\nu}p^{\mu}p^{\nu} = p_{\nu}p^{\nu}$ is an invariant]. Below we shall limit ourselves to an analysis of the photon case for which $p^2 = 0$, $p^0 > 0$.

Let us now consider the *Wigner little group* [271], a subgroup consisting of Lorentz transformations W^{μ}_{ν} that leave p^{μ} invariant

$$W^{\mu}_{\nu}p^{\nu} = p^{\mu}. \tag{7.65}$$

Detailed discussions of the little group structure for massive and massless particles have been given by Wigner [271] and Weinberg [261], e.g.

Consider now two four-momenta: (i) A standard four-momentum $k^{\mu} = (1,0,0,1)k$, which describes a photon moving along the z-axis with a three-vector momentum magnitude k, $\sim \exp[i(k^3 x_3 - k^0 x_0)] = \exp[ik(x_3 - x_0)]$. (ii) A time-like momentum four-vector $\{t^{\mu}\} = (1,0,0,0)t$. With no loss of generality we can set $k = t = 1$ in working out the structure of the photon little group. From Eq. (7.65) we obtain

$$(Wt)^{\mu}(Wt)_{\mu} = t^{\mu}t_{\mu} = -1, \tag{7.66}$$
$$(Wt)^{\mu}k_{\mu} = t^{\mu}k_{\mu} = -1. \tag{7.67}$$

Any four-vector that satisfy Eq. (7.67) has the form

$$(Wt)^{\mu} = (1 + \gamma, \alpha, \beta, \gamma), \tag{7.68}$$

and Eq. (7.66) then allows one to express γ in terms of α and β. Thus

$$\gamma = \frac{1}{2}(\alpha^2 + \beta^2). \tag{7.69}$$

A straightforward calculation shows that the Lorentz transformation

$$S^{\mu}_{\nu}(\alpha, \beta) = \begin{bmatrix} 1+\gamma & \alpha & \beta & -\gamma \\ \alpha & 1 & 0 & -\alpha \\ \beta & 0 & 1 & -\beta \\ \gamma & \alpha & \beta & 1-\gamma \end{bmatrix} \tag{7.70}$$

satisfies the relation

$$S^{\mu}_{\nu}t^{\nu} = W^{\mu}_{\nu}t^{\nu} = \begin{bmatrix} 1+\gamma \\ \alpha \\ \beta \\ \gamma \end{bmatrix}. \tag{7.71}$$

This implies that

$$S^{-1}Wt = t. \tag{7.72}$$

The transformation $S^{-1}W$ therefore leaves the time-like four-vector $\{t^\mu\}/t = (1,0,0,0)$ invariant, and $S^{-1}W$ must be a spatial rotation by some angle θ around the 3-axis because also the standard four-vector $\{k^\mu\}/k$ satisfies

$$S^\mu_\nu k^\nu = W^\mu_\nu k^\nu = \begin{bmatrix} 1 \\ 0 \\ 0 \\ 1 \end{bmatrix}. \tag{7.73}$$

In conclusion, it appears that the *general* element of the little group has the form

$$W(\theta, \alpha, \beta) = S(\alpha, \beta)R(\theta), \tag{7.74}$$

where

$$R^\mu_\nu(\theta) = \begin{bmatrix} 1 & 0 & 0 & 0 \\ 0 & \cos\theta & \sin\theta & 0 \\ 0 & -\sin\theta & \cos\theta & 0 \\ 0 & 0 & 0 & 1 \end{bmatrix}. \tag{7.75}$$

In order to determine the Lie algebra of the photon little group, which general group element is

$$W^\mu_\nu = S^\mu_\rho R^\rho_\nu$$
$$= \begin{bmatrix} 1+\gamma & \alpha\cos\theta - \beta\sin\theta & \alpha\sin\theta + \beta\cos\theta & -\gamma \\ \alpha & \cos\theta & \sin\theta & -\alpha \\ \beta & -\sin\theta & \cos\theta & -\beta \\ \gamma & \alpha\cos\theta - \beta\sin\theta & \alpha\sin\theta + \beta\cos\theta & 1-\gamma \end{bmatrix}, \tag{7.76}$$

we consider the form of the $W^\mu_\nu(\theta, \alpha, \beta)$ for infinitesimal θ, α and β. Keeping only lowest order terms

$$W^\mu_\nu(\theta, \alpha, \beta) = \begin{bmatrix} 1 & \alpha & \beta & 0 \\ \alpha & 1 & \theta & -\alpha \\ \beta & -\theta & 1 & -\beta \\ 0 & \alpha & \beta & 1 \end{bmatrix}$$
$$= \delta^\mu_\nu + \omega^\mu_\nu(\theta, \alpha, \beta), \tag{7.77}$$

and hence

$$\omega_{\nu\mu}(\theta, \alpha, \beta) = \eta_{\mu\rho}\omega^\rho_\nu(\theta, \alpha, \beta)$$
$$= \begin{bmatrix} 0 & -\alpha & -\beta & 0 \\ \alpha & 0 & \theta & -\alpha \\ \beta & -\theta & 0 & -\beta \\ 0 & \alpha & \beta & 0 \end{bmatrix}. \tag{7.78}$$

From the antisymmetric tensor $[J^{\nu\mu} = -J^{\mu\nu}]$

$$J^{\mu\nu} = \begin{bmatrix} 0 & J^{01} & J^{02} & J^{03} \\ J^{10} & 0 & J^{12} & J^{13} \\ J^{20} & J^{21} & 0 & J^{23} \\ J^{30} & J^{31} & J^{32} & 0 \end{bmatrix}, \tag{7.79}$$

and Eq. (7.78) one obtains

$$\frac{1}{2}\omega_{\mu\nu}J^{\mu\nu} = \alpha(J^{10} - J^{13}) + \beta(J^{20} - J^{23}) + \theta J^{12}. \tag{7.80}$$

Expressed in terms of the angular-momentum three-vector [Eq. (7.53)]

$$\mathbf{J} = (J_1, J_2, J_3) \equiv (J^{23}, J^{31}, J^{12}), \tag{7.81}$$

and the boost three-vector [Eq. (7.54)]

$$\mathbf{K} = (K_1, K_2, K_3) \equiv (J^{10}, J^{20}, J^{30}), \tag{7.82}$$

$$\frac{i}{2}\omega_{\mu\nu}J^{\mu\nu} = i\alpha(K_1 + J_2) + i\beta(K_2 - J_1) + i\theta J_3. \tag{7.83}$$

The corresponding Hilbert space operator [Eq. (7.43)] is

$$U(W(\theta, \alpha, \beta)[= U(1 + \omega, \varepsilon = 0)] = 1 + i\alpha A + i\beta B + i\theta J_3, \tag{7.84}$$

where

$$A = K_1 + J_2, \tag{7.85}$$
$$B = K_2 - J_1, \tag{7.86}$$

are the Hermitian operators. The generators A, B and J_3 satisfy the commutation relations

$$[A, B] = 0, \tag{7.87}$$
$$[J_3, A] = iB, \tag{7.88}$$
$$[J_3, B] = -iA. \tag{7.89}$$

As Wigner observed in 1939 [271], the closed set of commutation relations is just like that for generators of the two-dimensional Euclidean group [EU(2)]: One rotation (around the 3-axis) and two translations (along the 1- and 2-axes). We shall see in Section 8.1 that the part of the photon little group parametrized by α and β relates to the electromagnetic gauge invariance.

From the form of the little group element Eq. (7.74) [with explicit expression given in Eq. (7.76)] it is readily shown that the transformations with $\theta = 0$ form a subgroup

$$S(\overline{\alpha}, \overline{\beta})S(\alpha, \beta) = S(\overline{\alpha} + \alpha, \overline{\beta} + \beta). \tag{7.90}$$

The pure rotations ($\alpha = \beta = 0$) also form a subgroup

$$R(\bar{\theta})R(\theta) = R(\bar{\theta} + \theta). \tag{7.91}$$

Both subgroups are *Abelian*. With the help of the transformation

$$R(\theta)S(\alpha, \beta)R^{-1}(\theta) = S(\alpha \cos \theta + \beta \sin \theta, -\alpha \sin \theta + \beta \cos \theta), \tag{7.92}$$

which the reader may show to herself, the multiplication rules for the photon little group can be worked out, and recognized as those of the group EU(2).

7.6   REMARKS ON PHOTON HELICITY

Since A and B [Eqs. (7.85) and (7.86)] are commuting Hermitian operators that can be simultaneously diagonalized. Thus

$$A \left|\psi_{k,a,b}\right\rangle = a \left|\psi_{k,a,b}\right\rangle, \tag{7.93}$$

$$B \left|\psi_{k,a,b}\right\rangle = b \left|\psi_{k,a,b}\right\rangle, \tag{7.94}$$

where in the eigenstate $\left|\psi_{k,a,b}\right\rangle$ k refers to a three-vector momentum in the z-direction. Let us assume that the real eigenvalues a and b are non-zero. It appears from Eqs. (7.84) and (7.92) that the association

$$S(\alpha, \beta) \Rightarrow U = 1 + i\alpha A + i\beta B \tag{7.95}$$

implies

$$\begin{aligned} R(\theta)S(\alpha, \beta)R^{-1}(\theta) =& S(\alpha \cos \theta + \beta \sin \theta, -\alpha \sin \theta + \beta \cos \theta) \\ \Rightarrow U =& 1 + i(\alpha \cos \theta + \beta \sin \theta)A + i(-\alpha \sin \theta + \beta \cos \theta)B \\ =& 1 + i\alpha(A \cos \theta - B \sin \theta) + \beta(A \sin \theta + B \cos \theta). \end{aligned} \tag{7.96}$$

The transformations of A and B hence must be

$$U(R(\theta)AU^{-1}(R(\theta)) = A \cos \theta - B \sin \theta, \tag{7.97}$$

$$U(R(\theta)BU^{-1}(R(\theta)) = A \sin \theta + B \cos \theta. \tag{7.98}$$

By letting A act on the rotated eigenstate

$$\left|\psi_{k,a,b}(\theta)\right\rangle \equiv U^{-1}(R(\theta)) \left|\psi_{k,a,b}\right\rangle, \tag{7.99}$$

one obtains

$$\begin{aligned} A \left|\psi_{k,a,b}(\theta)\right\rangle =& U^{-1}UAU^{-1} \left|\psi_{k,a,b}\right\rangle \\ =& U^{-1}(A \cos \theta - B \sin \theta) \left|\psi_{k,a,b}\right\rangle \\ =& U^{-1}(a \cos \theta - b \sin \theta) \left|\psi_{k,a,b}\right\rangle, \end{aligned} \tag{7.100}$$

and then

$$A \left|\psi_{k,a,b}(\theta)\right\rangle = (a \cos \theta - b \sin \theta) \left|\psi_{k,a,b}(\theta)\right\rangle. \tag{7.101}$$

An analogous calculation gives

$$B \left| \psi_{k,a,b}(\theta) \right\rangle = (a \sin \theta + b \cos \theta) \left| \psi_{k,a,b}(\theta) \right\rangle . \tag{7.102}$$

The results in Eqs. (7.101) and (7.102) show that the commuting operators A and B have a whole continuum of eigenvalues *provided a and b are different from zero.* Since no experimental observations on photons have shown that there belongs to a continuum of eigenvalues to a given photon momentum $(\mathbf{k})$, we conclude that $a = b = 0$. The eigenstates $\left| \psi_{k,\sigma} \right\rangle$ with

$$A \left| \psi_{k,\sigma} \right\rangle = B \left| \psi_{k,\sigma} \right\rangle = 0, \tag{7.103}$$

now characterized by the *discrete* index σ (and k), are distinguished by the eigenvalue (σ) of the J_3 generator

$$J_3 \left| \psi_{k,\sigma} \right\rangle = \sigma \left| \psi_{k,\sigma} \right\rangle . \tag{7.104}$$

The quantity σ, named the *photon helicity* therefore gives the component of the angular momentum in the three-momentum direction (here $\mathbf{k}$). The helicity is Lorentz-invariant: A photon of a given helicity "looks the same" in all inertial systems, aside from its four-momentum. The eigenvalues of the photon helicity is $\pm\hbar$.

It follows from Eq. (7.103) that

$$U(W) \left| \psi_{k,\sigma} \right\rangle = \exp[i(\alpha A + \beta B + \theta J_3)] \left| \psi_{k,\sigma} \right\rangle = \exp(i\sigma\theta) \left| \psi_{k,\sigma} \right\rangle . \tag{7.105}$$

The helicity dependence of the phase factor $\exp(i\sigma\theta)$ implies that a state formed by a linear superposition of one-photon states with opposite helicities (+ and -), say

$$\left| \psi_{k,\sigma} \right\rangle = \alpha_+ \left| \psi_{k,+} \right\rangle + \alpha_- \left| \psi_{k,-} \right\rangle \tag{7.106}$$

where

$$\left| \alpha_+ \right|^2 + \left| \alpha_- \right|^2 = 1 \tag{7.107}$$

can have different polarizations: $|\alpha_+| \neq |\alpha_-| \neq 0$, elliptic; either $\alpha_+ = 0$ or $\alpha_- = 0$, circular; $|\alpha_+| = |\alpha_-|$ linear. A general analysis of the Lorentz transformation rule for a photon of arbitrary helicity can be found in [261].

CHAPTER 8

Wave-Packet Photons and Bi-Photons

8.1 PHOTON LITTLE GROUP AND GAUGE INVARIANCE

Let us consider the four-potential as the wave function of the photon. A monochromatic plane-wave photon propagating along the 3-axis (z-axis) thus has a wave function

$$A^\mu(z, t) = \begin{pmatrix} A^0(\mathbf{q}, \omega) \\ A^1(\mathbf{q}, \omega) \\ A^2(\mathbf{q}, \omega) \\ A^3(\mathbf{q}, \omega) \end{pmatrix} \exp[iq(z - ct)]. \tag{8.1}$$

Application of the Lorentz transformation given in Eq. (7.70) to the electromagnetic potential four-vector gives

$$S^\mu_\nu(\alpha, \beta) A^\nu(\mathbf{q}, \omega)$$
$$= \begin{pmatrix} 1 & \alpha & \beta & 0 \\ 0 & 1 & 0 & 0 \\ 0 & 0 & 1 & 0 \\ 0 & \alpha & \beta & 1 \end{pmatrix} \begin{pmatrix} A^0 \\ A^1 \\ A^2 \\ A^3 \end{pmatrix} + (A^0 - A^3) \begin{pmatrix} \gamma \\ \alpha \\ \beta \\ \gamma \end{pmatrix}. \tag{8.2}$$

If the four-vector satisfies the Lorenz gauge condition, $A^0 - A^3 = 0$, the last term in Eq.(8.2) vanishes, and the first part can rewritten as follows:

$$S^\mu_\nu(\alpha, \beta) A^\nu(\mathbf{q}, \omega)$$
$$= \begin{pmatrix} A^0(\mathbf{q}, \omega) \\ A^1(\mathbf{q}, \omega) \\ A^2(\mathbf{q}, \omega) \\ A^3(\mathbf{q}, \omega) \end{pmatrix} + f(\alpha, \beta | \mathbf{q}, \omega) \begin{pmatrix} 1 \\ 0 \\ 0 \\ 1 \end{pmatrix}, \tag{8.3}$$

where

$$f(\alpha, \beta | \mathbf{q}, \omega) = \alpha A^1(\mathbf{q}, \omega) + \beta A^2(\mathbf{q}, \omega). \tag{8.4}$$

DOI: 10.1201/9781003029458-8

The second part of Eq. (8.3) adds the *same* quantity $f(\alpha, \beta | \mathbf{q}, \omega)$ to the longitudinal and time-like components of the potential four-vector, and can therefore be eliminated by a gauge transformation within the Lorenz gauge. The transverse components of the four-potential is left invariant. Hence, it appears without loss of generality, that the photon four-potential may satisfy the photon little group criterion $(\theta = 0)$

$$S^\mu_\nu(\alpha, \beta) A^\nu(\mathbf{q}; \omega) = W^\mu_\nu(\alpha, \beta) A^\nu(\mathbf{q}, \omega) = A^\mu(\mathbf{q}, \omega). \tag{8.5}$$

8.2 POLYCHROMATIC PHOTONS

In most textbooks single photons are associated with the quanta of monochromatic plane-wave components of the electromagnetic field In this approach the Planck-Einstein relations [75, 76, 211, 212]

$$E = \hbar\omega, \ \mathbf{p} = \hbar\mathbf{q} \tag{8.6}$$

make the bridge between particle [energy E, momentum $\mathbf{p}$] and wave [angular frequency ω, wave vector $\mathbf{q}$] aspects of the photon. These relations were established prior to the development of the quantum theory for the electromagnetic field [44, 45, 69, 84], and to uphold the fundamental status of the Planck-Einstein relations it is speciously convincing to conclude that the photon *is* an entity associated with a monochromatic plane-wave quantum excitation of the electromagnetic field. Quantum field theory tells us that this conventional wisdom is too narrow. In the following we shall discuss the polychromatic single-photon concept, well-established but perhaps not well known,

The analysis below is carried out in the Schrödinger Picture, where the state vectors in Hilbert space evolve in time, and the operators are fixed. For the *free fields* the calculational labour needed to reach the conclusions essentially has the same size in the Heisenberg and Schrödinger Pictures. Although our subject here is *the polychromatic photon*, we start from a plane-wave mode functions. In the box (L^3)-mode description, the transverse vector potential operator is given by $[\omega_i \equiv cq_i]$

$$\hat{\mathbf{A}}_T(\mathbf{r}) = \left(\frac{\hbar}{2\varepsilon_0 L^3}\right)^{1/2} \sum_i \frac{1}{\omega_i^{1/2}} \left[\hat{a}_i \boldsymbol{\varepsilon}_i e^{i\mathbf{q}_i \cdot \mathbf{r}} + \hat{a}_i^\dagger \boldsymbol{\varepsilon}_i^* e^{-i\mathbf{q}_i \cdot \mathbf{r}}\right], \tag{8.7}$$

where the destruction $(\hat{a}_i)$ and creation $(\hat{a}_i^\dagger)$ operators are time independent (Schrödinger Picture), and satisfy the commutation rules

$$[\hat{a}_i, \hat{a}_j] = [\hat{a}_i^\dagger, \hat{a}_j^\dagger] = 0, \tag{8.8}$$

$$[\hat{a}_i, \hat{a}_j^\dagger] = \delta_{ij}. \tag{8.9}$$

In Eq. (8.7) the index $i = (\mathbf{q}_i, s)$, where $s = 1, 2$ refers to two linearly independent polarization states by the unit vector $\boldsymbol{\varepsilon}_i(\mathbf{q}/q)$.

8.2.1 Hilbert subspace

Let us consider a superposition

$$|\Phi\rangle(t) = \sum_i c_i(t)\hat{a}_i^\dagger |0\rangle \tag{8.10}$$

of plane-wave single-photon states, viz.,

$$|1_i\rangle = \hat{a}_i^\dagger |0\rangle, \tag{8.11}$$

where $|0\rangle$ is the global vacuum state of transverse modes. The $|0\rangle$-state is defined by

$$\hat{a}_i |0\rangle = 0, \quad \forall \ i = (\mathbf{q}_i, s). \tag{8.12}$$

The $|\Phi\rangle(t)$ state is normalized provided

$$\langle \Phi | \Phi \rangle = 1, \tag{8.13}$$

or equivalently

$$\sum_i |c_i(t)|^2 \left[= \sum_i |c_i(0)|^2 \right] = 1. \tag{8.14}$$

We have indicated that the absolute square of the individual c_i-coefficients is fixed in time $[|c_i(t)|^2 = |c_i(0)|^2, \forall i]$, so that the normalization condition [Eq. (8.13)] holds at all times. Although superimposed [Eq. (8.10)], the individual single-photon states, $|1_i\rangle$, are uncoupled in free space: A coupling requires interaction via matter. The time dependence of the $c_i(t)$'s is obtained from the Schrödinger equation

$$i\hbar \frac{d}{dt} |\Phi\rangle(t) = \hat{\mathcal{H}}_{\mathrm{ph}} |\Phi\rangle(t). \tag{8.15}$$

The time-independent photon (subscript: ph) Hamilton operator is obtained by subtracting the zero-point (vacuum) part, $\hat{\mathcal{H}}_0$, from the total Hamiltonian, $\hat{\mathcal{H}} \equiv \hat{\mathcal{H}}_T$ of the transverse field subspace, i.e.,

$$\hat{\mathcal{H}}_{ph} = \hat{\mathcal{H}} - \hat{\mathcal{H}}_0 = \sum_i \hbar\omega_i \hat{a}_i^\dagger \hat{a}_i. \tag{8.16}$$

Utilizing the commutator relation in Eq. (8.9), and the vacuum definition [Eq. (8.12)] one gets

$$\hat{\mathcal{H}}_{\mathrm{ph}} |\Phi\rangle(t) = \sum_{ij} \hbar\omega_j \hat{a}_j^\dagger \hat{a}_j [c_i(t)\hat{a}_i^\dagger |0\rangle] = \sum_i c_i(t)\hbar\omega_i \hat{a}_i^\dagger |0\rangle, \tag{8.17}$$

and from the Schrödinger equation in Eq. (8.15)

$$\sum_i \left[i\hbar \frac{dc_i(t)}{dt} - \hbar\omega_i c_i(t) \right] \hat{a}_i^\dagger |0\rangle = 0. \tag{8.18}$$

The independence of the i-modes implies that

$$c_i(t) = c_i(0)e^{-i\omega_i t}, \tag{8.19}$$

confirming that $|c_i(t)|^2 = |c_i(0)|^2$.

It appears from Eq. (8.17) that $|\Phi\rangle$ [Eq. (8.10)] is not an eigenstate for the photon energy unless $\omega_i = cq_i$ is independent of i. In the special case where the superposition in Eq. (8.10) contains only modes with the same magnitude of the wave vector, say q [$\omega_i = cq \equiv \omega \ \forall\ i$] $|0\rangle$ is of course an eigenstate for the photon energy, $\mathcal{H}_{\mathrm{ph}} |\Phi\rangle = \hbar\omega |\Phi\rangle$, $|\Phi\rangle (t) = |\Phi\rangle (0) \exp(-i\omega t)$.

Here comes the important point: *In order to qualify as a genuine single-photon state $|\Phi\rangle$ must be an eigenstate for the global number operator with eigenvalue 1.* The global number operator

$$\hat{N} = \sum_i \hat{a}_i^\dagger \hat{a}_i, \tag{8.20}$$

a fixed operator in the Schrödinger Picture, is the sum of the number operators for the individual plane-wave modes, $\hat{N}_i = \hat{a}_i^\dagger \hat{a}_i$. We have of course

$$\hat{N}_i |1_i\rangle = \hat{a}_i^\dagger \hat{a}_i \hat{a}_i^\dagger |0\rangle = \hat{a}_i^\dagger |0\rangle = 1 |1_i\rangle , \tag{8.21}$$

but most important also

$$\hat{N} |\Phi\rangle = \sum_{ij} c_i(t)\hat{a}_j^\dagger \hat{a}_j \hat{a}_i^\dagger |0\rangle = \sum_{ij} c_i(t)\delta_{ij}\hat{a}_j^\dagger |0\rangle , \tag{8.22}$$

and hence

$$\hat{N} |\Phi\rangle = 1 |\Phi\rangle . \tag{8.23}$$

Thus $|\Phi\rangle$ [Eq. (8.10)] is an eigenstate for the global number operator with eigenvalue 1. The state $|\Phi\rangle$ is called a *polychromatic single-photon state*. In the simplest case $|\Phi\rangle$ consist of only two superimposed plane-wave states (with $\omega_1 \neq \omega_2$). When only few plane-wave states are superimposed one may prefer to name $|\Phi\rangle$ a non-monochromatic single-photon state. An apparatus allowing one (in principle) to measure plane-wave photons of a specific polarization, shows us that (repeated) measurements on the single-photon state $|\Phi\rangle$ with probability $|c_i(0)|^2$ will find the photon in the plane wave state $|1_i\rangle$.

Although the single-photon state $|\Phi\rangle$ in general describes a *non-stationary state* all moments of the photon energy are time independent. In particular this holds for the mean value and the variance of the photon energy. Starting from Eq. (8.17), one readily realizes that

$$\hat{\mathcal{H}}_{\mathrm{ph}}^n |\Phi\rangle (t) = \sum_i c_i(t)(\hbar\omega_i)^n \hat{a}_i^\dagger |0\rangle . \tag{8.24}$$

The mean value of the nth order moment of the photon Hamilton operator therefore becomes

$$\langle\Phi| \hat{\mathcal{H}}_{\mathrm{ph}}^n |\Phi\rangle = \sum_i |c_i(0)|^2 (\hbar\omega_i)^n . \tag{8.25}$$

The mean value ($n = 1$) and variance ($n = 2$) of the energy thus are fixed in time. From a physical point of view it is obvious that all the moments of the photon energy must be time independent in free space, where the individual modes do not interact.

8.2.2 Direct space

In Section 3.3 we introduced a kind of mean position state, viz, $|\mathcal{R}\rangle$, in relation to an analysis of transverse-photon correlations. An acception of the position state construction in Eq. (3.21) implies that the photon wave function in direct space will be given by

$$\mathbf{\Phi}(\mathbf{r}, t) = \langle \mathcal{R}|\Phi\rangle\,(\mathbf{r}, t) = \left(\frac{2\varepsilon_0 c}{\hbar}\right)^{1/2} \langle 0|\,\hat{\mathbf{A}}_T^{(+)}(\mathbf{r}, t)\,|\Phi\rangle. \tag{8.26}$$

Let us now use the Heisenberg Picture form for $\hat{\mathbf{A}}_T^{(+)}$ in its plane-wave box (L^3) expansion, namely

$$\hat{\mathbf{A}}_T^{(+)}(\mathbf{r}, t) = \left(\frac{\hbar}{2\varepsilon_0 L^3}\right)^{1/2} \sum_i \frac{1}{\omega_i^{1/2}} \boldsymbol{\varepsilon}_i e^{i(\mathbf{q}_i \cdot \mathbf{r} - \omega_i t)} \hat{a}_i, \tag{8.27}$$

$\hat{a}_i \equiv \hat{a}_i(0)$, remembering $\hat{a}_i(t) = \hat{a}_i(0)\exp(-i\omega_i t)$ in free space. A calculation of the matrix element $\langle 0|\,\hat{\mathbf{A}}_T(\mathbf{r}, t)\,|\Phi\rangle$ in the Heisenberg Picture implies that the fixed single-photon state vector form, viz.,

$$|\Phi\rangle = \sum_i c_i(0)\hat{a}_i^\dagger\,|0\rangle, \tag{8.28}$$

has to be used. A combination of Eqs. (8.26)–(8.28) gives

$$\begin{aligned}
\mathbf{\Phi}(\mathbf{r}, t) &= L^{-3/2} \sum_{ij} \frac{1}{q_i^{1/2}} \boldsymbol{\varepsilon}_i e^{i(\mathbf{q}_i \cdot \mathbf{r} - \omega_i t)} c_j(0)\,\langle 0|\,\hat{a}_i \hat{a}_j^\dagger\,|0\rangle \\
&= L^{-3/2} \sum_i c_i(0) \frac{1}{q_i^{1/2}} \boldsymbol{\varepsilon}_i e^{i(\mathbf{q}_i \cdot \mathbf{r} - \omega_i t)}
\end{aligned} \tag{8.29}$$

since $\langle 0|\,\hat{a}_i \hat{a}_j^\dagger\,|0\rangle = \delta_{ij}$. By introduction of the scalar quantity

$$\Phi_i(\mathbf{q}_i) \equiv L^{3/2} q_i^{1/2} c_i(0), \tag{8.30}$$

the photon wave function takes the form

$$\mathbf{\Phi}(\mathbf{r}, t) = \frac{1}{L^3} \sum_i q_i^{-1} \Phi_i(\mathbf{q}_i) \boldsymbol{\varepsilon}_i e^{i(\mathbf{q}_i \cdot \mathbf{r} - \omega_i t)}. \tag{8.31}$$

In the continuous mode quantization Eq. (8.31) is replaced by

$$\mathbf{\Phi}(\mathbf{r}, t) = \sum_{s=1,2} \int_{-\infty}^{\infty} \Phi_s(\mathbf{q}) \boldsymbol{\varepsilon}_s(\boldsymbol{\kappa}) e^{i(\mathbf{q} \cdot \mathbf{r} - cqt)} \frac{d^3 q}{(2\pi)^3 q}. \tag{8.32}$$

The wave function in Eq. (8.32) can be divided into two parts relating to the two linearly independent polarization species that is

$$\Phi(\mathbf{r}, t) = \sum_{s=1,2} \Phi_s(\mathbf{r}, t) \tag{8.33}$$

with

$$\Phi_s(\mathbf{r}, t) = \int_{-\infty}^{\infty} \Phi_s(\mathbf{q}) \varepsilon_s(\boldsymbol{\kappa}) e^{i(\mathbf{q}\cdot\mathbf{r} - cqt)} \frac{\mathrm{d}^3 q}{(2\pi)^3 q} \tag{8.34}$$

The two vectors $\Phi_s(\mathbf{r}, t)$ [$s = 1, 2; +, -$ in the helicity basis] are the space-time single-photon wave functions of the two polarization species. Since $q^{-1}\mathrm{d}^3 q$ is the relativistically invariant volume element in $\mathbf{q}$-space [158], the form in Eq. (8.34) indicates that $\Phi_s(\mathbf{q}) \varepsilon_s(\boldsymbol{\kappa}) \exp(-icqt)$ is the vectorial photon wave function for species in the wave-vector ($\sim$ momentum) representation.

With the limitation forced upon us by the inability to localize a transverse photon completely in space (at any time) we may turn our perspective and consider the argument $\mathbf{r}$ in $\Phi(\mathbf{r}, t)$ as a sort of mean position coordinate for the photon in direct space. This point of view is general in relativistic wave mechanics. Hence it is known from the work of Foldy and Wouthuysen [87] that also the argument $\mathbf{r}$ in the Dirac wave function $\psi(\mathbf{r}, t)$, say for an electron, "only" qualifies as a mean position coordinate. Although the electron and the transverse photon share the "position coordinate problem" the length range involved in the mean-position calculation for a tightly localized electron is several orders of magnitude smaller than that for a strongly localized T-photon, at least in general.

As shown in my book "LIGHT -The Physics of the Photon" [158] the photon wave function in direct space $\Phi(\mathbf{r}, t) = \langle \mathcal{R}|\Phi \rangle (\mathbf{r}, t)$, satisfies a dynamical equation of the Hamiltonian-like form, viz.,

$$i\hbar \frac{\partial}{\partial t} \langle \mathcal{R}|\Phi \rangle (\mathbf{r}, t) = \langle \mathcal{R}| \hat{\mathcal{H}}_{ph} |\Phi \rangle (\mathbf{r}, t). \tag{8.35}$$

Furthermore, it turns out that the normalization condition for the wave function $\Phi(\mathbf{r}, t)$ takes the form [158]

$$\int_{-\infty}^{\infty} \Phi^{\dagger}(\mathbf{r}, t) \cdot \left[\frac{i}{c} \frac{\partial}{\partial t} \Phi(\mathbf{r}, t) \right] \mathrm{d}^3 r = 1. \tag{8.36}$$

The reader may recognize this form as the usual one for relativistic wave functions in direct space [95, 106, 285]. The dynamical equation allows one to rewrite the normalization condition as follows:

$$\frac{1}{\hbar c} \int_{-\infty}^{\infty} \langle \Phi|\mathcal{R} \rangle \langle \mathcal{R}| \hat{\mathcal{H}}_{ph} |\Phi \rangle \mathrm{d}^3 r = 1 \tag{8.37}$$

For comparison, the normalization condition in wave-vector space can be given in the form

$$\sum_{s=1,2} \int_{-\infty}^{\infty} |\Phi_s(\mathbf{q})|^2 \frac{d^3 q}{(2\pi)^3 q} = 1, \tag{8.38}$$

as one sees by combining Eqs. (8.14) and (8.30), and proceeding from box to continuous mode quantization.

8.2.3 Wave-packet basis

Notwithstanding the fact that polychromatic single-photon states in Hilbert Space, $|\Phi\rangle\,(t)$, conveniently (at least from a mathematical point of view) can be expressed as a linear superposition of monochromatic plane (P) one-photon states

$$|P_i\rangle\,[\equiv |1_i\rangle] = \hat{a}_i^\dagger |0\rangle,\tag{8.39}$$

with $i = (\mathbf{q}, s)$. The set $\{|P_i\rangle\}$ satisfies (in box quantization) the orthonormality

$$\langle P_j|P_i\rangle = \delta_{ij},\tag{8.40}$$

and completeness

$$\sum_i |P_i\rangle\,\langle P_i| = \mathbb{1}\tag{8.41}$$

relations. Although a *single-photon phenomenon* (in the Bohr sense of the word phenomenon [41, 268]) never can be associated with a single monochromatic plane-wave mode, many experimental results can be handled in an approximate sense starting from plane-wave states. This approach for instance is common in interference and scattering studies. From a cosmological point of view the possible "existence" of a monochromatic plane-wave photon raises deep unanswered questions related to the space-time structure of our universe.

In photon wave mechanics wave packets enter essentially all analyses, and to bridge the gap between QED and wave mechanics it can be convenient to start from a one-photon wave-packet basis in Hilbert space. Also from a fundamental point of view such a generalization is of importance conceptually. A set of one-photon wave-packet (W) states, $\{|W_i\rangle\}$, is introduced by a certain superposition of a complete set of plane single-photon states

$$|W_i\rangle = \sum_j t_{ij}\,|P_j\rangle.\tag{8.42}$$

The coefficient matrix $\mathbf{t} = \{t_{ij}\}$ must be unitary if one requires that the $\{|W_i\rangle\}$-set is orthonormalized and complete. The inner product of states i and j is given by

$$\langle W_j|W_i\rangle = \sum_{m,n} t_{jn}^* t_{im}\,\langle 0|\,\hat{a}_n\hat{a}_m^\dagger\,|0\rangle = \sum_{m,n} t_{jn}^* t_{im}\delta_{mn} = \sum_m t_{jm}^* t_{im}.\tag{8.43}$$

If $\mathbf{t}$ is unitary the last sum in Eq. (8.43) equals δ_{ij}, and the one-photon wave-packets hence satisfy the orthonormality condition

$$\langle W_j|W_i\rangle = \delta_{ij}.\tag{8.44}$$

The unitarity of the $\mathbf{t}$ matrix also ensures that the $|W_i\rangle$-set is complete. Hence,

$$\sum_i |W_i\rangle\,\langle W_i| = \sum_{i,m,n} t_{im}t_{in}^*\hat{a}_m^+\,|0\rangle\,\langle 0|\,\hat{a}_n = \sum_{n,m} \delta_{mn}\hat{a}_m^+\,|0\rangle\,\langle 0|\,\hat{a}_n = \sum_m |P_m\rangle\,\langle P_m|,$$

$$\tag{8.45}$$

and because it was assumed that the set $\{|P_m\rangle\}$ is complete, the one-photon wave packets also satisfy the completeness relation

$$\sum_i |W_i\rangle \langle W_i| = \mathbb{1}. \tag{8.46}$$

Since, for fixed i, $|W_i\rangle$ is just a superposition of monochromatic plane-wave modes, and the coefficients t_{ij} in Eq. (8.42), for running j, satisfy the normalization condition

$$\sum_j t_{ij}^* t_{ij} = 1, \tag{8.47}$$

it appears that the photon wave function belonging to each of the wave-packet basis states in direct space, viz.,

$$\mathbf{\Phi}_i^W(\mathbf{r}, t) = \langle \mathbf{R}|W_i\rangle (\mathbf{r}, t) \tag{8.48}$$

satisfies the dynamical equation

$$i\hbar \frac{\partial}{\partial t} \langle \mathbf{R}|W_i\rangle (\mathbf{r}, t) = \langle \mathbf{R}| \hat{H}_{ph} |W_i\rangle (\mathbf{r}, t), \tag{8.49}$$

and is normalized, that is

$$\int_{-\infty}^{\infty} \left[\mathbf{\Phi}_i^W(\mathbf{r}, t)\right]^\dagger \cdot \left[\frac{i}{c}\frac{\partial}{\partial t}\mathbf{\Phi}_i^W(\mathbf{r}, t)\right] \mathrm{d}^3 r = 1. \tag{8.50}$$

8.2.4   Multi-photon wave packets

As we shall see shortly, it is possible to introduce annihilation ($\hat{b}_i(0)$) and creation ($\hat{b}_i^\dagger(0)$) operators which relate to elementary excitations of the individual wave-packet modes, denoted by a subscript i as before [Eq. (8.42)].

Let the general polychromatic single-photon state $|\Phi\rangle (t)$ in the Schrödinger Picture be expanded after the wave-packet basis $\{|W_i\rangle\}$:

$$|\Phi\rangle (t) = \sum_i c_i(t) |W_i\rangle, \tag{8.51}$$

with

$$c_i(t) = \langle W_i|\Phi\rangle (t). \tag{8.52}$$

Using the plane-wave expansion of $|W_i\rangle$, given in Eq. (8.42 [with Eq. (8.39) inserted], Eq. (8.51) can be written in the form

$$|\Phi\rangle (t) = \sum_i c_i(t)\hat{b}_i^\dagger(0) |0\rangle, \tag{8.53}$$

where

$$\hat{b}_i^\dagger(0) \equiv \sum_j t_{ij}\hat{a}_j^\dagger(0). \tag{8.54}$$

The creation operator $\hat{b}_i^\dagger(0)$ hence generates the one-photon wave-packet state $|W_i\rangle$ by acting on the (global) vacuum state, that is

$$|W_i\rangle = \hat{b}_i^\dagger(0)\,|0\rangle. \tag{8.55}$$

The creation operator $\hat{b}_i^\dagger(0)$ is a linear superposition of the plane-mode operators $\hat{a}_i^\dagger(0)$, the coefficients being elements of the unitary $\mathbf{t}$-matrix; see Eq. (8.54).

The commutation relation between the wave-packet annihilation and creation operators follows from Eq. (8.9) and the unitary $\mathbf{t}$-matrix. Hence

$$[\hat{b}_i(0), \hat{b}_j^\dagger(0)] = \sum_{m,n} t_{im}^* t_{jn} [\hat{a}_m(0), \hat{a}_n^\dagger(0)] = \sum_{m,n} t_{im}^* t_{jn}\delta_{mn} = \sum_m t_{im}^* t_{jm} \tag{8.56}$$

and therefore

$$[\hat{b}_i(0), \hat{b}_j^\dagger(0)] = \delta_{ij}. \tag{8.57}$$

Furthermore, one obtains immediately by use of Eq. (8.8)

$$[\hat{b}_i(0), \hat{b}_j(0)] = [\hat{b}_i^\dagger(0), \hat{b}_j^\dagger(0)] = 0. \tag{8.58}$$

The commutator relations in Eqs. (8.57) and (8.58) refer to time $t = 0$. Since the elements of the $\mathbf{t}$-matrix are time independent [a time dependent $\{t_{ij}\}$ choice in Eq. (8.42) would complicate applications of the formalism unnecessarily. In the Heisenberg Picture the free-space time evolution $\hat{a}_i(t) = \hat{a}_i(0)\exp(-i\omega_i t)$ [$\hat{a}_i^\dagger(t) = \hat{a}_i^\dagger(0)\exp(i\omega_i t)$] leads to the following equal-time commutator relations:

$$[\hat{b}_i(t), \hat{b}_j^\dagger(t)] = \rho_{ij}, \tag{8.59}$$

$$[\hat{b}_i(t), \hat{b}_j(t)] = [\hat{b}_i^\dagger(t), \hat{b}_j^\dagger(t)] = 0. \tag{8.60}$$

A wave-packet number operator for mode i is defined as

$$\hat{\mathcal{N}}_i \equiv \hat{b}_i^\dagger(t)\hat{b}_i(t)[= \hat{b}_i^\dagger(0)\hat{b}_i(0)], \tag{8.61}$$

and the related eigenvalue problem

$$\hat{\mathcal{N}}_i\,|\mathcal{N}_i\rangle = \mathcal{N}_i\,|\mathcal{N}_i\rangle \tag{8.62}$$

can be solved in the same manner as for other boson operators [with algebra analogous to that of Eqs. (8.59) and (8.60)]. The ith ground state, $|0_i\rangle$, satisfies

$$\hat{b}_i\,|0_i\rangle = 0, \tag{8.63}$$

and the eigenvalues for the number operator are integers

$$\mathcal{N}_i = 0, 1, 2, \dots, \forall i. \tag{8.64}$$

The various number states eigenvectors can be generated from the ground state as follows

$$|\mathcal{N}_i\rangle = \frac{(\hat{b}_i^\dagger)^{\mathcal{N}_i}}{\sqrt{\mathcal{N}_i!}}\,|0_i\rangle. \tag{8.65}$$

The factor $(\mathcal{N}_i!)^{1/2}$ guarantees that the orthogonal wave-packet eigenstates are normalized, i.e.,

$$\langle \mathcal{N}_i | \mathcal{N}_i \rangle = 1, \tag{8.66}$$

at all times.

8.3   TWO-PHOTON STATES

Let us consider two plane-wave modes $i \equiv (\mathbf{q}_i, s_i)$ $j \equiv (\mathbf{q}_j, s_j)$, and let us form a *particular* two-photon state vector

$$|P_{ij}\rangle = \frac{1}{\sqrt{2!}} \hat{a}_i^\dagger \hat{a}_j^\dagger |0\rangle , \tag{8.67}$$

where

$$|0\rangle = |0_i, 0_j\rangle \otimes \Pi_{k \neq i,j} |0_k\rangle \tag{8.68}$$

is the global vacuum state, and

$$|0_i, 0_j\rangle \equiv |0_i\rangle \otimes |0_j\rangle . \tag{8.69}$$

The notation used in Eqs. (8.68) and (8.69) emphasizes our interest in just the modes i and j. The state $|P_{ij}\rangle$ holds precisely two photons For $i \neq j$, there is one in each of the states, $|P_{ij}\rangle = |1_i, 1_j\rangle$, but for $i = j$, the two photons are either in state i, $|P_{ii}\rangle = |2_i, 0_j\rangle$ or state j, $|P_{jj}\rangle = |0_i, 2_j\rangle$.

The inner product is given by

$$\langle P_{kl} | P_{ij} \rangle = \frac{1}{2} \langle 0| \hat{a}_k \hat{a}_l \hat{a}_i^\dagger \hat{a}_j^\dagger |0\rangle = \frac{1}{2}(\delta_{il}\delta_{jk} + \delta_{ik}\delta_{jl}), \tag{8.70}$$

a result readily obtained using the commutation relation for the boson operators [Eq. (8.9)] and the vacuum definition [Eq. (8.12)]. Since $|P_{ij}\rangle = |P_{ji}\rangle$, one next obtains

$$\langle P_{kl} | P_{ij} \rangle = \delta_{ik}\delta_{jl}. \tag{8.71}$$

The set $\{|P_{ij}\rangle\}$ thus is orthonormalized. The set is also complete because

$$\sum_{ij} |P_{ij}\rangle \langle P_{ij}| = \mathbb{1}, \tag{8.72}$$

where $\mathbb{1}$ is the unit-operator. In order to prove this we act with the operator $\sum_{kl} |P_{kl}\rangle \langle P_{kl}|$ on the *general* two-photon state

$$|\Phi_2\rangle = \frac{1}{\sqrt{2}} \sum_{ij} c_{ij} \hat{a}_i^\dagger \hat{a}_j^\dagger |0\rangle = \sum_{ij} c_{ij} |P_{ij}\rangle . \tag{8.73}$$

Since the inner product of $|P_{kl}\rangle$ and $|\Phi_2\rangle$ becomes

$$\langle P_{kl} | \Phi_2 \rangle = \sum_{i,j} c_{ij} \langle P_{kl} | P_{ij} \rangle = \sum_{i,j} c_{ij} \delta_{ik} \delta_{jl} = c_{kl}, \tag{8.74}$$

one finds

$$\sum_{k,l} |P_{kl}\rangle \langle P_{kl}|\Phi_2\rangle = \sum_{k,l} c_{kl} |P_{kl}\rangle = |\Phi_2\rangle \,, \tag{8.75}$$

a result which proves the completeness of the set $\{|P_{ij}\rangle\}$; Eq. (8.72).
   Utilizing Eq. (8.70)

$$\langle \Phi_2|\Phi_2\rangle = \frac{1}{2} \sum_{i,j,k,l} c_{kl}^* c_{ij} \langle 0| \hat{a}_k \hat{a}_l \hat{a}_i^\dagger \hat{a}_j^\dagger |0\rangle = \frac{1}{2} \sum_{i,j,k,l} c_{kl}^* c_{ij}(\delta_{il}\delta_{jk} + \delta_{ik}\delta_{jl})$$

$$= \frac{1}{2} \sum_{i,j}(c_{ji}^* c_{ij} + c_{ij}^* c_{ij}) = \sum_{i,j} |c_{ij}|^2, \tag{8.76}$$

The last member of Eq. (8.76) follows from the circumstance that the choice $c_{ji} = c_{ij}$ can be made without loss generality; cf. Eq. (8.67). The two-photon state $|\Phi_2\rangle$ thus is normalized provided

$$\sum_{i,j} |c_{ij}|^2 = 1. \tag{8.77}$$

It is the introduction of the factor $2^{-1/2}$ in Eq. (8.67) which results in the convenient condition in Eq. (8.77).
   The general two-photon state is an eigenstate for the global number operator $\hat{N} = \sum_j \hat{a}_i^\dagger \hat{a}_i$ with eigenvalue 2, i.e.

$$\hat{N} |\Phi_2\rangle = 2 |\Phi_2\rangle \,, \tag{8.78}$$

as expected. The proof is simple. Starting from

$$\hat{a}_i^\dagger \hat{a}_i \hat{a}_j^\dagger \hat{a}_k^\dagger |0\rangle = (\delta_{ij}\hat{a}_i^\dagger \hat{a}_k^\dagger + \delta_{ki}\hat{a}_i^\dagger \hat{a}_j^\dagger) |0\rangle \,, \tag{8.79}$$

one obtains

$$\hat{N} |\Phi_2\rangle = \frac{1}{\sqrt{2}} \sum_{i,j,k} c_{jk}\hat{a}_i^\dagger \hat{a}_i \hat{a}_j^\dagger \hat{a}_k^\dagger |0\rangle = \frac{1}{\sqrt{2}}(\sum_{i,k} c_{ik}\hat{a}_i^\dagger \hat{a}_k^\dagger + \sum_{i,j} c_{ji}\hat{a}_i^\dagger \hat{a}_j^\dagger) |0\rangle \,. \tag{8.80}$$

Replacing the dummy index k by j one gets

$$\hat{N} |\Phi_2\rangle = \frac{1}{\sqrt{2}} \sum_{i,j}(c_{ij} + c_{ji})\hat{a}_i^\dagger \hat{a}_j^\dagger |0\rangle \,. \tag{8.81}$$

The symmetric choice $c_{ji} = c_{ij}$ then gives Eq. (8.78); q.e.d.
   If wished (convenient in a given analysis), a general two-photon state can be expanded after a wave-packet basis, $\{|W_{ij}\rangle\}$. Thus,

$$|\Phi_2\rangle = \sum_{i,j} c_{ij} |W_{ij}\rangle \,, \tag{8.82}$$

where

$$|W_{ij}\rangle = \frac{1}{\sqrt{2!}} \hat{b}_i^\dagger \hat{b}_j^\dagger |0\rangle. \tag{8.83}$$

The boson commutation rules in Eq. (8.59) guarantees the orthonormality

$$\langle W_{kl}|W_{ij}\rangle = \delta_{ik}\delta_{jk}, \tag{8.84}$$

and completeness

$$\sum_{i,j} |W_{ij}\rangle \langle W_{ij}| = \mathbb{1} \tag{8.85}$$

of the $\{|W_{ij}\rangle\}$ set.

8.4 TWO-MODE TWO-PHOTON WAVE-PACKET STATES

Without losing sight of essential generality, one may simplify the two-photon study by assuming that the free electromagnetic field has been prepared in a state where only two modes, denoted 1 and 2, can contain photons.

The state vector of the field hence has the form

$$|\Psi\rangle = |\Psi_{12}\rangle \otimes \Pi_{i\neq 1,2} |0_i\rangle. \tag{8.86}$$

In the wave-packet basis the most general form of the state vector $|\Psi_{12}\rangle$ is a linear superposition

$$|\Psi_{12}\rangle = \sum_{\mathcal{N}_1,\mathcal{N}_2} c_{\mathcal{N}_1,\mathcal{N}_2} |\mathcal{N}_1,\mathcal{N}_2\rangle \tag{8.87}$$

of wave-packet basis states

$$|\mathcal{N}_1,\mathcal{N}_2\rangle = |\mathcal{N}_1\rangle \otimes |\mathcal{N}_2\rangle = \frac{1}{\sqrt{2!}} \frac{(\hat{b}_1^\dagger)^{\mathcal{N}_1}}{\sqrt{\mathcal{N}_1!}} |0_1\rangle \otimes \frac{(\hat{b}_2^\dagger)^{\mathcal{N}_2}}{\sqrt{\mathcal{N}_2!}} |0_2\rangle. \tag{8.88}$$

Among these states we find the single-photon wave-packet state

$$|\Phi_1\rangle = c_{10} |1,0\rangle + c_{01} |0,1\rangle = (c_{10}\hat{b}_1^\dagger + c_{01}\hat{b}_2^\dagger) |0,0\rangle, \tag{8.89}$$

normalized for $|c_{10}|^2 + |c_{01}|^2 = 1$.

In a slightly changed notation $[c_{10} = \cos\theta,\ c_{01} = \sin\theta,\ |0,0\rangle = |0\rangle]$ a particular (PAR) n-photon state $|\Phi_n^{\text{PAR}}\rangle$, can be obtained from the one-photon state

$$|\Phi_1\rangle = \frac{\hat{c}^\dagger}{\sqrt{1!}} |0\rangle \tag{8.90}$$

by repeated action with the operator

$$\hat{c}^\dagger = \hat{b}_1^\dagger \cos\theta + \hat{b}_2^\dagger \sin\theta. \tag{8.91}$$

Thus

$$\left|\Phi_n^{\mathrm{PAR}}\right\rangle = \frac{(\hat{c}^\dagger)^n}{\sqrt{n!}}\,|0\rangle. \tag{8.92}$$

In the two-photon case

$$\left|\Phi_2^{\mathrm{PAR}}\right\rangle = \frac{1}{\sqrt{2}}(\hat{c}^\dagger)^2\,|0\rangle = \frac{1}{\sqrt{2}}[(\hat{b}_1^\dagger)^2\cos^2\theta + 2\sin\theta\cos\theta\,\hat{b}_1^\dagger\hat{b}_2^\dagger + (\hat{b}_2^\dagger)^2\sin^2\theta]\,|0\rangle$$

$$= \frac{1}{\sqrt{2}}[\sqrt{2}\cos^2\theta\,|2,0\rangle + 2\sin\theta\cos\theta\,|1,1\rangle + \sqrt{2}\sin^2\theta\,|0,2\rangle], \tag{8.93}$$

remembering that $(\hat{b}_1^\dagger)^2\,|0,0\rangle = \sqrt{2}\,|2,0\rangle$ [and $(\hat{b}_2^\dagger)^2\,|0,0\rangle = \sqrt{2}\,|0,2\rangle$]. By construction [Eq. (8.92)] the particular two-photon wave-packet state is normalized, $\langle\Phi_2^{\mathrm{PAR}}|\Phi_2^{\mathrm{PAR}}\rangle = 1$.

8.5 THE TWO-PHOTON CONCEPT: TWO PHOTONS OR BI-PHOTONS

8.5.1 Factored and entangled states

Let us consider two single-photon wave packets, possibly well-localized in different regions of space at a given time. In chosen wave-packet bases, the most general states of the two particles (denoted by 1 and 2) are

$$|\Phi(1)\rangle = \sum_i a_i\,|W_i(1)\rangle \tag{8.94}$$

and

$$|\Phi(2)\rangle = \sum_j b_j\,|W_j(2)\rangle \tag{8.95}$$

In the union ($\mathcal{H}_1 \otimes \mathcal{H}_2$) of the two single-particle Hilbert spaces ($\mathcal{H}_1$ and $\mathcal{H}_2$) the basis vectors are

$$|W_{ij}\rangle\left[= \frac{1}{\sqrt{2!}}\hat{b}_i^\dagger\hat{b}_j^\dagger\,|0\rangle\right] = \frac{1}{\sqrt{2!}}\,|W_i(1)\rangle \otimes |W_j(2)\rangle, \tag{8.96}$$

see Eq. (8.88). The two photons can be considered as individual (independent) particles provided the two-photon state is a factored (FAC) state, i.e.,

$$\left|\Psi^{\mathrm{FAC}}(1,2)\right\rangle = |\Phi(1)\rangle \otimes |\Phi(2)\rangle, \tag{8.97}$$

or equivalently

$$\left|\Psi^{\mathrm{FAC}}(1,2)\right\rangle = \sum_{i,j} a_i b_j\,|W_i(1)\rangle \otimes |W_j(2)\rangle. \tag{8.98}$$

The two-photon state $\left|\Psi^{\mathrm{FAC}}(1,2)\right\rangle$ hence is the combined state of two photons.

The most general state in $\mathcal{H} = \mathcal{H}_1 \otimes \mathcal{H}_2$, $|\Psi(1,2)\rangle$, can be expanded as follows:

$$|\Psi(1,2)\rangle = \sum_{i,j} c_{ij} |W_i(1)\rangle \otimes |W_j(2)\rangle . \tag{8.99}$$

Since the elements of the arbitrary matrix $\{c_{ij}\}$, yet restricted by a normalization condition, cannot be expressed as the elements $\{a_i b_j\}$ of a dyadic product of two vectors with components $\{a_i\}$ and $\{b_j\}$, the state $|\Psi(1,2)\rangle$ cannot be factored, in general. A two-photon state which cannot be factored we call a bi-photon, a single unit also called an entangled state of the two photons.

The perhaps simplest entanglement is the entangled polarization state of a two-photon system, e.g., [185],

$$|\Psi(1,2)\rangle = \frac{1}{\sqrt{2}}(|R(1)\rangle \otimes |R(2)\rangle + |L(1)\rangle \otimes |L(2)\rangle) \tag{8.100}$$

where

$$|R(i)\rangle = \hat{a}_+^\dagger(i) |0\rangle , \quad |L(i)\rangle = \hat{a}_-^\dagger(i) |0\rangle , \tag{8.101}$$

refer to single-photon states of positive (R) and negative (L) helicity, respectively; $i = 1, 2$. The most general product state has the form

$$\left|\Psi^{\mathrm{FAC}}(1,2)\right\rangle = a_1 b_1 |R(1)\rangle \otimes |R(2)\rangle + a_1 b_2 |R(1)\rangle \otimes |L(2)\rangle \tag{8.102}$$

$$+ a_2 b_1 |L(1)\rangle \otimes |R(2)\rangle + a_2 b_2 |L(1)\rangle \otimes |L(2)\rangle .$$

The state in Eq. (8.100) is *not* among the factored states in Eq. (8.102) since the requirement $a_1 b_2 = a_2 b_1 = 0$ is incompatible with $a_1 b_1 = a_2 b_2 (= 2^{-1/2})$.

In [124, 125] microscopic single-photon diffraction from one and two mesoscopic holes is treated, and it is indicated [125] that such diffraction studies can be extended to two-photon diffraction. It is pointed out that two-photon that two-photon diffraction observations may provide us with information on bi-photons.

8.5.2 Two-photon wave mechanics

The wave-packet formalism described above in Hilbert space can be transferred to wave mechanics with the help of the two-photon "mean" position state, given as

$$|\mathcal{R}_2\rangle (\mathbf{r}, \mathbf{r}', t) \equiv \frac{2\varepsilon_0 c}{\hbar} \hat{\mathbf{A}}_T^{(-)}(\mathbf{r}, t) \hat{\mathbf{A}}_T^{(-)}(\mathbf{r}', t) |0\rangle . \tag{8.103}$$

In configuration space the two-point $(\mathbf{r}, \mathbf{r}')$ two-photon wave function now is defined as follows:

$$\boldsymbol{\Phi}_2(\mathbf{r}, \mathbf{r}', t) \equiv \langle \mathcal{R}_2 | \Phi_2 \rangle (\mathbf{r}, \mathbf{r}', t). \tag{8.104}$$

As shown in Ref. [158], the dyadic quantity $\boldsymbol{\Phi}_2(\mathbf{r}, \mathbf{r}', t)$ has in the helicity basis a plane-wave expansion (in the continuum limit) of the Lorentz-invariant integral form

$$\boldsymbol{\Phi}_2(\mathbf{r}, \mathbf{r}', t) = \sum_{s,s'} \int_{-\infty}^{\infty} \Phi_{ss'}(\mathbf{q}, \mathbf{q}') \mathbf{e}_s(\boldsymbol{\kappa}) \mathbf{e}_{s'}(\boldsymbol{\kappa}') \exp[i(\mathbf{q} \cdot \mathbf{r} + \mathbf{q}' \cdot \mathbf{r}')]$$

$$\times \exp[-i(\omega_q + \omega_{q'})t] \frac{\mathrm{d}^3 q}{(2\pi)^3 q} \frac{\mathrm{d}^3 q'}{(2\pi)^3 q'}. \tag{8.105}$$

Furthermore, it turns out that the two-photon wave function satisfies a dynamical equation

$$i\hbar \frac{\partial}{\partial t} \langle \mathcal{R}_2 | \Phi_2 \rangle (\mathbf{r}, \mathbf{r}', t) = \langle \mathcal{R}_2 | \hat{\mathcal{H}}_{ph} | \Phi_2 \rangle (\mathbf{r}, \mathbf{r}', t) \tag{8.106}$$

in the two-coordinate configuration space.

III

Conservation Laws and Lagrangians. Geometric Phase and Topology

From Gauge-Photon Wave Mechanics to the Magnetostatic Aharonov-Bohm Vector Potential

9.1 ON THE CORRECT IDENTIFICATION OF FIELD MOMENTUM AND ENERGY IN CLASSICAL ELECTRODYNAMICS

The description of the time development of a physical system is usually greatly simplified by a knowledge of those physical quantities that do not change during this development. Such quantities are said to be conserved, and the laws which state under what conditions a given quantity is conserved are known as *conservation laws*. In electrodynamics these laws are perhaps less evident than in mechanics.

In classical electrodynamics the differential forms of the conservation law for the field momentum and energy can be found in most textbooks. Thus, in terms of what usually is known as the electromagnetic energy density [118]

$$W = \frac{\varepsilon_0}{2}(E^2 + c^2 B^2), \tag{9.1}$$

the energy flux density (Poynting vector)

$$\mathbf{S} = \frac{1}{\mu_0}\mathbf{E} \times \mathbf{B}, \tag{9.2}$$

and the dyadic Maxwell stress tensor

$$\mathbf{T} = \varepsilon_0(\mathbf{E}\mathbf{E} + c^2\mathbf{B}\mathbf{B}) - \frac{\varepsilon_0}{2}(E^2 + c^2 B^2)\mathbf{1}, \tag{9.3}$$

DOI: 10.1201/9781003029458-9

the differential momentum and energy conservation laws can be written as [118]

$$\nabla \cdot \mathbf{T} - \frac{1}{c^2}\frac{\partial \mathbf{S}}{\partial t} = \mathbf{0}, \tag{9.4}$$

$$\nabla \cdot \mathbf{S} + \frac{\partial W}{\partial t} = 0. \tag{9.5}$$

The deeper physical meaning of these conservation laws only appears via the related integral forms. The integration over a finite volume in space is always necessary because it is not possible to measure the field in a specific point neither in classical electrodynamics nor in quantum electrodynamics. The question of the measurability of electromagnetic quantities in quantum physics was carefully studied by Bohr and Rosenfeld, who took as a starting point the integration of the various field commutation relations over a fixed space-time volume [42, 43]. As in the classical case, the measurement of a field average over a specific space-time volume demands the control of the total momentum transferred from this volume to a system of movable test bodies. Furthermore, the mere impossibility of localizing a photon precisely in space-time necessitates integration over a finite space-time volume in all situations [153].

In textbooks one proceeds from the differential forms in Eqs. (9.4) and (9.5) to integral forms by integration over a fixed spatial volume. However, relativistic invariance might be lost since the integral of a tensor over a spatial volume is not, in general, a tensor. The world line of particle determines uniquely its now-plane at every instant and as seen by any inertial observer. The surface element of such a plane is given by the vector $d\sigma^\mu = n^\mu d\sigma$, where $d\sigma$ is the invariant area element, and

$$\{n^\mu\} = (\gamma, \hat{\mathbf{n}}\sqrt{\gamma^2 - 1}), \tag{9.6}$$

where $\gamma = [1 - (v/c)^2]^{-1/2}$, and $\hat{\mathbf{n}}$ is the directional unit vector for the three vector frame velocity, $\mathbf{v} = v\hat{\mathbf{n}}$. In the frame S_0, in which $\{n^\mu\} = (1, 0, 0, 0)$, $d\sigma = dx^{(0)}dy^{(0)}dz^{(0)}$. By working with manifestly covariant expressions, the conservation laws take the integral form

$$\{P^\mu\} = \left(\frac{1}{c}W_{\text{em}}, \mathbf{P}_{\text{em}}\right), \tag{9.7}$$

where

$$W_{\text{em}} = \gamma \int W d\sigma - \frac{1}{c}\sqrt{\gamma^2 - 1} \int \mathbf{S} \cdot \hat{\mathbf{n}} d\sigma, \tag{9.8}$$

$$\mathbf{P}_{\text{em}} = \frac{\gamma}{c^2} \int \mathbf{S} d\sigma + \frac{1}{c}\sqrt{\gamma^2 - 1} \int \mathbf{T} \cdot \hat{\mathbf{n}} d\sigma. \tag{9.9}$$

Only in the special Lorentz frame S_0, in which $\gamma = 1$ do Eqs. (9.8) and (9.9) reduce to the textbook integrals. The designation "energy density" for W, and "momentum density" for $\mathbf{S}/c^2$, hence are meaningful only relative to S_0 [in the rest frame of a charge], where one regains the familiar textbook results.

9.2 DYNAMIC STRUCTURAL VACUUM AND PHOTON WAVE MECHANICS

Let us return to the expression for W_{em} in S_0 $[\gamma = 1]$, viz.,

$$W_{\mathrm{em}} = \frac{\varepsilon_0}{2} \int_{-\infty}^{\infty} (E_T^2 + c^2 B^2) d^3 r \tag{9.10}$$

in vacuum where $\mathbf{E} = \mathbf{E}_T$ since $\nabla \cdot \mathbf{E} = 0$. By introduction of $\mathbf{F}_{\pm} = (\varepsilon_0/2)^{1/2}[\mathbf{E}_T \pm ic\mathbf{B}]$, W_{em} can be written as

$$W_{\mathrm{em}} = \int_{-\infty}^{\infty} \mathbf{F}_+ \cdot \mathbf{F}_- d^3 r. \tag{9.11}$$

The positive frequency part of the $\mathbf{F}$'s indicated by a superscript $(+)$, namely

$$\mathbf{F}_{\pm}^{(+)} = \sqrt{\frac{\varepsilon_0}{2}}[\mathbf{E}_T^{(+)} \pm ic\mathbf{B}^{(+)}], \tag{9.12}$$

enters a two-component spinor

$$\mathbf{\Phi} = \begin{pmatrix} \mathbf{F}_+^{(+)} \\ \mathbf{F}_-^{(+)} \end{pmatrix}, \tag{9.13}$$

known as the energy wave function in photon wave mechanics. In terms of $\mathbf{\Phi}$ (and its Hermitian conjugate $\mathbf{\Phi}^{\dagger}$) one obtains the following expression for the field energy:

$$W_{\mathrm{em}} = \int_{-\infty}^{\infty} \mathbf{\Phi}^{\dagger} \cdot \mathbf{\Phi} d^3 r. \tag{9.14}$$

Proper normalization of $\mathbf{\Phi}$, implies that W_{em} Eq. (9.14) is identified as the energy of the photon.

Photon wave mechanics based on the gauge invariant potentials $\mathbf{A}_{T,\pm}^{(+)}$ [see Section 1.2] leads to the dynamical free-space evolution equations

$$i\frac{\partial}{\partial t}\mathbf{A}_{T,\pm}^{(+)} = \pm c\nabla \times \mathbf{A}_{T,\pm}^{(+)} \tag{9.15}$$

for the two helicity species. Expressed in terms of the *auxiliary fields*, defined as

$$\mathcal{E}_{T,\pm}^{(+)} \equiv -\frac{\partial}{\partial t}\mathbf{A}_{T,\pm}^{(+)}, \tag{9.16}$$

$$\mathcal{B}_{T,\pm}^{(+)} \equiv \nabla \times \mathbf{A}_{T,\pm}^{(+)}, \tag{9.17}$$

Eq. (9.15) can be written as

$$\sqrt{\frac{2}{\varepsilon_0}}\mathcal{F}_{\pm}^{(+)} \equiv \mathcal{E}_{T,\pm}^{(+)} \mp ic\mathcal{B}_{T,\pm}^{(+)} = \mathbf{0}. \tag{9.18}$$

Inspired by the magnetostatic structural vacuum introduced in relation to the Aharonov-Bohm phenomenon, we say that Eqs. (9.18) defines what I have called *dynamic structural vacuum.* It has been shown [see Chapter 2] that the $\mathcal{F}_{\pm}^{(+)}$'s satisfy the dynamic evolution equations

$$i\frac{\partial}{\partial t}\mathcal{F}_{\pm}^{(+)} = \pm c\boldsymbol{\nabla} \times \mathcal{F}_{\pm}^{(+)}. \tag{9.19}$$

It follows from the free-space Maxwell equations that the photon energy wave functions for the two helicity species, $\mathcal{F}_{\pm}^{(+)}$ [Eq. (9.12)] satisfy evolution equations form-identical to those in Eqs. (9.19); replacement $\mathcal{F}_{\pm}^{(+)} \Rightarrow \mathbf{F}_{\pm}^{(+)}$.

9.3 DYNAMICAL AHARONOV-BOHM EFFECT DERIVED FROM THE DYNAMICAL EQUATIONS FOR $\mathbf{A}_{T,\pm}^{(+)}$

Recent theoretical studies [130] have shown that there exists a close connection between the *dynamical* structural vacuum concept, outlined in Section 9.2, and the magnetostatic Aharonov-Bohm phenomenon [3] showing a quantum-field effect in regions where $\mathbf{E}(\mathbf{r};\omega \to 0) = c\mathbf{B}(\mathbf{r};\omega \to 0) = \mathbf{0}$, a domain called *static* structural vacuum. Rigorously, speaking, magnetostatic and electrostatic fields are abstractions, though extremely useful for many experimental studies where the limit $\omega \to 0$ is a useful approximation.

Let us for simplicity consider a linear circular cylindrical magnetic string of radius r_0 and with a rotationally symmetric time-dependent magnetization

$$\mathbf{M}(\mathbf{r};\omega) = M(\rho;\omega)\hat{\mathbf{z}}, \tag{9.20}$$

which is independent of z (the coordinate along the string). The related string current density is given by

$$\mathbf{J}_T = \boldsymbol{\nabla} \times \mathbf{M}(\mathbf{r};\omega) = -\hat{\boldsymbol{\phi}}\frac{\partial M(\rho;\omega)}{\partial \rho}. \tag{9.21}$$

As indicated, the current density is transverse $[\mathbf{J} = \mathbf{J}_T]$ since $\boldsymbol{\nabla} \cdot \boldsymbol{\nabla} \times \mathbf{M} = 0$. In Eqs. (9.20) and (9.21) plane polar coordinates (ρ,ϕ) are used, and $\hat{\boldsymbol{\phi}}(\phi) = \hat{\mathbf{y}}\cos\phi - \hat{\mathbf{x}}\sin\phi$ is the usual tangential unit vector. limiting ourselves to homogeneous string magnetization, $M(\rho;\omega) = M_0(\omega)$, one obtains

$$\mathbf{J}_T(\rho,\phi;\omega) = M_0(\omega)\hat{\boldsymbol{\phi}}\delta(\rho - r_0). \tag{9.22}$$

The Fourier integral transform of the surface current density $\mathbf{J}_T(\rho,\phi;\omega)$ is given by

$$\mathbf{J}_T(\mathbf{q},\omega) = i(2\pi)^2\delta(q_z)r_0 M_0(\omega)J_1(q_\parallel r_0)(\hat{\mathbf{q}}_\parallel \times \hat{\mathbf{z}}), \tag{9.23}$$

where $J_1(q_\parallel r_0)$ is the first-order Bessel function. The wave vector $\mathbf{q}$ of $\mathbf{J}_T(\mathbf{q};\omega)$ has no z-component, i.e., $\mathbf{q} = q_z\hat{\mathbf{z}} + \mathbf{q}_\parallel$ with $q_z = 0$. The direction of $\mathbf{J}_T(\mathbf{q},\omega)$ is given by the cross product $\hat{\mathbf{q}}_\parallel \times \hat{\mathbf{z}}$ [$\hat{\mathbf{q}}_\parallel = \mathbf{q}_\parallel/q_\parallel$], as expected.

By insertion of Eqs. (9.23) into the fundamental dynamical evolution equations, given in Eq. (1.16) [and here used in its field-unquantized version], it can be shown that the transverse vector potentials in the $(\mathbf{q},\omega)$-domain are given by

$$\mathbf{A}_{T,\pm}^{(+)}(\mathbf{q},\omega) = \alpha(\mathbf{q},\omega)\mathbf{e}_\pm, \tag{9.24}$$

where $\mathbf{e}_\pm$ are the positive (+) and negative (-) helicity unit vectors, and

$$\alpha(\mathbf{q},\omega) = \frac{i(2\pi)^2}{2\sqrt{2}\varepsilon_0 cq} r_0 M_0(\omega) \frac{J_1(q_\parallel r_0)}{\omega - cq} \delta(q_z). \tag{9.25}$$

For the dynamic Aharonov-Bohm effect it is sufficient to add the opposite helicity components to the transverse gauge potential, giving

$$\mathbf{A}_T^{(+)}(\mathbf{q},\omega) = \frac{i(2\pi)^2}{2\varepsilon_0 cq} r_0 M_0(\omega) \frac{J_1(q_\parallel r_0)}{\omega - cq} \delta(q_z)\hat{\mathbf{z}} \times \boldsymbol{\kappa}, \tag{9.26}$$

with $\boldsymbol{\kappa} \equiv \hat{\mathbf{q}}_\parallel$.

In the space $(\mathbf{r})$-frequency (ω) domain, Eq. (9.26) takes the form

$$\mathbf{A}_T^{(+)}(\mathbf{r}_\parallel,\omega) = \frac{ir_0 M_0(\omega)}{4\pi\varepsilon_0 c} \int_0^\infty \int_0^{2\pi} \frac{J_1(q_\parallel r_0)}{\omega - cq_\parallel} e^{iq_\parallel r_\parallel \cos\phi} \hat{\boldsymbol{\phi}}(\phi) d\phi dq_\parallel, \tag{9.27}$$

where

$$\hat{\boldsymbol{\phi}} = (\hat{\mathbf{z}} \times \hat{\mathbf{q}}_\parallel)\cos\phi - \hat{\mathbf{q}}_\parallel \sin\phi, \tag{9.28}$$

and $\mathbf{r}_\parallel = \mathbf{r} - z\hat{\mathbf{z}}\ [r_\parallel = |\mathbf{r}_\parallel|]$. The result in Eq. (9.27) is the basic one.

Superposition of different frequency components of $M_0(\omega)$ allows one to obtain a related wave packet $\mathbf{A}_T^{(+)}(\mathbf{r},t)$ signal. The driving current density, $\mathbf{J}_T(\mathbf{r},t)$, will have finite support in time, and when for, say $t > t_0$, $\mathbf{J}_T(\mathbf{r},t(> t_0)) = \mathbf{0}$, the emitted helicity components $\mathbf{A}_{T,\pm}^{(+)}(\mathbf{r},t)$ satisfy the dynamic structural vacuum condition in Eqs. (9.18).

In the magnestostatic limit $(\omega \to 0)$ one obtains

$$\mathbf{A}_T^{(+)}(\mathbf{r}_\parallel;\omega \to 0^+) = \frac{\mu_0}{2} r_0 M_0(0)\hat{\boldsymbol{\phi}}(\phi) \int_0^\infty \frac{J_1(q_\parallel r_0)J_1(q_\parallel r_\parallel)}{q_\parallel} dq_\parallel, \tag{9.29}$$

where $\hat{\boldsymbol{\phi}}(\phi) = \hat{\mathbf{y}}\cos\phi - \hat{\mathbf{x}}\sin\phi$ is the tangential unit vector in configuration space. Note that *three* length parameters $[r_0, r_\parallel, q_\parallel^{-1}]$ enter Eq. (9.29). If a small radius approximation $J_1(q_\parallel r_0) = q_\parallel r_0$ is used $[r_\parallel \gg r_0]$, one obtains

$$\mathbf{A}_T^{(+)}(\mathbf{r}_\parallel;\omega \to 0^+) \equiv \mathbf{A}_T(\mathbf{r}_\parallel) = \frac{\mu_0}{4\pi} \frac{r_0^2}{r_\parallel} \hat{\boldsymbol{\phi}}(\phi) M_0(\omega = 0), \tag{9.30}$$

precisely the Aharonov-Bohm transverse vector potential, giving

$$\mathbf{B}(\mathbf{r}_\parallel) = \boldsymbol{\nabla} \times \mathbf{A}_T(\mathbf{r}_\parallel) = \mathbf{0}. \tag{9.31}$$

Thus, with Aharonov-Bohm, we conclude $\mathbf{E} = c\mathbf{B} = \mathbf{0}$ in the simplest magnestotatic approach. A detailed account of the basis of the Aharonow-Bohm phenomenon [3] is given in Chapter 11.

The role of the Aharonov-Bohm effect in Bohm's causal interpretation of quantum mechanics [38] is described in a broader framework in Bohm and Hiley's book entitled *The undivided universe* [39].In philosophical terminology [39] presents an *ontology* for quantum systems. Ontology is concerned with that which IS, and not with how we obtain our knowledge about quantum systems, the point of view of Bohr [43]. Bohm's theory is to some extent related to de Broglie's *pilot-wave theory*, developed essentially to its final form in [68] as discussed e.g. in [9].

CHAPTER 10

Invariance Properties and Conservation Laws in Electrodynamics

The understanding of the time development of a physical system is aided by the knowledge (identification) of those quantities that do not change during the development. The so-called conservation laws state under what conditions the relevant quantities are time independent. In the following we shall realize that there is a close connection between the symmetry and general invariance properties of a physical system. In particular, we shall study the conservation laws in electrodynamics The general form of the conservation laws for energy, momentum and angular momentum are less evident (and perhaps less familiar) in (classical) electrodynamics than in mechanics.

10.1 NÖTHER'S THEOREM

10.1.1 Derivation of Emmy Nöther's first theorem. Simplest form

The connection between symmetry properties and conservation laws is provided by Nöther's theorem [192] in combination with Hamilton's principle of least action. In Chapter 7, the invariance properties of a physical (quantum) system under the symmetry transformations of the Poincare group were studied.

We now derive Nöther's (first) theorem in its simplest form, i.e. under a 1-parameter infinitesimal group of transformations. The generalization to p (> 1)-parameter groups is straightforward. Let $\{u_k(x)\}$, ($k = 1, 2, ..., n$), be a set of differentiable functions of the independent variable x, with first derivatives $\{v_k(x) \equiv du_k(x)/dx\}$, and let

$$\mathcal{L}[x] \equiv \mathcal{L}(\{u_k\}, \{v_k\}; x) \tag{10.1}$$

be a function of x and the 2n functions u_k and v_k. An infinitesimal transformation T

$$x \to x' = x + \delta x \tag{10.2}$$

DOI: 10.1201/9781003029458-10

with $\delta x = \varepsilon\eta(x)$. Hence, ε is infinitesimal, $\eta(x)$ being an arbitrary differentiable function of x. Under T

$$u_k(x) \to u'_k(x) = u_k(x) + \delta u_k(x), \tag{10.3}$$
$$v_k(x) \to v'_k(x) = v_k(x) + \delta v_k(x). \tag{10.4}$$

Let us assume the following: (i) $\mathcal{L}[x]$ is form-invariant under T, i.e.,

$$\mathcal{L}'[x'] = \mathcal{L}[x'] \equiv \mathcal{L}(\{u'_k(x')\}, \{v'_k('x)\}; x'), \tag{10.5}$$

and (ii) the integral

$$I = \int_X \mathcal{L}[x]\mathrm{d}x \tag{10.6}$$

is invariant under T, so that

$$\delta I = \int_{X'} \mathcal{L}'[x']\mathrm{d}x' - \int_X \mathcal{L}[x]\mathrm{d}x = 0. \tag{10.7}$$

The consequences of $\delta I = 0$ can be derived using the first-order identities

$$\mathrm{d}x' = \left(1 + \frac{d}{dx}\delta(x)\right)\mathrm{d}x, \tag{10.8}$$

and

$$\delta v_k(x)\left[= \delta\left(\frac{du_k(x)}{dx}\right)\right] = \frac{d}{dx}\delta u_k(x) - u_k(x)\frac{d}{dx}\delta(x). \tag{10.9}$$

To first order in ϵ we have

$$\mathcal{L}'[x'] = \mathcal{L}[x] + \frac{\partial\mathcal{L}}{\partial x}\delta x + \frac{\partial\mathcal{L}}{\partial u_k}\delta u_k + \frac{\partial\mathcal{L}}{\partial v_k}\delta v_k \tag{10.10}$$

and with

$$\frac{d\mathcal{L}}{dx} = \frac{\partial\mathcal{L}}{\partial x} + \frac{\partial\mathcal{L}}{\partial u_k}v_k + \frac{\partial\mathcal{L}}{\partial v_k}\left(\frac{dv_k}{dx}\right), \tag{10.11}$$

and use of Eq. (10.8), one obtains

$$\delta I = \int_X \left\{\frac{d}{dx}(\mathcal{L}\delta x) + \frac{\partial\mathcal{L}}{\partial u_k}(\delta u_k - v_k\delta x) + \frac{\partial\mathcal{L}}{\partial v_k}\left[\delta v_k - \left(\frac{dv_k}{dx}\right)\delta x\right]\right\}\mathrm{d}x. \tag{10.12}$$

By means of Eq. (10.9) the last term in the integrand of Eq. (10.12) can be rewritten as

$$\begin{aligned}
\frac{\partial\mathcal{L}}{\partial v_k}\left[\delta v_k - \left(\frac{dv_k}{dx}\right)\delta x\right] &= \frac{\partial\mathcal{L}}{\partial v_k}\frac{d}{dx}(\delta u_k - v_k\delta x) \\
&= \frac{d}{dx}\left[\frac{\partial\mathcal{L}}{\partial v_k}(\delta u_k - v_k\delta x)\right] - (\delta u_k - v_k\delta x)\frac{d}{dx}\frac{\partial\mathcal{L}}{\partial v_k}.
\end{aligned} \tag{10.13}$$

The final result for δI is therefore

$$\delta I = \int_X \left\{ \frac{d}{dx} \left[\mathcal{L}\delta x + \frac{\partial \mathcal{L}}{\partial v_k}(\delta u_k - v_k \delta x) \right] + \left[\frac{\partial \mathcal{L}}{\partial u_k} - \frac{d}{dx}\frac{\partial \mathcal{L}}{\partial v_k} \right] (\rho u_k - v_k \delta x) \right\} dx.$$

(10.14)

From the requirement $\delta I = 0$ it follows that

$$\left[\frac{d}{dx}\frac{\partial \mathcal{L}}{\partial v_k} - \frac{\partial \mathcal{L}}{\partial u_k} \right] (\delta u_k - v_k \delta x) = \frac{d}{dx}\left[\mathcal{L}\delta x + \frac{\partial \mathcal{L}}{\partial v_k}(\delta u_k - v_k \delta x) \right].$$

(10.15)

This is Nöther's theorem in its simplest form. At this point the reader may appreciate that we considered Eq. (10.14) as our final result for δI: The quantity $d(\partial\mathcal{L}/\partial v_k)/dx - \partial\mathcal{L}/\partial u_k$ entering Eq. (10.14) [Eq. (10.15)] has the characteristic form of Lagrangian dynamics in *a generalized form*, $\mathcal{L}$ being a Lagrangian density. Looking back at all the manipulations leading from Eq. (10.7) to Eq. (10.14), the guideline right from the initial step was to reach an expression in which the Lagrangian form enters explicitly. For physics the main lesson of Nöther's first theorem in its simplest form is as follows: If $I(X)$ [Eq. (10.6)] is invariant under the infinitesimal one-parameter group of transformations T, *and* it is required that

$$\frac{d}{dx}\left(\frac{\partial \mathcal{L}}{\partial v_k} \right) - \frac{\partial \mathcal{L}}{\partial u_k} = 0, \ k = 1, 2, ..., n$$

(10.16)

then Eq. (10.15) implies that

$$\mathcal{L}\delta x + \frac{\partial \mathcal{L}}{\partial v_k}(\delta u_k - v_k \delta x) = K,$$

(10.17)

where K is a constant; independent of x. Eq. (10.17) holds for each k, and with the summation convention in force for the whole k-set.

10.1.2 Hamilton's principle of least action

The principle of least action [46, 64, 189, 223, 246] states that

$$\delta \int_{t_0}^{t_1} L dt = 0.$$

(10.18)

For a system of (classical) point masses the solutions for the set of generalized coordinates $\{q_k(t)\}$ will be those functions for which the integral in Eq. (10.18) is an extremum compared with neighbouring functions integrated between the *fixed* instants of time t_0 and t_1. The calculus of variations leads from Hamilton's principle to the set of Lagrange equations of motions, i.e., [64]

$$\frac{d}{dt}\left(\frac{\partial L}{\partial \dot{q}_k} \right) - \frac{\partial L}{\partial q_k} = 0, \ k = 1, 2, ..., n$$

(10.19)

The principle of least action formally is closely related to Nöther's theorem. Thus, with fixed endpoints (t_0, t_1), Eq. (10.14) gives the Lagrange equations of motions in Eq. (10.16) [since the perfect-integral part vanishes].

A glimpse of the power of Nöther's theorem appears when one applies Eq. (10.17) $[\mathcal{L} \to L]$ to nonrelativistic mechanics: (i) Time transformations. These are charaterized by $\delta t \neq 0$ and $\delta q_k = 0$ and give

$$H \equiv \frac{\partial L}{\partial \dot{q}_k} \dot{q}_k - L = \text{const.} \tag{10.20}$$

The Hamilton H (defined in Eq. (10.20)) thus is a constant of the motion. (ii) Space translations, charaterized by $\delta \mathbf{r} \neq \mathbf{0}$ and $\delta t = 0$, and leads to $(\partial L/\partial \dot{q}_k)\delta \mathbf{r} \cdot \boldsymbol{\nabla} q_k = $ const, and hence since $\delta \mathbf{r}$ is an arbitrary infinitesimal

$$\mathbf{P} \equiv \sum_k \frac{\partial L}{\partial \dot{q}_k} \boldsymbol{\nabla} q_k = \mathbf{const.}, \tag{10.21}$$

The total linear momentum $\mathbf{P}$ (defined by Eq. (10.21)) thus is conserved.

10.1.3 Generalizations

Certain generalizations of the Nöther theorem lessons [Eqs. (10.16) and (10.17)] are of particular interest in electrodynamics [and for physics as such], namely the following: (i) from 1 to ν independent variables $x^1, x^2, ..., x^\nu$ and (ii) for 1 to p parameters $\varepsilon_1, \varepsilon_2, ..., \varepsilon_p$. Thus,

$$x^\mu \to x'^\mu = x^\mu + \delta x^\mu, \ \mu = 1, 2, ..., \nu \tag{10.22}$$

with

$$\delta x^\mu = \sum_{i=1}^p \delta_i x^\mu, \tag{10.23}$$

and

$$\delta u_k = \sum_{i=1}^p \delta_i u_k; \ \delta v_k^\mu = \sum_{i=1}^p \delta_i v_k^\mu. \tag{10.24}$$

The invariance of $\mathcal{L}(\{u_k\}, \{v_k^\mu\}; \{x^\mu\})$ under the p parameter infinitesimal group transformations T_p implies that set

$$\partial^\mu \left(\frac{\partial \mathcal{L}}{\partial v_k^\mu} \right) - \frac{\partial \mathcal{L}}{\partial u_k} = 0, \ \ k = 1, 2, .., n \tag{10.25}$$

implies p conservation laws

$$\partial_\mu F_i^\mu = 0, \ i = 1, 2, ..., p, \tag{10.26}$$

$$F_i^\mu \equiv \mathcal{L}\delta_i x^\mu + \frac{\partial \mathcal{L}}{\partial v_{k,u}}(\delta_i u_k - v_{k,\nu}\delta_i x^\nu). \tag{10.27}$$

Above, the summation convention (over repeated lower and upper indices) is implied.

10.2   ELECTRODYNAMICS AND NÖTHER'S THEOREM

In Chapter 7, we discussed the group theory for inhomogeneous Lorentz transformations, and established the related Poincare group algebra. The Poincare group is a 10-parameter Lie group $[i = 1, 2, ..., 10]$ and Nöther's theorem therefore furnishes one with 10 conservation laws. In the forthcoming Sections 10.3 and 10.4, we establish and analyze these laws in electrodynamics. The general framework for the connection between invariance properties and conservation laws in physical systems was summed up in Eqs. (10.25)–(10.27). In order to apply the scheme in these equations to electrodynamics we must identify the field variables $\{u_k\}$ and give their first derivatives, $\{v_k\}$. The number of independent variables is four, viz. $\{x^\mu\}$, $\mu = 0 - 3$ [the space-time coordinates: $\mu = 1 - 3$ the (Cartesian) space coordinates, $\mu = 0$ the time coordinate]. In the electrodynamic case the field variables become the potentials $A_\alpha(\alpha = 0 - 3)$, and the v_k^μ becomes $\partial^\mu A_\alpha$.

For what follows below, we shall be interested in free electromagnetic fields. Different equivalent choices can be made for the Lagrangian density of the field. Such choices, as well as so-called extended and nonequivalent Lagrangian densities, are studied in Chapter 12 Here we use the so-called standard Lagrangian density for the free field, viz. [63],

$$\mathcal{L}_F = -\frac{\varepsilon_0 c^2}{4} F_{\mu\nu} F^{\mu\nu}, \tag{10.28}$$

where $F_{\mu\nu}$ is defined as

$$F_{\mu\nu} = \partial_\mu A_\nu - \partial_\nu A_\mu, \tag{10.29}$$

Eq. (2.8). In three-vector notation

$$\mathcal{L}_F = \frac{\varepsilon_0}{2} \left[\dot{\mathbf{A}}^2 - c^2 (\boldsymbol{\nabla} \times \mathbf{A})^2 \right] = \frac{\varepsilon_0}{2} (\mathbf{E}^2 - c^2 \mathbf{B}^2), \tag{10.30}$$

where $\mathbf{A} = \mathbf{A}_T$ in free space. With $\mu_k \Rightarrow A_\alpha$ and $v_{k\mu} \Rightarrow \partial_\alpha A_\mu$, one obtains from Eq. (10.25) the Lagrange equations

$$\partial_\alpha \left(\frac{\partial \mathcal{L}_F}{\partial (\partial_\alpha A_\mu)} \right) - \frac{\partial \mathcal{L}_F}{\partial A_\mu} = 0, \ \mu = 0 - 3. \tag{10.31}$$

These, as we know, become the free-field wave equations for the four-potential, i.e., $\Box A^\mu = 0$. A moment of reflection shows that Nöther's theorem gives the following 10 conservation laws; cf. Eqs. (10.26) and (10.27):

$$\partial_\mu F_i^\mu = 0, \ i = 1, 2, ..., 10, \tag{10.32}$$

with

$$F_i^\mu = \frac{\partial \mathcal{L}_F}{\partial (\partial_\mu A_\alpha)} \delta_i A_\alpha - \left[\frac{\partial \mathcal{L}_F}{\partial (\partial_\mu A_\alpha)} \partial_\nu A_\alpha - \delta_\nu^\mu \mathcal{L}_F \right] \delta_i x^\nu. \tag{10.33}$$

For the Poincare group the group elements infinitely close to the identity elements are associated with infinitesimal inhomogeneous Lorentz transformations of the forms given in Eq. (7.40). For pure translations in space and time the parameters are $\delta x^\mu = \varepsilon^\mu$, and for rotations $\delta x^\mu = \omega^\mu_\nu x^\nu$.

In Sections 10.3 and 10.4 we shall study the electrodynamic conservation laws associated with translations and rotations in space-time, respectively.

10.3 INVARIANCE UNDER SPACE-TIME TRANSLATIONS

10.3.1 Conservation laws in covariant form

Let us return to Eq. (10.33). For translations the variations in A_α vanish; $\delta_\nu A_\alpha = 0$, so that we are left with the last term in Eq. (10.33). Since $\delta x^\mu = \varepsilon^\mu$ is an arbitrary infinitesimal, only the term in [...] of Eq. (10.33) enters the conservation law. Hence,

$$\partial_\mu T^{\mu\nu} = 0, \tag{10.34}$$

where

$$T^{\mu\nu} \equiv \frac{\partial \mathcal{L}_F}{\partial(\partial_\mu A_\alpha)} \partial^\nu A_\alpha - \eta^{\mu\nu} \mathcal{L}_F. \tag{10.35}$$

Note that $\eta_{\mu\nu}$ is the metric tensor, related to its reciprocal $\eta^{\mu\nu}$ [in Eq. (10.35)] by $\eta^{\mu\nu}\eta_{\mu\nu} = \delta^\mu_\nu$, where δ^μ_ν is the Kronecker delta [entering Eq. (10.33)]; $\delta^\mu_\nu = 1$ for $\mu = \nu$, and 0 for $\mu \neq \nu$. Because $T^{\mu\nu}$ is the coefficient of ϵ_ν, $T^{\mu\nu}$ is the generator of infinitesimal translations in space-time.

By re-introducing the explicit summation sign one has

$$\frac{\partial}{\partial(\partial_\mu A_\alpha)} \sum_{\gamma,\delta} F_{\gamma\delta} F^{\gamma\delta} = \frac{\partial}{\partial(\partial_\mu A_\alpha)} [F_{\mu\alpha}F^{\mu\alpha} + F_{\alpha\mu}F^{\alpha\mu}] = 2\frac{\partial}{\partial(\partial_\mu A_\alpha)}(F_{\mu\alpha}F^{\mu\alpha})$$

$$= 2\left[F^{\mu\alpha}\frac{\partial F_{\mu\alpha}}{\partial(\partial_\mu A_\alpha)} + F_{\mu\alpha}\frac{\partial F^{\mu\alpha}}{\partial(\partial_\mu A_\alpha)}\right]$$

$$= 2\left[F^{\mu\alpha} + F_{\mu\alpha}\eta^{\mu\mu}\eta^{\alpha\alpha}\right] = 4F^{\mu\alpha}. \tag{10.36}$$

By inserting this result into Eq. (10.35) one obtains

$$T^{\mu\nu} = -\frac{\varepsilon_0 c^2}{4}(4F^{\mu\alpha}\partial^\nu A_\alpha - \eta^{\mu\nu}F_{\alpha\beta}F^{\alpha\beta}). \tag{10.37}$$

Because [cf. the definition in Eq. (2.8)]

$$\partial^\nu A_\alpha = -F^\nu_\alpha + \delta_\alpha A^\nu, \tag{10.38}$$

we also have

$$T^{\mu\nu} = -\frac{\varepsilon_0 c^2}{4}\left[-4F^{\mu\alpha}F^\nu_\alpha - \eta^{\mu\nu}F_{\alpha\beta}F^{\alpha\beta}\right] - \varepsilon_0 c^2 F^{\mu\alpha}\partial_\alpha A^\nu. \tag{10.39}$$

If the last term in Eq. (10.39) could be omitted, $T^{\mu\nu}$, would be replaced by the *manifest gauge invariant tensor*

$$\Theta^{\mu\nu} \equiv \frac{\varepsilon_0 c^2}{4} \left[4 F^{\mu\alpha} F^{\nu}_{\alpha} - \eta^{\mu\nu} F_{\alpha\beta} F^{\alpha\beta} \right]. \tag{10.40}$$

By remembering that it is the four-divergence $\partial_\mu T^{\mu\nu}$ which is relevant in the conservation law [Eq. (10.34)]. The replacement $T^{\mu\nu} \Rightarrow \Theta^{\mu\nu}$ is justified if the divergence of $F^{\mu\alpha}\partial_\alpha A^\nu$ vanishes. Let us see if this is correct. From the transformation

$$\partial_\mu(F^{\mu\alpha}\partial_\alpha A^\nu) = \partial_\mu \partial_\alpha (F^{\mu\alpha} A^\nu) - \partial_\mu(A^\nu \partial_\alpha F^{\mu\alpha}) \tag{10.41}$$

it indeed follows that

$$\partial_\mu(F^{\mu\alpha}\partial_\alpha A^\nu) = 0, \tag{10.42}$$

since the first term on the right-hand side of Eq. (10.41) vanishes because it is a product of a symmetric tensor $(\partial_\mu \partial_\alpha)$ and an antisymmetric on $(F^{\mu\alpha})$. The second term vanishes for free fields: $\partial_\alpha F^{\mu\alpha} = 0$, as is well known [63, 106, 186].

The tensor $\Theta^{\mu\nu}$ is manifest symmetric, and it is known as the *symmetrical electromagnetic energy-momentum tensor* [although 9 of the 16 components of $\Theta^{\mu\nu}$ refer to stress]. For brevity we often call $\Theta^{\mu\nu}$ simply the energy tensor. It is easy to show that the trace of $\Theta^{\mu\nu}$ vanishes, i.e.,

$$\mathrm{Tr}\{\Theta^{\mu\nu}\} = 0. \tag{10.43}$$

Expressed in three-vector notation $\theta^{\mu\nu}$, and the conservation law

$$\partial_\mu \theta^{\mu\nu} = 0, \tag{10.44}$$

take forms appearing in many textbooks, as we shall see below. For readers not being experienced in the details of covariant calculations it may be useful to present the steps leading from $\theta^{\mu\nu}$ and $\partial_\mu \theta^{\mu\nu}$ to the three-vector forms of these.

10.3.2 Conservation laws in three-vector form

Let $\mathbf{E} = (E_1, E_2, E_3)$ and $\mathbf{B} = (B_1, B_2, B_3)$ be the electric and the magnetic three-vectors, related to the scalar and vector potentials Φ and $\mathbf{A} = (A_1, A_2, A_3)$ in the usual manner [Eq. (4.9) and (4.10)], viz., $\mathbf{E} = -\nabla\Phi - \partial\mathbf{A}/\partial t$ and $\mathbf{B} = \nabla \times \mathbf{A}$. From the covariant from of these relations, with components given in Eq. (2.8), it appears that the covariant electromagnetic field tensor has the matrix form

$$\{F_{\mu\nu}\} = \frac{1}{c} \begin{pmatrix} 0 & -E_1 & -E_2 & -E_3 \\ E_1 & 0 & cB_3 & -cB_2 \\ E_2 & -cB_3 & 0 & cB_1 \\ E_3 & cB_2 & -cB_1 & 0 \end{pmatrix}. \tag{10.45}$$

The field tensor's contravariant ($\{F^{\mu\nu}\}$) and mixed ($\{F^\nu_\mu\}$) forms are calculated raising and lowering indices with the identical contra- and covariant metric tensors of special relativity [here chosen with signature $(-1,1,1,1)$], viz.,

$$\{\eta_{\mu\nu}\} = \{\eta^{\mu\nu}\} = \begin{pmatrix} -1 & 0 & 0 & 0 \\ 0 & 1 & 0 & 0 \\ 0 & 0 & 1 & 0 \\ 0 & 0 & 0 & 1 \end{pmatrix}. \tag{10.46}$$

Thus,

$$\{F^{\mu\nu}\} = \{\eta^{\mu\alpha}\eta^{\nu\beta}F_{\alpha\beta}\} = \frac{1}{c}\begin{pmatrix} 0 & E_1 & E_2 & E_3 \\ -E_1 & 0 & cB_3 & -cB_2 \\ -E_2 & -cB_3 & 0 & cB_1 \\ -E_3 & cB_2 & -cB_1 & 0 \end{pmatrix}. \tag{10.47}$$

and

$$\{F^\nu_\mu\} = \{F_{\mu\alpha}\eta^{\alpha\nu}\} = \frac{1}{c}\begin{pmatrix} 0 & -E_1 & -E_2 & -E_3 \\ -E_1 & 0 & cB_3 & -cB_2 \\ -E_2 & -cB_3 & 0 & cB_1 \\ -E_3 & cB_2 & -cB_1 & 0 \end{pmatrix}. \tag{10.48}$$

Tedious matrix multiplications give the three-vector forms of $F^{\mu\alpha}F^\nu_\alpha$ (tensor) and $F_{\alpha\beta}F^{\alpha\beta}$ (scalar), and from Eq. (10.40) the three-vector elements of $\{\theta^{\mu\nu}\}$ are obtained. As the reader may verify, one obtains

$$\theta^{00} = -\frac{\varepsilon_0}{2}(E^2 + c^2B^2) = -W, \tag{10.49}$$

where W is known as the electromagnetic energy density, and ($i = 1 - 3$)

$$\theta^{0i} = \theta^{i0} = -\varepsilon_0 c(\mathbf{E} \times \mathbf{B})_i = -\frac{1}{c}(\mathbf{S})_i, \tag{10.50}$$

$\mathbf{S} = \mu_0^{-1}\mathbf{E} \times \mathbf{B}$ being the electromagnetic Poynting vector. Finally ($i, j = 1 - 3$)

$$\{\theta^{ij}\} \equiv \mathbf{T} = \varepsilon_0(\mathbf{EE} + c^2\mathbf{BB}) - \frac{\varepsilon_0}{2}(E^2 + c^2B^2)\mathbb{1}, \tag{10.51}$$

is the Maxwell stress density written in dyadic form ($\mathbb{1}$ is the dyadic form of the unit tensor). Eqs. (10.49)–(10.51) show that $\{\theta^{\mu\nu}\}$ is a symmetric tensor. The conservation law in Eq. (10.44) now gives (chosing $\nu = 0$)

$$\boldsymbol{\nabla} \cdot \mathbf{S} + \frac{\partial W}{\partial t} = 0, \tag{10.52}$$

and (using in turn $\nu = 1 - 3$)

$$\boldsymbol{\nabla} \cdot \mathbf{T} - \frac{1}{c^2}\frac{\partial \mathbf{S}}{\partial t} = \mathbf{0}. \tag{10.53}$$

Eqs. (10.52) and (10.53) express free field energy and momentum conservation in differential form. It must be remembered that the deeper physical meaning of these conservation laws only appears via proper integral forms of the conservation laws; c.f. our inability to localize a photon completely in space-time.

10.3.3 Extension to matter-coupled electromagnetic fields

In the presence of field-matter coupling the conservation law in Eq. (10.44) is no longer valid. The symmetrical electromagnetic energy-momentum tensor is still given (defined) by Eq. (10.39), but its four-divergence no longer vanishes. Thus,

$$\partial_\mu \theta^{\mu\nu} = \varepsilon_0 c^2 (\partial_\mu F^{\mu\alpha}) F^\nu_\alpha. \tag{10.54}$$

If the covariant form of the inhomogeneous Maxwell-Lorentz equations, viz. [63, 106, 158, 186],

$$\partial_\mu F^{\mu\alpha} = -\mu_0 J^\alpha, \tag{10.55}$$

where $\{J^\alpha\} = (c\rho, \mathbf{J})$ is the contravariant four-current density, is inserted on the right-hand side of Eq.(10.54), one obtains

$$\partial_\mu \theta^{\mu\nu} = -J^\alpha F^\nu_\alpha = -J^\alpha \eta_{\alpha\beta} F^{\beta\nu} = -J_\beta F^{\beta\nu}. \tag{10.56}$$

The antisymmetry of the contravariant field tensor finally gives

$$\partial_\mu \theta^{\mu\nu} = F^{\nu\alpha} J_\alpha. \tag{10.57}$$

The three-vector form of the term $F^{\nu\alpha} J_\alpha$ is determined as the matrix product of $\{F^{\nu\alpha}\}$ [Eq. (10.47)] and $\{J_\alpha\} = (-c\rho, \mathbf{J})$. For $\nu = 0$ one has

$$F^{0\alpha} J_\alpha = \mathbf{J} \cdot \mathbf{E}, \tag{10.58}$$

and from the equations for $\nu = i (i = 1 - 3)$ we get

$$F^{i\alpha} J_\alpha = \rho E_i + (\mathbf{J} \times \mathbf{B})_i. \tag{10.59}$$

In three-vector notation we hence obtain the following balance equations for the field energy and momentum:

$$\nabla \cdot \mathbf{S} + \frac{\partial W}{\partial t} = -\mathbf{J} \cdot \mathbf{E}, \tag{10.60}$$

and

$$\nabla \cdot \mathbf{T} - \frac{1}{c^2} \frac{\partial \mathbf{S}}{\partial t} = \rho \mathbf{E} + \mathbf{J} \times \mathbf{B}. \tag{10.61}$$

The reader may recognize the vector on the right-hand side of Eq. (10.61) as the Lorentz force density.

10.4 INVARIANCE UNDER BOOSTS AND SPATIAL ROTATIONS

For a proper homogeneous Lorentz transformation it follows from Eq. (7.40) that we are concerned with the invariance of the theory under the infinitesimal transformations

$$x'^\mu = x^\mu + \omega^\mu_\nu x^\nu \tag{10.62}$$

that is

$$\delta x^\mu = \omega^\mu_\nu x^\nu. \tag{10.63}$$

The four potential, transformed in the general case as

$$A'^\mu(x') = \Lambda^\mu_\nu A^\nu(x), \tag{10.64}$$

hence is also affected by ω^μ_ν. Thus

$$\delta A^\mu(x) = \omega^\mu_\nu A^\nu(x). \tag{10.65}$$

In the view of the analysis in Section 10.3, and with the help of the free-field result

$$\frac{\partial \mathcal{L}}{\partial(\partial_\mu A_\alpha)} = -\varepsilon_0 c^2 F^{\mu\alpha}, \tag{10.66}$$

the conserved quantity in Eq. (10.27) is in a first step given by

$$\frac{1}{2} M^{\mu\alpha\nu}\omega_{\alpha\nu} \equiv -\varepsilon_0 c^2 F^{\mu\alpha}\omega_{\alpha\nu} A^\nu - T^{\mu\alpha}\omega_{\alpha\nu} x^\nu. \tag{10.67}$$

Making use of the antisymmetry of $\omega_{\alpha\nu}$, and a relabeling $\mu \to \alpha$, $\alpha \to \mu$, one has

$$M^{\alpha\mu\nu} = -(T^{\alpha\mu}x^\nu - T^{\alpha\nu}x^\mu) - \varepsilon_0 c^2 (F^{\alpha\mu}A^\nu - F^{\alpha\nu}A^\mu). \tag{10.68}$$

Since the infinitesimal $\omega_{\mu\nu} = -\omega_{\nu\mu}$ is arbitrary, the six independent conservation laws are

$$\partial_\alpha M^{\alpha\mu\nu} = 0. \tag{10.69}$$

It may appear attempting to associate the quantity

$$L^{\alpha\mu\nu} \equiv -(T^{\alpha\mu}x^\nu - T^{\alpha\nu}x^\mu) \tag{10.70}$$

with an *orbital angular momentum density* related to four-dimensional rotations. Likewise,

$$S^{\alpha\mu\nu} \equiv -\varepsilon_0 c^2 (F^{\alpha\mu}A^\nu - F^{\alpha\nu}A^\mu) \tag{10.71}$$

perhaps might be a *spin angular momentum density*. In case

$$\tilde{J}^{\alpha\mu\nu} = L^{\alpha\mu\nu} + S^{\alpha\mu\nu} \tag{10.72}$$

would represent the generalized *total angular momentum density* of the free electromagnetic field. The suggested interpretation runs into troubles for several reasons, however. Thus, the quantity

$$\partial_\alpha L^{\alpha\mu\nu} = T^{\nu\mu} - T^{\mu\nu}(\neq 0) \tag{10.73}$$

does not vanish because the tensor $T^{\mu\nu}$ is not symmetric, and furthermore $T^{\mu\nu}$ is neither gauge invariant (see Eqs. (10.38) and (10.39)]. In free space there are no possibilities for transferring angular momentum between the "external" ($L^{\alpha\mu\nu}$) and "internal" ($S^{\alpha\mu\nu}$) contributions, so alone for this reason the above division of $\tilde{J}^{\alpha\mu\nu}$ into space- and spin-parts fails because $\partial_\alpha L^{\alpha\mu\nu} \neq 0$. It is suggestive to define instead of $L^{\alpha\mu\nu}$ the gauge invariant tensor

$$J^{\alpha\mu\nu} = -(\theta^{\alpha\mu}x^\nu - \theta^{\alpha\nu}x^\mu). \tag{10.74}$$

Although the inspiration to define $J^{\alpha\mu\nu}$ came from $L^{\alpha\mu\nu}$, a closer analysis shows that $J^{\alpha\mu\nu}$ is a fruitful replacement for the "total angular momentum" $\tilde{J}^{\alpha\mu\nu}$, given in Eq. (10.72). Utilizing Eq. (10.44), and the symmetry of $\theta^{\mu\nu}$, it is obvious that $J^{\alpha\mu\nu}$ satisfy the conservation law

$$\partial_\alpha J^{\alpha\mu\nu} = 0. \tag{10.75}$$

We know that *density* quantities (for energy, momentum and angular momentum), make no deeper sense for electromagnetic fields, only integrals of these local quantities over a sufficiently large space-time volume can be related to physical observations. If $d\sigma_\alpha$ denotes a space-like surface in Minkowski space [a three-dimensional surface on which the distance between any two points $\{x^\mu\}$ and $\{y^\mu\}$ is space-like, i.e., $(x_\mu - y_\mu)(x^\mu - y^\mu) > 0$], the integrated angular momentum

$$J^{\mu\nu} = \int J^{\alpha\mu\nu} d\sigma_\alpha \tag{10.76}$$

is a gauge-invariant tensor. We know from the analysis in Section 7.4 [and from elementary field theory] that only the space part $\{J^{ij}\}$, $i = 1 - 3$, describes the total angular momentum [conserved in free space due to invariance under usual rotations]. The conservation of mixed components $\{J^{ok}\}$, $k = 1 - 3$, expresses the center-of-mass theorem for free electromagnetic fields.

Rigorous three-vector discussions of the angular momentum conservation and our general inability to separate the angular momentum into orbital and spin parts can be found, e.g. in [63, 153, 157].

10.5 FOUR-MOMENTUM FIELD CONSERVATION ON COVARIANT INTEGRAL FORM

The physical interpretation of the differential conservation laws in Eqs. (10.52) and (10.53) requires an integration over an appropriate space-like plane, $d\sigma_\nu$. This is so because an instant of an inertial observer is characterized by such a plane. The unit normal vector related to a space-like plane, $\{n^\mu\}$, is necessarily time-like and given by

$$\{n^\mu\} = \frac{\{v^\mu\}}{c} \tag{10.77}$$

where $\{v^\mu\} = (c\gamma, \gamma\mathbf{v})$ is the four-velocity, $\mathbf{v}$ being the velocity three-vector, and $\gamma = (1 - \beta^2)^{-1/2}$ [with $\beta = \mathbf{v}/c$]. Since $v_\mu v^\mu = -c^2$, one has

$$n_\mu n^\mu = -1. \tag{10.78}$$

With the choice $n^0 > 0$, the unit vector points into the future. The surface element is given by

$$d\sigma^\mu = n^\mu d\sigma \tag{10.79}$$

where $d\sigma$ is the invariant area element.

Let us consider now the vectorial quantity

$$P^\mu \equiv \frac{1}{c} \int \Theta^{\mu\nu} n_\nu d\sigma. \tag{10.80}$$

Since $\{\Theta^{\mu\nu}\}$ is divergence free [Eq. (10.44)], $\{P^\mu\}$ is divergence free $[\partial_\mu P^\mu = 0]$. To make the bridge to the integral form of the conservation laws for the momentum four-vector of the electromagnetic field, we choose the specific Lorentz frame in which the surface normal is $\{n^\mu\} = (1, 0, 0, 0)$. For this choice the surface is a volume of three-space, with invariant area element

$$d\sigma = -n_\mu d\sigma^\mu = dx^{(0)} dy^{(0)} dz^{(0)}. \tag{10.81}$$

The superscript (0) on the three-space line element, is to remind the reader of the specific choice $\{n^\mu\} = (1, 0, 0, 0)$; $[n_0 = -n^0 = -1]$. The components

$$P^{\mu(0)} = -\frac{1}{c} \int \Theta^{\mu 0} d\sigma, \ \mu = 0 - 3, \tag{10.82}$$

lead, using three-vector notation [use Eq. (10.49) and Eq. (10.50)], to the textbook [118] integral expression

$$\{P^{\mu(0)}\} = \left(\frac{1}{c} W_{\text{em}}^{(0)}, \mathbf{P}_{\text{em}}^{(0)} \right) = \left(\frac{1}{c} \int W d^3x, \frac{1}{c^2} \int \mathbf{S} d^3x \right), \tag{10.83}$$

where W and $\mathbf{S}$ are given by Eqs. (10.49) and (10.50). To refer to $W^{(0)}$ and $\mathbf{P}^{(0)}$ as *the* energy and momentum of the electromagnetic field is from a basic point of view incorrect since these quantities belong to a specific Lorentz frame.

In order to obtain the expression for $\{P^\mu\}$ in an arbitrary inertial frame we chose an arbitrary $\{n^\mu\}$, viz.,

$$\{n^\mu\} = \frac{1}{c}\{v^\mu\} = \left(\gamma, \frac{\gamma}{c}\mathbf{v} \right) = \left(\gamma, \frac{\gamma}{c} v\hat{\mathbf{n}} \right) \tag{10.84}$$

where $\mathbf{v} = v\hat{\mathbf{n}}$ is the three-vector velocity, with directional unit vector $\hat{\mathbf{n}}$. Written in the form

$$\{n^\mu\} = (\gamma, \sqrt{\gamma^2 - 1}\hat{\mathbf{n}}), \tag{10.85}$$

Eq. (10.80) gives

$$P^\mu = \frac{1}{c} \int [\gamma \Theta^{\mu 0} + \sqrt{\gamma^2 - 1}\, \Theta^{\mu i} n_i] d\sigma. \tag{10.86}$$

The general integral form thus is

$$\{P^\mu\} = \left(\frac{1}{c} W_{\text{em}}, \mathbf{P}_{\text{em}} \right), \tag{10.87}$$

where

$$W_{\text{em}} = \gamma \int W d\sigma - \frac{1}{c}\sqrt{\gamma^2 - 1} \int \mathbf{S} \cdot \hat{\mathbf{n}} d\sigma, \tag{10.88}$$

$$\mathbf{P}_{\text{em}} = \frac{\gamma}{c^2} \int \mathbf{S} d\sigma + \frac{1}{c}\sqrt{\gamma^2 - 1} \int \mathbf{T} \cdot \hat{\mathbf{n}} d\sigma. \tag{10.89}$$

The *invariant* area element can in any case be taken from Eq. (10.81) if wished. It appears from Eqs. (10.87)–(10.89) that the *integral expressions* for W_{em} and $\mathbf{P}_{\text{em}}$, giving what an initial observer takes as the (free) electromagnetic fields energy and momentum, is frame-dependent. In the rest frame ($\gamma = 1$) one regains the familiar expression in Eq. (10.83).

CHAPTER 11

Topology and Geometric Phase

11.1 FORM-INVARIANCE OF THE FIELD-COUPLED DIRAC EQUATION

The requirement that local arbitrary phase transformations of the wave function $\psi(x)$ of a massive particle with charge e have no observable physical significance lead us in Section 2.1 to the conclusion that a gauge field (the electromagnetic four potential) necessarily must be present. Expressed in three vector notation we came to the conclusion that a gauge transformation of the vector potential $(\mathbf{A})$ and the scalar potential (V) viz.,

$$\mathbf{A}'(\mathbf{r}, t) = \mathbf{A}(\mathbf{r}, t) + \boldsymbol{\nabla}\chi(\mathbf{r}, t), \tag{11.1}$$

$$V'(\mathbf{r}, t) = V(\mathbf{r}, t) - \frac{\partial\chi(\mathbf{r}, t)}{\partial t}, \tag{11.2}$$

where $\chi(\mathbf{r}, t)$ is the so-called gauge function, necessarily must be accompanied by a local phase shift of the particle wave function, i.e.,

$$\psi'(\mathbf{r}, t) = \exp\left(i\frac{e}{\hbar}\chi(\mathbf{r}, t)\right)\psi(\mathbf{r}, t). \tag{11.3}$$

The above scheme can only be sensible if the dynamical equation describing the time evolution of the wave function is form-invariant under the transformation given by Eqs. (11.1)–(11.3). Let us show this for the relativistic Dirac equation [spin-1/2 particle of rest mass m and charge e] coupled to an electromagnetic field. In terms of the four Dirac matrices $\boldsymbol{\alpha} = (\alpha_x, \alpha_y, \alpha_z)$ and β, the Dirac equation has the Hamiltonian form [63, 69, 186]

$$i\hbar\frac{\partial\psi(\mathbf{r}, t)}{\partial t} = \hat{H}\psi(\mathbf{r}, t), \tag{11.4}$$

where

$$\hat{H} = c\boldsymbol{\alpha}\cdot\left(\frac{\hbar}{i}\boldsymbol{\nabla} - e\mathbf{A}(\mathbf{r}, t)\right) + eV(\mathbf{r}, t) + \beta mc^2 \tag{11.5}$$

is the Hamilton operator [the unit matrix ($\mathbb{1}$) on the scalar potential is kept implicit, $V\mathbb{1} = V$]. Inserting $\mathbf{A} = \mathbf{A}' - \boldsymbol{\nabla}\chi$, $V = V' + \partial\chi/\partial t$, and $\psi = \exp[-i(e/\hbar)\chi]\psi'$ into Eqs. (11.4) and (11.5) it is easy to show that the coupled wave equation for ψ' in the potential $(\mathbf{A}', V')$ has the same *form* as given in Eqs. (11.4) and (11.5). The conclusion we have reached of course also holds in the weakly relativistic (Pauli equation) and nonrelativistic (Schrödinger equation) regimes, and for a quantized electromagnetic field.

In a slightly different perspective, one may start from the free Dirac equation,

$$i\hbar\frac{\partial\psi(\mathbf{r}, t)}{\partial t} = \left(\frac{c\hbar}{i}\boldsymbol{\alpha}\cdot\boldsymbol{\nabla} + \beta mc^2\right)\psi(\mathbf{r}, t), \tag{11.6}$$

and then make the local phase shift given in Eq. (11.3). The resulting wave equation for $\psi'(\mathbf{r}, t)$ [renamed $\psi(\mathbf{r}, t)$ below] takes the form

$$i\hbar\frac{\partial\psi(\mathbf{r}, t)}{\partial t} = \left[c\boldsymbol{\alpha}\cdot\left(\frac{\hbar}{i}\boldsymbol{\nabla} - e\boldsymbol{\nabla}\chi(\mathbf{r}, t)\right) - e\frac{\partial\chi(\mathbf{r}, t)}{\partial t}\right]\psi(\mathbf{r}, t). \tag{11.7}$$

The demand of local phase invariance then dictates the presence of an electromagnetic field with potentials $\mathbf{A} = \boldsymbol{\nabla}\chi$ and $V = -\partial\chi/\partial t$.

11.2   THE DIRAC PHASE FACTOR IN A FIXED TIME LOOP

Among the closed curves (C) entering the Dirac phase factor given in Eq. (2.11) we shall in the following be interested in those for which time is fixed. For $dx^0 = 0$, $\Phi(C)$ takes the form

$$\Phi(C) = \exp\left[\frac{ie}{\hbar}\oint_C (A_x dx + A_y dy + A_z dz)\right] = \exp\left[i\frac{e}{\hbar}\oint_C \mathbf{A}(\mathbf{r}, t)\cdot d\mathbf{r}\right]. \tag{11.8}$$

In the present case Stokes' theorem [Eq. (2.7)] is reduced to

$$\oint_C \mathbf{A}(\mathbf{r}, t)\cdot d\mathbf{r} = \int_S [\boldsymbol{\nabla}\times\mathbf{A}(\mathbf{r}, t)]\cdot\hat{\mathbf{n}}(\mathbf{r})dS, \tag{11.9}$$

where $d\mathbf{S}(\mathbf{r}) = \hat{\mathbf{n}}(\mathbf{r})dS$ is the oriented surface area element on an open surface (S) bounded by C, and $\hat{\mathbf{n}}$ is the elements' outwards (c.f. Gauss' theorem) directed unit normal vector. If the curl of the vector potential ($\boldsymbol{\nabla}\times\mathbf{A}$) is zero everywhere throughout the region (Ω) of interest at all relevant times, the integral of $\mathbf{A}(\mathbf{r}, t)$ taken along any closed curve C lying within Ω is zero.

One may sharpen the physical perspective of Eq. (11.9) a bit. Thus, from the PRINCIPLE OF LOCAL PHASE INVARIANCE (Section 2.1) it readily follows that only the gauge invariant (transverse) part, $\mathbf{A}_T(\mathbf{r}, t)$, of the vector potential, $\mathbf{A}(\mathbf{r}, t)$ contributes to the Dirac phase factor. Hence

$$\Phi(C) = \exp\left[i\frac{e}{\hbar}\oint_C \mathbf{A}_T(\mathbf{r}, t)\cdot d\mathbf{r}\right]. \tag{11.10}$$

The local phase invariance principle is not in conflict with Stoke's theorem since $\boldsymbol{\nabla} \times \mathbf{A}_L = \mathbf{0}$ (per definition) allows us to write the theorem in the form

$$\oint_C \mathbf{A}_T(\mathbf{r}, t) \cdot d\mathbf{r} = \int_S [\boldsymbol{\nabla} \times \mathbf{A}_T(\mathbf{r}, t)] \cdot \hat{\mathbf{n}}(\mathbf{r}) dS, \tag{11.11}$$

in the chosen inertial system.

An utmost interesting situation appears if

$$\boldsymbol{\nabla} \times \mathbf{A}_T(\mathbf{r}, t) = \mathbf{0}, \quad \mathbf{r} \in \Omega, \tag{11.12}$$

in a certain spatial domain Ω. The conditions $\boldsymbol{\nabla} \times \mathbf{A}_T(\mathbf{r}, t) = \mathbf{0}$ and $\boldsymbol{\nabla} \cdot \mathbf{A}_T(\mathbf{r}, t) = \mathbf{0}$ (per definition) do not imply that $\mathbf{A}_T(\mathbf{r}, t) = \mathbf{0}$ identically since Ω is assumed to cover only a finite region of infinite space:

$$\int_\Omega (...) d^3 r = \int_{-\infty}^{\infty} (..) d^3 r - \int_{\tilde{\Omega}} (...) d^3 r, \tag{11.13}$$

where $\tilde{\Omega}$ is the domain outside Ω. From the Helmholtz identity [7,158,190] (with time kept implicit)

$$\mathbf{A}(\mathbf{r}) = \boldsymbol{\nabla} \times \int_{-\infty}^{\infty} \frac{\boldsymbol{\nabla}' \times \mathbf{A}(\mathbf{r}')}{4\pi |\mathbf{r} - \mathbf{r}'|} d^3 r' - \boldsymbol{\nabla} \int_{-\infty}^{\infty} \frac{\boldsymbol{\nabla}' \cdot \mathbf{A}(\mathbf{r}')}{4\pi |\mathbf{r} - \mathbf{r}'|} d^3 r', \tag{11.14}$$

it follows that for the points inside Ω $(\mathbf{r} \in \Omega)$

$$\mathbf{A}(\mathbf{r} | \mathbf{r} \in \Omega) = \boldsymbol{\nabla} \times \int_{\tilde{\Omega}} \frac{\boldsymbol{\nabla}' \times \mathbf{A}(\mathbf{r}')}{4\pi |\mathbf{r} - \mathbf{r}'|} d^3 r' - \boldsymbol{\nabla} \int_{\tilde{\Omega}} \frac{\boldsymbol{\nabla}' \cdot \mathbf{A}(\mathbf{r}')}{4\pi |\mathbf{r} - \mathbf{r}'|} d^3 r', \tag{11.15}$$

and specifically

$$\mathbf{A}_T(\mathbf{r} | \mathbf{r} \in \Omega) = \boldsymbol{\nabla} \times \int_{\tilde{\Omega}} \frac{\boldsymbol{\nabla}' \times \mathbf{A}_T(\mathbf{r}')}{4\pi |\mathbf{r} - \mathbf{r}'|} d^3 r'. \tag{11.16}$$

11.3 MAGNETIC (DIPOLE) STRING: THE AHARONOV-BOHM PHENOMENON

11.3.1 Vector potential of a magnetic string with curvature and torsion

In the Maxwell-Lorentz electrodynamics the microscopic charge-particle ($\sim$ electron) current density, $\mathbf{J}(\mathbf{r}, t)$, plays a central role, and for the present purpose it is useful to divide it into two parts in the following manner:

$$\mathbf{J}(\mathbf{r}, t) \equiv \frac{\partial \mathbf{P}(\mathbf{r}, t)}{\partial t} + \boldsymbol{\nabla} \times \mathbf{M}(\mathbf{r}, t). \tag{11.17}$$

The microscopic vector fields $\mathbf{P}(\mathbf{r}, t)$ and $\mathbf{M}(\mathbf{r}, t)$ are called the generalized polarization and magnetization [157], using words borrowed from macroscopic electrodynamics. In the macroscopic theory $\mathbf{P}_{\mathrm{macro}}$ and $\mathbf{M}_{\mathrm{macro}}$ refer to the electric and magnetic *dipole moment densities*, quantities which on a small ($\sim$ mesoscopic) length scale lose a precise physical meaning. In contrast the microscopic quantities $\mathbf{P}$ and $\mathbf{M}$ are good physical vector fields also on the atomic length scale.

In a sense, these microscopic concepts are physically well-defined down to a length scale where it is necessary to take into account the fact that the $\mathbf{r}$-coordinate in the Dirac wave function $\psi(\mathbf{r}, t)$ is just a sort of mean position coordinate, c.f. the Foldy-Wouthausen theory [87].

The division in Eq. (11.17) has a certain arbitrariness in the transverse parts $\mathbf{P}_T$ and $\nabla \times \mathbf{M}(= \nabla \times \mathbf{M}_T)$. Hence, if $\mathbf{N}_T(\mathbf{r}, t)$ denotes the arbitrary (yet differentiable in space and time) transverse vector field, the transformation

$$\mathbf{P}_T(\mathbf{r}, t) \Rightarrow \mathbf{P}'_T(\mathbf{r}, t) = \mathbf{P}_T(\mathbf{r}, t) - \nabla \times \mathbf{N}_T(\mathbf{r}, t), \qquad (11.18)$$

$$\mathbf{M}(\mathbf{r}, t) \Rightarrow \mathbf{M}'(\mathbf{r}, t) = \mathbf{M}(\mathbf{r}, t) + \frac{\partial}{\partial t}\mathbf{N}_T(\mathbf{r}, t), \qquad (11.19)$$

is seen to leave the (physical) microscopic current density invariant, $\mathbf{J}'(\mathbf{r}, t) = \mathbf{J}(\mathbf{r}, t)$. The choice $\nabla \times \mathbf{N}_T = \mathbf{P}_T$ removes the transverse part from the polarization, i.e., $\mathbf{P}'(\mathbf{r}, t) = \mathbf{P}_L(\mathbf{r}, t)$. It appears from the microscopic equation of continuity, $\nabla \cdot \mathbf{J} + \partial\rho/\partial t = 0$, that the longitudinal part $\mathbf{P}_L$ is related to the charge density by $\nabla \cdot \mathbf{P}_L = -\rho$. Since $\nabla \times \mathbf{P}_L = \mathbf{0}$, it follows from Helmholtz theorem that $\mathbf{P}_L(\mathbf{r}, t)$ (up to a physically unimportant constant) is fixed by charge density $\rho(\mathbf{r}, t)$. With no loss of generality we take

$$\mathbf{J}(\mathbf{r}, t) = \frac{\partial}{\partial t}\mathbf{P}_L(\mathbf{r}, t) + \nabla \times \mathbf{M}(\mathbf{r}, t). \qquad (11.20)$$

For good reasons $\mathbf{J}_M(\mathbf{r}, t) \equiv \nabla \times \mathbf{M}(\mathbf{r}, t)$ now is called the microscopic magnetization current density.

In the following we shall study the vector potential in the surrounding of a so-called magnetic string, and apply the general considerations to a discussion of the Aharonov-Bohm effect. The magnetic string concept is an idealization from a microscopic point of view representing a magnetized tube (see Fig. 11.1) in the limit where quantum mechanical interstate excitations in the transverse coordinates can be neglected. We shall assume additionally that the magnetization in the tube (of the string) is unaffected by the created vector potential. Hence, the string magnetization is considered as a prescribed quantity.

Geometrically, it is useful here to employ a parametrization $\mathbf{r} = \mathbf{r}(s)$, of the string characterized by its local curvature, $\kappa(s)$, and torsion, $\tau(s)$. We take the parameter s as the arclength on the string. A local coordinate system with orthogonal unit vectors $\hat{\mathbf{t}}$, $\hat{\mathbf{b}}$ and $\hat{\mathbf{n}} = \hat{\mathbf{b}} \times \hat{\mathbf{t}}$, is introduced next. With the notation $\mathbf{r}'(s) = d\mathbf{r}(s)ds$, $\mathbf{r}''(s) = d^2\mathbf{r}(s)ds^2$, and $\mathbf{r}'''(s) = d^3\mathbf{r}(s)ds^3$,

$$\hat{\mathbf{t}}(s) = \frac{\mathbf{r}'(s)}{|\mathbf{r}'(s)|} \qquad (11.21)$$

is the tangent unit vector,

$$\hat{\mathbf{b}}(s) = \frac{\mathbf{r}'(s) \times \mathbf{r}''(s)}{|\mathbf{r}'(s) \times \mathbf{r}''(s)|} \qquad (11.22)$$

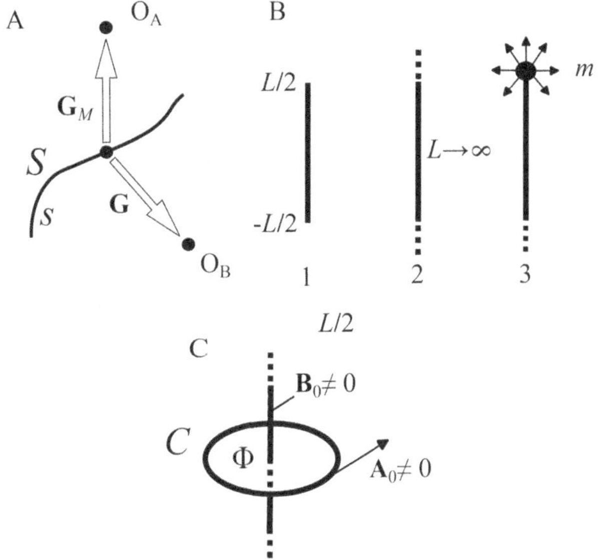

Figure 11.1 Schematic illustrations relating to the electrodynamics of a magnetic dipole string of length L. A: The position vector $\mathbf{r} = \mathbf{r}(s)$ of a field source point S is parametrized in terms of the arc length s on the string. The local magnetization (per unit length) is $\mathbf{M}(s;\omega) = \mathbf{M}(s;-\omega)d\mathbf{r}(s)/ds$, and the local curvature and torsion are denoted by $\kappa(s)$ and $\tau(s)$, respectively [see Subsection 11.3.1] The propagation of the vector potential $(\mathbf{A} = \mathbf{A_T})$ from S to an observation point $\mathcal{O}_A$ is described by the magnetic field propagator, $\mathbf{G}_M = \mathbf{G}_M(\mathbf{r} - \mathbf{r}(s);\omega)$. Likewise, the propagation of the magnetic field $(\mathbf{B})$ from S to the point $\mathcal{O}_B$ involves the dyadic Green function for the electic field, $\mathbf{G} = \mathbf{G}(\mathbf{r} - \mathbf{r}(s);\omega)$. B: B1. Linear magnetic string $(\kappa(s) = \tau(s) = 0 \; \forall \; s)$ of finite length L $[-L/2 \le s \le L/2]$. B2. In the ideal Aharonov-Bohm case it is assumed that $L \to \infty$ and that the magnetization is homogeneous $[M(s;\omega) = M(\omega)]$. B3. A linear Dirac string with an attached magnetic monopole (m) located at $\mathbf{r} = \mathbf{r}(s = 0)$. The semiinfinite string stretches out to infinity $[-\infty < s < 0]$. C: In the magnetostatic limit $[M(\omega \to 0) \equiv M_0]$, and magnetic field $(\mathbf{B}_0)$ is zero everywhere in the vacuum outside for an Aharonov-Bohm string, and infinite on the string. The vector potential $\mathbf{A}(\rho, \omega \to 0) \equiv A_0(\rho)\hat{\boldsymbol{\phi}}(\phi)$ is nonvanishing everywhere in the vacuum domain [see Eq. (11.42)]. The magnetic flux through an open surface (grey-toned) bounded by the circular contour C is $\Phi = \mu_0 M(\omega \to 0)$.

is the so-called binormal unit vector, and

$$\hat{\mathbf{n}}(s) = \frac{(\mathbf{r}'(s) \times \mathbf{r}''(s)) \times \mathbf{r}'(s)}{|\mathbf{r}'(s) \times \mathbf{r}''(s)||\mathbf{r}'(s)|} \tag{11.23}$$

is, what is named, the normal unit vector. The geometrical characterization presented above is particularly convenient in theoretical studies of the electrodynamics of Möbius bands, as we shall see in Chapter 32.

It is now assumed that the string magnetization per unit length everywhere is directed along the string, i.e.,

$$\mathbf{M}(s;\omega) = M(s;\omega)\hat{\mathbf{t}}(s) \tag{11.24}$$

in the frequency domain. The vector potential at the observation point $\mathbf{r}$ from the infinitesimal element

$$d\mathbf{M}(s;\omega) = M(s;\omega)ds\hat{\mathbf{t}}(s) = M(s;\omega)d\mathbf{r} \tag{11.25}$$

located on the string at $\mathbf{r} = \mathbf{r}(s)$ is given by [157]

$$d\mathbf{A}(\mathbf{r};\omega) = i\mu_0 q_0 \mathbf{G}_M(\mathbf{R};\omega) \cdot d\mathbf{M}(s;\omega), \tag{11.26}$$

where [with $\mathbf{R} = \mathbf{r} - \mathbf{r}(s)$ and, as before, $q_0 = \omega/c$]

$$\mathbf{G}_M(\mathbf{R};\omega) = \frac{iq_0}{4\pi}\left[\frac{1}{iq_0 R} - \frac{1}{(iq_0 R)^2}\right]\mathbf{U} \times \hat{\mathbf{R}}e^{iq_0 R} \tag{11.27}$$

is the magnetic field propagator [153, 157]. The vector potential from a magnetic string of arclength L, hence is given by the line integral

$$\mathbf{A}(\mathbf{r};\omega) = i\mu_0 q_0 \int_{-L/2}^{L/2} M(s;\omega)\mathbf{G}_M(\mathbf{r} - \mathbf{r}(s);\omega) \cdot \hat{\mathbf{t}}(s)ds. \tag{11.28}$$

If wished, the vector $(\mathbf{U} - \hat{\mathbf{R}}\hat{\mathbf{R}}) \cdot \hat{\mathbf{t}} = \hat{\mathbf{t}} - R_t\hat{\mathbf{R}}$ can be expressed in the local $(\hat{\mathbf{t}},\hat{\mathbf{n}}, \hat{\mathbf{b}})$-coordinate system, where $\hat{\mathbf{R}} = (R_t(s)\hat{\mathbf{t}} + R_n(s)\hat{\mathbf{n}} + R_b(s)\hat{\mathbf{b}})/[R_t^2(s) + R_n^2(s) + R_b^2(s)]^{1/2}$.

As the reader may show $\nabla \cdot \mathbf{G}_M(\mathbf{R};\omega) = 0$ and $\nabla \times \mathbf{G}_M(\mathbf{R};\omega) = -iq_0\mathbf{G}(\mathbf{R};\omega)$, where

$$\mathbf{G}(\mathbf{R};\omega) = \frac{iq_0}{4\pi}\left\{\frac{1}{iq_0 R}(\mathbf{U} - \hat{\mathbf{R}}\hat{\mathbf{R}}) - \left[\frac{1}{(iq_0 R)^2} - \frac{1}{(iq_0 R)^3}\right](\mathbf{U} - 3\hat{\mathbf{R}}\hat{\mathbf{R}})\right\}e^{iq_0 R} \tag{11.29}$$

is the so-called Green function (dyadic) for the electric field [157]. Its transverse part $\mathbf{G}_T(\mathbf{R};\omega)$ is the photon propagator given in the space-time domain in Eq. (3.31). From the divergence and the curl of $\mathbf{G}_M$, it follows that the vector potential in Eq. (11.28) is divergence-free,

$$\nabla \cdot \mathbf{A}(\mathbf{r};\omega) = 0, \tag{11.30}$$

and has a curl

$$\nabla \times \mathbf{A}(\mathbf{r};\omega) = \mu_0 q_0^2 \int_{-L/2}^{L/2} M(s;\omega)\mathbf{G}(\mathbf{r} - \mathbf{r}(s);\omega) \cdot \hat{\mathbf{t}}(s)ds, \tag{11.31}$$

which in general does not vanish in every space point.

In retrospect, the result in Eq. (11.30) could have been foreseen since the magnetic current density $\mathbf{J}_M \equiv \nabla \times \mathbf{M}$ is divergence-free ($\mathbf{J}_M = \mathbf{J}_{T,M}$). In turn, the derived vector potential becomes gauge invariant (transverse).

11.3.2 Rectilinear magnetic string

In the approach to the Aharonov-Bohm effect one considers a linear magnetic string ($\kappa = \tau = 0$), with a magnetization per unit length which is independent of s, i.e., $M(s;\omega) = M(\omega)$. Finally, the length of the string is assumed to be much larger than all other relevant lengths in the problem; see the microscopic approach described in Chapter 32. Thus, we take the limit $L \to \infty$ in Eq. (11.28). The line integral for $\mathbf{A}(\mathbf{r};\omega)$, given in Eq. (11.28), is reduced to a convenient form using cylindrical coordinates [local orthogonal unit vectors: $\hat{\mathbf{z}}$, $\hat{\boldsymbol{\rho}}$, $\hat{\boldsymbol{\phi}}$], with the tangential unit vector directed along the z-axis. Since $\mathbf{R} = (z - z')\hat{\mathbf{z}} + \rho\hat{\boldsymbol{\rho}}$, $\boldsymbol{\rho} = \rho\hat{\boldsymbol{\rho}}$ being the radial vector in the xy-plane, one has

$$\mathbf{U} \times \hat{\mathbf{R}} \cdot \hat{\mathbf{z}} = -\frac{\rho}{R}\hat{\boldsymbol{\phi}}(\phi), \tag{11.32}$$

where $R = [(z - z')^2 + \rho^2]^{1/2}$, and $\hat{\boldsymbol{\phi}}(\phi) = \hat{\mathbf{z}} \times \hat{\boldsymbol{\rho}}$. The result in Eq. (11.32) is readily obtained using the dyadic expansion $\mathbf{U} = \hat{\mathbf{z}}\hat{\mathbf{z}} + \hat{\boldsymbol{\rho}}\hat{\boldsymbol{\rho}} + \hat{\boldsymbol{\phi}}\hat{\boldsymbol{\phi}}$.

The integral expression for the vector potential surrounding the linear string hence takes the form

$$\mathbf{A}(\mathbf{r};\omega) = \frac{\mu_0}{4\pi}M(\omega)\rho\hat{\boldsymbol{\phi}}(\phi)\int_{-\infty}^{\infty}\frac{1}{R^3}(1 - iq_0 R)e^{iq_0 R}dz'. \tag{11.33}$$

Obviously, the vector potential cannot depend on z in this ideal case. The substitution $z' = z - u$ finally leads to

$$\mathbf{A}(\rho, \phi;\omega) = \frac{\mu_0}{2\pi}M(\omega)\rho\hat{\boldsymbol{\phi}}(\phi)\int_{0}^{\infty}\frac{1}{R^3}(1 - iq_0 R)e^{iq_0 R}du, \tag{11.34}$$

now with $R = (\rho^2 + u^2)^{1/2}$. In terms of the function

$$f(\rho; q_0) = \int_{0}^{\infty}\frac{1}{R^3}(1 - iq_0 R)e^{iq_0 R}du, \tag{11.35}$$

we have

$$\mathbf{A}(\rho, \phi;\omega) = \frac{\mu_0}{2\pi}M(\omega)\rho^{-1}f(\rho; q_0)\hat{\boldsymbol{\phi}}(\phi). \tag{11.36}$$

Since $\mathbf{A}(\rho, \phi;\omega)$ is independent of z the divergence and the curl of $\mathbf{A}$ can be obtained replacing the operator $\boldsymbol{\nabla}$ by

$$\boldsymbol{\nabla}_{\perp} = \hat{\boldsymbol{\rho}}\frac{\partial}{\partial\rho} + \hat{\boldsymbol{\phi}}\frac{1}{\rho}\frac{\partial}{\partial\phi} \tag{11.37}$$

in the calculations. The reader may show that

$$\boldsymbol{\nabla} \cdot \mathbf{A}(\rho, \phi;\omega) = \mathbf{0}, \tag{11.38}$$

thus confirming the general result for a string with nonvanishing curvature and torsion. Since $\boldsymbol{\nabla}\times\mathbf{G}_M(\mathbf{r};\omega) = \mathbf{0}$, the vector potential is divergence-free also for magnetic source regions of finite volume. For the curl of $\mathbf{A}(\rho, \phi;\omega)$ one obtains

$$\boldsymbol{\nabla} \times \mathbf{A}(\rho, \phi;\omega) = \frac{\mu_0}{2\pi}M(\omega)\left[\hat{\boldsymbol{\rho}} \times \hat{\boldsymbol{\phi}}\frac{d}{d\rho}(\rho f) + \hat{\boldsymbol{\phi}} \times \frac{d\hat{\boldsymbol{\phi}}}{d\phi}f\right] = \frac{\mu_0}{2\pi}M(\omega)\hat{\mathbf{z}}\left[2f + \rho\frac{df}{d\rho}\right]. \tag{11.39}$$

The magnetic field hence is directed along the direction of the string, and given by

$$\mathbf{B}(\rho, \phi; \omega) = \frac{\mu_0}{2\pi} M(\omega)\hat{\mathbf{t}}\left[2f(\rho; \omega) + \rho\frac{df(\rho; \omega)}{d\rho}\right]. \tag{11.40}$$

In the magnetostatic limit where

$$f(\rho; q_0 \to 0) = \int_0^\infty \frac{du}{(\rho^2 + u^2)^{3/2}} = \frac{1}{\rho^2}, \tag{11.41}$$

the vector potential takes the simple form

$$\mathbf{A}(\rho, \phi; \omega \to 0)[\equiv \mathbf{A}_0(\rho, \phi)] = \frac{\mu_0 M_0}{2\pi}\rho^{-1}\hat{\boldsymbol{\phi}}(\phi), \tag{11.42}$$

$M_0 = M(\omega \to 0)$ being the static string magnetization. By combining Eqs. (11.40) and (11.41) it appears that the magnetostatic magnetic field vanishes everywhere outside the string

$$\mathbf{B}(\rho, \phi; \omega \to 0) = [\equiv \mathbf{B}_0(\rho, \phi)] = \mathbf{0}, \quad \rho \neq 0. \tag{11.43}$$

The closed line integral of the vector potential in Eq. (11.42) now is calculated using a circular path centered on the string and lying in a plane perpendicular to $\hat{\mathbf{t}}(= \hat{\mathbf{z}})$. Hence,

$$\int_0^{2\pi} \mathbf{A}_0(\rho, \phi) \cdot \hat{\boldsymbol{\phi}}(\phi)\rho d\phi = \mu_0 M_0, \tag{11.44}$$

independent of ρ (of course).

Since the transverse part of the current density in Eq. (11.20) is given by $\mathbf{J}_T(\mathbf{r}; \omega) = \nabla \times \mathbf{M}(\mathbf{r}; \omega)$ in the general case, one obtains from the magnetostatic Maxwell equation $\nabla \times \mathbf{B}(\mathbf{r}; \omega) = \mu_0 \nabla \times \mathbf{M}(\mathbf{r}; \omega)$, the result $\mathbf{B}(\mathbf{r}; \omega) = \mu_0 \mathbf{M}(\mathbf{r}; \omega)$. A gradient term $\nabla\alpha(\mathbf{r}; \omega) = \mathbf{B}(\mathbf{r}; \omega) - \mu_0 \mathbf{M}(\mathbf{r}; \omega)$ cannot be present since $\mathbf{B}$ and $\mathbf{M}$ are divergence-free vector fields. In the magnetostatic limit we therefore reach the conclusion

$$\mathbf{B}_0(\rho, \phi) = \mu_0 M_0 \hat{\mathbf{t}}\delta(\rho) \tag{11.45}$$

for the rectilinear string. Hence, the magnetic flux through an open surface bounded by C is given by

$$\Phi = \mu_0 M_0 \tag{11.46}$$

in our example. For studies of persistent currents and angular momentum photon drag in mesoscopic rings and cylinders [144, 145, 167], it is physically attractive to introduce the fundamental (elementary) flux quantum

$$\Phi_0 = \frac{h}{|e|}. \tag{11.47}$$

A combination of Eqs. (11.8), (11.44), (11.46) and (11.47) finally gives the following expression for the Dirac phase factor:

$$\Phi(C) = \exp\left[-2\pi i \frac{\Phi}{\Phi_0}\right].$$
(11.48)

The Dirac phase factor is a periodic function of the magnetic flux in the string, the period being just $h/|e|$. Thus, starting from $\Phi_\star$ ($0 \leq \Phi_\star < \Phi_0$), the fluxes $\Phi_\star$ and

$$\Phi = \Phi_\star + M\frac{h}{|e|}, \quad M = 0, \pm 1, \pm 2, \ldots$$
(11.49)

give the same Dirac phase factor, $\Phi(C)$.

11.4 MAGNETIC FLUX TUBES AND CURRENT SOLENOIDS FOR THE AHARONOV-BOHM EFFECT

The study of the infinitely long rectilinear magnetostatic string has opened the door a little to the structural vacuum concept, and shown that a charged particle in quantum mechanics may "feel" a quantum electrodynamic effect in regions of space where there are no TIME INDEPENDENT magnetic and electric fields. A somewhat deeper insight in the Aharonov-Bohm mechanics potential for probing the physics of structural vacuum appears via a step-by-step generalization of the example analyzed in Section 11.3.

An infinitesimally thin string is a physical abstraction so let us assume we have a rectilinear string (wire) of finite circular cross-section which carries a rotational symmetric static magnetization $\mathbf{M}_0(\mathbf{r}) = M_0(\rho)\hat{\mathbf{t}}$ directed everywhere along the string ($\hat{\mathbf{t}} = \hat{\mathbf{z}}$). The radius of the wire is r_0. Under the assumption that the wire can be taken as infinitely long, symmetry dictates that the magnetostatic vector potential has the form

$$\mathbf{A}_0(\rho, \phi) = A_0(\rho)\hat{\boldsymbol{\phi}}(\phi).$$
(11.50)

The magnetostatic field is given by

$$\mathbf{B}_0(\mathbf{r}) = \mu_0 M_0(\rho)\hat{\mathbf{z}}\theta(r_0 - \rho),$$
(11.51)

where $\theta(r_0 - \rho)$ is the Heaviside unit step function. Application of Stokes theorem [Eq. (11.9)] to a circular path placed perpendicular to the z-axis and centered on the wire axis ($\rho = 0$) now gives

$$\mathbf{A}(\rho, \phi) = \hat{\boldsymbol{\phi}}(\phi)\rho^{-1}\mu_0 \int_0^\rho M_0(\rho')\rho'd\rho', \quad \rho \leq r_0$$
(11.52)

and

$$\mathbf{A}(\rho, \phi) = \hat{\boldsymbol{\phi}}(\phi)\rho^{-1}\mu_0 \int_0^{r_0} M_0(\rho')\rho'd\rho', \quad \rho > r_0.$$
(11.53)

The use of a rectilinear magnetic wire in the Aharonow-Bohm effect is just one possibility. Another possibility immediately appears if one considers the magnetization current density, $\mathbf{J}_M = \boldsymbol{\nabla} \times \mathbf{M}$, which in the present case is given by

$$\mathbf{J}_M(\rho, \phi) = \hat{\boldsymbol{\phi}}(\phi) \left(-\frac{dM_0(\rho)}{d\rho} \right). \tag{11.54}$$

If we assume that the magnetization is uniform in the bulk of the wire, the magnetization current density will be nonvanishing only in a (thin) surface layer where $M_0(\rho)$ changes rapidly with ρ from its bulk value to zero. In a quantum mechanical context the surface layer is characterized by the "spill-out" of the electron many-body wave function, a circumstance of utmost importance in microscopic electrodynamics [10,82,83,88,132,136,142]. Although the solenoidal surface current density in the example above is related to a (ferro)magnetic wire, the Aharonov-Bohm effect can also arise via a solenoidal microscopic polarization current density, $\mathbf{J}_L = \partial \mathbf{P}_L/\partial t$ c.f. Eq. (11.20). In a particular simple model one may associate the Aharonov-Bohm phenomenon to the time-independent vector potential from a mesoscopic cylindrical shell in which the dc electron dynamics is free-electron like.

One may illustrate the Aharonov-Bohm phenomenon by surrounding the magnetic tube (current solenoid) by a mescoscopic circular ring placed in the xy-plane and with its center at $(x, y) = (0, 0)$; see Fig. 11.2. Without lose of essential generality for the present purpose we assume that the electron dynamics along the ring is free-electron-like and that the ring cross section is so small that the electron dynamics in the transversal directions can be ignored. For definiteness, we consider the ring as an infinitely cylindrical shell with a height approaching zero. In the absence of a magnetic $\mathbf{B}_0$-flux in the rectilinear tube, and with use of cylindrical coordinates, (r, ϕ, z), the stationary states of the ring have the generic form

$$\Psi_0(r, \phi, z) = w(r, z)\psi_0(\phi), \tag{11.55}$$

with $|w(r, z)|^2 = \delta(r - r_0)\delta(z)/r_0$, r_0 being the ring radius. The ϕ-part of the wave function satisfies the free-electron (time independent) Schrödinger equation

$$-\frac{\hbar^2}{2mr_0^2}\frac{d^2\psi_0(\phi)}{d\phi^2} = E\psi_0(\phi), \tag{11.56}$$

where E is the yet undetermined eigenenergy. In the presence of a magnetic flux, the wave function $\psi_0(\phi)$ acquires a change in phase; c.f. Eq. (2.5), here with $dx^0 = 0$. If the reference point (P) is taken at $\phi = 0$, the phase change in Q $(\phi = \phi)$ is given by $\exp[-i(\Phi/\Phi_0)\phi]$. The new wave function, $\psi(\phi)$, therefore is related to the old ψ_0 as follows:

$$\psi(\phi) = \exp[-i(\Phi/\Phi_0)\phi]\psi_0(\phi). \tag{11.57}$$

The general solution to Eq. (11.56) can be written as a linear superposition of solutions of the form

$$\psi_0(\phi) = (2\pi)^{-1/2} \exp\left[\pm i\frac{r_0}{\hbar}(2mE)^{1/2}\phi\right]. \tag{11.58}$$

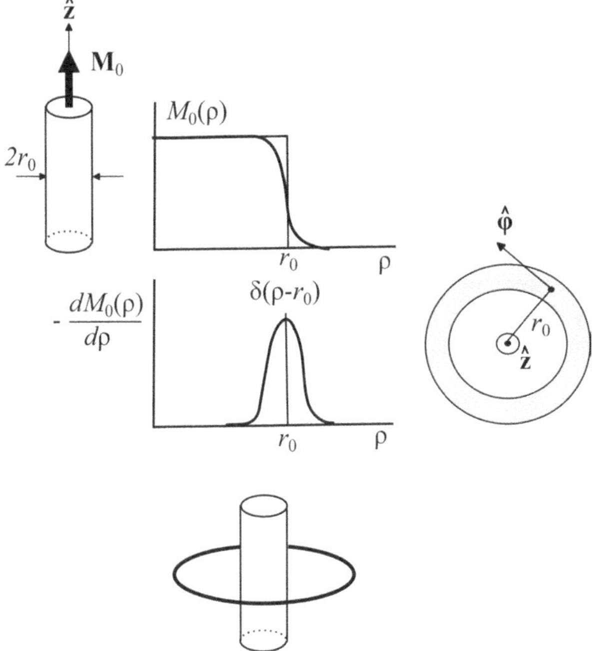

Figure 11.2 A magnetic circular cylindrical flux tube with a static magnetization $\mathbf{M}_0(\mathbf{r}) = M_0(\rho)\hat{\mathbf{z}}$ has a microscopic radial "spill-out" surface layer. In this layer (grey-toned domain in the top view) a tangential ($\hat{\boldsymbol{\phi}}$) magnetization current density of magnitude $-dM_0(\rho)/d\rho$ flows. In a sharp-boundary approximation $-dM_0(\rho)/d\rho = \delta(\rho - r_0)$, the Aharonov-Bohm analysis becomes equivalent to the standard macroscopic solenoidal approach. Bottom figure: Mesoscopic flux tube surrounded by a mesoscopic circular jellium ring. It has been suggested [145] that this configuration might be used to confirm the Dirac phase factor, $\Phi(C) = \exp[2\pi ie/|e|\Phi/|\Phi_0|]$, periodicity, i.e., $\Phi = \Phi_* + M\Phi_0$ [See Sections 11.3 and 11.4], where $\Phi_0 = h/|e|$ is the fundamental flux quantum.

These two solution types represent clockwise and counter clockwise running waves. Starting from these states, the $\psi(\phi)$-solutions are of the types

$$\psi(\phi) = (2\pi)^{-1/2} \exp\left\{ i\left[-\frac{\Phi}{\Phi_0} \pm \frac{r_0}{\hbar}(2mE)^{1/2} \right] \right\}. \tag{11.59}$$

The possible eigenenergies are determined from the requirement

$$\psi(\phi + 2\pi) = \psi(\phi). \tag{11.60}$$

The normalized eigenfunctions hence are

$$\psi(\phi) = (2\pi)^{-1/2} \exp(im\phi), \quad m = 0, \pm 1, \pm 2, \ldots \tag{11.61}$$

with associated eigenenergies

$$E_m(\Phi) = \frac{\hbar^2}{2mr_0^2} \left(m + \frac{\Phi}{\Phi_0} \right)^2. \tag{11.62}$$

The magnetic flux periodicity of the Dirac phase factor [see Eqs. (11.48) and (11.49)] gives the following relation between the eigenenergies:

$$E_{m+M}(\Phi_\star) = E_m(\Phi). \tag{11.63}$$

In Section V fundamental aspects of microscopic electrodynamics in mesoscopic media are studied, and for such media one can demonstrate that the Aharonov-Bohm effect gives rise to a persistent current in the circular ring [144]. This current is related to the so-called diamagnetic part of the steady-state current density operator, and can be expressed in terms of the changes in the electron Fermi-Dirac distribution caused by the magnetic flux ration $|e|\Phi/h$.

11.5 ADIABATIC GEOMETRIC PHASE

11.5.1 Evolution operator

The principle of the local phase invariance is a cornerstone in electrodynamics, and when properly extended also in the electroweak interaction [Part IX] and beyond. The local phase shift $\exp\{ie\chi(\mathbf{r}, t)/\hbar\}$, [Eq. (11.3)] makes contact to observationable physical quantities via the Dirac phase factor, given by $\Phi(C) = \exp[i(e/\hbar) \oint_C \mathbf{A}_T(\mathbf{r}, t) \cdot \mathbf{dr}]$ for a fixed-time loop. In relation to the Aharonov-Bohm effect we have realized that the local phase factor for a rectlinear magnetic flux tube, viz., $\exp[-i(\Phi/\Phi_0)\phi]$, where Φ is the $\mathbf{B}_0$-flux and $\Phi_0 = h/|e|$ is the fundamental flux quantum, gives rise to a (periodic) flux dependence of the energy eigenstates in a circular mesoscopic ring.

In the present section we shall study phase shifts in relation to the time evolution of a quantum state when the Hamiltonian of the system depends on some time dependent external parameter. We start the analysis with preliminary remarks on the linear and unitary time evolution operator $\hat{U}(t, t_0)$ $[\hat{U}^\dagger = \hat{U}^{-1}]$ which relates the state vectors at times t_0 and t, i.e.

$$|\psi(t)\rangle = \hat{U}(t, t_0) |\psi(t_0)\rangle, \tag{11.64}$$

with $\hat{U}(t_0, t_0) = \mathbb{1}$. Substituting Eq. (11.64) into the Schrödinger equation

$$i\hbar \frac{d}{dt} |\psi(t)\rangle = \hat{H}(t) |\psi(t)\rangle. \tag{11.65}$$

One obtains, since the state vector $|\psi(t_0)\rangle$ is arbitrary, the operator equation

$$i\hbar \frac{d}{dt} \hat{U}(t, t_0) = \hat{H}(t)\hat{U}(t, t_0). \tag{11.66}$$

One might be tempted to believe that the solution to Eq. (11.66) is given by

$$\hat{U}(t, t_0) = \exp\left[-\frac{i}{\hbar} \int_{t_0}^{t} \hat{H}(t')d't\right] \mathbb{1}, \quad ? \tag{11.67}$$

with the exponential of the operator defined by its power series expansion. However, this is not correct in general since the derivative $d\exp\left[\hat{A}(t)\right]/dt$ is not equal to $(d\hat{A}(t)/dt)\exp\left[\hat{A}(t)\right]$ unless $\hat{A}(t)$ and $d\hat{A}(t)/dt$ commute (use the power series expansion of $\exp\left[\hat{A}(t)\right]$ to prove this assertion). For a conservative system, where $\hat{H}$ is time independent,

$$\hat{U}(t,t_0) = \exp\left[-\frac{i}{\hbar}\hat{H}(t-t_0)\right]\mathbb{1}, \tag{11.68}$$

obviously. For an energy eigenstate $|\psi(t_0)\rangle = |n\rangle$, satisfying the time-independent Schrödinger equation

$$\hat{H}|n\rangle = E_n|n\rangle, \tag{11.69}$$

we get

$$|\psi(t)\rangle = \exp\left[-\frac{i}{\hbar}E_n(t-t_0)\right]|n\rangle, \tag{11.70}$$

a result is known from elementary quantum mechanics [62].

Returning to Eq. (11.66), it appears that this can be condensed into an integral equation,

$$\hat{U}(t,t_0) = \mathbb{1} + \frac{1}{i\hbar}\int_{t_0}^{t}\hat{H}(t')\hat{U}(t',t_0)dt', \tag{11.71}$$

satisfying the initial condition $\hat{U}(t_0,t_0) = \mathbb{1}$. By successive iterations, Eq. (11.71) leads to the well-known perturbative expansion of the evolution operator [64, 85, 184, 186]

$$\hat{U}(t,t_0) = \mathbb{1} + \sum_{n=1}^{\infty}\hat{U}_n(t,t_0), \tag{11.72}$$

where

$$\hat{U}_n(t,t_0) = \left(\frac{1}{i\hbar}\right)^n\int\hat{H}(\tau_n)...\hat{H}(\tau_2)\hat{H}(\tau_1)d\tau_n...d\tau_2 d\tau_1 \tag{11.73}$$

with $t_0 \leq \tau_1 \leq \tau_2... \leq \tau_n \leq t$. We have implicitly *assumed* that the series converges. In the special (unusual) case where the Hamilton operator commutes at (all) different times, i.e.,

$$[\hat{H}(t),\hat{H}(t')] = 0 \quad \forall\, t,t' \tag{11.74}$$

the order of the operators is not important. In case, one obtains

$$\hat{U}(t,t_0) = \sum_{n=0}^{\infty}\frac{1}{n!}\left(\frac{1}{i\hbar}\right)^n\int_{t_0}^{t}dt_1...\int_{t_0}^{t}dt_n\hat{H}(t_1)...\hat{H}(t_n)\mathbb{1} = \exp\left[-\frac{i}{\hbar}\int_{t_0}^{t}\hat{H}(t')dt'\right]\mathbb{1}. \tag{11.75}$$

The reader may show the result in Eq.(11.75) writing down the $n!$ equivalent permutations of the $\hat{H}$'s in Eq. (11.73) [start possible with the $n = 2$ term], or go back to Eq. (11.66), utilizing that $\hat{H}(t)$ and $d\hat{H}(t)/dt = \lim_{\Delta t\to 0}\{[\hat{H}(t+\Delta t) - \hat{H}(t)]/\Delta t\}$ commutes for $[\hat{H}(t),\hat{H}(t+\Delta t)] = 0$. Hence, the result in Eq.(11.67) *is* correct provided Eq. (11.74) holds.

11.5.2 The adiabatic theorem

We now consider the time evolution of a quantum state when the Hamiltonian depends on a time-dependent *external* parameter $R = R(t)$, or if needed several parameters $R(t) = (R_1(t), R_2(t), ...)$ [37]. The Hamilton operator is denoted

$$\hat{H}(R(t)) \equiv \hat{H}(R)[\equiv \hat{H}(t)]. \tag{11.76}$$

For a pure state, $|\psi(t)\rangle$, the evolution is in Hilbert space described by the Schrödinger equation

$$i\hbar\frac{d}{dt}|\psi(t)\rangle = \hat{H}(R)|\psi(t)\rangle, \tag{11.77}$$

whereas a mixed state is described by the Liouville equation

$$i\hbar\frac{d}{dt}\left|\hat{W}(t)\right\rangle = [\hat{H}(R), \hat{W}(t)] \tag{11.78}$$

for the density matrix (statistical) operator, $\hat{W}(t)$. The time evolution of the environmental system determines the way in which $R = R(t)$ evolves in time.

An *instantaneous* energy eigenstate

$$|n[R(t)]\rangle \equiv |n; R(t)\rangle \, [\equiv |n; R\rangle] \tag{11.79}$$

of time-dependent energy $E_n[R(t)] \equiv E_n(R)$ is *defined* by

$$\hat{H}(R)|n; R\rangle = E_n(R)|n; R\rangle, \tag{11.80}$$

where the state vectors form an orthonormal basis,

$$\langle m; R|n; R\rangle = \delta_{mn}. \tag{11.81}$$

The completeness relation

$$\sum_n |n; R\rangle \langle n; R| = \mathbb{1} \tag{11.82}$$

gives the instantaneous spectral resolution of the Hamiltonian, viz.,

$$\hat{H}(R) = \sum_n E_n(R)|n; R\rangle \langle n; R|. \tag{11.83}$$

The projection operator on the state $|n; R\rangle$, i.e.,

$$\hat{P}_n(R) = |n; R\rangle \langle n; R| \tag{11.84}$$

in general change in time, and

$$\sum_n \hat{P}_n(R) = \mathbb{1}, \tag{11.85}$$

c.f.

$$|\psi\rangle = \left[\sum_n \hat{P}_n(R)\right]|\psi\rangle = \sum_n \langle n; R|\psi\rangle\, |n; R\rangle. \tag{11.86}$$

In the following we focus our analysis on the case where $R(t)$ describes a closed path loop, C, of period T:

$$C : R(0) \to R(t) \to R(T) = R(0). \tag{11.87}$$

Thus

$$\hat{H}[R(T)] = \hat{H}[R(0)], \tag{11.88}$$

and for all n

$$E_n[R(T)] = E_n[R(0)], \tag{11.89}$$
$$|n; R(T)\rangle\langle n; R(T)| = |n; R(0)\rangle\langle n; R(0)|. \tag{11.90}$$

The condition in Eq. (11.90) *does not* imply that $|n; R(T)\rangle = |n; R(0)\rangle$, but only

$$|n; R(T)\rangle = e^{i\alpha_n^C}\, |n; R(0)\rangle, \tag{11.91}$$

where $\exp(i\alpha_n^C)$ is a global phase factor for a complete loop. The reason originates in circumstance that the eigenvectors in Eq. (11.80) are determined only up to a phase factor. Thus, instead of the set $|n; R\rangle$ one may use a new set

$$|n; R\rangle^{\mathrm{NEW}} = e^{i\alpha_n(R)}\, |n; R\rangle, \tag{11.92}$$

giving

$$\alpha_n^C = \alpha_n[R(0)] - \alpha_n[R(T)]. \tag{11.93}$$

If one inserts the ansatz

$$|\psi(t)\rangle = \sum_m c_m(t)\, |n; R(t)\rangle \tag{11.94}$$

into the Schrödinger equation [Eq. (11.77)], and in turn use the orthonormality of the instantaneous eigenstates, one finds

$$\frac{d}{dt}c_n(t) + \frac{iE_n[R(t)]}{\hbar}c_n(t) + \langle n[R(t)]|\frac{d}{dt}|n[R(t)]\rangle c_n(t)$$
$$= -\sum_{m \neq n} \langle n[R(t)]|\frac{d}{dt}|m[R(t)]\rangle c_m(t). \tag{11.95}$$

The adiabatic theorem is the approximation one obtains by neglecting the right-hand-side of Eq. (11.95), i.e.,

$$\frac{d}{dt}c_n(t) + \frac{iE_n[R(t)]}{\hbar}c_n(t) + \langle n[R(t)]|\frac{d}{dt}|n[R(t)]\rangle c_n(t) = 0, \quad [\text{adiabatic}]. \tag{11.96}$$

It appears from Eqs. (11.95) and (11.96) that the adiabatic theorem relates to an approximation where the transition rate from the level (n) in considerations to all other levels $(m \neq n)$ can be neglected. One may quantify this somewhat by the following calculation (leaving out the reference to $R = R(t)$ for brevity):

$$\langle n| \frac{d\hat{H}}{dt} |m\rangle = \langle n| \frac{d}{dt}[\hat{H}\,|m\rangle] - \langle n| \hat{H} \frac{d}{dt} |m\rangle = \langle n| \frac{d}{dt}[E_m\,|m\rangle] - E_n \langle n| \frac{d}{dt} |m\rangle =$$

$$\langle n|m\rangle \frac{d}{dt} E_m + E_m \langle n| \frac{d}{dt} |m\rangle - E_n \langle n| \frac{d}{dt} |m\rangle .$$

$$(11.97)$$

To reach the result in Eq. (11.97), we have used the hermiticity of the Hamiltonian, $\hat{H} = \hat{H}^\dagger$. For $m \neq m$, the first term in the last expression of Eq. (11.97) is zero. Summation of the result in Eq. (11.97) over m finally gives for the non-adiabatic factor to $c_m(t)$ [in Eq. (11.95)]

$$\sum_{m \neq n} \langle n| \frac{d}{dt} |m\rangle = \sum_{m \neq n} \frac{\langle n| \frac{d\hat{H}}{dt} |m\rangle}{E_m - E_n} = \left[\sum_{m \neq n} \frac{\langle n| \frac{d\hat{H}}{dR} |m\rangle}{E_m - E_n} \right] \frac{dR}{dt}. \qquad (11.98)$$

Roughly speaking, the non-adiabatic term is negligible if the time derivative dR/dt is small enough; hence the name adiabatic. Thus, the rate dR/dt must be compared with a characteristic (intrinsic) frequency scale for the system [rate R is dimensionless, and so is the prefactor to dR/dt in Eq. (11.98).]

11.5.3 Berry's adiabatic geometric phase

When the Hamilton operator commutes at different times [Eq. (11.74)] the common eigenvectors (and eigenspaces) of $\hat{H}(t)$ and $\hat{H}(t')$ are the same, and necessarily time independent. The now time-independent vectors $|n\rangle$ give one a simple spectral resolution of the evolution operator in Eq. (11.75). Taking a reference time $t_0 = 0$, one obtains

$$\hat{U}(t,0) = \sum_n \hat{U}(t,0) |n\rangle \langle n| = \sum_n \exp\left[-\frac{i}{\hbar} \int_0^t \hat{H}(t')dt'\right] |n\rangle \langle n| , \qquad (11.99)$$

and thus a spectrum

$$\hat{U}(t,0) = \sum_n \exp\left[-\frac{i}{\hbar} \int_0^t E_n[R(t')]\right] |n\rangle \langle n| . \qquad (11.100)$$

Note that for a time-independent Hamiltonian, $\hat{H}(t) = \hat{H}$, we regain the standard expression for conservative systems, viz.,

$$\hat{U}(t,0) = \exp\left[-\frac{i}{\hbar}\hat{H}t\right] \mathbb{1} = \sum_n \exp\left[-\frac{i}{\hbar}E_n t\right] |n\rangle \langle n| . \qquad (11.101)$$

Inspired by the result in Eq. (11.100), it appears natural to extract the so-called *dynamical phase*

$$\phi_n^{\mathrm{DYN}}(t) \equiv \frac{1}{\hbar} \int_0^t E_n[R(t')]dt' \qquad (11.102)$$

from the adiabatic time evolution, Eq. (11.96). Hence with

$$c_n(t) = b_n(t) \exp\left[-i\phi_n^{\text{DYN}}(t)\right], \tag{11.103}$$

one obtains

$$\frac{d}{dt} b_n(t) + \langle n; R(t)| \frac{d}{dt} |n; R(t)\rangle b_n(t) = 0. \tag{11.104}$$

with the initial condition $c_n(0)[= b_n(0)] = 1$ the solution of Eq. (11.104) is

$$b_n(t) = \exp\left[-i\phi_n^{\text{GEO}}(t)\right], \tag{11.105}$$

where

$$\phi_n^{\text{GEO}}(t) = \int_0^t i \langle n; R(t')| \frac{d}{dt'} |n; R(t')\rangle \, dt' \tag{11.106}$$

is the so-called *geometric (adiabatic) phase* also named the Berry phase [21–23]. The quantity $\phi_n^{\text{GEO}}(t)$ *is* a phase factor since

$$0 = \frac{d}{dt} \langle n|n\rangle = \frac{d\langle n|}{dt} |n\rangle + \langle n| \frac{d}{dt} |n\rangle = \langle n| \frac{d}{dt} |n\rangle^* + \langle n| \frac{d}{dt} |n\rangle \tag{11.107}$$

shows that $\langle n; R(t)| \frac{d}{dt} |n; R(t)\rangle$ is a purely imaginary quantity.

11.5.4 Geometrical phase as a flux

After one cycle in parameter space,

$$c_n(T) = \exp\{-i[\phi_n^{\text{DYN}}(T) + \phi_n^{\text{GEO}}(T)]\}c_n(0). \tag{11.108}$$

One thus obtains a total phase shift given by the sum of the dynamical shift

$$\phi_n^{\text{DYN}}(T) = \frac{1}{\hbar} \int_0^T E_n[R(t)]dt, \tag{11.109}$$

and a geometrical shift

$$\phi_n^{\text{GEO}}(T) = \int_0^T i \langle n; R(t)| \frac{d}{dt} |n; R(t)\rangle \, dt. \tag{11.110}$$

We may consider $\phi_n^{\text{GEO}}(t)$ [Eq. (11.106)] as a (curve) integral, $\mathbf{R}$ being a three-vector,i.e.,

$$\phi_n^{\text{GEO}}(T) = \int_{\mathbf{R}(t=0)}^{\mathbf{R}(t)} i \langle n; R(t)| \boldsymbol{\nabla}_{\mathbf{R}} |n; R(t)\rangle \cdot d\mathbf{R}, \tag{11.111}$$

For a closed loop $[\mathbf{R}(T) = \mathbf{R}(0)]$, one gets

$$\phi_n^{\text{GEO}}(T) = \oint_C i \langle n; R| \boldsymbol{\nabla}_{\mathbf{R}} |n; R\rangle \cdot d\mathbf{R}. \tag{11.112}$$

By application of Stokes' theorem we obtain

$$\phi_n^{\text{GEO}}(T) = \int_S \boldsymbol{\mathcal{B}}_n(\mathbf{R}) \cdot d\mathbf{S} \tag{11.113}$$

where

$$\boldsymbol{\mathcal{B}}_n(\mathbf{R}) = \boldsymbol{\nabla}_{\mathbf{R}} \times [i\,|n;R\rangle\,\boldsymbol{\nabla}_{\mathbf{R}}\,|n;R\rangle]. \tag{11.114}$$

The geometrical phase, $\phi_n^{\text{GEO}}(T)$ hence may be considered as a flux of the field $\boldsymbol{\mathcal{B}}_n(\mathbf{R})$ through a surface S bounded by the closed curve C. The geometrical adiabatic phase therefore appears similar to the Aharonov-Bohm phase. However, there is an important difference: *The geometrical adiabatic phase is based on a slow rate of change of* $\mathbf{R}(t)$ [and neglect of interlevel transitions], *whereas the Aharonov-Bohm effect does not rely on an adiabatic approximation.* Yet, we shall see in the following that an adiabatic change is not necessary for an introduction of a geometrical phase concept.

11.6   AHARONOV-ANANDAN PHASE

Although the Berry phase has been introduced and discussed on the basis of the adiabatic approximation this restriction is not essential. Below we shall realize that the geometrical phase can be defined in a general framework as shown by Aharonov and Anandan [2, 37]

11.6.1   Derivation of the non-adiabatic Berry phase

Among the solutions to the Schrödinger equation

$$i\hbar \frac{d}{dt}\,|\psi(t)\rangle = \hat{H}(t)\,|\psi(t)\rangle \tag{11.115}$$

we focus the attention on a special kind of quantum state $|\psi(t)\rangle$ namely one which after a time evolution from 0 to T $[\psi(o) \to \psi(t) \to \psi(T)]$ is identical to the original state up to a phase factor η, that is

$$|\psi(T)\rangle = e^{i\eta}\,|\psi(0)\rangle. \tag{11.116}$$

In the special case where $[\hat{H}(t), \hat{H}(t')] = 0$ [Eq. (11.74)] it apperas from Eqs. (11.64) and (11.75) that

$$\eta = \frac{1}{i\hbar} \int_0^T \hat{H}(t)dt. \tag{11.117}$$

We can obtain information on the general structure of η by seeking to construct a state vector $\left|\tilde{\psi}(t)\right\rangle$ which has the same phase at $t = 0$ and $t = T$, i.e.,

$$\left|\tilde{\psi}(T)\right\rangle = \left|\tilde{\psi}(0)\right\rangle. \tag{11.118}$$

The state $\left|\tilde{\psi}(t)\right\rangle$ is defined by

$$\left|\tilde{\psi}(t)\right\rangle = e^{-i\phi(t)}\left|\psi(t)\right\rangle. \tag{11.119}$$

In a sense the phase $\phi(t)$ compensates in a running (continuous) manner the phase shift of $|\psi(t)\rangle$ which ends up as η at time $t = T$. To determine the rate of change of $\phi(t)$, we look at the time evolution of $\left|\tilde{\psi}(t)\right\rangle$, namely,

$$\frac{d}{dt}\left|\tilde{\psi}(t)\right\rangle = -i\frac{d\phi(t)}{dt}\left|\tilde{\psi}(t)\right\rangle + e^{-i\phi(t)}\frac{d}{dt}\left|\phi(t)\right\rangle = -i\frac{d\phi(t)}{dt}\left|\tilde{\psi}(t)\right\rangle + \frac{1}{i\hbar}e^{-i\phi(t)}\hat{H}(t)\left|\phi(t)\right\rangle. \tag{11.120}$$

By multiplication of this equation with $\left\langle\tilde{\psi}(t)\right|$ one obtains

$$\frac{d\phi(t)}{dt} = i\left\langle\tilde{\psi}(t)\right|\frac{d}{dt}\left|\tilde{\psi}(t)\right\rangle - \frac{1}{\hbar}\langle\psi(t)|\hat{H}(t)|\psi(t)\rangle. \tag{11.121}$$

To reach this result we have utilized the Hermitian conjugate of Eq. (11.119), and $\left\langle\tilde{\psi}(t)\big|\tilde{\psi}(t)\right\rangle = 1$ [assuming $\langle\psi(t)|\psi(t)\rangle = 1$]. The rate of change of $\phi(t)$ hence consists of a dynamical part [last term] and a geometrical part [first term]. Integration over the time interval $(0\,|\,T)$ gives

$$\phi(T) - \phi(0) = i\int_0^T \left\langle\tilde{\psi}(t)\right|\frac{d}{dt}\left|\tilde{\psi}(t)\right\rangle dt - \frac{1}{\hbar}\int_0^T \langle\psi(t)|\hat{H}(t)|\psi(t)\rangle\, dt. \tag{11.122}$$

Utilizing that

$$\left|\tilde{\psi}(T)\right\rangle = e^{-i\phi(T)}|\psi(T)\rangle, \tag{11.123}$$

$$\left|\tilde{\psi}(0)\right\rangle = e^{-i\phi(0)}|\psi(0)\rangle, \tag{11.124}$$

Eq. (11.116) together with the requirement in Eq. (11.118) lead to the result

$$\eta = \phi(T) - \phi(0). \tag{11.125}$$

The state $|\psi\rangle$ thus has acquired a phase η which consists of two parts associated with dynamical time evolution and geometry; c.f. Eq. (11.122).

11.6.2 Geometric and dynamical phase of a harmonic oscillator

One may illustrate the Aharonov-Anandan theory by applying it to a harmonic oscillator with angular frequency Ω. Let the initial oscillator state $|\psi(0)\rangle$ be expanded into the oscillator spectrum of energy eigenstates $|m\rangle$, $m = 0, 1, 2, ...$, that is

$$|\psi(0)\rangle = \sum_m |m\rangle\langle m|\psi(0)\rangle = \sum_m \psi_m |m\rangle. \tag{11.126}$$

The time evolution of this state is given by

$$|\psi(t)\rangle = \sum_{m=0}^{\infty} \psi_m \exp\left[-i(m + \frac{1}{2})\Omega t\right] |m\rangle . \tag{11.127}$$

By inspection, it appears that

$$\left|\tilde{\psi}(t)\right\rangle = \exp\left(\frac{i}{2}\Omega t\right) |\psi(t)\rangle \tag{11.128}$$

is periodic and satisfies the requirement $\left|\tilde{\psi}(T)\right\rangle = \left|\tilde{\psi}(0)\right\rangle$ where $T = 2\pi/\Omega$. By inserting Eqs. (11.127), (11.128) and $\hat{H} = \sum_m (m + \frac{1}{2})\hbar\Omega$ into Eq. (11.122) one obtains

$$\eta = 2\pi \sum_m m|\psi_m|^2 - 2\pi \sum_m (m + \frac{1}{2})|\psi_m|^2 = -\pi. \tag{11.129}$$

The quantum mechanical mean value of the occupation number, $< m >= \sum_m m|\psi_m|^2$ multiplied by 2π hence is the oscillator's geometrical phase, whereas the dynamical phase is $-2\pi(< m > +1/2)$. Note that the sum of the two parts only relates to the zero point energy $\hbar\Omega/2$. *The oscillator's zero point energy cannot enter the geometrical phase.*

CHAPTER 12

Lagrangian Approach to Electrodynamics

12.1 LAGRANGIAN FORMALISM FOR A CONTINUUM OF DEGREES OF FREEDOM

The electromagnetic field is defined by its values at all points in space. The associated Lagrangian description therefore must be based on a continuous set of dynamical variables.We thus consider generalized "coordinates" which depend on a *continuous index*, denoted by $\mathbf{r}$ (a point in three-dimensional space), and a *discrete index j* [running over the chosen set of field coordinates ($j = 0 - 3$)]. Let the coordinates be named $A_j(\mathbf{r})$. Although not necessary for the moment, the reader may consider $A_j(\mathbf{r})$ for $j = 1 - 3$ as the jth component of the vector potential and $A_0(\mathbf{r})$ as the scalar potential. The coordinates $A_j(\mathbf{r})$ and the velocities

$$\dot{A}_j(\mathbf{r})[= \dot{A}_j(\mathbf{r}, t)] = \frac{\partial}{\partial t} A_j(\mathbf{r}, t). \tag{12.1}$$

The Lagrangian L can have a large variety of structures. For particles interacting with the electromagnetic field it is useful to divide L into three parts, i.e.,

$$L = L_P + L_F + L_I, \tag{12.2}$$

L_F and L_I being the field (F) and field-matter interaction (I) parts. The part L_P is the Lagrangian associated with the particles (P). In nonrelativistic quantum theory the Schrödinger matter field $\psi(\mathbf{r})$ and its conjugate $\psi^*(\mathbf{r})$ can be taken as the dynamical coordinates. Alternatively, one may use the real ($\psi_r(\mathbf{r})$) and imaginary ($\psi_i(\mathbf{r})$) parts of $\psi(\mathbf{r})$ as dynamical coordinates. Together with their time derivatives ("velocities") $\dot{\psi}_r(\mathbf{r})$ and $\dot{\psi}_i(\mathbf{r})$, the particle Lagrangian contains an excess of dynamical variables. Thus, it turns out that the imaginary quantities $\psi_i(\mathbf{r})$ and $\dot{\psi}_i(\mathbf{r})$ can be considered as redundant dynamic variables which can be eliminated from L_P. With focus on a later application to the electromagnetic field it is sufficient here only to study a Lagrangian formalism where $\{A_j(\mathbf{r})\}$ and $\{\dot{A}_J(\mathbf{r})\}$ form an ensemble of dynamical variables.

DOI: 10.1201/9781003029458-12

Let us assume that we can take

$$L = \int \mathcal{L}(A_j, \dot{A}_j, \partial_i A_j) \mathrm{d}^3 r. \tag{12.3}$$

The integration in Eq. (12.3) runs over the entire space, and $\mathcal{L}$ is called the Lagrangian density. $\mathcal{L}$ is a real function not only of the coordinates $(A_j(\mathbf{r}))$ and velocities $(\dot{A}_j(\mathbf{r}))$ but also of the *spatial derivatives* (denoted $\partial_i A_j(\mathbf{r})$ with $\partial_i = \partial_x, \partial_y, \partial_z$. Although spatial derivatives enter $\mathcal{L}$, these are not new (extra) dynamical variables but (so to speak) linear combinations of the generalized coordinates $A_j(\mathbf{r})$. Since Maxwell equations contain spatial derivatives, electrodynamics involves field coupling between neighboring points in space. If the $\partial_i A_j$-dependence was absent from $\mathcal{L}$ no nonlocal coupling would exist, and the field would (wrongly) evolve in time independently at each point in space. In $\mathcal{L}(A_j, \dot{A}_j, \partial_i A_j)$ we have kept implicitly a possible *explicit* dependence on $\mathbf{r}$ and t.

The *action S* is the time integral of the Lagrangian between, say, times t_1 and t_2, that is

$$S = \int_{t_1}^{t_2} \left[\int \mathcal{L}(A_j, \dot{A}_j, \partial_i A_j) \mathrm{d}^3 r \right] dt. \tag{12.4}$$

The fundamental *Principle of Least Action* says that S is an extremum when $A_j(\mathbf{r})$ corresponds to the actual "motion" of the field between t_1 and t_2. Hence

$$\delta S = \int_{t_1}^{t_2} \delta L dt = 0, \tag{12.5}$$

where

$$\delta L = \int \sum_{i,j} \left[\frac{\partial \mathcal{L}}{\partial A_j} \delta A_j(\mathbf{r}) + \frac{\partial \mathcal{L}}{\partial \dot{A}_j} \delta \dot{A}_j(\mathbf{r}) + \frac{\partial \mathcal{L}}{\partial(\partial_i A_j)} \delta(\partial_i A_j(\mathbf{r})) \right] d^3 r. \tag{12.6}$$

The idea now is to express the variation $\delta(\partial_i A_j(\mathbf{r}))$ as a function of the variation $\delta A_j(\mathbf{r})$ of the dynamical position variable. Assuming that the fields, $A_j(\mathbf{r})$, vanish at infinity (what else can we do?), and utilizing that $\delta(\partial_i A_j(\mathbf{r})) = \partial_i(\delta A_j(\mathbf{r}))$, a partial integration of the last term in Eq. (12.6) gives

$$\delta L = \int \sum_{i,j} \left\{ \left[\frac{\partial \mathcal{L}}{\partial A_j} - \partial_i \left(\frac{\partial \mathcal{L}}{\partial(\partial_i A_j)} \right) \right] \delta A_j(\mathbf{r}) + \frac{\partial \mathcal{L}}{\partial \dot{A}_j} \delta \dot{A}_j(\mathbf{r}) \right\} d^3 r. \tag{12.7}$$

To obtain the variation in the action, Eq. (12.7) is integrated over time. Since $\delta \dot{A}_j(\mathbf{r}) = d(\delta A_j)/dt$, a partial integration of the last term in Eq. (12.7) with respect to time altogether leads to

$$\delta S = \int_{t_1}^{t_2} \sum_j \left[\frac{\partial \mathcal{L}}{\partial A_j} - \sum_i \partial_i \left(\frac{\partial \mathcal{L}}{\partial(\partial_i A_j)} \right) - \frac{d}{dt} \left(\frac{\partial \mathcal{L}}{\partial \dot{A}_j} \right) \right] \delta A_j(\mathbf{r}) d^3 r dt, \tag{12.8}$$

remembering that the *Principle of Least Action is used with zero variations* δA_j *of the dynamical variables at* t_1 *and* t_2. The stationarity of the action with respect to infinitesimal variations of all the independent dynamical variables $[\delta S = 0]$ then gives the Lagrange equations

$$\frac{d}{dt}\left(\frac{\partial \mathcal{L}}{\partial \dot{A}_j}\right) = \frac{\partial \mathcal{L}}{\partial A_j} - \sum_{i=x,y,z} \partial_i \left(\frac{\partial \mathcal{L}}{\partial(\partial_i A_j)}\right). \tag{12.9}$$

Applied to electrodynamics $A_j(\mathbf{r})$, $j = 1 - 3$, are the components of the vector potential, and since the particle Lagrangian density is independent of the $A_j(\mathbf{r})$'s, the effective Lagrangian density in Eq.(12.9) is the sum of the free-field and the interaction parts, $\mathcal{L} = \mathcal{L}_F + \mathcal{L}_I$. In Section 3.5 the Lagrange equations in Eq. (12.9) was applied in a study of the dynamics of the transverse gauge photons; compare Eqs. (3.51) and (12.9).

The Lagrangian density entering the calculation of the variation of the action, δS, is not unique. Thus, a new density given by

$$\mathcal{L}' = \mathcal{L} + \frac{d}{dt}f_0(A_j(\mathbf{r}),\mathbf{r},t) + \sum_{i=x,y,z} \partial_i f_i(A_j(\mathbf{r}),\mathbf{r},t), \tag{12.10}$$

will result in a Lagrangian L' which(physically) is equivalent to L. This is so because the integral of $\sum_i \partial_i f_i = \mathbf{\nabla} \cdot \mathbf{f}$ is transformed to a surface integral of infinity (which vanishes) and $\int_{t_1}^{t_2} dt = f_0(t_2) - f_0(t_1) = 0$.

12.2 STANDARD AND COVARIANT LAGRANGIAN DENSITIES

For what follows it is sufficient to consider the field (L_F) and field-matter (L_I) parts of the total Lagrangian $(L = L_P + L_F + L_I)$. The standard (ST) Lagrangian (L_{F+I}) generally used in classical electrodynamics, namely $[63, 118, 157, 185, 186, 223, 230]$

$$L_{F+I}^{\text{ST}} = \int \mathcal{L}_F^{\text{ST}} d^3r + \int \mathcal{L}_I^{\text{ST}} d^3r \tag{12.11}$$

has a Lagrangian field density given by

$$\mathcal{L}_F^{\text{ST}} = \frac{\varepsilon_0}{2}(\mathbf{E}^2 - c^2\mathbf{B}^2) = \frac{\varepsilon_0}{2}[(-\dot{\mathbf{A}} - \mathbf{\nabla}\Phi)^2 - c^2(\mathbf{\nabla} \times \mathbf{A})^2], \tag{12.12}$$

and an interaction part of the form

$$\mathcal{L}_I^{\text{ST}} = \mathbf{J} \cdot \mathbf{A} - \rho\Phi. \tag{12.13}$$

In Eqs. (12.12) and (12.13) $\Phi = cA^0$ is the scalar potential. The three-vector particle current density is denoted by $\mathbf{J}$, and ρ is the related charge density. A comparison to the Lagrangian density $\mathcal{L}(A_j, \dot{A}_j, \partial_i A_j)$, $j = 0 - 3$, studied in Section 12.1, shows that there is no $\dot{A}_0 = \dot{\Phi}/c$ dependence in the electrodynamic case. In consequence the scalar potential has no conjugate momentum $[\partial\mathcal{L}^{\text{ST}}/\partial\dot{A}_0 = c\partial\mathcal{L}_{F+I}^{\text{ST}}/\partial\dot{\Phi} = 0]$. When a velocity (here $\dot{A}_0$) associated with a generalized coordinate (A_0) does not appear in the Lagrangian this coordinate can be eliminated by expressing it as a function of the other dynamical variables. This elimination brings in also the particle dynamical variables.

A gauge transformation of the vector and scalar potentials [Eqs.(11.1) and (11.2); $V = \Phi$ (in this section)] does not change the Lagrangian of the field, which depends only on the electric and magnetic fields, nor is the particle Lagrangian changed, obviously. The change in the interaction Lagrangian density, viz.,

$$(\mathcal{L}_I^{\mathrm{ST}})' - \mathcal{L}_I^{\mathrm{ST}} = \mathbf{J} \cdot \boldsymbol{\nabla}\chi + \rho\frac{\partial\chi}{\partial t} = \boldsymbol{\nabla} \cdot (\mathbf{J}\chi) + \frac{\partial}{\partial t}(\rho\chi) - (\boldsymbol{\nabla} \cdot \mathbf{J} + \frac{\partial\rho}{\partial t})\chi, \qquad (12.14)$$

has a form which makes the two Lagrangian densities equivalent. This is so because the equation of charge continuity implies that the last term in Eq. (12.14) is zero. The first two terms have the forms given in Eq. (12.10) and will therefore vanish upon integration.

In nonrelativistic electrodynamics it is often convenient to work in the Coulomb gauge, $\boldsymbol{\nabla} \cdot \mathbf{A} = 0$, where $\mathbf{A} = \mathbf{A}_T$ [63, 157, 223]. In this gauge the *total* Lagrangian takes the form

$$L^{\mathrm{ST}} = L_P - V_{\mathrm{Coul}} + \int \mathcal{L}_{\mathrm{Coul}}^{\mathrm{ST}}d^3r, \qquad (12.15)$$

where

$$\mathcal{L}_{\mathrm{coul}}^{\mathrm{ST}}[= \mathcal{L}_F^{\mathrm{ST}} + \mathcal{L}_{I,\mathrm{coul}}^{\mathrm{ST}}] = \frac{\varepsilon_0}{2}[\dot{\mathbf{A}}_T^2 - c^2(\boldsymbol{\nabla} \times \mathbf{A}_T)^2] + \mathbf{J} \cdot \mathbf{A}_T. \qquad (12.16)$$

In this well-known formulation, the Coulomb interaction energy [stemming from the $\rho\Phi$-term in Eq. (12.13)] can be expressed solely in terms of the particle position coordinates [63, 157]. Due to this fact, V_{Coul} is considered as a correction to the original particle Lagrangian L_P, that is $L_P^{\mathrm{Coul}} = L_P - V_{\mathrm{Coul}}$. For many-particle systems V_{Coul} makes the Schrödinger equation derived from L_P^{Coul} extremely difficult (impossible) to solve in general.

Let us now shift to covariant notation in which covariant and contravariant components of a general four-vector $F_\mu = \eta_{\mu\nu}F^\nu$ are related via the diagonal metric tensor with non vanishing elements $\eta_{00} = -1$, $\eta_{11} = \eta_{22} = \eta_{33} = +1$. With

$$\partial_\mu = \left(\frac{1}{c}\frac{\partial}{\partial t}, \boldsymbol{\nabla}\right); \quad A^\mu = (A^0, \mathbf{A}); \quad J^\mu = (J^0, \mathbf{J}), \qquad (12.17)$$

$[A^0 = \Phi/c,\ J^0 = c\rho]$, the field and interaction Lagrangian density of the standard description take the following forms:

$$\mathcal{L}_F^{\mathrm{ST}} = -\frac{\varepsilon_0 c^2}{4}F_{\mu\nu}F^{\mu\nu}, \qquad (12.18)$$

$F^{\mu\nu}$ being the electromagnetic field tensor, and

$$\mathcal{L}_I^{\mathrm{ST}} = J_\mu A^\mu. \qquad (12.19)$$

Above, and for what follows the summation convention over repeated upper and lower indices is in force.

In relativistic (quantum) electrodynamics [63,106,118,186,223] as well as in near-field nonrelativistic electrodynamics [128,151,156–158,162] it is very often convenient to employ a *manifest covariant* formalism (below just called covariant). Despite the name covariant one should remember that the Coulomb gauge description *is* relativistically correct, although the Lorentz transformations of the fields are cumbersome to work with in general.

We consider then first a manifest covariant (cov) field Lagrangian density

$$\mathcal{L}_F^{\text{cov}} = -\frac{\varepsilon_0 c^2}{2}(\partial_\mu A^\nu)(\partial^\mu A_\nu). \tag{12.20}$$

The reader may make a comparison to the standard field Lagrangian density by writing Eq. (12.20) in three-vector notation:

$$\mathcal{L}_F^{\text{cov}} = \frac{\varepsilon_0}{2}\left[\dot{\mathbf{A}}^2 - \left(\frac{\dot{\Phi}}{c}\right)^2 - c^2\sum_{i,j}(\partial_i A_j)^2 + (\boldsymbol{\nabla}\Phi)^2\right]. \tag{12.21}$$

It should be noted that $\mathcal{L}_F^{\text{cov}}$ has a (quadratic) term in $\dot{\Phi}$, so that Φ now has a conjugate momentum.

In the covariant description one retains the same interaction Lagrangian density ($\mathcal{L}_I^{\text{cov}}$) as in the standard formulation, that is

$$\mathcal{L}_I^{\text{cov}} = \mathcal{L}_I^{\text{ST}} = J_\mu A^\mu. \tag{12.22}$$

It is important to note that the two Lagrangian field densities $\mathcal{L}_I^{\text{ST}}$ and $\mathcal{L}_I^{\text{cov}}$ are *not equivalent*. In order to demonstrate this, it is useful to start from the so-called Fermi (F) Lagrangian density for the field ($\mathcal{L}_F^{\text{F}}$), given explicitly as [63]

$$\mathcal{L}_F^{\text{F}} = -\varepsilon_0 c^2\left[\frac{1}{4}F_{\mu\nu}F^{\mu\nu} + \frac{1}{2}(\partial_\mu A^\mu)^2\right]. \tag{12.23}$$

This Lagrangian density is obviously not equivalent to the standard field Lagrangian in Eq. (12.18), *unless* the Lorenz gauge condition

$$\partial_\mu A^\mu = 0 \tag{12.24}$$

is satisfied. However the Fermi Lagrangian density is equivalent to the covariant field Lagrangian density. Hence,

$$\mathcal{L}_F^{\text{cov}} - \mathcal{L}_F^{\text{F}} = \frac{\varepsilon_0 c^2}{2}\partial_\mu[A^\mu\partial_\nu A^\nu - A^\nu\partial_\nu A^\mu], \tag{12.25}$$

as a simple calculation shows. Since $\mathcal{L}_F^{\text{cov}}$ and $\mathcal{L}_F^{\text{F}}$ differ only by a four-divergence term, which we know disappears upon space/time integration; cf. Eq. (12.10), $\mathcal{L}_F^{\text{cov}}$ and $\mathcal{L}_F^{\text{F}}$ are equivalent. We can therefore conclude as follows: The standard ($\mathcal{L}_F^{\text{ST}}$) and covariant ($\mathcal{L}_F^{\text{cov}}$) field Lagrangian densities are not equivalent, unless the Lorenz gauge condition is put on $\mathcal{L}_F^{\text{cov}}$ as a subsidiary demand. In applying the two sets of Lagrange equations, $\mathcal{L}_{F+I}^{\text{ST}}$ leads to the Maxwell-Lorentz equations in the potential description, but $\mathcal{L}_{F+I}^{\text{cov}}$ gives the field equations only in the Lorenz gauge.

12.3 EXTENDED LAGRANGIANS: OHMURA'S SCALAR FIELD

Let us now ask the following question: In what sense is the Maxwell-Lorentz electrodynamics modified if one replaces the Fermi Lagrangian density for the field by

$$\mathcal{L}_F(\lambda) = -\frac{\varepsilon_0 c^2}{2}\left[\frac{1}{2}F_{\mu\nu}F^{\mu\nu} + \lambda(\partial_\mu A^\mu)^2\right], \tag{12.26}$$

where λ is a real parameter? Before studying this question, we note that $\lambda = 0$ gives us the standard Lagrangian density, $\mathcal{L}_F(\lambda = 0) = \mathcal{L}_F^{\text{ST}}$, and for $\lambda = 1$ one regains the Fermi Lagrangian density, $\mathcal{L}_F(\lambda = 1) = \mathcal{L}_F^F$. Thus, it follows that the Lagrangian densities for $\lambda = 0$ and $\lambda = 1$ are not equivalent. This circumstance can be generalized immediately: Two Lagrangians field densities corresponding to different values of λ (λ_1 and λ_2) are not equivalent, since the difference $(\lambda_1 - \lambda_2)(\partial_\mu A^\mu)^2$ is not a four-divergence. The Lagrangian in Eq. (12.26) is called an *extented field Lagrangian density*. Most importantly, one obtains for $\lambda \neq 0, 1$ an extension of the Maxwell-Lorentz electrodynamics. Whether such an extension relates to some kind of physical phenomena is not known. However, it has been suggested that the extended formalism may be needed to account for certain observations in General Relativity [160,277,278]. With the same field-matter interaction Lagrangian density for $\lambda = 0, 1$, the general description ($\lambda = 0$) of the potential formulation of electrodynamics is in agreement with the $\lambda = 1$ case in the Lorenz gauge. A "minimal" extension of electrodynamics thus appears if one keeps the interaction Lagrangian density in its original form, i.e.

$$\mathcal{L}_I(\lambda) = \mathcal{L}_I(0) \,[= \mathcal{L}_I], \quad \forall\lambda. \tag{12.27}$$

Starting form the generic form of the Euler-Lagrange equation, viz.,

$$\partial_\alpha\left(\frac{\partial\mathcal{L}}{\partial_\alpha(\partial A^\beta)}\right) = \frac{\partial\mathcal{L}}{\partial A^\beta}, \tag{12.28}$$

which just us the covariant form of Eq. (12.9), and with $\mathcal{L}(\lambda) \equiv \mathcal{L}_F(\lambda) + \mathcal{L}_I$, one obtains the following inhomogeneous wave equations in a potential formulation:

$$\partial_\mu\partial^\mu A^\nu - (1-\lambda)\partial^\nu(\partial_\mu A^\mu) = -\mu_0 J^\nu, \quad \nu = 0 - 3. \tag{12.29}$$

All covariant gauges are characterized by

$$\partial_\mu A^\mu = K, \tag{12.30}$$

where K is a constant that is independent of space and time. In the Lorenz gauge $K = 0$. For $\lambda = 0, 1$ the well-known wave equations are regained.

In three-vector notation Eq. (12.29) yields the following set of wave equations [160]:

$$\Box\mathbf{A}_T(\mathbf{r}, t) = -\mu_0\mathbf{J}_T(\mathbf{r}, t), \tag{12.31}$$

$$\Box\mathbf{A}_L(\mathbf{r}, t) - (1-\lambda)\boldsymbol{\nabla}C(\mathbf{r}, t) = -\mu_0\mathbf{J}_L(\mathbf{r}, t), \tag{12.32}$$

$$\Box\Phi(\mathbf{r}, t) + (1-\lambda)\frac{\partial}{\partial t}C(\mathbf{r}, t) = -\frac{1}{\varepsilon_0}\rho(\mathbf{r}, t), \tag{12.33}$$

where

$$C(\mathbf{r}, t) \equiv \boldsymbol{\nabla} \cdot \mathbf{A}_L(\mathbf{r}, t) + \frac{1}{c^2} \frac{\partial}{\partial t} \Phi(\mathbf{r}, t). \tag{12.34}$$

It appears from Eqs. (12.31)–(12.33) that the transverse (gauge photon) dynamics is unaffected by the extension. With the help of Eqs. (12.32) and (12.34) one obtains for the new scalar field $C(\mathbf{r}, t)$ the inhomogeneous wave equation

$$\lambda \square C(\mathbf{r}, t) = -\mu_0 [\boldsymbol{\nabla} \cdot \mathbf{J}_L(\mathbf{r}, t) + \frac{\partial}{\partial t} \rho(\mathbf{r}, t)]. \tag{12.35}$$

In the (accepted) non-extended form of electrodynamics, charge conservation ($\sim$ equation of continuity) is a consequence of the form of the set of the Maxwell-Lorentz equations. It appears from Eq. (12.35) that if one insists on upholding charge conservation, the scalar field $C(\mathbf{r}, t)$ satisfies the homogeneous wave equation $\square C(\mathbf{r}, t) = 0$ ($\lambda \neq 0$).

In General Relativity the extended Lagrangian ($\mathcal{L}(\lambda) = \mathcal{L}_F(\lambda) + \mathcal{L}_I$) takes the form

$$\mathcal{L} = \sqrt{-g} \left\{ -\frac{\varepsilon_0 c^2}{2} \left[\frac{F_{\mu\nu} F^{\mu\nu}}{2} + \lambda (\nabla_\mu A^\mu)^2 \right] + J_\mu A^\mu \right\}, \tag{12.36}$$

where ∇_μ is the covariant derivative [157, 223, 258] replacing ∂_μ, and $F^{\mu\nu} = \nabla^\mu A^\nu - \nabla^\nu A^\mu$. Indices are lowered with the metric tensor of General Relativity ($g_{\mu\nu}$) [and raised by its inverse]. The quantity g (< 0) is the determinant of the metric tensor. The extended theory opens the doorway for longitudinal wave propagation in General Relativity, a suggested possibility related to the curvature of space-time. A detailed discussion of extended electrodynamics in Special and General Relativity has been given in [160]; see also [278]. In the extension made by Ohmura in 1956 a scalar field $C^{\mathrm{Ohmura}} = -\lambda C$ is used [195]. However, Ohmura does not discuss in his short letter in any detail the electrodynamic consequences of the extension.

IV

Photon Scattering Theory. Density Operator Formalism

CHAPTER 13

Standard Light Scattering Theory, Ante Omnia

13.1 SCALAR THEORY OF SCATTERING: MACROSCOPIC APPROACH

In macroscopic electrodynamics the optical properties of a medium usually are characterized at a given angular frequency (ω) by the macroscopic polarization, $\mathbf{P}(\mathbf{r};\omega)$, and magnetization, $\mathbf{M}(\mathbf{r};\omega)$. From the Maxwell equations it then follows that the macroscopic electric field, $\mathbf{E}(\mathbf{r};\omega)$ satisfies the inhomogeneous wave equation (no free charges and currents present)

$$(\nabla^2 + q_0^2)\mathbf{E}(\mathbf{r};\omega) - \boldsymbol{\nabla}\boldsymbol{\nabla} \cdot \mathbf{E}(\mathbf{r};\omega) = -i\mu_0\omega[\boldsymbol{\nabla} \times \mathbf{M}(\mathbf{r};\omega) - i\omega\mathbf{P}(\mathbf{r};\omega)], \qquad (13.1)$$

where $q_0 = \omega/c$.

In the *scalar theory* of light scattering the term $\boldsymbol{\nabla}\boldsymbol{\nabla} \cdot \mathbf{E}(\mathbf{r};\omega)$ in Eq. (13.1) is assumed to be negligible [154, 193]. Roughly speaking this will be the case of the polarization varies sufficiently slowly with position. With the help pf the Huygens scalar propagator

$$g(R;\omega) = \frac{\exp(iq_0 R)}{4\pi R} \qquad (13.2)$$

one can obtain the following integral expression for the electric field in the scalar theory:

$$\mathbf{E}(\mathbf{r};\omega) = \mathbf{E}^i(\mathbf{r};\omega) + i\mu_0\omega \int g(|\mathbf{r} - \mathbf{r}'|;\omega)[\boldsymbol{\nabla}' \times \mathbf{M}(\mathbf{r}';\omega) - i\omega\mathbf{P}(\mathbf{r}';\omega)]d^3r', \quad (13.3)$$

where $\mathbf{E}^i(\mathbf{r};\omega)$ is the incident (i) field [in later chapters often called the external (ext) field $\mathbf{E}^{\text{ext}}(\mathbf{r};\omega)$]. The integral in Eq. (13.3) extends over the domain occupied by the medium in considerations. Eq. (13.3) is not an explicit solution for the electric field since $\mathbf{M}(\mathbf{r};\omega)$ and $\mathbf{P}(\mathbf{r};\omega)$ here are the magnetization and polarization induced by the prevailing (selfconsistent) electric field. For nonmagnetic, isotropic media, characterized by a refractive index $n(\mathbf{r};\omega)$, Eq. (13.3) leads to the textbook integral equation [46]

$$\mathbf{E}(\mathbf{r};\omega) = \mathbf{E}^i(\mathbf{r};\omega) + q_0^2 \int g(\mathbf{r} - \mathbf{r}';\omega)[n^2(\mathbf{r}';\omega) - 1]\mathbf{E}(\mathbf{r}';\omega)d^3r' \qquad (13.4)$$

DOI: 10.1201/9781003029458-13

in linear response theory. Provided that the medium scatters light rather weakly, an iterative procedure can be used to solve Eq. (13.4). This so-called Born series approach enables one to study multiple scattering processes in an elegant manner. In the first order Born approximation one puts $\mathbf{E}(\mathbf{r}';\omega) = \mathbf{E}^i(\mathbf{r}';\omega)$ under the integral sign. Thereby one obtains the explicit scalar theory solution

$$\mathbf{E}(\mathbf{r}';\omega) = \mathbf{E}^i(\mathbf{r}';\omega) + q_0^2 \int g(\mathbf{r} - \mathbf{r}';\omega)[n^2(\mathbf{r}';\omega) - 1]\mathbf{E}^i(\mathbf{r}';\omega)d^3r' \tag{13.5}$$

roughly speaking correct if the refractive index differs only slightly from unity. The result in Eq. (13.5) is extremely useful for the problem of inverse scattering; i.e. the problem of obtaining information on the physical properties of a body from measurements of the field scattered by it [46].

13.2   BEYOND THE SCALAR THEORY OF SCATTERING

It is possible to obtain an integral equation for the electric field without neglecting the term $\boldsymbol{\nabla}\boldsymbol{\nabla} \cdot \mathbf{E}(\mathbf{r};\omega)$ in Eq. (13.1). From a mathematical point of view the light scattering problem becomes much more complicated, but from a fundamental physical point of view much more information now may be obtained, and the nagging uncertainties one is left with from the scalar theory disappear. Thus, from Eq. (13.1) one is led to the following integral equation for the electric field:

$$\begin{aligned}\mathbf{E}(\mathbf{r};\omega) =&\mathbf{E}^i(\mathbf{r};\omega) + \mathbf{E}^{\mathrm{SF}}(\mathbf{r};\omega)\\ &+ i\mu_0\omega \int_{\varepsilon \to 0} \mathbf{G}(\mathbf{r} - \mathbf{r}';\omega) \cdot [\boldsymbol{\nabla}' \times \mathbf{M}(\mathbf{r}';\omega) - i\omega\mathbf{P}(\mathbf{r}';\omega)]d^3r'.\end{aligned} \tag{13.6}$$

The quantity $\mathbf{E}^{\mathrm{SF}}(\mathbf{r};\omega)$, named the self-field term, represents the local $(\mathbf{r}' = \mathbf{r})$ contributions to the scattered field in the various space points. The contributions stemming from all other points $(\mathbf{r}' \neq \mathbf{r})$ in the scattering medium are accounted for by the integral in Eq. (13.5). This is so because a small exclusion volume ε is excluded from the integral volume. By the end of a given calculation the exclusion volume is made infinitesimally small [as indicated by $\varepsilon \to 0$ at the foot of the integral sign]. Only the sum of the self-field an integral terms is of physical importance. This is so because each of the two terms (but not their sum) depends on the form of the contracting exclusion volume. The dyadic Green function $\mathbf{G}(\mathbf{R};\omega) = \mathbf{G}(\mathbf{r} - \mathbf{r}';\omega)$ can be obtained from a knowledge of the Huygens scalar propagator. Thus,

$$\mathbf{G}(\mathbf{R};\omega) = \left(\mathbf{U} + \frac{1}{q_0^2}\boldsymbol{\nabla}\boldsymbol{\nabla}\right) g(R;\omega). \tag{13.7}$$

It is obvious that $\mathbf{G}(\mathbf{R};\omega)$ is singular at $\mathbf{r}' = \mathbf{r}$, and the reader may show that the singularity is of the type $|\mathbf{r} - \mathbf{r}'|^{-3}$, and hence so strong that the integral in Eq. (13.5) is only conditionally convergent. Although only the sum of the self-field and the integral terms in Eq. (13.5) has physical importance, the division of the scattered field $\mathbf{E}(\mathbf{r};\omega) - \mathbf{E}^i(\mathbf{r};\omega)$, between the two parts depends on the chosen form of the contraction volume. In many cases it is preferable to employ a spherical exclusion volume [157].

It is possible to generalize the macroscopic light scattering theory to the microscopic regime. With emphasis on the scattering from mesoscopic media such a theory is established in [152].

The unpleasantness associated with the form of the singularity in $\mathbf{G}(\mathbf{R};\omega)$ is further increased by the fact that it is impossible to convert the dyadic Green function to the space-time domain. This fact in itself makes it difficult to unravel the physics hidden in $\mathbf{G}(\mathbf{R};\omega)$ [148]. In order to relate the expression in Eq. (13.5) to photon physics, it is necessary that a part of $\mathbf{G}(\mathbf{R};\omega)$ qualifies as an electromagnetic field *propagator*. A rigorous field propagator must vanish for $\omega \to 0$. It is clear that $\mathbf{G}(\mathbf{R};\omega)$ takes the nonvanishing asymptotic form $q_0^{-2}\boldsymbol{\nabla}\boldsymbol{\nabla}g(R;0)$ for $\omega \to 0$. Let us therefore subtract this form from $\mathbf{G}(\mathbf{R};\omega)$. Thus, we obtain a dyadic tensor

$$\mathbf{G}_T(\mathbf{R};\omega) \equiv \mathbf{G}(\mathbf{R};\omega) - \frac{1}{q_0^2}\boldsymbol{\nabla}\boldsymbol{\nabla}g(R;\omega) = g(R;\omega)\mathbf{U} + \frac{1}{q_0^2}\boldsymbol{\nabla}\boldsymbol{\nabla}[g(R;\omega) - g(R;0)].$$

$$(13.8)$$

It can be shown [157] that $\mathbf{G}_T(\mathbf{R};\omega)$ is a *genuine transverse (divergence-free) dyad*; indicated by the subscript T. Thus,

$$\mathbf{G}_T(\mathbf{R};0) = \mathbf{0} \tag{13.9}$$

for all $\mathbf{R}$, including $\mathbf{R} = \mathbf{0}$. The remaining part of $\mathbf{G}(\mathbf{R};\omega)$, viz.,

$$\mathbf{G}_L(\mathbf{R};\omega) = \frac{1}{q_0^2}\boldsymbol{\nabla}\boldsymbol{\nabla}g(\mathbf{R};\omega) \tag{13.10}$$

is a *longitudinal* (subscript L) dyadic quantity [$\boldsymbol{\nabla} \times \mathbf{G}_L(\mathbf{R};\omega) = \mathbf{0}$ obviously], and $\mathbf{G}_L(\mathbf{R};\omega)$ plays an important role in near-field electrodynamics [157]. The self-field $\mathbf{E}^{\mathrm{SF}}$ has both T and L-parts.

Our inability to localize a photon precisely in space-time is closely related to the presence of the transverse self-field, $\mathbf{E}_T^{\mathrm{SF}}$ [147], and the description of photon emission from the various points in the current density source distribution, given by $\mathbf{J} = \boldsymbol{\nabla} \times \mathbf{M} - i\omega\mathbf{P}$, is inherent in the transverse photon propagator as this appears in the space-time domain. The explicit expression for $\mathbf{G}_T(\mathbf{R} = \mathbf{r} - \mathbf{r}', \tau = t - t')$ was introduced already in Chapter 3 [Eq. (3.31)], where the properties of the gauge photon in structural vacuum was discussed in detail.

CHAPTER 14

Semiclassical Light Scattering

Over the years integral equation descriptions of classical electromagnetic scattering from macroscopic and mesoscopic objects have taken up a prominent position. Depending on the actual application, the starting points for the formulations can be quite different, and a survey of the literature shows that the various approaches seldom are mutually consistent. From a fundamental point of view, it appears natural to seek to formulate the semiclassical light scattering theory in such a manner that it links "smoothly" to the photon scattering theory; c.f. the Bohr Correspondence Principle. This is the idea for what follows below.

14.1 SPATIALLY NONLOCAL SCATTERING

14.1.1 Integral relation for the gauge invariant potential

We have realized in previous chapters that the transverse gauge photon, the fundamental entity in photon physics, is connected intimately to the quantization of the transverse vector potential, $\mathbf{A}_T(\mathbf{r}, t)$. In classical electrodynamics, $\mathbf{A}_T(\mathbf{r}, t)$ satisfies the inhomogeneous wave equation given in Eq. (3.41). The wave equation can be converted to an integral equation of the following form:

$$\mathbf{A}_T(\mathbf{r}, t) = \mathbf{A}_T^{\text{ext}}(\mathbf{r}, t) + \mu_0 \int_{-\infty}^{\infty} \int_{V_T} g(R, \tau) \mathbf{J}_T(\mathbf{r}', t') d^3 r' dt', \qquad (14.1)$$

where the spatial integral extends over the transverse current density domain, V_T. In Section 16.1, we shall show how this integral equation can be derived in an elegant manner on the basis of the Ewald-Oseen extinction theorem. The quantity $\mathbf{A}_T^{\text{ext}}(\mathbf{r}, t)$ is the transverse vector potential stemming from the external (ext) source exciting the medium (object) under study. We shall assume that the external source is so far from the object that the transverse current densities of the two do not overlap in space. In this case only the transverse part of the total external vector potential, $\mathbf{A}^{\text{ext}} = \mathbf{A}_T^{\text{ext}} + \mathbf{A}_L^{\text{ext}}$, enters the description. We shall assume also that the dynamics of the external source develops independently of the scattering from the object (no

DOI: 10.1201/9781003029458-14

feedback coupling, c.f. the discussion in Section 4.4). The quantity $g(R, \tau)$ is the retarded Huygens scalar propagator which explicit form is given in Eq. (3.42).

The transverse current density $\mathbf{J}_T(\mathbf{r}, t)$ entering Eq. (14.1) is related to the current density (itself) $\mathbf{J}(\mathbf{r}, t)$, by a *spatially nonlocal* relation

$$\mathbf{J}_T(\mathbf{r}, t) = \int_{-\infty}^{\infty} \boldsymbol{\delta}_T(\mathbf{r} - \mathbf{r}') \cdot \mathbf{J}(\mathbf{r}', t) d^3 r'. \tag{14.2}$$

For a macroscopic object, the difference between the volumes V and V_T of $\mathbf{J}$ and $\mathbf{J}_T$, respectively, is in general negligible. However, this circumstance does not mean that $\mathbf{J}_T$ can be replaced by $\mathbf{J}$ in Eq. (14.1), since the integral equation relates divergence-free vector fields. For light scattering from mesoscopic (and microscopic) objects, the difference $\Delta V = V_T - V$ cannot be neglected. From the point of view of photon physics the fact that the relation in Eq. (14.2) is *local* in time has a deeper significance. As we have realized in Sections 3.3 and 3.4, the transverse gauge photon emitted from a $\mathbf{J}$-source domain cannot be localized better than the $\mathbf{J}_T$-domain. This turns out to be a satisfactory conclusion, which, in a certain sense also makes the lack complete spatial confinement uniform for massive and massless elementary particles, albeit in a qualitative manner only. In [143] the difference between the $\mathbf{J}$ and $\mathbf{J}_T$-domains as exemplified by a quantitative calculation for the hydrogen $1s \Leftrightarrow 2p_z$ transition. A comprehensive discussion of the source domain issue is given in [153, 157], e.g.

By inserting Eq. (14.2) in Eq. (14.1), one obtains an integral relation between $\mathbf{A}_T(\mathbf{r}, t)$ and $\mathbf{J}(\mathbf{r}', t')$, viz.,

$$\mathbf{A}_T(\mathbf{r}, t) = \mathbf{A}_T^{\text{ext}}(\mathbf{r}, t) + \mu_0 \int_{-\infty}^{\infty} \int_{V_T} \mathbf{G}_T(\mathbf{R}, \tau) \cdot \mathbf{J}(\mathbf{r}', t') d^3 r' dt', \tag{14.3}$$

where $\mathbf{R} = \mathbf{r} - \mathbf{r}'$ and $\tau = t - t'$. The transverse Feynman propagator

$$\mathbf{G}_T(\mathbf{R}, \tau) = \frac{1}{4\pi R} \delta\left(\frac{R}{c} - \tau\right)(\mathbf{U} - \mathbf{e_R e_R}) - \frac{c^2 \tau}{4\pi R^3} \Theta(\tau) \Theta\left(\frac{R}{c} - \tau\right)(\mathbf{u} - 3\mathbf{e_R e_R}), \tag{14.4}$$

already introduced in Section 3.3 [Eq. (3.31)], hold the spatial localization issue in its near field ($\sim R^{-3}$) part. The two formulations of the semiclassical light scattering approaches given above are illustrated graphically in Fig. 14.1.

To proceed from here it is usual to transfer the scattering problem to the wave vector $(\mathbf{q})$ -f frequency (ω) domain. Since the integral over V in Eq. (14.3) can be extended to an integral over the entire space, the integrals in Eq. (14.3) are folding integrals of $\mathbf{G}_T$ and $\mathbf{J}$ in space and time. In the $(\omega, \mathbf{q})$-domain we therefore get the algebraic relation

$$\mathbf{A}_T(\mathbf{q}, \omega) = \mathbf{A}_T^{\text{ext}}(\mathbf{q}, \omega) + \mu_o \mathbf{G}_T(\mathbf{q}, \omega) \cdot \mathbf{J}(\mathbf{q}, \omega). \tag{14.5}$$

It can be shown [157] that the transverse propagator in the $(\omega, \mathbf{q})$-domain is given by the dyadic form

$$\mathbf{G}_T(\mathbf{q}, \omega) = \frac{\mathbf{U} - \boldsymbol{\kappa}\boldsymbol{\kappa}}{q^2 - q_0^2 + i\varepsilon}, \tag{14.6}$$

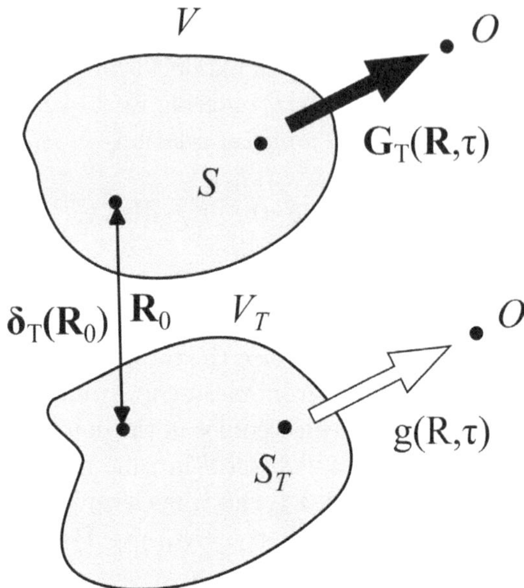

Figure 14.1 Schematic illustration of the two physically equivalent approaches to semi-classical light scattering. "Electron-eye view": Scattering from current density domain, V. The field propagation from a source point S to an observation point O is described by the dyadic transverse Feynman propagator, $\mathbf{G}_T(\mathbf{R}, \tau)$. By means of the transverse delta function, $\boldsymbol{\delta}_T(\mathbf{R}_0)$, a timely local operator, the current density distribution in V is projected onto the transverse current density occupying the domain V_T. "Photon-eye view": The field propagation from a source point S_T to an observation point O is described by the scalar Huygens propagator $g(|\mathbf{R}|, \tau)$.

where $\boldsymbol{\kappa} = \mathbf{q}/q$, $q^2 = \mathbf{q} \cdot \mathbf{q}$ and $q_0 = \omega/c$. An infinitesimal imaginary quantity $i\varepsilon$ ($\varepsilon > 0$) has been added to $q^2 - q_0^2$ to remind the reader that integrals involving the singular form in Eq. (14.6) are carried out as contour integrals (usually in a complex ω-plane). The transverse propagator hence is the product of the Huygens scalar propagator, $g(q, \omega) = (q^2 - q_0^2 + i\varepsilon)^{-1}$, and the transverse part, $\mathbf{U} - \boldsymbol{\kappa}\boldsymbol{\kappa}$, of the unit dyad that is

$$\mathbf{G}_T(\mathbf{q}, \omega) = g(q, \omega)(\mathbf{U} - \boldsymbol{\kappa}\boldsymbol{\kappa}). \tag{14.7}$$

We know that the Huygens form $g(q, \omega)$ leads to retarded field propagation with the speed of light in the space-time domain. Despite the simple connection given in Eq. (14.7) between $g(q, \omega)$ and $\mathbf{G}_T(\mathbf{q}, \omega)$, the "harmless" $(\mathbf{U} - \boldsymbol{\kappa}\boldsymbol{\kappa})$-factor modifies the field "propagation" in such a manner that only the far-field R^{-1} part of the scattered $\mathbf{A}_T$-field exhibits genuine Einsteinian retardation.

14.1.2 Linear constitutive relations. Causal response tensor

Let us assume that the current density vanishes everywhere inside the object under study in the absence of the external (incoming) transverse vector potential

$[\mathbf{A}_T^{\text{ext}}(\mathbf{r}, t) = \mathbf{0}, \forall \mathbf{r} \in V]$. This assumption certainly is an approximation because it is impossible to decouple our medium from the surroundings. In some cases one may treat the surroundings as a heatbath. Although the mean value of the current density may vanish in each point $\mathbf{r} \in V$, classical and quantum mechanical fluctuations will be present. The influence of such fluctuations on the electrodynamics of mesoscopic objects has until now not been studied in detail.

If the incident field, $\mathbf{A}_T^{\text{ext}}$, is sufficiently weak, the field-induced current density, $\mathbf{J}(\mathbf{r}, t)$, usually will be related in a linear fashion to the prevailing transverse field, $\mathbf{A}_T(\mathbf{r}, t)$, acting on the charge particles.

The most general linear relation between $\mathbf{J}$ and $\mathbf{A}_T$ will be nonlocal in both space and time that is of the form

$$\mathbf{J}(\mathbf{r}, t) = \int_{-\infty}^{\infty} \mathbf{S}(\mathbf{r}, \mathbf{r}', t, t') \cdot \mathbf{A}_T(\mathbf{r}', t') d^3 r' dt'. \tag{14.8}$$

The reader may wonder why contributions from the longitudinal part of the vector potential, $\mathbf{A}_L(\mathbf{r}, t)$, and the gradient of the scalar potential, $\nabla \Phi(\mathbf{r}, t)$, is not included as driving mechanisms for the induced current density. First of all, it must be remembered that $\mathbf{A}_L$ and Φ are gauge-dependent quantities, in contrast to $\mathbf{J}$. So, even if one included contributions from $\mathbf{A}_L$ and $\nabla \Phi$, the final result for $\mathbf{J}$ *must be* gauge invariant. A proof of this gauge invariance has been given comparing the particle charge density, ρ induced by the scalar potential Φ to the longitudinal current density, $\mathbf{J}_L$, induced by $\mathbf{A}_L$ and using additionally the equation of continuity, $\nabla \cdot \mathbf{J}_L + \partial \rho / \partial t = 0$ by Bagchi [10] for one-electron systems. The proof holds even in the many-body case, as shown by the present author [144]. An early review of local-field electrodynamics was given in a book dedicated to the memory of Prof. A. O. Barut [146], one of the pioneers of the self-field approach to quantum electrodynamics. Imagine next that we work in the Coulomb gauge, where $\mathbf{A}_L = \mathbf{0}$ $[\nabla \cdot \mathbf{A} = 0]$. In this gauge the scalar potential is (or can be) eliminated in favour of the particle position coordinates, and hence included in the particle part of the Hamiltonian. From a quantum physical point of view the form given in Eq. (14.8) therefore appears basal in the framework of linear response theory, We call Eq. (14.8) a microscopic constitutive equation.

The local field $\mathbf{A}_T(\mathbf{r}, t)$ inside the medium is not a measurable quantity and the constitutive relation does not relate $\mathbf{J}(\mathbf{r}, t)$ in a causal manner to $\mathbf{A}_T(\mathbf{r}, t)$. However, if one inserts Eq.(14.8) into Eq. (14.3), a complicated (!) integro-differential equation for $\mathbf{A}_T(\mathbf{r}, t)$ follows. Since the transverse Feynman propagator, $\mathbf{G}_T(\mathbf{R}, \tau)$, is a known quantity (function of $\mathbf{R} = \mathbf{r} - \mathbf{r}'$ and $\tau = t - t'$), *in principle* the solution to the integro-differential equation gives us a relation between $\mathbf{A}_T$ and $\mathbf{A}_T^{\text{ext}}$. If this relation, which is nonlocal in both space and time, is inserted into Eq. (14.8) one obtains a linear constitutive relation of the form

$$\mathbf{J}(\mathbf{r}, t) = \int_{-\infty}^{\infty} \mathbf{S}^{\text{CAU}}(\mathbf{r}, \mathbf{r}', t, t') \cdot \mathbf{A}_T^{\text{ext}}(\mathbf{r}', t') d^3 r' dt'. \tag{14.9}$$

Despite all the technical difficulties involved in order to reach Eq. (14.9) [that is to determine $\mathbf{S}^{\text{CAU}}(\mathbf{r}, \mathbf{r}', t, t')$] which in general is an insuperable task unless approximations are made, the constitutive relation in Eq. (14.9) has from a fundamental theoretical point of view a very satisfactory form. Let me emphasize four advantages: (i)

The linear response tensor $\mathbf{S}^{\mathrm{CAU}}(\mathbf{r}, \mathbf{r}', t, t')$ obeys microcausality (superscript CAU), since the contribution to the current density at space point $\mathbf{r}$ from the external field at the point $\mathbf{r}'$ cannot exist before there is an external field at $\mathbf{r}'$ to act on the charged particles here. Thus,

$$\mathbf{S}^{\mathrm{CAU}}(\mathbf{r}, \mathbf{r}', t(< t'), t') = \mathbf{0}, \quad \forall \ (\mathbf{r}, \mathbf{r}'). \tag{14.10}$$

(ii) Technically, this implies that a genuine Kramers-Krönig relation exists between the cause ($\mathbf{A}_T^{\mathrm{ext}}$) and the response ($\mathbf{J}$). (iii) $\mathbf{A}_T^{\mathrm{ext}}$ is a vectorial field which, at least in principle, can be controlled by the experimenter. (iv) A later field quantization of $\mathbf{A}_T^{\mathrm{ext}}$ provides us with a clear roadmap for the extension of the theory to the quantum electrodynamic (QED) level. In a sense, the Bohr Correspondence Principle connects QED for globally coherent photon distributions with a large mean number of gauge photons to the present semiclassical theory, which in turn leads to classical electrodynamics in the macroscopic limit.

14.1.3 Elastic scattering

In the space ($\mathbf{r}$)-frequency (ω) domain the causal constitutive equation [Eq. (14.9)] takes the form

$$\mathbf{J}(\mathbf{r}; \omega) = \frac{1}{2\pi} \int_{-\infty}^{\infty} \mathbf{S}^{\mathrm{CAU}}(\mathbf{r}, \mathbf{r}', \omega, -\Omega) \cdot \mathbf{A}_T^{\mathrm{ext}}(\mathbf{r}', \Omega) d\Omega d^3 r'. \tag{14.11}$$

Written in this form parts of $\mathbf{S}^{\mathrm{CAU}}(\mathbf{r}, \mathbf{r}', \omega, -\Omega)$ select various physical processes, e.g. elastic scattering, and in Raman and Brillouin scattering Stokes and antiStokes processes.

Part of the induced current density $\mathbf{J}(\mathbf{r}; \omega)$ stems from the external field of the same frequency, $\mathbf{A}_T^{\mathrm{ext}}(\mathbf{r}'; \omega)$. The related scattering is called elastic with a word borrowed from the particle kinematics of photons; see Section 14.2. We select the elastic scattering processes setting

$$\mathbf{S}^{\mathrm{CAU}}(\mathbf{r}, \mathbf{r}'; \omega, -\Omega) = 2\pi\delta(\omega - \Omega)\mathbf{S}^{\mathrm{CAU}}(\mathbf{r}, \mathbf{r}'; \Omega), \tag{14.12}$$

where $\mathbf{S}^{\mathrm{CAU}}(\mathbf{r}, \mathbf{r}'; \Omega)$ is a new causal response tensor associated with these processes. Retaining only the elastic current density

$$\mathbf{J}(\mathbf{r}; \omega) = \int_{-\infty}^{\infty} \mathbf{S}^{\mathrm{CAU}}(\mathbf{r}, \mathbf{r}'; \omega) \cdot \mathbf{A}_T^{\mathrm{ext}}(\mathbf{r}'; \omega) d^3 r', \tag{14.13}$$

one obtains from Eq. (14.3)

$$\mathbf{A}_T(\mathbf{r}; \omega) = \mathbf{A}_T^{\mathrm{ext}}(\mathbf{r}; \omega) + \int_{-\infty}^{\infty} \mathbf{R}^{\mathrm{CAU}}(\mathbf{r}, \mathbf{r}'; \omega) \cdot \mathbf{A}_T^{\mathrm{ext}}(\mathbf{r}'; \omega) d^3 r' \tag{14.14}$$

where

$$\mathbf{R}^{\mathrm{CAU}}(\mathbf{r}, \mathbf{r}'; \omega) = \mu_0 \int_{-\infty}^{\infty} \mathbf{G}_T(\mathbf{r} - \mathbf{r}''; \omega) \cdot \mathbf{S}^{\mathrm{CAU}}(\mathbf{r}'', \mathbf{r}'; \omega) d^3 r''. \tag{14.15}$$

To make oneself familiar with the physics of the formal approach offered by Eqs. (14.14) and (14.15), let us consider two borderline cases:

(i) In scattering from mesoscopic and microscopic objects the *external* field will be slowly varying in space across the object. Also the transverse propagator given by [157]

$$\mathbf{G}_T(\mathbf{R};\omega) = \frac{iq_0}{4\pi}\left\{\frac{e^{iq_0 R}}{iq_0 R}(\mathbf{U}-\mathbf{e_R}\mathbf{e_R}) - \left[\frac{e^{iq_0 R}}{(iq_0 R)^2} - \frac{e^{iq_0 R}-1}{(iq_0 R)^3}\right](\mathbf{U}-3\mathbf{e_R}\mathbf{e_R})\right\},$$

(14.16)

is slowly varying across the object. For $q_0 R \to 0$, $\mathbf{G}_T$ is proportional to R^{-1}. Placing the origo of the coordinate system at a point inside the mesoscopic object, one obtains

$$\mathbf{A}_T(\mathbf{r};\omega) = \mathbf{A}_T^{\text{ext}}(\mathbf{r};\omega) + \mu_0\mathbf{G}_T(\mathbf{r};\omega)\cdot\left[\int_{-\infty}^{\infty}\mathbf{S}^{\text{CAU}}(\mathbf{r},\mathbf{r}';\omega)d^3r'd^3r\right]\cdot\mathbf{A}_T^{\text{ext}}(\mathbf{0};\omega)$$

(14.17)

if the variations of $\mathbf{G}_T$ and $\mathbf{A}_T^{\text{ext}}$ across the object is neglected. The double integral

$$\mathbf{S}_{\text{ED-ED}}^{\text{CAU}}(\omega) = \int_{-\infty}^{\infty}\mathbf{S}^{\text{CAU}}(\mathbf{r},\mathbf{r}';\omega)d^3r'd^3r$$

(14.18)

is the so-called ED-ED response tensor, a frequency-dependent quantity describing the mesoscopic (microscopic) object as both an electric dipole (ED) absorber and emitter [141, 142].

(ii) It may sometimes be justified to make the displacement invariance approximation

$$\mathbf{S}^{\text{CAU}}(\mathbf{r},\mathbf{r}';\omega) = \mathbf{S}^{\text{CAU}}(\mathbf{r}-\mathbf{r}';\omega),$$

(14.19)

for instance in free-electron-like media, if the elastic scattering from the surface region is negligible. The approximation in Eq. (14.19) implies that $\mathbf{S}^{\text{CAU}}$ can be transformed to the wave-vector ($\mathbf{q}$) domain. Media for which $\mathbf{S}^{\text{CAU}}(\mathbf{q},\omega)$ can represent the causal response we characterize as *spatially dispersive* media. Remembering the division of the microscopic Maxwell-Lorentz current density, given in Eq. (11.17), and in the $(\omega,\mathbf{q})$-domain by

$$\mathbf{J}(\mathbf{q},\omega) = -i\omega\mathbf{P}(\mathbf{q},\omega) + i\mathbf{q}\times\mathbf{M}(\mathbf{q},\omega),$$

(14.20)

it appears that the response to magnetic fields is included as part of $\mathbf{S}^{\text{CAU}}(\mathbf{q}\neq\mathbf{0},\omega)$. In the general case the magnetic response appears via spatial derivatives of $\mathbf{S}^{\text{CAU}}(\mathbf{r},\mathbf{r}';\omega)$ [first- and higher-derivatives in both $\mathbf{r}$ and $\mathbf{r}'$]. In the spatially local limit, $\mathbf{S}^{\text{CAU}}(\mathbf{q}\to\mathbf{0},\omega)$ all magnetic effects disappear.

The induced elastic current density $\mathbf{J}(\mathbf{r},\omega)$ also can be represented as a response to the local field, $\mathbf{A}_T(\mathbf{r},\omega)$ c.f. Eq. (14.8) for $\mathbf{S}(\mathbf{r},\mathbf{r}',t,t') \Rightarrow \mathbf{S}(\mathbf{r},\mathbf{r}',t-t')$. In this approach Eq. (14.14) is replaced by

$$\mathbf{A}_T(\mathbf{r};\omega) = \mathbf{A}_T^{\text{ext}}(\mathbf{r};\omega) + \int_{-\infty}^{\infty}\mathbf{R}(\mathbf{r},\mathbf{r}';\omega)\cdot\mathbf{A}_T(\mathbf{r}';\omega)d^3r',$$

(14.21)

where $\mathbf{R}(\mathbf{r}, \mathbf{r}'; \omega)$ has the form given in Eq. (14.15) with $\mathbf{S}^{\mathrm{CAU}}(\mathbf{r}'', \mathbf{r}'; \omega)$ replaced by $\mathbf{S}(\mathbf{r}'', \mathbf{r}'; \omega)$. For a given $\mathbf{R}(\mathbf{r}, \mathbf{r}'; \omega)$, Eq. (14.21) is a 3D integral equation for the vector field $\mathbf{A}_T(\mathbf{r}; \omega)$. For a mesoscopic (or microscopic) object the spatially slow variation of $\mathbf{A}_T^{\mathrm{ext}}(\mathbf{r}'; \omega)$ across the object makes it in most cases sufficient to use a Taylor expansion of $\mathbf{A}_T$ (around $\mathbf{r} = \mathbf{0}$) in which only lowest order terms are kept will be sufficient in most cases [in Eq. (14.17) only the zeroth order was kept]. The local field, $\mathbf{A}_T(\mathbf{r}; \omega)$, in Eq. (14.21) varies rapidly across the scattering medium, and therefore an iteration procedure [with many iterations in general] usually is needed to solve Eq. (14.21). Physically, one may characterize the procedure as a (microscopic) multiple scattering scheme, as exemplified below for spatially dispersive media. An analogous procedure is well known in macroscopic elastic scattering theory in inhomogeneous media [46, 165, 166, 193].

14.1.4   Spatial dispersion subfield

In the framework of the spatial dispersion approximation all relevant relations take algebraic form in the $(\omega, \mathbf{q})$-domain. The algebraic relation in Eq. (14.5) [with $\mathbf{G}_T(\mathbf{q}, \omega)$ given by Eq. (14.6)] holds in general. The two versions of the *elastic* and *linear* constitutive relations now are

$$\mu_0 \mathbf{J}(\mathbf{q}, \omega) = \mathbf{S}(\mathbf{q}, \omega) \cdot \mathbf{A}_T(\mathbf{q}, \omega), \tag{14.22}$$

and

$$\mu_0 \mathbf{J}(\mathbf{q}, \omega) = \mathbf{S}^{\mathrm{CAU}}(\mathbf{q}, \omega) \cdot \mathbf{A}_T^{\mathrm{ext}}(\mathbf{q}, \omega), \tag{14.23}$$

where for the notational simplicity an extra factor μ_0 has been included in the definitions of the response tensor [but only in the remaining part of this subsection]. By insertion of respectively Eq. (14.22) and Eq. (14.23) in Eq. (14.5), one obtains

$$\mathbf{A}_T(\mathbf{q}, \omega) = [\mathbf{U} - \mathbf{G}_T(\mathbf{q}, \omega) \cdot \mathbf{S}(\mathbf{q}, \omega)]^{-1} \cdot \mathbf{A}_T^{\mathrm{ext}}(\mathbf{q}, \omega), \tag{14.24}$$

$$\mathbf{A}_T(\mathbf{q}, \omega) = [\mathbf{U} + \mathbf{G}_T(\mathbf{q}, \omega) \cdot \mathbf{S}^{\mathrm{CAU}}(\mathbf{q}, \omega)] \cdot \mathbf{A}_T^{\mathrm{ext}}(\mathbf{q}, \omega), \tag{14.25}$$

By combining Eqs. (14.24) and (14.25) we can express $\mathbf{S}^{\mathrm{CAU}}$ in terms of $\mathbf{S}$. Thus,

$$\mathbf{S}^{\mathrm{CAU}}(\mathbf{q}, \omega) = [\mathbf{U} - \mathbf{S}(\mathbf{q}, \omega) \cdot \mathbf{G}_T(\mathbf{q}, \omega)]^{-1} \cdot \mathbf{S}(\mathbf{q}, \omega). \tag{14.26}$$

Further insight is gained by dividing $\mathbf{S}$ and $\mathbf{S}^{\mathrm{CAU}}$ into their geometrically transverse (subscript: T) and longitudinal (subscript: L) parts, viz.,

$$\mathbf{S}^{\mathrm{CAU}}(\mathbf{q}; \omega) = S_T^{\mathrm{CAU}}(q; \omega)(\mathbf{U} - \boldsymbol{\kappa}\boldsymbol{\kappa}) + S_L^{\mathrm{CAU}}(q, \omega)\boldsymbol{\kappa}\boldsymbol{\kappa}, \tag{14.27}$$

and

$$\mathbf{S}(\mathbf{q}; \omega) = S_T(q; \omega)(\mathbf{U} - \boldsymbol{\kappa}\boldsymbol{\kappa}) + S_L(q, \omega)\boldsymbol{\kappa}\boldsymbol{\kappa}. \tag{14.28}$$

The four scalar quantities S_T^{CAU}, S_L^{CAU}, S_T and S_L are those of central importance in the subfield of spatial dispersion. It is obvious that only the transverse parts S_T^{CAU}

and S_T enter the scattering analysis presented here [see Eqs. (14.24) and (14.25)]. Since

$$\mathbf{S}(\mathbf{q},\omega) \cdot \mathbf{G}_T(\mathbf{q},\omega) = g(q,\omega)S_T(q,\omega)(\mathbf{U} - \boldsymbol{\kappa\kappa}), \tag{14.29}$$

one obtains from Eq. (14.26) the relation

$$S_T^{\text{CAU}}(q,\omega) = \frac{S_T(q,\omega)}{1 - g(q,\omega)S_T(q,\omega)}, \tag{14.30}$$

and hereafter from Eq. (14.25) the scalar relation

$$A_T(q,\omega) = \frac{A_T^{\text{ext}}(q,\omega)}{1 - g(q,\omega)S_T(q,\omega)} \tag{14.31}$$

between $A_T = (\mathbf{U}-\boldsymbol{\kappa\kappa})\cdot\mathbf{A}_T$ and $A_T^{\text{ext}} = (\mathbf{U}-\boldsymbol{\kappa\kappa})\cdot\mathbf{A}_T^{\text{ext}}$. The result in Eq. (14.31) describes the modification one obtains of the transverse external vector potential by the scattering process in the entire $(\omega,\mathbf{q})$-domain. Self-sustaining modes $(A_T^{\text{ext}}(q,\omega) \to 0)$ are obtained from the resonance condition $1 = gS_T$, giving via

$$q^2 - \left(\frac{\omega}{c}\right)^2 + i\varepsilon - S_T(q,\omega) = 0, \tag{14.32}$$

possible $\omega = \omega(q)$ relations (dispersion relations) in a complex $(\omega,\mathbf{q})$-domain.

14.2 SCATTERING MATRIX

In Section 14.1, we discussed general aspects of the semiclassical light scattering theory, paying particular attention to linear elastic scattering processes. The elastic scattering of incident $\mathbf{A}_T^{\text{ext}}(\mathbf{r},t)$ wave packets, composed of a given frequency $[\mathbf{A}(\mathbf{r};\omega)]$ distribution, can be studied by suitable superposition of the results obtained for monochromatic external fields in Subsection 14.1.3. In the wave-packet case the interaction with the charged particles of the object under study lasts for a finite time. In the light scattering process an initially free (incoming) wave packet will collide with the object, and after the process is finished a scattered free wave packet flies away. The light scattering matrix, also called the S-matrix, which here connects scattered and incident wave packet of essentially arbitrary forms, constitutes a particular (simple) part of the S-matrix in relativistic quantum electrodynamics [106, 186, 223, 261] describes the manner in which particles ($\sim$ electrons, positrons and photons) can be scattered, created and destroyed. In essence, the S-matrix relates to the solution, say in the Lorenz gauge, of the inhomogeneous wave equations for $\mathbf{A}_T$, $\mathbf{A}_L$ and Φ with the Dirac (or charged Klein-Gordon) current density as source term. The solution of the extremely difficult system of coupled non-linear field equations can in general at best be solved in perturbation theory.

In the Interaction Picture (in the following denoted by a $\sim$ on top of state vectors and operators), the state vector $\left|\tilde{\Psi}(t)\right\rangle$ satisfies the dynamical equation

$$i\hbar\frac{d}{dt}\left|\tilde{\Psi}(t)\right\rangle = \tilde{H}_I(t)\left|\tilde{\Psi}(t)\right\rangle, \tag{14.33}$$

where $\tilde{H}_I(t)[\equiv \hat{\tilde{H}}_I(t)]$ is the interaction Hamiltonian (subscript I). In the absence of interaction the state vector is constant in time. The Hermiticity of the interaction Hamiltonian ($\tilde{H}_I = \tilde{H}_I^+$) implies that the normalization of the state $\left|\tilde{\Psi}(t)\right\rangle$ is preserved over time, i.e.,

$$\left\langle \tilde{\Psi}(t) \middle| \tilde{\Psi}(t) \right\rangle = 1. \tag{14.34}$$

During the interaction the state

$$\left|\tilde{\Psi}(t)\right\rangle = \left|\tilde{\mathcal{O}}, \tilde{\Gamma}\right\rangle (t), \tag{14.35}$$

is an entangled object ($\mathcal{O}$)-field (Γ) state. Before the interaction [which we technically take as $t \to -\infty$], the state is a tensor product of the initial (i) object [$\left|\tilde{A}\right\rangle$] and field [$\left|\tilde{\Phi}_i\right\rangle$] states, that is

$$\left|\tilde{\Psi}(-\infty)\right\rangle = \left|\tilde{A}\right\rangle \otimes \left|\tilde{\Phi}_i\right\rangle. \tag{14.36}$$

After the interaction [$t \to \infty$], one has the product state

$$\left|\tilde{\Psi}(\infty)\right\rangle = \left|\tilde{B}\right\rangle \otimes \left|\tilde{\Phi}_f\right\rangle, \tag{14.37}$$

$\left|\tilde{B}\right\rangle$ and $\left|\tilde{\Phi}_f\right\rangle$ being the final (f) object and field state, respectively. The general S-matrix (operator), denoted by $\tilde{S}$, defined by

$$\left|\tilde{\Psi}(\infty)\right\rangle = \tilde{S}\left|\tilde{\Psi}(-\infty)\right\rangle, \tag{14.38}$$

thus relates $\left|\tilde{\Psi}(\infty)\right\rangle$ to $\left|\tilde{\Psi}(-\infty)\right\rangle$. If the light scattering process is elastic the object state is the same before and after the scattering process, $\left|\tilde{A}\right\rangle = \left|\tilde{B}\right\rangle$, at least up to a physically unimportant global phase factor, ignored here.

In elastic scattering, where

$$\left|\tilde{A}\right\rangle \otimes \left|\tilde{\Phi}_f\right\rangle = \tilde{S}(\left|\tilde{A}\right\rangle \otimes \left|\tilde{\Phi}_i\right\rangle), \tag{14.39}$$

a reduced (R) S-matrix

$$\tilde{S}_R = \left\langle \tilde{A}\middle| \tilde{S}\middle|\tilde{A}\right\rangle \tag{14.40}$$

describes all transitions, $\tilde{\Phi}_i \Rightarrow \tilde{\Phi}_f$, in collision theory. Thus,

$$\left|\tilde{\Phi}_f\right\rangle = \tilde{S}_R\left|\tilde{\Phi}_i\right\rangle \tag{14.41}$$

relates to scattering between many different initial, $\{\left|\tilde{i}\right\rangle\}$, and final, $\{\left|\tilde{f}\right\rangle\}$ photon states. Preparation of a specific initial state $\left|\tilde{i}\right\rangle$, and the transition amplitude to a selected final state $\left|\tilde{f}\right\rangle$ is given by the matrix element

$$S_{R,fi} = \left\langle \tilde{f}\right| \tilde{S}_R \left|\tilde{i}\right\rangle. \tag{14.42}$$

The completeness of the set of final states allows one to expand $\left|\tilde{\Phi}_f\right\rangle$ as follows:

$$\left|\tilde{\Phi}_f\right\rangle = \sum_f \left|\tilde{f}\right\rangle \left\langle \tilde{f}\middle|\tilde{\Phi}_f\right\rangle = \sum_f \left|\tilde{f}\right\rangle S_{R,fi}. \tag{14.43}$$

The inner product of Eq. (14.41) with itself, viz.,

$$\left\langle \tilde{\Phi}_f\middle|\tilde{\Phi}_f\right\rangle = \left\langle \tilde{\Phi}_i\right| \tilde{S}_R^\dagger \tilde{S}_R \left|\tilde{\Phi}_i\right\rangle \tag{14.44}$$

shows that the reduced S-matrix operator must by unitary, i.e.,

$$\tilde{S}_R^\dagger = \tilde{S}_R^{-1}, \tag{14.45}$$

in order to conserve the number of photons in elastic collision processes. This conservation of probability can be expressed as

$$\sum_f |S_{R,fi}|^2 = 1. \tag{14.46}$$

In order to calculate the general S-matrix operator one must solve Eq. (14.33), describing the time evolution of $\left|\tilde{\Psi}(t)\right\rangle$. Given $\left|\tilde{\Psi}(-\infty)\right\rangle$, the dynamical equation can be turned into the integral equation

$$\left|\tilde{\Psi}(t)\right\rangle = \left|\tilde{\Psi}(-\infty)\right\rangle - \frac{i}{\hbar} \int_{-\infty}^{t} \tilde{H}_I(t_1)\left|\tilde{\Psi}(t_1)\right\rangle dt_1. \tag{14.47}$$

Solving Eq. (14.47) by iteration gives

$$\left|\tilde{\Psi}(t)\right\rangle = \left|\tilde{\Psi}(-\infty)\right\rangle - \frac{i}{\hbar} \int_{-\infty}^{t} \tilde{H}_I(t_1)\left|\tilde{\Psi}(-\infty)\right\rangle dt_1 +$$
$$\left(\frac{-i}{\hbar}\right)^2 \int_{-\infty}^{t} \int_{-\infty}^{t_1} \tilde{H}_I(t_1)\tilde{H}_I(t_2)\left|\tilde{\Psi}(t_2)\right\rangle dt_2 dt_1 + \dots. \tag{14.48}$$

Thus, we obtain (in the limit $t \to \infty$), the following formal expression for the S-matrix

$$\tilde{S} = \sum_{n=0}^{\infty} \left(\frac{-i}{\hbar}\right)^n \int_{-\infty}^{\infty} \int_{-\infty}^{t_1} \dots \int_{-\infty}^{t_{n-1}} \tilde{H}_I(t_1)\tilde{H}_I(t_2)\dots\tilde{H}_I(t_n) dt_n \dots dt_1. \tag{14.49}$$

Although the elastic scattering S-matrix, S_R, relates "only" photon states, its calculation involves complicated time-dependent correlations between intermediate electron-photon states.

14.3 WEAKLY RELATIVISTIC INTERACTION HAMILTONIAN. ELECTRON CURRENT DENSITY

The physics behind the field-matter interaction Hamiltonian in relativistic electrodynamics is closely related to the electron localization problem, and to the role of photon localization in measurement theory, as analyzed in [158]. With special reference to photon physics in mesoscopic media (Part V) we shall here focus our attention on the interaction Hamiltonian in the weakly relativistic domain. Starting from relativistic wave mechanics Foldy and Wouthuysen [87] have provided us with a deeper insight in the physics of the electron spin and given a rigorous derivation of the weakly relativistic interaction Hamiltonian. The Foldy-Wouthuysen transformation thus leads to the following one-body expression for the electron-field interaction Hamiltonian in the weakly relativistic region, and the Coulomb gauge, [$\mathbf{r}$-representation]:

$$\hat{H}_I = -\frac{e}{2m}\left(\hat{\mathbf{p}}\cdot\hat{\mathbf{A}}_T + \hat{\mathbf{A}}_T\cdot\hat{\mathbf{p}}\right) + \frac{e^2}{2m}\hat{\mathbf{A}}_T^2 + \frac{e\hbar}{4m^2c^2}\boldsymbol{\nabla}V\times\hat{\boldsymbol{\sigma}}\cdot\hat{\mathbf{A}}_T - \frac{e\hbar}{2m}\hat{\boldsymbol{\sigma}}\cdot\boldsymbol{\nabla}\times\hat{\mathbf{A}}_T,$$

$$(14.50)$$

where $\hat{\mathbf{p}}) = (\hbar/i)\boldsymbol{\nabla}$. The two terms with the Pauli spin operator, $\hat{\boldsymbol{\sigma}}$, are absent in the nonrelativistic limit [the electron spin *is* a relativistic quantity]. The term proportional to the gradient of the Coulomb energy, $\boldsymbol{\nabla}V$, is the so-called spin-orbit term [62], and the last term involving the magnetic field operator, $\hat{\mathbf{B}} = \boldsymbol{\nabla}\times\hat{\mathbf{A}}_T$ is known as the Pauli interaction term. The role of the Pauli term has been studied in certain mesoscopic systems, e.g., in cases where the *jellium* approximation appears justified (to some extent). Despite the fact that the gradient of the Coulomb potential can be large in the surface region of a mesoscopic medium, not least in the electron tail (spill-out region), it appears that the importance of the spin-orbit coupling has not been appreciated. In bulk matter the spin-orbit coupling has been included in numerous studies.

In nonrelativistic, as well as in weakly relativistic electrodynamics it is convenient to use the Coulomb gauge because (i) the gauge photon alone enters the quantized vector potential, and (ii) the Coulomb interaction between the electrons becomes a part of the particle Hamiltonian. It should be noted, however, that the use of the Lorenz gauge with its three-photon types is convenient for the understanding of certain fundamental aspects in near-field electrodynamics [128, 151, 162].

As we have come to conclude, it is the gauge photon potential operator, $\hat{\mathbf{A}}_T$, which is the fundamental quantity in photon physics. The gauge photon dynamics is driven by the prevailing charged-particle current density operator. Thus, the dynamical equations for the two helicity species of the part of the gauge potential operator, $\hat{\mathbf{A}}_{T,t}^{(+)}$ satisfy Eqs. (3.79) in the space-time domain. The transverse current density operator, $\hat{\mathbf{J}}_T$, and the current density operator itself, $\hat{\mathbf{J}}$, are in the space-time domain related via the (time independent) transverse delta function, $\boldsymbol{\delta}_T$; see Eq. (3.40). The one-body current density operator is a Fermion [Dirac equation] or Boson [Klein-Gordon] operator, or a combination of the two. In the Coulomb gauge, where the transverse set of Maxwell operator equations is given by Eqs. (4.25)–(4.28), it is the symmetrized current density operator, $\hat{\mathbf{J}}_T = \hat{\mathbf{J}}_T^{\dagger}$, which is the source term.

Underlying the driving current density operator in the space-time domain, $\hat{\mathbf{J}}(\mathbf{r}, t)$, there is a Hermitian one-body current density operator given in the Fermion case by

$$\hat{\mathbf{j}}(\mathbf{r}) = \frac{e}{2m}[\hat{\mathbf{p}}\delta(\mathbf{r} - \mathbf{r}_e) + \delta(\mathbf{r} - \mathbf{r}_e)\hat{\mathbf{p}}] - \frac{e^2}{m}\hat{\mathbf{A}}_T\delta(\mathbf{r} - \mathbf{r}_e)$$
$$+ \frac{ie}{2m}[\delta(\mathbf{r} - \mathbf{r}_e)\hat{\mathbf{p}} \times \hat{\boldsymbol{\sigma}} - \hat{\mathbf{p}} \times \hat{\boldsymbol{\sigma}}\delta(\mathbf{r} - \mathbf{r}_e)], \tag{14.51}$$

where $\mathbf{r}_e$ denotes the Fermion ($\sim$ electron $(e < 0)$] position vector. The current density $\mathbf{J}(\mathbf{r})$ from $\hat{\mathbf{j}}(\mathbf{r})$ is gauge invariant, and the operator is here divided into a part $-(e^2/m)\hat{\mathbf{A}}_T\delta(\mathbf{r} - \mathbf{r}_e)$ and the rest. This division is particularly useful because the part containing $\hat{\mathbf{A}}_T$ in itself leads to a gauge-invariant contribution [named the diamagnetic (dia)], $\mathbf{J}^{\text{dia}}$ [157]. The expression for $\hat{\mathbf{j}}(\mathbf{r})$ follows in the usual manner from an extension of the classical Hamiltonian to the operator level. A comprehensive discussion of the gauge transformation related to the mechanical momentum operator $\hat{\boldsymbol{\Pi}} = (\hbar/i)\boldsymbol{\nabla} - e\hat{\mathbf{A}}$, and this transformation's influence on the current density operator in view of unitary transformations of the state vector, is given in [157]. In the nonrelativistic limit the spin contribution to $\hat{\mathbf{j}}(\mathbf{r})$ vanishes, of course.

The one-body operators in Eqs. (14.50) and (14.51) are of great importance also for many-body systems. Let us now show this in the fermion case, using the second-quantized formalism. We start form the general one-body fermion operator (matrix)

$$\hat{\mathcal{O}}(\mathbf{r}_1, ..., \mathbf{r}_N, t) = \sum_{i=1}^{N} \hat{\mathcal{O}}(\mathbf{r}_i, t) \tag{14.52}$$

belonging to a system of N Fermions having spatial coordinates $\mathbf{r}_i$ $(i = 1 - N)$. As indicated $\hat{\mathcal{O}}$ is allowed to depend on time, and for our spin-1/2 system it can be represented by a 2 x 2 matrix. In second quantized form the operator takes the form

$$\hat{\mathcal{O}}(t) = \int \hat{\boldsymbol{\Psi}}^{\dagger}(\mathbf{r}) \cdot \hat{\mathcal{O}} \cdot \hat{\boldsymbol{\Psi}}(\mathbf{r})d^3r, \tag{14.53}$$

where $\hat{\boldsymbol{\Psi}}$ and $\hat{\boldsymbol{\Psi}}^{\dagger}$ are the Fermion field operator and its Hermitian conjugate. In the present case, the field operator is a two-component spinor i.e.,

$$\hat{\boldsymbol{\Psi}}(\mathbf{r}) = \begin{pmatrix} \hat{\Psi}_{\uparrow}(\mathbf{r}) \\ \hat{\Psi}_{\downarrow}(\mathbf{r}) \end{pmatrix}, \tag{14.54}$$

where the arrows denote the spin-up ($\uparrow$) and spin-down ($\downarrow$) states. In component form, Eq. (14.53) is given by

$$\hat{\mathcal{O}}(t) = \sum_{\sigma,\sigma'} \int \hat{\Psi}_{\sigma'}^{\dagger}(\mathbf{r}) \cdot \hat{\mathcal{O}}_{\sigma',\sigma} \cdot \hat{\Psi}_{\sigma}(\mathbf{r})d^3r, \tag{14.55}$$

where σ and σ' are the spin state labels. In terms of the single-particle wave functions $\psi_{\mathbf{k},\sigma}$ [with space and spin quantum numbers $\mathbf{k}$ and σ] the spinor components of the

Fermion field operators are

$$\hat{\Psi}_\sigma(\mathbf{r}) = \sum_{\mathbf{k}} \psi_{\mathbf{k},\sigma}\hat{b}_{\mathbf{k},\sigma}, \quad \sigma = \uparrow \text{ or } \downarrow, \tag{14.56}$$

$$\hat{\Psi}^\dagger_\sigma(\mathbf{r}) = \sum_{\mathbf{k}} \psi^*_{\mathbf{k},\sigma}\hat{b}^\dagger_{\mathbf{k},\sigma}, \quad \sigma = \uparrow \text{ or } \downarrow. \tag{14.57}$$

The Fermion annihilation $(\hat{b}_{\mathbf{k},\sigma})$ and creation $(\hat{b}^\dagger_{\mathbf{k},\sigma})$ operators satisfy the anti-commutator relation

$$\left\{\hat{b}_{\mathbf{k},\sigma}, \hat{b}^\dagger_{\mathbf{k}',\sigma'}\right\} = \delta_{\mathbf{k},\mathbf{k}'}\delta_{\sigma,\sigma'}, \tag{14.58}$$

the remaining anti-commutator relations being zero. In a number of important cases (exemplified in Part V, e.g.) the single-particle states are conveniently chosen as free electron states (plane-wave states).

By combining Eqs. (14.55)–(14.57) the general one-body Fermion operator takes the form

$$\hat{O}(t) = \sum_{\mathbf{k},\mathbf{k}',\sigma,\sigma'} \langle \mathbf{k}',\sigma'|\hat{O}_{\sigma',\sigma}|\mathbf{k},\sigma\rangle \hat{b}^\dagger_{\mathbf{k}',\sigma'}\hat{b}_{\mathbf{k},\sigma}. \tag{14.59}$$

The expression in Eq. (14.59) can be applied to $\hat{H}_I$ [Eq. (14.50)] and $\hat{\mathbf{j}}$ [Eq. (14.51)]. In these cases various combinations of $\{\hat{a}_{\mathbf{q},s}\}$, $\{\hat{a}^\dagger_{\mathbf{q},s}\}$, $\{\hat{b}_{\mathbf{k},s}\}$ and $\{\hat{b}^\dagger_{\mathbf{k},s}\}$ appear in relation to the dynamical evolution of the electromagnetic field.

CHAPTER 15

Density Matrix Formalism. Resolvent Operator

15.1 STATISTICAL MIXTURE OF STATES. ENSEMBLE AVERAGE

A quantum state of a system is seldom perfectly determined (known), and the preparation of a wished pure state constitutes in a general a complicated task. In fact, a preparation of such a state is newer perfect. The *incomplete information* we possess about the state of a system enters the quantum formalism in a convenient manner in terms of the density operator, $\hat{\rho}(t)$. Let us assume that the system with probability P_i is in the quantum state $|\psi_i\rangle$ of a set $\{|\psi_i\rangle\}$. Then, the hermitian density operator is defined by

$$\hat{\rho} \equiv \sum_i P_i |\psi_i\rangle \langle\psi_i| \ [= \hat{\rho}^\dagger]. \tag{15.1}$$

The various states, $|\psi_i\rangle$, need not constitute a complete set, nor is it necessary that the assumed normalized states are mutually orthogonal. The probabilities satisfy

$$0 \leq P_i \leq 1, \ \sum_i P_i = 1 \ [\Rightarrow \sum_i P_i^2 \leq 1]. \tag{15.2}$$

If the system is in a specific state, say $|\psi_j\rangle$, we have $P_j = 1$, and the density operator simplifies to

$$\hat{\rho} = |\psi_j\rangle \langle\psi_j|. \tag{15.3}$$

A system satisfying Eq. (15.3) is said to be in a *pure* state. *If* the set $\{|\psi_i\rangle\}$ consists of orthonormal states $[\langle\psi_i|\psi_j\rangle = \delta_{ij}]$ one has $\hat{\rho}|\psi_i\rangle = P_i|\psi_i\rangle$, and the off diagonal matrix elements $\langle\psi_j|\hat{\rho}|\psi_i\rangle$, $i \neq j$ obviously vanish.

   In a given orthonormal basis $\{|n\rangle\}$, always satisfying the closure relation $\sum_n |n\rangle \langle n| = \mathbb{1}$, and with

$$|\psi_i\rangle = \sum_n |n\rangle \langle n|\psi_i\rangle \equiv \sum_n c_{ni} |n\rangle, \tag{15.4}$$

DOI: 10.1201/9781003029458-15

the density operator has the following expression:

$$\hat{\rho} = \sum_i P_i \sum_{n,m} |n\rangle \langle n|\psi_i\rangle \langle\psi_i|m\rangle \langle m| = \sum_i P_i \sum_{n,m} c_{ni} c_{mi}^* |n\rangle \langle m|. \tag{15.5}$$

The often used name "density matrix" refers to the circumstance that in applications one works with the various matrix elements

$$\langle n| \hat{\rho} |m\rangle = \sum_i P_i c_{ni} c_{mi}^*, \tag{15.6}$$

in the chosen basis. In Section V we shall exemplify the use of the density matrix in various cases. For the trace (Tr) of the density matrix one has (in an arbitrary basis)

$$\mathrm{Tr}\hat{\rho} \equiv \sum_n \langle n| \hat{\rho} |n\rangle = \sum_{n,i} P_i \langle n|\psi_i\rangle \langle\psi_i|n\rangle = \sum_i P_i < \psi_i \left(\sum_n |n\rangle \langle n| \right) \psi_i >$$

$$= \sum_i P_i \langle\psi_i|\psi_i\rangle = \sum_i P_i = 1. \tag{15.7}$$

Since $\hat{\rho}$ is Hermitian, there must exist a basis, say $\{|q\rangle\}$, in which $\hat{\rho}$ is diagonal [This is the case if the set $\{|\psi_i\rangle\}$ defining $\hat{\rho}$ in Eq. (15.1) is a complete orthonormal set, as we known]. The trace of $\hat{\rho}$ is of course independent of the chosen basis. In the diagonal basis for $\hat{\rho}$, $\hat{\rho}|q\rangle = P_q|q\rangle$, and thus $\hat{\rho}^2|q\rangle = P_q^2|q\rangle$. In turn,

$$\mathrm{Tr}\{\hat{\rho}^2\} = \sum_q \langle q| \hat{\rho}^2 |q\rangle = \sum_q P_q^2 \le 1. \tag{15.8}$$

For a pure state $\mathrm{Tr}\{\hat{\rho}^2\} = \mathrm{Tr}\{\hat{\rho}\} = 1$. It is easy to show that the relation $\hat{\rho}^2 = \hat{\rho}$ also is sufficient to ensure that the state in question is a pure state.

If $\hat{A}$ is an observable [Hermitian operator, which eigenspectrum is complete], the *ensemble average* of $\hat{A}[= \hat{A}^\dagger]$ is defined by

$$< \hat{A} > \equiv \sum_i P_i \langle\psi_i| \hat{A} |\psi_i\rangle, \tag{15.9}$$

the expectation value of $\hat{A}$ being $\langle\psi_i| \hat{A} |\psi_i\rangle$ in state $|\psi_i\rangle$. Let us now consider the trace of the operator

$$\hat{\rho}\hat{A} = \sum_i P_i |\psi_i\rangle \langle\psi_i| \hat{A}. \tag{15.10}$$

In the arbitrary basis $\{|n\rangle\}$ one obtains using the closure theorem $[\sum |n\rangle \langle n| = \mathbb{1}]$

$$\mathrm{Tr}\{\hat{\rho}\hat{A}\} = \sum_{i,n} P_i \langle n|\psi_i\rangle \langle\psi_i| \hat{A} |n\rangle = \sum_i P_i < \psi_i \left(\sum_n |n\rangle \langle n| \right) \hat{A}\psi_i >= \sum_i P_i \langle\psi_i| \hat{A} |\psi_i\rangle.$$

$$\tag{15.11}$$

A comparison of Eqs. (15.9) and (15.11) gives the important result

$$< \hat{A} >= \mathrm{Tr}\{\hat{\rho}\hat{A}\}[= \mathrm{Tr}\{\hat{A}\hat{\rho}\}]. \tag{15.12}$$

For a pure state $|\psi\rangle$, $\mathrm{Tr}\{\hat{\rho}\hat{A}\} = \langle\psi| \hat{A} |\psi\rangle$, the usual expectation value of a quantum state.

15.2 LIOUVILLE EQUATION AND ITS INTEGRAL FORM. TIME DEVELOPMENT OF ENSEMBLE AVERAGE

15.2.1 Integral equation for the evolution operator in the interaction representation

The Interaction Representation is particularly suited to study the solutions of the time-dependent Schrödinger, Dirac and Klein-Gordon equations when the Hamiltonian is of the form

$$\hat{H}(t) = \hat{H}_0 + \hat{H}_I(t), \tag{15.13}$$

where $\hat{H}_I(t)$ is regarded as a perturbation, which in general depends on time. In weakly relativistic electrodynamics $\hat{H}_I(t)$ has the explicit form given in Eq. (14.50). In the problems in the present book it is assumed that $\hat{H}_0$ is independent of time, and that the eigenfunctions *in principle* are known. In practice, exact eigenfunctions and energy eigenvalues to $\hat{H}_0$ are known only for simple models. Nevertheless, extremely many analyses pivots on the division in Eq. (15.13). For a closed system of assumed noninteracting photons and electrons ($\sim$ charged particles)

$$\hat{H}_0 = \hat{H}_P + \hat{H}_F, \tag{15.14}$$

where both the particle ($\hat{H}_P$) and field ($\hat{H}_F$) parts are time independent.

State vectors, $\left|\tilde{\psi}(t)\right\rangle$, and operators, $\tilde{A}(t)$, in the Interaction Representation, denoted by a tilde ($\sim$), are obtained by the transformations

$$\left|\tilde{\psi}(t)\right\rangle = \exp\left[\frac{i}{\hbar}\hat{H}_0(t - t_0)\right]\left|\psi(t)\right\rangle, \tag{15.15}$$

$$\tilde{A}(t) = \exp\left[\frac{i}{\hbar}\hat{H}_0(t - t_0)\right]\hat{A}\exp\left[-\frac{i}{\hbar}\hat{H}_0(t - t_0)\right]. \tag{15.16}$$

At t_0 it has been required that $\left|\tilde{\psi}(t_0)\right\rangle = |\psi(t_0)\rangle$ and $\tilde{A}(t_0) = \hat{A}$. The transformations above ensure the invariance of the matrix elements and commutation relations in the Schrödinger $[|\psi(t)\rangle, \hat{A}]$ and Interaction $[\left|\tilde{\psi}(t)\right\rangle, \tilde{A}(t)]$ Pictures. The time t_0 is an arbitrary fixed reference time; c.f. the discussion of the evolution operator $\hat{U}(t, t_0)$ given in Subsection 11.5.1. By inserting the inverse transformations to those in Eqs. (15.15) and (15.16) into the time-dependent Hilbert Schrödinger equation

$$i\hbar\frac{d}{dt}|\psi(t)\rangle = (\hat{H}_0 + \hat{H}_I(t))|\psi(t)\rangle, \tag{15.17}$$

one obtains the dynamical equation

$$i\hbar\frac{d}{dt}\left|\tilde{\psi}(t)\right\rangle = \tilde{H}_I(t)\left|\tilde{\psi}(t)\right\rangle \tag{15.18}$$

in the Interaction Representation. The time evolution of the $\tilde{A}(t)$-operator is

$$i\hbar\frac{d}{dt}\tilde{A}(t) = [\tilde{A}(t), \hat{H}_0]. \tag{15.19}$$

Note that the Hamilton operator is identical in the two representations, $\tilde{H}_0 = \hat{H}_0$, since $\hat{H}_0$ and $\exp\left[\pm(i/\hbar)\hat{H}_0(t - t_0)\right]$ commute.

In the context of the time evolution of the density operator in the Interaction Representation, $\tilde{\rho}(t)$, the time evolution operator $\tilde{U}(t, t_0)$ plays a central role. In analogy with Eq. (11.64), one describes the time evolution of the state vector, $\left|\tilde{\psi}(t)\right\rangle$, from t_0 to t formally as

$$\left|\tilde{\psi}(t)\right\rangle = \tilde{U}(t, t_0)\left|\tilde{\psi}(t_0)\right\rangle. \tag{15.20}$$

By utilizing Eqs. (15.15) and (15.16) [for $\tilde{A}(t) = \tilde{U}(t, t_0)$] it appears that

$$\tilde{U}(t, t_0) = \exp\left[\frac{i}{\hbar}\hat{H}_0 t\right]\hat{U}(t, t_0)\exp\left[-\frac{i}{\hbar}\hat{H}_0 t_0\right], \tag{15.21}$$

and $\tilde{U}(t_0, t_0)[= \hat{U}(t_0, t_0)] = \hat{\mathbb{1}}$. The time evolution of $\tilde{U}(t, t_0)$ is obtained by insertion of Eq. (15.20) into Eq. (15.18). This yields the differential equation

$$i\hbar\frac{d}{dt}\tilde{U}(t, t_0) = \tilde{H}_I(t)\tilde{U}(t, t_0), \tag{15.22}$$

and hereafter an equivalent integral equation

$$\tilde{U}(t, t_0) = \mathbb{1} + \frac{i}{\hbar}\int_{t_0}^{t}\tilde{H}_I(t')\tilde{U}(t', t_0)dt', \tag{15.23}$$

with initial condition $\tilde{U}(t_0, t_0) = \mathbb{1}$. The perturbative expansion of the evolution operator next takes the form

$$\tilde{U}(t, t_0) = \mathbb{1} + \sum_{n=1}^{\infty}\tilde{U}_n(t, t_0), \tag{15.24}$$

where [Eq. (11.73) with the replacement $\hat{H} \Rightarrow \tilde{H}_I$]

$$\tilde{U}_n(t, t_0) = \left(\frac{1}{i\hbar}\right)^n\int\tilde{H}_I(\tau_n)...\tilde{H}_I(\tau_2)\tilde{H}_I(\tau_1)d\tau_n...d\tau_2 d\tau_1, \tag{15.25}$$

with $t_0 \leq \tau_1 \leq \tau_2... \leq \tau_n \leq t$.

15.2.2 Dynamical equations for the density operator. Integral equation for ensemble average

Let us return to the definition of the density operator [Eq. (15.1)], and calculate its time derivative (multiplied by $i\hbar$). Thus, with help of the time-dependent Schrödinger equation [Eq. (15.17)], we obtain

$$i\hbar\frac{d}{dt}\hat{\rho}(t) = i\hbar\sum_i P_i\left[\left(\frac{d}{dt}|\psi_i(t)\rangle\right)\langle\psi_i(t)| + |\psi_i(t)\rangle\frac{d}{dt}\langle\psi_i(t)|\right]$$

$$= \sum_i P_i\left[\hat{H}(t)|\psi_i(t)\rangle\langle\psi_i(t)| - |\psi_i(t)\rangle\langle\psi_i(t)|\hat{H}(t)\right] = \hat{H}(t)\hat{\rho}(t) - \hat{\rho}(t)\hat{H}(t),$$

$$\tag{15.26}$$

or

$$i\hbar\frac{d}{dt}\hat{\rho}(t) = [\hat{H}(t), \hat{\rho}(t)] \equiv \hat{L}(t)\hat{\rho}(t), \tag{15.27}$$

where

$$\hat{L}(t) = \hat{H}(t) - \hat{\rho}(t)\hat{H}(t)\hat{\rho}^{-1}(t). \tag{15.28}$$

The operator $\hat{L}(t)$ is known as the *Liouville operator* [263], and Eq. (15.27) is the quantum mechanical *Liouville equation*, describing the time development of the density operator in the Schrödinger representation, yet with a time-dependent Hamilton operator. The Liouville equation may be converted into an equivalent integral equation

$$\hat{\rho}(t) = \hat{\rho}(t_0) + \frac{1}{i\hbar}\int_{t_0}^{t}\hat{L}(t')\hat{\rho}(t')dt', \tag{15.29}$$

here expressed in terms of the Liouville operator. In a sense, it is obvious that the Liouville equation in the Interaction Representation must be

$$i\hbar\frac{d}{dt}\tilde{\rho}(t) = [\tilde{H}_I(t), \tilde{\rho}(t)], \tag{15.30}$$

and its integral equation form

$$\tilde{\rho}(t) = \tilde{\rho}(t_0) + \frac{1}{i\hbar}\int_{t_0}^{t}\tilde{L}(t')\tilde{\rho}(t')dt', \tag{15.31}$$

where

$$\tilde{L}(t) = \tilde{H}_I(t) - \tilde{\rho}(t)\tilde{H}_I(t)\tilde{\rho}^{-1}(t), \tag{15.32}$$

and, of course, $\tilde{\rho}(t_0) = \hat{\rho}(t_0)$. As before, a formal iterative solution of Eq. (15.31) may be set up, if wished.

In relation to the studies in Section V, where the time development of the ensemble average, $< \hat{\mathbf{j}} > (t)$ [Eq. (14.50)] is of central importance, let us finally consider the time evolution of the ensemble average of the operator $\hat{A}(t)$ in the Interaction Representation. Since the mean value of an operator is (and must be) the same in all representations, for a quantum system in a pure state, it is obvious that the same is the case for a mixed state. In consequence

$$< \tilde{A}(t) > = \mathrm{Tr}\{\tilde{\rho}(t)\tilde{A}(t)\}. \tag{15.33}$$

By inserting Eq. (15.29) into Eq. (15.33), one obtains the exact formula

$$< \tilde{A}(t) > = < \hat{A}(t_0) > + \frac{i}{\hbar}\int_{t_0}^{t}\mathrm{Tr}\{\tilde{L}(t')\tilde{\rho}(t')\tilde{A}(t)\}dt'. \tag{15.34}$$

The compact result given in Eq. (15.34) is convenient for perturbative studies of quantum electrodynamic interactions in atomic and mesoscopic systems, e.g.

15.3 THERMAL EQUILIBRIUM. KUBO FORMALISM

15.3.1 Reservoir for electron system

One of the most important applications of the density matrix formalism arises when a system is in contact with a heatbath, i.e., a reservoir in thermal equilibrium, at least in the absence of coupling to our system. A system in contact with a reservoir is what is called an open system. Is it important to include heatbath considerations when studying the (electron)dynamics of a system? Yes and no, because the importance of a heatbath depends on what we chose to consider as our system, and on the temperature of the reservoir. Consider for example a many-body solid-state system of electrons brought out of equilibrium by an external electromagnetic field. The electron system tends to relax towards equilibrium by a coupling to phonons, impurities, etc. If, say, the coupling to the phonons is the strongest coupling mechanism one may include the phonon spectrum as a part of the system. In case, this reduces the importance of the heatbath coupling. However, one must realize that an enlargement of the system's degrees of freedom makes actual system dynamical calculations difficult to carry out, in general. At low temperatures, reservoir couplings may or may not be of importance, depending on the system studied.

Let us consider a system of uncoupled photons and electrons coupled to a common reservoir. In some cases the coupling of the photons to the heatbath is most important, e.g. in single atom electrodynamic studies (e.g. in spontaneous emission), and in other cases (e.g. in solid-state and mesoscopic electrodynamics) the coupling of the electrons to the reservoir is the dominating part. In the following we shall consider the electron-reservoir coupling, assuming a fixed number of system electrons.

Let us excite the system (object) by a prescribed external electromagnetic field, $\mathbf{A}_T^{\mathrm{ext}}(\mathbf{r}, t)$, possibly described as a distribution of photon modes, and let us ignore all other photon (reservoir) modes. In this case the total Hamiltonian needed is

$$\hat{H} = \hat{H}_P + \hat{H}_I(t), \tag{15.35}$$

where $\hat{H}_I(t)$ is given by the weakly relativistic Hamiltonian in Eq. (14.50). In the absence of the external photon field, the density operator in thermal equilibrium ($\hat{\rho}_0$) is given by [12, 63, 108, 182, 185]

$$\hat{\rho}_0 = Z^{-1} \exp\left(-\beta \hat{H}_P\right), \tag{15.36}$$

where

$$Z = \mathrm{Tr}\{\exp\left(-\beta \hat{H}_P\right)\}, \tag{15.37}$$

and $\beta = (kT)^{-1}$, k being the Boltzmann constant. The operator expression given in Eq. (15.36) leads to the correct Boltzmann distribution for the P_i's of Eq. (15.1). To prove this, we take as a starting point, the energy eigenvalue equation for $\hat{H}_P$, viz.,

$$\hat{H}_P |I\rangle = E_I |I\rangle, \tag{15.38}$$

where $|I\rangle$ and E_I are the many-body electron eigenstate and eigenenergy for state I. Utilizing the completeness relation $\sum_I |I\rangle\langle I| = \mathbb{1}$ one obtains

$$\hat{\rho}_0 = \sum_I \hat{\rho}_0 |I\rangle\langle I| = \frac{1}{Z}\sum_I \exp\left(-\beta\hat{H}_P\right)|I\rangle\langle I| = \frac{1}{Z}\sum_I \exp(-\beta E_I)|I\rangle\langle I| \quad (15.39)$$

with

$$Z = \mathrm{Tr}\{\exp\left(-\beta\hat{H}_P\right)\} = \sum_I \langle I|\exp\left(-\beta\hat{H}_P\right)|I\rangle = \sum_I \exp(-\beta E_I). \quad (15.40)$$

Hence,

$$\hat{\rho}_0 = \sum_I P_I |I\rangle\langle I|, \quad (15.41)$$

where

$$P_I = \frac{\exp(-\beta E_I)}{\sum_I \exp(-\beta E_I)} \quad (15.42)$$

is the correct Boltzmann distribution (probability) factor [189].

15.3.2 Kubo formalism for the time-dependent density matrix

In the presence of $\hat{H}_I(t)$, the density operator satisfies the Liouville equation

$$i\hbar\frac{d}{dt}\hat{\rho}(t) = [\hat{H}_P + \hat{H}_I(t), \hat{\rho}(t)]. \quad (15.43)$$

The density operator, $\hat{\rho}(t)$, now deviates from $\hat{\rho}_0$ by a time-dependent amount $\hat{f}(t)$:

$$\hat{\rho}(t) = \hat{\rho}_0 + \hat{f}(t). \quad (15.44)$$

The thermal equilibrium density operator is time independent, as its Liouville equation shows, viz.,

$$i\hbar\frac{d}{dt}\hat{\rho}_0 = [\hat{H}_P, \hat{\rho}_0] = \left[\hat{H}_P, Z^{-1}\exp\left(-\beta\hat{H}_P\right)\right] = 0. \quad (15.45)$$

The Liouville equation hence take the form

$$i\hbar\frac{d}{dt}\hat{f}(t) = [\hat{H}_P, \hat{f}(t)] + [\hat{H}_I(t), \hat{\rho}_0] + i\hbar\hat{C}(t), \quad (15.46)$$

with

$$\hat{C}(t) \equiv \frac{1}{i\hbar}[\hat{H}_I(t), \hat{f}(t)]. \quad (15.47)$$

When the influence of $\hat{H}_I(t)$ on the electron system is sufficiently weak, and hence the change $\hat{\rho}_0 \Rightarrow \hat{\rho}_0 + \hat{f}(t)$ small, the second-order term $\hat{C}(t)$ may be negligible. This is the regime of linear electrodynamics [118, 157]; see also Section 15.1.

The dynamical equation for the density matrix element (from state $|J\rangle$ to the state $|I\rangle$)

$$f_{IJ}(t) = \langle I| \hat{f}(t) |J\rangle \tag{15.48}$$

is given by

$$i\hbar\frac{d}{dt}f_{IJ}(t) = \langle I| [\hat{H}_P, \hat{f}(t)] |J\rangle + \langle I| [\hat{H}_I(t), \hat{\rho}_0] |J\rangle + i\hbar C_{IJ}(t), \tag{15.49}$$

where

$$C_{IJ}(t) = \langle I| \hat{C}(t) |J\rangle. \tag{15.50}$$

A simplification of Eq, (15.49) is obtained utilizing that

$$\langle I| [\hat{H}_P, \hat{f}(t)] |J\rangle = (E_I - E_J)f_{IJ}(t), \tag{15.51}$$

[remembering that $\hat{H}_P = \hat{H}_P^\dagger$], and

$$\langle I| [\hat{H}_I(t), \hat{\rho}_0] |J\rangle = (P_J - P_I) \langle I| \hat{H}_I(t) |J\rangle, \tag{15.52}$$

since also $\hat{H}_I$ is Hermitian. Thus,

$$\frac{d}{dt}f_{IJ}(t) = \frac{i}{\hbar}(E_J - E_I)f_{IJ}(t) + \frac{i}{\hbar}(P_I - P_J) \langle I| \hat{H}_{IJ}(t) |J\rangle + C_{IJ}(t). \tag{15.53}$$

In a notation where the subscript index (I) is changed to a superscript index, we introduce the abbreviation

$$H_{I,IJ}(t) \equiv H_{IJ}^I(t) = \langle I| \hat{H}_I(t) |J\rangle. \tag{15.54}$$

With this, and the definition of $\hat{C}(t)$ [Eq. (15.47)] one obtains in two steps, first

$$\begin{aligned}
C_{IJ}(t) &= \frac{1}{i\hbar} \langle I| [\hat{H}_I(t), \hat{f}(t)] |J\rangle \\
&= \frac{1}{i\hbar} \sum_K [\langle I| \hat{H}_I(t) |K\rangle \langle K| \hat{f} |J\rangle - \langle I| \hat{f}(t) |K\rangle \langle K| \hat{H}_I(t) |J\rangle]
\end{aligned} \tag{15.55}$$

utilizing the completeness relation $\sum_K |K\rangle \langle K| = \mathbb{1}$, and then

$$C_{IJ}(t) = \frac{1}{i\hbar} \sum_K [H_{IK}^I(t)f_{KJ}(t) - f_{IK}(t)H_{KJ}^I(t)]. \tag{15.56}$$

By insertion of this expression into Eq. (15.53), one finally gets the following set of linear differential equations among the density matrix elements of the perturbation $\hat{f}(t)$:

$$\begin{aligned}
i\hbar\frac{d}{dt}f_{IJ}(t) =& (E_I - E_J)f_{IJ}(t) + (P_J - P_I)H_{IJ}^I(t) \\
&+ \sum_K [H_{IK}^I(t)f_{KJ}(t) - f_{IK}(t)H_{KJ}^I(t)].
\end{aligned} \tag{15.57}$$

In Section V, Eq. (15.57) will be employed in calculations of the mean value of the one-body current density operator $(\hat{\mathbf{j}})$ in selected mesoscopic systems, that is

$$< \hat{\mathbf{J}} > = \mathrm{Tr}\{\hat{\rho}\hat{\mathbf{j}}\}. \tag{15.58}$$

A perturbation scheme is used to obtain various linear and nonlinear (in lowest order) electrodynamic response tensors.

The Kubo formalism presented in this subsection essentially was established by Kubo in 1957-59 [172,173], in a study dealing with the linear electrical conductivity in solids. Kubo formulas are the name applied to linear correlation functions. Formulas of the Kubo-like type were first proposed by Green [103,104] for transport in liquids.

15.4 RESOLVENT PROPAGATOR

15.4.1 Mathematical digression: Analytical part of delta function

In electrodynamics one often comes across the so-called *analytical* part $[\delta^{(+)}(x)]$ of the delta $[\delta(x)]$. It is defined by [for $x \geq 0$]

$$\delta^{(+)}(x) = \frac{1}{2\pi} \lim_{\eta \to 0^+} \left[\int_0^\infty e^{(ix-\eta)t} dt \right], \quad \eta > 0, \tag{15.59}$$

In the case where $x = \omega$, we call $\delta^{(+)}(\omega)$ the positive-frequency part, or analytical part, of the monochromatic signal, $\sim \exp(i\omega t)$. The concept already has been used in many places in the previous part of this book [first in Eq. (2.26)]. When $x = k$, one speaks of a positive wave-number signal, $\sim \exp(ikx)$. By carrying out the integration one obtains

$$\delta^{(+)}(x) = \frac{i}{2\pi} \lim_{\eta \to 0^+} \left[\frac{1}{x + i\eta} \right] = \frac{i}{2\pi} \lim_{\eta \to 0^+} \left[\frac{x}{x^2 + \eta^2} - \frac{i\eta}{x^2 + \eta^2} \right] \tag{15.60}$$

The first part of Eq. (15.60) [times $2\pi/i$] is the Principal (P) value of x^{-1}, viz.,

$$\lim_{\eta \to 0^+} \left[\frac{x}{x^2 + \eta^2} \right] = P\left(\frac{1}{x}\right) \equiv \begin{cases} x^{-1} \text{ for } x \neq 0 \\ 0 \text{ for } x = 0 \end{cases}, \tag{15.61}$$

and the last part is proportional to

$$\lim_{\eta \to 0^+} \left[\frac{\eta}{x^2 + \eta^2} \right] = \pi \delta(x). \tag{15.62}$$

Hence,

$$\delta^{(+)}(x) = \frac{i}{2\pi} P(x) + \frac{1}{2}\delta(x), \quad x \geq 0. \tag{15.63}$$

15.4.2 Retarded and advanced propagators

For what follows, let us introduce the evolution operator associated with the time independent $\hat{H}_0$-operator [Eq. (15.13)], viz,

$$\hat{U}_0(t, t_0)\ [= \hat{U}_0(t - t_0)] = \exp\left[-\frac{i}{\hbar}\hat{H}_0(t - t_0)\right]\mathbb{1}, \tag{15.64}$$

[compare to Eq. (11.68), which holds when $\hat{H}$ is time independent (conservative system), and take $\hat{H} = \hat{H}_0$]. In the following we use the Schrödinger Picture, in which all dynamical observables are time independent [that is both $\hat{H}_0$ and $\hat{H}_I$]. The differential equation for the general evolution operator,

$$i\hbar\frac{d}{dt}\hat{U}(t,t_0) = [\hat{H}_0 + \hat{H}_I]\hat{U}(t,t_0), \tag{15.65}$$

can now be converted into an integral equation in which $\hat{U}_0(t,t_0)$ enters in an explicit manner. Thus,

$$\hat{U}(t,t_0) = \hat{U}_0(t,t_0) + \frac{1}{i\hbar}\int_{t_0}^{t}\hat{U}_0(t,t')\hat{H}_I\hat{U}(t',t_0)dt', \tag{15.66}$$

as the reader may verify by substitution of Eq. (15.66) into Eq. (15.65), and use afterwards use of Eq. (15.64).

The evolution operator describes forward ($t > t_0$) as well as backward ($t < t_0$) time developments of the state function $|\psi(t)\rangle$, from its reference, $|\psi(t_0)\rangle$, at the *arbitrary* time $t = t_0$. The evolution forward in time is given by the *retarded* (superscript $+$) propagator (operator)

$$\hat{K}_+(t - t_0) = \hat{U}(t - t_0)\theta(t - t_0), \tag{15.67}$$

and in the absence of $\hat{H}_I(t)$ by

$$\hat{K}_{0+}(t - t_0) = \hat{U}_0(t - t_0)\theta(t - t_0), \tag{15.68}$$

In the two equations above, $\theta(t-t_0)$ is the Heaviside unit step function. The retarded propagator obeys the differential equation

$$\left[i\hbar\frac{d}{dt} - \hat{H}\right]\hat{K}_+(t,t_0) = i\hbar\delta(t - t_0)\mathbb{1}, \tag{15.69}$$

as it is easy to verify via Eq. (15.65), and use of the fact that $d\theta(\tau)d\tau = \delta(\tau)$, and $\hat{U}(t_0,t_0) = \mathbb{1}$. Therefore $\hat{K}_+(t,t_0)$ is sometimes called a (retarded) Green function. To simplify the notation we leave out the unit operator $\mathbb{1}$ in all relevant equations (expressions) in the remaining part of Subsection 15.4.2. By multiplying Eq. (15.66) by $\theta(t - t_0)$, and since $\hat{U}(t',t_0)$ under the integral sign can be replaced by $\hat{U}(t',t_0)\theta(\theta(t'-t_0)$, remembering that $t' > t_0$, it appears that the retarded propagator satisfies the integral equation

$$\hat{K}_+(t,t_0) = \hat{K}_{0+}(t - t_0) + \frac{1}{i\hbar}\int_{-\infty}^{\infty}\hat{K}_{0+}(t - t')\hat{H}_I\hat{K}_+(t',t_0)dt'. \tag{15.70}$$

The bonus going from Eq. (15.66) [for $t > t_0$] to Eq. (15.70) is associated with the circumstance that the limits of integration are extended to $\pm\infty$. This allows one to use the folding theorem on the convolution product of $\hat{K}_{0+}(t - t')$ and $\hat{H}_I\hat{K}_+(t',t_0)$.

Before employing the folding theorem, we introduce the Fourier integral transform of $\hat{K}_+(\tau)$ [and $\hat{K}_{0+}(\tau)$], $\tau = t - t_0$. It follows from the integral of the folding type in Eq. (15.70) that $\hat{K}_+$ actually depends only on τ; $\hat{K}_+(t,t_0) = \hat{K}_+(t-t_0)$. With a convenient prefactor $i/(2\pi)$, let us take

$$\hat{K}_+(\tau) = \frac{i}{2\pi} \int_{-\infty}^{\infty} \hat{G}_+(E) \exp\left(-\frac{i}{\hbar}E\tau\right) dE, \tag{15.71}$$

a form which in a sense gives the most simple explicit expression for $\hat{G}_+(E)$, as one can see starting by inverting Eq. (15.71). With $\omega = E/\hbar$,

$$\hat{K}_+(\tau) = \frac{1}{2\pi} \int_{-\infty}^{\infty} i\hbar\hat{G}_+(E)e^{-i\omega t}d\omega, \tag{15.72}$$

inversion gives

$$\hat{G}_+(E) = \frac{1}{i\hbar} \int_{-\infty}^{\infty} \hat{K}_+(\tau) \exp\left(\frac{i}{\hbar}E\tau\right) d\tau. \tag{15.73}$$

Since

$$\hat{K}_+(\tau) = \exp\left[-\frac{i}{\hbar}\hat{H}\tau\right] \theta(\tau), \tag{15.74}$$

it appears that formally

$$\hat{G}_+(E) = \frac{1}{i\hbar} \int_0^{\infty} \exp\left[\frac{i}{\hbar}(E-\hat{H})\tau\right] d\tau. \tag{15.75}$$

The mathematical digression on the analytical part of the delta function [Subsection 15.4.1] now tells us that the integral in Eq. (15.75) has to be calculated by regularization, that is

$$\begin{aligned}
\hat{G}_+(E) &= \lim_{\eta\to 0^+} \left[\frac{1}{i\hbar} \int_0^{\infty} \exp\left[\frac{i}{\hbar}(E-\hat{H}+i\eta)\tau\right] d\tau\right] \\
&= \lim_{\eta\to 0^+} \frac{1}{E-\hat{H}+i\eta} \left[= \frac{2\pi}{i}\delta^{(+)}(E-\hat{H})\right]
\end{aligned} \tag{15.76}$$

Advanced propagators (with subscript $-$) are introduced as follows:

$$\hat{K}_-(t,t_0) = -\hat{U}(t,t_0)\theta(t_0-t), \tag{15.77}$$

$$\hat{K}_{0-}(t,t_0) = -\hat{U}_0(t,t_0)\theta(t_0-t). \tag{15.78}$$

The $\hat{K}_-(t,t_0) = \hat{K}_-(t-t_0)$ operator obeys the same evolution equation as $\hat{K}_+(t,t_0)$, viz,

$$\left[i\hbar\frac{d}{dt} - \hat{H}\right] \hat{K}_-(t,t_0) = i\hbar\delta(t-t_0). \tag{15.79}$$

With the Fourier transform $\hat{G}_-(E)$ defined by

$$\hat{K}_-(\tau) = \frac{i}{2\pi} \int_{-\infty}^{\infty} \hat{G}_-(E) \exp\left[-\frac{i}{\hbar}E\tau\right] dE, \tag{15.80}$$

one ends up with

$$\hat{G}_-(E) = \lim_{\eta \to 0^+} \frac{1}{E - \hat{H} - i\eta}, \tag{15.81}$$

as similar calculations as those leading to $\hat{G}_+(E)$ [Eq. (15.76)] show. Note the different signs in front of η for $\hat{G}_+(E)$ and $\hat{G}_-(E)$.

Remembering at this point that the integral in Eq. (15.70) is a folding integral [with $\hat{H}_I$ independent of τ: Schrödinger Picture], the Fourier transform of the integral equation for $\hat{K}_+(t, t_0)$ becomes an algebraic operator equation

$$\hat{G}_+(E) = \hat{G}_{0+}(E) + \hat{G}_{0+}(E)\hat{H}_I\hat{G}_+(E). \tag{15.82}$$

The "free" retarded propagator $\hat{G}_{0+}(E)$ is given by

$$\hat{G}_{0+}(E) = \lim_{\eta \to 0^+} \frac{1}{E - \hat{H}_0 + i\eta}; \tag{15.83}$$

c.f. Eq. (15.76).

15.4.3 Resolvent of the Hamiltonian. Iterative solution for resolvent

The evolution operator $\hat{U}(t, t_0) = \hat{U}(t - t_0) \equiv \hat{U}(\tau)$, can be expressed in terms of $\hat{G}_+(E)$ and $\hat{G}_-(E)$. Since $\theta(\tau) + \theta(-\tau) = 1$, one immediately sees from Eqs. (15.67) and (15.77) that

$$\hat{U}(\tau) = \hat{K}_+(\tau) - \hat{K}_-(\tau). \tag{15.84}$$

Hence one yields the Fourier integral decomposition

$$\hat{U}(\tau) = \frac{1}{2\pi i} \int_{-\infty}^{\infty} [\hat{G}_-(E) - \hat{G}_+(E)] \exp\left(-\frac{i}{\hbar}E\tau\right) dE. \tag{15.85}$$

The simple forms of

$$\hat{G}_\pm(E) \equiv \lim_{\eta \to 0^+} \hat{G}_\pm(E \pm i\eta) = \lim_{\eta \to 0^+} \frac{1}{E - \hat{H} \pm i\eta} \tag{15.86}$$

suggest that further insight may be obtained by contour integration technique [in a complex z-plane, say]. Let us therefore introduce the so-called *resolvent* of the Hamiltonian by the definition

$$\hat{G}(z) = \frac{1}{z - \hat{H}}. \tag{15.87}$$

With reference to the contour in Fig. 15.1, it appears that the evolution operator can be represented in the compact form

$$\hat{U}(\tau) = \frac{1}{2\pi i} \int_{C_+ + C_-} \hat{G}(z) \exp\left(-\frac{i}{\hbar}z\tau\right) dz. \tag{15.88}$$

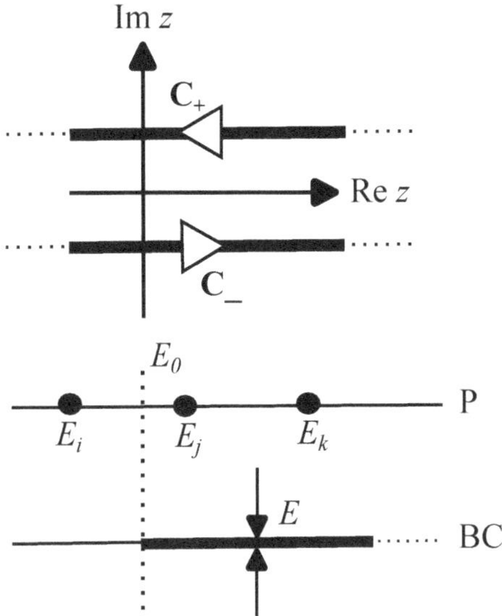

Figure 15.1 Top figure: Integration contour $(C_+ + C_-)$ for $\hat{U}(\tau)$ in a complex z-plane. For $\tau > 0$ $(\tau < 0)$ the contribution of $C_-(C_+)$ is zero. Bottom figure; Discrete eigenvalues $(...E_i, E_j, E_k, ...)$, [Poles (P)], and continuous eigenvalue spectrum, [Branch cut (BC) starting from E_0], of the Hamilton operator $(\hat{H})$. As indicated by the vertical arrows, the eigenvalues E are different on the two sides of the cut.

The straight lines C_+ and C_- are located immediately $(\eta \to 0^+)$ above (C_+) and below (C_-) the real axis, and "running" right $\to$ left (C_+) and left $\to$ right (C_-), as shown in Fig. 15.1. For $\tau > 0$ $(\tau < 0)$ one must use only C_+ (C_-) in contour analyses. The singularities of the matrix elements of $\hat{G}(z)$ are all on the real axis. To exemplify this consider an arbitrary normalized state $|w\rangle$; $\langle w|w \rangle = 1$. In this state

$$\langle w| \hat{G}(z) |w\rangle = \langle w| \frac{1}{z - \hat{H}} |w\rangle, \tag{15.89}$$

and if $|u\rangle$ is an eigenstate of $\hat{H}$ with eigenvalue E_u, we have

$$\langle u| \hat{G}(z) |u\rangle = \frac{1}{z - E_u}. \tag{15.90}$$

The singularities of $\hat{G}(z)$, all located on the real axis, are poles (the discrete eigenvalues of $\hat{H}$) and branch cuts (the continuous spectrum of $\hat{H}$ eigenvalue); see Figure 15.1. Outside the real axis, all matrix elements of $\hat{G}(z)$ are analytic functions of z.

15.5 CORRELATED SYSTEMS: REDUCED DENSITY OPERATORS

An important application of the density operator formalism relates to the description of the interaction between two physical systems A and B. Let the Hamiltonian of the composed system (AB) be

$$\hat{H}_{AB} = \hat{H}_A + \hat{H}_B + \hat{V}, \tag{15.91}$$

where $\hat{V}$ represents the interaction between A and B. As a typical case one may think of a situation where the two systems are uncorrelated [with Hamiltonians $\hat{H}_A$ and $\hat{H}_B$] for times $t < t_0$. At $t = t_0$ the interaction [with coupling Hamiltonian $\hat{V}$] is turned on, and the systems become correlated. In line with the general definition given in Eq. (15.1), let the density operator of the correlated AB system be

$$\hat{\rho}_{AB} = \sum_{\{AB\}} P_{AB} \, |\psi_{AB}\rangle \, \langle\psi_{AB}| \,, \tag{15.92}$$

where P_{AB} is the probability that the correlated system is in the quantum state $|\psi_{Ab}\rangle$, the summation ($\{AB\}$) being over all composed system states. When the systems are uncorrelated [superscript: uncorr]

$$|\psi_{AB}^{uncorr}\rangle = |\psi_A\rangle \otimes |\psi_B\rangle \,, \tag{15.93}$$

that is, the spectrum [taking as the energy eigenvalue spectrum] is tensor product states of states belonging to A and B, $|\psi_A\rangle$ and $|\psi_B\rangle$. The probability P_{AB}^{uncorr} is the product of the probabilities for the uncorrelated systems to be in the states $|\psi_A\rangle$ and $|\psi_B\rangle$:

$$P_{AB}^{\text{uncorr}} = P_A P_B. \tag{15.94}$$

In the absence of correlations

$$\hat{\rho}_{AB} = \sum_{A,B} P_A P_B [|\psi_A\rangle \otimes |\psi_B\rangle][\langle\psi_A| \otimes \langle\psi_B|] = \hat{\rho}_A \otimes \hat{\rho}_B. \tag{15.95}$$

In the correlated case $\hat{\rho}_{AB} \neq \hat{\rho}_A \otimes \hat{\rho}_B$.

In the presence of interactions between A and B it is useful often to introduce so-called reduced operators by the definitions

$$\hat{\rho}_A \equiv \text{Tr}_B \hat{\rho}_{AB}; \quad \hat{\rho}_B \equiv \text{Tr}_A \hat{\rho}_{AB}, \tag{15.96}$$

where the Tr_A (Tr_B) stands for the trace over the states of the uncorrelated A (B) systems. For uncorrelated systems

$$\hat{\rho}_A = \text{Tr}_B \{\hat{\rho}_A \otimes \hat{\rho}_B\} = \hat{\rho}_A \otimes \text{Tr}_B \hat{\rho}_B, \tag{15.97}$$

and since $\text{Tr}_B \hat{\rho}_B = 1$ [Eq. (15.7)], one regains the density operator for A. An analogous conclusion holds for $\hat{\rho}_B = \text{Tr}_A \hat{\rho}_{AB}$, of course.

Let us finally study the time development of the reduced density operator $\hat{\rho}_A(t)$. Starting with the Liouville equation for the composite system, i.e.,

$$i\hbar \frac{d}{dt} \hat{\rho}_{AB} = [\hat{H}_A + \hat{H}_B + \hat{V}, \, \hat{\rho}_{AB}], \tag{15.98}$$

we have

$$i\hbar \frac{d}{dt} \hat{\rho}_A = i\hbar \text{Tr}_B \left(\frac{d}{dt} \hat{\rho}_{AB} \right) = \text{Tr}_B [\hat{H}_A + \hat{H}_B + \hat{V}, \hat{\rho}_{AB}]. \tag{15.99}$$

Now we use the following results:

$$\mathrm{Tr}_B[\hat{H}_A, \hat{\rho}_{AB}] = [\hat{H}_A, \mathrm{Tr}_B\hat{\rho}_{AB}] = [\hat{H}_A, \hat{\rho}_A], \tag{15.100}$$

$$\mathrm{Tr}_B[\hat{H}_B, \hat{\rho}_{AB}] = \sum_B \langle\psi_B| \hat{H}_B\hat{\rho}_{AB} - \hat{\rho}_{AB}\hat{H}_B |\psi_B\rangle = 0 \tag{15.101}$$

To obtain the last equation, the eigenvalue equation $\hat{H}_B |\psi_B\rangle = E_B |\psi_B\rangle$ was employed. Therefore,

$$i\hbar\frac{d}{dt}\hat{\rho}_A = [\hat{H}_A, \hat{\rho}_A] + \mathrm{Tr}_B[\hat{V}, \hat{\rho}_{AB}], \tag{15.102}$$

and by the same procedure as above, an analogous Liouville equation for $\hat{\rho}_B(t)$ is obtained. In the interaction representation Eq. (15.102) is transformed into

$$i\hbar\frac{d}{dt}\tilde{\rho}_A(t) = \mathrm{Tr}_B[\tilde{V}(t), \tilde{\rho}_{AB}(t)], \tag{15.103}$$

emphasizing explicitly the time dependence of the various operators. If wished Eq. (15.103) can be written in integral form, and solved iteratively, noting that $\hat{\rho}_{AB}(t_0) = \hat{\rho}_A(t_0) \otimes \hat{\rho}_B(t_0)$.

CHAPTER 16

Microscopic Extinction Theorems

16.1 EWALD-OSEEN FIELD EXTINCTION IN ELECTRODYNAMICS

We have already met a simplified version of the vacuum extinction theorem in electromagnetic in Section 6.1, and made use of the theorem in relation to the study of the mean-field operator for the transverse vector potential operator (Section 6.2). In the present section we shall extend the theorem for the case where microscopic current density distributions are present in parts of space. We shall limit ourselves to the establishment of the microscopic extinction theorem in semiclassical electrodynamics, although its extension to the field-quantized level poses no insurmountable difficulties. The theorem was first formulated by Ewald [79, 80] in 1912 in his basic investigations on the foundations of crystal optics. Starting from molecular optics Oseen [199] gave in 1915 a form of the extinction theorem which in a sense is quite close to its form in microscopic electrodynamics [46]. A generalized extinction theorem was established by Wolf in a paper in which he put forward a new hypothesis as to the true meaning of the theorem: A nonlocal boundary condition for the solution of the equation of motion for the interior scattering problem [276]. The formulation established below is called the *microscopic Ewald-Oseen extinction theorem*. The microscopic form of the theorem was established by the present author in a work dealing with both linear and nonlinear aspects of the theorem [133]. In [157] a detailed account of the microscopic extinction theorem is presented, and the theorem's relation to a so-called coupled-antenna description of microscopic electrodynamics [152] is emphasized. Molecular optics goes beyond the macroscopic Maxwell theory, and may be considered as a simple microscopic classical field theory. The microscopic extinction theorem for the transverse part of the vector potential is derived starting from the inhomogeneous wave equation

$$(\nabla^2 + q_0^2)\mathbf{A}_T(\mathbf{r};\omega) = -\mu_0 \mathbf{J}_T(\mathbf{r};\omega) \qquad (16.1)$$

and the differential equation for the Huygens propagator [Eq. (6.2)]

$$(\nabla^2 + q_0^2)g(R;\omega) = -\delta(\mathbf{R}). \qquad (16.2)$$

DOI: 10.1201/9781003029458-16

A calculation analogous to the one given in Section 6.1 shows that Eqs. (6.10) and (6.12) [semiclassical version] are replaced by

$$\mathbf{A}_T(\mathbf{r};\omega) = \mu_0 \int_{V_T} g(R;\omega)\mathbf{J}_T(\mathbf{r}';\omega)d^3r' + \boldsymbol{\Sigma}_T(\mathbf{r};\omega),\ \ \mathbf{r} \in V_T,\ \ \mathbf{r}' \in V_T \qquad (16.3)$$

and

$$\mathbf{0} = \mu_0 \int_{V_T} g(R;\omega)\mathbf{J}_T(\mathbf{r}';\omega)d^3r'\ \ \mathbf{r} \notin V_T,\ \ \mathbf{r}' \in V_T. \qquad (16.4)$$

The volume of integration, V_T, has such a (minimum) size that the entire transverse current density distribution is contained inside V_T. The point of observation ($\mathbf{r}$) is either inside [Eq. (16.3)] or outside [Eq. (16.4)] V_T. The closed surface of V_T is denoted by Σ_T and, as in Section 6.1

$$\boldsymbol{\Sigma}_T(\mathbf{r};\omega) = \int^{\Sigma_T} \left[g(R;\omega)\frac{\partial \mathbf{A}_T(\mathbf{r}';\omega)}{\partial n'} - \mathbf{A}_T(\mathbf{r}';\omega)\frac{\partial g(R;\omega)}{\partial n'}\right] dS'. \qquad (16.5)$$

In a typical experimental situation a source with a prescribed transverse current density distribution $\mathbf{J}_T^0(\mathbf{r};\omega)$ contained in a volume V_T^0, excites a system contained in a volume V_T and gives rise to a *self-consistent* transverse current density distribution $\mathbf{J}_T(\mathbf{r};\omega)$. The vacuum domain outside V_T and V_T^0 is denoted by $\tilde{V}_T$, for what follows. The surfaces enclosing V_T and V_T^0 are named Σ_T and Σ_T^0, respectively. The non-simply connected vacuum domain is closed by an imaginary outer surface Σ^∞, see Fig. 16.1. We let Σ^∞ recede towards infinity ["observed" from every point in the V_T and V_T^0 domains]. In the limit "$\Sigma^\infty \to \infty$" there is no surface integral contribution from Σ^∞ [$\int^{\Sigma^\infty}(...)dS' \to 0$]. The outwards directed unit normal vectors from V_T and V_T^0 are denoted by $\hat{\mathbf{n}}(\mathbf{r})$ so that $\hat{\mathbf{n}} \cdot \boldsymbol{\nabla} \equiv \partial/\partial n$. If Green's theorem is applied to the vacuum domain, two equations analogous to Eqs. (16.3) and (16.4) are obtained. Remembering that we now have two surfaces Σ_T and Σ_T^0 in play and that the *outwards directed* unit normal vector from $\tilde{V}_T$ is $-\mathbf{n}(\mathbf{r})$, the explicit relations take the forms

$$\mathbf{A}_T(\mathbf{r};\omega) = -\boldsymbol{\Sigma}_T(\mathbf{r};\omega) - \boldsymbol{\Sigma}_T^0(\mathbf{r};\omega),\ \ \mathbf{r} \in \tilde{V}_T, \mathbf{r}' \in \tilde{V}_T \qquad (16.6)$$

$$\mathbf{0} = -\boldsymbol{\Sigma}_T(\mathbf{r};\omega) - \boldsymbol{\Sigma}_T^0(\mathbf{r};\omega),\ \ \mathbf{r} \notin \tilde{V}_T, \mathbf{r}' \in \tilde{V}_T. \qquad (16.7)$$

A current density contribution, similar to the one in Eq. (16.3) does not appear in Eq. (16.6) since $\mathbf{J}(\mathbf{r};\omega) = \mathbf{0}$ in the transverse vacuum region. Integration over the source domain immediately gives

$$\mathbf{A}_T(\mathbf{r};\omega) = \mu_0 \int_{V_t^0} g(R;\omega)\mathbf{J}_T^0(\mathbf{r}';\omega)d^3r' + \boldsymbol{\Sigma}_T^0(\mathbf{r};\omega),\ \ \mathbf{r} \in V_T^0,\ \ \mathbf{r}' \in V_T^0, \qquad (16.8)$$

$$\mathbf{0} = \mu_0 \int_{V_t^0} g(R;\omega)\mathbf{J}_T^0(\mathbf{r}';\omega)d^3r' + \boldsymbol{\Sigma}_T^0(\mathbf{r};\omega),\ \ \mathbf{r} \notin V_T^0,\ \ \mathbf{r}' \in V_T^0. \qquad (16.9)$$

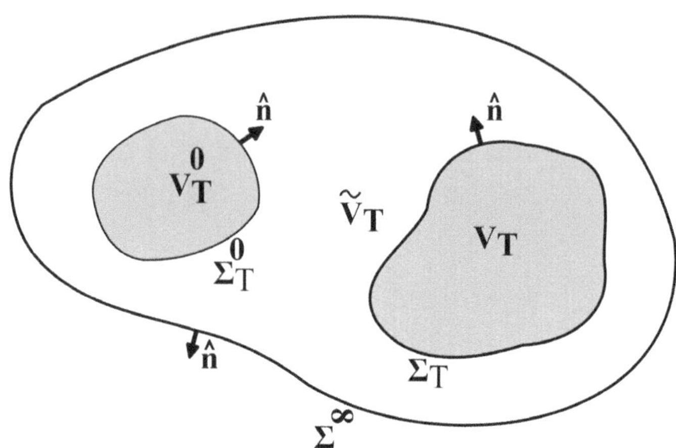

Figure 16.1 Schematic illustration of the three-domain structure used in a microscopic extinction theorem analysis of the transverse vector potential ($\mathbf{A}_T$). The prescribed source current density ($\mathbf{J}_T^0$) distribution and the self-consistent system current density ($\mathbf{J}_T$) distribution are contained in the volumes V_T^0 and V_T, respectively. The surfaces enclosing these volumes are denoted by Σ_T^0 and Σ_T. The non-simply connected vacuum domain is denoted by $\tilde{V}_T$. The outer surface Σ^∞ closes the vacuum domain. Letting this surface recede towards infinity the surface integral contribution of Σ^∞ tends to zero. The relevant normal unit vectors ($\hat{\mathbf{n}}$) used are indicated by black arrows.

When the system in V_T is excited by the field from a prescribed source current density (located in V_T^0), the central problem (goal) in all cases is a determination of the self-consistent vector potential $\mathbf{A}_T(\mathbf{r};\omega)$ in V_T. How does this problem appear in the microscopic extension of the Ewald-Oseen approach? With focus on the field in V_T, Eqs. (16.3), (16.7) and (16.9) are the relevant ones. The term

$$\mathbf{A}_T^0(\mathbf{r};\omega) = \mu_0 \int_{V_T^0} g(R;\omega)\mathbf{J}_T^0(\mathbf{r};\omega)d^3r', \quad \mathbf{r} \notin V_T^0, \tag{16.10}$$

obviously can be identified as the transverse source field outside the source region. From Eq. (16.9) it then appears that this field can be obtained from a knowledge of $\mathbf{A}_T^0$ on the Σ_T^0 surface, i.e.,

$$\mathbf{A}_T^0(\mathbf{r};\omega) = -\Sigma_T^0(\mathbf{r};\omega) = \int^{\Sigma_T^0} \left[\mathbf{A}_T^0(\mathbf{r}';\omega)\frac{\partial g(R;\omega)}{\partial n'} - g(R;\omega)\frac{\partial \mathbf{A}_T^0(\mathbf{r}';\omega)}{\partial n'} \right] dS', \quad \mathbf{r} \notin V_T^0. \tag{16.11}$$

By means of Eq. (16.7) it therefore follows that the transverse field form the *prescribed* source current density at points *inside* the system under study is given by

$$\Sigma_T(\mathbf{r};\omega) = \mathbf{A}_T^0(\mathbf{r};\omega), \quad \mathbf{r} \in V_T, \tag{16.12}$$

where

$$\boldsymbol{\Sigma}_T(\mathbf{r};\omega) = \int^{\Sigma_T} \left[g(R;\omega)\frac{\partial \mathbf{A}_T^0(\mathbf{r}';\omega)}{\partial n'} - \mathbf{A}_T^0(\mathbf{r}';\omega)\frac{\partial g(R;\omega)}{\partial n'} \right] dS', \quad \mathbf{r} \in V_T, \quad (16.13)$$

Eq. (16.13) is an example of the microscopic Ewald-Oseen extinction theorem. The word *extinction* dates back to the formulation of the theorem in molecular optics, where it was asserted that the incident field is extinguished by those molecular dipoles that are situated on the surface (boundary) of the dielectric medium in considera-tion. From a physical point view, this historical interpretation cannot be upheld in a rigorous microscopic framework.

In the case of elastic scattering Eq. (16.3) is equivalent to Eq. (14.21) [with $\mathbf{A}_T^{\text{ext}} \equiv \mathbf{A}_T^0$]. With $\mathbf{A}_T^0(\mathbf{r}';\omega)$ given by Eq. (16.11), the basic integral equation for the prevailing transverse vector potential inside the medium under study takes the following form:

$$\mathbf{A}_T(\mathbf{r};\omega) = \int_{V_T} \mathbf{R}(\mathbf{r},\mathbf{r}';\omega) \cdot \mathbf{A}_T(\mathbf{r}';\omega)d^3r'$$
$$+ \int^{\Sigma_T} \left[g(R,\omega)\frac{\partial \mathbf{A}_T^0(\mathbf{r}';\omega)}{\partial n'} - \mathbf{A}_T^0(\mathbf{r}';\omega)\frac{\partial g(R;\omega)}{\partial n'} \right] dS'. \qquad (16.14)$$

Let us denote an arbitrary point on the Σ_T-surface by $\mathbf{r}_S$. In the limit where the point of observation $(\mathbf{r})$ approach $\mathbf{r}_s$ $[\mathbf{r} \to \mathbf{r}_S]$ $\mathbf{A}_T(\mathbf{r} \to \mathbf{r}_S;\omega) = \mathbf{A}_T^0(\mathbf{r}_S;\omega)$, i.e.,

$$\mathbf{A}_T^0(\mathbf{r}_S;\omega) = \int_{V_T} \mathbf{R}(\mathbf{r}_S,\mathbf{r}';\omega) \cdot \mathbf{A}_T(\mathbf{r};\omega)d^3r'$$
$$+ \int^{\Sigma_T} \left[g(|\mathbf{r}_s - \mathbf{r}'|;\omega)\frac{\partial \mathbf{A}_T^0(\mathbf{r}';\omega)}{\partial n'} - \mathbf{A}_T^0(\mathbf{r}';\omega)\frac{\partial g(|\mathbf{r}_S - \mathbf{r}'|;\omega)}{\partial n'} \right] dS'.$$
$$(16.15)$$

Seen in the Ewald-Oseen perspective, the extinction principle appears as a kind of spatially nonlocal boundary condition on the homogeneous integral equation

$$\mathcal{A}_T^0(\mathbf{r}_S;\omega) = \int_{V_T} \mathbf{R}(\mathbf{r}_S,\mathbf{r}';\omega) \cdot \mathbf{A}_T(\mathbf{r}';\omega)d^3r', \qquad (16.16)$$

where

$$\mathcal{A}_T^0(\mathbf{r}_S;\omega) = \mathbf{A}_T^0(\mathbf{r}_S;\omega) + \int^{\Sigma_T} \left[\mathbf{A}_T^0(\mathbf{r}';\omega)\frac{\partial g(|\mathbf{r}_S - \mathbf{r}'|;\omega)}{\partial n'} - g(|\mathbf{r}_S - \mathbf{r}'|;\omega)\frac{\partial \mathbf{A}_T^0(\mathbf{r}';\omega)}{\partial n'} \right].$$
$$(16.17)$$

In the framework of a macroscopic approach, in which polarization and mag-netization vector fields $\mathbf{P}$ and $\mathbf{M}$ appear, a generalization of the original Ewald-Oseen extinction (and its role in scattering theory) has been given by Pattanayak and Wolf [202]; see also [276]. Due to the fact that macroscopic jump (boundary) conditions are needed at the surface of the medium being investigated, the quantity $\boldsymbol{\Sigma}_T(\mathbf{r};\omega)$ depends on from which side the limiting value of the field is taken In the rigorous microscopic approach, where $\mathbf{P}$ and $\mathbf{M}$ are microscopic quantities related to $\mathbf{J}$ via Eq. (11.17) there is no jump in $\boldsymbol{\Sigma}_T$ when Σ_T is crossed.

16.2 EXTINCTION THEOREM FOR THE COMPLEX KLEIN-GORDON FIELD

The extinction theorem we have obtained for electromagnetic scattering has an analogue in quantum-mechanical potential scattering. In the nonrelativistic regime the extinction theorem related to the Schrödinger equation has been discussed by Wolf [276], and recently employed in a study of electron scattering from a mesoscopic hole in a jellium screen [120]. In the present section we shall establish an extinction theorem for the complex Klein-Gordon field. This field describes the wave mechanics of electrically charged spin-0 particles, and below we shall concentrate on the coupling of those scalar particles to the electromagnetic field.

16.2.1 Free Klein-Gordon equation

The relativistic wave equation for a spinless scalar particle of mass m is obtained from the (squared) energy (E) and momentum $(\mathbf{p})$ relation,

$$E^2 = (cp)^2 + (mc^2)^2, \tag{16.18}$$

by substituting differential operators for E and $\mathbf{p}$ in the manner standard in quantum theory, viz.,

$$E \to i\hbar\frac{\partial}{\partial t}, \quad \mathbf{p} \to -i\hbar\boldsymbol{\nabla}. \tag{16.19}$$

The resulting second-order differential equation for the scalar wave function, $\phi(x)$, is the Klein-Gordon equation. Written in covariant notation it has the form

$$(\partial_\mu\partial^\mu - q_C^2)\phi(x) = 0, \tag{16.20}$$

where $q_C = mc/\hbar$ is the Compton wave number, and $\partial_\mu\partial^\mu = \square$ [Eq. (2.33)].

Let us write (in three-vector notation) the Klein-Gordon equation in the symbolic form

$$\left(\frac{i}{c}\frac{\partial}{\partial t} + \sqrt{-\nabla^2 + q_C^2}\right)\left(\frac{i}{c}\frac{\partial}{\partial t} - \sqrt{-\nabla^2 + q_C^2}\right)\phi(\mathbf{r}, t) = 0. \tag{16.21}$$

The operator $\sqrt{-\nabla^2 + q_C^2}$ is defined by the manner in which it acts in wave-vector $(\mathbf{q})$ space. Thus,

$$\sqrt{-\nabla^2 + q_C^2}\phi(\mathbf{r}, t) \equiv (2\pi)^{-3}\int_{-\infty}^{\infty}(q^2 + q_C^2)^{1/2}\phi(\mathbf{q}, t)e^{i\mathbf{q}\cdot\mathbf{r}}d^3q. \tag{16.22}$$

Note that for $q_C = 0$, the definition in Eq. (16.22) is identical to the one used in our discussion of the massless scalar and longitudinal photon wave mechanics; see Sections 2.4 and 2.5 [in particular Eq. (2.71)]. The form in Eq. (16.21) shows that every solution to the first-order equation

$$\left(\frac{i}{c}\frac{\partial}{\partial t} - \sqrt{-\nabla^2 + q_C^2}\right)\phi(\mathbf{r}, t) = 0 \tag{16.23}$$

is a solution also to the Klein-Gordon equation. In the $(\mathbf{q}; t)$-domain Eq. (16.23) takes the form

$$\left[\frac{i}{c}\frac{\partial}{\partial t} - (q^2 + q_C^2)^{1/2}\right]\phi(\mathbf{q}; t) = 0, \tag{16.24}$$

and by insertion of a stationary-state solution, $\phi(\mathbf{q}; t) = \phi(\mathbf{q})\exp(-iE_+t/\hbar)$, into this equation it appears that the eigenstate energy is positive and given by

$$E_+ = +\sqrt{(c\hbar q)^2 + (mc^2)^2}. \tag{16.25}$$

Since $\hbar q = p$, E_+ is the energy obtained by Eq. (16.18) using a plus sign in front of square root.

If one writes the symbolic operator factors in Eq. (16.21) in opposite order, it appears that all solutions to

$$\left(\frac{i}{c}\frac{\partial}{\partial t} + \sqrt{-\nabla^2 + q_C^2}\right)\phi(\mathbf{r}, t) = 0 \tag{16.26}$$

are solutions also to the Klein-Gordon equation. Furthermore, it is obvious that the energy E of the eigenstate $\phi(\mathbf{q}; t) = \phi(\mathbf{q})\exp(-iE_-t/\hbar)$ is negative and given by

$$E_- = -\sqrt{(c\hbar q)^2 + (mc^2)^2}, \tag{16.27}$$

i.e. precisely the solution to Eq. (16.18) with a minus sign in front of the square root.

The considerations lead to the conclusion that the solutions to the free Klein-Gordon equation can be divided into two groups in which the members have positive and negative eigenenergies, respectively.

In order to obtain further insight in the physical interpretation of the Klein-Gordon equation, Eq. (16.20) is multiplied by $\phi^*(x)$, and the resulting equation thereafter rewritten in the form

$$\partial^\mu(\phi^*\partial_\mu\phi) - (\partial^\mu\phi^*)(\partial_\mu\phi) - q_C^2\phi^*\phi = 0, \tag{16.28}$$

remembering that $\partial_\mu\partial^\mu = \partial^\mu\partial_\mu$. Similarly, the complex conjugate Klein-Gordon equation is multiplied by $\phi(x)$ giving

$$\partial^\mu(\phi\partial_\mu\phi^*) - (\partial^\mu\phi)(\partial_\mu\phi^*) - q_C^2\phi\phi^* = 0, \tag{16.29}$$

Subtracting Eq. (16.29) from Eq. (16.28) one obtains

$$\partial^\mu j_\mu(x) = 0, \tag{16.30}$$

where

$$j_\mu(x) = i[\phi^*(x)\partial_\mu\phi(x) - \phi(x)\partial_u\phi^*(x)]. \tag{16.31}$$

[we remind the reader that $\{\partial_\mu\} = \{\partial/\partial x^\mu\} = (\frac{1}{c}\frac{\partial}{\partial t}, \boldsymbol{\nabla})$ $\{j_\mu\} = (-j_0, \mathbf{j})$ with the choice used in this book for the metric tensor]. The multiplicate i-factor in Eq.(16.31) makes $j_\mu(x)$ real.

The form of Eq. (16.30) *might* suggest that $\{j_\mu(x)\}$ is interpreted as a *probability* four-current density However,

$$j^0[=?c\rho] = i(\phi^*\partial^0\phi - \phi\partial^0\phi^*), \tag{16.32}$$

divided by c cannot be considered as a probability density because ρ is not necessarily positive definite, because the Klein-Gordon equation is not of first order in the time derivative. This means that one must know both ϕ and $\partial\phi/\partial t$ at a particular time in order to determine the wave function at later times, However, as we shall soon see, this is *to some extent* only an apparent problem and will lead to a satisfactory reinterpretation of the wave function in the framework of *relativistic quantum mechanics*. A problem is still left: Negative energy states (as for the Dirac equation). If the Klein-Gordon is used as basis for a *quantum field theory*, then the negative energy states can be reinterpreted as positive energy states of antiparticles. Below we shall stick to the first-quantized form of the theory.

In any case, we know that the differential form in Eq. (16.30) *is* correct. We now integrate Eq. (16.30 over all space (below taken as a box of volume $V = L^3$, with periodic boundary conditions), and use Gauss' theorem

$$\frac{\partial}{\partial t}\int_V j_0 d^3r = -\int_V \boldsymbol{\nabla}\cdot\mathbf{j}d^3r = -\int_{S\to``\infty"}\mathbf{j}\cdot d\mathbf{S} = 0. \tag{16.33}$$

The surface integral vanishes at $S \to$ "∞". The quantity

$$Q \equiv i\int_V (\phi^*\partial_0\phi - \phi\partial_0\phi^*)d^3r, \tag{16.34}$$

which one *defines* as the charge of the given Klein-Gordon particle, is a conserved quantity. As we shall see, the positive energy solutions have positive norms, $Q > 0$, and the negative energy solutions have negative norms, $Q < 0$. The *complex Klein-Gordon equation* hence describes the quantum mechanics of charged spin-0 particles, and the wave function relates to the *charge* density and current density. A real wave function ($\phi^* = \phi$) describes the relativistic wave mechanics of electrically neutral spin-0 particles. For such particles $\{j_\mu\} = 0$.

The free Klein-Gordon equation obviously has plane-wave solutions and these form a complete set. In box normalization ($V = L^3$) with periodic boundary conditions these solutions are

$$\phi_n^\pm(\mathbf{r}, t) = N\exp\left[i\left(\mathbf{q}_n\cdot\mathbf{r} \mp \frac{E_n}{\hbar}t\right)\right], \tag{16.35}$$

where $E_n \equiv E_{+,n} = +\sqrt{(c\hbar q_n)^2 + (mc^2)^2}$ is always positive. The superscript $(\pm)$ refers to the positive $(+)$ and negative $(-)$ energy solutions. The wave vector can take on one of the values

$$\mathbf{q}_n = \frac{2\pi}{L}(n_x, n_y, n_z), \quad n_i = 0, \pm1, \pm2, ... \tag{16.36}$$

Since

$$\frac{i}{c}\int_{L^3}\left[\phi_n^{(\pm)*}\frac{\partial}{\partial t}\phi_m^{(\pm)} - \phi_m^{(\pm)}\frac{\partial}{\partial t}\phi_n^{(\pm)*}\right]d^3r = \pm\frac{E_n}{c\hbar}L^3|N|^2\delta_{nm}, \qquad (16.37)$$

it appears that the norms $(n = m)$ of the positive $[+ : Q > 0]$ and negative $[- : Q < 0]$ solutions have opposite signs. With the choice $N = [\hbar c/(2E_n L^3)]^{1/2}$ the two types of solutions each will be orthonormalized:

$$\phi_n^{(\pm)}(\mathbf{r}, t) = \left(\frac{\hbar c}{2E_n L^3}\right)^{1/2}\exp\left[i\left(\mathbf{q}_n\cdot\mathbf{r}\mp\frac{E_n}{\hbar}t\right)\right]. \qquad (16.38)$$

Note also (i)

$$\frac{E_n}{\hbar c} = [q_n^2 + q_C^2]^{1/2}, \qquad (16.39)$$

and (ii)

$$\frac{i}{c}\int_{L^3}\left[\phi_n^{(+)*}\frac{\partial}{\partial t}\phi_m^{(-)} - \phi_m^{(-)}\frac{\partial}{\partial t}\phi_n^{(+)*}\right]d^3r = 0. \qquad (16.40)$$

Eq. (16.40) shows that the invariant scalar products (defined using the zeroth component of the conserved current) of members from different groups are orthogonal. The quantity in Eq. (16.39) appears in integrations over $\mathbf{q}$-space, where (in continuum quantization) $d^3q/\sqrt{q^2 + q_C^2}$ is the relativistically invariant volume element [158,261].

16.2.2 Electromagnetic coupling

The charged Klein-Gordon equation is modified in its interaction with an electromagnetic field. The coupled Klein-Gordon equation is obtained making the minimal coupling substitution [Eq. (2.56)]

$$\partial_\mu \to D_\mu(\equiv \nabla_\mu^G) = \partial_\mu - \frac{ie}{\hbar}A_\mu \qquad (16.41)$$

in Eq, (16.20), This gives

$$(\partial_\mu\partial^\mu - q_C^2)\phi(x) = \hat{V}(x)\phi(x) \qquad (16.42)$$

where

$$\hat{V}(x) = \frac{ie}{\hbar}[A_\mu(x)\partial^\mu + \partial_\mu A^\mu(x)] + \left(\frac{e}{\hbar}\right)^2 A_\mu(x)A^\mu(x) \qquad (16.43)$$

is the "generalized potential operator" [or "squared Compton operator"]. Remembering that $\partial_\mu A^\mu$ operates as follows: $\partial_\mu(A^\mu\phi) = (\partial_\mu A^\mu)\phi + A^\mu\partial_\mu\phi$, one can write $\hat{V}(x)$ in the alternative form

$$\hat{V}(x) = \frac{ie}{\hbar}[2A_\mu(x)\partial^\mu + (\partial^\mu A_\mu(x))] + \left(\frac{e}{c\hbar}\right)^2 A_\mu(x)A^\mu(x). \qquad (16.44)$$

In the Lorenz gauge $\partial^\mu A_\mu = 0$.

The minimal coupling principle changes the free current density [Eq. (16.31)] to

$$j_\mu(x) = i[\phi^*(x)D_\mu\phi(x) - \phi(x)D_\mu^*\phi^*(x)]. \tag{16.45}$$

Note that the presence of D_μ^* guarantees that $j_\mu(x)$ is real. If wished, the reader may verify Eq. (16.45) starting from $\phi^*D_\mu D^\mu\phi - \phi D_\mu^* D^{\mu*}\phi^* = 0$, and following analogous steps to those leading to Eq. (16.31). By insertion of the expression for D_μ [Eq. (16.41)], Eq. (16.45) becomes

$$j_\mu(x) = i[\phi^*(x)\partial_\mu\phi(x) - \phi(x)\partial_\mu\phi^*(x)] + \frac{2e}{\hbar}A_\mu(x)\phi(x)\phi^*(x). \tag{16.46}$$

The presence of the $A_\mu(x)$-dependent part of $j_\mu(x)$ ensures that the current density is gauge invariant [for a detailed discussion of this aspect, see e.g. [4]].

Since $\partial^\mu j_\mu(x) = 0$, it follows in analogy to Eq. (16.34) that

$$\int_{L^3} \left\{ i\left[\phi^*(x)\partial_0\phi(x) - \phi(x)\partial_0\phi^*(x)\right] + \frac{2e}{\hbar}A_0(x)|\phi(x)|^2 \right\} d^3r = \text{constant.} \tag{16.47}$$

Thus, the conserved norm involves the time component of the four-potential.

16.3 ELECTROMAGNETIC $\phi^{(+)} \Leftrightarrow \phi^{(-)}$ COUPLING

16.3.1 Meson propagator

The coupling between positive and negative solutions to the Klein-Gordon equation, as well as the extinction theorem for the complex field, we shall now study on the basis of propagator theory. For the Klein-Gordon equation the meson propagator (also called the scalar Feynman (F) propagator), $\Delta_F(x - x')$, is defined to satisfy the equation [157, 186]

$$(\partial_\mu\partial^\mu - q_C^2)\Delta_F(x - x') = -\delta(x - x'). \tag{16.48}$$

To investigate the structure of this propagator, it is useful to look for Δ_F's wave-vector ($\sim$ momentum) – space representation. The formal Fourier transform of the configuration space propagator is (with $X = x - x'$)

$$\Delta_F(X) = \frac{1}{(2\pi)^4} \int_{-\infty}^{\infty} \Delta_F(q)\exp(iq_\mu x^\mu)d^4q. \tag{16.49}$$

Substituting this Fourier integral into Eq. (16.48), and remembering that the transform of $\delta(X)$ equals 1, give

$$\Delta_F(q) = [q_\mu q^\mu + q_C^2]^{-1} = \left[\mathbf{q}\cdot\mathbf{q} + q_C^2 - \left(\frac{\omega}{c}\right)^2\right]^{-1}. \tag{16.50}$$

It appears that $\Delta_F(q)$ is singular for [with $q^2 = \mathbf{q}\cdot\mathbf{q}$]

$$E_\pm[= \hbar\omega_\pm] = \pm\sqrt{(c\hbar q)^2 + m^2c^4} = \pm E, \tag{16.51}$$

which means that the integral in Eq. (16.49) only makes sense when evaluated as a contour integral In three-vector notation $[q_0 \equiv \omega/c]$, $\mathbf{R} = \mathbf{r} - \mathbf{r}'$, $\tau = t - t'$]

$$\Delta_F(\mathbf{R}, \tau) = \frac{1}{(2\pi)^4} \int_{-\infty}^{\infty} \left\{ \int_C \frac{\exp[i(\mathbf{q} \cdot \mathbf{R} - cq_0\tau]}{q^2 + q_C^2 - q_0^2} dq_0 \right\} d^3q, \qquad (16.52)$$

where we (for physical reasons) have chosen a contour (C) in the complex q_0 plane. We perform the q_0-integration by closing the contour with a semicircle at infinity either in the upper $(\tau = t - t' < 0)$ or lower (for $\tau = t - t' > 0$) half plane. The residue calculation gives, remembering that

$$q^2 + q_C^2 - q_0^2 = \left(\frac{E}{c\hbar} - q_0\right)\left(\frac{E}{c\hbar} + q_0\right), \qquad E > 0, \qquad (16.53)$$

the following result:

$$\frac{1}{2\pi} \int_C \frac{e^{-icq_0\tau}}{q^2 + q_C^2 - q_0^2} dq_0 = \frac{c\hbar i}{2E} \left[\theta(\tau) \exp\left(-\frac{iE}{\hbar}\tau\right) + \theta(-\tau) \exp\left(\frac{iE}{\hbar}\tau\right)\right]. \qquad (16.54)$$

The meson propagator

$$\Delta_F(\mathbf{R}, \tau) = \theta(\tau)\Delta_F^{(+)}(\mathbf{R}, \tau) + \theta(-\tau)\Delta_F^{(-)}(\mathbf{R}, \tau), \qquad (16.55)$$

hence is the sum of a positive energy ($\sim$ frequency) part

$$\Delta_F^{(+)}(\mathbf{R}, \tau) = \frac{ic\hbar}{2} \int_{-\infty}^{\infty} \frac{1}{E} \exp\left[i\left(\mathbf{q} \cdot \mathbf{R} - \frac{E}{\hbar}\tau\right)\right] \frac{d^3q}{(2\pi)^3}, \qquad (16.56)$$

for propagation *forward* in time, $t > t'$, and a negative energy ($\sim$ frequency) part

$$\Delta_F^{(-)}(\mathbf{R}, \tau) = \frac{ic\hbar}{2} \int_{-\infty}^{\infty} \frac{1}{E} \exp\left[i\left(\mathbf{q} \cdot \mathbf{R} + \frac{E}{\hbar}\tau\right)\right] \frac{d^3q}{(2\pi)^3}, \qquad (16.57)$$

for propagation *backwards* in time. For $q_C \to 0$, $\Delta_F(\mathbf{R}, \tau)$ is the massless Feynman photon propagator. Note that $(c\hbar/E)d^q3 = [q^2 + q_C^2]^{-1/2}d^3q$ is the Lorentz-invariant volume element in $\mathbf{q}$-space [[158, 261]]. *In quantum field theory, the present negative-energy particle moving backwards in time, becomes a positive-energy antiparticle moving forward in time.*

16.3.2 Integral equations for $\phi^{(+)}$ and $\phi^{(-)}$ for particles

A combination of Eqs. (16.42) and (16.48) allows one to obtain the following integral for the complex Klein-Gordon field:

$$\phi(x) = \phi_0(x) - \int \Delta_F(X)\hat{V}(x')\phi(x')d^4x', \qquad (16.58)$$

where $X = x - x'$. The integration extends over the space-time domain where the electromagnetic coupling is nonvanishing. The function $\phi_0(x)$ is an arbitrary solution to the free Klein-Gordon equation When the scattering problem under investigation is specified the physically relevant $\phi_0(x)$-field can be selected.

Let us assume that a particle with positive charge, and given by a wave packet $\phi_0^{(+)}(x)$ is incident on the scattering domain. The analytic part of Eq. (16.58), viz,

$$\phi^{(+)}(x) = \phi_0^{(+)}(x) - \int \Delta_F^{(+)}(X)\hat{V}(x')[\phi^{(+)}(x') + \phi^{(-)}(x')]d^4x', \qquad (16.59)$$

describes this scattering problem, remembering that $\Delta_F^{(+)}(X)$ is the positive-frequency part of the meson propagator. Although $\Delta_F^{(+)}$ selects the analytic part of $\hat{V}(x')\phi(x')$, this does not imply that only the positive-frequency part of $\phi(x')$ appears in the integral of Eq. (16.59). As the form of the interaction operator $\hat{V}(x)$ [Eq. (16.43)] shows, there will be a cross-coupling from negative to positive frequency wave-packets unless $\hat{V}(x)\phi^{(-)}(x) = 0$. The cross-coupling vanishes if $\hat{V}(x)$ is time independent since

$$c\int \Delta_F^{(+)}(\mathbf{r}, t - t')\hat{V}(\mathbf{r})\phi^{(-)}(t')dt' = 0, \qquad (16.60)$$

as the folding theorem shows. However, even in the perhaps simplest case, viz. so-called Coulomb scattering from a fixed scalar potential, $\Phi = cV^0$, where

$$\hat{V} = -\frac{2ie\hbar}{c^2}\Phi\frac{\partial}{\partial t} - \left(\frac{e}{\hbar c}\right)^2 \Phi^2, \qquad (16.61)$$

one has a time dependence of the $\hat{V}$-operator. The form in Eq.(16.61) is used in studies of pair creation [106]. In the absence of an incident $\phi^{(-)}$-wave packet, one obtains

$$\phi^{(-)}(x) = -\int \Delta_F^{(-)}(X)\hat{V}(x')[\phi^{(+)}(x') + \phi^{(-)}(x')]d^4x'. \qquad (16.62)$$

Eqs. (16.61) and (16.62) constitute a set of coupled integral equations for $\phi^{(+)}$ and $\phi^{(-)}(x)$, equations to be solved for a *prescribed* incident Klein-Gordon field $\phi^{(0)}(x)$. In *symbolic* notation, the solutions are given as

$$\phi^{(+)} = \hat{\mathcal{O}}\phi_0^{(+)}, \qquad (16.63)$$

and

$$\phi^{(-)} = -[\hat{1} + (\Delta_F^{(-)}\hat{V})^{-1}]^{-1}\hat{\mathcal{O}}\phi_0^{(+)}, \qquad (16.64)$$

where

$$\hat{\mathcal{O}} = \left\{\hat{1} + (\Delta_F^{(+)}\hat{V}) - (\Delta_F^{(+)}\hat{V})[\hat{1} + (\Delta_F^{(-)}\hat{V})]^{-1}\right\}^{-1}. \qquad (16.65)$$

In the so-called Klein paradox, [95, 106], the Coulomb interaction, with $\hat{V}$ given by the form in Eq. (16.63), one studies the reflection and transmission of a $\phi_0^{(+)}$ wave packet [e.g., a meson $(\pi^{(+)})$ particle composed of a quark (q) and antiquark $(\bar{q})$ pair, surrounded by a "sea" of gluons and other $q\vec{q}$ pairs] from a Coulomb step. If the barrier is sufficiently high $[e\Phi > E+mc^2]$, $\phi^{(+)}$ (particle)-$\phi^{(-)}$ (antiparticle) pairs are created at the barrier edge. As a consequence, it turns out the amount of positive charge (and energy) of the reflected $\phi^{(+)}$-packet exceeds the positive charge (and energy) of the incoming $\phi_0^{(+)}$-packet. This apparent paradox dissolves itself due to the fact that the created $\phi^{(-)}$-packet propagates forwards *inside* the barrier with negative charge and energy. Both charge and energy are thus conserved in this example studied by Klein [170].

16.4   SCALAR EXTINCTION THEOREM

The technique used to establish the Ewald-Oseen extinction theorem for the transverse part of the vector potential, can be used also for the Klein-Gordon field. Thus, from Eqs. (16.42) and (16.48), one obtains the following extinction theorem in the $(\mathbf{r};\omega)$-domain:

$$\psi_0(\mathbf{r};\omega) + \Sigma(\mathbf{r};\omega) = 0, \quad \mathbf{r} \in V, \tag{16.66}$$

for points inside the scattering volume (V). The incident field, $\psi_0(\mathbf{r};\omega)$, *prescribed* by the primary source is extinguished by the contribution from the surface integral

$$\Sigma(\mathbf{r};\omega) = \int^{\Sigma} \left[\Delta_F(\mathbf{R};\omega)\frac{\partial\phi(\mathbf{r}';\omega)}{\partial n'} - \phi(\mathbf{r}';\omega)\frac{\partial\Delta_F(\mathbf{R};\omega)}{\partial n'} \right] dS', \tag{16.67}$$

with $\mathbf{R} = \mathbf{r} - \mathbf{r}'$. In the type of study, which as textbook paradigm has the Klein example, the extinction theorem for the positively charge field takes the form

$$\psi_0^{(+)}(\mathbf{r};\omega) + \int^{\Sigma} \left\{ \Delta_F^{(+)}(\mathbf{R};\omega)\frac{\partial}{\partial n'}[\phi^{(+)}(\mathbf{r}';\omega) + \phi^{(-)}(\mathbf{r}';\omega)] \right.$$
$$\left. -[\phi^{(+)}(\mathbf{r}';\omega) + \phi^{(-)}(\mathbf{r}';\omega)]\frac{\partial}{\partial n'}\Delta_F^{(+)}(\mathbf{R};\omega) \right\} dS' = 0, \ \mathbf{r} \in V. \tag{16.68}$$

In the absence of an incoming negatively charged field part one obtains

$$\int^{\Sigma} \left\{ \Delta_F^{(-)}(\mathbf{R};\omega)\frac{\partial}{\partial n'}[\phi^{(+)}(\mathbf{r}';\omega) + \phi^{(-)}(\mathbf{r}';\omega)] \right.$$
$$\left. -[\phi^{(+)}(\mathbf{r}';\omega) + \phi^{(-)}(\mathbf{r}';\omega)]\frac{\partial}{\partial n'}\Delta_F^{(-)}(\mathbf{R};\omega) \right\} dS' = 0, \ \mathbf{r} \in V. \tag{16.69}$$

In cases where the particle-antiparticle generation is of importance, the extinction theorem has the form of a set of two coupled equations for $\phi^{(+)}$ and $\phi^{(-)}$.

In the case of a sufficiently high Coulomb barrier at the surface of the scattering volume, one may use the approximation $\phi^{(+)}(\mathbf{r};\omega) = 0$ in the scattering domain $[\mathbf{r} \in V]$. Thus, Eq. (16.68) is reduced to

$$\psi_0^{(+)}(\mathbf{r};\omega) + \int^{\Sigma} \left[\Delta_F^{(+)}(\mathbf{R};\omega)\frac{\partial\phi^{(-)}(\mathbf{r}';\omega)}{\partial n'} - \phi^{(-)}(\mathbf{r}';\omega)\frac{\partial\Delta_F^{(+)}(\mathbf{r}';\omega)}{\partial n'} \right] dS' = 0. \ \mathbf{r} \in V \tag{16.70}$$

The (in time) forward propagating positive-energy incident field (wave packet) now is extinguished inside the scattering volume by the forward propagating negative-energy wave packet. The $\phi^{(-)}(\mathbf{R};\omega)$ field *inside* the Coulomb scattering domain must satisfy

$$\int^{\Sigma} \left[\Delta_F^{(-)}(\mathbf{R};\omega) \frac{\partial \phi^{(-)}(\mathbf{r}';\omega)}{\partial n'} - \phi^{(-)}(\mathbf{r}';\omega) \frac{\partial \Delta_F^{(-)}(\mathbf{r}';\omega)}{\partial n'} \right] dS', \quad \mathbf{r} \in V \qquad (16.71)$$

in every $\mathbf{r}$-point. It is obvious that an application of the extinction theorem to the domain outside V $[\mathbf{r} \in \tilde{V}]$ in the Coulomb scattering case leads to the conclusion that there is no negative-energy field (antiparticle) backscattered. In the somewhat idealized case where the scattering volume is a semi-infinite medium with a flat and sharp surface, the extinction theorem for an incoming $\phi^{(+)}$-field gives one the result calculated by Klein [170].

The extinction theorem for the Schrödinger equation [120, 276] is simpler than for the Klein-Gordon equation since there is no antiparticle generation in the non-relativistic domain. Following the Klein-Gordon approach, one may establish an extinction theorem for the electromagnetic field $(\mathbf{A}_T)$ in the high-field limit, where electron-positron pairs are "pulled out" of the vacuum in a certain region of space.

V

Basics of Mesoscopic Electrodynamics

CHAPTER 17

Notes on Nonlocal Electrodynamics

17.1   TWO KINDS OF SPATIAL NONLOCALITY

In the special theory of relativity an essential distinction must be made between uniform and nonuniform particle motion relative to an inertial frame. This distinction is 'transferred' to the electromagnetic fields produced by the various charges. Thus, if a charge is moving uniformly one speaks of *velocity fields*, while for nonuniform motion the related fields are denoted as *radiation fields*, also called *acceleration fields*.

In mesoscopic electrodynamics, where one most often are dealing with a (large) number of particles it is common to separate electromagnetic fields into *solenoidal* [also called transverse (T)] and *irrotational* [also called longitudinal (L)] *fields*. However, it must be remembered that the division of the field into T- and L-parts is not relativistically invariant. Thus, in the wave vector $(\mathbf{q})$- frequency (ω) domain the longitudinal electric fields (i.e. fields along the wave vector direction) are related as [158]

$$\mathbf{q}' \cdot \mathbf{E}'(\mathbf{q}', \omega') = \gamma \mathbf{q} \cdot \mathbf{E}(\mathbf{q}, \omega) + i\mu_0 c_0 \boldsymbol{\beta} \cdot \mathbf{J}(\mathbf{q}, \omega), \tag{17.1}$$

in inertial frames K' and K [assuming, for simplicity, that the Cartesian axes are parallel and that the coordinate origins are coincident at time $t = t' = 0$]. Viewed from K, the frame K' is assumed to move with a velocity $\mathbf{v}$. The usual abbreviations $\boldsymbol{\beta} = \mathbf{v}/c$ and $\gamma = (1 - \beta^2)^{-1/2}$. For free fields, where the current density $\mathbf{J}(\mathbf{r}, t) = \mathbf{0}$ in every space-time point, no longitudinal fields can exist in any inertial system. This is obvious from the Maxwell-Lorentz equation $\mathbf{q} \times \mathbf{B}(\mathbf{q}, \omega) = i\mu_0 \mathbf{J}(\mathbf{q}, \omega) - (\omega/c^2)\mathbf{E}(\mathbf{q}, \omega)$ which for $\mathbf{J}(\mathbf{q}, \omega) = \mathbf{0}$ gives $\mathbf{q} \cdot \mathbf{E}(\mathbf{q}, \omega) = -(c^2/\omega)\mathbf{q} \cdot \mathbf{q} \times \mathbf{B}(\mathbf{q}, \omega) = 0$.

The transverse part of the electric field, $\mathbf{E}_T(\mathbf{r}, t)$, which is so important in photon physics, satisfies the inhomogeneous wave equation

$$\Box \mathbf{E}_T(\mathbf{r}, t) = \mu_0 \frac{\partial \mathbf{J}_T}{\partial t}, \tag{17.2}$$

in the framework of classical Maxwell-Lorentz electrodynamics. If one insists on the orthodox point of view, that the current density itself, $\mathbf{J}(\mathbf{r}, t)$, is the quantity of

DOI: 10.1201/9781003029458-17

physical importance, and not its transverse part, $\mathbf{J}_T(\mathbf{r}, t)$, Eq. (17.2) exhibits, at first sight, a peculiar kind of spatial nonlocality, not related to a retarded (with speed c) field response to $\mathbf{J}(\mathbf{r}, t)$. Thus, from the formula [7, 158, 190]

$$\mathbf{J}_T(\mathbf{r}, t) = \nabla \times \left[\nabla \times \int_{-\infty}^{\infty} \frac{\mathbf{J}(\mathbf{r}', t) d^3 r'}{4\pi |\mathbf{r} - \mathbf{r}'|} \right]. \tag{17.3}$$

it appears that $\mathbf{J}_T(\mathbf{r}, t)$ and $\mathbf{J}(\mathbf{r}, t)$ are related nonlocally in space, but locally in time. The spatial nonlocality has a range characterized qualitatively by the profile form $|\mathbf{r} - \mathbf{r}'|^{-3}$, generally.

It has turned out, however that also $\mathbf{J}_T(\mathbf{r}, t)$ qualify as a genuine physical quantity. We shall discuss this point in Chapter 27 in relation to a study of transverse fields tied to massive particles. Hence the $\mathbf{J}_T(\mathbf{r}, t)$ distribution characterizes the best confinement region for a photon emitted from the electronic current density distribution $\mathbf{J}(\mathbf{r}, t)$ at time t.

The other kind of spatial nonlocality is associated with the constitutive relation giving the relation between the induced current density, $\mathbf{J}(\mathbf{r}, t)$, at a given space-time point and the driving (prevailing) field, $\mathbf{E}(\mathbf{r}', t')$, in surrounding space-time points. This second kind of nonlocality is of particular importance in mesoscopic electrodynamics, as we shall indicate in the following.

17.2   LINEAR NONLOCAL ELECTRODYNAMICS IN A FORMAL SENSE

Formally, the most general linear relation between the microscopic current density and the prevailing electric field is given by

$$\mathbf{J}(\mathbf{r}, t) = \int_{-\infty}^{\infty} \boldsymbol{\sigma}(\mathbf{r}, \mathbf{r}', t, t') \cdot \mathbf{E}(\mathbf{r}', t') dt' d^3 r', \tag{17.4}$$

where $\boldsymbol{\sigma}(\mathbf{r}, \mathbf{r}', t, t')$ is the so-called two space-time microscopic conductivity tensor. The constitutive relation in Eq. (17.4) does not obey the principle of causality. However, let us assume that we can eliminate the prevailing electric field in favour of the external (ext) electric field, $\mathbf{E}_{\text{ext}}(\mathbf{r}, t)$ acting on the medium in consideration. In linear electrodynamics the relation between $\mathbf{E}(\mathbf{r}, t)$ and $\mathbf{E}_{\text{ext}}(\mathbf{r}, t)$ is nonlocal in space-time. Hence,

$$\mathbf{E}(\mathbf{r}, t) = \int_{-\infty}^{\infty} \mathbf{L}(\mathbf{r}, \mathbf{r}', t, t') \cdot \mathbf{E}_{\text{ext}}(\mathbf{r}', t') dt' d^3 r' \tag{17.5}$$

where $\mathbf{L}(\mathbf{r}, \mathbf{r}', t, t')$ is known as the local-field tensor. By inserting Eq. (17.5) into Eq. (17.4) one obtains

$$\mathbf{J}(\mathbf{r}, t) = \int_{\infty}^{\infty} \boldsymbol{\sigma}_{\text{CAU}}(\mathbf{r}, \mathbf{r}', t, t') \cdot \mathbf{E}_{\text{ext}}(\mathbf{r}', t') dt' d^3 r', \tag{17.6}$$

with

$$\boldsymbol{\sigma}_{\text{CAU}}(\mathbf{r}, \mathbf{r}', t, t') = \int_{-\infty}^{\infty} \boldsymbol{\sigma}(\mathbf{r}, \mathbf{r}'', t, t'') \cdot \mathbf{L}(\mathbf{r}'', \mathbf{r}', t'', t') dt'' d^3 r''. \tag{17.7}$$

The tensor $\boldsymbol{\sigma}_{\text{CAU}}(\mathbf{r}, \mathbf{r}', t, t')$ is a causal (CAU) response function because the cause, $\mathbf{E}_{\text{ext}}(\mathbf{r}', t')$, precedes the *effect*, $\mathbf{J}(\mathbf{r}, t)$, in time:

$$\boldsymbol{\sigma}_{\text{CAU}}(\mathbf{r}, \mathbf{r}', t, t'(> t)) = \mathbf{0} \tag{17.8}$$

for all pairs $(\mathbf{r}, \mathbf{r}')$ of space points. The real and imaginary parts of $\boldsymbol{\sigma}_{\text{CAU}}$ thus constitute a pair of Hilbert transforms [46, 157, 185, 194, 250].

Let us assume that we can make the approximation

$$\boldsymbol{\sigma}_{\text{CAU}}(\mathbf{r}, \mathbf{r}', t, t') = \boldsymbol{\sigma}_{\text{CAU}}(\mathbf{r}, \mathbf{r}', t - t'). \tag{17.9}$$

Physically, this means that the medium exhibits translation invariance in time. On a microscopic time scale fluctuations in the properties of the medium are always present, but if these fluctuations vary rapidly on the characteristic time scale of variations in $\mathbf{E}_{\text{ext}}(\mathbf{r}', t')$ Eq. (17.9) may appear as a reasonable approximation. Using the folding theorem, the causal constitutive relation in Eq. (17.6) [with the approximation in Eq. (17.9) takes the following form in the frequency (ω) domain:

$$\mathbf{J}(\mathbf{r}; \omega) = \int_{-\infty}^{\infty} \boldsymbol{\sigma}_{\text{CAU}}(\mathbf{r}, \mathbf{r}', \omega) \cdot \mathbf{E}_{\text{ext}}(\mathbf{r}'; \omega) d^3 r'. \tag{17.10}$$

If the external field is slowly varying in space in the range where $\boldsymbol{\sigma}_{\text{CAU}}(\mathbf{r}, \mathbf{r}', \omega)$ effectively is nonvanishing, one may expand $\mathbf{E}_{\text{ext}}(\mathbf{r}'; \omega)$ in a Taylor series around the point $\mathbf{r}$. In lowest order, the medium appears as a inhomogeneous (inh) medium, that is

$$\mathbf{J}(\mathbf{r}, \omega) = \boldsymbol{\sigma}_{\text{CAU}}^{\text{inh}}(\mathbf{r}; \omega) \cdot \mathbf{E}_{\text{ext}}(\mathbf{r}; \omega), \tag{17.11}$$

where

$$\boldsymbol{\sigma}_{\text{CAU}}^{\text{inh}}(\mathbf{r}; \omega) = \int_{-\infty}^{\infty} \boldsymbol{\sigma}_{\text{CAU}}(\mathbf{r}, \mathbf{r}'; \omega) d^3 r'. \tag{17.12}$$

17.3 MICROSCOPIC "INHOMOGENEOUS" RESPONSE

The elimination of the prevailing field $\mathbf{E}(\mathbf{r}, t)$ in favour of the external field is an extremely difficult task to carry out, and in microscopic electrodynamics it turns out to be favourable to return to Eq. (17.4). With the adoption of translational invariance in time, Eq. (17.4) takes the form

$$\mathbf{J}(\mathbf{r}; \omega) = \int_{-\infty}^{\infty} \boldsymbol{\sigma}(\mathbf{r}, \mathbf{r}'; \omega) \cdot \mathbf{E}(\mathbf{r}'; \omega) d^3 r'. \tag{17.13}$$

Now, the linear conductivity tensor can in the nonrelativistic regime be calculated on the basis of the Schrödinger equation. It turns out then that $\boldsymbol{\sigma}(\mathbf{r}, \mathbf{r}'; \omega)$ has the following structure if only bound-state transitions between the various (many-body) states are included [152, 157]:

$$\boldsymbol{\sigma}(\mathbf{r}, \mathbf{r}'; \omega) = \sum_{I, J} A_{IJ}(\omega) \mathbf{J}_{I \to J}(\mathbf{r}) \mathbf{J}_{J \to I}(\mathbf{r}'). \tag{17.14}$$

The conductivity tensor hence is seen to be a weighted [weight factor: $A_{IJ}(\omega)$] sum of dyadic tensor products, $\mathbf{J}_{I\to J}(\mathbf{r})\mathbf{J}_{J\to I}(\mathbf{r}')$, where the first vector, $\mathbf{J}_{I\to J}(\mathbf{r})$ is a function of $\mathbf{r}$ alone, and the second vector $\mathbf{J}_{J\to I}(\mathbf{r}')$ depends only on $\mathbf{r}'$. The two vectors are *transition current density* between the (many-body) states I and J: From I to J $[\mathbf{J}_{I\to J}(\mathbf{r})]$ and from J to I $[\mathbf{J}_{I\to J}(\mathbf{r})]$. The frequency-dependent factor $A_{IJ}(\omega)$ is given by

$$A_{IJ}(\omega) = \frac{2i\hbar}{E_I - E_J}\frac{P_J - P_I}{\hbar\omega + E_J - E_I} \tag{17.15}$$

where $E_I(E_J)$ is the stationary-state energy of state $I(J)$, and $P_I(P_J)$ the probabilities that this state is occupied. Note that resonance is obtained for $\hbar\omega + E_J = E_I$, the Bohr formula.

By combining Eqs. (17.13) and (17.14) one obtains

$$\mathbf{J}(\mathbf{r};\omega) = \sum_{I,J} A_{IJ}(\omega)N_{J\to I}(\omega)\mathbf{J}_{I\to J}(\mathbf{r}), \tag{17.16}$$

where the number $N_{J\to I}(\omega)$ is given by

$$N_{J\to I}(\omega) = \int_{-\infty}^{\infty} \mathbf{J}_{J\to I}(\mathbf{r}') \cdot \mathbf{E}(\mathbf{r}';\omega)d^3r'. \tag{17.17}$$

The microscopic current density in Eq. (17.16) hence appears as a sum of *inhomogeneous* (but local) contributions, where the individual terms in the sum relate to a given transition $(I \to J)$. In the electrodynamics of mesoscopic (or atomic) structures, where the number of needed transitions is small, Eq. (17.16) is of great value for theoretical studies and numerical calculations.

CHAPTER 18

Quantum-Field Formalism

18.1 INTERACTION HAMILTONIAN AND CURRENT DENSITY

The complete set of single-particle ($\sim$ electron) wave functions, $\psi_{\mathbf{k},\sigma}$, entering as coefficients in the Fermi-field operator expansion [Eqs. (14.55) and (14.56)], can be chosen in an indefinitely number of ways. The optimal choice depends on the actual application.

The reader may appreciate that an analogous scheme is used in the second-quantization of the Maxwell theory. For instance, if one considers the Riemann-Silberstein $\rightarrow$ Oppenheimer $\rightarrow$ Bialynicki bispinor

$$\boldsymbol{\Psi}_i(\mathbf{r}, t) = \begin{pmatrix} \boldsymbol{\Psi}_{i,+}(\mathbf{r}, t) \\ \boldsymbol{\Psi}_{i,-}(\mathbf{r}, t) \end{pmatrix} \equiv \begin{pmatrix} \boldsymbol{F}_+^{(+)}(\mathbf{r}, t) \\ \boldsymbol{F}_-^{(+)}(\mathbf{r}, t) \end{pmatrix}, \tag{18.1}$$

here written in helicity basis, as single-state wave functions for members of the i-set, the related expansions of the second-quantized vectorial field operators have the forms

$$\hat{\boldsymbol{\mathcal{F}}}_+(\mathbf{r}, t) = \sum_i \boldsymbol{\Psi}_{i,+}(\mathbf{r}, t)\hat{a}_{i,+}, \tag{18.2}$$

$$\hat{\boldsymbol{\mathcal{F}}}_-(\mathbf{r}, t) = \sum_i \boldsymbol{\Psi}_{i,+}^*(\mathbf{r}, t)\hat{a}_{i,+}^\dagger. \tag{18.3}$$

In a number of important cases the coefficients conveniently are chosen as members of a complete set of monochromatic plane waves, in other studies a complete set of wave-packet modes is used. The present author has advocated the use of wave-packet modes in single (or few)-photon diffraction studies from mesoscopic holes [124, 158].

In the following we shall for the electron system make use of a basis set of plane-wave wave functions [wave vector: $\mathbf{k}$, spin index: s]. In the electron subspace of Hilbert space these have the generic form

$$|\mathbf{k}, s\rangle = |\mathbf{k}\rangle \otimes |s\rangle, \tag{18.4}$$

i.e., a tensor product ($\otimes$) of states in the direct ($|\mathbf{k}\rangle$) and spin ($|s\rangle$) subspaces. Any state, $|\psi\rangle$, can be expanded after the $\{|\mathbf{k}, s\rangle\}$-set as follows:

$$|\psi\rangle = \sum_s \int_{\mathbf{k}} |\mathbf{k}, s\rangle \langle \mathbf{k}, s|\psi\rangle, \tag{18.5}$$

DOI: 10.1201/9781003029458-18

and represented by the numbers

$$\psi_{\mathbf{k},s} = \langle \mathbf{k}, s | \psi \rangle . \tag{18.6}$$

These numbers are the wave functions appearing in the field-operator expansion in Eq. (14.55).

In the plane-wave basis, the second-quantized form of the weakly relativistic interaction Hamiltonian [Eq. (14.50)] thus becomes

$$\mathcal{H}_I(t) = \sum_{\mathbf{k},\mathbf{k}',s,s'} \langle \mathbf{k}', s' | -\frac{e}{2m}(\hat{\mathbf{p}} \cdot \hat{\mathbf{A}}_T + \hat{\mathbf{A}}_T \cdot \hat{\mathbf{p}}) + \frac{e^2}{2m}\hat{\mathbf{A}}_T^2 - \frac{e\hbar}{4m^2c^2}\boldsymbol{\nabla}V \times \hat{\mathbf{A}}_T \cdot \hat{\boldsymbol{\sigma}}$$

$$- \frac{e\hbar}{2m}\hat{\boldsymbol{\sigma}} \cdot \boldsymbol{\nabla} \times \hat{\mathbf{A}}_T |\mathbf{k}, s\rangle \, \hat{b}_{\mathbf{k}',s'}^{\dagger}\hat{b}_{\mathbf{k},s}. \tag{18.7}$$

The electron spin operator ($\hat{\boldsymbol{\sigma}}$) enters the spin-orbit and Pauli part of the interaction Hamiltonian in a linear fashion, and the tensor-product form in Eq. (18.4) makes it easy to calculate the matrix elements $\langle s' | \hat{\boldsymbol{\sigma}} | s\rangle$, and reduce the double sum $\sum_{s,s'}(...)$ to a single sum over the chosen basis states for the spin. Below, we shall use as basis states the eigenstates $|1\rangle$, $|-1\rangle$ for the z-component of the spin operator, i.e., $\hat{\sigma}_z |s\rangle = s|s\rangle$, where $s = +1$ and $s = -1$ for, respectively, the spin-up ($|1\rangle \equiv |\uparrow\rangle$) and spin-down ($|-1\rangle \equiv |\downarrow\rangle$) states. The Cartesian components of the (dimensionless) Pauli spin operator are

$$\hat{\sigma}_x = \begin{pmatrix} 0 & 1 \\ 1 & 0 \end{pmatrix}, \; \hat{\sigma}_y = \begin{pmatrix} 0 & -i \\ i & 0 \end{pmatrix}, \; \hat{\sigma}_z = \begin{pmatrix} 1 & 0 \\ 0 & -1 \end{pmatrix}. \tag{18.8}$$

By utilizing the orthonormality of the spin eigenstates [$\langle s'|s\rangle = \delta_{ss'}$] and Eq. (18.8) one obtains

$$\sum_{s,s'} \langle s' | \hat{\boldsymbol{\sigma}} | s\rangle \, \hat{b}_{\mathbf{k}',s'}^{\dagger}\hat{b}_{\mathbf{k},s} = \sum_{s=+,-} [(\mathbf{e}_x + is\mathbf{e}_y)\hat{b}_{\mathbf{k}',-s}^{\dagger}\hat{b}_{\mathbf{k},s} + \mathbf{e}_z s\hat{b}_{\mathbf{k}',s}^{\dagger}\hat{b}_{\mathbf{k},s}], \tag{18.9}$$

where $\mathbf{e}_i$ ($i = x, y, z$) denote the unit vectors along the axes of the Cartesian coordinate system used. The reader may notice that a spin flip ($s \leftrightarrow -s$) appears in the x- and y-components, but not in the z-component ($s \leftrightarrow s$). Furthermore, the vectorial parts of the spin-flip terms are proportional to the helicity unit vectors

$$\mathbf{e}_s = \frac{1}{\sqrt{2}}(\mathbf{e}_x + is\mathbf{e}_y), \quad s = 1, -1, \tag{18.10}$$

for internal rotations around $\mathbf{e}_z$ [$\mathbf{e}_+ \equiv \mathbf{e}_1$, $\mathbf{e}_- \equiv \mathbf{e}_{-1}$].

The interaction Hamiltonian next is divided into three parts

$$\hat{\mathcal{H}}_I(t) = \hat{\mathcal{H}}_{1+2}(t) + \hat{\mathcal{H}}_1^{\sigma}(t) + \hat{\mathcal{H}}_1^{so}(t). \tag{18.11}$$

The operator $\hat{\mathcal{H}}_{1+2}(t)$ is the orbital (spin-independent) part of the Hamiltonian [no superscript "o" for orbital is put on this part], $\hat{\mathcal{H}}_1^{\sigma}(t)$ is the Pauli part (superscript: σ), and $\hat{\mathcal{H}}_1^{so}(t)$ is the spin-orbit part (superscript: so). The subscript on the various parts indicates in which order the transverse vector potential (operator) enters. The orbital part of $\hat{\mathcal{H}}_I(t)$ is subdivided into parts proportional to $\hat{\mathbf{A}}_T$ and $\hat{\mathbf{A}}_T^2$, that is

$$\hat{\mathcal{H}}_{1+2}(t) = \hat{\mathcal{H}}_1(t) + \hat{\mathcal{H}}_2(t). \tag{18.12}$$

In explicit form, the different parts are as follows:

$$\hat{\mathcal{H}}_1(t) = -\frac{e}{2m} \sum_{\mathbf{k},\mathbf{k}',s} = \langle \mathbf{k}'| \,\hat{\mathbf{p}} \cdot \hat{\mathbf{A}}_T + \hat{\mathbf{A}}_T \cdot \hat{\mathbf{p}} \,|\mathbf{k}\rangle \, \hat{b}_{\mathbf{k}',s}^\dagger \hat{b}_{\mathbf{k},s}, \tag{18.13}$$

$$\hat{\mathcal{H}}_2(t) = \frac{e^2}{2m} \sum_{\mathbf{k},\mathbf{k}',s} \langle \mathbf{k}'| \,\hat{\mathbf{A}}_T^2 \,|\mathbf{k}\rangle \, \hat{b}_{\mathbf{k}',s}^\dagger \hat{b}_{\mathbf{k},s}, \tag{18.14}$$

$$\hat{\mathcal{H}}_1^\sigma(t) = -\frac{e\hbar}{2m} \sum_{\mathbf{k},\mathbf{k}',s} [\sqrt{2}e_s \hat{b}_{\mathbf{k}',-s}^\dagger \hat{b}_{\mathbf{k},s} + e_z s \hat{b}_{\mathbf{k}',s}^\dagger \hat{b}_{\mathbf{k},s}] \cdot \langle \mathbf{k}'| \,\boldsymbol{\nabla} \times \hat{\mathbf{A}}_T \,|\mathbf{k}\rangle, \tag{18.15}$$

and

$$\hat{\mathcal{H}}_1^{\mathrm{so}}(t) = -\frac{e\hbar}{4m^2c^2} \sum_{\mathbf{k},\mathbf{k}',s} [\sqrt{2}e_s \hat{b}_{\mathbf{k}',-s}^\dagger \hat{b}_{\mathbf{k},s} + e_z s \hat{b}_{\mathbf{k}',s}^\dagger \hat{b}_{\mathbf{k},s}] \cdot \langle \mathbf{k}'| \,\boldsymbol{\nabla} V \times \hat{\mathbf{A}}_T \,|\mathbf{k}\rangle. \tag{18.16}$$

In order to apply the density matrix formalism in a second-quantized calculation of various field-induced microscopic current densities, we write the one-body current density operator [Eq. (14.51)] in second quantized form, and again in the plane-wave basis. Hence,

$$\hat{\boldsymbol{\mathcal{J}}}(t) = \sum_{\mathbf{k},\mathbf{k}',s,s'} \langle \mathbf{k}', s'| \,\frac{e}{2m}[\hat{\mathbf{p}}\delta(\mathbf{r} - \mathbf{r}_e) + \delta(\mathbf{r} - \mathbf{r}_e)\hat{\mathbf{p}}] - \frac{e^2}{m} \hat{\mathbf{A}}_T \delta(\mathbf{r} - \mathbf{r}_e)$$

$$+ \frac{ie}{2m}[\delta(\mathbf{r} - \mathbf{r}_e)\hat{\mathbf{p}} \times \hat{\boldsymbol{\sigma}} - \hat{\mathbf{p}} \times \hat{\boldsymbol{\sigma}}\delta(\mathbf{r} - \mathbf{r}_e)] \,|\mathbf{k}, s\rangle \, \hat{b}_{\mathbf{k}',s'}^\dagger \hat{b}_{\mathbf{k},s}. \tag{18.17}$$

Also the current density operator is divided into parts, viz.,

$$\hat{\boldsymbol{\mathcal{J}}}(t) = \hat{\boldsymbol{\mathcal{J}}}_F + \hat{\boldsymbol{\mathcal{J}}}_F^\sigma + \hat{\boldsymbol{\mathcal{J}}}_1(t), \tag{18.18}$$

where

$$\hat{\boldsymbol{\mathcal{J}}}_F = \frac{e}{2m} \langle \mathbf{k}'| \,\hat{\mathbf{p}}\delta(\mathbf{r} - \mathbf{r}_e) + \delta(\mathbf{r} - \mathbf{r}_e)\hat{\mathbf{p}} \,|\mathbf{k}\rangle \, \hat{b}_{\mathbf{k}',s}^\dagger \hat{b}_{\mathbf{k},s}, \tag{18.19}$$

is the field-unperturbed, also called the free (subscript: F), part of the spin-independent current density,

$$\hat{\boldsymbol{\mathcal{J}}}_F^\sigma = \frac{ie}{2m} \sum_{\mathbf{k},\mathbf{k}',s} \langle \mathbf{k}'| \,\delta(\mathbf{r} - \mathbf{r}_e)\hat{\mathbf{p}} - \hat{\mathbf{p}}\delta(\mathbf{r} - \mathbf{r}_e) \,|\mathbf{k}\rangle \times \left[\sqrt{2}e_s \hat{b}_{\mathbf{k}',-s}^\dagger \hat{b}_{\mathbf{k},s} + e_z \hat{b}_{\mathbf{k}',s}^\dagger \hat{b}_{\mathbf{k},s} \right], \tag{18.20}$$

is the spin part of the free current density operator [note $\times$ stands for vector product], and

$$\hat{\boldsymbol{\mathcal{J}}}_1(t) = -\frac{e^2}{m} \sum_{\mathbf{k},\mathbf{k}',s} \langle \mathbf{k}'| \,\hat{\mathbf{A}}_T \delta(\mathbf{r} - \mathbf{r}_e) \,|\mathbf{k}\rangle \, \hat{b}_{\mathbf{k}',s}^\dagger \hat{b}_{\mathbf{k},s}, \tag{18.21}$$

is the field-induced part, linear in the transverse vector potential (operator).

18.2 FUNDAMENTAL PHOTON-ELECTRON QUANTUM INTERACTION PROCESSES

By inserting the general expression for the transverse vector potential operator, given in box quantization in Eq. (3.63), in Eqs. (18.13)–(18.16), the interaction Hamiltonian takes a form in which the photon and electron fields both are second quantized. Myriads of quantal photon-electron interaction processes flourish in this garden. To obtain an overview of the structure of these processes it is sufficient to consider the interaction of the electron system with a single plane-wave mode of the quantized electromagnetic field. The transverse vector potential operator hence is given by

$$\hat{\mathbf{A}}_T(\mathbf{r}, t) = \left(\frac{\hbar}{2\varepsilon_0 c q L^3}\right)^{1/2} [\hat{a}_{\mathbf{q},\sigma}(t)\boldsymbol{\varepsilon}_{\mathbf{q},\sigma}e^{i\mathbf{q}\cdot\mathbf{r}} + \hat{a}^\dagger_{\mathbf{q},\sigma}(t)\boldsymbol{\varepsilon}^*_{\mathbf{q},\sigma}e^{-i\mathbf{q}\cdot\mathbf{r}}], \tag{18.22}$$

where $\mathbf{q}$ is the mode wave vector, and σ the index characterizing the type of polarization. The (complex) polarization unit vector is denoted by $\boldsymbol{\varepsilon}_{\mathbf{q},\sigma}$. The annihilation $[\hat{a}_{\mathbf{q},\sigma}(t)]$ and creation $[\hat{a}^\dagger_{\mathbf{q},\sigma}(t)]$ are time dependent here, and they satisfy the Heisenberg equation in Eq. (3.62) and its Hermitian conjugate.

The elementary photon-electron interactions are different for the four parts of $\hat{\mathcal{H}}_I(t)$ [Eqs. (18.13)–(18.16)], so let us therefore study them separately, starting with $\hat{\mathcal{H}}_1(t)$ [Eq. (18.13)]. Since $\nabla \cdot \hat{\mathbf{A}}_T = 0$, per definition, one has

$$\langle \mathbf{k}'| \hat{\mathbf{p}} \cdot \hat{\mathbf{A}}_T + \hat{\mathbf{A}}_T \cdot \hat{\mathbf{p}} |\mathbf{k}\rangle = 2 \langle \mathbf{k}'| \hat{\mathbf{A}}_T \cdot \hat{\mathbf{p}} |\mathbf{k}\rangle = 2\hbar\mathbf{k} \langle \mathbf{k}'| \hat{\mathbf{A}}_T |\mathbf{k}\rangle . \tag{18.23}$$

The matrix element $\langle \mathbf{k}'| \hat{\mathbf{A}}_T |\mathbf{k}\rangle$ involves spatial integrals of two types, viz.,

$$\int_{-\infty}^{\infty} \exp[i(\mathbf{k} \pm \mathbf{q} - \mathbf{k}') \cdot \mathbf{r}]d^3r = (2\pi)^3\delta(\mathbf{k} \pm \mathbf{q} - \mathbf{k}') \tag{18.24}$$

in the large box (continuum) limit ($L \to \infty$). The related fundamental photon-electron processes at $t = t$ therefore have the structures

$$\hat{\mathcal{H}}_1(t) = \begin{cases} \hat{a}_{\mathbf{q},\sigma}\hat{b}^\dagger_{\mathbf{k}+\mathbf{q},s}\hat{b}_{\mathbf{k},s} \\ \hat{a}^\dagger_{\mathbf{q},\sigma}\hat{b}^\dagger_{\mathbf{k}-\mathbf{q},s}\hat{b}_{\mathbf{k},s} \end{cases} . \tag{18.25}$$

In the upper panel an electron in state $|\mathbf{k}, s\rangle$ is scattered to state $|\mathbf{k} + \mathbf{q}, s\rangle$ by *absorption* of a photon in state $|\mathbf{q}, \sigma\rangle$. In the lower panel an electron is scattered from $|\mathbf{k}, s\rangle$ to $|\mathbf{k} - \mathbf{q}, s\rangle$ and *emitting* a photon in state $|\mathbf{q}, s\rangle$. In both processes momentum is conserved. The discussion of the manner in which the conservation laws for energy and angular momentum enter the analysis will be postponed to later.

The structure of the Pauli interaction term $\hat{\mathcal{H}}_1^\sigma(t)$, again gives terms of the types $\exp(i\mathbf{q} \cdot \mathbf{r})$ and $\exp(-i\mathbf{q} \cdot \mathbf{r})$ for the parts of $\nabla \times \mathbf{A}_T$ proportional to $\hat{a}_{\mathbf{q},\sigma}$ and $\hat{a}^\dagger_{\mathbf{q},s}$, respectively. For the $\mathbf{e}_z$-component of the spin contribution [Eq. (18.9)] one thus obtains fundamental processes as those given in Eq. (18.25). Additionally, the $\mathbf{e}_s$-component leads to processes involving a spin flip of the electron state. Altogether, the fundamental photon-electron interactions of the Pauli term are

$$\hat{\mathcal{H}}_1^\sigma(t) \sim \begin{cases} \hat{a}_{\mathbf{q},\sigma}\hat{b}^\dagger_{\mathbf{k}+\mathbf{q},s}\hat{b}_{\mathbf{k},s}; & \hat{a}_{\mathbf{q},\sigma}\hat{b}^\dagger_{\mathbf{k}+\mathbf{q},-s}\hat{b}_{\mathbf{k},s} \\ \hat{a}^\dagger_{\mathbf{q},\sigma}\hat{b}^\dagger_{\mathbf{k}-\mathbf{q},s}\hat{b}_{\mathbf{k},s}; & \hat{a}^\dagger_{\mathbf{q},\sigma}\hat{b}^\dagger_{\mathbf{k}-\mathbf{q},-s}\hat{b}_{\mathbf{k},s} \end{cases} . \tag{18.26}$$

The $\hat{\mathcal{H}}_2(t)$-interaction [Eq. (18.14)] involves simultaneously $(t = t)$ two photons, since the $\hat{\mathbf{A}}_T^2$-operator has four different parts proportional to $\hat{a}_{\mathbf{q},\sigma}^2 \exp(2i\mathbf{q}\cdot\mathbf{r})$, $\hat{a}_{\mathbf{q},\sigma}\hat{a}_{\mathbf{q},\sigma}^\dagger$, $\hat{a}_{\mathbf{q},\sigma}^\dagger\hat{a}_{\mathbf{q},\sigma}$ and $(\hat{a}_{\mathbf{q},\sigma}^\dagger)^2 \exp(-2i\mathbf{q}\cdot\mathbf{r})$. In turn, this leads to the following fundamental photon-electron processes:

$$\hat{\mathcal{H}}_2(t) \sim \begin{cases} \hat{a}_{\mathbf{q},\sigma}\hat{a}_{\mathbf{q},\sigma}\hat{b}_{\mathbf{k}+2\mathbf{q},s}^\dagger\hat{b}_{\mathbf{k},s} \\ \hat{a}_{\mathbf{q},\sigma}^\dagger\hat{a}_{\mathbf{q},\sigma}^\dagger\hat{b}_{\mathbf{k}-2\mathbf{q},s}^\dagger\hat{b}_{\mathbf{k},s} \\ \hat{a}_{\mathbf{q},\sigma}^\dagger\hat{a}_{\mathbf{q},\sigma}\hat{b}_{\mathbf{k},s}^\dagger\hat{b}_{\mathbf{k},s} \\ \hat{a}_{\mathbf{q},\sigma}\hat{a}_{\mathbf{q},\sigma}^\dagger\hat{b}_{\mathbf{k},s}^\dagger\hat{b}_{\mathbf{k},s} = (\hat{a}_{\mathbf{q},\sigma}^\dagger\hat{a}_{\mathbf{q},\sigma} + 1)\hat{b}_{\mathbf{k},s}^\dagger\hat{b}_{\mathbf{k},s} \end{cases} \tag{18.27}$$

The interpretation of the structures of the two upper panels in Eq. (18.27) is obvious: An electron is scattered from the state $|\mathbf{k},s\rangle$ to $|\mathbf{k}+2\mathbf{q},s\rangle$ by simultaneous absorption of two photons from $|\mathbf{q},\sigma\rangle$, *or* scattered from $|\mathbf{k},s\rangle$ to $|\mathbf{k}-2\mathbf{q},s\rangle$ by emission of two photons from $|\mathbf{q},\sigma\rangle$. In the two lower panels the electron stays in $|\mathbf{k},s\rangle$ due to simultaneous absorption and emission of photons from state $|\mathbf{q},\sigma\rangle$. The number of photons in the $|\mathbf{q},\sigma\rangle$-mode is unaltered in the two lower panels, but if the $|\mathbf{q},\sigma\rangle$ *initially* is without photons (vacuum: $|0\rangle$) the matrix element $\langle 0|\,\hat{a}_{\mathbf{q},\sigma}^\dagger\hat{a}_{\mathbf{q},\sigma}\,|0\rangle = 0$ but $\langle 0|\,\hat{a}_{\mathbf{q},\sigma}\hat{a}_{\mathbf{q},\sigma}^\dagger\,|0\rangle = 1$. A schematic illustration of the fundamental photon-electron interaction processes related to $\hat{\mathcal{H}}_1$, $\hat{\mathcal{H}}_1^\sigma$ and $\hat{\mathcal{H}}_2$ is shown in Fig. 18.1.

The spin-orbit part of the interaction Hamiltonian, $\hat{\mathcal{H}}_1^{\text{so}}(t)$, has the same spin structure as the Pauli part, $\hat{\mathcal{H}}_1^\sigma(t)$. The presence of the gradient of the single-particle potential energy, $\boldsymbol{\nabla}V(\mathbf{r})$, implies that the matrix element $\langle\mathbf{k}'|\,\boldsymbol{\nabla}V\times\hat{\mathbf{A}}_T\,|\mathbf{k}\rangle$ involves spatial integrals more complicated than those in Eq. (18.24). The matrix element in $\hat{\mathcal{H}}_1^{\text{so}}(t)$ instead can be expressed in terms of the Fourier integral vector of $\boldsymbol{\nabla}V(\mathbf{r})$, viz.,

$$\mathbf{I}(\mathbf{Q}) = \int_{-\infty}^{\infty} \boldsymbol{\nabla}V(\mathbf{r})e^{-i\mathbf{Q}\cdot\mathbf{r}}d^3r. \tag{18.28}$$

Remembering that $\hat{\mathbf{A}}_T$ contains terms of the types $\exp(\pm i\mathbf{q}\cdot\mathbf{r})$, spatial integrals of the form

$$\mathbf{I}(\mathbf{k}'-\mathbf{k}\mp\mathbf{q}) = \int_{-\infty}^{\infty} \boldsymbol{\nabla}V(\mathbf{r})\exp[-i(\mathbf{k}'-\mathbf{k}\mp\mathbf{q})\cdot\mathbf{r}]d^3r \tag{18.29}$$

appear.

18.3   CURRENT DENSITY OPERATOR: QUANTUM-FIELD FORM

The current density operator, given in Eq. (18.17) is a local quantity varying from point to point in space. Its two free parts [Eqs. (18.19) and (18..20)] are easily calculated, utilizing that $\hat{\mathbf{p}}\,|\mathbf{k}\rangle = \hbar\mathbf{k}\,|\mathbf{k}\rangle$, and $\langle\mathbf{k}'|\,\hat{\mathbf{p}}^\dagger = \langle\mathbf{k}'|\,\hat{\mathbf{p}} = \hbar\mathbf{k}'\,\langle\mathbf{k}'|$. Hence, one obtains

$$\hat{\boldsymbol{\mathcal{J}}}_F(\mathbf{r}) = \frac{e\hbar}{2m}\sum_{\mathbf{k},\mathbf{k}',s}(\mathbf{k}+\mathbf{k}')e^{i(\mathbf{k}-\mathbf{k}')\cdot\mathbf{r}}\hat{b}_{\mathbf{k}',s}^\dagger\hat{b}_{\mathbf{k},s}, \tag{18.30}$$

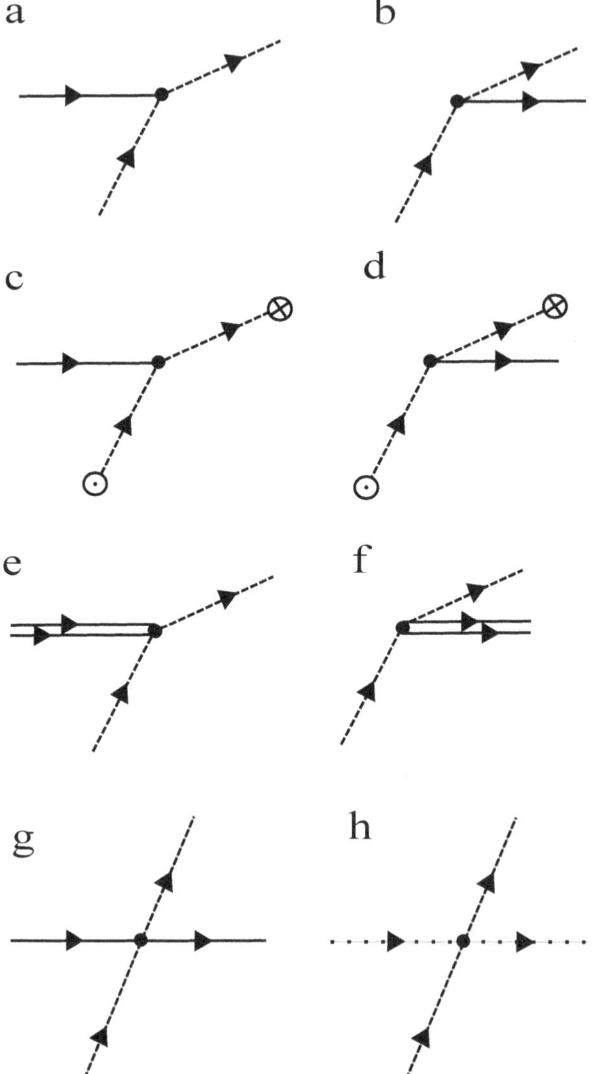

Figure 18.1 Schematic diagrams illustrating the fundamental momentum conservation photon-electron interaction processes in the framework of the Pauli theory, assuming the presence of only a single photon mode $(\mathbf{q}, \sigma)$. Unbroken line: PHOTON. Dashed line: ELECTRON [modes: $((\mathbf{k}, s), (\mathbf{k} \pm \mathbf{q}, s), (\mathbf{k} \pm \mathbf{q}, -s), (\mathbf{k} \pm 2\mathbf{q}, s)]$. Dotted line: PHOTON VACUUM. a: Single-photon absorption. b: Single-photon emission. c and d: Single-photon absorption and emission with electron spin flip $[\odot \Rightarrow \otimes$, schematically]. e. Two-photon absorption. f: Two-photon emission. g: Photon scattering into its own state. h: Photon vacuum scattering [in mode $(\mathbf{q}, \sigma)$].

and

$$\hat{\mathcal{J}}_F^\sigma(\mathbf{r}) = \frac{ie}{2m} \sum_{\mathbf{k},\mathbf{k}',s} (\mathbf{k} - \mathbf{k}') e^{i(\mathbf{k}-\mathbf{k}')\cdot\mathbf{r}} \times [\sqrt{2} \mathbf{e}_S \hat{b}^\dagger_{\mathbf{k}',-s} \hat{b}_{\mathbf{k},s} + \mathbf{e}_z \hat{b}^\dagger_{\mathbf{k}',s} \hat{b}_{\mathbf{k},s}]. \qquad (18.31)$$

The field-induced part, $\hat{\boldsymbol{\mathcal{J}}}_1(\mathbf{r}, t)$, we calculate in the presence of one field mode, only [Eq. (18.22)]. The generalization to many field modes is straightforward. The results for $\hat{\boldsymbol{\mathcal{J}}}_1(\mathbf{r}, t)$, which is time dependent because of the photon field, is given by

$$\hat{\boldsymbol{\mathcal{J}}}_1(\mathbf{r}, t) = -\frac{e^2}{m} \left(\frac{\hbar}{2\varepsilon_0 c q L^3} \right)^{1/2} \sum_{\mathbf{k}, \mathbf{k}', s} [\hat{a}_{\mathbf{q},s}(t) \boldsymbol{\varepsilon}_{\mathbf{q},\sigma} e^{i(\mathbf{k}+\mathbf{q}-\mathbf{k}')\cdot\mathbf{r}} \qquad (18.32)$$
$$+ \hat{a}^\dagger_{\mathbf{q},\sigma}(t) \boldsymbol{\varepsilon}^*_{\mathbf{q},\sigma} e^{i(\mathbf{k}-\mathbf{q}-\mathbf{k}')\cdot\mathbf{r}}] \hat{b}^\dagger_{\mathbf{k}',s} \hat{b}_{\mathbf{k},s}.$$

Although we in Section 18.1 carried out all calculations by means of a plane-wave basis set, $\{|\mathbf{k}, s\rangle\} = \{|k\rangle\} \otimes \{|s\rangle\}$, one may instead use other complete sets of single-particle state vectors characterized by a triple set of spatial quantum numbers $\mathbf{k}(\mathbf{k}')$ and the spin-state label $s(s')$. The Fermion annihilation and creation operators belonging the general single-particle state $|\mathbf{k}, s\rangle$ may still be denoted by $\hat{a}_{\mathbf{k},s}$ and $\hat{a}^\dagger_{\mathbf{k},s}$, respectively. In the Random-Phase-Approximation (RPA) description, where the light unperturbed many-electron Hamiltonian, $\hat{\mathcal{H}}_p$, is composed of direct products of single-particle-(like) states, it is convenient to use single-particle wave functions, $\psi_\mathbf{k}(\mathbf{r})$, satisfying the time-independent single-electron Schrödinger equation

$$\left[-\frac{\hbar^2}{2m} \nabla^2 + V(\mathbf{r}) \right] \psi_\mathbf{k}(\mathbf{r}) = \varepsilon_\mathbf{k} \psi_\mathbf{k}(\mathbf{r}), \qquad (18.33)$$

where $\varepsilon_\mathbf{k}$ denotes the single-particle eigenenergy. If the free Schrödinger equation [as in Eq. (18.33)] is spin independent the orbital states of the spin-up and spin-down components are identical, so that the tensor product decomposition given in Eq. (18.4) can be used. The complications arising from the presence of spin-orbit couplings will not be addressed here.

18.4 SEMICLASSICAL CURRENT DENSITY: MANY-ELECTRON THEORY

18.4.1 Liouville equation: Monochromatic driving field

In the semiclassical approach $\hat{\mathbf{A}}_T$ is a c-number quantity, $\hat{\mathbf{A}}_T \Rightarrow \mathbf{A}_T$. With this replacement, the Hamiltonian operator $\hat{\mathcal{H}}(t) = \hat{\mathcal{H}}_1(t) + \hat{\mathcal{H}}_1^\sigma(t) + \hat{\mathcal{H}}_1^{so}(t) + \hat{\mathcal{H}}_2(t)$ [Eqs. (18.13)–(18.16)] is a many-electron operator in the particle subspace.

We consider the situation where a monochromatic electromagnetic field, here given as

$$\hat{A}_T(\mathbf{r}, t) = \frac{1}{2} [\mathbf{A}_T(\mathbf{r}; \omega) e^{-i\omega t} + \mathbf{A}^*_T(\mathbf{r}; \omega) e^{i\omega t}], \qquad (18.34)$$

drives the many-electron system. Starting from Eq. (18.22), Eq. (18.34) is reached (i) by replacing $\hat{a}_{\mathbf{q},\sigma}(t)$ and $\hat{a}^\dagger_{\mathbf{q},\sigma}(t)$ by c-numbers $\alpha_{\mathbf{q},\sigma} \exp(-i\omega t)$ and $\alpha^*_{\mathbf{q},\sigma} \exp(i\omega t)$, (ii) by forcing ω to be the same for all modes (by *brute force*; prescribed frequency) and (iii) by making a summation over wave vectors.

Let us now turn the attention towards the dynamics of a many-electron system driven by a monochromatic electromagnetic field, given by the transverse vector potential $\mathbf{A}_T(\mathbf{r}, t)$ [Eq. (18.34)]. In the Liouville equation for the many-body density operator, $\hat{\rho}(t)$, viz.,

$$i\hbar \frac{d}{dt}\hat{\rho}(t) = [\hat{\mathcal{H}}(t), \hat{\rho}(t)], \tag{18.35}$$

the many-electron Hamilton operator has three parts,

$$\hat{\mathcal{H}}(t) = \hat{\mathcal{H}}_0(t) + \hat{\mathcal{H}}_I(t) + \hat{\mathcal{H}}_\tau(t). \tag{18.36}$$

The time independent particle part, here denoted by $\hat{\mathcal{H}}_0(t)$, gives the many-body eigenstates and eigenenergies. Thus, for the state I [Eq. (15.38) repeated]

$$\hat{\mathcal{H}}_0 \left| I \right\rangle = E_I \left| I \right\rangle. \tag{18.37}$$

The part $\hat{\mathcal{H}}_\tau(t)$ describes couplings to a thermal reservoir. The explicit form of $\hat{\mathcal{H}}_\tau(t)$ can only be given once the most important couplings have been identified. In a jellium the coupling can be to phonons, impurities etc. In Chap. 19, we shall consider a case where $\hat{\mathcal{H}}_0$ relates to a spectrum of surface plasmons and $\hat{\mathcal{H}}_\tau(t)$ describes these plasmons relaxation by single-particle (electron-hole excitations in the Fermi sea. Without specifying $\hat{\mathcal{H}}_\tau(t)$, let us assume that the interaction with the reservoir has the effect of relaxing the dynamic density operator, $\hat{\rho}(t)$, towards its time-independent thermal equilibrium value, $\hat{\rho}_0$, with a phenomenological time constant, τ. *Even if* such a relaxation form makes physical sense, heavy theory is needed to calculate τ, in general. In mesoscopic quantum well films with in-plane jellium dynamics a 2D-calculation of τ often is sufficient, and much easier to carry to its conclusion. In the Liouville equation we thus assume an ansatz of the form

$$[\hat{\mathcal{H}}_\tau(t), \hat{\rho}(t)] = -i\hbar \frac{\hat{\rho}(t) - \hat{\rho}_0}{\tau}. \tag{18.38}$$

From Eqs. (15.41) and (15.42) it appears that

$$\hat{\rho}_0 \left| I \right\rangle = P_I \left| I \right\rangle. \tag{18.39}$$

The arbitrary many-body state $\left| I \right\rangle$, hence is an eigenstate for the equilibrium density matrix, $\hat{\rho}_0$, with the many-body Boltzmann distribution probability, P_I, as eigenvalue.

The monochromatic driving of the many-electron system by the vector potential $\mathbf{A}_T(\mathbf{r}, t)$, given in Eq. (18.34), suggests that one divides the interaction Hamiltonian and the density operator into harmonic series in frequencies. The interaction Hamiltonian has the general form

$$\hat{\mathcal{H}}_I(t) = \frac{1}{2}\sum_{\alpha=1}^{2}(\hat{\mathcal{H}}_\alpha e^{-i\alpha\omega t} + \hat{\mathcal{H}}_\alpha^\dagger e^{i\alpha\omega t}) + \hat{\mathcal{H}}_0^{\mathrm{NL}} \tag{18.40}$$

In $\hat{\mathcal{H}}_1$ now we have gathered all the parts which are proportional to $\mathbf{A}_T(\mathbf{r}; \omega)$. In the first quantized description $\hat{\mathcal{H}}_1 \Rightarrow \hat{H}_1$ [See Eq. (18.7)],

$$\hat{H}_1 = -\frac{e}{2m}(\hat{\mathbf{p}} \cdot \mathbf{A}_T + \mathbf{A}_T \cdot \hat{\mathbf{p}}) - \frac{e\hbar}{4m^2c^2}\hat{\boldsymbol{\sigma}} \cdot \boldsymbol{\nabla} V \times \mathbf{A}_T - \frac{e\hbar}{2m}\hat{\boldsymbol{\sigma}} \cdot \boldsymbol{\nabla} \times \mathbf{A}_T. \tag{18.41}$$

In $\hat{\mathcal{H}}_2$ we only have one term $\hat{\mathcal{H}}_2 \Rightarrow \hat{H}_2$, namely

$$\hat{H}_2 = \frac{e^2}{4m} \mathbf{A}_T \cdot \mathbf{A}_T. \tag{18.42}$$

Finally, a frequency independent term, $\hat{\mathcal{H}}_0^{\mathrm{NL}} \Rightarrow \hat{H}_0^{\mathrm{NL}}$, with first-quantized form

$$\hat{H}_0^{\mathrm{NL}} = \frac{e^2}{4m} \mathbf{A}_T \cdot \mathbf{A}_T^*, \tag{18.43}$$

is present. We have added a superscript NL to this term to indicate that it is nonlinear in $\mathbf{A}_T$ (like $\hat{H}_2$). The term gives a field-dependent correction to the time-independent Hamiltonian, i.e., $\hat{\mathcal{H}}_0 \Rightarrow \hat{\mathcal{H}}_0 + \hat{\mathcal{H}}_0^{\mathrm{NL}}$.

Although the interaction Hamiltonian in general has only ω and 2ω frequency components, the density operator must be expanded in a complete series of harmonics. Hence,

$$\hat{\rho}(t) = \hat{\rho}_0 + \hat{\rho}_0^{\mathrm{NL}} + \frac{1}{2} \sum_{m=1}^{\infty} (\hat{\rho}_m e^{-im\omega t} + \hat{\rho}_m^{\dagger} e^{im\omega t}). \tag{18.44}$$

The part $\hat{\rho}_0^{\mathrm{NL}}$ accounts for a field-induced nonlinear contribution to the dc part of $\hat{\rho}$.

18.4.2 Iterative expansion of the Liouville equation

In many cases the mesoscopic medium under study is not driven far away from thermal equilibrium by the prescribed external field. In such situations an iterative approach can be employed to solve the Liouville equation. In the following we indicate this approach in the frequency domain.

A combination of Eqs. (18.38), (18.40) and (18.44) gives in first order the Liouville equation for $\hat{\rho}_1$, viz.,

$$\hbar\left(\omega + \frac{i}{\tau}\right)\hat{\rho}_1 = [\hat{\mathcal{H}}_0, \hat{\rho}_1] + [\hat{\mathcal{H}}_1, \hat{\rho}_0]. \tag{18.45}$$

It appears that $\hat{\rho}_1$, as determined from Eq. (18.45), is linear in the vector potential. As we shall realize below, this circumstance leads to a many-body expression for the microscopic linear conductivity tensor.

To second order in $\mathbf{A}_T(\mathbf{r};\omega)$ one obtains the following equation for $\hat{\rho}_2$:

$$\hbar\left(2\omega + \frac{i}{\tau}\right)\hat{\rho}_2 = [\hat{\mathcal{H}}_0, \hat{\rho}_2] + [\hat{\mathcal{H}}_2, \hat{\rho}_0] + \frac{1}{2}[\hat{\mathcal{H}}_1, \hat{\rho}_1] \tag{18.46}$$

Starting from Eq. (18.46) a microscopic theory for parametric electromagnetic second-harmonic generation can be obtained, exemplified in [134] for a semi-infinite jellium system. The 2ω-generation obviously has contributions from $\mathbf{p} \cdot \mathbf{A}_T + \mathbf{A}_T \cdot \mathbf{p}$ [via $\hat{\mathcal{H}}_1$]. Since $\hat{\rho}_1$ is linear in $\mathbf{A}_T$, $\hat{\rho}_2$ becomes a quadratic function of $\mathbf{A}_T$ [last term in Eq. (18.46)]. The commutator $[\hat{\mathcal{H}}_2, \hat{\rho}_0]$ involves the $\mathbf{A}_T \cdot \mathbf{A}_T$ term [see Eq. (18.42)] and this term therefore gives a quadratic term in $\mathbf{A}_T$ to $\hat{\rho}_2$, as it must do.

In second order one also obtains a Liouville equation for the nonlinear part of the density operator, $\hat{\rho}_0^{\text{NL}}$, namely,

$$\frac{i\hbar}{\tau}\hat{\rho}_0^{\text{NL}} = [\hat{\mathcal{H}}_0, \hat{\rho}_0^{\text{NL}}] + [\hat{\mathcal{H}}_0^{\text{NL}}, \hat{\rho}_0] + \frac{1}{4}[\hat{\mathcal{H}}_1, \hat{\rho}_1^{\dagger}] + \frac{1}{4}[\hat{\mathcal{H}}_1^{\dagger}, \hat{\rho}_1]. \tag{18.47}$$

In the derivation of Eq. (18.47) we have made use of the fact that

$$[\hat{\mathcal{H}}_0, \hat{\rho}_0] = 0. \tag{18.48}$$

That $\hat{\mathcal{H}}_0$ and $\hat{\rho}_0$ commute is a consequence of Eqs. (18.37) and (18.39). From these it follows immediately that $[\hat{\mathcal{H}}_0, \hat{\rho}_0]\,|I\rangle = 0$. The arbitrariness of the state $|I\rangle$ implies Eq. (18.48). Eq. (18.47) is the starting point for studies of electromagnetic rectification processes. A particular interesting situation appears in BCS-superconductors, where the rectification process gives rise to a nonlinear Meissner effect [138, 140]. We shall discuss the rectification process in superconductors in Chapter 25.

To go beyond the parametric approximation for 2ω-generation, and calculate the feedback of the generated second-harmonic field on the first-harmonic field one must carry out a third-order calculation. The related Liouville equation has the form

$$\hbar\left(\omega + \frac{i}{\tau}\right)\hat{\rho}_1 = [\hat{\mathcal{H}}_0, \hat{\rho}_1] + [\hat{\mathcal{H}}_1, \hat{\rho}_0] + [\hat{\mathcal{H}}_0^{\text{NL}}, \hat{\rho}_1] + [\hat{\mathcal{H}}_1, \hat{\rho}_0^{\text{NL}}] + \frac{1}{2}[\hat{\mathcal{H}}_1^{\dagger}, \hat{\rho}_2] + \frac{1}{2}[\hat{\mathcal{H}}_2, \hat{\rho}_1^{\dagger}]. \tag{18.49}$$

The last four commutators in Eq. (18.49) are responsible for the feedback process. In third order other physical phenomena can be described, e.g., third-harmonic generation, dynamical inelastic diffraction and degenerate four-wave mixing (dynamical holography).

18.4.3 Field-induced current density

It was realized in Section 15.3, that the Kubo formalism leads to the following expression for the mean value of the current density operator $(\hat{\mathbf{j}})$:

$$< \hat{\mathbf{J}} >= \text{Tr}\{\hat{\rho}\hat{\boldsymbol{\mathcal{J}}}\}, \tag{18.50}$$

where the second-quantized form of the current density operator is given by Eqs. (18.18)–(18.21). To determine the ensemble average, $< \hat{\mathbf{J}} >\equiv \mathbf{J}$, of the field-induced current density via the iterative procedure described in Subsection 18.4.2, we put together the two field-independent parts, $\hat{\boldsymbol{\mathcal{J}}}_F$ [Eq. (18.19)] and $\hat{\boldsymbol{\mathcal{J}}}_F^{\sigma}$ [Eq. (18.20] in one term

$$\hat{\boldsymbol{\mathcal{J}}}_0 = \hat{\boldsymbol{\mathcal{J}}}_F + \hat{\boldsymbol{\mathcal{J}}}_F^{\sigma}. \tag{18.51}$$

As for the density operator, we write the induced current density in a frequency expansion:

$$\mathbf{J}(\mathbf{r}, t) = \mathbf{J}_0^{\text{NL}}(\mathbf{r}) + \frac{1}{2}\sum_{m=1}^{\infty}[\mathbf{J}_m(\mathbf{r}; m\omega)e^{-im\omega t} + \mathbf{J}_m^{*}(\mathbf{r}; m\omega)e^{im\omega t}]. \tag{18.52}$$

The fact that there is no average current density in thermal equilibrium, expressed as

$$\mathbf{J}_0(\mathbf{r}) = \mathrm{Tr}\{\hat{\rho}_0 \hat{\boldsymbol{\mathcal{J}}}_0(\mathbf{r})\} = \sum_{I,J} \langle I| \hat{\rho}_0 |J\rangle \langle J| \hat{\boldsymbol{\mathcal{J}}}(\mathbf{r}) |I\rangle) = \sum_I P_I \langle I| \hat{\boldsymbol{\mathcal{J}}}_0(\mathbf{r}) |I\rangle = \mathbf{0},$$

(18.53)

is the reason that only a nonlinear dc current density, $\mathbf{J}_0^{\mathrm{NL}}(\mathbf{r})$, appears in Eq. (18.52). Since the current density operator $\hat{\boldsymbol{\mathcal{J}}}_1$ is proportional to $\mathbf{A}_T$, it is easy to see that the linear (ω-) part of the induced current density is to be calculated from the expression

$$\mathbf{J}_1(\mathbf{r};\omega) = \mathrm{Tr}\{\hat{\rho}_0 \hat{\boldsymbol{\mathcal{J}}}_1\} + \mathrm{Tr}\{\hat{\rho}_1 \hat{\boldsymbol{\mathcal{J}}}_0\}$$

(18.54)

in the frequency domain. For the 2ω-part one obtains

$$\mathbf{J}_2(\mathbf{r};2\omega) = \mathrm{Tr}\{\hat{\rho}_2 \hat{\boldsymbol{\mathcal{J}}}_0\} + \mathrm{Tr}\{\hat{\rho}_1 \hat{\boldsymbol{\mathcal{J}}}_1\},$$

(18.55)

and the nonlinear (dynamic) dc current density is determined as

$$\mathbf{J}_0^{\mathrm{NL}}(\mathbf{r}) = \mathrm{Tr}\{\hat{\rho}_0^{\mathrm{NL}} \hat{\boldsymbol{\mathcal{J}}}_0\} + \frac{1}{4}\mathrm{Tr}\{\hat{\rho}_1 \hat{\boldsymbol{\mathcal{J}}}_1^\dagger\} + \frac{1}{4}\mathrm{Tr}\{\hat{\rho}_1^\dagger \hat{\boldsymbol{\mathcal{J}}}_1\}.$$

(18.56)

In Eqs. (18.54)–(18.56) appear in second-quantized form the single-particle current-density operator in the frequency domain, viz.,

$$\hat{\mathbf{j}}_0(\mathbf{r};\omega) = -\frac{e^2}{m}\mathbf{A}_T(\mathbf{r}_e;\omega)\delta(\mathbf{r} - \mathbf{r}_e).$$

(18.57)

In Eq. (18.56) also the Hermitian conjugate $\hat{\mathbf{j}}_0^\dagger(\mathbf{r};\omega)$ enters. In the field-unquantized formalism $\hat{\mathbf{j}}_0^\dagger = \hat{\mathbf{j}}_0^*$.

18.4.4 Single-particle excitations in the occupation number basis

In the occupation number formalism the preferred basis states often are tensor products of eigenstates, $|n_i\rangle$, of the number operator $\hat{n}_i = \hat{b}_i^\dagger \hat{b}_i$, for each mode ($i$). Thus

$$|n_1, n_2, ..., n_\infty\rangle = |n_1\rangle \otimes |n_2\rangle \otimes ... \otimes |n_\infty\rangle.$$

(18.58)

The anticommutation rules for fermions

$$\{\hat{b}_i, \hat{b}_j^\dagger\} = \delta_{ij},$$

(18.59)

$$\{\hat{b}_i, \hat{b}_j\} = \{\hat{b}_i^\dagger, \hat{b}_j^\dagger\} = 0,$$

(18.60)

produces the correct statistics:

1. $\hat{b}_i^2 = (\hat{b}_i^\dagger)^2$; therefore $\hat{b}_i^\dagger \hat{b}_i^\dagger |0_i\rangle = 0$, which prevents two particles from occupying the same quantum state.

2. From the anticommutation rules the following relations are obtained:

$$\hat{b}_i^\dagger \left|0_i\right\rangle = \left|1_i\right\rangle, \quad \hat{b}_i \left|1_i\right\rangle = \left|0_i\right\rangle, \tag{18.61}$$

$$\hat{b}_i^\dagger \left|1_i\right\rangle = 0, \quad \hat{b}_i \left|0\right\rangle = 0. \tag{18.62}$$

3. Since (from point 2) $\hat{b}_i^\dagger \hat{b}_i \left|0_i\right\rangle = 0$, and $\hat{b}_i^\dagger \hat{b}_i \left|1_i\right\rangle = \left|1_i\right\rangle$ the number operator for mode i has eigenvalues zero and one, as required.

In order to use the occupation number formalism to calculate the various microscopic conductivity tensors (linear as well as all nonlinear types) it is apparent from Eqs. (18.13)–(18.16), and (18.19)–(18.21) that one needs to know the action of the annihilation and creation operators on a given state. One finds

$$\hat{b}_i \left|...n_i...\right\rangle = (-1)^{S_i} n_i \left|...n_i-1, ...\right\rangle, \tag{18.63}$$

and

$$\hat{b}_i^\dagger \left|...n_i...\right\rangle = (-1)^{S_i} (1 - n_i) \left|...n_i+1, ...\right\rangle, \tag{18.64}$$

where

$$S_i = n_1 + n_2 + ... + (n_i - 1). \tag{18.65}$$

The results obtained in Eqs. (18.63) and (18.64) follow directly from the fermion anticommutation rules. By a combination of Eqs. (18.63) and (18.64) one gets

$$\hat{b}_i^\dagger \hat{b}_i \left|..., n_i, ...\right\rangle = n_i \left|..., n_i, ...\right\rangle, \tag{18.66}$$

the expected result for the action of the single-mode number operator on a global fermion basis state.

CHAPTER 19

Surface Plasmons, Plasmaritons and Jellions: Classical Electrodynamics

In solid-state physics the so-called *jellium* model [85, 132, 184, 210, 213] has played an important role in the development of our understanding of the physics of many-body electron systems. In jellium it is assumed that the positive charge of the ions is spread uniformly through the solid. The mobile electrons collectively behave like an electron gas, in which the particles interact mutually by Coulomb's law ($\sim e^2/r$). Although the jellium model certainly is a very naive model, which does not allow to address important issues in condensed matter physics, it has a number of attractive aspects for theorists. Thus, exchange and correlation effects can be studied in detail in jellium, e.g. A detailed account of these effects is presented in [184].

When the jellium is excited by the field from a prescribed (also called "external") current density distribution excitations of the ground state of the electron gas occur. Roughly speaking one observes two main types of excitations: (i) Single-particle excitations of electron-hole pairs in the Fermi sea, and (ii) collective excitations of many electrons [88, 94, 178]. The two types interact, however. Hence, the collective modes may decay by transfer of energy and momentum to single-particle excitations. The collective modes excited by the longitudinal (L) part of the external electromagnetic field are called *plasmons*, and those excited by the transverse (T) part of the external field we call *plasmaritons*, a name dating back to Patel and Slusher [203], at least. In the majority of the literature the collecitve T-excitations are referred to as *polaritons* [61, 169, 280]. However, I prefer to reserve the name polariton to excitations involving induced polarizations of the ion-lattice. The parts plasm-, and plasmarit-, underline that one is dealing with excitations in the electron plasma (charged gas).

In mescoscopic electrodynamics collective modes excited at a condensed matter surface play a particular role, and these modes are called surface plasmons (SPL) and surface plasmaritons (SPT). In constrast to what is most often believed these modes are seldom eigenmodes (resonances) related to excitations by external sources. The SPL- and SPT-modes interact and form new eigenmodes. *I have given these*

DOI: 10.1201/9781003029458-19

more general eigenmodes the new name jellions, and the physics of the jellions is a central issue for what follows. In Chapter 20, the description culminates with a second-quantized theory for collective fields in jellium. Surface jellions are interface entangled plasmon-plasmariton modes, and a rigorous theory for these has appeared only recently [129]. Outside the region of entanglement one regains the characteristics of surface plasmons and surface plasmaritons.

19.1 SCREENED GREEN FUNCTION APPROACH FOR COLLECTIVE JELLIUM EXCITATIONS

19.1.1 Green function of an infinite jellium with nonlocal linear response

In the framework of the microscopic classical Maxwell-Lorentz theory (equations) the local electric field, $\mathbf{E}(\mathbf{r};\omega)$, satisfies in the space $(\mathbf{r})$-frequency (ω) domain of the wave equation

$$(\mathbf{U}\Box(\mathbf{r};\omega) - \boldsymbol{\nabla}\boldsymbol{\nabla}) \cdot \mathbf{E}(\mathbf{r};\omega) = -i\mu_0\omega\mathbf{J}(\mathbf{r};\omega), \qquad (19.1)$$

where

$$\Box(\mathbf{r};\omega) = \nabla^2 + \left(\frac{\omega}{c}\right)^2 \qquad (19.2)$$

is the d' Alembertian operator. An infinitely extended jellium possesses in its ground state translational invariance in space and time, and its spatially nonlocal linear response to a local field therefore is described by a microscopic constitutive equation of the form

$$\mathbf{J}(\mathbf{r};\omega) = \int_{-\infty}^{\infty} \boldsymbol{\sigma}^{\infty}(\mathbf{r} - \mathbf{r}';\omega) \cdot \mathbf{E}(\mathbf{r}';\omega)d^3r'. \qquad (19.3)$$

The conductivity tensor $\boldsymbol{\sigma}^{\infty}(\mathbf{r} - \mathbf{r}';\omega)$ of the infinite (superscript ∞) jellium relates the current density, $\mathbf{J}(\mathbf{r};\omega)$, at the point $\mathbf{r}$ to the local electric field, $\mathbf{E}(\mathbf{r}';\omega)$ in surrounding points $\mathbf{r}'$. Together Eqs. (19.1)–(19.3) show that $\mathbf{E}(\mathbf{r};\omega)$ satisfies the integro-differential equation

$$\left\{\mathbf{U}\left[\nabla^2 + \left(\frac{\omega}{c}\right)^2\right] - \boldsymbol{\nabla}\boldsymbol{\nabla}\right\} \cdot \mathbf{E}(\mathbf{r};\omega) + i\mu_0\omega \int_{-\infty}^{\infty} \boldsymbol{\sigma}^{\infty}(\mathbf{r} - \mathbf{r}';\omega) \cdot \mathbf{E}(\mathbf{r}';\omega)d^3r' = \mathbf{0}.$$

$$(19.4)$$

A screened (also called a pseudo-vacuum) dyadic Green function $\mathbf{G}(\mathbf{r} - \mathbf{r}';\omega)$ now is introduced (indirectly) as a solution to the integro-differential equation

$$\left\{\mathbf{U}\left[\nabla^2 + \left(\frac{\omega}{c}\right)^2\right] - \boldsymbol{\nabla}\boldsymbol{\nabla}\right\} \cdot \mathbf{G}(\mathbf{r} - \mathbf{r}';\omega)$$

$$+ i\mu_0\omega \int_{-\infty}^{\infty} \boldsymbol{\sigma}^{\infty}(\mathbf{r} - \mathbf{r}'';\omega) \cdot \mathbf{G}(\mathbf{r}'' - \mathbf{r}';\omega)d^3r'' = \mathbf{U}\delta(\mathbf{r} - \mathbf{r}'). \qquad (19.5)$$

For practical reasons (and in line with the original article [135,137] on pseudo-vacuum Green functions), we have used the form $+\mathbf{U}\delta(\mathbf{r} - \mathbf{r}')$ on the right side of Eq. (19.5) and not $-\mathbf{U}\delta(\mathbf{r} - \mathbf{r}')$ as is most often seen when vacuum $[\boldsymbol{\sigma}^\infty = 0]$ Green functions are introduced. By inserting the Fourier integrals $[\mathbf{R} \equiv \mathbf{r} - \mathbf{r}']$

$$\mathbf{G}(\mathbf{R};\omega) = (2\pi)^{-3} \int_{-\infty}^{\infty} \mathbf{G}(\mathbf{q},\omega) e^{i\mathbf{q}\cdot\mathbf{R}} d^3q, \tag{19.6}$$

$$\boldsymbol{\sigma}^\infty(\mathbf{R};\omega) = (2\pi)^{-3} \int_{-\infty}^{\infty} \boldsymbol{\sigma}^\infty(\mathbf{q},\omega) e^{i\mathbf{q}\cdot\mathbf{R}} d^3q, \tag{19.7}$$

and the plane-wave expansion of the delta function, $\delta(\mathbf{R}) = (2\pi)^{-3} \int_{-\infty}^{\infty} e^{i\mathbf{q}\cdot\mathbf{R}} d^3q$, in Eq. (19.5), it follows that the screened Green function in the wave vector $(\mathbf{q})$-frequency (ω) representation satisfies the algebraic equation $[\mathbf{q} \cdot \mathbf{q} = q^2]$

$$\left\{ \left[\left(\frac{\omega}{c}\right)^2 - q^2 \right] \mathbf{U} + \mathbf{q}\mathbf{q} + i\mu_0\omega\boldsymbol{\sigma}^\infty(\mathbf{q},\omega) \right\} \cdot \mathbf{G}(\mathbf{q},\omega) = \mathbf{U}. \tag{19.8}$$

The rotational symmetry of the jellium around a given wave-vector direction $\boldsymbol{\kappa} = \mathbf{q}/q$, implies that the conductivity tensor's most general form is

$$\boldsymbol{\sigma}^\infty(\mathbf{q},\omega) = \sigma_T(q,\omega)(\mathbf{U} - \boldsymbol{\kappa}\boldsymbol{\kappa}) + \sigma_L(q,\omega)\boldsymbol{\kappa}\boldsymbol{\kappa}, \tag{19.9}$$

where $\sigma_T(q,\omega)$ and $\sigma_L(q,\omega)$ are the conductivities related to the transverse (T) and longitudinal (L) current densities generated by transverse $(\mathbf{E}_T)$ and longitudinal $(\mathbb{E}_L)$ local electric fields. These microscopic conductivities are scalar quantities which depend on the magnitude (q) of the wave vector and on the frequency. By insertion of Eq. (19.9) in Eq. (19.8), and by writing the unit tensor as $\mathbf{U} = (\mathbf{U} - \boldsymbol{\kappa}\boldsymbol{\kappa}) + \boldsymbol{\kappa}\boldsymbol{\kappa}$, one obtains

$$[N_T(q,\omega)(\mathbf{U} - \boldsymbol{\kappa}\boldsymbol{\kappa}) + N_L(q,\omega)\boldsymbol{\kappa}\boldsymbol{\kappa}] \cdot \mathbf{G}(\mathbf{q},\omega) = \mathbf{U}, \tag{19.10}$$

where

$$N_T(q,\omega) = \left(\frac{\omega}{c}\right)^2 \left[1 + \frac{i}{\varepsilon_0\omega}\sigma_T(q,\omega)\right] - q^2, \tag{19.11}$$

and

$$N_L(q,\omega) = \left(\frac{\omega}{c}\right)^2 \left[1 + \frac{i}{\varepsilon_0\omega}\sigma_L(q,\omega)\right]. \tag{19.12}$$

The quantities

$$\varepsilon_T(q,\omega) = 1 + \frac{i}{\varepsilon_0\omega}\sigma_T(q,\omega), \tag{19.13}$$

and

$$\varepsilon_L(q,\omega) = 1 + \frac{i}{\varepsilon_0\omega}\sigma_L(q,\omega), \tag{19.14}$$

the reader may recognize as the relative dielectric functions of transverse (ε_T) and longitudinal (ε_L) dynamics. These must be determined by a quantum mechanical calculation.

The solution for the screened Green tensor satisfying Eq. (19.10) obviously is given by

$$\mathbf{G}(\mathbf{q}, \omega) = \frac{\mathbf{U} - \kappa\kappa}{N_T(q, \omega)} + \frac{\kappa\kappa}{N_L(q, \omega)}. \tag{19.15}$$

Resonances occur in the dyadic Green function for

$$N_T(q, \omega) = \left(\frac{\omega}{c}\right)^2 \varepsilon_T(q, \omega) - q^2 = 0, \tag{19.16}$$

and

$$N_L(q, \omega) = \left(\frac{\omega}{c}\right)^2 \varepsilon_L(q, \omega) = 0. \tag{19.17}$$

The relations Eqs. (19.16) and (19.17) are the dispersion relations for transverse and longitudinal bulk excitations in jellium. For small q-values, $N_L(q, \omega) = 0$ and $N_T(q, \omega) = 0$ reduce to the dispersion relation for *bulk plasmons* and *bulk plasmaritons*.

19.1.2 Adjacent jellium-vacuum half-spaces: The SCIB model

Independent of any details, the external source will always be located outside the jellium, and this means that one must consider types of dyadic Green functions more general than the one for an infinitely extended jellium, viz.,

$$\mathbf{G}(\mathbf{R}, \omega) = \frac{1}{(2\pi)^3} \int_{-\infty}^{\infty} \left[\frac{\mathbf{U} - \kappa\kappa}{N_T(q, \omega)} + \frac{\kappa\kappa}{N_L(q., \omega)}\right] e^{i\mathbf{q}\cdot\mathbf{R}} d^3q. \tag{19.18}$$

The perhaps simplest generalization is for adjacent jellium-vacuum half-spaces, but even in this case the construction of the dyadic Green function constitutes an insurmountable theoretical problem if the density profile (spill-out) at the surface is to be incorporated in the Green function including spatially nonlocal effects. However, if one assumes (*i*) that the jellium electrons are scattered specularly when hitting the surface (located at $z = 0$) and (*ii*) that the wave functions of the incoming and scattered electron do not interfere it is possible to construct a pseudo-vacuum Green function adequate for studies of the dynamics of surface jellions, plasmaritons and plasmons. The model based on the assumptions (*i*) and (*ii*) is called the semiclassical infinite-barrier model (SCIB-model). Translation symmetry parallel ($\|$) to the surface allows one to Fourier analyze the half-space conductivity tensor,

$$\sigma(\mathbf{r}, \mathbf{r}'; \omega) = \sigma(\mathbf{r}_\| - \mathbf{r}'_\|, z, z'; \omega), \tag{19.19}$$

in the coordinates parallel [$\mathbf{r}_\| = (x, y)$] to the surface. Hence,

$$\sigma(\mathbf{r}_\| - \mathbf{r}'_\|, z, z'; \omega) = (2\pi)^{-2} \int_{-\infty}^{\infty} \sigma(z, z'; \mathbf{q}_\|, \omega) e^{i\mathbf{q}_\|\cdot(\mathbf{r}_\| - \mathbf{r}'_\|)} d^2q_\|. \tag{19.20}$$

It has been shown in the literature [88, 92, 171] that the conductivity tensor of the SCIB-model, can be expressed in terms of the infinite-space conductivity tensor as follows:

$$\sigma_{ij}(z, z'; \mathbf{q}_\parallel, \omega) = \Theta(z)\Theta(z')[\sigma_{ij}^\infty(z - z'; \mathbf{q}_\parallel, \omega) + \xi_j \sigma_{ij}^\infty(z + z'; \mathbf{q}_\parallel, \omega)] \qquad (19.21)$$

where $\xi_j = 1$ for $j = x, y$ and $\xi_j = -1$ for $j = z$.

A theoretical method going far beyond the scope of the present chapter allows one to construct the screened Green function of the SCIB model [135]. The result appears in the form of a plane-wave expansion over the surface plane, a so-called Weyl expansion [185, 266]. Thus,

$$\mathbf{G}(\mathbf{r}, \mathbf{r}'; \omega) = (2\pi)^{-2} \int_{-\infty}^{\infty} \mathbf{S}^{-1} \cdot \mathbf{G}(z, z'; q_\parallel, \omega) \cdot \mathbf{S} e^{i\mathbf{q}_\parallel \cdot (\mathbf{r}_\parallel - \mathbf{r}_\parallel')} d^2 q_\parallel', \qquad (19.22)$$

where

$$\mathbf{S} = \frac{1}{q_\parallel} \begin{pmatrix} q_{\parallel,x} & q_{\parallel,y} & 0 \\ -q_{\parallel,y} & q_{\parallel,x} & 0 \\ 0 & 0 & q_\parallel \end{pmatrix} \qquad (19.23)$$

is a transformation which rotates the vector $(q_{\parallel,x}, q_{\parallel,y}, 0)$ into $(q_\parallel, 0, 0)$. The Green function in the integrand of Eq. (19.22) has the general matrix form

$$\mathbf{G}(z, z'; q_\parallel, \omega) = \begin{pmatrix} G_{xx} & 0 & G_{xz} \\ 0 & G_{yy} & 0 \\ G_{zx} & 0 & G_{zz} \end{pmatrix} \qquad (19.24)$$

The explicit forms of the tensor components of $\mathbf{G}(z, z'; q_\parallel, \omega)$ depend on which half-space the source $(z' < 0, z' > 0)$ and observation $(z < 0, z > 0)$ points are located.

19.1.3 Green tensor $\mathbf{G}^{<<}(z, z'; q_\parallel, \omega)$

Experimental investigations are carried out under conditions where the source and observation points both are located in vacuum $(z' < 0, z < 0)$. The related Green tensor is given the superscript $<<$, and it has three parts, i.e.

$$\mathbf{G}^{<<}(z, z'; q_\parallel, \omega) = \mathbf{D}^{<<}(z - z'; q_\parallel, \omega) + \mathbf{I}^{<<}(z + z'; q_\parallel, \omega) + \mathbf{g}^{<<}(z - z'; q_\parallel, \omega).$$
$$(19.25)$$

The tensor $\mathbf{D}^{<<}$ is responsible for the direct (D) field propagation between the source and the observation points, and $\mathbf{I}^{<<}$ connects these points via reflection from the surface; so-called indirect (I) propagation. For $z = z'$ the Green tensor also has a self-field part $g^{<<}$. The interested reader is referred to my paper "Tensor-product structure of a new electromagnetic propagator for nonlocal surface optics of metals" [137] for a detailed analysis of the structure of $\mathbf{G}(z, z'; q_\parallel, \omega)$ in the various domains $[z, z' < 0; z, z' > 0; z' < 0, z > 0; z' > 0, z < 0]$.

The dynamics of the various collective surface excitations are hidden in the indirect part of the propagator, which is given by

$$\mathbf{I}^{<<}(z + z'; q_{\parallel}, \omega) = \frac{\exp\left[-iq_{\perp}^0(z + z')\right]}{2iq_{\perp}^0}\left[r_s(q_{\parallel}, \omega)\mathbf{e}_y\mathbf{e}_y + r_p(q_{\parallel}, \omega)\mathbf{e}_r\mathbf{e}_i\right], \qquad (19.26)$$

where $\mathbf{e}_y = (0, 1, 0)$, $\mathbf{e}_i = (q_{\perp}^0, 0, -q_{\parallel})c/\omega$ and $\mathbf{e}_r = (-q_{\perp}^0, 0, -q_{\parallel})c/\omega$ are unit vectors. The quantity $q_{\perp}^0$ is given by

$$q_{\perp}^0 = \left[\left(\frac{\omega}{c}\right)^2 - q_{\parallel}^2\right]^{1/2}. \qquad (19.27)$$

In the Weyl expansion of the propagator in Eq.(19.22) over the entire surface $(\mathbf{q}_{\parallel})$ plane in wave-vector space, $q_{\perp}^0$ is either real $(q_{\parallel} < \omega/c)$ or purely imaginary $(q_{\parallel} > \omega/c)$. The quantities $r_s(q_{\parallel}, \omega)$ and $r_p(q_{\parallel}, \omega)$ are s- and p-polarized GENERALIZED amplitude reflection coefficients for the plane-wave mode characterized by $(q_{\parallel}, \omega)$. These coefficients depend on collective and single-particle excitations of the jellium. For $q_{\parallel} << q_F$, where q_F is the electron Fermi wave number, the low-energy collective excitations determine $r_s(q_{\parallel}, \omega)$ and $r_p(q_{\parallel}, \omega)$, essentially. In the collective-mode approximation it turns out [137] that

$$r_s(q_{\parallel}, \omega) = \frac{q_{\perp}^0 - \kappa_{\perp}^T}{q_{\perp}^0 + \kappa_{\perp}^T} \qquad (19.28)$$

and

$$r_p(q_{\parallel}, \omega) = \frac{q_{\perp}^0 \varepsilon_T(\kappa_T, \omega) - \kappa_{\perp}^T - (q_{\parallel}/\kappa_L)^2 \kappa_{\perp}^L \varepsilon_T(\kappa_T, \omega)[\varepsilon_L^{-1}(q_{\parallel}, \omega) - 1]}{q_{\perp}^0 \varepsilon_T(\kappa_T, \omega) + \kappa_{\perp}^T + (q_{\parallel}/\kappa_L)^2 \kappa_{\perp}^L \varepsilon_T(\kappa_T, \omega)[\varepsilon_L^{-1}(q_{\parallel}, \omega) - 1]}. \qquad (19.29)$$

The quantities $\kappa_{\perp}^T(q_{\parallel}, \omega)$ and $\kappa_{\perp}^L(q_{\parallel}, \omega)$ are determined by the dispersion relations [cf. Eqs. (19.16) and (19.17)]

$$N_T(q_{\parallel}, \kappa_{\perp}^T, \omega) = 0, \qquad (19.30)$$

and

$$N_L(q_{\parallel}, \kappa_{\perp}^L, \omega) = 0. \qquad (19.31)$$

In the following section we shall see that an analysis of the resonance structure of the generalized p-polarized reflection coefficient gives one the dispersion relations $q_{\parallel} = q_{\parallel}(\omega)$, for surface plasmons, surface plasmaritons and surface jellions.

19.2 DISPERSION RELATIONS FOR COLLECTIVE BULK AND SURFACE MODES

The transverse $[\varepsilon_T(\mathbf{q}, \omega)]$ and longitudinal $[\varepsilon_L(\mathbf{q}, \omega)]$ dielectric functions appearing in Eqs. (19.16) and (19.17) can be calculated on the basis of the Kubo formalism.

A many-electron version is needed in some cases, e.g., for studies of the electromagnetic properties of BCS-superconductors. Two types of excitations enter $\varepsilon_L(\mathbf{q}, \omega)$ and $\varepsilon_T(\mathbf{q}, \omega)$: ($i$) single-particle excitations in the Fermi sea. These generate electron-hole pairs, in particular in the vicinity of the Fermi surface, and (ii) collective excitations, viz. plasmons [in $\varepsilon_L(\mathbf{q}, \omega)$] and plasmaritons [in $\varepsilon_T(\mathbf{q}, \omega)$]. The excitation energy needed from the external source is (much) lower for collective excitations than for single-particle excitations. Furthermore, it is known [94, 178, 210, 213] that the collective-mode contributions to the dielectric functions can be calculated quite accurately within the framework of the Random-Phase-Approximation (RPA) [40, 178, 209]. The RPA approach is a type of one-electron theory in which the eigenstates of the jellium Hamiltonian ($\mathcal{H}_0$) are direct products of single-particle-like states. In RPA, the jellium ground state, $|G\rangle$, is given by [10, 144]

$$|G\rangle = \Pi_{k \leq k_F, s}\, \hat{b}^{\dagger}_{\mathbf{k}, s} |0\rangle \tag{19.32}$$

at $T = 0K$. The global vacuum is denoted by $|0\rangle$. The excited single-particle eigenstates are of the particle-hole type, i.e.,

$$|\mathbf{k}', s'; \mathbf{k}, s\rangle = \hat{b}^{\dagger}_{\mathbf{k}', s'} \hat{b}_{\mathbf{k}, s} |G\rangle\,, \tag{19.33}$$

for an electron in $(\mathbf{k}', s')$ and a hole in $(\mathbf{k}, s)$. In the notation $|\mathbf{k}', s'; \mathbf{k}, s\rangle$ only the two states involved in the given excitation $(\mathbf{k}, s) \Rightarrow (\mathbf{k}', s')$ are kept explicit in the global state vector.

The collective excitations are related to the pole structure of the Green tensor. In the infinite medium case [Eq. (19.15)], the poles are obtained from $N_T(q, \omega) = 0$ and $N_L(q, \omega) = 0$ [see Eqs. (19.16) and (19.17)]. The plasmon dispersion relation is obtained by calculating $\varepsilon_L(q, \omega)$ to order q^2 in a series expansion from $q = 0$. In this expansion there is no first-order term because the jellium possesses inversion symmetry [$\mathbf{q}$ and $-\mathbf{q}$ give the same $\varepsilon_L(q, \omega)$]. Let us write $\varepsilon_L(q, \omega)$ in the usual "hydrodynamic" form [92]

$$\varepsilon_L(q, \omega) = 1 - \frac{\omega_p^2}{\omega^2 - Dq^2}, \tag{19.34}$$

where $\omega_p = [n_0 e^2/(m\varepsilon_0)]^{1/2}$ is the bulk plasma frequency, and $D \sim (3/5)v_f^2$ is the so-called diffusion coefficient. The solid-state plasma is characterized by the equilibrium electron density n_0 and the Fermi speed, v_F [which can be expressed in terms of n_0, if wished]. Since $D^{1/2}q << \omega$, the form in Eq. (19.34) effectively is a series expansion to order q^2. According to Eq. (19.17), the bulk plasma dispersion is given by $\varepsilon_L(\kappa_L, \omega) = 0$ in implicit form, and explicitly as

$$D^{1/2}\kappa_L = (\omega^2 - \omega_p^2)^{1/2}, \tag{19.35}$$

where the plasmon wave number has been denoted by κ_L.

The presence of the (large) q^2-term in Eq. (19.11) implies that it is sufficient to use the long-wavelength ($q \to 0$) expression for the transverse dielectric function in the following, viz,

$$\varepsilon_T(q \to 0, \omega) \equiv \varepsilon_T(\omega) = 1 - \left(\frac{\omega_p}{\omega}\right)^2. \tag{19.36}$$

In the $q \to 0$ limit, the L- and T-dielectric functions are equal, $\varepsilon_T(q \to 0, \omega) = \varepsilon_L(q \to 0, \omega)$. Obviously this must be so because one cannot distinguish transverse and longitudinal displacement fields from each other for $q = 0$. From, the condition $N_T(\kappa_T, \omega) = 0$, one obtains the following dispersion relation (wave number: κ_T) for the bulk plasmariton mode:

$$c\kappa_T = (\omega^2 - \omega_p^2)^{1/2}. \tag{19.37}$$

It should be noted that the L- and T-dispersion relations apart from respective $D^{1/2}(\sim v_F)$ and c factors are identical. The speed of light cannot enter the plasmon dispersion relation since this is associated solely with the Coulomb field between the electrons [and of course also the Pauli principles for fermions].

The dispersion relation for the collective surface modes can be obtained in an elegant manner with the help of the so-called causal (CAU) conductivity tensor, $\boldsymbol{\sigma}^{\mathrm{CAU}}$, relating the microscopic current density in the jellium to the *incoming (external)* prescribed electric field, $\mathbf{E}^{\mathrm{ext}}$. Hence,

$$\mathbf{J}(z, \mathbf{q}_\parallel, \omega) = \int \boldsymbol{\sigma}^{\mathrm{CAU}}(z, z'; \mathbf{q}_\parallel, \omega) \cdot \mathbf{E}^{\mathrm{ext}}(z'; \mathbf{q}_\parallel, \omega) dz', \tag{19.38}$$

for the case of a jellium half-space. The reflected electric field, $\mathbf{E}_r$, in turn is given by

$$\mathbf{E}_r(z; \mathbf{q}_\parallel, \omega) = \mathbf{E}^{\mathrm{ext}}(z; \mathbf{q}_\parallel, \omega)$$
$$- i\mu_0\omega \int \mathbf{I}^{<<}(z + z'; q_\parallel, \omega) \cdot \boldsymbol{\sigma}^{\mathrm{CAU}}(z', z''; \mathbf{q}_\parallel, \omega) \cdot \mathbf{E}^{\mathrm{ext}}(z''; \mathbf{q}_\parallel, \omega) dz'' dz'.$$
$$\tag{19.39}$$

The self-sustaining solutions are obtained letting $\mathbf{E}^{\mathrm{ext}} \to \mathbf{0}$. In order to obtain a solution for the reflected field in this situation one must require that $r_p(q_\parallel, \omega) \to \infty$, or $r_s(q_\parallel, \omega) \to \infty$. Since $q_\perp^0 + \kappa_\perp^T \neq 0$, obviously, there is never a pole in the s-polarized reflection coefficient, given in Eq. (19.28). Self-sustaining solutions *can* be found for p-polarized light, as it is well known from elementary studies [1, 33]. The resonance condition is obtained by setting the denominator in Eq. (19.29) equal to zero, that is

$$q_\perp^0 \varepsilon_T(\kappa_T, \omega) + \kappa_\perp^T + \left(\frac{q_\parallel}{\kappa_L}\right)^2 \kappa_\perp^L \varepsilon_T(\kappa_T, \omega)[\varepsilon_L^{-1}(q_\parallel, \omega) - 1] = 0. \tag{19.40}$$

The condition in Eq. (19.40) gives one the dispersion relation for the collective surface modes, $q_\parallel = q_\parallel(\omega)$, using for instance the expression for ε_L and ε_T given in Eqs. (19.34) and (19.36), for simplicity.

For the elementary quanta associated with the dispersion relation I suggest the new name JELLIONS. These quanta are fundamental in the sense that they represent the *true* surface bosons following from a solution of the eigenvalue problem for coupled photon, surface plasmon and surface polaritons. The surface plasmon and plasmariton are massive boson fields, whereas the photon is a massless boson field. When the surface plasmon and plasmariton fields are strongly coupled through the photon field the dispersion relation in Eq. (19.40) has three branches [129, 163], as shown in Fig. 19.1–19.3. The numerical calculation was carried out using for ε_L and ε_T the expressions in Eqs. (19.34) and (19.36). Outside the strong coupling regime the dispersion has only two branches. These branches of the dispersion relation are those of the surface plasmon (L-mode) and surface plasmariton (T-mode).

The plasmon mode disappears for $D \to 0$, and from Eq. (19.40) on then obtains

$$q_\perp^0 \varepsilon_T(\omega) + \kappa_\perp^T = 0. \tag{19.41}$$

By inserting $q_\perp^0 = [(\omega/c)^2 - q_\parallel^2]^{1/2}$ and $\kappa_\perp^T = [(\omega/c)^2 \varepsilon_T(\omega) - q_\parallel^2]^{1/2}$ in Eq. (19.41) the dispersion relation of the surface plasmariton takes the form

$$q_\parallel = \frac{\omega}{c} \left(\frac{\varepsilon_T(\omega)}{1 + \varepsilon_T(\omega)} \right)^{1/2}, \tag{19.42}$$

which is well known from textbooks on the optical properties of solids. With $\varepsilon_T(\omega)$ given by Eq. (19.36) one obtain

$$q_\parallel = \frac{\omega}{c} \left(\frac{\omega^2 - \omega_p^2}{2\omega^2 - \omega_p^2} \right)^{1/2}. \tag{19.43}$$

As we shall realize in Section 21.2 the form of the dispersion relation for the surface plasmariton with its two branches has it routes in the first-quantized (wave mechanical) picture. The non-retarded ($c \to \infty$) surface plasmon dispersion relation is obtained inserting the limiting forms $q_\perp^0 \to iq_\parallel$, $\kappa_\perp^T \to iq_\parallel$, and $\varepsilon_T(\kappa_T, \omega) \to \varepsilon(\omega)$ in Eq. (19.40). Thus, one obtains

$$1 + \varepsilon(\omega) + i \frac{q_\parallel \kappa_\perp^L}{\kappa_L^2} \varepsilon(\omega)[1 - \varepsilon_L^{-1}(q_\parallel, \omega)] = 0. \tag{19.44}$$

This somewhat complicated expression can be reduced to

$$q_\parallel = \frac{\omega}{\sqrt{D}} \left[1 - \left(\frac{\omega_{pS}}{\omega} \right)^2 \right], \tag{19.45}$$

where $\omega_{pS} = \omega_p/\sqrt{2}$ is the surface plasma frequency. The reader may verify Eq. (19.45) using Eq. (19.34) [with the replacement $q \to q_\parallel$] and (19.35), and $\kappa_\perp^L = (\kappa_L^2 - q_\parallel^2)^{1/2}$. The form given in Eq. (19.45) is the so-called Ritchie dispersion relation [222]; see also [33].

19.3 GENERAL STRUCTURE OF JELLIUM SURFACE MODES

It is possible to calculate the s- and p-polarized reflection coefficients in the general case where both collective and single-particle excitations in the jellium contribute

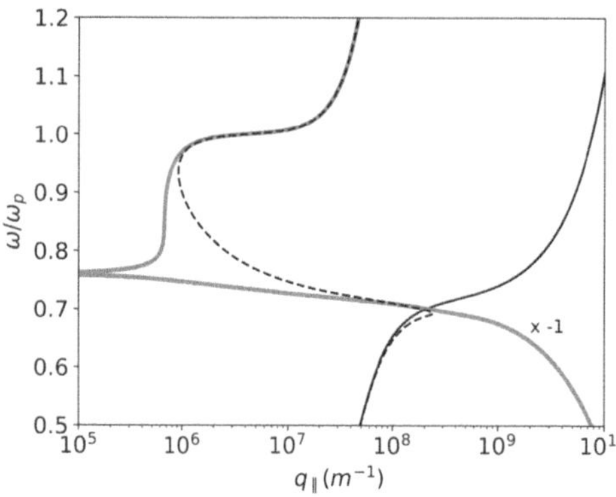

Figure 19.1 Dispersion relation $\omega = \omega(\mathrm{Re}\,q_{\parallel})$ [normalized to ω_p]. Three branches are indicated by thick, thin, and dashed lines. Note that the thick-line branch has $\mathrm{Re}\,q_{\parallel} < 0$ for $\omega \leqslant \omega_p^s = \omega_p/\sqrt{2}$. The numerical calculations are for an Al (jellium) electron density $n_0 = 1.81 \times 10^{29}\,\mathrm{m}^{-3}$. The relaxation time used is $\tau = 7.5 \times 10^{-15}\mathrm{s}$. Qualitatively significant single-particle excitations occur for $q_{\parallel} \geqslant k_F \sim 1.8 \times 10^{10}\mathrm{m}^{-1}$.

Figure 19.2 Three-branch eigenmode ($\tau \to \infty$) dispersion relation $\omega = \omega(q_{\parallel})$ [normalized to ω_p]. The white domain in the (ω/ω_p)-$q_{\parallel}$ plane indicates the region in which the eigenmodes are exponentially confined, which means that all three underlying modes decay exponentially with the distance from the surface.

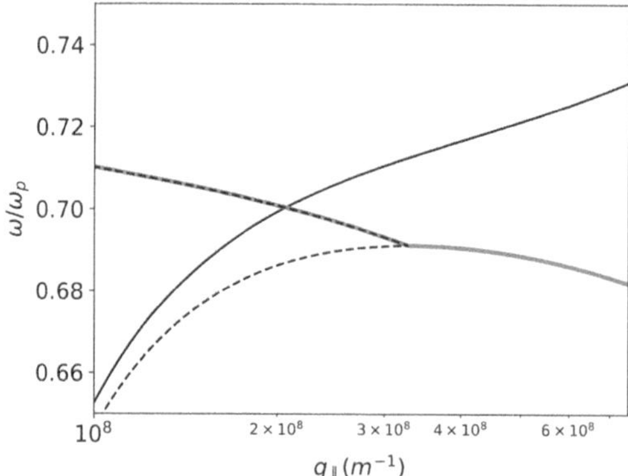

Figure 19.3 Enlarged plot of the three branches of the eigenmode dispersion relation in the range in which the surface plasmon-plasmariton entanglement is pronounced. For a detailed account of the interface entangled plasmon-plasmariton modes see [129].

to the field-matter interaction. The Green function formalism for the general case is quite complicated, and here we shall just cite the final result [137]. For the s-polarized amplitude reflection coefficient one obtains

$$r_s(q_\parallel,\omega) = \frac{q_\perp^0 L(q_\parallel,\omega) - 1}{q_\perp^0 L(q_\parallel,\omega) + 1}, \tag{19.46}$$

where

$$L(q_\parallel,\omega) = \frac{2i}{\pi} \int_0^\infty \frac{\cos(q_\perp 0^+)}{N_T(q)} dq_\perp, \tag{19.47}$$

in the notation

$$\lim_{z\to 0^+} \int_0^\infty F(q_\perp,z) dq_\perp \equiv \int_0^\infty F(q_\perp,0^+) dq_\perp. \tag{19.48}$$

Obviously, the reflection of an s-polarized field can only give rise to transversely polarized excitations in the jellium. The function $N_T(q)$ contains the transverse dielectric function $\varepsilon_T(q,\omega)$; see Eqs. (19.11) and (19.13). The integral in Eq. (19.47) is calculated by contour integration in a complex $q_\perp$-plane. On the basis of the Lindhard-Mermin transverse dielectric function [RPA approach] the collective and single-particle excitations can be identified by studying the analytical properties of $\varepsilon_T(q_\parallel,q_\perp,\omega)$. The collective excitations are related to the pole(s) and the single-particle excitation spectrum is contained in the branch cuts. In the nonretarded limit $(c \to \infty)$ one has $N_T(q) \to -q^2$ and $q_\perp^0 \to iq_\parallel$. Contour integration in the upper half-plane around the pole at $q_\perp = iq_\parallel$ gives $L(q_\parallel,\omega) = (iq_\parallel)^{-1}$. In consequence $r_s = 0$, as one might have expected.

The p-polarized amplitude reflection coefficient is given by

$$r_p(q_\parallel,\omega) = \frac{q_\perp^0 - (\omega/c)^2 M(q_\parallel,\omega)}{q_\perp^0 + (\omega/c)^2 M(q_\parallel,\omega)}, \tag{19.49}$$

where

$$M(q_\parallel,\omega) = \frac{2i}{\pi} \int_0^\infty \left[\frac{q_\perp^2}{N_T(q)} + \frac{q_\parallel^2}{N_L(q)} \right] \frac{\cos(q_\perp 0^+)}{q^2} dq_\perp. \tag{19.50}$$

The p-polarized reflection hence depends on both the transverse $(\sim N_T)$ and longitudinal $(\sim N_L)$ excitation spectrum of the jellium, and each of the spectra has contributions from both collective and single-particle processes.

Unlike $L(q_\parallel,\omega)$, which contains a contribution from the pole at $q_\perp = iq_\parallel$ in the nonretarded limit, $M(q_\parallel,\omega)$'s integral representation does not have poles at $q_\perp = \pm iq_\parallel (q = 0)$. This follows from the fact that

$$\lim_{q_\perp \to iq_\parallel} \left[\frac{q_\perp^2}{N_T(q)} + \frac{q_\parallel^2}{N_L(q)} \right] = q_\parallel^2 \lim_{q_\perp \to iq_\parallel} \left[\frac{1}{N_L(q)} - \frac{1}{N_T(q)} \right] \propto q_\parallel^2 \lim_{q_\perp \to iq_\parallel} [q^2]. \tag{19.51}$$

This asymptotic from divided by q^2 [see Eq. (19.50)] proves that $M(q_\parallel,\omega)$ has no pole at $q_\perp = iq_\parallel$.

In certain limiting cases it is justified to neglect the transverse dynamics completely. In the language of photons, these are cases where only virtual ($\mathbf{\nabla} \times \mathbf{E} = 0$) photons are present. In these nonretarded ($c \to \infty$) situations, and with the neglect of the T-part of the $M(q_{\parallel}, \omega)$-integral, the p-polarized reflection coefficient takes the form

$$r_p(q_{\parallel}, \omega) = \frac{\pi - q_{\parallel} \displaystyle\int_{-\infty}^{\infty} \frac{\exp(iq_{\perp}0^+)dq_{\perp}}{q^2 \varepsilon_L(q, \omega)}}{\pi + q_{\parallel} \displaystyle\int_{-\infty}^{\infty} \frac{\exp(iq_{\perp}0^+)dq_{\perp}}{q^2 \varepsilon_L(q, \omega)}}. \tag{19.52}$$

When single-particle excitations are included in the analysis surface eigenmodes also may exist in the s-polarized case. The dispersion relation for these modes is given in implicit form by

$$\left[\left(\frac{\omega}{c} \right)^2 - q_{\parallel}^2 \right]^{1/2} L(q_{\parallel}, \omega) + 1 = 0. \tag{19.53}$$

Without single-particle excitations $L(q_{\parallel}, \omega) = 1/\kappa_{\perp}^T$, and these modes disappear; cf. the discussion in Section 19.2. The dispersion relation for p-polarized modes is in the general case given by

$$\left[\left(\frac{\omega}{c} \right)^2 - q_{\parallel}^2 \right]^{1/2} + \left(\frac{\omega}{c} \right)^2 M(q_{\parallel}, \omega) = 0, \tag{19.54}$$

and it is reduced to the jellion eigenmode condition when only collective jellium excitations contribute.

CHAPTER 20

Quantum Theory of Collective Jellium Fields

20.1 BULK PLASMONS

In this section we shall establish and discuss the first- and second-quantized theories of bulk plasmons in jellium. Starting from the plasmon dispersion relation a complex plasmon field satisfying a Klein-Gordon type wave equation is introduced. A rigorous Lagrangian formalism is set up for the plasmon field, and via the minimal coupling principle, the plasmon's interaction with external fields is derived. The plasmon Hamiltonian is upgraded to the field-quantized level, and the plasmon quasiparticle concept is introduced following the procedure well known for the photon and electron fields, e.g. The plasmon quasiparticle is a boson, and the commutation relations among the plasmon annihilation and creation mode operators display this in the usual manner.

20.2 KLEIN-GORDON EQUATION AND EINSTEIN-DE BROGLIE RELATIONS FOR PLASMONS

Let us return to the bulk plasmon dispersion relation in Eq. (19.35), use the notation $\kappa_L \equiv q$, and introduce an effective plasmon phase velocity as

$$a \equiv D^{1/2} \tag{20.1}$$

where $D = (3/5)v_F^2$ is the diffusion coefficient. This plasmon phase velocity is of the order of the jellium Fermi velocity, and the precise value, $(3/5)^{1/2}$ here, is not of qualitative importance for the following. From the squared dispersion relation

$$\omega^2 = (aq)^2 + \omega_p^2 \tag{20.2}$$

a wave equation for a SCALAR PLASMON FIELD, $\phi(\mathbf{r}, t)$, is obtained via the usual connection's $\partial/\partial t \leftrightarrow -i\omega$, $\nabla \leftrightarrow i\mathbf{q}$. Thus,

$$\left(\nabla^2 - \frac{1}{a^2} \frac{\partial^2}{\partial t^2} \right) \phi(\mathbf{r}, t) - Q_C^2 \phi(\mathbf{r}, t) = 0, \tag{20.3}$$

DOI: 10.1201/9781003029458-20

where Q_C is a plasmon Compton wave number defined by

$$Q_C \equiv \frac{\omega_p}{a}. \tag{20.4}$$

The second-order wave equation for the plasmon field is of the Klein-Gordon type, but is must be remembered that it is *not* a relativistic wave equation, since it originates in the many-electron Schröinger equation for jellium. This fact is manifest from Eqs. (20.3) and (20.4), since $a \sim v_f$ appears instead of the vacuum speed of light. The inherent nonrelativistic character of Eq. (20.3) implies that we need not worry about negative frequency ($\sim$ energy) solutions: The plasmon quasiparticle has no anti-quasiparticle, as such. The scalar plasmon field, $\phi(\mathbf{r}, t)$, is a complex quantity since the plasmon field carries a charge (the electron charge, here). In the correct interpretation of the complex one-plasmon wave function, $\phi(\mathbf{r}, t)$ therefore is a CHARGE WAVE FUNCTION, so to speak. We shall return to this point when we introduce an equation of continuity related to $\phi(\mathbf{r}, t)$. On a heuristic level one can introduce an antiplasmon quasiparticle in for instance free-electron-like semiconductors with jellium-like dynamics of electrons and holes, since these have opposite effective charges. The analogy is to the complex relativistic Klein-Gordon equations for electrons and positrons. Inspired by the relativistic energy-momentum relation, we take the square root of Eq. (20.2) [with the choice $\omega > 0$] and multiply by $\hbar$. Hence, we reach the form

$$\hbar\omega = [(a\hbar q)^2 + (Ma^2)^2]^{1/2}, \tag{20.5}$$

where

$$M \equiv \frac{\hbar\omega_p}{a^2} \tag{20.6}$$

plays the role of a PLASMON REST MASS. This is understandable since one has no plasmon wave propagation in the long wavelength limit ($q \to 0$). The PLASMON REST ENERGY

$$E_0 \equiv Ma^2 = \hbar\omega_p \tag{20.7}$$

then must be identified with $\hbar$ times the plasma frequency. In a sense, $E_0 \equiv \hbar\omega_p$ represents the SELF-FIELD ENERGY of the scalar field. The Einstein-de Broglie plasmon energy (E) and momentum ($\mathbf{p}$) are $E = \hbar\omega$ and $\mathbf{p} = \hbar\mathbf{q}$, where $\mathbf{q} = q\boldsymbol{\kappa}$ is the wave vector in the plasmon propagation direction, $\boldsymbol{\kappa}$. In the PLASMON KINETIC ENERGY, $a\hbar q$, the characteristic speed is $a \sim v_F$ [in the relativistic case it is $c\hbar q$]. The plasmon Compton wavelength is

$$\Lambda_C = \frac{h}{Ma}. \tag{20.8}$$

20.2.1 Plasmon "four-current" density. Plasmon charge conservation

For what follows it is convenient to make use of a sort of "covariant" notation. Thus, we introduce the "covariant and contravariant derivatives"

$$\left\{\frac{\partial}{\partial x^\mu}\right\} \equiv \{\partial_\mu\} \equiv \left(\frac{1}{a}\frac{\partial}{\partial t}, \nabla\right), \tag{20.9}$$

$$\left\{\frac{\partial}{\partial x_\mu}\right\} \equiv \{\partial^\mu\} \equiv \left(\frac{1}{a}\frac{\partial}{\partial t}, -\nabla\right). \tag{20.10}$$

Above I have put relevant names between quotation marks, because the quantities above are *not relativistic four-vectors* [remember the theory underlying the plasmon description is nonrelativistic]. Below, where other "four-vectors" are introduced, we shall leave out the quotation marks for simplicity. Note that in this chapter we use a metric signature $(1,-1,-1,-1)$.

In our formally covariant notation the wave equation for the plasmon field reads

$$(\partial_\mu\partial^\mu + Q_C^2)\phi(x) = 0. \tag{20.11}$$

In analogy to the treatment of the relativistic Klein-Gordon equation [Section 16.2.1], we introduce a contravariant PLASMON CHARGE FOUR-CURRENT DENSITY ($\{J\}$) with the μth component

$$J^\mu = i(\phi^*\partial^\mu\phi - \phi\partial^\mu\phi^*), \ \mu = 0 - 3. \tag{20.12}$$

The presence of the factor i makes J^μ real. The equation of continuity

$$\partial_\mu J^\mu = 0, \tag{20.13}$$

obviously is satisfied by the plasmon field, and the charge (Q) of the plasmon is given by

$$Q = i\int_V (\phi^*\partial_0\phi - \phi\partial_0\phi^*)d^3r, \tag{20.14}$$

in a box of volume V.

20.2.2 Lagrangian formalism for free plasmons

Starting from a Lagrangian density,

$$\mathcal{L}_P = (\partial^\mu\phi)(\partial_\mu\phi^*) - Q_C^2\phi\phi^*, \tag{20.15}$$

for the free plasmon field [subscript P for plasmon or particle, as wished], the Euler-Lagrange equation for ϕ^*, i.e.,

$$\partial_\mu\left[\frac{\partial\mathcal{L}_P}{\partial(\partial_\mu\phi^*)}\right] - \frac{\partial\mathcal{L}_P}{\partial\phi^*} = 0, \tag{20.16}$$

immediately gives the Klein-Gordon equation in Eq. (20.11). In passing, notice that ϕ and ϕ^* are regarded as independent fields.

The second quantization of the plasmon field is based on the plasmon Hamiltonian density, $\mathcal{H}_p$. In order to set up the explicit expression one needs the canonical momenta π_ϕ and π_{ϕ^*}. From their definitions we obtain

$$\pi_\phi \equiv \frac{\partial \mathcal{L}_P}{\partial(\partial_0 \phi^*)} = \partial^0 \phi, \qquad (20.17)$$

and

$$\pi_{\phi^*} \equiv \frac{\partial \mathcal{L}_P}{\partial(\partial_0 \phi)} = \partial^0 \phi^*. \qquad (20.18)$$

The Hamiltonian density, given by

$$\mathcal{H}_P = \pi_\phi \dot{\phi}^* + \pi_{\phi^*} \dot{\phi} - \mathcal{L}_P \qquad (20.19)$$

with $\dot{\phi} \equiv \partial_0 \phi [= \partial^0 \phi]$, hence becomes

$$\mathcal{H}_P = (\partial^0 \phi)(\partial_0 \phi^*) + (\boldsymbol{\nabla}\phi) \cdot (\boldsymbol{\nabla}\phi^*) + Q_C^2 \phi\phi^*, \qquad (20.20)$$

since

$$\pi_\phi \dot{\phi}^* = \pi_{\phi^*} \dot{\phi} = (\partial^0 \phi)(\partial_0 \phi^*). \qquad (20.21)$$

The total Hamiltonian, viz.

$$H_P = \int_V \mathcal{H}_P(\mathbf{r}, t) d^3 r \qquad (20.22)$$

is obtained by integration of the Hamiltonian density over the quantization volume, V.

20.2.3 Plasmon quasiparticles and antiparticles

It is convenient for the extension of the plasmon formalism given in Subsections 20.1.1–20.1.3 to second quantization to give an alternative form for the total energy, H_p. Thus, if we integrate the term $(\boldsymbol{\nabla}\phi) \cdot (\boldsymbol{\nabla}\phi^*)$ by parts (dropping surface terms), and use the Klein-Gordon equation we obtain

$$H_P = \int_V [(\partial^0 \phi)(\partial_0 \phi^*) - \phi^*(\nabla^2 \phi - Q_C^2 \phi)] d^3 r = \int_V [(\partial^0 \phi)(\partial_0 \phi^*) - \phi^* \partial_0 \partial^0 \phi] d^3 r. \qquad (20.23)$$

The plasmon Hamilton operator, $\hat{H}_P$, is obtained by upgradation of the fields ϕ and ϕ^* to operator status, i.e., $\phi \Rightarrow \hat{\phi}$ and $\phi^* \Rightarrow \hat{\phi}^\dagger$. Next, we expand these operators in complete plane-wave sets as follows:

$$\hat{\phi}(\mathbf{r}, t) = \sum_n [\hat{a}_n \phi_n^{(+)}(\mathbf{r}, t) + \hat{b}_n^\dagger \phi_{-n}^{(-)}(\mathbf{r}, t)], \qquad (20.24)$$

$$\hat{\phi}^\dagger(\mathbf{r}, t) = \sum_n [\hat{a}_n^\dagger [\phi_n^{(+)}(\mathbf{r}, t)]^* + \hat{b}_n [\phi_{-n}^{(-)}(\mathbf{r}, t)]^*], \qquad (20.25)$$

where the mode functions are those of Eq. (16.38) yet with the plasmon energy given by

$$E_n(= \hbar\omega_n) = [(a\hbar q_n)^2 + (\hbar\omega_p)^2]^{1/2}, \tag{20.26}$$

that is Eq. (16.39) with the replacement $c \to a$. The annihilation $[\hat{a}_n, \hat{b}_n]$ and creation $[\hat{a}_n^\dagger, \hat{b}_n^\dagger]$ operators satisfy the commutation relations

$$[\hat{a}_n, \hat{a}_m^\dagger] = [\hat{b}_n, \hat{b}_m^\dagger] = \delta_{nm}, \tag{20.27}$$

$$[\hat{a}_n, \hat{a}_m] = [\hat{a}_n^\dagger, \hat{a}_m^\dagger] = [\hat{a}_n, \hat{b}_n] = [\hat{a}_n^\dagger, \hat{b}_m^\dagger] = [\hat{b}_n, \hat{b}_m] = [\hat{b}_n^\dagger, \hat{b}_m^\dagger] = 0. \tag{20.28}$$

By insertion of the mode expressions given in Eqs. (20.24) and (20.25) into the quantum form of the Hamiltonian given in Eq. (20.23) one obtains

$$\hat{H}_p = \sum_n \hbar\omega_n[\hat{a}_n^\dagger \hat{a}_n + \hat{b}_n^\dagger \hat{b}_n] \tag{20.29}$$

omitting the vacuum energy. The expression for $\hat{H}_p$ is more general than needed for jellium plasmon quasiparticles, because it includes their antiparticles, via the $\hat{b}(\hat{b}^\dagger)$-operators. The Hamilton and momentum operators of the bulk plasmon field thus are

$$\hat{H}_P = \sum_{\mathbf{q}} \hbar[(aq)^2 + \omega_p^2]^{1/2} \left[\hat{N}(\mathbf{q}) + \frac{1}{2}\right], \tag{20.30}$$

and

$$\hat{\mathbf{P}}_p = \sum_{\mathbf{q}} \hbar\mathbf{q}\hat{N}(\mathbf{q}), \tag{20.31}$$

where

$$\hat{N}(\mathbf{q}) = \hat{a}^\dagger(\mathbf{q})\hat{a}(\mathbf{q}) \tag{20.32}$$

is the mode ($\mathbf{q}$) number operator. The charge of the plasmon, given by Eq. (20.14), is to be identical with the total charge of the jellium electrons.

The formalism described briefly above allows one to think about possibilities for generating coherent, squeezzed, entangled states, etc., following the lines of contemporary developments in quantum optics.

20.2.4 Plasmon-gauge field interaction

How does one describe the interaction of the plasmon with an electromagnetic field? We suggest here that the coupling is effectuated via a minimum coupling principle for $\phi(\mathbf{r}, t)$. Thus, a "covariant derivative (D_μ)" is assumed to act on ϕ as follows:

$$D_\mu\phi = \left(\partial_\mu + \frac{iQ}{\hbar}A_\mu\right)\phi. \tag{20.33}$$

Since $\{\partial_\mu\} = (\partial_0, \nabla)$ $\{A_\mu\} = (A_0, -\mathbf{A})$, the space part the substitution rule is $\nabla \Rightarrow \nabla - i(e/\hbar)\mathbf{A}$, and hence in agreement with the minimal coupling principle for the canonical momentum, $\mathbf{p} \Rightarrow \mathbf{p} - e\mathbf{A}$. The new feature is that the substitution for $\mu = 0$ involves the characteristic plasmon speed a ($\sim v_F$) instead of the speed of light:

$$\partial_0 = \frac{1}{a}\frac{\partial}{\partial t} \Rightarrow \frac{1}{a}\frac{\partial}{\partial t} + \frac{ie}{\hbar}A_0. \tag{20.34}$$

The sum of the plasmon ($\mathcal{L}_P$) and interaction ($\mathcal{L}_I$) Lagrangian densities is obtained using additionally the covariant action

$$D_\mu\phi^* = \left(\partial_\mu - \frac{iQ}{\hbar}A_\mu\right)\phi^* \tag{20.35}$$

on ϕ^*. The replacements given in Eqs. (20.33) and (20.35) transform Eq. (20.15) into

$$\mathcal{L}_P + \mathcal{L}_I = \left(\partial^\mu\phi + \frac{iQ}{\hbar}A^\mu\phi\right)\left(\partial_\mu\phi^* - \frac{iQ}{\hbar}A_\mu\phi^*\right)) - Q_C^2\phi\phi^*, \tag{20.36}$$

The expression in Eq. (20.36) shows that the suggested interaction Lagrangian density is

$$\mathcal{L}_I = \frac{iQ}{\hbar}[(A^\mu\phi)(\partial_\mu\phi^*) - (\partial^\mu\phi)(A_\mu\phi^*)] + \left(\frac{Q}{\hbar}\right)^2 A_\mu A^\mu\phi\phi^*. \tag{20.37}$$

With the replacement $a \to c$, $\mathcal{L}_I$ takes the usual form for a complex scalar field ϕ; c.f. Subsection 16.2.2, Eq. (16.43) in particular.

The Euler-Lagrange equation

$$\partial_\mu\left[\frac{\partial(\mathcal{L}_P + \mathcal{L}_I)}{\partial(\partial_\mu\phi^*)}\right] - \frac{(\mathcal{L}_P + \mathcal{L}_I)}{\partial\phi^*} = 0 \tag{20.38}$$

gives the field-induced modification of the free Klein-Gordon equation for the plasmon, viz.,

$$(\partial_\mu\partial^\mu + Q_C^2)\phi = -\frac{iQ}{\hbar}[(\partial_\mu A^\mu)\phi + 2A^\mu\partial_\mu\phi] + \left(\frac{Q}{\hbar}\right)^2 A_\mu A^\mu\phi, \tag{20.39}$$

where we have utilized $\partial_\mu(A^\mu\phi) = (\partial_\mu A^\mu)\phi + A^\mu\partial_\mu\phi$.

In the presence of plasmon-field interaction, the minimal coupling principle modifies the current density in Eq. (20.12) to

$$J^\mu = i\left[\phi^*\left(\partial^\mu\phi + \frac{iQ}{\hbar}A^\mu\phi\right) - \phi\left(\partial^\mu\phi^* - \frac{iQ}{\hbar}A^\mu\phi^*\right)\right]. \tag{20.40}$$

This expression for J^μ may be written also as

$$J^\mu(A^\mu) = J^\mu(A^\mu = 0) - \frac{2Q}{\hbar}A^\mu\phi\phi^*, \tag{20.41}$$

where $J^\mu(A^\mu = 0)$ is the paramagnetic part and $-(2Q/\hbar)A^\mu\phi\phi^*$ is the diamagnetic part.

The reader should note that the usual interaction Hamiltonian density form, $-J^\mu A_\mu$, is not found in the Klein-Gordon case. The reason stems from the fact that the canonical momentum for ϕ, viz.,

$$\pi_\phi = \frac{\partial(\mathcal{L}_P + \mathcal{L}_I)}{\partial(\partial_0\phi^*)} = \partial^0\phi + \frac{iQ}{\hbar}A^0\phi, \tag{20.42}$$

together with a corresponding one for ϕ^*, lead to an extra term, $(iQ/\hbar)A^0[\phi(\partial_0\phi^*) - \phi^*(\partial_0\phi)]$, in the Hamiltonian density $\mathcal{H}_P + \mathcal{H}_I$. Hence, one can conclude that it is the time dependence $\partial_0\phi[\partial_0\phi^*]$ which prevents one from having $\mathcal{H}_I = -\mathcal{L}_I$ for the Klein-Gordon equation. Only when $\mathcal{H}_I = -\mathcal{L}_I$ does one reach the $-J^\mu A_\mu$-form. The "complication" above does not imply that a Hamiltonian formalism cannot be used, if wished.

One may illustrate the plasmon's coupling to a gauge field by a simple example, viz., the coupling to a space-time constant scalar potential $\Phi = cA^0$. From Eq. (20.39), one obtains the inhomogeneous plasmon Klein-Gordon equation

$$(\partial_\mu\partial^\mu + Q_C^2)\phi + \frac{2i}{\hbar c}Q\Phi\frac{\partial\phi}{\partial t} + \frac{1}{\hbar^2}(Q\Phi)^2\phi = 0. \tag{20.43}$$

By inserting a monochromatic plane-wave ansatz ($\sim \exp[i(\mathbf{q}\cdot\mathbf{r} - \omega t)]$) into Eq. (20.43), one gets (for $\omega > 0$) the following dispersion relation:

$$\omega = \frac{a}{c}\frac{aQ\Phi}{\hbar} + \left\{(aq)^2 + \omega_p^2 + \left(\frac{aQ\Phi}{\hbar}\right)^2\left[1 + \left(\frac{a}{c}\right)\right]\right\}^{1/2}. \tag{20.44}$$

Since $a/c << 1$, the dispersion relation may ($\sim$ always) be reduced to

$$\omega \simeq \left[(aq)^2 + \omega_p^2 + \left(\frac{aQ\Phi}{\hbar}\right)^2\right]^{1/2}. \tag{20.45}$$

The term $(aQ\Phi/\hbar)^2$ originates in the nonlinear part of the coupling, and this displaces the plasma frequency to

$$\omega_p(\Phi) = \omega_p(\Phi = 0)\left[1 + \left(\frac{Q\Phi}{\hbar Q_C}\right)^2\right]^{1/2}. \tag{20.46}$$

One may conclude that it is a simultaneous two-photon scalar photon interaction with the plasmon which is responsible for the upwards displacement $[\omega_p(\Phi) > \omega_p(\Phi = 0)]$ of the plasma frequency. As always, the results for constant potential are to be used when regions of different potential appear in the given problem.

20.3 BULK PLASMARITONS

To some extent the quantum theory of bulk plasmaritons can be established following the scheme presented for bulk plasmons. However, there is an important difference between the two types of modes. Thus, the characteristic phase velocity of the plasmon, $a = \sqrt{3/5}v_F$, is essentially the Fermi velocity of the jellium gas, whereas the phase velocity of the plasmariton equals the vacuum speed of light, c. This difference can give rise to resonant plasmariton-photon interactions because the characteristic phase velocity of the photon in vacuum is $c(!)$. Photons and plasmaritons can be strongly coupled at a jellium/vacuum surface in a certain frequency range. At the kinematic level the resonant coupling shows up in the two characteristic branches of the dispersion relation for surface plasmaritons.

20.3.1   Klein-Gordon equation and Einstein-de Broglie kinematics

As we shall realize in Section 21.1 it is an extremely good approximation to neglect the wave-number dependence of the transverse dielectric function, $\varepsilon_T(q,\omega)$ in the establishment of the plasmariton dispersion relation. Roughly speaking this is so because $(D_T/c^2)(\omega_p/\omega)^2 << 1$, for frequencies near (above) the bulk plasma frequency. The quantity $D_T = (1/5)v_F^2$ is the so-called transverse diffusion coefficient. Squaring the plasmariton dispersion relation, given in Eq. (19.37), and using the notation $\kappa_T \equiv q$, one has

$$\omega^2 = (cq)^2 + \omega_p^2. \tag{20.47}$$

A wave equation for a VECTORIAL PLASMARITON FIELD, $\boldsymbol{\Phi}(\mathbf{r},t)$, is obtained via the connections $\partial/\partial t \leftrightarrow -i\omega$ and $\boldsymbol{\nabla} \leftrightarrow iq$. Hence, one gets the following Klein-Gordon-like equation:

$$\left(\nabla^2 - \frac{1}{c^2}\frac{\partial^2}{\partial t^2}\right)\boldsymbol{\Phi}(\mathbf{r},t) - Q_T^2\boldsymbol{\Phi}(\mathbf{r},t) = \mathbf{0}, \tag{20.48}$$

where

$$Q_T \equiv \frac{\omega_p}{c} \tag{20.49}$$

is the characteristic Compton wave number for the transverse (T) plasmariton. At this point we realize that the plasmariton wave equation is a *genuine relativistic wave equation for a boson*. The relativistic PLASMARITON REST MASS is

$$M_T \equiv \frac{\hbar\omega_p}{c^2}, \tag{20.50}$$

and the rest energy $E_T[= E_0] = M_T c^2 = \hbar\omega_p$ is the same as for the plasmon. In a sense, this is obvious because the plasmon and plasmariton masses are connected to the energy in the long-wavelength $(q \to 0)$ limit. For $q \to 0$ it becomes impossible to distinguish "transverse" and "longitudinal" displacements in the jellium from each other [No $\mathbf{q}$-vector!]. The plasmariton Compton wavelength has the relativistic form, viz.,

$$\Lambda_T = \frac{h}{M_T c}. \tag{20.51}$$

20.3.2 Proca equation for massive photon

The Klein-Gordon equation for the plasmariton has a certain similarity to the Proca equation for a (hypothetical) massive photon. If one identifies the photon wave field with the four potential $\{A^\mu\}$, the standard wave equation for the free photon field is

$$\partial_\nu\partial^\nu A^\mu(x) - \partial^\mu(\partial_\nu A^\nu(x)) = 0, \ \mu = 0 - 3. \tag{20.52}$$

In the Proca theory one makes the replacement

$$\partial_\nu\partial^\nu \Rightarrow \partial_\nu\partial^\nu + Q^2_{\mathrm{PH}} \tag{20.53}$$

where

$$Q_{\mathrm{PH}} = \frac{M_{\mathrm{PH}}c}{\hbar} \tag{20.54}$$

is the Compton wave number of a photon with mass M_{PH}. The replacement in Eq. (20.53) transforms Eq. (20.52) to

$$(\partial_\nu\partial^\nu + Q^2_{\mathrm{PH}})A^\mu(x) - \partial^\mu(\partial_\nu A^\nu(x)) = 0. \tag{20.55}$$

On taking the four-divergence of this equation one obtains

$$Q^2_{\mathrm{PH}}(\partial_\mu A^\mu(x)) = 0. \tag{20.56}$$

From the requirement $M_{\mathrm{PH}} \neq 0$, it thus appears that the Lorenz condition

$$\partial_\mu A^\mu(x) = 0 \tag{20.57}$$

necessarily must be satisfied to reach a Klein-Gordon equation of the usual form for a photon, viz.,

$$(\partial_\nu\partial^\nu + Q^2_{\mathrm{PH}})A^\mu(x) = 0, \tag{20.58}$$

for metric signature (1,-1,-1,-1). Eq. (20.58) is the Proca equation of the photon. Conclusion: If the Lorenz condition is a requirement for a massive photon, the Klein-Gordon equations for the plasmon, plasmariton and photon are form-identical. Note that the plasmon wave function is a scalar, and the plasmariton wave function is a three-vector and the photon wave function a four-vector.

20.3.3 Lagrangian formalism for free plasmaritons

The second-quantized theory for the plasmariton can be established in analogy with what we did for the plasmon. Essentially, only the following two changes are needed: (i) The effective plasmon phase velocity, $a = \sqrt{3/5}v_F$, must be replaced by the vacuum speed of light, c, and (ii) the one-component plasmon field, $\phi(x)$, is replaced by the three-component field

$$\mathbf{\Phi}(x) = \sum_i \mathbf{e}_i \Phi_i(x), \ i = 1 - 3 \tag{20.59}$$

$\mathbf{e}_i$ being Cartesian unit vectors.

As Lagrangian density for the plasmariton we take

$$\mathcal{L}_P = \sum_i \mathcal{L}_{P,i} = \sum_i [(\partial^\mu \Phi_i)(\partial_\mu \Phi_i^*) - Q_T^2 \Phi_i \Phi_i^*] \tag{20.60}$$

and from the Euler-Lagrange equations

$$\partial_\mu \left[\frac{\partial \mathcal{L}_P}{\partial(\partial_\mu \Phi_i^*)} \right] - \frac{\partial \mathcal{L}_P}{\partial \Phi_i^*} = 0, \ i = 1 - 3 \tag{20.61}$$

we obtain the Klein-Gordon equations

$$(\partial_\mu \partial^\mu + Q_T^2)\Phi_i(x) = 0, \ i = 1 - 3 \tag{20.62}$$

for the three plasmariton field components. Taken together, Eqs. (20.59) and (20.62) lead back to the vectorial wave equation given in Eq. (20.48). The six canonical momenta

$$\Pi_{\Phi_i} \equiv \frac{\partial \mathcal{L}_P}{\partial(\partial_0 \Phi_i^*)} = \partial^0 \Phi_i, \tag{20.63}$$

$$\Pi_{\Phi_i^*} \equiv \frac{\partial \mathcal{L}_P}{\partial(\partial_0 \Phi_i)} = \partial^0 \Phi_i^*, \tag{20.64}$$

in turn, via

$$\Pi_{\Phi_i} \dot{\Phi}_i^* = \Pi_{\Phi_i^*} \dot{\Phi}_i = (\partial^0 \Phi_i)(\partial_0 \Phi_i^*), \tag{20.65}$$

give one a Hamiltonian density

$$\mathcal{H}_P = \sum_i [\Pi_{\Phi_i} \dot{\Phi}_i^* + \Pi_{\Phi_i^*} \dot{\Phi}_i - \mathcal{L}_{P,i}]. \tag{20.66}$$

Written in the "oscillator" form

$$\mathcal{H}_P = \sum_i [(\partial^0 \Phi_i)(\partial_0 \Phi_i^*) + (\boldsymbol{\nabla} \Phi_i) \cdot (\boldsymbol{\nabla} \Phi_i^*) + Q_T^2 \Phi_i \Phi_i^*], \tag{20.67}$$

and extended to the operator level with the associations

$$\Phi_i \Rightarrow \hat{\Phi}_i, \ \Phi_i^* \Rightarrow \hat{\Phi}_i^\dagger, \ i = 1 - 3 \tag{20.68}$$

one obtains from the integral

$$\hat{H}_P = \int_V \hat{\mathcal{H}}_P d^3 r \tag{20.69}$$

the plasmariton particle Hamiltonian operator. Thus,

$$\hat{H}_p = \sum_i \sum_{\mathbf{q}} \hbar [(cq)^2 + \omega_p^2]^{1/2} \left[\hat{N}_i(\mathbf{q}) + \frac{1}{2} \right], \tag{20.70}$$

where

$$\hat{N}_i(\mathbf{q}) = \hat{a}_i^\dagger(\mathbf{q})\hat{a}_i(\mathbf{q}), \;\; i = 1 - 3 \tag{20.71}$$

is the plasmariton number operator belonging for a given wave vector $(\mathbf{q})$ to each of the three polarization components. The momentum properties carried by a plasmariton in a given quantum state can be studied starting from the plasmariton momentum operator

$$\hat{P}_P = \sum_i \sum_{\mathbf{q}} \hbar \mathbf{q} \hat{N}_i(\mathbf{q}). \tag{20.72}$$

As a boson field, the mode annihilation $(\hat{a}_i(\mathbf{q}))$ and creation $(\hat{a}_i^\dagger(\mathbf{q}))$ operators satisfy the commutator relation

$$[\hat{a}_i(\mathbf{q}), \hat{a}_j^\dagger(\mathbf{q}')] = \delta_{ij}\delta_{\mathbf{q},\mathbf{q}'}, \tag{20.73}$$

all other commutators being zero.

20.3.4 Plasmariton-gauge field interaction

The coupling of the plasmariton to the photon field can quantitatively be *very* different from that of the plasmon. It is obvious that the three-vector form of the plasmariton wave field makes a difference, but the main cause originates in the fact that the plasmariton field propagates with the speed of light and not with a speed of the order of the Fermi velocity. Hence, it is clear *a priori* that there exists a potential possibility for resonant plasmariton-photon interaction. This possibility is realized in a manifest form in the so-called surface plasmaritons of a semiinfinite jellium.

The basic form of the plasmariton-photon interaction is derived starting from the covariant derivatives $[i = 1 - 3]$

$$D_\mu \Phi_i = \left(\partial_\mu + \frac{iQ}{\hbar}A_\mu\right)\Phi_i, \tag{20.74}$$

$$D_\mu \Phi_i^* = \left(\partial_\mu - \frac{iQ}{\hbar}A_\mu\right)\Phi_i^*, \tag{20.75}$$

In a sense the coupled theory of the plasmariton is easier than that of the plasmon, due to the fact that *all* derivatives appearing are genuine covariant derivatives, not a mixture of formally [Eqs. (20.9) and (20.10)] and genuine $[\{\partial_\mu\} = (c^{-1}\partial/\partial t, \nabla), \{\partial^\mu\} = (c^{-1}\partial/\partial t, -\nabla)]$ covariant derivatives. The genuine covariant derivatives have been used throughout the previous chapters of the book, up to Section 20.1.

The sum of the particle and interaction Lagrangian densities

$$\mathcal{L}_P + \mathcal{L}_I = \sum_i (\mathcal{L}_{P,i} + \mathcal{L}_{I,i}) = \sum_i [(D^\mu \Phi_i)(D_\mu \Phi_i^*) - Q_T^2 \Phi_i \Phi_i^*] \tag{20.76}$$

shows that the interaction part is given by

$$\mathcal{L}_I = \sum_i \mathcal{L}_{I,i} = \sum_i \left\{ \frac{iQ}{\hbar}[(A^\mu \Phi_i)(\partial_\mu \Phi_i^*) - (\partial^\mu \Phi_i)(A_\mu \Phi_i^*)] + \left(\frac{Q}{\hbar}\right)^2 A^\mu A_\mu \Phi_i \Phi_i^* \right\}. \tag{20.77}$$

As in the plasmon-gauge field interaction, we can also here use the Euler-Lagrange equations $[i = 1 - 3]$,

$$\partial_\mu \left[\frac{\partial(\mathcal{L}_P + \mathcal{L}_I)}{\partial(\partial_\mu \Phi_i^*)} \right] - \frac{\partial(\mathcal{L}_P + \mathcal{L}_I)}{\partial \Phi_i^*} = 0, \tag{20.78}$$

to obtain inhomogeneous plasmariton Klein-Gordon equations for the three Cartesian components. Not unexpectedly, the result is as follows:

$$(\partial_\mu \partial^\mu + Q_T^2)\Phi_i + \frac{iQ}{\hbar}[\partial_\mu(A^\mu \Phi_i) + A_\mu(\partial^\mu \Phi_i)] - \left(\frac{Q}{\hbar}\right)^2 A_\mu A^\mu \Phi_i = 0. \quad i = 1 - 3 \tag{20.79}$$

The Hamiltonian formalism in the first-quantized description is obtained starting from the canonical momenta $[i = 1 - 3]$

$$\Pi_{\Phi_i} \equiv \frac{\partial(\mathcal{L}_P + \mathcal{L}_I)}{\partial(\partial_0 \Phi_i^*)} = \partial^0 \Phi_i + \frac{iQ}{\hbar} A^0 \Phi_i, \tag{20.80}$$

and

$$\Pi_{\Phi_i^*} \equiv \frac{\partial(\mathcal{L}_P + \mathcal{L}_I)}{\partial(\partial_0 \Phi_i)} = \partial^0 \Phi_i^* - \frac{iQ}{\hbar} A^0 \Phi_i^*, \tag{20.81}$$

which imply that

$$\Pi_{\Phi_i} \dot{\Phi}_i^* = \Pi_{\Phi_i^*} \dot{\Phi}_i = (\partial^0 \Phi_i)(\partial_0 \Phi_i^*). \tag{20.82}$$

By means of the relations in Eq. (20.82) the explicit form of the Hamiltonian density

$$\mathcal{H}_P + \mathcal{H}_I = \sum_i [\Pi_{\Phi_i} \dot{\Phi}_i^* + \Pi_{\Phi_i^*} \dot{\Phi}_i - \mathcal{L}_{P,i} - \mathcal{L}_{I,i}], \tag{20.83}$$

is obtained immediately. Hence,

$$\mathcal{H}_P + \mathcal{H}_I = \sum_i \left\{ (\partial^0 \Phi_i)(\partial_0 \Phi_i^*) + (\boldsymbol{\nabla}\Phi_i) \cdot (\boldsymbol{\nabla}\Phi_i^*) + Q_T^2 \Phi_i \Phi_i^* \right.$$
$$\left. \frac{iQ}{\hbar}[(\mathbf{A}\Phi_i) \cdot (\boldsymbol{\nabla}\Phi_i^*) + (\mathbf{A}\Phi_i^*) \cdot (\boldsymbol{\nabla}\Phi_i)] - \left(\frac{Q}{\hbar}\right)^2 A_\mu A^\mu \Phi_i \Phi_i^* \right\}. \tag{20.84}$$

The reader may recognize the expression in the first line on the right-hand side of Eq. (20.84) as the free plasmariton Hamiltonian density. The expression in the first term in the second line relates to the linear interaction between the plasmariton and photon fields (in first-quantization). In the last term of the second line the manifest nonlinear coupling between the two fields appears.

20.4 ELEMENTS OF THE QED THEORY FOR A COUPLED PLASMARITON-PHOTON SYSTEM

Let us return to the Lagrangian formalism for the plasmariton field coupled to the electromagnetic field, and let us treat the field not as a prescribed quantity but as a dynamical entity, To $\mathcal{L}_P + \mathcal{L}_I$, given in terms of covariant derivatives in Eq. (20.76) one now must add the field Lagrangian density $\mathcal{L}_F$. In its standard form this is given by Eq. (10.28). For the total Lagrangian density ($\mathcal{L}$) one thus can make the following choice:

$$\mathcal{L} = \mathcal{L}_P + \mathcal{L}_I + \mathcal{L}_F = \sum_i [(D^\mu \Phi_i)(D_\mu \Phi_i^*) - Q_T^2 \Phi_I \Phi_i^*] - \frac{\varepsilon_0 c^2}{4} F_{\mu\nu} F^{\mu\nu}, \qquad (20.85)$$

with field tensor elements

$$F_{\mu\nu} = \partial_\mu A_\nu - \partial_\nu A_\mu. \qquad (20.86)$$

The description based on Eq. (20.85) is extended to the second-quantized level by the usual association

$$\Phi_i \Rightarrow \hat{\Phi}_i; \ \ \Phi_i^* \Rightarrow \hat{\Phi}_i^\dagger; \ \ A_\mu \Rightarrow \hat{A}_\mu. \qquad (20.87)$$

The QED theory starting from the quantized form of $\mathcal{L}$, although rigorous and general for the field interaction with the plasmariton becomes very complicated and in order to apply the formalism to specific cases approximations have to be made. A substantial simplification appears if one neglects the manifest nonlinear interaction term $(Q/\hbar)^2 A_\mu A^\mu \Phi_i \Phi_i^*$ [Eq. (20.77)]. Since the plasmariton theory is based on the Schrödinger equation, the nonrelativistic QED theory takes a convenient form in the Coulomb gauge where $\mathbf{A} = \mathbf{A}_T$. In the linearized approximation one thus may start from the following Hamiltonian operator density:

$$\hat{\mathcal{H}} = \hat{\mathcal{H}}_P + \hat{\mathcal{H}}_I + \hat{\mathcal{H}}_F = \sum_i \left\{ (\partial^0 \hat{\Phi}_i)(\partial_0 \hat{\Phi}_i^\dagger) + (\nabla \hat{\Phi}_i) \cdot (\nabla \hat{\Phi}_i^\dagger) + Q_T^2 \hat{\Phi}_i \hat{\Phi}_i^\dagger \right.$$

$$\left. + \frac{iQ}{\hbar} [(\hat{\mathbf{A}}_T \hat{\Phi}_i) \cdot (\nabla \hat{\Phi}_i^\dagger) + (\hat{\mathbf{A}}_T \hat{\Phi}_i^\dagger) \cdot (\nabla \hat{\Phi}_i)] \right\} + \frac{\varepsilon_0}{2} \left[\left(\frac{\hat{\mathbf{\Pi}}_T}{\varepsilon_0} \right)^2 + c^2 (\nabla \times \hat{\mathbf{A}}_T)^2 \right],$$

$$(20.88)$$

where $\hat{\mathbf{\Pi}}_T = \varepsilon_0 \partial \hat{\mathbf{A}}_T / \partial t$ is the canonical momentum of the transverse gauge photon.

Let us now turn to plane-wave expansions of the vector potential ($\hat{\mathbf{A}}_T$) and plasmariton ($\hat{\mathbf{\Phi}}$) fields, and consider a *common wave* of the two species. In box quantization expansion, and with the plasmariton mode annihilation and creation operators renamed as $\hat{b}_i(\mathbf{q})$ and $\hat{b}_i^\dagger(\mathbf{q})$, we have

$$\hat{\mathbf{A}}_T = \sum_{\mathbf{q},\lambda} \left(\frac{\hbar}{2\varepsilon_0 c V q} \right)^{1/2} \left[\varepsilon_{\mathbf{q}\lambda} \hat{a}_{\mathbf{q}\lambda} e^{i\mathbf{q}\cdot\mathbf{r}} + \varepsilon_{\mathbf{q}\lambda}^* \hat{a}_{\mathbf{q}\lambda}^\dagger e^{-i\mathbf{q}\cdot\mathbf{r}} \right], \qquad (20.89)$$

and

$$\hat{\mathbf{\Phi}} = \sum_{\mathbf{q},\lambda} \left(\frac{\hbar c}{2 E_q V} \right)^{1/2} \left[\boldsymbol{\varepsilon}_{\mathbf{q}\lambda} \hat{b}_{\mathbf{q}\lambda} e^{i\mathbf{q}\cdot\mathbf{r}} + \boldsymbol{\varepsilon}^*_{\mathbf{q}\lambda} \hat{b}^\dagger_{\mathbf{q}\lambda} e^{-i\mathbf{q}\cdot\mathbf{r}} \right] \tag{20.90}$$

where $\boldsymbol{\varepsilon}_{\mathbf{q}\lambda}$ is the common unit polarization vector and λ is the polarization index .The quantity $E_q = \hbar[(cq)^2 + \omega_p^2]^{1/2}$ is the plasmariton energy in the mode $\mathbf{q}$. The expressions in Eqs. (20.89) and (20.90) enter the $\mathcal{H}_I$-part of Eq. (20.88). Integration over the box volume gives a Hamilton operator

$$\hat{H} = \int_V \hat{\mathcal{H}} d^3 r, \tag{20.91}$$

with a free part

$$\hat{H}_F + \hat{H}_P = \sum_{\mathbf{q},\lambda} \hbar c q \left[\hat{a}^\dagger_{\mathbf{q}\lambda} \hat{a}_{\mathbf{q}\lambda} + \frac{1}{2} \right] + \sum_{\mathbf{q},\lambda} \hbar[(cq)^2 + \omega_p^2]^{1/2} \left[\hat{b}^\dagger_{\mathbf{q}\lambda} \hat{b}_{\mathbf{q},\lambda} + \frac{1}{2} \right]. \tag{20.92}$$

To obtain the eigenmodes of the coupled photon-plasmariton system one first has to determine the explicit form of $\hat{H}_I = \int_V \hat{\mathcal{H}}_I d^3 r$. Afterwards, one must seek to diagonalize the total Hamiltonian, $\hat{H} = \hat{H}_F + \hat{H}_P + \hat{H}_I$. We shall not here carry out this technically somewhat difficult program. At the end, the solution will lead to the dispersion relation for the photon-plasmariton mode; basically the eigensolution for two coupled oscillators. A quantum theory of the interaction of optical phonons with photons is given in [81, 116]. Elements of this theory is presented in tutorial form in [168]. In a series of comprehensive articles wave-mechanical and second-quantized theories have been established for bulk and surface plasmons [126], bulk plasmaritons [127] and surface plasmaritons [128].

CHAPTER 21

Heading for Jellion Quasiparticles

In this chapter we shall deepen our understanding of the coupling between the L-plasmon, the T-plasmon, and the photon. Particular attention is paid to the particle aspects, as these appear in particle wave mechanics (PaWM) and QED. Together, the three particles form a new quasiparticle, which I have named a JELLION. The plasmariton is a bound state of a T-plasmon and T-photon. In a sense this bound state is like that between two quarks because it is impossible to split the plasmariton into a free T-plasmon and a gauge photon. A SURFACE JELLION is an interface entangled plasmon (L)-plasmariton particle, analyzed in detail in [129]

21.1 PARTICLE WAVE MECHANICS OF THE L-PLASMON, T-PLASMON AND PHOTON TRIPLET: THE JELLION

21.1.1 Longitudinal and transverse Lindhard dielectric functions

An infinitely extended jellium exhibits translational (arbitrary displacements) and rotational (arbitrary rotations) invariance in space. Furthermore, the jellium has in an average sense ($\sim$ neglect of fluctuations) time-independent properties. If the jellium is subjected to a sufficiently weak external stimulus it responds in a linear fashion. In consequence of the above properties, the weak response can conveniently be studied in the wave vector ($\mathbf{q}$)- frequency (ω) domain. The linear microscopic conductivity tensor, $\boldsymbol{\sigma}(\mathbf{q}, \omega)$, is the central concept for jellium response theory [or equivalently the dielectric tensor, $\boldsymbol{\varepsilon}(\mathbf{q}, \omega)$]. The two are related as follows:

$$\boldsymbol{\varepsilon}(\mathbf{q}, \omega) = \mathbf{U} + \frac{i}{\varepsilon_0 \omega} \boldsymbol{\sigma}(\mathbf{q}, \omega). \tag{21.1}$$

It is obvious that the jellium response will be different for excitations parallel and perpendicular to $\mathbf{q}$. This fact. together with the jellium's rotational symmetry around $\mathbf{q}$ imply that the general form of $\boldsymbol{\sigma}(\mathbf{q}, \omega)$, viz.

$$\boldsymbol{\sigma}(\mathbf{q}, \omega) = (\mathbf{U} - \hat{\mathbf{q}}\hat{\mathbf{q}})\sigma_T(q, \omega) + \hat{\mathbf{q}}\hat{\mathbf{q}}\sigma_L(q, \omega) \tag{21.2}$$

DOI: 10.1201/9781003029458-21

can be characterized by two scalar quantities, named the transverse $(\sigma_T(q,\omega))$ and longitudinal $(\sigma_L(q,\omega))$ conductivities. From

$$\varepsilon(\mathbf{q},\omega) = (\mathbf{U} - \hat{\mathbf{q}}\hat{\mathbf{q}})\varepsilon_T(q,\omega) + \hat{\mathbf{q}}\hat{\mathbf{q}}\varepsilon_L(q,\omega), \tag{21.3}$$

and Eqs. (21.1) and (21.2), one obtains

$$\varepsilon_T(q,\omega) = 1 + \frac{i}{\varepsilon_0\omega}\sigma_T(q,\omega), \tag{21.4}$$

$$\varepsilon_L(q,\omega) = 1 + \frac{i}{\varepsilon_0\omega}\sigma_L(q,\omega). \tag{21.5}$$

Note that $\sigma_T(q,\omega)$ $[\varepsilon_T(q,\omega)]$ and $\sigma_L(q,\omega)$ $[\varepsilon_L(q,\omega)]$ depend on the magnitude of the wave vector only. Since the jellium also possesses inversion symmetry [same response functions for $\mathbf{q}$ and $-\mathbf{q}$] the T- and L-response functions can depend only on the square (q^2) of the wave vector.

In the framework of RPA explicit expressions can be derived for $\sigma_T(q,\omega)$ and $\sigma_L(q,\omega)$. It can be shown, that these so-called Lindhard conductivity response functions in the low-temperature limit $(T \to 0\mathrm{K})$ and with the neglect of relaxation mechanics $(\tau \to \infty)$ have the forms [178]

$$\sigma_T(q,\omega) = \frac{3ine^2}{8m\omega}\left(z^2 + 3u^2 + 1 \right.$$
$$\left. - \frac{1}{4z}\left\{[1 - (z-u)^2]^2 \ln\left|\frac{z-u+1}{z-u-1}\right| + [1 - (z+u)^2]^2 \ln\left|\frac{z+u+1}{z+u-1}\right|\right\}\right), \tag{21.6}$$

and

$$\sigma_L(q,\omega) = \frac{3ne^2u^2}{2im\omega}\left(1 \right.$$
$$\left. + \frac{1}{4z}\left\{[1 - (z-u)^2] \ln\left|\frac{z-u+1}{z-u-1}\right| + [1 - (z+u)^2] \ln\left|\frac{z+u+1}{z+u-1}\right|\right\}\right). \tag{21.7}$$

The $T \to 0\mathrm{K}$ approximation is sufficient for our purpose, and $\tau \to \infty$ makes the jellium a closed system and thus sufficient for the subsequent eigenmode analyses. The results in Eqs. (21.6) and (21.7) depend on two important nonlocality parameters, viz.,

$$z = \frac{q}{2k_F}, \tag{21.8}$$

and

$$u = \frac{\omega}{qv_F}. \tag{21.9}$$

The z-parameter (often called the "quantum" parameter) characterizes the ratio between the wave number of the excitation and the jellium's Fermi wave number. For long-wavelength excitations $z << 1$. The u-parameter (called the "classical" param-eter) gives the phase velocity of the excitation (ω/q) in comparison to the Fermi speed of the jellium. Single-particle excitations are related to the singularities in the ln-functions. [94, 178]

The collective jellium modes of interest here relate to expressions of $\sigma_T(q,\omega)$ and $\sigma_L(\mathbf{q},\omega)$ from $q = 0$ to lowest order (q^2) in q [85]. Thus, one obtains for the dielectric functions the long-wavelength results [164]

$$\varepsilon_T(q,\omega) = 1 - \left(\frac{\omega_p}{\omega}\right)^2 \left[1 + D_T \left(\frac{q}{\omega}\right)^2\right], \tag{21.10}$$

and

$$\varepsilon_L(q,\omega) = 1 - \left(\frac{\omega_p}{\omega}\right)^2 \left[1 + D_L \left(\frac{q}{\omega}\right)^2\right], \tag{21.11}$$

where

$$D_T = \frac{1}{5}v_F^2, \quad D_L = \frac{3}{5}v_F^2 \tag{21.12}$$

are the transverse (D_T) and longitudinal (D_L) diffusion coefficients $[D_L \equiv D$ in Chapts. 19 and 20].

Provided $D_L(q/\omega)^2 = 3D_T(q/\omega)^2 << 1$ [and this will always be correct for what follows] one can use the so-called hydrodynamical forms

$$\varepsilon_T(q,\omega) = 1 - \frac{\omega_p^2}{\omega^2 - D_T q^2}, \tag{21.13}$$

$$\varepsilon_L(q,\omega) = 1 - \frac{\omega_p^2}{\omega^2 - D_L q^2}. \tag{21.14}$$

The hydrodynamic form of $\varepsilon_L(q,\omega)$ already has been introduced in Section 19.2 [Eq. (19.34)]. The form in Eq. (21.13) is often called the *extended transverse* dielectric function. In most cases it is sufficient to use the limiting $(q \to 0)$ form for $\varepsilon_T(q,\omega)$, cf. Eq. (19.36), and its use in p-polarized reflection studies [see Section 19.1]. However, despite of the fact that

$$\frac{D_T^{1/2}}{c} << 1, \tag{21.15}$$

it may in certain cases be important to keep the $D_T q^2$ - quantity in Eq.(21.13).

21.1.2 The Klein-Gordon triplet

Let us consider the combination of the Maxwell-Lorentz equations and the constitu-tive equation for the jellium plasma, as this appears the wave vector $(\mathbf{q})$-frequency domain, and let us divide the combination into its L- and T-parts.

From the L-parts of the field equations $\nabla \cdot \mathbf{E} = \rho/\varepsilon_0$ and $\nabla \times \mathbf{B} = \mu_0 \mathbf{J} + c^{-2}\partial \mathbf{E}/\partial t$ one obtains

$$i\mathbf{q} \cdot \mathbf{E}_L(\mathbf{q}, \omega) = \varepsilon_0^{-1}\rho(\mathbf{q}, \omega), \tag{21.16}$$

$$\mathbf{J}_L(\mathbf{q}, \omega) = i\varepsilon_0\omega\mathbf{E}_L(\mathbf{q}, \omega). \tag{21.17}$$

These equations contain implicitly the equation of continuity $\mathbf{q} \cdot \mathbf{J}_L(\mathbf{q}, \omega) = \omega\rho(\mathbf{q}, \omega)$. The constitutive equation for the L-dynamics has the form

$$\mathbf{J}_L(\mathbf{q}, \omega) = \sigma_L(q, \omega)\mathbf{E}_L(\mathbf{q}, \omega), \tag{21.18}$$

c.f. Eq. (21.2). By combining Eqs. (21.17) and (21.18), and remembering the relation between $\sigma_L(q, \omega)$ and $\varepsilon_L(q, \omega)$ [Eq. (21.5)] one finds

$$\varepsilon_L(q, \omega) = 0; \tag{21.19}$$

the eigenmode condition for the jellium L-dynamics. The transverse field equation, $\nabla \times \mathbf{B} = \mu_0 \mathbf{J}_T + c^{-2}\partial \mathbf{E}_T/\partial t$, given by

$$i\mathbf{q} \times \mathbf{B}(\mathbf{q}, \omega) = \mu_0\mathbf{J}_T(\mathbf{q}, \omega) - \frac{i\omega}{c^2}\mathbf{E}_T(\mathbf{q}, \omega) \tag{21.20}$$

in the $(\omega, \mathbf{q})$-domain, taken together with the constitutive equation for the T-dynamics,

$$\mathbf{J}_T(\mathbf{q}, \omega) = \sigma_T(q, \omega)\mathbf{E}_T(\mathbf{q}, \omega), \tag{21.21}$$

and the relation in Eq. (21.4), give

$$\varepsilon_T(q, \omega)\mathbf{E}_T(\mathbf{q}, \omega) = -\frac{c^2}{\omega}\mathbf{q} \times \mathbf{B}(\mathbf{q}, \omega). \tag{21.22}$$

Since $\mathbf{q} \cdot \mathbf{B}(\mathbf{q}, \omega) = 0$ [from the field equation $\nabla \cdot \mathbf{B} = 0$], Eq. (21.22) can be transformed to

$$\varepsilon_T(q, \omega)\mathbf{q} \times \mathbf{E}_T(\mathbf{q}, \omega) = \frac{(cq)^2}{\omega}\mathbf{B}(\mathbf{q}, \omega). \tag{21.23}$$

Using the Maxwell equation $\nabla \times \mathbf{E}_T = -\partial\mathbf{B}/\partial t$ [$\mathbf{q} \times \mathbf{E}_T = \omega\mathbf{B}$], one obtains

$$\left[\left(\frac{\omega}{c}\right)^2\varepsilon_T(q, \omega) - q^2\right]\mathbf{B}(\mathbf{q}, \omega) = 0, \tag{21.24}$$

and since $\mathbf{B}(\mathbf{q}, \omega) \neq \mathbf{0}$, one finally gets the transverse dispersion relation

$$\left(\frac{\omega}{c}\right)^2\varepsilon_T(q, \omega) = q^2. \tag{21.25}$$

In the following the two L- and T-dispersion relations [Eqs. (21.19) and (21.25)] are considered together with that of the massless gauge photon in squared form,

$$\omega^2 = (cq)^2, \tag{21.26}$$

obtained as we know from the $(\omega,\mathbf{q})$-from of the wave equation for the transverse vector potential viz.,

$$[\omega^2 - (cq)^2]\mathbf{A}_T(\mathbf{q},\omega) = \mathbf{0}. \tag{21.27}$$

Before proceeding the general analysis let us insert the expression for $\varepsilon_T(q,\omega)$, given in Eq. (21.10), into the dispersion relation in Eq. (21.25), and rewrite this as follows:

$$\omega^2 - \omega_p^2 = (cq)^2 \left[1 + \frac{D_T}{c^2}\left(\frac{\omega_p}{\omega}\right)^2\right]. \tag{21.28}$$

The inequality in Eq. (21.15) justifies a neglect of the last term in the square bracket for frequencies in the vicinity of ω_p.

To sum up we now have in our hands the three important (squared) dispersion relations

$$(a_L q)^2 = \omega^2 - \omega_p^2, \quad [\text{Bulk } L\text{-plasmon}], \tag{21.29}$$

$$(cq)^2 = \omega^2 - \omega_p^2, \quad [\text{Bulk plasmariton}], \tag{21.30}$$

$$(cq)^2 = \omega^2, \quad [\text{Gauge photon}]. \tag{21.31}$$

$a_L = \sqrt{3/5}v_f$ being the characteristic phase ($\sim$ Fermi) velocity for the plasmon. The corresponding quantity for the plasmariton is the speed of light, and this fact indicates the possibility of a strong plasmariton-photon coupling.

The usual connections $\partial/\partial t \leftrightarrow -i\omega$ $\nabla \leftrightarrow i\mathbf{q}$ extend the formalism to the quantum mechanical level. The plasmon is described by a scalar wave function $\phi_L(\mathbf{r},t)$, and the plasmariton and the gauge photon by vectorial wave functions $\mathbf{\Phi}_T(\mathbf{r},t)$ and $\mathbf{A}_T(\mathbf{r},t)$, respectively. All three particles are bosons, and the related free-particle wave functions satisfy the following Klein-Gordon equations:

$$\left(\nabla^2 - \frac{1}{a_L^2}\frac{\partial^2}{\partial t^2} - Q_L^2\right)\phi_L(\mathbf{r},t) = 0, \tag{21.32}$$

$$\left(\nabla^2 - \frac{1}{c^2}\frac{\partial^2}{\partial t^2} - Q_T^2\right)\mathbf{\Phi}_T(\mathbf{r},t) = \mathbf{0}, \tag{21.33}$$

$$\left(\nabla^2 - \frac{1}{c^2}\frac{\partial^2}{\partial t^2}\right)\mathbf{A}_T(\mathbf{r},t) = \mathbf{0}. \tag{21.34}$$

The quantities $Q_L \equiv \omega_p/a$ and $Q_T \equiv \omega_p/c$ are Compton wave numbers of the plasmon and plasmariton.

21.2 THE "INNER" STRUCTURE OF THE PLASMARITON – A QUASIPARTICLE -

Let us return to the transverse and longitudinal dielectric functions of the electron gas as these appear in the hydrodynamic version [Eqs. (21.13) and (21.14)]. We have learned in Subsec 21.1.2 that the bulk plasmon dispersion relation originates in the condition $\varepsilon_L(q,\omega) = 0$ [Eq. (21.19)]. For what follows we rename the plasmon a L-plasmon. At first sight it may appear a bit odd that the dispersion relation for the transverse jellium is given by Eq. (21.25) and *not* by the condition

$$\varepsilon_T(q,\omega) = 0. \tag{21.35}$$

Going through the classical electrodynamic calculations the reason for reaching the dispersion $(\omega/c)^2\varepsilon_T(q,\omega) = q^2$ is obvious, of course. However, from a quantum physical point of view it is useful, and to a certain extend necessary, to take a closer look at this "problem". We have two aspects of the classical theories which help us to see the transverse dynamics in a deeper perspective: (i) The Klein-Gordon equation for the L-plasmon [wave function, $\Phi_L(\mathbf{r},t)$, [Eq. (21.32)] depends *alone* on plasma parameters $[a_L, Q_L]$, and the massless Klein-Gordon equation for the photon wave function $\mathbf{A}_T(\mathbf{r},t)$ only depends on the light parameter $[c]$. The Klein-Gordon equation for the plasmariton wave function, $\mathbf{\Phi}_T(\mathbf{r},t)$ given in Eq. (21.33), has *mixed* photon and plasma parameters $[c, Q_T]$, making one guess that the plasmariton has an "inner" structure. (ii) The bulk plasmariton has *one* branch (for $q \geq 0$) [see Eq. (21.30)], whereas the surface plasmariton has *two* branches [see Eq. (19.43)].

How does our understanding shift if we introduce a new transverse plasmon, T-plasmon, with a dispersion relation determined by $\varepsilon_T = 0$ [Eq. (21.35)]? Using the hydrodynamic expression for $\varepsilon_T(q,\omega)$ [Eq. (21.13)] one obtains the squared T-plasmon bulk dispersion relation

$$(a_Tq)^2 = \omega^2 - \omega_p^2, \quad [\text{Bulk } T\text{-plasmon}], \tag{21.36}$$

where

$$a_T = D^{1/2} = \sqrt{\frac{1}{5}}v_F. \tag{21.37}$$

The related *scalar* Klein-Gordon equation for the new T-plasmon wave function, $\phi_T(\mathbf{r},t)$, viz.,

$$\left(\nabla^2 - \frac{1}{a_T^2}\frac{\partial^2}{\partial t^2} - Q_T^2\right)\phi_T(\mathbf{r},t) = 0, \tag{21.38}$$

is quite satisfactory because it depends alone on plasma parameters $[a_T, Q_T]$. For each $\mathbf{q}$ component of transverse excitations one has two independent polarization modes, but in jellium the rotational symmetry around $\mathbf{q}$ makes it clear that one for the underlying structure only a single scalar wave function is needed.

In a sense, a free L-plasmon can exist, but not a free T-plasmon. This is due to the fact that the transverse dynamics must satisfy the transverse set of microscopic

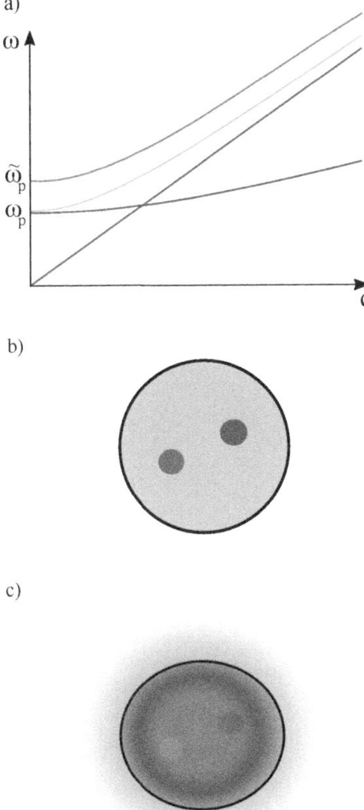

Figure 21.1 Schematic illustration of the "inner" bulk plasmariton structure. a): Light-line in red (straight line), T-plasmon in blue (lower curved line), plasmariton in green (middle curved line) and the renormalized in violet, $\tilde{\omega}_p = \sqrt{2}\omega_p$ (upper curved line) dispersion relations. b): A localized plasmariton quasiparticle (big circle) with its "internal" boson particles: T-photon, red (left) and T-plasmon, blue (right). c): Plasmariton quasiparticle with its tied T-photon cloud in violet (gray shaded); after [127]. The plasmariton quasiparticle description is discussed in Chapter 28.

Maxell-Lorentz equations. Thus, since $\mathbf{B} \neq \mathbf{0}$ it is clear from Eq. (21.24) that the dispersion relation in Eq. (21.25) is the eigenmode closest to observation through an external excitation of the jellium. In the perspective we may consider the plasmariton as a quasiparticle consisting of a T-plasmon-photon pair inevitably coupled. The dispersion diagram shown in Fig. 21.1 reflects this in a direct manner.

Note that there is only one branch of the bulk plasmariton dispersion relation. This stems from the fact that the inequality $(a_T q)^2 = \omega^2 - \omega_p^2 \geq 0$ must be satisfied. For real q this implies that $\omega \geq \omega_p$ For surface wave where $q_\| > \omega/c$ [cf. the Weyl expansion], one has

$$(a_T q_\|)^2 = \omega^2 - \omega_p^2 - (a_T q_\perp)^2 = \omega^2 - \omega_p^2 + (a_T |q_\perp|)^2 \geq 0, \tag{21.39}$$

an inequality which open the doorway for a potential plasmariton branch lying below ω_p! However, it takes a surface mode to achieve the condition $q_\parallel > \omega/c$. For surface modes $\omega_p \Rightarrow \omega_{pS}$. The upper (existing) bulk plasmariton branch hence goes into the radiative Brewster branch [1, 33] of the surface plasmariton, and the non-exsisting lower branch goes into the existing Fano branch [81] of the surface plasmariton.

21.3 JELLION QUASIPARTICLE

21.3.1 Kinematics

The three fundamental Klein-Gordon equations

$$\left(\nabla^2 - \frac{1}{a_L^2}\frac{\partial^2}{\partial t^2} - Q_L^2\right)\phi_L(\mathbf{r}, t) = 0, \tag{21.40}$$

$$\left(\nabla^2 - \frac{1}{c^2}\frac{\partial^2}{\partial t^2} - Q_T^2\right)\phi_T(\mathbf{r}, t) = \mathbf{0}, \tag{21.41}$$

$$\left(\nabla^2 - \frac{1}{c^2}\frac{\partial^2}{\partial t^2}\right)\mathbf{A}_T(\mathbf{r}, t) = \mathbf{0}. \tag{21.42}$$

describe in wave mechanics the free L-plasmon, T-plasmon [degeneracy $= 2$], and gauge photon. These free particles are abstractions of our mind since they are not observable. Experimentally, a jellium with these eigenmodes must *always* be excited and measured by *external* devices [$\sim$ source, detector]. Quantum mechanical phenomena are closed (cf. Bohr [41, 158]), and our observations relate to the physics of the total setup: Source + jellium + detector. An indirect glimpse of the L- and T-plasmons can be obtained by resonance excitation of the jellium. Thus, excitations by light and electrons typically lead to information on the T-plasmon and L-plasmon, respectively.

We have realized in Section 21.2 that the photon and the T-plasmon resonantly coupled form a quasiparticle, the plasmariton. The bulk plasmariton dispersion relation is at first sight expected to have two branches, one with $\omega > \omega_p$ and one with $\omega < \omega_p$. Only the upper branch can be resonantly excited in a bulk experiment. A finger-print of the lower branch appears in the form of surface plasmaritons. The two branches of the surface plasmaritons are separated by a frequency gap $\omega_{pS} < \omega < \omega_p$ [neglecting irreversible relaxation mechanisms].

Since the T-plasmon may interact strongly with the gauge photon one might expect that a quasiparticle can be formed by resonant photon$\leftrightarrow$ L-plasmon$\leftrightarrow$ T-plasmon interaction. I have named this *triplet quasiparticle* a *jellion*, as already mentioned in Chapter 19, where it was shown that its closely related partner the *surface jellion* can be resonantly excited with a p-polarized externally impressed electromagnetic field. If the three eigenmodes (photon, L-plasmon, plasmariton) are uncoupled the three dispersion relations can be combined into the following compact form:

$$\left[q^2 - \left(\frac{\omega}{c}\right)^2\right]\left[q^2 - \kappa_L^2\right]\left[q^2 - \kappa_T^2\right] = 0. \tag{21.43}$$

The characteristic L- and T (squared) wave numbers of the jellium are $\kappa_L^2 = (\omega^2 - \omega_p^2)/a_L^2$ [Eq. (19.35) with $D^{1/2} \equiv D_L^{1/2} = a_L$] and $\kappa_T^2 = (\omega^2 - \omega_p^2)/c^2$ [Eq. (19.37)].

It appear from Eq. (21.43) that the L-plasmon and the plasmariton dispersion relations each have one branch both located above ω_p. However, the photon branch, $0 < \omega < \infty$, holds the possibility that the coupled system's partner, the surface jellion, can possess solutions below ω_p. We know that the surface plasmariton lower branch is pushed below ω_{pS} [Upper branch: $\omega > \omega_p$; lower branch: $\omega < \omega_{pS}$]. Because of this one might anticipate that the surface jellion dispersion relation will behave qualitative different in the three frequency regions $\omega < \omega_{pS}$, $\omega_{pS} < \omega < \omega_p$, and $\omega_p < \omega$. Since Eq. (21.43) is of third degree in q^2, one expects that the surface triplet in general would posses three branches for $q_\parallel > 0$ [The solutions $\sqrt{q_\parallel^2} = \pm q_\parallel$ represent waves with physically identical properties propagating in opposite directions along the surface.

The surface jellion dispersion relation has already been given in Eq. (19.40), and its "universal" character established on the basis of screened Green function formalism for adjacent jellium/vacuum halfspaces [Chapter 19]. A closer formal connection to Eq. (21.43) appears by rewriting Eq. (19.40) as follows:

$$\left\{ q_\parallel^2 [\kappa_T^2 + \kappa_L^2(1 + \varepsilon(\omega))] - \kappa_T^2\kappa_L^2 \right\}^2 = 4\varepsilon(\omega)\kappa_L^2 q_\parallel^6, \tag{21.44}$$

where $\varepsilon(\omega) = 1 - (\omega_p/\omega)^2$. Above, the wave-vector dispersion in the transverse dielectric function has been neglected; see Eq. (21.28) , and remember the inequality $a_T/c << 1$.

In Section 19.2, the surface plasmariton and the Ritchie dispersion relations [Eq. (19.43) and (19.45)] were derived starting from Eq. (19.40). The reader may find it easier to derive these taking Eq. (21.44) as starting point. In the limit $a = D^{1/2} \to 0$, $\kappa_T^2/\kappa_L^2 \to 0$, and this reduces Eq. (21.44) to

$$[q_\parallel^2(1 + \varepsilon(\omega) - \kappa_T^2]^2 = 4\varepsilon\kappa_L^{-2}q_\parallel^6 \to 0, \tag{21.45}$$

so that

$$q_\parallel^2(1 + \varepsilon(\omega)) = \kappa_T^2 = \left(\frac{\omega}{c}\right)^2 \varepsilon(\omega). \tag{21.46}$$

The square root of Eq. (21.46) is Eq. (19.42), or equivalent the surface plasmariton dispersion relation in Eq. (19.43). In the nonretarded ($c \to \infty$) case $\kappa_T \to 0$, and Eq. (21.44) simplifies to

$$q_\parallel = \kappa_L \frac{1 + \varepsilon(\omega)}{2\varepsilon^{1/2}(\omega)}. \tag{21.47}$$

By inserting $\kappa_L = D^{-1/2}(\omega^2 - \omega_p^2)^{1/2}$ in Eq. (21.47) one obtains the Ritchie dispersion relation [Eq. (19.45)], q.e.d.

21.3.2 Role of the transverse self-field

In the Ritchie dispersion relation [Eq. (19.45)] a trace of deeper physics appears in disguise. To reveal the secret, let us return to the general expression of the p-polarized amplitude reflection coefficient, given in Eq. (19.29). If one neglects the L-plasmon part $(D \to 0)$, $r_p(q_\parallel,\omega)$ becomes

$$r_p(q_\parallel,\omega) = \frac{q_\perp^0 \varepsilon_T(\kappa_T,\omega) - \kappa_\perp^T}{q_\perp^0 \varepsilon_T(\kappa_T,\omega) + \kappa_\perp^T},\tag{21.48}$$

a well-known result in elementary physical optics. In the limit $c \to \infty$, Eq. (21.48) is reduced to

$$r_p(q_\parallel,\omega; c \to \infty) = \frac{\varepsilon(\omega) - 1}{\varepsilon(\omega) + 1}.\tag{21.49}$$

The result in Eq. (21.48) holds for the transverse (T) field dynamics, but it is not correct that the T-dynamics relates solely to gauge photons (T-photons), propagating with the speed of light. If this was the case, $r_p(q_\parallel,\omega)$ would approach zero for $c \to \infty$. The non-propagating part of $r_p(q_\parallel,\omega)$, which we are left with in Eq. (21.49), may be called the transverse self-field part. In the Ritchie dispersion relation, written in the form given in Eq. (19.44), the quantity $1 + \varepsilon(\omega)$ originates in the transverse self-field, obviously. The nonretarded Ritchie formula hence is a mixture of L- and T-dynamics. This conclusion holds beyond the collective-mode approach. Thus, with the inclusion of single-particle excitations, $r_p(q_\parallel,\omega)$ is given by Eq. (19.49), and the explicit expression for $M(q_\parallel,\omega)$ [Eq. (19.50)] shows the generality of the $T - L$ mixing. An explicit calculation of the T-part of $M(q_\parallel,\omega)$, $M_T(q_\parallel,\omega)$, gives by residue calculation

$$M_T(q_\parallel,\omega) = \frac{\kappa_\perp^T}{\kappa_T^2} \to iq_\parallel \,,\tag{21.50}$$

where the last expression is obtained for $c \to \infty$. If one neglects the T-part of the p-polarized reflection coefficient, one obtains from Eq. (19.29) the result

$$r_p(q_\parallel,\omega) = -1.\tag{21.51}$$

CHAPTER 22

Photon-Spin Interaction

22.1 LINEAR SPIN ELECTRODYNAMICS

22.1.1 General considerations

Let us return to the first-order Liouville equation for the density matrix operator [Eq. (18.45)]. To turn the attention to the spin we divide $\hat{\mathcal{H}}_1$ into its space- and spin-parts, i.e., $\hat{\mathcal{H}}_1 \Rightarrow \hat{\mathcal{H}}_1 + \hat{\mathcal{H}}_1^\sigma$, where in $\hat{\mathcal{H}}_1$ [Eq. (18.13)] and $\hat{\mathcal{H}}_1^\sigma$ [Eq. (18.15)]. It is assumed that the transverse vector-potential operator is a classical quantity. Also the first-harmonic density operator is split into space- and spin-parts, that $\hat{\rho}_1 \Rightarrow \hat{\rho}_1 + \hat{\rho}_1^\sigma$. The first-order perturbations of the density matrix obviously satisfy the equation

$$\hbar\omega \begin{bmatrix} \hat{\rho}_1 \\ \hat{\rho}_1^\sigma \end{bmatrix} = \left[\hat{\mathcal{H}}_0, \begin{bmatrix} \hat{\rho}_1 \\ \hat{\rho}_1^\sigma \end{bmatrix} \right] + \left[\begin{bmatrix} \hat{\mathcal{H}}_1 \\ \hat{\mathcal{H}}_1^\sigma \end{bmatrix}, \hat{\rho}_0 \right], \tag{22.1}$$

neglecting the irreversible coupling to the reservoir ($\tau \to \infty$), and spin-orbit couplings. The matrix element of Eq. (22.1) between the many-body states $|I\rangle$ and $|J\rangle$ is readily obtained utilizing Eqs. (18.37) and (18.39). Hence

$$\langle I| \begin{bmatrix} \hat{\rho}_1 \\ \hat{\rho}_1^\sigma \end{bmatrix} |J\rangle = \frac{P_J - P_I}{\hbar\omega + E_J - E_I} \langle I| \begin{bmatrix} \hat{\mathcal{H}}_1 \\ \hat{\mathcal{H}}_1^\sigma \end{bmatrix} |J\rangle. \tag{22.2}$$

It is important to emphasize that the *field-unperturbed* states ($|I\rangle$, $|J\rangle$, ...) may be spin-dependent. In many mesoscopic systems the unperturbed eigenstates depend on the electron spin via the spin-orbit part of the Hamiltonian, $[\hbar/(4m^2c^2)]\boldsymbol{\nabla} V(\mathbf{r}) \times \hat{\mathbf{p}} \cdot \hat{\boldsymbol{\sigma}}$, because large gradients of the Coulomb energy can occur (near surfaces and interfaces, e.g.). In ferromagnetic (antiferromagnetic) materials the spin plays a dominating role. The low-lying energy states of the spin systems coupled by exchange interactions are wavelike. These spin waves can couple to the electromagnetic field, and the elementary quantum events are magnon-photon interactions. A brief account of the ferromagnetic spin waves and their interaction with photons is given in Chapter 24.

The linear part of induced current density $[\mathbf{J}_1(\mathbf{r};\omega)]$ is given by Eq. (18.54) in the space-frequency domain. Remembering that $\hat{\mathcal{J}}_0$ is the sum of spin-independent ($\hat{\mathcal{J}}_F$) and spin-dependent ($\hat{\mathcal{J}}_F^\sigma$) parts [Eq. (18.51)], the formal expression for $\mathbf{J}_1(\mathbf{r};\omega)$ can be divided as follows:

$$\mathbf{J}_1(\mathbf{r};\omega) = \mathbf{J}_1^{\text{SPACE}}(\mathbf{r};\omega) + \mathbf{J}_1^{\text{SPACE-SPIN}}(\mathbf{r};\omega) + \mathbf{J}_1^{\text{SPIN}}(\mathbf{r};\omega), \tag{22.3}$$

DOI: 10.1201/9781003029458-22

where

$$\mathbf{J}_1^{\text{SPACE}}(\mathbf{r};\omega) = \text{Tr}\{\hat{\rho}_0\hat{\mathcal{J}}_1\} + \text{Tr}\{\hat{\rho}_1\hat{\mathcal{J}}_F\}, \tag{22.4}$$

$$\mathbf{J}_1^{\text{SPACE-SPIN}}(\mathbf{r};\omega) = \text{Tr}\{\hat{\rho}_1\hat{\mathcal{J}}_F^\sigma\} + \text{Tr}\{\hat{\rho}_1^\sigma\hat{\mathcal{J}}_F\}, \tag{22.5}$$

$$\mathbf{J}_1^{\text{SPIN}}(\mathbf{r};\omega) = \text{Tr}\{\hat{\rho}_1^\sigma\hat{\mathcal{J}}_F^\sigma\}. \tag{22.6}$$

In Eq.(22.5) the space and spin properties are mixed in the sense that in each of the involved operators, $\hat{\rho}_1\hat{\mathcal{J}}_F^\sigma$ and $\hat{\rho}_1^\sigma\hat{\mathcal{J}}_F$, a product of a space and spin operator appears. If one uses the plane-wave basis set with members $|\mathbf{k}, s\rangle = |\mathbf{k}\rangle \otimes |s\rangle$ [Eq. (18.4)] it can be shown [144] without difficulties that

$$\mathbf{J}_1^{\text{SPACE-SPIN}}(\mathbf{r};\omega) = \mathbf{0} \tag{22.7}$$

if the free Hamiltonian ($\hat{\mathcal{H}}_0$) is spin independent. Thus, if the spin-orbit term in $\hat{\mathcal{H}}_0$ is neglected ($\sim$ small), only Eq. (22.6) contributes to the linearized spin dynamics.

22.1.2   Space conductivity tensor

The many-body (MB) space conductivity tensor $\sigma_{\text{MB}}^{\text{SPACE}}$, can be derived from Eq. (22.4). Below we shall cite the result of the calculation [144, 157] because it put the final formula for the spin conductivity tensor, $\sigma_{\text{MB}}^{\text{SPIN}}$, in perspective. For what follows it is useful to write $\sigma_{\text{MB}}^{\text{SPACE}}$ as a sum of two parts:

$$\sigma_{\text{MB}}^{\text{SPACE}}(\mathbf{r}, \mathbf{r}';\omega) = \sigma_{\text{MB}}^{\text{dia}}(\mathbf{r}, \mathbf{r}';\omega) + \sigma_{\text{MB}}^{\text{para}}(\mathbf{r}, \mathbf{r}';\omega). \tag{22.8}$$

The diamagnetic (dia) part, given by

$$\sigma_{\text{MB}}^{\text{dia}}(\mathbf{r}, \mathbf{r}';\omega) = \frac{ie^2}{m\omega}N_0(\mathbf{r})\delta(\mathbf{r} - \mathbf{r}')\mathbf{U}, \tag{22.9}$$

where

$$N_0(\mathbf{r}) = \sum_{\mathbf{k},s} N_{\mathbf{k},s}|\psi_{\mathbf{k}}(\mathbf{r})|^2 \tag{22.10}$$

is the electron density at space point $\mathbf{r}$ in the field-unperturbed state. The quantity

$$N_{\mathbf{k},s} = \sum_I N_{\mathbf{k},s}^I P_I \tag{22.11}$$

gives the number of particles in the single-particle state $(\mathbf{k}, s)$ when the system is in the mixed many-body state characterized by the P_I's. The diamagnetic part plays a particularly important role in superconductors; see Chapter 25. The diamagnetic term originates in the $\text{Tr}\{\hat{\rho}_0\hat{\mathcal{J}}_1\}$-part of Eq. (22.5).

The paramagnetic (para) part of the many-body conductivity is given by

$$\sigma_{\mathrm{MB}}^{\mathrm{para}}(\mathbf{r},\mathbf{r}';\omega) = \sum_{\mathbf{k},\mathbf{k}'} A_{\mathbf{k},\mathbf{k}'}(\omega) \mathbf{j}_{\mathbf{k}'\to\mathbf{k}}^{\mathrm{SPACE}}(\mathbf{r}) \mathbf{j}_{\mathbf{k}\to\mathbf{k}'}^{\mathrm{SPACE}}(\mathbf{r}'). \tag{22.12}$$

The space transition current density from state $|\alpha\rangle$ to state $|\beta\rangle$, i.e. [with $e < 0$],

$$\mathbf{j}_{\alpha\to\beta}^{\mathrm{SPACE}}(\mathbf{r}) = \frac{e\hbar}{2mi}\left[\psi_\beta^*(\mathbf{r})\boldsymbol{\nabla}\psi_\alpha(\mathbf{r}) - \psi_\alpha(\mathbf{r})\boldsymbol{\nabla}\psi_\beta^*(\mathbf{r})\right], \tag{22.13}$$

enters Eq. (22.12) in two different space points $\mathbf{r}$ and $\mathbf{r}'$ and with opposite transitions $[\mathbf{k}' \to \mathbf{k},\ \mathbf{k} \to \mathbf{k}']$. The quantity $\mathbf{J}_{\mathbf{k}'\to\mathbf{k}}^{\mathrm{SPACE}}(\mathbf{r})\mathbf{J}_{\mathbf{k}\to\mathbf{k}'}^{\mathrm{SPACE}}(\mathbf{r}')$ is a dyadic product of transition current densities, and $\sigma_{\mathrm{MB}}^{\mathrm{para}}(\mathbf{r},\mathbf{r}';\omega)$ hence is weighted sum of dyadic products belonging to pairs $(\mathbf{k},\mathbf{k}')$ of one-electron states. The complex frequency-dependent weight factor is given by

$$A_{\mathbf{k},\mathbf{k}'}(\omega) = \frac{i}{\omega}\sum_{I,J,s}\frac{P_J - P_I}{\hbar\omega + E_J - E_I}\langle I|\,\hat{b}_{\mathbf{k}',s}^\dagger \hat{b}_{\mathbf{k},s}\,|J\rangle \langle J|\,\hat{b}_{\mathbf{k}',s}^\dagger \hat{b}_{\mathbf{k}',s}\,|I\rangle. \tag{22.14}$$

It is possible to rewrite the expression for $\sigma_{\mathrm{MB}}^{\mathrm{dia}}(\mathbf{r},\mathbf{r}';\omega)$ [Eq. (22.9)] in such a manner that when added to $\sigma_{\mathrm{MB}}^{\mathrm{para}}(\mathbf{r},\mathbf{r}';\omega)$ [Eq. (22.12)] the singularities in the two parts for $\omega = 0$ cancel [157]. This means that even without irreversible relaxation mechanics the space conductivity stays finite for $\omega \to 0$. There is an exception to this conclusion: In a superconductor the paramagnetic conductivity vanishes below the gab frequency. Left with the diamagnetic part, the spatial conductivity is infinite at $\omega = 0$; as we know this is the correct result.

22.1.3 Spin conductivity tensor

Let us write the linear spin current density in Eq. (22.6) in the explicit form

$$\mathbf{J}_1^{\mathrm{SPIN}}(\mathbf{r};\omega) = \sum_{I,J}\langle I|\,\hat{\rho}_1^\sigma\,|J\rangle \langle J|\,\hat{\boldsymbol{\mathcal{J}}}_F^\sigma\,|I\rangle. \tag{22.15}$$

By utilizing the relation given in Eq. (22.2) between $\hat{\rho}_1^\sigma$ and $\hat{\mathcal{H}}_1^\sigma$, one obtains

$$\mathbf{J}_1^{\mathrm{SPIN}}(\mathbf{r};\omega) = \sum_{I,J}\frac{P_J - P_I}{\hbar\omega + E_J - E_I}\langle I|\,\hat{\boldsymbol{\mathcal{J}}}_F^\sigma\,|J\rangle \langle J|\,\hat{\mathcal{H}}_1^\sigma\,|I\rangle. \tag{22.16}$$

It appears from Eq. (18.20) that it is convenient to introduce a one-electron spin transition current density $[e < 0]$,

$$\mathbf{j}_{\alpha\to\beta}^{\mathrm{SPIN}}(\mathbf{r}) = \frac{ie}{2m}\langle\beta|\,\delta(\mathbf{r}-\mathbf{r}_e)\hat{\mathbf{p}} - \hat{\mathbf{p}}\delta(\mathbf{r}-\mathbf{r}_e)\,|\alpha\rangle = \frac{e\hbar}{2m}[\psi_\beta^*(\mathbf{r})\boldsymbol{\nabla}\psi_\alpha(\mathbf{r}) + \psi_\alpha(\mathbf{r})\boldsymbol{\nabla}\psi_\beta^*(\mathbf{r})], \tag{22.17}$$

in relation to the calculation of the matrix element $\langle I|\,\hat{\boldsymbol{\mathcal{J}}}_F^\sigma\,|J\rangle$. In passing we notice that $i\mathbf{j}_{\alpha\to\beta}^{\mathrm{SPACE}}(\mathbf{r})$ [Eq. (22.13)] and $\mathbf{j}_{\alpha\to\beta}^{\mathrm{SPIN}}$ [Eq. (22.17)] deviate only through the sign between the two terms in the respective parentheses. We shall understand the physical importance of this difference when we analyze the electrodynamics of a mesoscopic circular ring (wire) in Section 22.2. If one writes $\mathbf{j}_{\alpha\to\beta}^{\mathrm{SPIN}}(\mathbf{r})$ in the form

$$\mathbf{j}_{\alpha\to\beta}^{\mathrm{SPIN}}(\mathbf{r}) = \frac{e\hbar}{2m}\boldsymbol{\nabla}[\psi_\alpha(\mathbf{r})\psi_\beta^*(\mathbf{r})], \tag{22.18}$$

it is manifest that $\boldsymbol{\nabla}\times\mathbf{J}_{\alpha\to\beta}^{\mathrm{SPIN}}(\mathbf{r}) = \mathbf{0}$. Hence, it follows that the spin transition current density is a rotational-free vector field.

It appears from Eq. (18.15) that $\mathbf{j}_{\alpha\to\beta}^{\mathrm{SPIN}}(\mathbf{r})$ also enters the calculation of the matrix element $\langle J|\,\hat{\mathcal{H}}_1^\sigma\,|I\rangle$ in Eq. (22.15). Thus,

$$\langle\mathbf{k}'|\,\boldsymbol{\nabla}\times\mathbf{A}_T\,|\mathbf{k}\rangle = -\frac{2m}{e\hbar}\int_{-\infty}^{\infty}\mathbf{j}_{\mathbf{k}\to\mathbf{k}'}^{\mathrm{SPIN}}(\mathbf{r}')\times\mathbf{A}_T(\mathbf{r}';\omega)d^3r'. \tag{22.19}$$

A tedious calculation gives the following results for the linear spin current density in the space-frequency domain:

$$\mathbf{J}_1^{\mathrm{SPIN}}(\mathbf{r};\omega) = \sum_{\mathbf{k}'\to\mathbf{k}}\mathbf{j}_{\mathbf{k}'\to\mathbf{k}}^{\mathrm{SPIN}}(\mathbf{r})\times\boldsymbol{\mathcal{B}}_{\mathbf{k},\mathbf{k}'}(\omega)\cdot\left[\int_{-\infty}^{\infty}\mathbf{j}_{\mathbf{k}\to\mathbf{k}'}^{\mathrm{SPIN}}(\mathbf{r}')\right]\times\mathbf{E}_T(\mathbf{r}';\omega)d^3r', \tag{22.20}$$

remembering that $\mathbf{A}_T(\mathbf{r},\omega) = \mathbf{E}_T(\mathbf{r};\omega)/(i\omega)$. The general expression for the tensor $\boldsymbol{\mathcal{B}}_{\mathbf{k},\mathbf{k}'}(\omega)$ is given in [144]. Under the assumption that the free Hamiltonian is spin independent (spin-orbit effects, etc. being neglected) the tensor takes the simplified form

$$\boldsymbol{\mathcal{B}}_{\mathbf{k},\mathbf{k}'}(\omega) = -\frac{g}{2}\mathcal{A}_{\mathbf{k},\mathbf{k}'}(\omega)\mathbf{U}. \tag{22.21}$$

Since the gyromagnetic factor $g \simeq 2$, $\boldsymbol{\mathcal{B}}_{\mathbf{k},\mathbf{k}'}(\omega)$ is just the unit tensor multiplied by $-\mathcal{A}_{\mathbf{k},\mathbf{k}'}(\omega)$, where $\mathcal{A}_{\mathbf{k},\mathbf{k}'}(\omega)$ is given by Eq. (22.14). The linear nonlocal many-body spin conductivity tensor $\boldsymbol{\sigma}_{\mathrm{MB}}^{\mathrm{SPIN}}(\mathbf{r},\mathbf{r}';\omega)$, defined via

$$\mathbf{J}_1^{\mathrm{SPIN}}(\mathbf{r};\omega) = \int_{-\infty}^{\infty}\boldsymbol{\sigma}_{\mathrm{MB}}^{\mathrm{SPIN}}(\mathbf{r},\mathbf{r}';\omega)\cdot\mathbf{E}_T(\mathbf{r}';\omega)d^3r', \tag{22.22}$$

now can be extracted from Eq. (22.20). Hence,

$$\boldsymbol{\sigma}_{\mathrm{MB}}^{\mathrm{SPIN}}(\mathbf{r},\mathbf{r}';\omega) = \sum_{\mathbf{k},\mathbf{k}'}\mathbf{j}_{\mathbf{k}'\to\mathbf{k}}^{\mathrm{SPIN}}(\mathbf{r})\times\boldsymbol{\mathcal{B}}_{\mathbf{k},\mathbf{k}'}(\omega)\times\mathbf{j}_{\mathbf{k}\to\mathbf{k}'}^{\mathrm{SPIN}}(\mathbf{r}). \tag{22.23}$$

From the dyadic relation $\boldsymbol{\alpha}\times\mathbf{U}\times\boldsymbol{\beta} = \boldsymbol{\beta}\boldsymbol{\alpha} - \mathbf{U}(\boldsymbol{\alpha}\cdot\boldsymbol{\beta})$, it is seen that the spin conductivity tensor can be written in the alternative form

$$\boldsymbol{\sigma}_{\mathrm{MB}}^{\mathrm{SPIN}}(\mathbf{r},\mathbf{r}';\omega) = \frac{g}{2}\sum_{\mathbf{k},\mathbf{k}'}\mathcal{A}_{\mathbf{k},\mathbf{k}'}(\omega)\left[\mathbf{U}\mathbf{j}_{\mathbf{k}'\to\mathbf{k}}^{\mathrm{SPIN}}(\mathbf{r})\cdot\mathbf{j}_{\mathbf{k}\to\mathbf{k}'}^{\mathrm{SPIN}}(\mathbf{r}') - \mathbf{j}_{\mathbf{k}\to\mathbf{k}'}^{\mathrm{SPIN}}(\mathbf{r})\mathbf{j}_{\mathbf{k}'\to\mathbf{k}}^{\mathrm{SPIN}}(\mathbf{r}')\right].$$

$$\tag{22.24}$$

22.2 SPIN ELECTRODYNAMICS IN RPA

22.2.1 Space-conductivity tensor

For many purposes it is sufficient to calculate the space- and spin-conductivity tensors, given in the many-body theory in Eqs. (22.9), (22.12) and (22.24) [for a spin-independent free particle Hamiltonian], in the framework of the RPA approach. In RPA it is assumed that the eigenstates of the free many-electron Hamiltonian $\hat{\mathcal{H}}_F$ are direct (tensor) products of single-particle-like states. If $|0\rangle$ denotes the global particle-vacuum state, the RPA ground state, $|G\rangle$, is given by

$$|G\rangle = \Pi_{k \leq k_F, s} \hat{b}^{\dagger}_{\mathbf{k},s} |0\rangle. \tag{22.25}$$

In the ground state only states at and below k_F are occupied, as indicated. The excited states are of the particle-hole type

$$|\text{PARTICLE, HOLE}\rangle = |\mathbf{k}'s'; \mathbf{k}, s\rangle = \hat{b}^{\dagger}_{\mathbf{k}',s'} \hat{b}_{\mathbf{k},s} |G\rangle. \tag{22.26}$$

To determine $\mathcal{A}_{\mathbf{k},\mathbf{k}'}(\omega)$ [Eq. (22.14)] in RPA, one has to calculate $P_J - P_I$ and $E_J - E_I$ for the cycle $|I\rangle \Rightarrow |J\rangle \propto \hat{b}^{\dagger}_{\mathbf{k},s} \hat{b}_{\mathbf{k}',s} |I\rangle \Rightarrow |I\rangle$. Clearly, $E_J - E_I = \varepsilon_\mathbf{k} - \varepsilon_{\mathbf{k}'}$, where $\varepsilon_\mathbf{k}$ and $\varepsilon_{\mathbf{k}'}$ are the energies of the single-electron-like states $(\mathbf{k}, s)$ and $(\mathbf{k}', s')$. Furthermore, for degenerate spin states, one obtains

$$P_J - P_I = P_I^{\mathbf{k},s}(1 - P_I^{\mathbf{k}',s}) - P_I^{\mathbf{k}',s}(1 - P_I^{\mathbf{k},s}) = P_I^{\mathbf{k},s} - P_I^{\mathbf{k}',s} \tag{22.27}$$

where $P_I^{\mathbf{k},s}$ $(P_I^{\mathbf{k}',s})$ is the probability that there *is* an electron in the one-electron state $(\mathbf{k}, s)$ $[\mathbf{k}', s]$. Next,

$$\sum_{I,s}(P_I^{\mathbf{k},s} - P_I^{\mathbf{k}',s}) = 2(f_\mathbf{k} - f_{\mathbf{k}'}), \tag{22.28}$$

$f_\mathbf{k}$ $(f_{\mathbf{k}'})$ being the Fermi-Dirac occupation factor for the single-particle state $\mathbf{k}$ $(\mathbf{k}')$.

Gathering the results above, it is realized that $\mathcal{A}_{\mathbf{k},\mathbf{k}'}(\omega)$ [Eq. (22.14)] takes the form

$$\mathcal{A}_{\mathbf{k},\mathbf{k}'}(\omega) = \frac{2i}{\omega} \frac{f_\mathbf{k} - f_{\mathbf{k}'}}{\hbar\omega + \varepsilon_\mathbf{k} - \varepsilon_{\mathbf{k}'}} \tag{22.29}$$

in RPA. The field-unperturbed electron density obviously is given by

$$N_0^{\text{RPA}} = 2 \sum_\mathbf{k} f_\mathbf{k} |\psi_\mathbf{k}(\mathbf{r})|^2 \tag{22.30}$$

in the RPA approach. By inserting Eqs. (22.29) and (22.30) into Eqs. (22.9) and (22.12), respectively, one obtains the following expressions for the dia- and para-magnetic parts of the space-conductivity tensor:

$$\sigma_{\text{RPA}}^{\text{dia}}(\mathbf{r}, \mathbf{r}'; \omega) = \frac{2ie^2}{m\omega} \delta(\mathbf{r} - \mathbf{r}') \sum_\mathbf{k} f_\mathbf{k} |\psi_\mathbf{k}(\mathbf{r})|^2, \tag{22.31}$$

and

$$\sigma_{\text{RPA}}^{\text{para}}(\mathbf{r}, \mathbf{r}'; \omega) = \frac{2i}{\omega} \sum_{\mathbf{k},\mathbf{k}'} \frac{f_\mathbf{k} - f_{\mathbf{k}'}}{\hbar\omega + \varepsilon_\mathbf{k} - \varepsilon_{\mathbf{k}'}} \mathbf{j}_{\mathbf{k}' \to \mathbf{k}}^{\text{SPACE}}(\mathbf{r}) \mathbf{j}_{\mathbf{k} \to \mathbf{k}'}^{\text{SPACE}}(\mathbf{r}'). \tag{22.32}$$

22.2.2 Addition of the dia- and para-magnetic parts of the space conductivity

Let us return to the diamagnetic part of the many-body conductivity tensor [Eq. (22.9)]. As shown in my book Quantum Theory of Near-Field Electrodynamics [157], Section 12. 3, it is possible to rewrite $\boldsymbol{\sigma}_{\text{MB}}^{\text{dia}}(\mathbf{r}, \mathbf{r}'; \omega)$ as follows:

$$\boldsymbol{\sigma}_{\text{MB}}^{\text{dia}}(\mathbf{r}, \mathbf{r}'; \omega) = \frac{1}{i\omega} \sum_{I,J} \frac{P_J - P_I}{E_J - E_I} \mathbf{J}_{I \to J}^{\text{SPACE}}(\mathbf{r}) \mathbf{J}_{J \to I}^{\text{SPACE}}(\mathbf{r}'), \qquad (22.33)$$

where $\mathbf{J}_{I \to J}^{\text{SPACE}}$ ($\mathbf{J}_{I \to J}^{\text{SPACE}}(\mathbf{r})$) is the many-body transition current density from state I to state J (vice versa). It is possible to prove that the term with $I = J$ does not contribute to $\boldsymbol{\sigma}_{\text{MB}}^{\text{dia}}$ [146]. The idea of writing the diamagnetic part of the space-conductivity tensor as a double sum $[\sum_{I,J}(...)]$ is convenient because this allows one to add it term by term to the paramagnetic part, given by

$$\boldsymbol{\sigma}_{\text{MB}}^{\text{para}}(\mathbf{r}, \mathbf{r}'; \omega) = \frac{i}{\omega} \sum_{I,J} \frac{P_J - P_I}{\hbar\omega + E_J - E_I} \mathbf{J}_{I \to J}^{\text{SPACE}}(\mathbf{r}) \mathbf{J}_{J \to I}^{\text{SPACE}}(\mathbf{r}'). \qquad (22.34)$$

In Eqs. (22.33) and (22.34) we have returned from second- to first-quantization utilizing that

$$\sum_{\mathbf{k},\mathbf{k}',s} \langle I | \hat{b}_{\mathbf{k}',s}^{\dagger} \hat{b}_{\mathbf{k},s} | J \rangle \langle J | \hat{b}_{\mathbf{k},s}^{\dagger} \hat{b}_{\mathbf{k}',s} | I \rangle \, \mathbf{j}_{\mathbf{k}' \to \mathbf{k}}^{\text{SPACE}}(\mathbf{r}) \mathbf{j}_{\mathbf{k} \to \mathbf{k}'}^{\text{SPACE}}(\mathbf{r}') = \mathbf{J}_{I \to J}^{\text{SPACE}}(\mathbf{r}) \mathbf{J}_{J \to I}^{\text{SPACE}}(\mathbf{r}').$$

$$(22.35)$$

By addition of Eqs. (22.33) and (22.34) one obtains the many-body result

$$\boldsymbol{\sigma}_{\text{MB}}^{\text{SPACE}}(\mathbf{r}, \mathbf{r}'; \omega) = \frac{\hbar}{i} \sum_{I,J} \frac{P_J - P_I}{E_J - E_I} \frac{1}{\hbar\omega + E_J - E_I} \mathbf{J}_{I \to J}(\mathbf{r}) \mathbf{J}_{J \to I}(\mathbf{r}'). \qquad (22.36)$$

Written in this form it appears manifestly correct that $\boldsymbol{\sigma}_{\text{MB}}^{\text{SPACE}}(\mathbf{r}, \mathbf{r}'; \omega)$ in general is not singular at $\omega = 0$, a fact already indicated in the text below Eq. (22.14). A BCS superconductor makes an exception because $\boldsymbol{\sigma}_{\text{MB}}^{\text{para}}$ vanishes below the gap frequency, and thus also for $\omega = 0$.

In RPA (here with spin degeneracy) one then obtains a space conductivity in the compact form

$$\boldsymbol{\sigma}_{\text{RPA}}^{\text{SPACE}}(\mathbf{r}, \mathbf{r}'; \omega) = \frac{2\hbar}{i} \sum_{\mathbf{k},\mathbf{k}'} \frac{f_{\mathbf{k}} - f_{\mathbf{k}'}}{\varepsilon_{\mathbf{k}} - \varepsilon_{\mathbf{k}'}} \frac{1}{\hbar\omega + \varepsilon_{\mathbf{k}} - \varepsilon_{\mathbf{k}'}} \mathbf{j}_{\mathbf{k}' \to \mathbf{k}}^{\text{SPACE}}(\mathbf{r}) \mathbf{j}_{\mathbf{k} \to \mathbf{k}'}^{\text{SPACE}}(\mathbf{r}') \qquad (22.37)$$

by addition of Eqs. (22.31) and (22.32).

22.2.3 Spin-conductivity tensor

On the basis of the considerations made in Subsection 22.2.1, the RPA expression for the linear spin conductivity can be obtained immediately by inserting the expression given for $\mathcal{A}_{\mathbf{k},\mathbf{k}'}(\omega)$ in Eq. (22.29) into Eq.(22.24). Hence, one obtains

$$\boldsymbol{\sigma}_{\text{RPA}}^{\text{SPIN}}(\mathbf{r}, \mathbf{r}'; \omega) = \frac{g}{i\omega} \sum_{\mathbf{k},\mathbf{k}'} \left\{ \frac{f_{\mathbf{k}} - f_{\mathbf{k}'}}{\hbar\omega + \varepsilon_{\mathbf{k}} - \varepsilon_{\mathbf{k}'}} \left[\mathbf{j}_{\mathbf{k} \to \mathbf{k}'}^{\text{SPIN}}(\mathbf{r}) \mathbf{j}_{\mathbf{k}' \to \mathbf{k}}^{\text{SPIN}}(\mathbf{r}') - \mathbf{U} \mathbf{j}_{\mathbf{k}' \to \mathbf{k}}^{\text{SPIN}}(\mathbf{r}) \cdot \mathbf{j}_{\mathbf{k} \to \mathbf{k}'}^{\text{SPIN}}(\mathbf{r}') \right] \right\}.$$

$$(22.38)$$

Due to the fact that the spin part of the interaction Hamiltonian is proportional to $\boldsymbol{\nabla} \times \mathbf{A}_T = (i\omega)^{-1}\boldsymbol{\nabla} \times \mathbf{E}_T$ it is obvious from the outset of the analysis that the spins contributes only to the transverse electrodynamics. Furthermore, the prefractor $(i\omega)^{-1}$ in Eq. (22.38) does not lead to any troublesome divergences in the theory for $\omega \to 0$, because physical results are obtained only after reinserting the relation between the electric and magnetic fields in the formalism $[\mathbf{A}_T(\mathbf{r};\omega) = (i\omega)^{-1}\mathbf{E}_T(\mathbf{r};\omega)]$.

22.3 SPIN ELECTRODYNAMICS IN A MESOSCOPIC RING

In order to obtain a qualitative understanding of the linear spin electrodynamics in mesoscopic systems, and to compare its importance relative to the linear space electrodynamics, we shall study the conductivity response functions in a circular mesoscopic ring (quantum wire). The following calculations are based on the RPA approach. We have already met such a ring in connection to an illustration of the Aharonov-Bohm phenomenon in Section 11.4. As a forerunner for the analysis below, a schematic illustration of the currents induced in the mesoscopic ring by a plane electromagnetic field is shown in Fig. 22.1

22.3.1 Stationary states and eigenenergies

Let us start by considering a circular cylinder shell having its center axis coincident with the z-axis in a cylindrical (r, ϕ, z)-coordinate system. The height of the cylinder is L, and its shell thickness is d. Inside the shell, i.e., here for $r_0 \leq r \leq r_0+d$ and $-L/2 \leq z \leq L/2$, we use the jellium approximation $(V(\mathbf{r}) = 0)$ for the electrons. Furthermore, we take the semiclassical infinite-barrier model as basis for the description of the electron scattering (reflection) at all surfaces.

Using the ansatz

$$\psi(r, \phi, z) = R(r)\Phi(\phi)Z(z), \tag{22.39}$$

the time-independent single-particle free Schrödinger equation,

$$-\frac{\hbar^2}{2m}\nabla^2\psi(\mathbf{r}) = \varepsilon\psi(\mathbf{r}), \tag{22.40}$$

breaks into three ordinary differential equations, viz.,

$$\left(\frac{d^2}{dz^2} + k_z^2\right) Z(z) = 0, \tag{22.41}$$

$$\left(\frac{d^2}{d\phi^2} + m^2\right) \Phi(\phi) = 0, \tag{22.42}$$

$$\left(\frac{d^2}{dr^2} + \frac{1}{r}\frac{d}{dr} + k^2 - k_z^2 - \frac{m^2}{r^2}\right) R(r) = 0, \tag{22.43}$$

where $k = (2m\varepsilon/\hbar^2)^{1/2}$. Since the wave function vanishes at the top and the bottom surfaces of the cylinder normalized standing solutions of Eq. (22.41) have wave

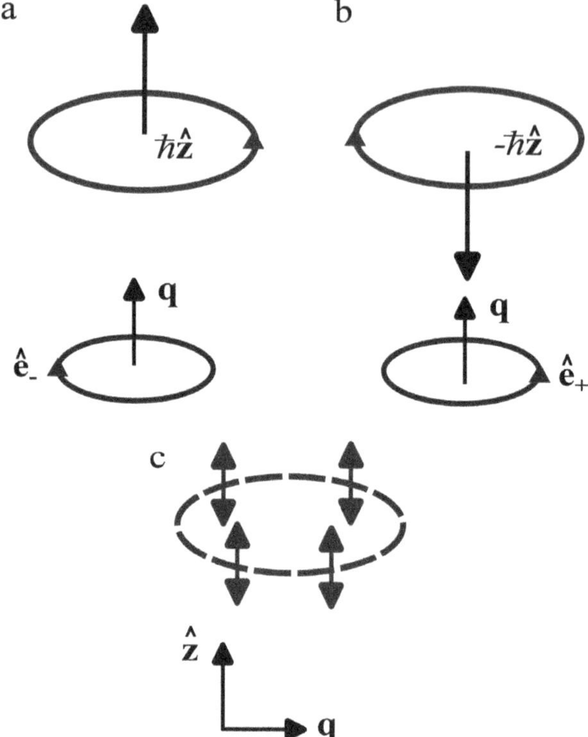

Figure 22.1 Schematic illustration of currents [in blue] induced in a mesoscopic circular ring by electromagnetic fields of different wave vectors **q** [in black] and polarizations [in red]. In figures a and b the field has in-plane positive ($\hat{\mathbf{e}}_-$) and negative ($\hat{\mathbf{e}}_+$) helicity, respectively. The associated induced electron angular momentum is $\pm\hbar\hat{\mathbf{z}}$, as in indicated [$\hat{\mathbf{z}}$ is a unit vector perpendicular to the plane of the ring]. In figure c the field is polarized in the $\hat{\mathbf{z}}$-direction. In this case the *internal* oscillating spin current [in blue double arrows] is induced perpendicular to the plane of the ring [dashed in blue].

numbers $k_z = \pi l/L$, and form a fundamental set

$$Z(z) = \left(\frac{2}{L}\right)^{1/2} \sin\left[\frac{\pi l}{L}\left(z + \frac{L}{2}\right)\right], \quad l = 1, 2, \dots \tag{22.44}$$

In order that $\Phi(\phi)$ be single valued, normalized running eigensolutions form a set

$$\Phi(\phi) = (2\pi)^{-1/2}\exp(im\phi), \quad m = 0 \pm 1 \pm 2, \dots \tag{22.45}$$

For the radial equation the Bessel function of the first kind (J_m) and Neumann function (N_m), both of order m form a pair of linearly independent solutions, i.e.,

$$R(r) = A J_m\left(\left[k^2 - \left(\frac{\pi l}{L}\right)^2\right]^{1/2} r\right) + B N_m\left(\left[k^2 - \left(\frac{\pi l}{L}\right)^2\right]^{1/2} r\right), \tag{22.46}$$

where A and B are so far arbitrary constants. Since $R(r_0) = R(r_0 + d) = 0$ for the SCIB model, the yet unknown quantity k can for a given set (l, m) take on values $k = k_{n,l,m}$ $(n = 1, 2, ..)$ determined by

$$J_m \left(\left[k_{n,l,m}^2 - \left(\frac{\pi l}{L} \right)^2 \right]^{1/2} r_0 \right) N_m \left(\left[k_{n,l,m}^2 - \left(\frac{\pi l}{L} \right)^2 \right]^{1/2} (r_0 + d) \right) =$$

$$J_m \left(\left[k_{n,l,m}^2 - \left(\frac{\pi l}{L} \right)^2 \right]^{1/2} (r_0 + d) \right) N_m \left(\left[k_{n,l,m}^2 - \left(\frac{\pi l}{L} \right)^2 \right]^{1/2} r_0 \right). \tag{22.47}$$

Based on the elementary considerations presented above, it is realized that one may chose energy eigenfunctions of the type

$$\psi_{n,l,m}(r, \phi, z) = A_{n,l,m} \sin \left[\frac{\pi l}{L} \left(z + \frac{L}{2} \right) \right] \exp(im\phi) \left\{ J_m \left(\left[k_{n,l,m}^2 - \left(\frac{\pi l}{L} \right)^2 \right]^{1/2} r \right) - \right.$$

$$\left. \frac{J_m \left(\left[k_{n,l,m}^2 - \left(\frac{\pi l}{L} \right)^2 \right]^{1/2} r_0 \right)}{N_m \left(\left[k_{n,l,m}^2 - \left(\frac{\pi l}{L} \right)^2 \right]^{1/2} r_0 \right)} N_m \left(\left[k_{n,l,m}^2 - \left(\frac{\pi l}{L} \right)^2 \right]^{1/2} r \right) \right\}. \tag{22.48}$$

The corresponding energy eigenvalues are

$$\varepsilon_{n,l,m} = \frac{\hbar^2}{2m} k_{n,l,m}^2. \tag{22.49}$$

The constants $A_{n,l,m}$ can be determined by normalization of $\psi_{n,l,m}$. The normalized energy eigenfunctions can be written as follows:

$$\psi_{n,l,m}(r, \phi, z) - (2\pi)^{-1/2} \exp(im\phi) \left(\frac{2}{L} \right)^{1/2} \sin \left[\frac{\pi l}{L} \left(z + \frac{L}{2} \right) \right] R_{n,l,m}(r), \tag{22.50}$$

where $R_{n,l,m}(r)$ is separately normalized, i.e.,

$$\int_{r_0}^{r_0+d} |R_{n,l,m}(r)|^2 r\, dr = 1. \tag{22.51}$$

Let us assume now that the shell is so thin and the cylinder height so small that electronic transitions between quantum states in the r- and $z-$directions can be neglected for the range of photon energies ($\hbar\omega$) of interest. Effectively, one is left with single wavefunctions $[R(r), Z(z)]$ related to r and z. Although the forms of these wavefunctions are of interest in certain studies, it is for the subsequent qualitative discussion sufficient to assume complete electron confinement, and thus take

$$|R(r)|^2 = \frac{\delta(r - r_0)}{r_0}, \tag{22.52}$$

and

$$|Z(z)|^2 = \delta(z). \tag{22.53}$$

In this idealized quantum ring, only the set

$$\Phi_\alpha(\phi) = (2\pi)^{-1/2} \exp(im_\alpha\phi), \quad m_\alpha = 0, \pm 1, \pm 2, , ... \tag{22.54}$$

of stationary left- and right-hand running free electron states is needed. The associated eigenenergies are

$$\varepsilon_\alpha = \frac{\hbar^2}{2mr_0^2} m_\alpha^2. \tag{22.55}$$

22.3.2 Scattering in first-order Born approximation

In Chapter 14, we discussed fundamental aspects of the semiclassical theory of light scattering paying particular attention to linear elastic scattering. Below we briefly describe a simplified heuristic approach which readers often come across in the literature, and which we consider sufficient for an elementary treatment of the electrodynamics of a mesoscopic ring with complete transverse electron confinement.

In the standard approach one converts the microscopic Maxwell-Lorentz wave equation for the electric field in the space-frequency domain, $\mathbf{E}(\mathbf{r};\omega)$, into a scattering integral equation of the form

$$\mathbf{E}(\mathbf{r};\omega) = \mathbf{E}^0(\mathbf{r};\omega) + i\mu_0\omega \int_{-\infty}^{\infty} \mathbf{G}(\mathbf{r},\mathbf{r}';\omega) \cdot \mathbf{J}(\mathbf{r}';\omega)d^3r'. \tag{22.56}$$

The prescribed external (incident) electric field $\mathbf{E}^0(\mathbf{r};\omega)$ induces a current density response, $\mathbf{J}(\mathbf{r}';\omega)$, in the (mesoscopic) medium. In linear electrodynamics one combines Eq. (22.56) with a spatially nonlocal constitutive equation

$$\mathbf{J}(\mathbf{r};\omega) = \int_{-\infty}^{\infty} \boldsymbol{\sigma}(\mathbf{r},\mathbf{r}';\omega) \cdot \mathbf{E}(\mathbf{r}',\omega)d^3r', \tag{22.57}$$

where $\boldsymbol{\sigma}(\mathbf{r},\mathbf{r}'\omega)$ is an unspecified microscopic linear conductivity tensor. Before proceeding, let me draw the readers attention to the limitations inherent in a scattering theory based on Eqs. (22.56) and (22.57). The standard Green function, $\mathbf{G}(\mathbf{r},\mathbf{r}';\omega)$, used in Eq. (22.56), is *not* a propagator, because it contains a non-propagating part. Technically, this manifests itself in the fact that $\mathbf{G}(\mathbf{r},\mathbf{r}';\omega)$ cannot be transformed to the space-time domain [118, 148]. The transverse part of the Green function $\mathbf{G}_T(\mathbf{r},\mathbf{r}';\omega)$, which as we have discussed in Part I, relates to the physics of the transverse gauge photon, *can be* transformed to the space-time domain, $\mathbf{G}_T(\mathbf{r},\mathbf{r}';\omega) \Rightarrow \mathbf{G}_T(\mathbf{r},\mathbf{r}',\tau)$. Although it might be useful to incorporate screening effects in $\mathbf{G}(\mathbf{r},\mathbf{r}';\omega)$, as we have seen in relation to the treatment of collective excitations in Chapter 19, in most cases one uses the vacuum form of $\mathbf{G}(\mathbf{r},\mathbf{r}';\omega) = \mathbf{G}(\mathbf{r}-\mathbf{r}';\omega)$. The physics of the transverse vacuum propagator $\mathbf{G}_T(\mathbf{R},\tau) = \mathbf{G}_T(\mathbf{r}-\mathbf{r}',t-t')$, was discussed in Sections 3.3, 3.4 and 14.1. In the constitutive relation in Eq. (22.57) we have not made the fundamental distinction between the transverse and longitudinal dynamics, as one needs to do in refined analyses. The microscopic conductivity tensor contains the space- and spin-parts, and if needed also a part associated with spin-orbit dynamics. The incident electric field, $\mathbf{E}^0(\mathbf{r};\omega)$, induces a certain motion in the charges of the medium, and this nonuniform motion generates an extra electric field inside and outside the medium. The induced current density $\mathbf{J}(\mathbf{r};\omega)$ then is assumed to be driven in a spatially nonlocal fashion by the local field $\mathbf{E}(\mathbf{r},\omega)$ in the medium.

By combination of Eqs. (22.56) and (22.57) one obtains an integral equation for the electric field, $\mathbf{E}(\mathbf{r};\omega)$, viz.,

$$\mathbf{E}(\mathbf{r};\omega) = \mathbf{E}^0(\mathbf{r};\omega) + i\mu_0\omega \int_{-\infty}^{\infty} \mathbf{G}(\mathbf{r}-\mathbf{r}'';\omega) \cdot \boldsymbol{\sigma}(\mathbf{r}'',\mathbf{r}';\omega) \cdot \mathbf{E}(\mathbf{r}';\omega)d^3r''d^3r'. \quad (22.58)$$

Let us assume that $\boldsymbol{\sigma}(\mathbf{r}'',\mathbf{r}';\omega)$ is a known quantity, the central problem then is a calculation of the local electric field *inside* the medium. Once $\mathbf{E}(\mathbf{r};\omega)$ has been determined in all points of the medium, the field *outside* the medium can be obtained by projection with the vacuum Green function $\mathbf{G}(\mathbf{R};\omega)$, for which an explicit expression is known [157]. Although more advanced methods exist the local field most often is calculated by iteration in a so-called Born series approach. In the first-order Born approximation one replaces $\mathbf{E}(\mathbf{r};\omega)$ by $\mathbf{E}^0(\mathbf{r};\omega)$ in the integral of Eq. (22.58). The approximate field inside as well as outside the medium, denoted by $\mathbf{E}^1(\mathbf{r};\omega)$, hence is given by

$$\mathbf{E}^1(\mathbf{r};\omega) = \mathbf{E}^0(\mathbf{r};\omega) + i\mu_0\omega \int_{-\infty}^{\infty} \mathbf{G}(\mathbf{r}-\mathbf{r}'';\omega) \cdot \boldsymbol{\sigma}(\mathbf{r}'',\mathbf{r}';\omega) \cdot \mathbf{E}^0(\mathbf{r}';\omega)d^3r''d^3r'.$$

$$(22.59)$$

The first-order Born approximation consequently corresponds to neglect of local-field effects in the constitutive equation, i.e.,

$$\mathbf{J}(\mathbf{r};\omega) = \int_{-\infty}^{\infty} \boldsymbol{\sigma}(\mathbf{r},\mathbf{r}';\omega) \cdot \mathbf{E}^0(\mathbf{r}',\omega)d^3r'. \quad (22.60)$$

However, one must be aware of the fact that the very determination of the conductivity tensor depends on the (approximate) "choice" made for the field acting on the charges from the outset. In a mesoscopic ring, where one adopts complete transverse confinement of the electrons [Eqs. (22.52) and (22.53)] as a premise for the calculation, the local-field correction in a given point of the ring alone arises from the 1D current density prevailing in all other points along the ring.

Information about the spin electrodynamics of a mesoscopic ring, for instance, can be obtained by light scattering studies. In the framework of the first-Born approximation the linearly scattered electric field, $\mathbf{E}^{\mathrm{scatt}}(\mathbf{r};\omega)$, is given by

$$\mathbf{E}^{\mathrm{scatt}}(\mathbf{r};\omega) \equiv \mathbf{E}(\mathbf{r};\omega) - \mathbf{E}^0(\mathbf{r};\omega)$$

$$= i\mu_0\omega \int_{-\infty}^{\infty} \mathbf{G}(\mathbf{r}-\mathbf{r}'';\omega) \cdot \boldsymbol{\sigma}(\mathbf{r}'',\mathbf{r}';\omega) \cdot \mathbf{E}^0(\mathbf{r}';\omega)d^3r''d^3r'. \quad (22.61)$$

It is obvious that it is an extremely complicated task, if not an impossible one, to determine the general structure of $\boldsymbol{\sigma}(\mathbf{r}, \mathbf{r}'; \omega)$ from scattering data. Even if this could be done one still has to filter out the spin part from the total conductivity tensor. Some information can be obtained rather easily in the case where the linear dimensions of the mesoscopic object are small in comparison to the characteristic wavelength spectrum of the incoming field. In this case one may assume that the spatial variations of the incoming field and the vacuum Green functions across the mesoscopic objects are small. If the origo of our coordinate system is placed inside the object, e.g. in the center for a circular ring, the scattered field becomes

$$\mathbf{E}^{\text{scatt}}(\mathbf{r}; \omega) = i\mu_0 \omega \mathbf{G}(\mathbf{r}, \mathbf{0}; \omega) \cdot \boldsymbol{\sigma}_{\text{ED}}(\omega) \cdot \mathbf{E}^0(\mathbf{0}; \omega), \tag{22.62}$$

where

$$\boldsymbol{\sigma}_{\text{ED}}(\omega) = \int_{-\infty}^{\infty} \boldsymbol{\sigma}(\mathbf{r}, \mathbf{r}'; \omega) d^3r' d^3r. \tag{22.63}$$

In this limit the mesoscopic object behaves like an electric dipole (ED) absorber and radiator in the scattering process. In this double ED approximation sometimes use the more precise notation $\boldsymbol{\sigma}_{\text{ED-ED}}(\omega)$ instead of $\boldsymbol{\sigma}_{\text{ED}}(\omega)$. Taylor series expansions of $\mathbf{G}$ and $\mathbf{E}^0$ around the origo allow one to include higher-order electric (E) and magnetic (B) moments of $\boldsymbol{\sigma}(\mathbf{r}, \mathbf{r}'; \omega)$ [ED-EQ, EQ-ED, ED-MD,MD-ED,...] in the scattering analysis. The moment expansion idea has recently been used to study the light scattering from a mesoscopic hole in a free-electron-like flat screen [119, 120, 122]. In this example the mesoscopic hole behaves electrodynamically much like a mesoscopic disk placed on a screen without a hole.

22.3.3 Integrated spin and space transition current densities

It appears from the RPA expression for the space [Eq. (22.37)] - and spin [Eq. (22.38)] conductivities that only integrated transition current densities $(i \to j)$

$$\mathbf{I}_{i \to j}^{\text{space}} = \int_{-\infty}^{\infty} \mathbf{j}_{i \to j}^{\text{space}}(\mathbf{r} d^3 r, \tag{22.64}$$

and

$$\mathbf{I}_{i \to j}^{\text{spin}} = \int_{-\infty}^{\infty} \mathbf{j}_{i \to j}^{\text{spin}}(\mathbf{r} d^3 r, \tag{22.65}$$

appear in $\boldsymbol{\sigma}_{\text{ED}}(\omega)$ [Eq. (22.63)]. The relevant transition current densities for a circular ring with complete transverse confinement are obtained by means of the reduction

$$\boldsymbol{\nabla} \Rightarrow \hat{\boldsymbol{\phi}} \frac{1}{r_0} \frac{\partial}{\partial \phi}, \quad \hat{\boldsymbol{\phi}} = -\hat{\mathbf{x}} \sin \phi + \hat{\mathbf{y}} \cos \phi. \tag{22.66}$$

Hence,

$$\mathbf{j}_{\alpha \to \beta}^{\text{spin}}(\mathbf{r}) = \frac{e\hbar i}{4\pi m r_0^2} \delta(r - r_0)\delta(z)(m_\alpha - m_\beta)e^{i(m_\alpha - m_\beta)\phi}\hat{\boldsymbol{\phi}} \tag{22.67}$$

$$\mathbf{j}_{\alpha \to \beta}^{\text{space}}(\mathbf{r}) = \frac{e\hbar i}{4\pi m r_0^2} \delta(r - r_0)\delta(z)(m_\alpha + m_\beta)e^{i(m_\alpha - m_\beta)\phi}\hat{\boldsymbol{\phi}} \tag{22.68}$$

The expressions in Eqs. (22.67) and (22.68) in turn give

$$\mathbf{I}^{\text{spin}}_{\alpha\to\beta} = \frac{e\hbar}{2\sqrt{2}mr_0}\left(\delta_{m_\beta,m_\alpha-1}\mathbf{e}_+ + \delta_{m_\beta,m_\alpha+1}\mathbf{e}_-\right),$$

(22.69)

$$\mathbf{I}^{\text{space}}_{\alpha\to\beta} = \frac{e\hbar}{2\sqrt{2}imr_0}\left[(2m_\alpha - 1)\delta_{m_\beta,m_\alpha-1}\mathbf{e}_+ - (2m_\alpha + 1)\delta_{m_\beta,m_\alpha+1}\mathbf{e}_-\right]$$

(22.70)

where

$$\mathbf{e}_\pm = \frac{1}{\sqrt{2}}(\hat{\mathbf{x}} \pm i\hat{\mathbf{y}})$$

(22.71)

are helicity unit vectors.

22.3.4 ED Spin- and Space-Conductivity Tensors

A combination of Eqs. (22.37), (22.38), (22.63)–(22.65), (22.69) and (22.70) results after a somewhat lengthy calculation the following final results [197]:

$$\boldsymbol{\sigma}^{\text{spin}}_{\text{ED}}(\omega) = \frac{ig}{\omega}\left(\frac{e\hbar}{2\sqrt{2}m_0r_0}\right)^2$$
$$\times \sum_m\left[\frac{f_{m-1} - f_m}{\hbar\omega + \frac{\hbar^2}{2m_0r_0^2}(1-2m)}(\hat{\mathbf{z}}\hat{\mathbf{z}} + \mathbf{e}_+\mathbf{e}_-) + \frac{f_{m+1} - f_m}{\hbar\omega + \frac{\hbar^2}{2m_0r_0^2}(1+2m)}(\hat{\mathbf{z}}\hat{\mathbf{z}} + \mathbf{e}_-\mathbf{e}_+)\right]$$

(22.72)

and

$$\boldsymbol{\sigma}^{\text{space}}_{\text{ED}}(\omega) = \frac{\hbar e^2}{2im_0}\sum_m\left[\frac{(1-2m)(f_{m-1} - f_m)}{\hbar\omega + \frac{\hbar^2}{2m_0r_0^2}(1-2m)}\mathbf{e}_+\mathbf{e}_- + \frac{(1+2m)(f_{m+1} - f_m)}{\hbar\omega + \frac{\hbar^2}{2m_0r_0^2}(1+2m)}\mathbf{e}_-\mathbf{e}_+\right],$$

(22.73)

with the relabeling $\sum_\alpha \Rightarrow \sum_m$, $m_\alpha \Rightarrow m$, and for the electron mass $m \Rightarrow m_0$. The chemical potential, $\mu(T)$, is obtained normalizing the sum (over m) of the Fermi factors to half the number of jellium electrons in the ring. A comprehensive discussion and comparison of the results in Eqs. (22.72) and (22.73) can be found in [197]. The electron excitation-deexcitation cycle needed for the calculation of the spin conductivity tensors is shown in Fig. 22.2.

22.3.5 Far-field scattering from the spins

In most cases the light scattering from the mesoscopic ED ring will be dominated by the space contribution, $\boldsymbol{\sigma}^{\text{SPACE}}_{\text{ED}}(\omega)$. There is an obvious exception, however. Thus if one assumes that the incident field on the ring is polarized perpendicular to the plane of the ring, $\mathbf{E}^0(\mathbf{0};\omega) = E^0(\omega)\hat{\mathbf{z}}$, the current density induced by the space dynamics vanishes, i.e.,

$$\mathbf{J}^{\text{space}}(\omega) = \boldsymbol{\sigma}^{\text{space}}_{\text{ED}}(\omega)\mathbf{E}^0(\mathbf{0};\omega) = \mathbf{0}.$$

(22.74)

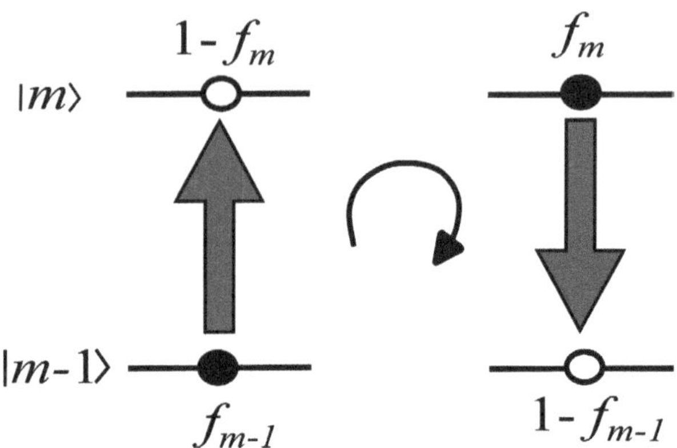

Figure 22.2 Schematic illustration of the electron excitation ($|m - 1\rangle \to |m\rangle$) deexcitation ($|m\rangle \to |m - 1\rangle$) cycle entering the calculation of the spin conductivity. The cycle involves transition between neighbouring angular momentum states only. The relevant finite-temperature Fermi factors are shown.

In a sense, this result is obvious because the complete transverse confinement forbids the electrons to achieve a field-induced motion in the direction perpendicular to the plane of the ring. The complete confinement assumption is an idealization of course, but if the height (L) of the ring is so small that only a single bound level exists for the z-part of the wavefunction in a potential with the finite barrier height, alone the diamagnetic part of the space conductivity tensor will be nonvanishing. This part in general gives a very small contribution to the mesoscopic electrodynamics, superconductivity being an exception, though.

For $\mathbf{E}^0(0;\omega) \parallel \hat{\mathbf{z}}$, Eq. (22.72) shows that only the $\hat{\mathbf{z}}\hat{\mathbf{z}}$-part of $\boldsymbol{\sigma}_{\mathrm{ED}}^{\mathrm{SPIN}}(\omega)$ contributes to the electrodynamics. Upon a number of algebraic manipulations one obtains in compact form

$$\sigma_{\mathrm{ED},zz}^{\mathrm{spin}}(\omega) \equiv \hat{\mathbf{z}} \cdot \boldsymbol{\sigma}_{\mathrm{ED}}^{\mathrm{spin}}(\omega) \cdot \hat{\mathbf{z}} = \frac{ig}{\omega} \frac{e^2 \hbar^4}{8m_0^3 r_0^4} \sum_m \frac{(2m-1)(f_{m-1} - f_m)}{(\hbar\omega)^2 - \left[\frac{\hbar^2(2m-1)}{2m_0 r_0^2}\right]^2}. \tag{22.75}$$

Why is it possible to obtain a finite spin contribution with complete transverse electron confinement, and an incident field polarized perpendicular to the plane of the ring? The answer is that the induced spin dynamics is *not* accompanied by a dynamic spin charge density, $\rho^{\mathrm{SPIN}}(\mathbf{r};\omega)$, not even in the general case. Thus it is possible to show, using Eq. (22.20) that

$$\boldsymbol{\nabla} \cdot \mathbf{J}_{\mathrm{MB}}^{\mathrm{SPIN}}(\mathbf{r};\omega) = \int_{-\infty}^{\infty} [\boldsymbol{\nabla} \cdot \boldsymbol{\sigma}_{\mathrm{MB}}^{\mathrm{SPIN}}(\mathbf{r},\mathbf{r}';\omega)] \cdot \mathbf{E}_T(\mathbf{r}';\omega) d^3 r' = 0. \tag{22.76}$$

From the equation of continuity for the spin dynamics, viz.,

$$\boldsymbol{\nabla} \cdot \mathbf{J}_{\mathrm{MB}}^{\mathrm{SPIN}}(\mathbf{r};\omega) = i\omega \rho^{\mathrm{SPIN}}(\mathbf{r};\omega) \tag{22.77}$$

it then follows that $\rho^{\mathrm{SPIN}}(\mathbf{r};\omega) = 0$. One might have guessed the conclusion reached above, because the spin is an *internal* degree of freedom. Although a spin flip is accompanied by a spin current flowing in *real* space, the flow is *always* incompressible [157], so that $\rho^{\mathrm{SPIN}}(\mathbf{r};\omega) = 0$.

Let us assume that a monochromatic field polarized perpendicular to the plane of the ring excites the spin system. In a typical light scattering experiment one measures the cycle-averaged Poynting vector of the scattered electromagnetic field, $< \mathbf{S}^{\mathrm{scatt}} >$, in the far field (FF). An elementary calculation gives

$$< \mathbf{S}^{\mathrm{scatt}}(\mathbf{r}, t)) >_{FF} = \frac{\mu_0 (E^0)^2}{32\pi^2 c} \left(\frac{\sin\theta}{r} \right)^2 \omega^2 \left| \sigma^{\mathrm{spin}}_{ED,zz}(\omega) \right|^2 \hat{\mathbf{r}}, \qquad (22.78)$$

in the scattering direction given by the unit vector $\hat{\mathbf{r}}$. The energy flow is independent of the azimuth angle (ϕ), as required by symmetry: $\mathbf{E}^0$ points in the z-direction, and the physical properties of the circular ring are invariant against arbitrary rotations around $\hat{\mathbf{z}}$. The polar angle (θ) and distance (r) dependencies of $< \mathbf{S}^{\mathrm{scatt}} >_{FF}$ are given by the factor $(\sin\theta/r)^2$. It appears from Eq. (22.75) that the frequency spectrum depends on the ring radius, the temperature and the number (N) of free electrons hold by the ring. A detailed analysis also shows that the spectra are qualitatively different from rings where half the number $(N/2)$ of electrons is even and uneven. The reader may find a number of numerical results in [197].

VI

Basics of Magnon and Superconductor Electrodynamics

CHAPTER 23

Spin-1/2 Current Density: Exponential Photon Source Confinement

The electron spin plays a conceptually important role in various subfields of photon physics, and in the chapters subsequent to this chapter we shall illustrate this by two examples from condensed-matter electrodynamics. In Chapter 22, we already discussed elements of the linear photon-spin interaction. Thus, in Chapter 24, a brief discussion of spin waves in ferromagnets is given paying particular attention to the magnon-photon interaction. The importance of the electron spin in linear BCS-superconductivity is well known and treated in many textbooks and reference works, so in Chapter 25, we take up a less well-known subject, namely the role of spin paring in nonlinear electrodynamics. We illustrate the nonlinear photon-electron coupling by studying a particular example, viz., the process of nonlinear electromagnetic rectification. Main emphasis is devoted to discussions of the nonlinear dynamical Meissner effect in an infinitely extended jellium, and the rectification process at a superconductor-vacuum interface.

We know that the extension of the photon source current distribution determines the best possible confinement of photons in space at a given time. Usually, the photon confinement exhibits algebraic decay. However, in a pure spin-flip process the photon source domain is exponentially confined, as we shall realize below.

23.1 RELATIVISTIC DIRAC CURRENT DENSITY

In relativistic quantum mechanics the spin-1/2 current density of a particle with charge q is given by [95, 106, 133, 158]

$$\mathbf{J}(\mathbf{r}, t) = cq\psi^{\dagger}(\mathbf{r}, t)\boldsymbol{\alpha}\psi(\mathbf{r}, t), \tag{23.1}$$

DOI: 10.1201/9781003029458-23

where the vector $\boldsymbol{\alpha} = (\alpha_1, \alpha_2, \alpha_3)$ in the Dirac 4x4-matrix realization has the components

$$\alpha_i = \begin{pmatrix} 0 & \sigma_i \\ \sigma_i & 0 \end{pmatrix}, \quad i = 1 - 3, \tag{23.2}$$

$\boldsymbol{\sigma} = (\sigma_1, \sigma_2, \sigma_3)$ being the Pauli 2x2-spin vector. The Pauli spin matrices have the explicit 2x2 form

$$\sigma_1 = \begin{pmatrix} 0 & 1 \\ 1 & 0 \end{pmatrix}, \quad \sigma_2 = \begin{pmatrix} 0 & -i \\ i & 0 \end{pmatrix}, \quad \sigma_3 = \begin{pmatrix} 1 & 0 \\ 0 & -1 \end{pmatrix}. \tag{23.3}$$

For what follows, it is convenient to write the four-component Dirac wave function $\Psi = (\Psi_1, \Psi_2, \Psi_3, \Psi_4)$ in the form of a pair of two-component spinors, i.e.,

$$\Psi(\mathbf{r}, t) = \begin{pmatrix} \phi(\mathbf{r}, t) \\ \chi(\mathbf{r}, t) \end{pmatrix} \tag{23.4}$$

where $\phi = (\psi_1, \psi_2)$ and $\chi = (\psi_3, \psi_4)$. The ansatz

$$\begin{pmatrix} \phi(\mathbf{r}, t) \\ \chi(\mathbf{r}, t) \end{pmatrix} = \begin{pmatrix} \phi_0(\mathbf{r}, t) \\ \chi_0(\mathbf{r}, t) \end{pmatrix} \exp\left(-i\frac{mc^2 t}{\hbar}\right), \tag{23.5}$$

where m is the electron mass, then shows that the Dirac current density can be written in the form

$$\mathbf{J}(\mathbf{r}, t) = cq \left[\phi_0^\dagger \boldsymbol{\sigma} \chi_0 + \chi_0^\dagger \boldsymbol{\sigma} \phi_0 \right]. \tag{23.6}$$

Although it is possible to approach the role of the spin in a general relativistic sense using the so-called Gordon decomposition of $\mathbf{J}(\mathbf{r}, t)$ [102, 105, 158], and from there derive the well-known Pauli spin contribution, it is sufficient here to proceed directly to the weakly relativistic limit.

23.2 WEAKLY RELATIVISTIC PAULI SPIN CURRENT DENSITY

In the weakly relativistic domain the bispinor component (ϕ_0, χ_0) are related by [95, 105, 106, 158]

$$\chi_0 = \frac{1}{2mc} \boldsymbol{\sigma} \cdot (\hat{\mathbf{p}} - q\mathbf{A}) \phi_0, \tag{23.7}$$

where $\hat{\mathbf{p}} = (\hbar/i)\boldsymbol{\nabla}$. The components χ_0 and ϕ_0 are often referred to as the small and large components of Ψ due to the fact that ϕ_0 is the dominating component when relativistic effects are small. By inserting Eq. (23.7) into Eq. (23.6) it appears that the current density is given by

$$\mathbf{J}(\mathbf{r}, t) = \frac{q}{2m} \left[\phi_0^\dagger \boldsymbol{\sigma}\boldsymbol{\sigma} \cdot (\hat{\mathbf{p}} - q\mathbf{A}) \phi_0 + h.c. \right] \tag{23.8}$$

in the weakly relativistic domain. In Eq (23.8), $h.c.$ stands for Hermitian conjugate.

Utilizing the easily proved identity

$$\sigma\sigma \cdot (\hat{\mathbf{p}} - q\mathbf{A}) = (\hat{\mathbf{p}} - q\mathbf{A}) - i\sigma \times (\hat{\mathbf{p}} - q\mathbf{A}), \tag{23.9}$$

the current density divides into space- and spin-parts:

$$\mathbf{J}(\mathbf{r}, t) = \mathbf{J}^{\mathrm{space}}(\mathbf{r}, t) + \mathbf{J}^{\mathrm{spin}}(\mathbf{r}, t), \tag{23.10}$$

where

$$\mathbf{J}^{\mathrm{space}}(\mathbf{r}, t) = \frac{q}{2m}\left[\phi_0^\dagger(\hat{\mathbf{p}} - q\mathbf{A})\phi_0 + h.c.\right] \tag{23.11}$$

and

$$\mathbf{J}^{\mathrm{spin}}(\mathbf{r}, t) = \frac{q}{2m}\left[i\phi_0^\dagger(\hat{\mathbf{p}} - q\mathbf{A}) \times \sigma\phi_0 + h.c.\right] = \frac{iq}{2m}\phi_0^\dagger\hat{\mathbf{p}} \times \sigma\phi_0 + h.c. \tag{23.12}$$

The last member of Eq. (23.12) follows from the fact that the Pauli spin matrices are Hermitian $[\sigma = \sigma^\dagger]$, and the two terms containing the factor $q\mathbf{A}$ therefore cancel each other.

23.3   SPIN SOURCE FOR PHOTONS

Let us now expand the Hilbert state $|\phi_0\rangle$ of the weakly relativistic spin-1/2 particle in the $\mathbf{r}$-representation and the (S^2, S_z)-spin basis. The two-component Pauli spinor hence is

$$\phi_0(\mathbf{r}, t) = \begin{pmatrix} \langle \mathbf{r}, \uparrow | \phi_0 \rangle \\ \langle \mathbf{r}, \downarrow | \phi_0 \rangle \end{pmatrix} = \begin{pmatrix} \psi_\uparrow(\mathbf{r}, t) \\ \psi_\downarrow(\mathbf{r}, t) \end{pmatrix}, \tag{23.13}$$

where $\psi_\uparrow(\mathbf{r}, t)$ and $\psi_\downarrow(\mathbf{r}, t)$ are the spin-up ($\uparrow$) and spin-down ($\downarrow$) components of the spinor.

The spin probability current density in Eq. (23.12) also appears as the quantum mechanical mean value of the one-particle Hermitian spin current density operator

$$\hat{\mathbf{j}}^{\mathrm{spin}}(\mathbf{r}, t) = \frac{iq}{2m}[\delta(\mathbf{r} - \mathbf{r}_e)\hat{\mathbf{p}}_e \times \sigma - \hat{\mathbf{p}}_e \times \sigma\delta(\mathbf{r} - \mathbf{r}_e)], \tag{23.14}$$

$\mathbf{r}_e$ and $\hat{\mathbf{p}}_e = (\hbar/i)\nabla_e$ denoting the particle [electron (e)] position and momentum operators. Thus,

$$\mathbf{J}^{\mathrm{spin}}(\mathbf{r}, t) = \int_{-\infty}^{\infty} \phi_0^\dagger(\mathbf{r}_e, t)\hat{\mathbf{j}}^{\mathrm{spin}}(\mathbf{r}, t)\phi_0(\mathbf{r}_e, t)d^3r_e, \tag{23.15}$$

where the integration is over the particle coordinate, $\mathbf{r}_e$. The spin current density, given in Eq. (23.12) can be written in terms of $\psi_\uparrow(\mathbf{r}, t)$ and $\psi_\downarrow(\mathbf{r}, t)$ making use of the operator matrix form

$$\sigma \times \nabla = \begin{pmatrix} \hat{\mathbf{z}} \times \nabla & \sqrt{2}\mathbf{e}_- \times \nabla \\ \sqrt{2}\mathbf{e}_+ \times \nabla & -\mathbf{z} \times \nabla \end{pmatrix}, \tag{23.16}$$

where

$$\mathbf{e}_\pm = \frac{1}{\sqrt{2}}(\hat{\mathbf{x}} \pm i\hat{\mathbf{y}}) \tag{23.17}$$

are specific helicity unit vectors. By combining Eqs. (23.12), (23.13) and (23.16) one obtains the following result:

$$\mathbf{J}^{\text{spin}}(\mathbf{r},t) = \frac{q\hbar}{2m}\left\{\hat{\mathbf{z}} \times \boldsymbol{\nabla}[|\psi_\downarrow(\mathbf{r},t)|^2 - |\psi_\uparrow(\mathbf{r},t)|^2]\right.$$
$$\left. - \sqrt{2}\mathbf{e}_+ \times \boldsymbol{\nabla}[\psi_\uparrow(\mathbf{r},t)\psi_\downarrow^*(\mathbf{r},t)] - \sqrt{2}\mathbf{e}_- \times \boldsymbol{\nabla}[\psi_\downarrow(\mathbf{r},t)\psi_\uparrow^*(\mathbf{r},t)]\right\} \tag{23.18}$$

If one utilizes that $\boldsymbol{\nabla} \cdot (\mathbf{v} \times \boldsymbol{\nabla} f(\mathbf{r})) = 0$ in the case where the vector $\mathbf{v}$ is independent of $\mathbf{r}$, it readily follows that

$$\boldsymbol{\nabla} \cdot \mathbf{J}^{\text{spin}}(\mathbf{r},t) = 0. \tag{23.19}$$

Hence, it appears that the Pauli spin current density is a transverse vector field, i.e.,

$$\mathbf{J}^{\text{spin}}(\mathbf{r},t) = \mathbf{J}_T^{\text{spin}}(\mathbf{r},t). \tag{23.20}$$

It appears from Eq. (23.18) that the extension of the source domain of a photon emitted from a spin flip, $\psi_\downarrow \leftrightarrow \psi_\uparrow$, has an extension given by the electronic orbitals $\psi_\uparrow(\mathbf{r},t)$ and $\psi_\downarrow(\mathbf{r},t)$.

23.4   EXPONENTIAL PHOTON LOCALIZATION

Although the spin source domain has an extension of the involved electronic orbitals, one must not forget that the spatial current density in general plays a dominant role for the limitation of the spatial photon confinement in an emission process. In the Pauli spinor formalism the space part of the current density [Eq. (23.11)] is the sum of spin-up and spin-down parts. Thus,

$$\mathbf{J}^{\text{space}}(\mathbf{r},t) = \mathbf{J}_\uparrow^{\text{space}}(\mathbf{r},t) + \mathbf{J}_\downarrow^{\text{space}}(\mathbf{r},t), \tag{23.21}$$

where

$$\mathbf{J}_{\uparrow(\downarrow)}^{\text{space}}(\mathbf{r},t) = \frac{q\hbar}{2m}\left[\psi_{\uparrow(\downarrow)}^*(\mathbf{r},t)\boldsymbol{\nabla}\psi_{\uparrow(\downarrow)}(\mathbf{r},t) - \psi_{\uparrow(\downarrow)}(\mathbf{r},t)\boldsymbol{\nabla}\psi_{\uparrow(\downarrow)}^*(\mathbf{r},t)\right]$$
$$- \frac{q^2}{m}\left|\psi_{\uparrow(\downarrow)}(\mathbf{r},t)\right|^2\mathbf{A}(\mathbf{r},t). \tag{23.22}$$

In what one may call a *pure* spin transition the spatial particle (electron) configuration does not change during the photon emission process. The paradigm of a pure spin-1/2 source appears in the ground state ($1s$) of a hydrogen atom [62, 158]. The $1s$ state has no fine structure, and the electron in this state has no orbital angular momentum. The magnetic interaction between the proton and the electron spins gives rise to a so-called hyperfine structure of the various electron levels. The hyperfine split $1s$ state thus is a pure spin-1/2 source.

When the spatial dynamics is frozen, the state space for the state vector at all times is a simple tensor product of the space- and spin-state spaces, and the Pauli spinor hence takes the form

$$\begin{pmatrix} \psi_\uparrow(\mathbf{r},t) \\ \psi_\downarrow(\mathbf{r},t) \end{pmatrix} = \psi(\mathbf{r}) \begin{pmatrix} \uparrow(t) \\ \downarrow(t) \end{pmatrix}. \tag{23.23}$$

In $1s$ state of hydrogen $\psi(\mathbf{r}) = \psi(r) = (\pi a_0^3)^{-1/2}\exp(-r/a_0)$, where a_0 is the Bohr radius. In the frozen state the spatial current density [choosing a real $\psi(\mathbf{r})$] is given by

$$\mathbf{J}^{\text{space}}(\mathbf{r},t) = -\frac{q^2}{2m}|\psi(\mathbf{r})|^2\left[|\uparrow(t)|^2 + |\downarrow(t)|^2\right]\mathbf{A}(\mathbf{r},t) = -\frac{q^2}{m}|\psi(\mathbf{r})|^2\mathbf{A}(\mathbf{r},t). \tag{23.24}$$

Hence, one cannot neglect $\mathbf{J}^{\text{space}}(\mathbf{r},t)$ even for the ansatz in Eq. (23.23). If one nevertheless neglects this usually small $\mathbf{A}$-dependent space part, one is left with a spin current density

$$\mathbf{J}^{\text{spin}}(\mathbf{r},t) = \frac{q\hbar}{2m}\left\{\left[|\downarrow(t)|^2 - |\uparrow(t)|^2\right]\hat{\mathbf{z}} - \sqrt{2}\uparrow(t)\downarrow^*(t)\mathbf{e}_+ - \sqrt{2}\downarrow(t)\uparrow^*(t)\mathbf{e}_-\right\}$$
$$\times \boldsymbol{\nabla}|\psi(\mathbf{r})|^2. \tag{23.25}$$

It appears from Eq. (23.25) that in a pure spin-flip process the photon source domain has the extension of the electronic orbital probability density $|\psi(\mathbf{r})|$. This, roughly speaking, implies that the source domain of the photon is exponentially confined; see also Fig. 23.1. Note, that the presence of the term $-(q^2/m)|\psi_{\uparrow(\downarrow)}(\mathbf{r},t)|^2\mathbf{A}(\mathbf{r},t)$ in Eq. (23.22) ensures that $\mathbf{J}^{\text{space}}_{\uparrow(\downarrow)}(\mathbf{r},t)$ is gauge independent [158]. The current density in Eq. (23.24) thus is not gauge invariant unless $\mathbf{A} = \mathbf{A}_T$.

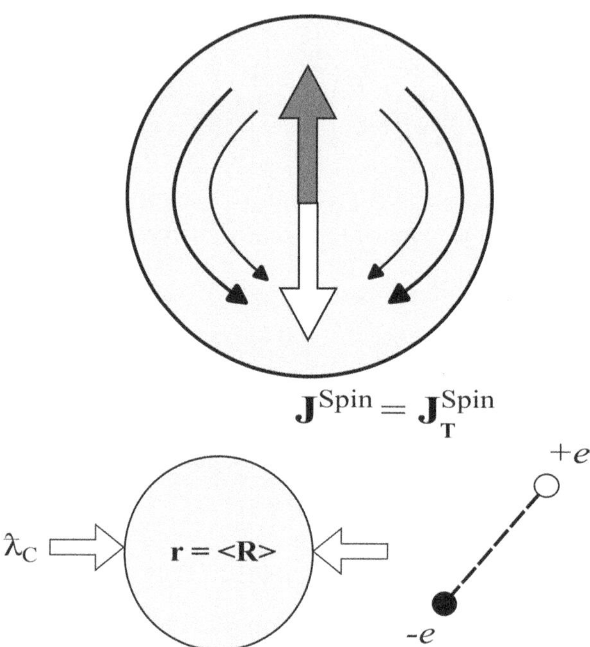

Figure 23.1 Schematic illustration of the source confinement problem for a transverse photon emitted in a pure spin transition. Top figure: In a spin flip a current is flowing in direct space around the point-particle, but since the flow is incompressible no rim zone exists, i.e. $\mathbf{J}^{\mathrm{spin}}_{T} = \mathbf{J}^{\mathrm{spin}}$. The photon source thus is confined to a region identical to the electronic size of the atom (or mesoscopic particle). Bottom figure: In relativistic quantum mechanics, even a spin source for photons cannot be truncated to a point, because the electron-position coupling makes the $\mathbf{r}$-vector in the Dirac equation a mean position vector, $< \mathbf{R} >$, in a region of linear extension comparable to the Compton wavelength (divided by 2π), $\lambdabar_C = \hbar/(mc_0)$.

CHAPTER 24

Spin Waves in Ferromagnets

24.1 HEISENBERG HAMILTONIAN. EFFECTIVE MAGNETIC FIELD

Our concern in the following is the phenomenological approach to ferromagnetics, in particular the collective spin waves, and their excitation by linear interaction with light. It is assumed that it is meaningful to consider a crystal lattice in which the only relevant degrees of freedom are the individual atomic spins. For simplicity we consider a case where there is only one magnetic atom in the unit cell. On the lattice cite $\mathbf{R}_j$, we let $\hat{\mathbf{S}}_j$ denote the spin operator, and we assume that the spin operators at different cites $(\mathbf{R}_j, \mathbf{R}_k, ...)$ commute. The commutation rules thus are

$$\hat{\mathbf{S}}_j \times \hat{\mathbf{S}}_k = i\hbar\delta_{jk}\hat{\mathbf{S}}_j. \tag{24.1}$$

The assumed particle Hamiltonian is

$$\hat{H}_P = -2\sum_{i>j} J_{ij}\hat{\mathbf{S}}_i \cdot \hat{\mathbf{S}}_j. \tag{24.2}$$

This Hamiltonian, introduced by Dirac in 1929 [70] and used in 1932 by van Vleck [254] and others in studies of magnetically ordered solids, goes under the name Heisenberg Hamiltonian (!). The quantities

$$J_{ij} = J_{ij}(|\mathbf{R}_i - \mathbf{R}_j|) \tag{24.3}$$

are referred to as *exchange parameters*. For a rigorous analysis of the many-body physics forming the basis for Eq. (24.3) the reader is referred to the elegant work of Herring [110], and Herring and Flicker [111]. The ground state will be ferromagnetic if all the J_{ij}'s are positive. For atoms with orbitally nondegenerate ground states $(L = 0, J = S$ in standard notation) the lowest-lying excited states are spin waves [in a certain formal analogy to the plasmons and plasmaritons in jellium]. The magnetic moment operator at cite $\mathbf{R}_i$, is related to $\hat{\mathbf{S}}_i$ as follows:

$$\hat{\boldsymbol{\mu}}_i = \Gamma_i\hat{\mathbf{S}}_i, \tag{24.4}$$

DOI: 10.1201/9781003029458-24

where Γ_i is the gyromagnetic ratio of the atom (spin part only), Without loss of qualitative generality for the present description, from here on it is assumed that all the atoms are identical [$\Gamma_i = \Gamma$, $\forall i$].

In some studies it appears sufficient to work with an effective magnetic-field concept, and with the restriction that a given atom alone is coupled to its nearest neighbours (Z in numbers). From Eq. (24.2) one sees that the single-atom Hamiltonian belonging to cite i can be written as

$$\hat{H}_{P,i} = -\Gamma \hat{\mathbf{S}}_I \cdot \hat{\mathbf{B}}_i, \tag{24.5}$$

where

$$\hat{\mathbf{B}}_i = \frac{2J}{\Gamma} \sum_{j=1}^{Z} \hat{\mathbf{S}}_j \tag{24.6}$$

is the magnetic field operator at the cite. The essential assumption now is that each $\hat{\mathbf{S}}_j$ can be replaced by its quantum mechanical mean value, $< \hat{\mathbf{S}}_j >$. This means that the magnetic field operator becomes a c-number, $\mathbf{B}_i$; the effective magnetic field at cite i. If N is the number of atoms per unit volume, the semi-macroscopic magnetization, $\mathbf{M}$, will be given by

$$\mathbf{M} = N\Gamma < \hat{\mathbf{S}}_j > . \tag{24.7}$$

The relation between $\mathbf{M}$ and the effective magnetic field,

$$\mathbf{B}_{\text{eff}} = \frac{2JZ}{\Gamma} < \hat{\mathbf{S}}_j >, \tag{24.8}$$

hence takes the form

$$\mathbf{B}_{\text{eff}} = \gamma \mathbf{M}, \tag{24.9}$$

with $\gamma = 2ZJ/(N\Gamma^2)$. The direction of the spin alignment, considered in the quantum mechanical sense is arbitrary, since nothing in the Heisenberg Hamiltonian ties the spins to the crystal axes (no anisotropy). If a space-time constant external magnetic field, $\mathbf{B}_{\text{ext}}$, is add a preferred alignment direction occurs. The total field is $\mathbf{B} = \mathbf{B}_{\text{ext}} + \mathbf{B}_{\text{eff}}$, and one may use $\mathbf{B}_{\text{ext}}$ to define the spin-quantization direction (usually the z-axis).

24.2   SPIN WAVES AND THEIR STATISTICS

In this section we consider the Heisenberg Hamiltonian from a more rigorous quantum physical point of view than offered by the so-called molecular field theory which is based on use of the effective magnetic field concept. Rather complete surveys of the wave theory have been given by Akhiezer *et al.* [6] and Keffer [131], e.g. See also [5,54].

24.2.1 Ground state of the Heisenberg ferromagnet

For what follows it is convenient to introduce the raising, $\hat{S}_i^{\dagger}$ and lowering, $\hat{S}_i^{-}$, operators

$$\hat{S}_i^{\pm} = \hat{S}_i^{x} \pm i\hat{S}_i^{y}. \tag{24.10}$$

These spins operators commute if they belong to different sites, and one has

$$[\hat{S}_i^{z}, \hat{S}_j^{\pm}] = \pm\delta_{ij}\hbar\hat{S}_i^{\pm}, \tag{24.11}$$

and

$$[\hat{S}_i^{+}, \hat{S}_j^{-}] = 2\delta_{ij}\hbar\hat{S}_i^{z}. \tag{24.12}$$

With the definition $J_{ii} = 0$, we can rewrite the Heisenberg Hamiltonian, [Eq. (24.2)], as follows:

$$\hat{H}_P = -\sum_{i,j} J_{ij}\hat{\mathbf{S}}_i \cdot \hat{\mathbf{S}}_j = \sum_{i,j} J_{ij}\left(\hat{S}_i^{z}\hat{S}_j^{z} + \frac{1}{2}\hat{S}_i^{+}\hat{S}_j^{-} + \frac{1}{2}\hat{S}_i^{-}\hat{S}_j^{+}\right). \tag{24.13}$$

Since

$$\frac{1}{2}\sum_{i,j}\left(\hat{S}_i^{-}\hat{S}_j^{+} - \hat{S}_i^{+}\hat{S}_j^{-}\right) = 0, \tag{24.14}$$

one finally obtains the form

$$\hat{H}_P = -\sum_{i,j} J_{ij}\left(\hat{S}_i^{z}\hat{S}_j^{z} + \hat{S}_i^{-}\hat{S}_j^{+}\right). \tag{24.15}$$

Let us consider the state in which each $\hat{S}_i^{z}$ attains its maximum value, denoted by $\hbar S$. We call the associated single spin state $|0\rangle$. Hence, for $\forall i$,

$$\hat{S}_i^{z}|0\rangle = \hbar S|0\rangle, \quad \hat{S}_i^{+}|0\rangle = 0. \tag{24.16}$$

Utilizing the expression for the Heisenberg Hamiltonian in Eq. (24.15), one gets

$$\hat{H}_P|0\rangle = -\hbar^2 S^2 \sum_{i,j} J_{ij}|0\rangle. \tag{24.17}$$

For positive J_{ij}'s, $|0\rangle$ certainly is the ground state of the spin system, with energy

$$E_0 = -\hbar^2 S^2 \sum_{i,j} J_{ij}. \tag{24.18}$$

With interaction only between nearest neighbours $E_0 = -NZ(\hbar S)^2 J$, where N is the total number of atoms in the crystal. In the absence of an external magnetic field there are $2NS + 1$ possible orientations of the total ground state spin, $N\hbar S$.

24.2.2 Excited spin states. Dispersion relation

Let

$$|n\rangle = (2\hbar^2 S)^{-1/2}\hat{S}_n^- |0\rangle, \tag{24.19}$$

in which the spin at site n has been lowered a single step. The states $\{|n\rangle\}$ are localized, but they are not eigenstates of $\hat{H}_P$, as the reader may show with the help of Eq. (24.17), Eq. (24.11) and Eq. (24.12). The excited states are orthonormal, however:

$$\langle m|n\rangle = (2\hbar^2 S)^{-1}\langle 0|\hat{S}_m^+ \hat{S}_n^- |0\rangle = (2\hbar^2 S)^{-1}\langle 0| 2\delta_{mn}\hbar\hat{S}_m^z + \hat{S}_n^- \hat{S}_m^+ |0\rangle$$
$$= (2\hbar^2 S)^{-1}(2\hbar^2 S)\delta_{mn} = \delta_{mn}. \tag{24.20}$$

The set $\{|n\rangle\}$ furnishes a convenient basis for the solution of determining the collective spin eigenstates of the Heisenberg ferromagnet. For the generic state $|E\rangle$, which satisfies

$$\hat{H}_P |E\rangle = E |E\rangle, \tag{24.21}$$

we use the expansion

$$|E\rangle = \sum_n \phi(\mathbf{R}_n) |n\rangle. \tag{24.22}$$

In order to find an equation satisfied by the expansion coefficients $\phi(\mathbf{R}_n)$, we use a familiar standard procedure in quantum physics. Thus, in the inner product

$$\sum_n \langle m|\hat{H}_P |n\rangle \phi(\mathbf{R}_n) = E\phi(\mathbf{R}_m) \tag{24.23}$$

we need to evaluate the matrix element $\langle m|\hat{H}_P |n\rangle$. By utilizing Eqs. (24.11) (24.12), (24.15) and (24.16), elementary manipulations of

$$\langle m|\hat{H}_P |n\rangle = -(2\hbar^2 S)^{-1}\sum_{i,j} J_{ij} \langle 0|\hat{S}_m^+ \left|\hat{S}_i^z \hat{S}_j^z + \hat{S}_i^- \hat{S}_j^+ \right\rangle \hat{S}_n^- |0\rangle, \tag{24.24}$$

give

$$\langle m|\hat{H}_P |n\rangle = -\delta_{mn}\hbar^2 \left[S^2 \sum_{i,j} J_{ij} - 2S \sum_j J_{mj}\right] - 2\hbar^2 S J_{mn}. \tag{24.25}$$

In the relation in Eq. (24.23) now one inserts Eq. (24.25), use the expression for the ground energy E_0 [Eq. (24.18)], and introduce the deviation $\epsilon = E - E_0$ from the ground state energy. Hence, one obtains

$$2\hbar^2 S \sum_n \left[\delta_{mn}\sum_j J_{nj} - J_{mn}\right] \phi(\mathbf{R}_n) = \varepsilon\Phi(\mathbf{R}_m). \tag{24.26}$$

If one assumes that the exchange integral coupling the spins on two arbitrary sites is a function of the numerical distance between the sites only, i.e.,

$$J_{mn} = J(|\mathbf{R}_m - \mathbf{R}_n|), \tag{24.27}$$

it is easy to see that the wavelike form

$$\phi(\mathbf{R}_n) \propto \exp(i\mathbf{k} \cdot \mathbf{R}_n) \tag{24.28}$$

yields a solution to Eq. (24.26) with $\varepsilon = \varepsilon(\mathbf{k})$ given by

$$\varepsilon(\mathbf{k}) = 2\hbar^2 S \sum_m \mathbf{J}(\mathbf{R}_m)[1 - \exp(i\mathbf{k} \cdot \mathbf{R}_m)]. \tag{24.29}$$

The relation in Eq. (24.29) is the spin-wave (magnon) dispersion relation. The wave vector $\mathbf{k}$ is restricted to the Brillouin zone: If $\mathbf{k}$ is replaced by $\mathbf{k} + \mathbf{K}$ where $\mathbf{K}$ is any reciprocal lattice vector, both ϕ and ε are unchanged. For a crystal with a center of inversion $\varepsilon(-\mathbf{k}) = \varepsilon(\mathbf{k})$, and

$$\varepsilon(\mathbf{k}) = 2\hbar^2 S \sum_m \mathbf{J}(\mathbf{R}_m)[1 - \cos(\mathbf{k} \cdot \mathbf{R}_m)]. \tag{24.30}$$

24.2.3 Spin waves of fixed wave vector

Let us introduce the wave vector $(\mathbf{k})$ into the designation of ϕ, and $\phi(\mathbf{R}_n) \equiv \phi(\mathbf{k}; \mathbf{R}_n)$ [see Eq. (24.28)]. Use of discrete orthonormalization

$$\sum_n \phi^*(\mathbf{k}, \mathbf{R}_n)\phi(\mathbf{q}, \mathbf{R}_n) = \delta_{\mathbf{k},\mathbf{q}}, \tag{24.31}$$

gives

$$\phi(\mathbf{k}, \mathbf{R}_n) = N^{-1/2} \exp(i\mathbf{k} \cdot \mathbf{R}_n), \tag{24.32}$$

where N is the number of unit cells in the crystal. A spin wave state of wave vector $\mathbf{k}$ may be constructed by the following superposition of single-spin deviation states [Eq. (24.19)]

$$|\mathbf{k}\rangle = N^{-1/2} \sum_n \exp(i\mathbf{k} \cdot \mathbf{R}_n) |n\rangle. \tag{24.33}$$

The $|\mathbf{k}\rangle$ state is an eigenstate of $\hat{\mathbf{S}}^2$ for the entire crystal;

$$\hat{\mathbf{S}}^2 = \left(\sum_i \hat{\mathbf{S}}_i\right)^2 = \sum_{i,j} \hat{\mathbf{S}}_i \cdot \hat{\mathbf{S}}_j \tag{24.34}$$

with eigenvalue $\hbar^2 \mathcal{S}(\mathcal{S} + 1)$:

$$\hat{\mathbf{S}}^2 |\mathbf{k}\rangle = \hbar^2 \mathcal{S}(\mathcal{S} + 1) |\mathbf{k}\rangle. \tag{24.35}$$

In order to prove Eq. (24.35), first one operates with $\hat{\mathbf{S}}^2$ on an excited single-spin state. As the reader may show, this gives

$$\hat{\mathbf{S}}^2 \left|n\right\rangle = (2\hbar^2 S)^{-1/2} \sum_{i,j} \left[\hat{S}_i^z \hat{S}_j^z + \frac{1}{2}\hat{S}_i^- \hat{S}_j^+ + \frac{1}{2}\hat{S}_i^+ \hat{S}_j^- \right] \hat{S}_n^- \left|0\right\rangle$$

$$= \hbar^2 NS(NS - 1)\left|n\right\rangle + 2\hbar^2 S \sum_j \left|j\right\rangle. \tag{24.36}$$

Next, then

$$\hat{\mathbf{S}}^2 \left|\mathbf{k}\right\rangle = \hbar^2 NS(NS - 1)\left|\mathbf{k}\right\rangle + \frac{2\hbar^2 S}{N^{1/2}} \left[\sum_n \exp(i\mathbf{k} \cdot \mathbf{R}_n) \right] \sum_j \left|j\right\rangle$$

$$= \hbar^2 [NS(NS - 1) + 2NS\delta_{\mathbf{k},\mathbf{0}}]\left|\mathbf{k}\right\rangle, \tag{24.37}$$

because

$$\sum_n \exp(i\mathbf{k} \cdot \mathbf{R}_n) = N\delta_{\mathbf{k},\mathbf{0}}; \quad \sum_j \left|j\right\rangle = N^{1/2}\left|\mathbf{k} = \mathbf{0}\right\rangle. \tag{24.38}$$

For $k = 0$, $\mathcal{S} = NS$, and for $k \neq 0$, $\mathcal{S} = NS - 1$. Note that a spin wave of infinite wavelength ($k = 0$) has the same value for $\mathcal{S}$ as the ground state. Thus

$$\left(\sum_i \hat{S}_i^z \right) \left|0\right\rangle = \hbar NS \left|0\right\rangle, \tag{24.39}$$

using Eq. (24.16) [first member].

24.2.4 Holstein-Primakoff correspondence

The three Cartesian components $(\hat{S}_j^x, \hat{S}_j^y, \hat{S}_j^z)$ of the spin operator $\hat{\mathbf{S}}_j$ at site j are not independent, but are connected by the operator identity

$$\hat{\mathbf{S}}_j \cdot \hat{\mathbf{S}}_j = \hbar^2 S(S + 1). \tag{24.40}$$

Spins are not exactly bosons or fermions, that is, the spin operators are not those of a boson or fermion field. Thus, the $\hat{S}_j^+$, $\hat{S}_j^-$ and $\hat{S}_j^z$, operators satisfy the commutator relations in Eqs. (24.11) and (24.12) [with $i = j$]. It turns out, as we shall see below, that at low temperatures where only a few spin waves are excited, spin waves may treated as independent bosons. The identity in Eq. (24.40) makes it worthwhile to seek to replace spin operators, below $\hat{S}_i^+$, $\hat{S}_i^-$ and $\hat{S}_i^z$, ($\forall i$), with boson annihilation $(\hat{a}_i)$ and creation $(\hat{a}_i^+)$ operators required to obey the commutation rules

$$[\hat{a}_i, \hat{a}_j^\dagger] = \delta_{ij}, \quad [\hat{a}_i^\dagger \hat{a}_j^\dagger] = [\hat{a}_i, \hat{a}_j] = 0. \tag{24.41}$$

In a first attempt we make the correspondence

$$\hat{S}_i^+ \rightarrow \hbar(2S)^{1/2}\hat{a}_i, \quad \hat{S}_i^- \rightarrow \hbar(2S)^{1/2}\hat{a}_i^\dagger, \quad \hat{S}_i^z \rightarrow \hbar(S - \hat{a}_i^\dagger \hat{a}_i). \tag{24.42}$$

In a mean-field picture, where $\hat{S}_i^+$ and $\hat{S}_i^-$ are replaced by their quantum mechanical mean values, $< \hat{S}_i^+ >$ and $< \hat{S}_i^- >$, the correspondence in Eq. (24.42) makes sense, because $< \hat{S}_i^z > = \hbar(S - < \hat{a}_i^\dagger \hat{a}_i >)$ relates to a reduction from the spin alignment (at site i) given by the mean number of boson excitations. With the help of the boson commutator relations, it is easy to show that the correspondence in Eq. (24.42) upholds the spin commutation rule in Eq. (24.11). Equally easy one can show that the commutator between $\hat{S}_i^+$ and $\hat{S}_j^-$ is not satisfied.

Guided by the considerations above, in a second attempt, we can essentially guess the correct form of the correspondence. In a sense it is obvious that one should keep the transformation for $\hat{S}_i^z$ as given in Eq. (24.42). With this in mind, it is clear that a correction of $\hat{S}_i^+$ and $\hat{S}_i^-$ by a square-root operator of the type $\hbar[\alpha S - \hat{a}^\dagger \hat{a}]^{1/2}$, α being a c-number, would allow one to satisfy $both$ Eqs. (24.11) and (24.12). A straightforward calculation shows that these operator equations are obeyed for $\alpha = 2$. Thus, the transformation

$$\hat{S}_i^+ = \hbar(2S - \hat{a}_i^+ \hat{a}_i)^{1/2}\hat{a}_i, \tag{24.43}$$

$$\hat{S}_i^- = \hbar\hat{a}_i^\dagger(2S - \hat{a}_i^+ \hat{a}_i)^{1/2}, \tag{24.44}$$

$$\hat{S}_i^z = \hbar(S - \hat{a}_i^\dagger \hat{a}_i), \tag{24.45}$$

provides the exact correspondence between the spin and boson operators at site i. Note that the square-root operator must be placed to the left (right) of $\hat{a}_i$ ($\hat{a}_i^\dagger$) to obtain $(\hat{S}_i^-)^\dagger = \hat{S}_i^\dagger$. The exact correspondence in Eqs. (24.43)–(24.45) was originally obtained in 1940 in a fundamental paper by Holstein and Primakoff [115]. Let us check that the Holstein-Primakoff transformation is exact. One has [dropping the subscript i]

$$\hat{S}^+\hat{S}^- = \hbar^2 \left[(2S - \hat{a}^\dagger\hat{a})^{1/2}\hat{a}\hat{a}^\dagger(2S - \hat{a}^\dagger\hat{a})^{1/2} \right]$$

$$= \hbar^2 \left[(2S - \hat{a}^\dagger\hat{a})^{1/2}(1 + \hat{a}^\dagger\hat{a})(2S - \hat{a}^\dagger\hat{a})^{1/2} \right]$$

$$= \hbar^2(2S - \hat{a}^\dagger\hat{a})(1 + \hat{a}^\dagger\hat{a}) = \hbar^2(2S - \hat{a}^\dagger\hat{a})\hat{a}\hat{a}^\dagger, \tag{24.46}$$

$$\hat{S}^-\hat{S}^+ = \hbar^2\hat{a}^\dagger(2S - \hat{a}^\dagger\hat{a})\hat{a}, \tag{24.47}$$

and thus

$$[\hat{S}^+, \hat{S}^-] = \hbar^2 \left\{ 2S - \left[\hat{a}^\dagger\hat{a}\hat{a}\hat{a}^\dagger - \hat{a}^\dagger\hat{a}^\dagger\hat{a}\hat{a} \right] \right\} = 2\hbar^2(S - \hat{a}^\dagger\hat{a}) = 2\hbar\hat{S}^z. \tag{24.48}$$

Hence, we have proved Eq. (24.11) for $i = j$. The result for $i \neq j$ is obviously correct. Furthermore,

$$[\hat{S}^z, \hat{S}^+] = \hbar^2 \left[(S - \hat{a}^\dagger\hat{a})(2S - \hat{a}^\dagger\hat{a})^{1/2}\hat{a} - (2S - \hat{a}^\dagger\hat{a})^{1/2}\hat{a}(S - \hat{a}^\dagger\hat{a}) \right]$$

$$= \hbar^2(2S - \hat{a}^\dagger\hat{a})^{1/2} \left[(S - \hat{a}^\dagger\hat{a})\hat{a} - \hat{a}(S - \hat{a}^\dagger\hat{a}) \right]$$

$$= \hbar^2(2S - \hat{a}^\dagger\hat{a})^{1/2} \left(\hat{a}\hat{a}^\dagger\hat{a} - \hat{a}^\dagger\hat{a}\hat{a} \right)$$

$$= \hbar^2(2S - \hat{a}^\dagger\hat{a})^{1/2}\hat{a} = \hbar\hat{S}^+, \tag{24.49}$$

I urge the reader to prove that $[\hat{S}^z, \hat{S}^-] = -\hbar\hat{S}^-$. Hence, Eq. (24.12) has also been verified.

24.2.5 Creation and destruction operators for spin waves

Let us define the "Fourier transform" of the set of local boson annihilation operators
as follows:

$$\hat{a}(\mathbf{k}) \equiv N^{-1/2} \sum_n \exp(i\mathbf{k} \cdot \mathbf{R}_n)\hat{a}_n, \tag{24.50}$$

with related creation operators

$$\hat{a}^\dagger(\mathbf{k}) = N^{-1/2} \sum_n \exp(-i\mathbf{k} \cdot \mathbf{R}_n)\hat{a}_n^\dagger. \tag{24.51}$$

As in Section 24.2.3, N is the number of unit cells in the crystal. The wave vector
$\mathbf{k}$ is confined to the Brillouin zone. The inverse relations to Eqs. (24.50) and (24.51)
are

$$\hat{a}_i = N^{-1/2} \sum_{\mathbf{k}} \exp(-i\mathbf{k} \cdot \mathbf{R}_i)\hat{a}(\mathbf{k}), \tag{24.52}$$

$$\hat{a}_i^\dagger = N^{-1/2} \sum_{\mathbf{k}} \exp(i\mathbf{k} \cdot \mathbf{R}_i)\hat{a}^\dagger(\mathbf{k}), \tag{24.53}$$

It follows from the commutation rules in Eq. (24.41) that

$$[\hat{a}(\mathbf{k}), \hat{a}^\dagger(\mathbf{q})] = \delta_{\mathbf{k},\mathbf{q}}, \quad [\hat{a}(\mathbf{k}), \hat{a}(\mathbf{q})] = [\hat{a}^\dagger(\mathbf{k}), \hat{a}^\dagger(\mathbf{q})] = 0. \tag{24.54}$$

The operators $\hat{a}^\dagger(\mathbf{k})$ and $\hat{a}(\mathbf{k})$ are bosonic creation and destruction operators for spin
waves. Thus, the single-spin wave state

$$|\mathbf{k}\rangle = \hat{a}^\dagger(\mathbf{k})\,|0\rangle \tag{24.55}$$

is precisely the single-spin deviation state given in Eq.(24.33). However, the states in
which more than one spin-wave state is excited, such as the two spin-wave state

$$|\mathbf{k}, \mathbf{q}\rangle = \hat{a}^\dagger(\mathbf{k})\hat{a}^\dagger(\mathbf{q})\,|0\rangle, \tag{24.56}$$

are not eigenstates of the Hamiltonian [Eq. (24.15)]. This follows from the fact
that the Holstein-Primakoff transformation [Eqs. (24.43)–(24.45)] is more complicated than the correspondence in Eq. (24.42). In turn, this implies that there
is an effective interaction between spin waves, as discussed by several authors
[49, 73, 74, 238, 281, 282]. Although not necessarily the most convenient in all applications, the interaction between spin waves may be handled in a perturbative approach
by expanding the square roots in the Holstein-Primakoff transformation in orders
of $\hat{a}_i^\dagger \hat{a}_i$. Roughly speaking this approach will be useful when mean fractional spin
reversal is small:

$$< \hat{a}_i \hat{a}_i > /S \leq 1. \tag{24.57}$$

We shall not discuss the theory of magnon-magnon interaction nor magnon-magnon
coupled states interaction with photons.

Not surprisingly, the operator $\hat{a}^\dagger(\mathbf{k})\hat{a}(\mathbf{k}) \equiv \hat{a}^\dagger_\mathbf{k}\hat{a}_\mathbf{k}$ turns out to be the occupation number operator for the magnon state $|\mathbf{k}\rangle$. In order to prove this let us return to the Holstein-Primakoff transformation for $\hat{S}^z_i = \hbar(S - \hat{a}^\dagger_i\hat{a}_i)$ [Eq. (24.45)]. By summation over the i's of the system one has

$$\hat{S}^z \equiv \sum_i \hat{S}^z_i = \hbar\left(NS - \sum_i \hat{a}^\dagger_i\hat{a}_i\right). \tag{24.58}$$

Since

$$\sum_i \hat{a}^\dagger_i\hat{a}_i = \frac{1}{N}\sum_{i,\mathbf{k},\mathbf{q}} e^{i(\mathbf{k}-\mathbf{q})\cdot\mathbf{R}_i}\hat{a}^\dagger_\mathbf{k}\hat{a}_\mathbf{q} = \sum_{\mathbf{k},\mathbf{q}} \delta_{\mathbf{k},\mathbf{q}}\hat{a}^\dagger_\mathbf{k}\hat{a}_\mathbf{q} = \sum_\mathbf{k} \hat{a}^\dagger_\mathbf{k}\hat{a}_\mathbf{k}, \tag{24.59}$$

we obtain

$$\hat{S}^z = \hbar\left(NS - \sum_\mathbf{k} \hat{a}^\dagger_\mathbf{k}\hat{a}_\mathbf{k}\right), \tag{24.60}$$

a result which shows that $\hat{a}^\dagger_\mathbf{k}\hat{a}_\mathbf{k}$ may be viewed as the number operator for the magnon state $|\mathbf{k}\rangle$.

24.3 SPIN WAVES IN EXTERNAL MAGNETIC FIELDS

24.3.1 Spin wave in a space-time constant magnetic field

The ground state of the Heisenberg ferromagnets is a highly degenerate as there is nothing to fix the direction of the z-axis used in Section 24.2. The states belonging to the $2NS+1$ possible orientations of the total spin all have the same energy. One may lift the degeneracy by means of an externally impressed space- and time-independent magnetic field, $\mathbf{B}_0 = [B_0\hat{\mathbf{z}}]$, lining up the atomic spins along $\mathbf{B}_0$, here chosen as the z-axis.

For the subsequent study of the interaction of spin waves with photons, it is convenient to rename the creation and destruction operators for the magnon quanta: $\hat{a}_\mathbf{k} \Rightarrow \hat{b}_\mathbf{k},\ \hat{a}^\dagger_\mathbf{k} \Rightarrow \hat{b}^\dagger_\mathbf{k}$. In second quantization the spin wave Hamiltonian operator thus reads

$$\hat{H}_P = \sum_\mathbf{k} \varepsilon(\mathbf{k})\left[\hat{b}^\dagger_\mathbf{k}\hat{b}_\mathbf{k} + \frac{1}{2}\right], \tag{24.61}$$

where, under the assumption that the crystal has inversion symmetry $[\varepsilon(-\mathbf{k}) = \varepsilon(\mathbf{k})]$, the mode energy is given by [Eq. (24.29)]

$$\varepsilon(\mathbf{k}) = 2\hbar^2 S \sum_m J(\mathbf{R}_m)[1 - \cos(\mathbf{k} \cdot \mathbf{R}_m)]. \tag{24.62}$$

The interaction Hamiltonian operator of the atomic magnetic moment of site i m_i, with the external magnetic field, $\mathbf{B}_0$ is given by

$$\hat{H}_{I,i} = -\hat{\mathbf{m}}_i \cdot \mathbf{B}_0. \tag{24.63}$$

For a single electron spin

$$\hat{\mathbf{m}}_i = \left(\frac{g}{2}\right) 2\frac{\mu_B}{\hbar}\hat{\mathbf{S}}_i \tag{24.64}$$

where $\mu_B = e\hbar/(2m)$, $e < 0$, is the signed Bohr magneton. With $\mathbf{B}_0 \parallel \hat{\mathbf{z}}$ one has

$$\hat{H}_{I,i} = -\frac{ge}{2m}B_0\hat{S}_i^z, \tag{24.65}$$

and hence for the whole crystal

$$\hat{H}_I = -\frac{ge}{2m}B_0\hat{S}^z, \tag{24.66}$$

where $\hat{S}^z$ is given by Eq. (24.60), with the notational renaming introduced above Eq. (24.61). The constant term $\hbar NS$ in Eq. (24.60) is of no physical importance here, so one can use an interaction Hamiltonian of the form

$$\hat{H}_I = \frac{g}{2}\left(\frac{e\hbar}{m}\right)B_0\sum_{\mathbf{k}}\hat{b}_{\mathbf{k}}^\dagger\hat{b}_{\mathbf{k}}. \tag{24.67}$$

With the addition of $\hat{H}_I$ to the particle Hamiltonian operator given in Eq. (24.61), it appears that the magnon dispersion relation in Eq. (24.62) is changed to

$$\varepsilon(\mathbf{k}) = 2\hbar^2 S\sum_{m} J(\mathbf{R}_m)\left[1 - \cos(\mathbf{k}\cdot\mathbf{R}_m)\right] + g\mu_B B_0. \tag{24.68}$$

While the static Zeeman effect can be included in the magnon dispersion, the nonlinear interaction of spin waves with photons can give rise to a photon drag phenomenon ($\sim$ nonlinear rectification process) of the photon-magnon interaction. A somewhat similar process is discussed for BCS superconductors in Chapter 25.

24.3.2 Magnon-photon interaction

Let us generalize our considerations to the situation where the spin waves interact with a space-time-varying microscopic magnetic quantum field. At site j the magnetic field operator in plane-wave expansion given by

$$\hat{\mathbf{B}}(\mathbf{R}_j,t) = \left(\frac{\hbar}{2\varepsilon_0 c^2 V}\right)^{1/2}\sum_{i} i\omega_i^{1/2}\left[\hat{a}_i(t)\boldsymbol{\kappa}_i \times \boldsymbol{\varepsilon}_i e^{i\mathbf{q}_i\cdot\mathbf{R}_j} - \hat{a}_i^\dagger(t)\boldsymbol{\kappa}_i \times \boldsymbol{\varepsilon}_i^* e^{-i\mathbf{q}_i\cdot\mathbf{R}_j}\right] \tag{24.69}$$

in usual notation; see Chapter 4. The magnon-photon interaction now is described by the interaction Hamiltonian

$$\hat{H}_I = -\left(\frac{g}{2}\right)\frac{e}{m}\sum_{j}\hat{\mathbf{S}}_j \cdot \hat{\mathbf{B}}(\mathbf{R}_j,t), \tag{24.70}$$

where the local Cartesian spin-operator components, expressed in terms of the magnon creation and destruction operators, take the forms

$$\hat{S}_j^x = \frac{\hbar}{2}(2S)^{1/2}(\hat{b}_j + \hat{b}_j^\dagger), \tag{24.71}$$

$$\hat{S}_j^y = \frac{\hbar}{2i}(2S)^{1/2}(\hat{b}_j - \hat{b}_j^\dagger), \tag{24.72}$$

$$\hat{S}_j^z = \hbar(S - \hat{b}_j^\dagger \hat{b}_j). \tag{24.73}$$

Since $\hat{\mathbf{B}}(\mathbf{r}, t)$ is a *microscopic* magnetic field operator one cannot conclude that it is slowly varying in space from one lattice site to the next. Although one may assume in quantum optics that the externally impressed (prescribed) electromagnetic field is slowly varying in space, local field corrections associated with magnetic dipole-dipole (and higher order) interactions among the local atomic spins, result in rapid variations in space of the microscopic magnetic field operator.

The starting point for a quantum electrodynamic description of the photon-spin-wave interaction is the total Hamiltonian operator

$$\hat{H} = \sum_i \hbar\omega_i \left[\hat{a}_i^\dagger(t)\hat{a}_i(t) + \frac{1}{2}\right] + \sum_\mathbf{k} \varepsilon(\mathbf{k})\left[\hat{b}_\mathbf{k}^\dagger\hat{b}_\mathbf{k} + \frac{1}{2}\right] - \left(\frac{g}{2}\right)\frac{e}{m}\sum_j \hat{\mathbf{S}}_j \cdot \mathbf{B}(\mathbf{R}_j, t),$$

$$\tag{24.74}$$

with $\varepsilon(\mathbf{k})$ given by Eq. (24.68) for spins aligned along the $\mathbf{B}_0 = B_0\hat{\mathbf{z}}$ direction in a constant magnetic field.

24.4 LOCAL-FIELD EFFECTS IN MAGNON-PHOTON INTERACTION

The form of the interaction Hamiltonian operator, $\hat{H}_I$ given in Eq. (24.70), raises immediately a complicated problem due to the fact that a prescribed external magnetic field, $\mathbf{B}_{\text{ext}}(\mathbf{r}, t)$, induces an extra microscopic magnetic moment in the spin system of the ferromagnets. In turn the sum of these induced magnetic moments gives rise to a local-field correction, $\mathbf{B}_{\text{ind}}(\mathbf{r}, t)$, to the external field. It is the sum $\mathbf{B}(\mathbf{r}, t) = \mathbf{B}_{\text{ext}}(\mathbf{r}, t) + \mathbf{B}_{\text{ind}}(\mathbf{r}, t)$ which drives the photon-magnon interaction. The magnetic field operator entering in Eq. (24.70) hence is

$$\hat{\mathbf{B}}(\mathbf{R}_j, t) = \hat{\mathbf{B}}_{\text{ext}}(\mathbf{R}_j, t) + \hat{\mathbf{B}}_{\text{ind}}(\mathbf{R}_j, t). \tag{24.75}$$

The form of $\hat{\mathbf{B}}_{\text{ext}}(\mathbf{R}_j, t)$ is determined by the external current density source distribution.

To understand the physical structure of the local-field correction let us consider the magnetic field emitted by a time-varying magnetic dipole moment $\mathbf{m}(\mathbf{0}, t)$ located at the origo of a spherical coordinate system. At the space-point $\mathbf{R}$ the magnetic field is in the space-frequency domain given by the following well-known result [157]

$$\mathbf{B}(\mathbf{R}; \omega) = \mu_0 q_0^2 \mathbf{G}(\mathbf{R}; \omega) \cdot \mathbf{m}(\mathbf{0}; \omega), \tag{24.76}$$

where $q_0 = \omega/c$. The dyadic electromagnetic propagator one now divides into transverse (T) and longitudinal (L) parts, viz. [157],

$$\mathbf{G}_T(\mathbf{R};\omega) = \frac{iq_0}{4\pi}\left\{\frac{e^{iq_0 R}}{iq_0 R}\left(\mathbf{U} - \hat{\mathbf{R}}\hat{\mathbf{R}}\right) - \left[\frac{e^{iq_0 R}}{(iq_0 R)^2} - \frac{e^{iq_0 R} - 1}{(iq_0 R)^3}\right]\left(\mathbf{U} - 3\hat{\mathbf{R}}\hat{\mathbf{R}}\right)\right\}$$

(24.77)

with $\hat{\mathbf{R}} = \mathbf{R}/R$, and

$$\mathbf{G}_L(\mathbf{R};\omega) = \frac{\mathbf{U} - 3\hat{\mathbf{R}}\hat{\mathbf{R}}}{4\pi q_0^2 R^3}.$$

(24.78)

The transverse propagator in Eq. (24.77) can be transformed to the space-time domain, $\mathbf{G}_T(\mathbf{R}, t)$. The related magnetic field, $\mathbf{B}_{T,\mathrm{ind}}(\mathbf{R}, t)$, is Einstein retarded, and contains in the near-field zone surrounding the magnetic dipole a transverse part originating in the circumstance that a transverse photon cannot be completely localized in space-time [153]. Unless the external source and the ferromagnet are in near-field contact, the external magnetic field inside the ferromagnet is a divergence vector-field, $\mathbf{B}_{\mathrm{ext}}(\mathbf{r}, t) = \mathbf{B}_{T,\mathrm{ext}}(\mathbf{r}, t)$.

Since $q_0^2 \mathbf{G}_L(\mathbf{R};\omega)$ is frequency independent, it appears that the longitudinal part of the magnetic field from the magnetic dipole is given by

$$\mathbf{B}_L(\mathbf{R}, t) = \frac{\mu_0}{4\pi R^3}\left(\mathbf{U} - 3\hat{\mathbf{R}}\hat{\mathbf{R}}\right) \cdot \mathbf{m}(\mathbf{0}, t)$$

(24.79)

in the time domain. The field $\mathbf{B}_L(\mathbf{R}, t)$ is nonvanishing only in the near-field $(\sim R^{-3})$ zone of the dipole, and the relation between $\mathbf{m}(\mathbf{0}, t)$ and $\mathbf{B}_L(\mathbf{R}, t)$ is nonretarded.

The result in Eq. (24.79) readily can be generalized to the case where the dipole is located at the space-point $\mathbf{R}_i$ (the position vector of a spin at site i in the crystal). At spin site j (position vector $\mathbf{R}_j$), the longitudinal magnetic field thus is given by

$$\mathbf{B}_L(\mathbf{R}_j, t) = \frac{\mu_0}{4\pi R_{ji}^3}\left(\mathbf{U} - 3\hat{\mathbf{R}}_{ji}\hat{\mathbf{R}}_{ji}\right) \cdot \mathbf{m}(\mathbf{R}_i, t)$$

(24.80)

where $\mathbf{R}_{ji} \equiv \mathbf{R}_j - \mathbf{R}_i$, $\hat{\mathbf{R}}_{ji} = \mathbf{R}_{ij}/|\mathbf{R}_{ij}|$.

The analysis above shows that the interaction Hamiltonian operator given in Eq. (24.70) contains a nonretarded magnetic dipole-dipole (MD-MD) part with the explicit form

$$\hat{H}_I^{\mathrm{MD\text{-}MD}} = \frac{\mu_0}{4\pi}\frac{\Gamma^2}{2}\sum_{i,j} R_{ji}^{-3}[\hat{\mathbf{S}}_i \cdot \hat{\mathbf{S}}_j - 3(\hat{\mathbf{S}}_i \cdot \hat{\mathbf{R}}_{ji})(\hat{\mathbf{R}}_{ji} \cdot \hat{\mathbf{S}}_j)].$$

(24.81)

In writing down Eq. (24.81), the relation $\hat{\mathbf{m}}_i = \Gamma\hat{\mathbf{S}}_i$ [Eq. (24.4] was used. The nonretarded MD-MD interaction depends alone on the single-spin operators at the various sites of the ferromagnet, and therefore must be included as part of the particle Hamiltonian that is

$$\hat{H}_P^{\mathrm{NEW}} = \sum_{\mathbf{k}} \varepsilon(\mathbf{k})\left[\hat{b}_{\mathbf{k}}^\dagger \hat{b}_{\mathbf{k}} + \frac{1}{2}\right] + \hat{H}_I^{\mathrm{MD\text{-}MD}}$$

(24.82)

The Cartesian components of $\hat{\mathbf{S}}_j(\hat{\mathbf{S}}_i)$ can be expressed in terms of the creation $(\hat{b}_\mathbf{k}^\dagger)$ and destruction $(\hat{b}_\mathbf{k})$ operators for the spin-wave quanta [see Eqs. (24.71)–(24.73)] when the mean fractional spin reversal is small. It was shown by Holstein and Primakoff [115] that an explicit expression can be obtained for the eigenenergies belonging to $\hat{H}_P^{\mathrm{NEW}}$. The calculation is rather long and the result will not be given here. The interested reader is referred also to the nice form given by Charap and Boyd [60]; see also [54].

A rigorous quantum electrodynamic study of many photon-spin-wave interaction including local-field effects thus may start from the total Hamiltonian

$$\hat{H} = \hat{H}_F + \hat{H}_P^{\mathrm{NEW}} - \Gamma \sum_j \hat{\mathbf{S}}_j \cdot \mathbf{B}_T(\mathbf{R}_j, t), \tag{24.83}$$

where

$$\hat{\mathbf{B}}_T(\mathbf{r}, t) = \hat{\mathbf{B}}_{T,\mathrm{ext}}(\mathbf{r}, t) + \hat{\mathbf{B}}_{T,\mathrm{ind}}(\mathbf{r}, t) \tag{24.84}$$

is the magnetic field operator belonging to the dynamic photon field.

24.5   NONLOCAL RESPONSE FORMALISM: SPIN-WAVE CORRELATION

Spin-wave propagation implies that the magnetic-field-induced microscopic magnetization field at given lattice site in the ferromagnets depends on the prevailing magnetic field at surrounding sites. In the framework of linear response theory the Kubo formalism allows one to obtain the quantum mechanical mean value of the magnetic field density (microscopic magnetization) operator, i.e.,

$$\hat{\mathbf{M}}(\mathbf{r}) = \sum_i \hat{\mathbf{m}}_i \delta(\mathbf{r} - \mathbf{R}_i) = \Gamma \sum_i \hat{\mathbf{S}}_i \delta(\mathbf{r} - \mathbf{R}_i), \tag{24.85}$$

using the procedure described in Chapter 18. Although local-field effects may be included in the particle Hamiltonian, as described in Section 24.4 we omit these in the following. The free-field Hamiltonian operator is not needed in the analysis below, and we consider the magnetic field as a classical (unquantized) quantity. Thus, our starting point is the Hamiltonian operator

$$\hat{H} = \sum_\mathbf{k} \varepsilon(\mathbf{k}) \left[\hat{b}_k^\dagger \hat{b}_k + \frac{1}{2} \right] + \hat{H}^{(1)}, \tag{24.86}$$

where

$$\hat{H}^{(1)} = -\Gamma \sum_i \hat{\mathbf{S}}_i \cdot \mathbf{B}(\mathbf{R}_i; \omega) \tag{24.87}$$

is the interaction operator in the space-frequency domain. Including a static uniform magnetic field $\mathbf{B}_0 = B_0\hat{\mathbf{z}}$, the magnon energy is given by Eq. (24.68), under the assumption that the crystal has inversion symmetry.

In the linearized Kubo formalism the quantum mechanical mean value of the microscopic magnetization is given by

$$< \hat{\mathbf{M}}(\omega) > = \text{Tr}\left\{ \hat{\rho}^{(1)}\hat{\mathbf{M}} \right\} = \sum_{\mathbf{k},\mathbf{k}'} \langle \mathbf{k}| \hat{\rho}^{(1)} |\mathbf{k}'\rangle \langle \mathbf{k}'| \hat{\mathbf{M}} |\mathbf{k}\rangle , \qquad (24.88)$$

where the matrix element of the first-order part of the density matrix operator, $\hat{\rho}^{(1)}$, between the spin-wave states $|\mathbf{k}\rangle$ and $|\mathbf{k}'\rangle$ takes the usual form [see Eq. (22.2)]

$$\langle \mathbf{k}| \hat{\rho}^{(1)} |\mathbf{k}'\rangle = \frac{f_0(\varepsilon_{\mathbf{k}'}) - f_0(\varepsilon_{\mathbf{k}})}{\hbar\omega + \varepsilon_{\mathbf{k}'} - \varepsilon_{\mathbf{k}}} \langle \mathbf{k}| \hat{H}^{(1)} |\mathbf{k}'\rangle . \qquad (24.89)$$

The quantity $f_0(\varepsilon_{\mathbf{k}})$ $[f_0(\varepsilon_{\mathbf{k}'})]$ is the probability for the bosonic spin-wave state $|\mathbf{k}\rangle$ ($|\mathbf{k}'\rangle$) to be occupied in thermal equilibrium.

In order to determine the matrix element

$$\langle \mathbf{k}| \hat{H}^{(1)} |\mathbf{k}'\rangle = -\Gamma \sum_{i} \langle \mathbf{k}| \hat{\mathbf{S}}_i |\mathbf{k}'\rangle \cdot \mathbf{B}(\mathbf{R}_i; \omega) \qquad (24.90)$$

the central point is a calculation of $\langle \mathbf{k}| \hat{\mathbf{S}}_i |\mathbf{k}'\rangle$. For the x-component we have

$$\langle \mathbf{k}| \hat{S}_i^x |\mathbf{k}'\rangle \propto \langle 0| \hat{b}_{\mathbf{k}}(\hat{b}_i + \hat{b}_i^\dagger)\hat{b}_{\mathbf{k}}^\dagger |0\rangle = 0 \qquad (24.91)$$

due to the fact that $\hat{b}_i(\hat{b}_i^\dagger)$ is a linear combination of magnon operators $\hat{a}_{\mathbf{q}}(\hat{a}_{\mathbf{q}}^\dagger)$ [see Eqs. (24.52) and (24.53)]. Matrix elements of the types $\langle 0| \hat{a}_{\mathbf{k}}\hat{a}_{\mathbf{q}}\hat{a}_{\mathbf{k}}^\dagger |0\rangle$ and $\langle 0| \hat{a}_{\mathbf{k}}\hat{a}_{\mathbf{q}}^\dagger\hat{a}_{\mathbf{k}}^\dagger |0\rangle$ hence appear. These are zero, obviously. The matrix element of $\hat{S}_i^x$ therefore is zero, and the same holds for $\hat{S}_i^y$:

$$\langle \mathbf{k}| \hat{S}_i^x |\mathbf{k}'\rangle = \langle \mathbf{k}| \hat{S}_i^y |\mathbf{k}'\rangle = 0. \qquad (24.92)$$

The matrix element in Eq. (24.89) in the view of Eq. (24.73) now takes the form

$$\langle \mathbf{k}| \hat{H}^{(1)} |\mathbf{k}'\rangle = -\Gamma \sum_{i} \langle \mathbf{k}| \hbar(S - \hat{b}_i^\dagger\hat{b}_i |\mathbf{k}'\rangle B_z(\mathbf{R}_i; \omega). \qquad (24.93)$$

The part of Eq. (24.93) proportional to

$$\langle \mathbf{k}| \hbar S |\mathbf{k}'\rangle = \hbar S \delta_{\mathbf{k},\mathbf{k}'} \qquad (24.94)$$

clearly does not contribute to $\langle \mathbf{k}| \hat{\rho}^{(1)} |\mathbf{k}'\rangle$, as one immediately sees from Eq. (24.89). In the remaining part we replace the product $\hat{b}_i^\dagger\hat{b}_i$ by a weighted double sum of magnon creation and annihilation operators using Eqs. (24.52) and (24.53). Thus

$$\hat{b}_i^\dagger\hat{b}_i = \frac{1}{N} \sum_{\mathbf{q},\mathbf{q}'} \exp[i(\mathbf{q}' - \mathbf{q}) \cdot \mathbf{R}_i]\hat{b}_{\mathbf{q}'}^\dagger\hat{b}_{\mathbf{q}}. \qquad (24.95)$$

Since,

$$\langle \mathbf{k}| \hat{b}_{\mathbf{q}'}^\dagger\hat{b}_{\mathbf{q}} |\mathbf{k}'\rangle = \langle 0| \hat{b}_{\mathbf{k}}\hat{b}_{\mathbf{q}'}^\dagger\hat{b}_{\mathbf{q}}\hat{b}_{\mathbf{k}'}^\dagger |0\rangle = \delta_{\mathbf{q},\mathbf{k}'}\delta_{\mathbf{q}',\mathbf{k}}, \qquad (24.96)$$

one obtains

$$\langle \mathbf{k} | \, \hat{H}^{(1)} \, | \mathbf{k}' \rangle = \frac{\Gamma \hbar}{N} \sum_i \exp[i(\mathbf{k} - \mathbf{k}') \cdot \mathbf{R}_i] \hat{\mathbf{z}} \cdot \mathbf{B}(\mathbf{R}_i; \omega), \qquad (24.97)$$

and finally

$$\langle \mathbf{k} | \, \hat{\rho}^{(1)} \, | \mathbf{k}' \rangle = \frac{\Gamma \hbar}{N} \frac{f_0(\varepsilon_{\mathbf{k}'}) - f_0(\varepsilon_{\mathbf{k}})}{\hbar \omega + \varepsilon_{\mathbf{k}'} - \varepsilon_{\mathbf{k}}} \sum_j \exp\left[i(\mathbf{k} - \mathbf{k}') \cdot \mathbf{R}_j\right] \hat{\mathbf{z}} \cdot \mathbf{B}(\mathbf{R}_j; \omega), \qquad (24.98)$$

where the running index i has been changed to j.

The matrix element of the microscopic magnetization, viz.,

$$\langle \mathbf{k}' | \, \hat{\mathbf{M}} \, | \mathbf{k} \rangle = \Gamma \sum_i \delta(\mathbf{r} - \mathbf{R}_i) \langle \mathbf{k}' | \, \hat{\mathbf{S}}_i \, | \mathbf{k} \rangle = \Gamma \sum_i \delta(\mathbf{r} - \mathbf{R}_i) \langle \mathbf{k}' | \, \hat{S}_i^z \, | \mathbf{k} \rangle \, \hat{\mathbf{z}}. \qquad (24.99)$$

The calculation of $\langle \mathbf{k}' | \, \hat{S}_i^z \, | \mathbf{k} \rangle$ proceeds as done in relation to the determination of $\langle \mathbf{k}' | \, \hat{H}^{(1)} \, | \mathbf{k} \rangle$. The result one obtains is

$$\langle \mathbf{k}' | \, \hat{\mathbf{M}} \, | \mathbf{k} \rangle = -\frac{\Gamma \hbar}{N} \hat{\mathbf{z}} \sum_i \delta(\mathbf{r} - \mathbf{R}_i) \exp[i(\mathbf{k}' - \mathbf{k}) \cdot \mathbf{R}_i]. \qquad (24.100)$$

By inserting Eqs. (24.98) and (24.100) into Eq. (24.88) one gets

$$< \hat{\mathbf{M}}(\omega) > = -\left(\frac{\Gamma \hbar}{N}\right)^2 \hat{\mathbf{z}}\hat{\mathbf{z}} \cdot \left\{ \sum_{i,j} \sum_{\mathbf{k},\mathbf{k}'} \delta(\mathbf{r} - \mathbf{R}_i) \frac{f_0(\varepsilon_{\mathbf{k}'}) - f_0(\varepsilon_{\mathbf{k}})}{\hbar \omega + \varepsilon_{\mathbf{k}'} - \varepsilon_{\mathbf{k}}} \right.$$
$$\left. \times \exp[i(\mathbf{k} - \mathbf{k}') \cdot (\mathbf{R}_i - \mathbf{R}_j)] \hat{\mathbf{z}} \cdot \mathbf{B}(\mathbf{R}_j; \omega) \right\}. \qquad (24.101)$$

The result in Eq. (24.101) tells one how the quantum mechanical mean value of the induced microscopic magnetization in the space point $\mathbf{r}$, $< \hat{\mathbf{M}}(\mathbf{r}; \omega) > [\equiv < \hat{\mathbf{M}}(\omega) >]$ is correlated, via spin-wave propagation, to the microscopic local magnetic fields on the various lattice sites $(\mathbf{R}_j)$.

CHAPTER 25

Nonlinear Electromagnetic Rectification in BCS Superconductors

Pioneering theoretical works in nonlinear optics were published by Bloembergen and coworkers [8,31,32]. A comprehensive monograph on nonlinear optics by Bloembergen was published in 1965 [30]. Many excellent books on nonlinear optics have been published over the years, e.g., [50, 52, 205, 232, 237]. In nonlinear electrodynamics (optics) studies of sum- and difference-frequency generation have taken a prominent position in the wake of the invention of the laser in 1960 and Franken and coworkers detection of second harmonic generation of light in 1961 [89]. This work was reviewed in a number of papers [29, 90, 206].

In brief, the lowest-order nonlinear mixing of electromagnetic signals of angular frequencies ω_1 and ω_2 reads

$$\cos \omega_1 t \cos \omega_2 t = \frac{1}{2} \left[\cos(\omega_1 + \omega_2)t + \cos(\omega_1 - \omega_2)t \right]. \tag{25.1}$$

As pointed out by the present book author in 1988-1989 [139] this nonlinear process may be of importance in the electrodynamics of BCS superconductors. Thus, mixing two (high-frequency) signals so close in frequencies that the difference-frequency $|\omega_1 - \omega_2|$ falls below the superconducting gap frequency, nonlinear electrodynamics might give one new insight in the fingerprints of superconductivity: The Meissner effect and the infinite electrical conductivity. In particular, the second harmonic generation process $[\omega_1 = \omega_2 = \omega]$, given as

$$\cos^2 \omega t = \frac{1}{2} \left[\cos 2\omega t + 1 \right], \tag{25.2}$$

implies that a dc current flowing without resistance might appear in the supercon-ductor.

Let us now turn the attention from this kindergarten description towards the microscopic physics, limiting ourselves to the electromagnetic rectification process

DOI: 10.1201/9781003029458-25

accompanying the second-harmonic generation process. The theoretical description given in this chapter is based on my two Phys. Rev. papers on the subject [138,140]. For a full account of the theory the reader is referred to these articles. In order to ease the reading of these papers, the notation in the remaining part of this chapter closely follows Refs. [138] and [140].

25.1 INTERACTION HAMILTONIAN. ONE-BODY CURRENT DENSITY OPERATOR

Let us consider the general one-body fermion operator

$$\hat{\mathbf{O}}(\mathbf{r}_1, \mathbf{r}_2, \ldots, \mathbf{r}_N, t) = \sum_{i=1}^{N} \hat{\mathbf{O}}(\mathbf{r}_i, t), \tag{25.3}$$

belonging to a system of N fermions ($\approx$ electrons) having the spatial coordinate position labels $\mathbf{r}_i$ $[i = 1 - N]$. For a spin-1/2 fermion system the one-body operator can be represented by a 2x2 matrix. Let the field operator be given by the two-component spinor

$$\hat{\boldsymbol{\Psi}}(\mathbf{r}) = \begin{pmatrix} \hat{\Psi}_\uparrow(\mathbf{r}) \\ \hat{\Psi}_\downarrow(\mathbf{r}) \end{pmatrix}, \tag{25.4}$$

where the arrow subscript denotes the spin-up ($\uparrow$) and spin-down ($\downarrow$) states. In second-quantization the one-body operator takes the form

$$\hat{\mathcal{O}}(t) = \int \hat{\boldsymbol{\Psi}}^{\dagger}(\mathbf{r}) \cdot \hat{\mathbf{O}}(\mathbf{r}, t) \cdot \hat{\boldsymbol{\Psi}}(\mathbf{r}) d^3 r, \tag{25.5}$$

If σ and σ' denote the spin state labels, Eq. (25.5) is in component form given by

$$\hat{\mathcal{O}}(t) = \sum_{\sigma,\sigma'} \int \hat{\Psi}_{\sigma'}^{\dagger}(\mathbf{r}) \cdot \hat{O}_{\sigma',\sigma}(\mathbf{r}, t) \cdot \hat{\Psi}_\sigma(\mathbf{r}) d^3 r. \tag{25.6}$$

In terms of the single-particle wave function $\psi_{\mathbf{k},\sigma}$ [quantum number $\{\mathbf{k}, \sigma\}$] and the related fermion annihilation $\{\hat{a}_{\mathbf{k},\sigma}\}$ and creation $\{\hat{a}_{\mathbf{k},\sigma}^{\dagger}\}$ operators the spinor component of the field operator is

$$\hat{\Psi}_\sigma(\mathbf{r}) = \sum_{\mathbf{k}} \psi_{\mathbf{k},\sigma} \hat{a}_{\mathbf{k},\sigma}, \quad \sigma = \uparrow \text{ or } \downarrow \tag{25.7}$$

with Hermitian conjugate

$$\hat{\Psi}_\sigma^{\dagger}(\mathbf{r}) = \sum_{\mathbf{k}} \psi_{\mathbf{k},\sigma}^{*} \hat{a}_{\mathbf{k},\sigma}^{\dagger}. \tag{25.8}$$

The one-body operator hence can be written in the convenient form

$$\hat{\mathcal{O}}(t) = \sum_{\mathbf{k},\mathbf{k}',\sigma,\sigma'} \langle \mathbf{k}', \sigma' | \hat{O}_{\sigma'\sigma} | \mathbf{k}, \sigma \rangle \hat{a}_{\mathbf{k}',\sigma'}^{\dagger} \hat{a}_{\mathbf{k},\sigma}. \tag{25.9}$$

As a starting point for the theory of the nonlinear electromagnetic response of a BCS superconductor to an externally impressed field, we consider the Pauli interaction Hamiltonian operator $\hat{H}_I$ in minimal coupling form, viz.,

$$\hat{H}_I = \left[\frac{e}{2m} (\hat{\mathbf{p}} \cdot \mathbf{A} + \mathbf{A} \cdot \hat{\mathbf{p}}) + \frac{e^2}{2m} \mathbf{A} \cdot \mathbf{A} \right] U + \frac{g}{2} \mu_B \hat{\boldsymbol{\sigma}} \cdot \boldsymbol{\nabla} \times \mathbf{A}, \tag{25.10}$$

leaving out the spin-orbit coupling, cf. Eq. (18.41). The quantities g and $\mu_B = e\hbar/(2m)$ are the g-factor and the Bohr magneton, respectively. The electron charge is taken as $-e$. Although the spin part of the Hamiltonian contributes to electromagnetic second-harmonic generation, as discussed in Chapter 18, it can be shown that it does not contribute in the electromagnetic rectification process. Hence, the spin part of the Hamiltonian is left out in the following.

Being interested in the effect of transverse electromagnetic fields we take $\mathbf{A} = \mathbf{A}_T$ in Eq. (25.10). Furthermore, we limit ourselves to unquantized electromagnetic fields $[\hat{\mathbf{A}} \Rightarrow \mathbf{A}]$, and assume that the self-consistent vector potential is a monochromatic plane-wave field. Hence,

$$\mathbf{A}(\mathbf{r}, t) = \frac{1}{2} \left[\mathbf{A}(\mathbf{q}, \omega) e^{i(\mathbf{q} \cdot \mathbf{r} - \omega t)} + \mathbf{A}^*(\mathbf{q}, \omega) e^{-i(\mathbf{q} \cdot \mathbf{r} - \omega t)} \right], \tag{25.11}$$

where $\mathbf{q}$ is the wave vector of the driving field.

In second-quantization notation our interaction Hamiltonian is given by

$$\mathcal{H}_I(t) = \sum_{\mathbf{k}, \mathbf{k}'} \langle \mathbf{k}' | \frac{e}{2m} (\hat{\mathbf{p}} \cdot \mathbf{A} + \mathbf{A} \cdot \hat{\mathbf{p}}) + \frac{e^2}{2m} \mathbf{A} \cdot \mathbf{A} | \mathbf{k} \rangle \sum_\sigma \hat{a}^\dagger_{\mathbf{k}',\sigma} \hat{a}_{\mathbf{k},\sigma}. \tag{25.12}$$

Now, for simplicity we treat the BCS superconductor as a Cooper-paired jellium by taking the single-particle states as free-electron states (in a confinement volume V), i.e.,

$$\langle \mathbf{r} | \mathbf{k} \rangle = V^{-1/2} e^{i\mathbf{k} \cdot \mathbf{r}}. \tag{25.13}$$

From Eq. (25.12) it then appears that the interaction Hamiltonian

$$\hat{\mathcal{H}}_I(t) = \frac{1}{2} \sum_{\alpha=1}^{2} \left(\hat{\mathcal{H}}_\alpha e^{-i\alpha\omega t} + \hat{\mathcal{H}}^\dagger_\alpha e^{i\alpha\omega t} \right) + \hat{\mathcal{H}}_0, \tag{25.14}$$

has the following parts:

$$\hat{\mathcal{H}}_1 = \frac{e\hbar}{2m} \mathbf{A}(\mathbf{q}, \omega) \cdot \left[\sum_{\mathbf{k}, \sigma} (2\mathbf{k} + \mathbf{q}) \hat{a}^\dagger_{\mathbf{k}+\mathbf{q}, \sigma} \hat{a}_{\mathbf{k}, \sigma} \right], \tag{25.15}$$

$$\hat{\mathcal{H}}_2 = \frac{e^2}{4m} \mathbf{A}(\mathbf{q}, \omega) \cdot \mathbf{A}(\mathbf{q}, \omega) \sum_{\mathbf{k}, \sigma} \hat{a}^\dagger_{\mathbf{k}+2\mathbf{q}, \sigma} \hat{a}_{\mathbf{k}, \sigma}, \tag{25.16}$$

$$\hat{\mathcal{H}}_0 = \frac{e^2}{4m} \mathbf{A}(\mathbf{q}, \omega) \cdot \mathbf{A}^*(\mathbf{q}, \omega) \sum_{\mathbf{k}, \sigma} \hat{a}^\dagger_{\mathbf{k}, \sigma} \hat{a}_{\mathbf{k}, \sigma}. \tag{25.17}$$

The $\hat{\mathcal{H}}_2$-part does not contribute to the rectification process.

In order to calculate the nonlinear dc current density generated in the superconductor one needs an expression for the one-body current density operator. It may be shown that the spin part of the current density cannot contribute to the rectification process. Hence, one obtains in second-quantized form the following result for the current density operator:

$$\hat{\boldsymbol{J}}(\mathbf{r},t) = \hat{\boldsymbol{J}}_F(\mathbf{r},t) + \frac{1}{2}\left[\hat{\boldsymbol{J}}_1(\mathbf{r},t)e^{-i\omega t} + \hat{\boldsymbol{J}}_1^\dagger(\mathbf{r},t)e^{i\omega t}\right], \tag{25.18}$$

where the free (F) and field perturbed parts are given by

$$\hat{\boldsymbol{J}}_F(\mathbf{r},t) = -\frac{e\hbar}{2mV}\sum_{\mathbf{k},\mathbf{k}',\sigma}(\mathbf{k}+\mathbf{k}')\hat{a}^\dagger_{\mathbf{k}',\sigma}\hat{a}_{\mathbf{k},\sigma}e^{i(\mathbf{k}-\mathbf{k}')\cdot\mathbf{r}}, \tag{25.19}$$

and

$$\hat{\boldsymbol{J}}_1(\mathbf{r},t) = -\frac{e^2}{mV}\mathbf{A}(\mathbf{q},\omega)e^{i\mathbf{q}\cdot\mathbf{r}}\sum_{\mathbf{k},\mathbf{k}',\sigma}\hat{a}^\dagger_{\mathbf{k}',\sigma}\hat{a}_{\mathbf{k},\sigma}e^{i(\mathbf{k}-\mathbf{k}')\cdot\mathbf{r}}. \tag{25.20}$$

The reader may appreciate that the establishment of the second-quantized forms of the interaction Hamiltonian and the one-body current density operator in broad outline follows the general quantum-field formalism presented in Chapter 18.

25.2 PARING APPROXIMATION. QUASIPARTICLE CONCEPT

In this section we give a short discussion of some of the relevant physical properties of the BCS superconductor in the electromagnetically unperturbed state. We limit ourselves to the most simple microscopic approach. In a number of good book references deeper aspects of the theory are described [67, 200, 220, 231].

Within the framework of the pairing approximation the ground state of a superconductor at zero temperature, $|G\rangle$, is given by [11]

$$|G\rangle = \left[\Pi_\mathbf{k}(u_\mathbf{k} + v_\mathbf{k}\hat{a}^\dagger_{\mathbf{k}\uparrow}\hat{a}^\dagger_{-\mathbf{k}\downarrow})|0\rangle\right] \tag{25.21}$$

in second quantization, the fermion vacuum state is denoted by $|o\rangle$. The quantities $u_\mathbf{k}$ and $v_\mathbf{k}$, which satisfy the relation

$$u_\mathbf{k}^2 + v_\mathbf{k}^2 = 1, \tag{25.22}$$

when the ground state is normalized, $\langle G|G\rangle = 1$, are the probability amplitudes for the Cooper-pair state $(\mathbf{k}\uparrow, -\mathbf{k}\downarrow)$ being empty and full respectively.

At finite temperatures *quasiparticles* are excited. To describe the quasiparticle excitation one introduces the *one-quasiparticle state*

$$|\mathbf{p}\uparrow\rangle = \left[\hat{a}^\dagger_{\mathbf{p}\uparrow}\Pi_{\mathbf{k}\neq\mathbf{p}}(u_\mathbf{k} + v_\mathbf{k}\hat{a}^\dagger_{\mathbf{k}\uparrow}\hat{a}^\dagger_{-\mathbf{k}\downarrow})\right]|0\rangle. \tag{25.23}$$

It appears from Eq. (25.23) that the state $|\mathbf{p}\uparrow\rangle$ has with certainty an electron in state $\mathbf{p}_\uparrow$ and its mate $-\mathbf{p}_\downarrow$ is empty. The states $\mathbf{p}_\uparrow$ and $-\mathbf{p}_\downarrow$ are blocked from participating in the pairing interaction [$\mathbf{k} \neq \mathbf{p}$ in Eq. (25.23)]. It follows from Eq. (25.21) that

$$\hat{a}^{\dagger}_{\mathbf{p}\uparrow} |G\rangle = u_{\mathbf{p}} |\mathbf{p} \uparrow\rangle , \tag{25.24}$$

$$\hat{a}_{-\mathbf{p}\downarrow} |G\rangle = - v_{\mathbf{p}} |\mathbf{p} \uparrow\rangle . \tag{25.25}$$

The reader readily may prove these relations by means of the anticommutation rules for fermions; see Subsection 18.4.4. For what follows it is convenient to introduce the quasiparticle operators

$$\hat{\gamma}^{\dagger}_{\mathbf{p}\uparrow} = u_{\mathbf{p}} \hat{a}^{\dagger}_{\mathbf{p}\uparrow} - v_{\mathbf{p}} \hat{a}_{-\mathbf{p}\downarrow}, \tag{25.26}$$

$$\hat{\gamma}_{-\mathbf{p}\downarrow} = u_{\mathbf{p}} \hat{a}_{-\mathbf{p}\downarrow} + v_{\mathbf{p}} \hat{a}^{\dagger}_{\mathbf{p}\uparrow}. \tag{25.27}$$

These operators create, respectively, the excited state $|\mathbf{p} \uparrow\rangle$ and null, i.e.,

$$\hat{\gamma}^{\dagger}_{\mathbf{p}\uparrow} |G\rangle = |\mathbf{p} \uparrow\rangle , \tag{25.28}$$

$$\hat{\gamma}_{-\mathbf{p}\downarrow} |G\rangle = 0. \tag{25.29}$$

I leave it to the reader to demonstrate the correctness of Eqs. (25.28) and (25.29) using the fermion anticommutation rules.

As shown in [11], the quasiparticle excitation energy is at temperature T given by

$$E_{\mathbf{k}} = +[\tilde{\varepsilon}^2_{\mathbf{k}} + \Delta^2_{\mathbf{k}}(T)]^{1/2} \tag{25.30}$$

for state $\mathbf{k}$ (or $-\mathbf{k}$), where $\Delta_{\mathbf{k}}(T)$ is the wave-number dependent gap parameter

$$\Delta_{\mathbf{k}} = - \sum_{\mathbf{k}'} W_{\mathbf{k},\mathbf{k}'} \frac{\Delta_{\mathbf{k}'}}{2E_{\mathbf{k}'}} \tanh \left(\frac{E_{\mathbf{k}'}}{2k_B T} \right) . \tag{25.31}$$

The quantity

$$W_{\mathbf{k},\mathbf{k}'} = \langle \mathbf{k} \uparrow, -\mathbf{k} \downarrow | W | \mathbf{k}' \uparrow, -\mathbf{k}' \downarrow \rangle \tag{25.32}$$

is the scattering matrix element between the pair states $(\mathbf{k}' \uparrow, -\mathbf{k}' \downarrow)$ and $(\mathbf{k} \uparrow, -\mathbf{k} \downarrow)$ in the two-body potential W. Furthermore,

$$\tilde{\varepsilon}_{\mathbf{k}} = \varepsilon_{\mathbf{k}} - \mu, \tag{25.33}$$

$\varepsilon_{\mathbf{k}} = \hbar^2 k^2/(2m)$ being the kinetic energy of the free electron and μ the chemical potential. The quasiparticle distribution is given by the Fermi-Dirac distribution

$$f_{\mathbf{k}}[\equiv f(E_k)] = \left[\exp \left(\frac{E_{\mathbf{k}}}{k_B T} \right) + 1 \right]^{-1} . \tag{25.34}$$

Above the transition temperature where $\Delta_{\mathbf{k}} = 0, \forall \mathbf{k}$, $f(E_k) = f(|\varepsilon_{\mathbf{k}} - \mu|)$, as expected.

The probability amplitudes are related to the quasiparticle excitation energy by [11]

$$u_{\mathbf{k}} = \frac{1}{\sqrt{2}} \left[1 + \frac{\tilde{\varepsilon}_{\mathbf{k}}}{E_{\mathbf{k}}} \right]^{1/2}, \tag{25.35}$$

$$v_{\mathbf{k}} = \frac{1}{\sqrt{2}} \left[1 - \frac{\tilde{\varepsilon}_{\mathbf{k}}}{E_{\mathbf{k}}} \right]^{1/2}, \tag{25.36}$$

giving the correct expression in the normal metallic state ($\Delta_{\mathbf{k}} = 0$).

With the summary of the field-unperturbed BCS superconductor, given in Section 25.2, we turn the attention towards the rectification process.

25.3 $\hat{\mathcal{H}}_1$ AND $\hat{\mathcal{J}}_1$ OPERATORS IN QUASIPARTICLE NOTATION

It is fruitful for the physical understanding of the rectification process in a Cooper-paired jellium to express $\hat{\mathcal{H}}_1$, $\hat{\mathcal{H}}_0$, $\hat{\mathcal{J}}_F$ and $\hat{\mathcal{J}}_1$ in terms of the quasiparticle operators. It appears from Eqs. (25.15), (25.17), (25.19) and (25.20) that one just needs the inverse of the transformations given in Eqs. (25.26) and (25.27). The relations in these equations are an example of the so-called Bogoliubov-Valatin transformations [34, 253], and their inverse are

$$\hat{a}^{\dagger}_{\mathbf{k}\uparrow} = u_{\mathbf{k}} \hat{\gamma}^{\dagger}_{\mathbf{k}\uparrow} + v_{\mathbf{k}} \hat{\gamma}_{-\mathbf{k}\downarrow}, \tag{25.37}$$

$$\hat{a}_{-\mathbf{k}\downarrow} = u_{\mathbf{k}} \hat{\gamma}_{-\mathbf{k}\downarrow} - v_{\mathbf{k}} \hat{\gamma}^{\dagger}_{\mathbf{k}\uparrow}. \tag{25.38}$$

By inserting Eqs. (25.37) and (25.38), and the Hermitian conjugate of these, into Eqs. (25.15) and (25.20) one obtains after some algebra, which the reader may find in the Appendix to [194], the following results:

$$\hat{\mathcal{H}}_1 = \frac{e\hbar}{2m} \mathbf{A}(\mathbf{q}, \omega) \cdot \left\{ \sum_{\mathbf{k}} (2\mathbf{k} + \mathbf{q}) \left[u_{\mathbf{k}} u_{\mathbf{k}+\mathbf{q}} + v_{\mathbf{k}} v_{\mathbf{k}+\mathbf{q}}) (\hat{\gamma}^{\dagger}_{\mathbf{k}+\mathbf{q}\uparrow} \hat{\gamma}_{\mathbf{k}\uparrow} + \hat{\gamma}_{-(\mathbf{k}+\mathbf{q})\downarrow} \hat{\gamma}^{\dagger}_{-\mathbf{k}\downarrow}) \right. \right.$$
$$\left. \left. + (u_{\mathbf{k}+\mathbf{q}} v_{\mathbf{k}} - v_{\mathbf{k}+\mathbf{q}} u_{\mathbf{k}}) (\hat{\gamma}^{\dagger}_{\mathbf{k}+\mathbf{q}\uparrow} \hat{\gamma}^{\dagger}_{-\mathbf{k}\downarrow} + \hat{\gamma}_{-(\mathbf{k}+\mathbf{q})\downarrow} \hat{\gamma}_{\mathbf{k}\uparrow}) \right] \right\}, \tag{25.39}$$

and

$$\hat{\mathcal{J}}_1(\mathbf{r}) = -\frac{e^2}{mV} \mathbf{A}(\mathbf{q}, \omega) e^{i\mathbf{q}\cdot\mathbf{r}} \left\{ \sum_{\mathbf{k},\mathbf{k}'} e^{i(\mathbf{k}-\mathbf{k}')\cdot\mathbf{r}} \left[(u_{\mathbf{k}'} u_{\mathbf{k}} - v_{\mathbf{k}'} v_{\mathbf{k}}) (\hat{\gamma}^{\dagger}_{\mathbf{k}'\uparrow} \hat{\gamma}_{\mathbf{k}\uparrow} + \hat{\gamma}^{\dagger}_{-\mathbf{k}\downarrow} \hat{\gamma}_{-\mathbf{k}'\downarrow}) \right. \right.$$
$$\left. \left. + (u_{\mathbf{k}'} v_{\mathbf{k}} + v_{\mathbf{k}'} u_{\mathbf{k}}) (\hat{\gamma}^{\dagger}_{\mathbf{k}'\uparrow} \hat{\gamma}^{\dagger}_{-\mathbf{k}\downarrow} - \hat{\gamma}_{\mathbf{k}\uparrow} \hat{\gamma}_{-\mathbf{k}'\downarrow}) + 2v_{\mathbf{k}}^2 \delta_{\mathbf{k},\mathbf{k}'} \right] \right\}. \tag{25.40}$$

The quasiparticle expressions for $\hat{\mathcal{H}}_0$ and $\hat{\mathcal{J}}_F$ can be found in [194], but these are not needed in the remaining part of the analysis.

Let us finish Section 25.3 by noting that the quasiparticle operators satisfy Fermi-Dirac statistics, i.e., the following anticommutator relations:

$$\{\hat{\gamma}^{\dagger}_{\mathbf{k}\uparrow}, \hat{\gamma}_{\mathbf{p}\uparrow}\} = \{\hat{\gamma}^{\dagger}_{-\mathbf{k}\downarrow}, \hat{\gamma}_{-\mathbf{p}\downarrow}\} = \delta_{\mathbf{k},\mathbf{p}}, \tag{25.41}$$

$$\{\hat{\gamma}^{\dagger}_{\mathbf{k}\uparrow}, \hat{\gamma}_{-\mathbf{p}\downarrow}\} = \{\hat{\gamma}_{\mathbf{k}\uparrow}, \hat{\gamma}_{\mathbf{p}\uparrow}\} = \{\hat{\gamma}^{\dagger}_{-\mathbf{k}\downarrow}, \hat{\gamma}_{-\mathbf{p}\downarrow}\} = \{\hat{\gamma}_{\mathbf{k}\uparrow}, \hat{\gamma}_{-\mathbf{p}\downarrow}\} = 0, \tag{25.42}$$

and the Hermitian conjugated relations. The reader may prove these relations using the anticommutator relations satisfied by the fermion annihilation ($\hat{a}_{\mathbf{k},\sigma}$) and creation ($\hat{a}^{\dagger}_{\mathbf{k},\sigma}$) operators, and the Bogoliubov-Valatin transformations [Eq. (25.37) and (25.38)].

25.4 NONLINEAR RECTIFICATION RESPONSE TENSOR

25.4.1 Many-body density-matrix approach

To determine the dc-current density induced in the superconductor by the transverse electromagnetic vector potential given in Eq. (25.11), we make use of the quantum-field formalism established in Chapter 18, in particular the many-body density-matrix approach.

In lowest order it was shown that the nonlinear (dynamic) dc current density was given by Eq. (18.56), in general. However, it turns out from an elaborate calculation of the present author [194] that

$$\mathrm{Tr}\{\hat{\rho}_0^{\mathrm{NL}}\hat{\boldsymbol{\mathcal{J}}}_0\} = 0. \tag{25.43}$$

Thus, it appears that the electromagnetically induced rectification current density ($\mathbf{J}_0$) in the BCS superconductor jellium can be calculated from the expression

$$\mathbf{J}_0 = \frac{1}{4}\left[\mathrm{Tr}\{\hat{\rho}_1\hat{\boldsymbol{\mathcal{J}}}_1^{\dagger}\} + \mathrm{Tr}\{\hat{\rho}_1^{\dagger}\hat{\boldsymbol{\mathcal{J}}}_1\}\right]. \tag{25.44}$$

Since

$$\mathrm{Tr}\{\hat{\rho}_1\hat{\boldsymbol{\mathcal{J}}}_1^{\dagger}\} = \sum_{I,J}\langle I|\,\hat{\rho}_1\,|J\rangle\,\langle J|\,\hat{\boldsymbol{\mathcal{J}}}_1^{\dagger}\,|I\rangle = \sum_{I,J}\frac{P_J - P_I}{\hbar\omega + E_J - E_I}\,\langle I|\,\hat{\mathcal{H}}_1\,|J\rangle\,\langle J|\,\hat{\boldsymbol{\mathcal{J}}}_1^{\dagger}\,|I\rangle, \tag{25.45}$$

one may conclude that only the quasiparticle expression for $\hat{\mathcal{H}}_1$ [Eq. (25.39)] and $\hat{\boldsymbol{\mathcal{J}}}_1$ [Eq. (25.40)] are needed to obtain $\mathbf{J}_0$, as said before.

25.4.2 Response tensor and its structure

Remembering that we here are dealing with transverse electromagnetic excitations we have $\mathbf{A} \equiv \mathbf{A}_T = \mathbf{E}_T/(i\omega) \equiv \mathbf{E}/(i\omega)$. The information in Eqs. (25.39), (25.40), (25.44) and (25.45) together tells us that the rectification current density ($\mathbf{J}_0$) must be proportional to the tensor product $\mathbf{A}(\mathbf{q},\omega)\mathbf{A}^*(\mathbf{q},\omega) = \mathbf{E}(\mathbf{q},\omega)\mathbf{E}^*(\mathbf{q},\omega)/\omega^2$. In consequence, the nonlinear relation between the rectification current density and the electric field, hence must take the general form

$$\mathbf{J}_0 = \frac{1}{2}\left[\mathbf{\Sigma}_0(\mathbf{q},\omega) : \mathbf{E}(\mathbf{q},\omega)\mathbf{E}^*(\mathbf{q},\omega) + c.c.\right]. \tag{25.46}$$

It is beyond the scope of our book concept to present the calculation of the nonlinear response tensor, so let us just cite the final result [194], viz.,

$$\mathbf{\Sigma}_0(\mathbf{q},\omega) = -\frac{\hbar e^3}{4m^2\omega^2 V}\mathbf{U}\mathbf{M}_0(\mathbf{q},\omega), \tag{25.47}$$

where

$$\mathbf{M}_0(\mathbf{q},\omega) = \sum_{\mathbf{k}}(2\mathbf{k}+\mathbf{q})\left\{(u_{\mathbf{k}}^2 u_{\mathbf{k}+\mathbf{q}}^2 - v_{\mathbf{k}}^2 v_{\mathbf{k}+\mathbf{q}}^2)\left[\frac{1}{\hbar\omega + E_{\mathbf{k}} - E_{\mathbf{k}+\mathbf{q}}} + \frac{1}{\hbar\omega - E_{\mathbf{k}} + E_{\mathbf{k}+\mathbf{q}}}\right]\right.$$
$$\times (f_{\mathbf{k}} - f_{\mathbf{k}+\mathbf{q}}) + (v_{\mathbf{k}}^2 u_{\mathbf{k}+\mathbf{q}}^2 - u_{\mathbf{k}}^2 v_{\mathbf{k}+\mathbf{q}}^2)$$
$$\left.\times \left[\frac{1}{\hbar\omega - E_{\mathbf{k}} - E_{\mathbf{k}+\mathbf{q}}} + \frac{1}{\hbar\omega + E_{\mathbf{k}} + E_{\mathbf{k}+\mathbf{q}}}\right](1 - f_{\mathbf{k}} - f_{\mathbf{k}+\mathbf{q}})\right\}. \quad (25.48)$$

Let us now take a brief look at the structure of the response tensor. It appears that

$$\mathbf{\Sigma}_0(\mathbf{q},\omega) = \mathbf{\Sigma}_0^*(\mathbf{q},\omega), \quad (25.49)$$

and that the tensorial form of $\mathbf{\Sigma}_0(\mathbf{q},\omega)$ is a triadic product of the unit tensor, $\mathbf{U}$ and the vector $\mathbf{M}_0(\mathbf{q},\omega)$. The physical implication of this is discussed in the following Section 25.5. Under the $\mathbf{k}$-summation sign one has terms proportional to $f_{\mathbf{k}} - f_{\mathbf{k}+\mathbf{q}}$ and $1 - f_{\mathbf{k}} - f_{\mathbf{k}+\mathbf{q}}$. The term proportional to $f_{\mathbf{k}} - f_{\mathbf{k}+\mathbf{q}}$ describes a *quasiparticle photon scattering cycle* in which a photon is scattered between the quasiparticle states $\mathbf{k}$ and $\mathbf{k}+\mathbf{q}$. The scattering cycle is at resonance for $\hbar\omega + E_{\mathbf{k}} = E_{\mathbf{k}+\mathbf{q}}$ and $\hbar\omega + E_{\mathbf{k}+\mathbf{q}} = E_{\mathbf{k}}$. The scattering cycle provides a contribution to the rectification process proportional to the coherence factor $u_{\mathbf{k}}^2 u_{\mathbf{k}+\mathbf{q}}^2 - v_{\mathbf{k}}^2 v_{\mathbf{k}+\mathbf{q}}^2$. The term under the $\mathbf{k}$-summation sign proportional to $1 - f_{\mathbf{k}} - f_{\mathbf{k}+\mathbf{q}}$ stems from a *two-quasiparticle photon excitation-deexcitation cycle* involving the quasiparticle states $\mathbf{k}$ and $\mathbf{k}+\mathbf{q}$. The excitation-deexcitation cycle is at resonance for $\hbar\omega = E_{\mathbf{k}} + \mathbf{E}_{\mathbf{k}+\mathbf{q}}$ and $-\hbar\omega = E_{\mathbf{k}} + E_{\mathbf{k}+\mathbf{q}}$. The coherence factor attached to the two-quasiparticle cycle is $v_{\mathbf{k}}^2 u_{\mathbf{k}+\mathbf{q}}^2 - u_{\mathbf{k}}^2 v_{\mathbf{k}+\mathbf{q}}^2$. A discussion of the many-body picture leading up to the result for $M_0(\mathbf{q},\omega)$ is given in [194]. In Fig. 25.1 the fundamental photon processes underlying the structure of $M_0(\mathbf{q},\omega)$ are shown.

For temperatures above the superconducting transition temperature, T_C, it may be shown that

$$\mathbf{\Sigma}_0(\mathbf{q},\omega; T > T_C) = -\frac{\hbar e^3}{2m^2\omega^2 V}\mathbf{U}\sum_{\mathbf{k}}(2\mathbf{k}+\mathbf{q})\frac{f_{\mathbf{k}}^{\mathrm{NS}} - f_{\mathbf{k}+\mathbf{q}}^{\mathrm{NS}}}{\hbar\omega + \varepsilon_{\mathbf{k}} - \varepsilon_{\mathbf{k}+\mathbf{q}}}, \quad (25.50)$$

where $f_{\mathbf{k}}^{\mathrm{NS}}$ ($f_{\mathbf{k}+\mathbf{q}}^{\mathrm{NS}}$) is the normal state (NS) Fermi-Dirac distribution function belonging to the free-electron energy $\varepsilon_{\mathbf{k}}$ ($\varepsilon_{\mathbf{k}+\mathbf{q}}$).

25.5 NONLINEAR DYNAMICAL MEISSNER EFFECT

A somewhat deeper insight into the physics of the rectification process can be obtained by a comparison to the processes of linear electrodynamics and second-harmonic generation. For such a comparison it is convenient to generalize the analysis to inhomogenous BCS superconductors, where the spatially nonlocal character of the electrodynamics appears in a manifest manner.

From the quantum-field formalism described in Chapter 18 it appears [Eq. (18.54)] that the linear part of the induced current density is obtained from

$$\mathbf{J}_1(\mathbf{r}) = \mathrm{Tr}\left\{\rho_F \hat{\mathcal{J}}_1\right\} + \mathrm{Tr}\left\{\rho_1 \hat{\mathcal{J}}_F\right\} \quad (25.51)$$

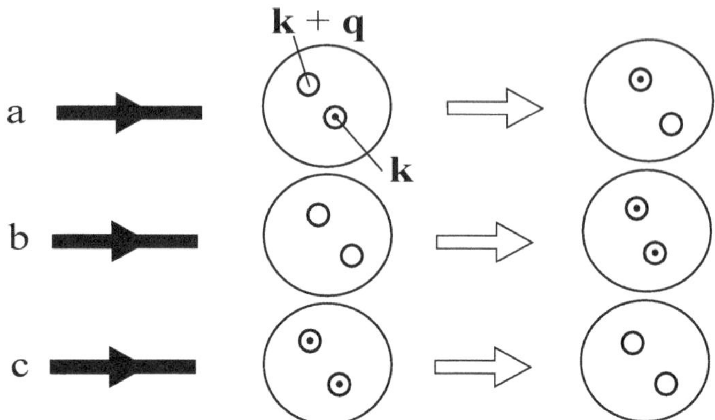

Figure 25.1 Schematic diagrams showing the three fundamental photon [black arrow] absorption processes in a Cooper-paired jellium. Diagram a: A photon absorption causes a quasiparticle scattering from state to $\mathbf{k}$ to state $\mathbf{k}+\mathbf{q}$ [⊙: filled state, O: empty state]. Diagram b: Photon absorption gives rise to creation of two quasiparticles. Diagram c: Photon absorption gives rise to destruction of two quasiparticles. This energy non-conserving process is omitted in a rotating wave approximation, The large circles indicate the Fermi surface. In general quasiparticles exist both inside and outside the Fermi surface. Three equivalent diagrams can be drawn for photon emission processes. In the calculation of the nonlinear response tensor for the rectification process, $\Sigma_0(\mathbf{q},\omega)$ [$\mathbf{M}_0(\mathbf{q},\omega)$], the photon excitation-deexcitation cycle is build from all combinations of a photon absorption process with a photon emission process.

using here the notation changes $\hat{\rho}_0 \Rightarrow \hat{\rho}_F$, $\mathcal{J}_0 \to \mathcal{J}_F$. In Eq. (25.51) and in the subsequent description the reference to the frequency (ω) is left out from the notation. As shown in schematic form in Fig. 25.2, the term $\mathrm{Tr}\left\{\rho_1\hat{\mathcal{J}}_F\right\}$ relates the current density [$\mathbf{J}_1(\mathbf{r})$] at a given space point $\mathbf{r}$ to the electric field [$\mathbf{E}(\mathbf{r}')$] in surrounding points, $\mathbf{r}'$. This is what we call a spatially nonlocal connection. The other term, $\mathrm{Tr}\left\{\rho_F\hat{\mathcal{J}}_1\right\}$, relates the current density at $\mathbf{r}$ to the electric field in the same point ($\mathbf{r}' = \mathbf{r}$). This is the local London response [179–181], also called the diamagnetic response. In the low-temperature limit ($T \to 0K$) only the diamagnetic coupling survives. The famous Meissner effect [187] originates in the diamagnetic response.

The induced second-harmonic current density, $\mathbf{J}_2(\mathbf{r};2\omega) \equiv \mathbf{J}_2(\mathbf{r})$ is obtained from

$$\mathbf{J}_2(\mathbf{r}) = \mathrm{Tr}\left\{\rho_2\hat{\mathcal{J}}_F\right\} + \mathrm{Tr}\left\{\rho_1\hat{\mathcal{J}}_1\right\} \tag{25.52}$$

and it relates the current density at $\mathbf{r}$ [$\mathbf{J}_2(\mathbf{r})$] to the electric-field dyad [$\mathbf{E}(\mathbf{r}')\mathbf{E}(\mathbf{r}'')$] at two neighbouring points, $\mathbf{r}'$ and $\mathbf{r}''$. It is known that the term $\mathrm{Tr}\{\rho_2\hat{\mathcal{J}}_F\}$ is associated with two types of nonlocal couplings. Thus, the induced nonlinear current density at $\mathbf{r}$ [$\mathbf{J}_2(\mathbf{r})$] have contributions (i) from two different neighbouring field points $\mathbf{r}' \neq \mathbf{r}''$ [$\mathbf{E}(\mathbf{r}')\mathbf{E}(\mathbf{r}'')$], and ($ii$) from coincident field points, $\mathbf{r}' = \mathbf{r}''$ [$\mathbf{E}(\mathbf{r}')\mathbf{E}(\mathbf{r}')$]. The term $\mathrm{Tr}\{\rho_1\hat{\mathcal{J}}_1\}$ is particularly interesting for us because it is semilocal in structure, which means that the nonlinear current density at $\mathbf{r}$ [$\mathbf{J}_2(\mathbf{r})$] relates to the fundamental

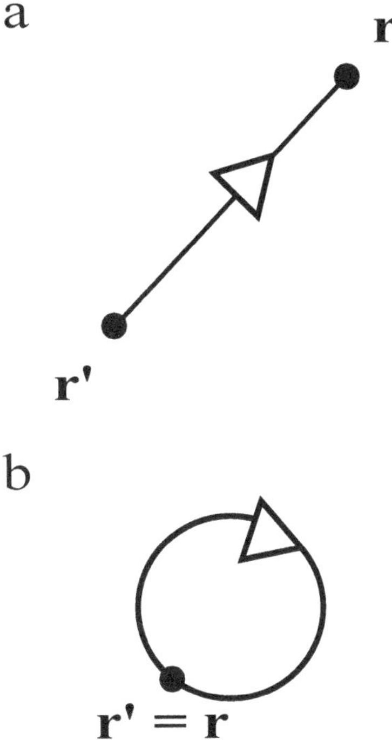

Figure 25.2 Schematic diagrams showing the two basic local and nonlocal processes associated with the linear electromagnetic response of a superconductor. To obtain the induced current density at space point $\mathbf{r}$, the electric field at $\mathbf{r}'$ must be known. (a) The nonlocal paramagnetic response. (b) the local London response.

electric field in neighbouring points $\mathbf{r}'$ and in $\mathbf{r}$, the dyadic field product thus being $\mathbf{E}(\mathbf{r})\mathbf{E}(\mathbf{r}')$. The $\mathrm{Tr}\left\{\rho_1\hat{\mathcal{J}}_1\right\}$-contribution hence is a product of a spatially nonlocal effect and a Meissner (local London) – like effect.

A moment of reflection makes it clear that the optical rectification process given by Eq (25.44), and repeated here,

$$\mathbf{J}_0 = \frac{1}{4}\left[\mathrm{Tr}\{\hat{\rho}_1\hat{\mathcal{J}}_1^\dagger\} + c.c.\right],\tag{25.53}$$

always depends on the Meissner effect [the ring structure in the schematic illustration of Fig. 25.3]. Because of this, I have called the nonlinear electromagnetic rectification phenomenon a *nonlinear dynamical Meissner effect*. The diagrams in Fig. 25.3 show the three basic nonlocal processes entering a second-order nonlinear response.

As a consequence of Eq. (25.53) the rectification current density has the general form

$$\mathbf{J}_0(\mathbf{r}) = \frac{1}{2}\left[\mathbf{E}^*(\mathbf{r};\omega)\int_{-\infty}^{\infty}\mathbf{R}_0(\mathbf{r},\mathbf{r}';\omega)\cdot\mathbf{E}(\mathbf{r}';\omega)d^3r' + c.c.\right],\tag{25.54}$$

where $\mathbf{R}_0(\mathbf{r},\mathbf{r}';\omega)$ is a spatially nonlocal vectorial response function relating the nonlinear dc-current density at point $\mathbf{r}$ to the fundamental field in neighbouring points

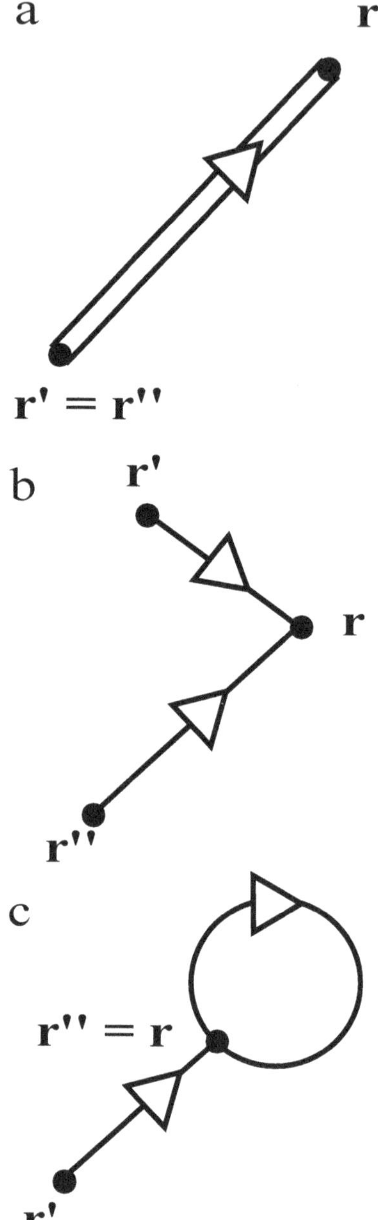

Figure 25.3 Schematic diagrams showing the three basic nonlocal processes associated with the second-order nonlinear response tensor. (a) The simultaneous two-photon excitation process. (b) The double-nonlocal excitation process. (c) The semilocal excitation process. The induced nonlinear current density at space point $\mathbf{r}$, has contributions from the fundamental electric field at $\mathbf{r}'$ and $\mathbf{r}''$. In second-harmonic generation from a Cooper-paired superconductor all the three basis processes contribute, whereas only the semilocal process gives rise to nonlinear electromagnetic rectification.

$\mathbf{r}'$. In a medium exhibiting translation invariance in space

$$\mathbf{R}_0 = \mathbf{R}_0(\mathbf{r} - \mathbf{r}'; \omega). \tag{25.55}$$

In order to link the general expression in Eq. (25.54) to the response formalism presented in Section 25.4, we write $\mathbf{J}_0(\mathbf{r})$ as

$$\mathbf{J}_0(\mathbf{r}) = \frac{1}{2}\left[\mathbf{J}_0^{(+)}(\mathbf{r}) + \mathbf{J}_0^{(-)}(\mathbf{r})\right], \tag{25.56}$$

where $\mathbf{J}_0^{(-)}(\mathbf{r}) = [\mathbf{J}_0^{(+)}(\mathbf{r})]^*$. The spatial Fourier integrals

$$\mathbf{E}(\mathbf{r}; \omega) = (2\pi)^{-3}\int_{-\infty}^{\infty}\mathbf{E}(\mathbf{Q}, \omega)e^{i\mathbf{Q}\cdot\mathbf{r}}d^3Q, \tag{25.57}$$

$$\mathbf{J}_0^{(+)}(\mathbf{r}; \omega) = (2\pi)^{-3}\int_{-\infty}^{\infty}\mathbf{J}_0^{(+)}(\mathbf{Q}, \omega)e^{i\mathbf{Q}\cdot\mathbf{r}}d^3Q, \tag{25.58}$$

then are used to represent the rectification current density [Eq. (25.54)] for a homogeneous medium [Eq. (25.55)] in Fourier space as follows:

$$\mathbf{J}_0(\mathbf{q}) = \frac{1}{2}\left\{\mathbf{J}_0^{(+)}(\mathbf{q}) + [\mathbf{J}_0^{(+)}(\mathbf{q})]^*\right\}, \tag{25.59}$$

where

$$\mathbf{J}_0^{(+)}(\mathbf{q}) = (2\pi)^{-3}\int_{-\infty}^{\infty}\mathbf{E}^*(\mathbf{Q} - \mathbf{q}, \omega)\mathbf{R}_0(\mathbf{Q}, \omega)\cdot\mathbf{E}(\mathbf{Q}, \omega)d^3Q. \tag{25.60}$$

In Section 25.4 the analysis was limited to the case where the fundamental field only has one spatial Fourier component, here denoted by $\mathbf{K}$. For a single plane-wave excitation, i.e.,

$$\mathbf{E}(\mathbf{Q}, \omega) = \mathbf{E}(\mathbf{K}, \omega)\delta(\mathbf{Q} - \mathbf{K}), \tag{25.61}$$

Eq. (25.60) first is reduced to

$$\mathbf{J}_0^{(+)}(\mathbf{q}) = (2\pi)^{-3}\mathbf{E}^*(\mathbf{K} - \mathbf{q}, \omega)\mathbf{R}_0(\mathbf{K}, \omega)\cdot\mathbf{E}(\mathbf{K}, \omega). \tag{25.62}$$

Next, in order for $\mathbf{J}_0^{(+)}(\mathbf{q})$ to be nonvanishing, one must require that $\mathbf{K} - \mathbf{q} = \mathbf{K}$, and thus $\mathbf{q} = \mathbf{0}$. A comparison of the form

$$\mathbf{J}_0(\mathbf{0}) = \frac{(2\pi)^{-3}}{2}[\mathbf{E}^*(\mathbf{K}, \omega)\mathbf{R}_0(\mathbf{K}, \omega)\cdot\mathbf{E}(\mathbf{K}, \omega) + c.c.], \tag{25.63}$$

to Eq. (25.46) [with Eq. (25.47)] leads for a normalization volume V $[(2\pi)^3 \to V]$ to the following relation between $\mathbf{R}_0(\mathbf{Q}, \omega)$ and $\mathbf{M}_0(\mathbf{Q}, \omega)$:

$$\mathbf{R}_0(\mathbf{Q}, \omega) = -\frac{\hbar e^3}{4m^2\omega^2 V}\mathbf{M}_0(\mathbf{Q}, \omega), \tag{25.64}$$

with $\mathbf{M}_0(\mathbf{Q}, \omega)$ given by Eq. (25.48).

25.6 RECTIFICATION AT A SUPERCONDUCTING SURFACE

25.6.1 Free Meissner current density

In Section 25.5 we have studied the rectification current density arising from nonlinear dynamic Meissner effects. We may characterize this process as a *forced interaction* arising from the impressed fundamental electric field (calculated in a self-consistent manner if needed). We have seen that the forced rectification process is semilocal (SL) and in the following we shall rename the forced rectification current density $\mathbf{J}_0^{\mathrm{SL}}(\mathbf{r})$.

The forced dc-current density will give rise to a dc magnetic field inside the superconductor. Having said *inside* we have implicitly indicated that we now aim at a description where surface effects are included in the formalism. It is known (from textbook descriptions, e.g. [61]) that the impressed dc magnetic field will be partly screened by the magnetic arising from a *linear* dc current density induced in the vicinity of the surface. This so-called Meissner current density is responsible for the famous Meissner effect [187] in linear electromagnetics of superconductors. Below we shall denote the free Meissner (M) current density by $\mathbf{J}_0^{M}(\mathbf{r})$. The Meissner current density is related in a spatially linear and nonlocal manner to the self-consistent dc vector potential $\mathbf{A}_0(\mathbf{r})$. Thus,

$$\mathbf{J}_0^{M}(\mathbf{r}) = \int_{-\infty}^{\infty} \mathbf{S}_0(\mathbf{r}, \mathbf{r}') \cdot \mathbf{A}_0(\mathbf{r}') d^3 r'. \tag{25.65}$$

In the case where the superconductor exhibits infinitesimal translational invariance parallel to the plane $z = 0$, the response tensor has the form $\mathbf{S}_0(\mathbf{r}, \mathbf{r}') = \mathbf{S}_0(\mathbf{r}_\parallel - \mathbf{r}'_\parallel, z, z')$. In the mixed Fourier representation

$$\mathbf{J}_0^{M}(z; \mathbf{Q}_\parallel) = \int_{-\infty}^{\infty} \mathbf{S}_0(z, z'; \mathbf{Q}_\parallel) \cdot \mathbf{A}_0(z'; \mathbf{Q}_\parallel) dz'. \tag{25.66}$$

In the most simple case, namely that of complete translational invariance, $\mathbf{S}_0(\mathbf{r}, \mathbf{r}') = \mathbf{S}_0(\mathbf{r} - \mathbf{r}')$, one obtains in Fourier space ($\mathbf{Q}$-space), the algebraic relation

$$\mathbf{J}_0^{M}(\mathbf{Q}) = \mathbf{S}_0(\mathbf{Q}) \cdot \mathbf{A}_0(\mathbf{Q}). \tag{25.67}$$

Within the pairing approximation, the linear response tensor $\mathbf{S}_0(\mathbf{Q})$ is given by the following well-known expression [134]:

$$\mathbf{S}_0(\mathbf{Q}) = -\frac{ne^2}{m}\mathbf{U} + \left(\frac{e\hbar}{2m}\right)^2 \frac{2}{V} \sum_{\mathbf{k}} (2\mathbf{k} + \mathbf{Q})(2\mathbf{k} + \mathbf{Q})$$

$$\times \left[(u_{\mathbf{k}} u_{\mathbf{k}+\mathbf{Q}} + v_{\mathbf{k}} v_{\mathbf{k}+\mathbf{Q}})^2 \frac{f_{\mathbf{k}} - f_{\mathbf{k}+\mathbf{Q}}}{E_{\mathbf{k}+\mathbf{Q}} - E_{\mathbf{k}}} + (u_{\mathbf{k}+\mathbf{Q}} v_{\mathbf{k}} - v_{\mathbf{k}+\mathbf{Q}} u_{\mathbf{k}})^2 \frac{1 - f_{\mathbf{k}} - f_{\mathbf{k}+\mathbf{Q}}}{E_{\mathbf{k}+\mathbf{Q}} + E_{\mathbf{k}}} \right]. \tag{25.68}$$

25.6.2 Fundamental integro-differential equation for the static vector potential

In order to determine the Meissner screening of the forced part of the rectification current density a self-consistent solution for the vector potential $\mathbf{A}_0(\mathbf{r})$ is needed, cf.

Eq. (25.65). Once $\mathbf{A}_0(\mathbf{r})$ is obtained, the prevailing dc current density

$$\mathbf{J}_0(\mathbf{r}) = \mathbf{J}_0^{\mathrm{SL}}(\mathbf{r}) + \mathbf{J}_0^{M}(\mathbf{r}) \tag{25.69}$$

inside the superconductor can be found. In silence, we have assumed that the fundamental electric field $\mathbf{E}(\mathbf{r};\omega)$ inside the superconductor is known. The field is of course the self-consistent field associated with a prescribed externally impressed field. The reader may find an account of this part of the problem in [134].

From the magnetostatic Maxwell equation $\boldsymbol{\nabla} \times \mathbf{B}_0(\mathbf{r}) = \mu_0 \mathbf{J}_0(\mathbf{r})$, where $\mathbf{B}_0(\mathbf{r}) = \boldsymbol{\nabla} \times \mathbf{A}_0(\mathbf{r})$, one thus obtains the following inhomogeneous integro-differential equation for the static vector potential

$$\boldsymbol{\nabla} \times (\boldsymbol{\nabla} \times \mathbf{A}_0(\mathbf{r})) - \mu_0 \int_{-\infty}^{\infty} \mathbf{S}_0(\mathbf{r},\mathbf{r}') \cdot \mathbf{A}_0(\mathbf{r}')d^3r' = \mu_0 \mathbf{J}_0^{\mathrm{SL}}(\mathbf{r}). \tag{25.70}$$

Let us now consider the case where the superconductor occupies the halfspace $z > 0$, the rest of the space being vacuum. Furthermore, we shall assume that the superconductor exhibits translationally invariance properties parallel to the surface. Finally, we shall limit ourselves to the situation where the electrons interact with a plane, monochromatic electric field incident from vacuum at an oblique angle $\Theta = \arcsin\left(cK_{\parallel}/\omega\right)$ where $\mathbf{K}_{\parallel}$ is the component of the vector parallel to the surface. The fundamental electric field inside the superconductor thus is given by

$$\mathbf{E}(\mathbf{r};\omega) = \frac{(2\pi)^{-2}}{2}\left[\mathbf{E}(z;\mathbf{K}_{\parallel},\omega)e^{i\mathbf{K}_{\parallel}\cdot\mathbf{r}_{\parallel}} + c.c.\right]. \tag{25.71}$$

In the mixed Fourier representation the self-consistent vector potential $\mathbf{A}_0(z;\mathbf{K}_{\parallel} = \mathbf{0}) \equiv \mathbf{A}_0(z,\mathbf{0})$ satisfies the integro-differential equation

$$(\mathbf{U}-\mathbf{e}_z\mathbf{e}_z) \cdot \frac{\partial^2 \mathbf{A}_0(z;\mathbf{0})}{\partial z^2} + \mu_0 \int_0^{\infty} \mathbf{S}_0(z,z',\mathbf{0}) \cdot \mathbf{A}_0(z';\mathbf{0})dz'$$
$$+ \frac{\mu_0}{2}\left[\mathbf{J}_0^{(+)}(z;\mathbf{0}) + (\mathbf{J}_0^{(+)}(z;\mathbf{0}))^*\right] = \mathbf{0}, \tag{25.72}$$

where

$$\mathbf{J}_0^{(+)}(z;\mathbf{0}) = (2\pi)^{-2}\mathbf{E}^*(z;\mathbf{K}_{\parallel},\omega) \int_0^{\infty} \mathbf{R}_0(z,z';\mathbf{K}_{\parallel},\omega) \cdot \mathbf{E}(z';\mathbf{K}_{\parallel},\omega)dz'. \tag{25.73}$$

25.6.3 Specular reflection model

Facing the fact that the surface-sensitive response functions $\mathbf{R}_0(z,z';\mathbf{K}_{\parallel},\omega)$ and $\mathbf{S}_0(z,z';\mathbf{0})$ are extremely complicated entities, it is beyond our ability to make analytical progress in the solution of the integro-differential equation for $\mathbf{A}_0(z;\mathbf{0})$ unless a model simplification is made. We know pretty well from our study of surface plasmons, plasmaritons and jellions in the normal metallic state [Chapter 19] that the SCIB model [Subsection 19.1.2], despite its physical shortcomings, offers a good basis for our understanding. The SCIB model is based on two premises namely, (i) specular reflection of the electon at the surface, and (ii) neglect of the quantum interference between the incoming and reflected parts of the electron wave function. The response functions entering the SCIB model are bulk response functions. Thus, in applying the SCIB model to a BCS superconductor jellium one only needs the linear and nonlinear bulk response functions $\mathbf{S}_0(\mathbf{Q})$ [Eq. (25.68)] and $\mathbf{R}_0(\mathbf{Q}, \omega) = -[\hbar e^3/(4m^2\omega^2 V)]\mathbf{M}_0(\mathbf{Q}, \omega)$ with $\mathbf{M}_0(\mathbf{Q}, \omega)$ given by Eq. (25.48).

In the framework of the SCIB model, for which

$$\mathbf{S}_0(z, z'; \mathbf{0}) = \mathbf{S}_0(z - z', \mathbf{0}) + \mathbf{S}_0(z + z'; \mathbf{0}) \cdot \boldsymbol{\alpha}, \tag{25.74}$$

and

$$\mathbf{R}_0(z, z'; \mathbf{K}_\parallel, \omega) = \mathbf{R}_0(z - z'; \mathbf{K}_\parallel, \omega) + \boldsymbol{\alpha} \cdot \mathbf{R}_0(z + z'; \mathbf{K}_\parallel, \omega), \tag{25.75}$$

where

$$\boldsymbol{\alpha} = \begin{pmatrix} 1 & 0 & 0 \\ 0 & 1 & 0 \\ 0 & 0 & -1 \end{pmatrix}, \tag{25.76}$$

it is possible to solve Eq. (25.72) for $\mathbf{A}_0(z; \mathbf{0})$ in a manner described in my Phys. Rev. B43, 10293 (1991) article.

The final result for the prevailing magnetostatic vector potential inside ($z > 0$) the superconductor is given by the following integral expression:

$$\mathbf{A}_0(z; \mathbf{0}) = (2\pi)^{-1} \int_{-\infty}^{\infty} \left[\frac{\mathbf{U} - \mathbf{e}_z\mathbf{e}_z}{\mathcal{N}_0^T(q_\perp)} + \frac{\mathbf{e}_z\mathbf{e}_z}{\mathcal{N}_0^L(q_\perp)} \right] \cdot \left[\mathbf{g}_0 - \mu_0 \mathbf{J}_0^{\mathrm{SL}}(q_\perp\mathbf{e}_z) \right] e^{iq_\perp z} dq_\perp. \ z > 0. \tag{25.77}$$

Before taking a look at the various quantities entering the integrand of Eq. (25.77), let us return to the tensor $\mathbf{S}_0(z - z'; \mathbf{0}) \equiv \mathbf{S}_0(z - z'; \mathbf{Q}_\parallel = \mathbf{0})$ which Fourier decomposition in $Z = z - z'$ is

$$\mathbf{S}_0(Z, \mathbf{0}) = (2\pi)^{-1} \int_{-\infty}^{\infty} \mathbf{S}_0(q_\perp, \mathbf{0}) e^{iq_\perp Z} dq_\perp. \tag{25.78}$$

It is a straightforward matter to show from Eq. (25.68) that the tensor $\mathbf{S}_0(q_\perp, \mathbf{Q}_\parallel = \mathbf{0}) \equiv \mathbf{S}_0(q_\perp\mathbf{e}_z)$ has only nonvanishing diagonal components in our Cartesian coordinate system. Thus,

$$\mathbf{S}_0(q_\perp\mathbf{e}_z) = S_0^T(q_\perp)(\mathbf{U} - \mathbf{e}_z\mathbf{e}_z) + S_0^L(q_\perp)\mathbf{e}_z\mathbf{e}_z, \tag{25.79}$$

where $S_0^T(q_\perp)$ and $S_0^L(q_\perp)$ are the transverse (T) and longitudinal (L) response functions of the scalar quantities $q_\perp$, only. Now in Eq. (25.77)

$$\mathcal{N}_0^T(q_\perp) = \mu_0 S_0^T(q_\perp) - q_\perp^2, \tag{25.80}$$

$$\mathcal{N}_0^L(q_\perp) = \mu_0 S_0^L(q_\perp). \tag{25.81}$$

A comparison to the normal state expression in Eqs. (19.15)–(19.17) shows that

$$\mathbf{G}(q_\perp\mathbf{e}_z) = \frac{\mathbf{U} - \mathbf{e}_z\mathbf{e}_z}{\mathcal{N}_0^T(q_\perp)} + \frac{\mathbf{e}_z\mathbf{e}_z}{\mathcal{N}_0^L(q_\perp)} \tag{25.82}$$

is the Green tensor giving the $\mathbf{A}_0(z;\mathbf{0})$ response to an effective current density $\mathbf{g}_0/\mu_0 - \mathbf{J}_0^{\mathrm{SL}}(q_\perp \mathbf{e}_z)$, in Eq. (25.77).

The quantity

$$\mathbf{J}_0^{\mathrm{SL}}(q_\perp \mathbf{e}_z) = \int_{-\infty}^{\infty} \mathbf{J}_0^{\mathrm{SL}}(z\mathbf{e}_z)e^{iq_\perp z}dz \qquad (25.83)$$

is the Fourier integral transform of the forced semilocal current density

$$\mathbf{J}_0^{\mathrm{SL}}(z\mathbf{e}_z) = \frac{1}{2}\left\{(2\pi)^{-2}\mathbf{E}^*(z;\mathbf{K}_\|,\omega)\int_0^{\infty}\mathbf{R}_0(z,z';\mathbf{K}_\|,\omega|\mathrm{SR})\cdot\mathbf{E}(z',\mathbf{K}_\|,\omega)dz' + c.c.\right\} \qquad (25.84)$$

calculated in the framework of the specular-reflection (SR) model, as the SR label added in the argument of $\mathbf{R}_0$ indicates, see Eq. (25.75). The quantity $\mathbf{g}_0$ is given by

$$\mathbf{g}_0 = \begin{pmatrix} \frac{\partial A_{0,x}}{\partial z}(z\to 0^+;\mathbf{0}) \\ \frac{\partial A_{0,y}}{\partial z}(z\to 0^+;\mathbf{0}) \\ 0 \end{pmatrix}, \qquad (25.85)$$

and represents (apart from a trivial constant) a fictitious surface current density. The $\mathbf{g}_0$ surface current density, which has no component perpendicular to the vacuum/superconductor interface, originates in the use of the specular-reflection model.

The Green tensor $\mathbf{G}(q_\perp \mathbf{e}_z)$ given by Eq. (25.82) is a *screened propagator* taking into account the free Meissner current density via $S_0^T(q_\perp)$ and $S_0^L(q_\perp)$. Therefore, the driving current density distribution for $A_0(z;\mathbf{0})$ consists alone of the nonlinear forced current density part ($\mathbf{J}_0^{\mathrm{SL}}$) and the surface part ($\sim \mathbf{g}_0$).

25.6.4 Selfconsistency requirement

Let us divide the forced current density into its components parallel ($\|$) and perpendicular $\perp$ to the surface, i.e.,

$$\mathbf{J}_\|^{\mathrm{SL}} = (\mathbf{U} - \mathbf{e}_z\mathbf{e}_z)\cdot\mathbf{J}_0^{\mathrm{SL}}; \quad \mathbf{J}_\perp^{\mathrm{SL}} = \mathbf{e}_z(\mathbf{e}_z\cdot\mathbf{J}_0^{\mathrm{SL}}). \qquad (25.86)$$

In terms of these parts the magnetostatic vector potential in Eq. (25.77) can be written in the following form;

$$\mathbf{A}_0(z;\mathbf{0}) = (2\pi)^{-1}\int_{-\infty}^{\infty}\left[\frac{\mathbf{g}_0 - \mu\mathbf{J}_\|^{\mathrm{SL}}(q_\perp \mathbf{e}_z)}{\mathcal{N}_0^T(q_\perp)} - \frac{\mu_0\mathbf{J}_\perp^{\mathrm{SL}}(q_\perp \mathbf{e}_z)}{\mathcal{N}_0^L(q_\perp)}\right]e^{iq_\perp z}dq_\perp. \qquad (25.87)$$

For the theory to be selfconsistent it must be required that $\partial\mathbf{A}_0(z;\mathbf{0})/\partial z$ taking in the limit $z\to 0^+$ gives one $\mathbf{g}_0$. A careful analysis [134] leads to the conclusion that the selfconsistency requirement can be expressed in the form

$$\lim_{q_\perp\to\infty}\mathbf{J}_\|^{\mathrm{SL}}(q_\perp \mathbf{e}_z) = \mathbf{0}. \qquad (25.88)$$

This condition on the semilocal current density involves only its component parallel to the surface. This should not surprise the reader since $\mathbf{g}_0$ is a vector parallel to the surface plane.

25.6.5 Magnetic field of the rectification process

The self-consistent nonlinear dc magnetic field ($\mathbf{B}_0$) inside the BCS superconductor only depends on the distance (z) from the surface, and can readily be calculated from $\mathbf{B}_0 = \mathbf{\nabla} \times \mathbf{A}_0(z; \mathbf{0})$. Thus,

$$\mathbf{B}_0(z) = \frac{i}{2\pi}\mathbf{e}_z \times \left\{ \int_{-\infty}^{\infty} [\mathbf{g}_0 - \mu_0 \mathbf{J}_{\|}^{\mathrm{SL}}(q_\perp \mathbf{e}_z)] \frac{q_\perp}{\mathcal{N}_0^T(q_\perp)} e^{iq_\perp z} dq_\perp \right\}. \tag{25.89}$$

It was shown in my Phys. Rev. B43, 10293 (1991) article that the total nonlinear current density, $\mathbf{J}_0(z; \mathbf{0}) = \mathbf{J}_0^M(z; \mathbf{0}) + \mathbf{J}_0^{\mathrm{SL}}(z; \mathbf{0})$ everywhere is directed parallel to the surface plane, i.e.,

$$\mathbf{e}_z \cdot \mathbf{J}_0(z; \mathbf{0}) = 0. \tag{25.90}$$

This is in agreement with the fact that $\mathbf{e}_z \cdot \mathbf{B}_0(z) = 0$. It was also shown in the aforementioned article that

$$\mathbf{B}_0(z \to 0^+) = \frac{1}{2}\mathbf{e}_z \times \mathbf{g}_0 = \mathbf{B}_0^V \tag{25.91}$$

where $\mathbf{B}_0^V$ is the z-independent nonlinear magnetic field in the vacuum (V) halfspace.

VII

Photons Tied to Massive Particles

CHAPTER 26

Radiation Reaction in the Self-Field Approach

In atomic physics the radiative transition process of an electron between two atomic electronic levels has been studied for many years using various theories. The approaches fall into two main categories, viz., (i) the perturbative QED family [188, 219, 264, 265] and (ii) the self-field method, also called the semiclassical method [18, 229, 249]. In the semiclassical approach one views radiative corrections as arising from radiation reaction effects due to the interaction of the atom with its own self-field [13–17, 28, 201]. The self-field approach (and also the QED approach) predicts that an atom will decay "spontaneously" from an excited state with a characteristic time constant equal to the phenomenological Einstein A coefficient [77] for the transition. The self-field method also predicts that the light radiated during the transition will have a frequency slightly different from the electronic transition frequency (the Bohr frequency). The frequency shift, called the Lamb shift, was first observed by Lamb and Retherford [174–176]. Bethe [24] succeeded in calculating the Lamb shift quite accurately, and his nonrelativistic calculation, in a broader context, indicated that "the problem of infinities" that was a nightmare for quantum electrodynamics for two decades might be solved. For the $2S_{1/2}$ and $2P_{1/2}$ states in Hydrogen, the measured Lamb shift was ≈ 1.057 MHz and Bethe obtained a value ≈ 1.040 MHz. A detailed account of the birth of quantum field theory has been given by Weinberg [261].

26.1 SPONTANEOUS DECAY AND LAMB SHIFT: THE ASSOCIATED TRANSVERSE VECTOR POTENTIAL

The transverse part of the atomic current density, $\mathbf{J}_T(\mathbf{r}, t)$, creates a classical transverse electromagnetic field, given below by the transverse vector potential, $\mathbf{A}_T(\mathbf{r}, t) - \mathbf{A}_T^{\mathrm{ext}}(\mathbf{r}, t)$, where $\mathbf{A}_T^{\mathrm{ext}}(\mathbf{r}, t)$ is the externally impressed field. Thus,

$$\mathbf{A}_T(\mathbf{r}, t) = \mathbf{A}_T^{\mathrm{ext}}(\mathbf{r}, t) + \frac{\mu_0}{4\pi} \int_{-\infty}^{\infty} \frac{\mathbf{J}_T(\mathbf{r}', t - |\mathbf{r} - \mathbf{r}'|/c)}{|\mathbf{r} - \mathbf{r}'|} d^3 r'. \tag{26.1}$$

DOI: 10.1201/9781003029458-26

For source ($\mathbf{r'}$) and observation ($\mathbf{r}$) points located inside or in the vicinity of the atom's transverse current density domain the retardation $|\mathbf{r} - \mathbf{r'}|/c$ is so small in comparison to the characteristic time variation in $\mathbf{J}_T(\mathbf{r}, t)$ that Eq. (26.1) can be approximated by

$$\mathbf{A}_T(\mathbf{r}, t) = \mathbf{A}_T^{\text{ext}}(\mathbf{0}, t) + \frac{\mu_0}{4\pi} \int_{-\infty}^{\infty} \frac{\mathbf{J}_T(\mathbf{r}', t)}{|\mathbf{r} - \mathbf{r'}|} d^3 r' - \frac{\mu_0}{4\pi c} \frac{d}{dt} \int_{-\infty}^{\infty} \mathbf{J}_T(\mathbf{r}', t) d^3 r', \qquad (26.2)$$

assuming that the external field essentially is constant across the atom [with its nucleus at the origo, $\mathbf{r} = \mathbf{0}$].

Let us consider an atom in a state described by the one-electron wave function

$$\Psi(\mathbf{r}, t) = \sum_j a_j(t) \psi_j(\mathbf{r}), \qquad (26.3)$$

where $\psi_j(t)$ is the stationary-state eigenfunction of the field-unperturbed Hamiltonian. By neglecting the $\mathbf{A}$-dependent term in the associated current density, one obtains

$$\mathbf{J}(\mathbf{r}, t) = \sum_{\alpha, \beta} \rho_{\beta\alpha}(t) \mathbf{j}_{\beta\alpha}(\mathbf{r}), \qquad (26.4)$$

where $\rho_{\beta\alpha}(t) = a_\beta(t) a_\alpha^*(t)$ is the $(\beta\alpha)$-matrix element of the density matrix.

The spontaneous (spon) emission is related to the transverse vector potential

$$\mathbf{A}_T^{\text{spon}} = -\frac{\mu_0}{4\pi c} \frac{d}{dt} \int_{-\infty}^{\infty} \mathbf{J}_T(\mathbf{r}, t) d^3 r = -\frac{\mu_0}{6\pi c} \frac{d}{dt} \int_{-\infty}^{\infty} \mathbf{J}(\mathbf{r}, t) d^3 r \qquad (26.5)$$

and since

$$\int_{-\infty}^{\infty} \mathbf{j}_{\alpha\beta}(\mathbf{r}) d^3 r = i\omega_{\alpha\beta} \mathbf{P}_{\beta\alpha} \qquad (26.6)$$

where $\omega_{\alpha\beta} = (\varepsilon_\alpha - \varepsilon_\beta)/\hbar$ is the Bohr transitions frequency [from β to α], and (with $e > 0$)

$$\mathbf{P}_{\alpha\beta} = -e \int \psi_\beta^*(\mathbf{r}) \mathbf{r} \psi_\alpha(\mathbf{r}) d^3 r \qquad (26.7)$$

is the electric-dipole transition matrix element between the energy eigenstates α and β, one obtains

$$\mathbf{A}_T^{\text{spon}} = \frac{\mu_0}{6\pi c} \sum_{\alpha, \beta} \rho_{\beta\alpha}(t) \omega_{\beta\alpha}^2 \mathbf{P}_{\beta\alpha}. \qquad (26.8)$$

In order to reach Eq. (26.8) we have used the approximation $d\rho_{\beta\alpha}(t)/dt \approx -i\omega_{\beta\alpha}\rho_{\beta\alpha}(t)$.

The vector potential associated with the Lamb shift is technically a bit more complicated to derive [66], and here we cite only the final result. Hence,

$$\mathbf{A}_T^{\text{Lamb}}(\mathbf{r}, t) = \mu_0 \sum_{\alpha, \beta} \rho_{\beta\alpha}(t) \mathbf{I}_{\beta\alpha}^T(\mathbf{r}), \qquad (26.9)$$

where

$$\mathbf{I}_{\beta\alpha}^T(\mathbf{r}) = \int_{-\infty}^{\infty} q^{-2}\mathbf{j}_{\beta\alpha}^T(\mathbf{q})e^{i\mathbf{q}\cdot\mathbf{r}}\frac{d^3q}{(2\pi)^3}, \tag{26.10}$$

$\mathbf{j}_{\beta\alpha}^T(\mathbf{q})$ being the Fourier transform of the transverse part of $\mathbf{j}_{\beta\alpha}(\mathbf{r})$. Altogether, it appears that

$$\mathbf{A}_T(\mathbf{r},t) = \mathbf{A}_T^{\text{ext}}(\mathbf{0},t) + \mu_0 \sum_{\alpha,\beta} \rho_{\beta\alpha}(t)\left[\mathbf{I}_{\beta\alpha}^T(\mathbf{r}) + \frac{\omega_{\beta\alpha}^2}{6\pi c}\mathbf{P}_{\beta\alpha}\right]. \tag{26.11}$$

In order to complete the calculation of the spontaneous decay rate and the Lamb shift, one has to calculate $\rho_{\beta\alpha}(t)$ from the equation of motion for the density matrix, using the interaction Hamiltonian $(e/2m)(\hat{\mathbf{p}}\cdot\mathbf{A}_T + \mathbf{A}_T\cdot\hat{\mathbf{p}})$ [the small nonlinear term $(e^2/2m)\mathbf{A}_T^2$ being neglected].

26.2 BETHE'S THEORY: A BRIEF REVIEW

The interaction of a free electron and a quantized electromagnetic field gives a correction of the bare electron mass (here denoted by m_0). The total mass of the electron (denoted by m) is given by the bare mass plus a small contribution (Δm) from the electromagnetic field. Thus,

$$\frac{1}{m} = \frac{1}{m_0 + \Delta m} \approx \frac{1}{m_0} - \frac{\Delta m}{m_0^2}. \tag{26.12}$$

From second-order perturbation theory Bethe [24] found for a free electron $[E = \hbar c q]$

$$\Delta m = \frac{e^2}{3\pi^2\varepsilon_0\hbar c^3}\int_0^{m_0 c^2} dE = \frac{e^2 q_M}{3\pi^2\varepsilon_0 c^2}, \tag{26.13}$$

where $q_M = m_0 c/\hbar$ is the Compton wave number. Note that the integral in this non-relativistic calculation is cut off at $E = m_0 c^2$. Using the Pauli-Fierz transformation a comprehensive discussion of the tied photons' correction of the bare particle (electron) mass is undertaken in Chapter 27. The result in Eq. (26.13) reappears in Eq. (27.44).

Let us now consider the Hamiltonian operator for an electron in a potential $V_0(\mathbf{r})$ and with a bare mass m_0. The unobservable (unobs) Hamiltonian operator is

$$\hat{H}_0^{\text{unobs}} = \frac{\hat{\mathbf{p}}^2}{2m_0} + V_0(\mathbf{r}) \approx \frac{\hat{\mathbf{p}}^2}{2m} + V_0(\mathbf{r}) + \frac{\hat{\mathbf{p}}^2}{2m_0^2}\Delta m. \tag{26.14}$$

To $\hat{H}_0^{\text{unobs}}$ we now add the unobservable second-quantized interaction Hamiltonian in the Coulomb gauge, namely

$$\hat{H}_I^{\text{unobs}} = \frac{e}{m_0}\hat{\mathbf{A}}_T(\mathbf{r})\cdot\hat{\mathbf{p}} + \frac{e^2}{2m_0}\hat{\mathbf{A}}_T^2(\mathbf{r}), \tag{26.15}$$

with electron charge $-e$. The total Hamiltonian operator,

$$\hat{H} = \hat{H}_0^{\text{unobs}} + \hat{H}_I^{\text{unobs}} \approx \frac{\hat{\mathbf{p}}^2}{2m} + V_0(\mathbf{r}) + \hat{H}_I^{\text{unobs}} + \frac{\hat{\mathbf{p}}^2}{2m_0^2}\Delta m, \tag{26.16}$$

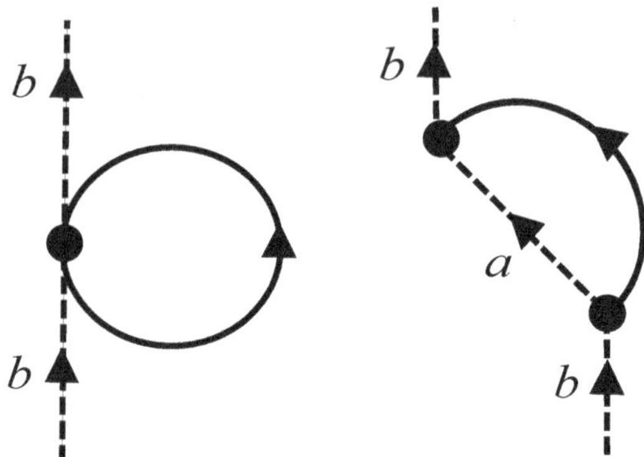

Figure 26.1 Feynman diagrams for the radiative corrections entering the Lamb shift calculation in atoms and mesoscopic rings. Diagram to the left: An electron (dashed line) in state b instantaneously emits and reabsorbs a photon (unbroken line). Diagram to the right: An electron initially in state b makes a transition to state a by emitting a photon, Subsequently the electron reabsorbs the photon and returns to state b.

Bethe then regards it as a sum of two *observable* parts: (*i*) A field-unperturbed part, $\hat{\mathbf{p}}^2/(2m) + V_0(\mathbf{r})$, with the observable mass m, and (*ii*) an interaction part, $\hat{H}_I^{\mathrm{unobs}} + [\hat{\mathbf{p}}^2/(2m_0^2)]\Delta m$.

Let $|b;0\rangle = |b\rangle \otimes |0\rangle$ describe an atomic state $|b\rangle$ and $|0\rangle$ the global vacuum state of the field [both normalized, i.e., $\langle b|b\rangle = 1$ and $\langle 0|0\rangle = 1$]. A state $|a;\mathbf{q}\varepsilon\rangle = |a\rangle \otimes |\mathbf{q}\varepsilon\rangle$, with the atom in state $|a\rangle$ and a photon field $|\mathbf{q}\varepsilon\rangle$ with just one photon of wave vector $\mathbf{q}$ and polarization index ε, serves as an intermediate state for the second-order perturbative calculation of the Lamb shift in state $|b\rangle$. The relevant Feynman diagrams for the calculation are shown in Fig. 26.1. The diagram (*a*) gives the same shift of all atomic energy levels, viz., $[e^2/(2m_0)]\langle 0|\hat{\mathbf{A}}_T^2|0\rangle$, and is therefore not important for the Lamb shift calculation. From diagram (*b*) one obtains the following energy shift of state $|b\rangle$:

$$W_b = \sum_{a,\varepsilon} \int \frac{\left|\langle a;\mathbf{q}\varepsilon|\frac{e}{m_0}\hat{\mathbf{A}}_T(0)\cdot\hat{\mathbf{p}}|0;b\rangle\right|^2}{E_b - E_a - \hbar\omega}d^3q + \frac{\Delta m}{2m_0^2}\langle b|\hat{\mathbf{p}}^2|b\rangle, \qquad (26.17)$$

where we have used the long-wavelength approximation $\hat{\mathbf{A}}_T(\mathbf{r}) \approx \hat{\mathbf{A}}_T(0)$. Using the closure relation on the electron states, namely $\sum_a |a\rangle\langle a| = \mathbb{1}$, it follows after some straightforward calculations that

$$\begin{aligned}
W_b &= \frac{e^2}{6\pi^2\varepsilon_0\hbar c^3 m_0^2}\sum_a |\mathbf{p}_{ba}|^2 \int_0^{m_0c^2} \frac{E\,dE}{E_b - E_a - E}dE + \frac{\Delta m}{2m_0^2}\sum_a |\mathbf{p}_{ba}|^2 \\
&= \frac{e^2}{6\pi^2\varepsilon_0\hbar c^3 m_0^2}\sum_a |\mathbf{p}_{ba}|^2 (E_a - E_b)\ln\left|\frac{E_b - E_a - m_0c^2}{E_b - E_a}\right|,
\end{aligned} \qquad (26.18)$$

with the matrix element $\mathbf{p}_{ba} = \langle b| \hat{\mathbf{p}} |a\rangle$. Since the energy difference between the two electron states a and b satisfies $|E_a - E_b| << m_0 c^2$, the Lamb shift for an electron in state b finally according to Bethe becomes

$$W_b = \frac{e^2}{6\pi^2 \varepsilon_0 \hbar c^3 m_0^2} \sum_a |\mathbf{p}_{ba}|^2 (E_a - E_b) \ln \frac{m_0 c^2}{|E_b - E_a|}. \tag{26.19}$$

It is convenient to write the result in Eq. (26.19) in terms of an average

$$K_b = \frac{\sum_a |\mathbf{p}_{ba}|^2 (E_a - E_b) \ln \frac{m_0 c^2}{|E_b - E_a|}}{\sum_a |\mathbf{p}_{ba}|^2 (E_a - E_b)} \tag{26.20}$$

of $\ln[m_0 c^2/|E_b - E_a|]$ with weight function $|\mathbf{p}_{ba}|^2 (E_a - E_b)$, giving

$$W_b = \frac{e^2 K_b}{6\pi^2 \varepsilon_0 \hbar c^3 m_0^2} \sum_a |\mathbf{p}_{ba}|^2 (E_a - E_b). \tag{26.21}$$

26.3   LAMB-SHIFT THEORY FOR A NANO-SIZED CONDUCTING RING

By means of the commutator relation $[\hat{p}_x, [\hat{p}_x, \hat{H}_0^{\text{obs}}]] = -\hbar^2 \partial^2 V_0(\mathbf{r})/\partial x^2$, and similar ones with $\hat{p}_y$ and $\hat{p}_z$, it can be shown [108] that

$$\sum_a |\mathbf{p}_{ba}|^2 (E_a - E_b) = \frac{\hbar^2}{2} \langle b| \nabla^2 V_0(\mathbf{r}), |b\rangle . \tag{26.22}$$

where from the Poisson equation

$$\nabla^2 V_0(\mathbf{r}) = -e\nabla^2 \phi(\mathbf{r}) = \frac{e}{\varepsilon_0}\rho(\mathbf{r}), \tag{26.23}$$

$\rho(\mathbf{r})$ being the charge density. By combining Eqs. (26.21)–(26.23) one obtains

$$W_b = \frac{e^3 \hbar}{12\pi^2 \varepsilon_0^2 c^3 m_0^2} K_b \langle b| \rho(\mathbf{r} |b\rangle . \tag{26.24}$$

The formula in Eq. (26.24) is an adequate starting point for calculating Lamb shifts in mesoscopic conducting rings, as we shall indicate below.

In passing it should be noted that Eq. (26.24) was the starting point for Bethe's calculation of the $2s_{1/2}$, Lamb shift ($\sim$ 1040 MHz) in the hydrogen atom, where $V_0(r) = -[e^2/(4\pi\varepsilon_0 r)]$.

In Section 22.3 the space- and spin-electrodynamics in a mesoscopic circular ring were studied starting from an analysis of the stationary states and eigenenergies of the ring (Subsection 22.3.1). For the ring the charge density consists of the sum of the contributions from the positively charged ion cores, $\rho_+(\mathbf{r})$ [which we already have assumed to be uniform in Section 22.3], and the negatively charged electrons $\rho_-(\mathbf{r})$. The contribution to the negative charge density comes from all the conduction electrons in the ring except for the one in the state, $|b\rangle$, for which the Lamb shift is to be calculated Hence, in terms of the wavefunctions, $\psi_a(\mathbf{r}) = \langle \mathbf{r}|a\rangle$, one has

$$\rho(\mathbf{r}) = \rho_+ + 2\sum_{a \neq b}(-e)f_a|\psi_a(\mathbf{r})|^2, \qquad (26.25)$$

where $f_a = \{\exp[(E_a - \mu)/(k_B T)] + 1\}^{-1}$ is the Fermi-Dirac distribution in usual notation. The degeneration factor 2 originates in the electron spin summation. The chemical potential (μ) can be found from the relation $N/2 = \sum_a f_a$, where $N = n_e V$ is the number of ring electrons, n_e being the conduction electron density and $V = \pi L(d^2 + 2r_0 d)$ is the ring volume. If one neglects the (often small) curvature term, $r^{-1}dR(r)/dr$ in Eq. (22.43) the radial wave function is given by

$$R(r) = \frac{2}{\sqrt{d(d + 2r_0)}} \sin\left[\frac{\pi n}{d}(r - r_0 - d)\right], \quad n = 1, 2, \ldots \qquad (26.26)$$

in the notation of Section 22.3. With the omission of the curvature term, the energy eigenvalues [Eq. (22.49)] are

$$\varepsilon_{n,l,m} = \frac{\hbar^2}{2m_0}\left[\left(\frac{\pi n}{d}\right)^2 + \left(\frac{\pi l}{L}\right)^2 + \left(\frac{m}{r_0}\right)^2\right]. \qquad (26.27)$$

The Lamb shift of an electron in state $|b\rangle$ hence is given by

$$W_b = \frac{e^3 \hbar K_b}{12\pi^2 \varepsilon_0^2 c^3 m_0^2} \langle b| \rho_+ + 2\sum_{a \neq b}(-e)f_a|\psi_a(\mathbf{r})|^2 |b\rangle. \qquad (26.28)$$

The Lamb shift phenomenon manifests itself through a change $\Delta W = W_f - W_i$ in the pure electron transition energy between the initial $i \equiv [n_i, l_i, m_i]$ and final $f = [n_f, l_f, m_f]$ states, a change which may be measured via resonance absorption of light $[|\Delta W| = |W_f - W_i|]$, e.g.

With the abbreviation

$$C_b = \frac{e^3 \hbar}{6\pi^2 \varepsilon_0^2 c^3 m_0^2} K_b, \qquad (26.29)$$

$$\Delta W = h\Delta\nu = C_i \sum_{a \neq i} f_a \int_{-\infty}^{\infty} |\psi_a(\mathbf{r})|^2|\psi_i(\mathbf{r})|^2 d^3 r - C_f \sum_{a \neq f} f_a \int_{-\infty}^{\infty} |\psi_a(\mathbf{r})|^2|\psi_f(\mathbf{r})|^2 d^3 r. \qquad (26.30)$$

With the definition

$$I_{j,j'} = 2 + \delta_{jj'} \qquad (26.31)$$

we obtain

$$\int_{-\infty}^{\infty} |\psi_{n,l,m}(\mathbf{r})|^2|\psi_{n',l',m'}(\mathbf{r})|^2 d^3 r = \frac{I_{l,l'} I_{n,n'}}{4V}. \qquad (26.32)$$

Assume now that $K_i \approx K_f = K$, then

$$\frac{\Delta \nu}{K} = \frac{C_0}{V}\left[\sum_{n,l,m} f_{n,l,m}(I_{l,l_i} I_{n,n_i} - I_{l,l_f} I_{n,n_f}) - 9(f_{n_i,l_i,m_i} - f_{n_f,l_f,m_f})\right], \qquad (26.33)$$

with

$$C_0 = \frac{e^4}{48\pi^3 \varepsilon_0^2 c^3 m_0^2}. \qquad (26.34)$$

In [196], a number of numerical results based on Eq. (26.33) were carried out for mesoscopic GaAs and Al rings.

CHAPTER 27

Particle Dressed by a Cloud of Transverse Photons

27.1   TRANSVERSE FIELD TIED TO A CLASSICAL PARTICLE

27.1.1   An unspoken lesson of classical electrodynamics

Let us consider a charged (charge: q) classical particle (P) in uniform motion with velocity $\mathbf{v}_q$. The accompanying field can conveniently be derived from the retarded scalar (Φ_P) and vector ($\mathbf{A}_P$) potentials. Thus, if $\rho(\mathbf{r}',t')$ and $\mathbf{j}(\mathbf{r}',t')$ denoted the charge and current densities associated with the particle classical electrodynamics gives the general result

$$\Phi_P(\mathbf{r},t) = \frac{1}{4\pi\varepsilon_0} \int_{-\infty}^{\infty} \frac{\rho\left(\mathbf{r}',t - \frac{|\mathbf{r}-\mathbf{r}'|}{c}\right)}{|\mathbf{r} - \mathbf{r}'|} d^3 r' \tag{27.1}$$

$$\mathbf{A}_P(\mathbf{r},t) = \frac{\mu_0}{4\pi} \int_{-\infty}^{\infty} \frac{\mathbf{j}\left(\mathbf{r}',t - \frac{|\mathbf{r}-\mathbf{r}'|}{c}\right)}{|\mathbf{r} - \mathbf{r}'|} d^3 r', \tag{27.2}$$

for the potentials at the space-time point $(\mathbf{r},t)$. The electric field then is obtained from

$$\mathbf{E}(\mathbf{r},t) = -\frac{\partial \mathbf{A}_P(\mathbf{r},t)}{\partial t} - \boldsymbol{\nabla}\Phi_P(\mathbf{r},t). \tag{27.3}$$

At first sight, it appears that the electric field originating in the source point $(\mathbf{r}',t')$ arrives at the point of observation $(\mathbf{r},t)$ with a time delay given by $|\mathbf{r} - \mathbf{r}'|/c$. Accordingly, the electrodynamic interaction over finite distance, i.e., for $|\mathbf{r} - \mathbf{r}'| \neq 0$ always appears retarded as dictated by the vacuum speed of light (c). This "obvious" conclusion is the wisdom reached in all textbooks on classical electrodynamics. But note, there is an unspoken condition usually overlooked: In order to obtain the electric field, $\mathbf{E}(\mathbf{r},t)$, one has to know *both* the current density $\mathbf{j}(\mathbf{r}',t - |\mathbf{r} - \mathbf{r}'|/c)$ and the charge density, $\rho(\mathbf{r}',t - |\mathbf{r} - \mathbf{r}'|/c)$, at $(\mathbf{r}',t')$. Since the charge and current densities are not independent, but related via the equation of continuity,

$$\boldsymbol{\nabla} \cdot \mathbf{j}_L(\mathbf{r},t) + \frac{\partial}{\partial t}\rho(\mathbf{r},t) = 0, \tag{27.4}$$

DOI: 10.1201/9781003029458-27

where $\mathbf{j}_L(\mathbf{r}, t)$ is the longitudinal part of $\mathbf{j}(\mathbf{r}, t)$, the current density must be known in the neighborhood of $(\mathbf{r}, t)$ in order to determine the charge density at $(\mathbf{r}, t)$. To clarify the consequences of this circumstance, we transform Eq. (27.3) into the space-frequency domain. With the help of Eqs. (27.1) and (27.2) one obtains

$$\mathbf{E}(\mathbf{r}; \omega) = \frac{i\omega}{\varepsilon_0 c^2} \int_{-\infty}^{\infty} g(|\mathbf{r} - \mathbf{r}'|; \omega)\mathbf{j}(\mathbf{r}'; \omega)d^3r' - \frac{1}{\varepsilon_0}\nabla \int_{-\infty}^{\infty} g(|\mathbf{r} - \mathbf{r}'|; \omega)\rho(\mathbf{r}'; \omega)d^3r',$$

$$(27.5)$$

where

$$g(|\mathbf{r} - \mathbf{r}'|; \omega) = \frac{\exp\left[i\frac{\omega}{c}|\mathbf{r} - \mathbf{r}'|\right]}{4\pi|\mathbf{r} - \mathbf{r}'|} \tag{27.6}$$

is the retarded Huygens scalar propagator. Use of the equation of continuity, $\rho(\mathbf{r}; \omega) = \nabla \cdot \mathbf{j}_L(\mathbf{r}; \omega)/(i\omega)$, the relation $\nabla g(|\mathbf{r} - \mathbf{r}'|; \omega) = -\nabla' g(|\mathbf{r} - \mathbf{r}'|; \omega)$ followed by a partial integration over a domain enclosing the $\mathbf{j}_L$-domain (mathematically the entire space), enables one to rewrite the gradient of the last integral in Eq. (27.5) in the form

$$\nabla \int_{-\infty}^{\infty} g(|\mathbf{r} - \mathbf{r}'|; \omega)\rho(\mathbf{r}'; \omega)d^3r' = \frac{i}{\omega} \int_{-\infty}^{\infty} \nabla' g(|\mathbf{r} - \mathbf{r}'|; \omega)\nabla' \cdot \mathbf{j}_L(\mathbf{r}'; \omega)d^3r'$$

$$= \frac{1}{i\omega} \int_{-\infty}^{\infty} [\nabla' \cdot \nabla' g(|\mathbf{r} - \mathbf{r}'|; \omega)]\mathbf{j}_L(\mathbf{r}'; \omega)d^3r'. \tag{27.7}$$

By utilizing the differential equation satisfied by the Huygens propagator [Eq. (6.2)] one finally obtains

$$\nabla \int_{-\infty}^{\infty} g(|\mathbf{r} - \mathbf{r}'|; \omega)\rho(\mathbf{r}'; \omega)d^3r' = \frac{i}{\omega}\mathbf{j}_L(\mathbf{r}; \omega) - \left(\frac{\omega}{c}\right)^2 \int_{-\infty}^{\infty} g(|\mathbf{r} - \mathbf{r}'|; \omega)\mathbf{j}_L(\mathbf{r}'; \omega)d^3r'.$$

$$(27.8)$$

Because $\mathbf{j} = \mathbf{j}_T + \mathbf{j}_L$, a combination of Eqs. (27.5) and (27.8) gives the following result for the electric field:

$$\mathbf{E}(\mathbf{r}; \omega) = i\mu_0\omega \int_{-\infty}^{\infty} g(|\mathbf{r} - \mathbf{r}'|; \omega)\mathbf{j}_T(\mathbf{r}'; \omega)d^3r' + \frac{1}{i\varepsilon_0\omega}\mathbf{j}_L(\mathbf{r}; \omega). \tag{27.9}$$

By rewriting the classical result for the electric field [Eqs. (27.1)–(27.3)] in *propagator* form, the physical structure of the field is clear. From each space-time point $(\mathbf{r}', t')$ in the *transverse current density domain* retarded wavelets reach the point of observation, $(\mathbf{r}, t)$. But, note that we have found a "hidden" longitudinal self-field contribution, $\mathbf{j}_L(\mathbf{r}; \omega)/(i\varepsilon_0\omega)$, *nonvanishing outside* $\mathbf{j}(\mathbf{r}, t)$. This contribution is not subjected to Einsteinian retardation, but links, as we shall realize later on, to (*i*) the "transverse photon tied to the particle"-concept and (*ii*) the spatial localization problem for the transverse photon. The result obtained in Eq. (27.9) was derived originally in my paper entitled *Propagator picture of the spatial confinement of quantized light emitted from an atom* [147].

27.1.2 Expansion of Lienard-Wiechert potentials

For a moving point particle of charge q one can derive the well-known Lienard-Wiechert potentials [118, 157], viz.

$$\Phi_P(\mathbf{r}, t) = \frac{1}{4\pi\varepsilon_0} \left[\frac{q}{R(1 - \boldsymbol{\beta} \cdot \hat{\mathbf{R}})} \right]_{\text{ret}} \tag{27.10}$$

$$\mathbf{A}_P(\mathbf{r}, t) = \frac{\mu_0}{4\pi} \left[\frac{qc\boldsymbol{\beta}}{R(1 - \boldsymbol{\beta} \cdot \hat{\mathbf{R}})} \right]_{\text{ret}} \tag{27.11}$$

where $R = |R| = |\mathbf{r} - \mathbf{r}'|$, $\hat{\mathbf{R}} = \mathbf{R}/R$, and

$$\boldsymbol{\beta} = \frac{\mathbf{v}_q}{c}. \tag{27.12}$$

In Eqs. (27.10) and (27.11) use have been made of the compact notation

$$\mathcal{F}(t_r \equiv t - |\mathbf{r} - \mathbf{r}'|/c) \equiv [\mathcal{F}]_{\text{ret}}. \tag{27.13}$$

These potentials were derived by Lienard in 1898 and independently by Wiechert two years later, as mentioned by Whittaker [269] and Sommerfeld [243].

In the non-relativistic regime, $|\boldsymbol{\beta}| << 1$, Eqs. (27.10) and (27.11) reduce to the well-known quasistatic forms

$$\Phi_P(\mathbf{r}, t) = \frac{q}{4\pi\varepsilon_0 |\mathbf{r} - \mathbf{r}_q(t)|} + \dots, \tag{27.14}$$

$$\mathbf{A}_P(\mathbf{r}, t) = \frac{q\mathbf{v}_q}{4\pi\varepsilon_0 c^2 |\mathbf{r} - \mathbf{r}_q(t)|} + \dots, \tag{27.15}$$

where $\mathbf{r}_q(t)$ is the position of the point particle at the time t. The $+\dots$ indicates the presence of higher-order terms in $\mathbf{v}_q = \dot{\mathbf{r}}_q$. For a motionless particle Eq.(27.14) is the Coulomb potential and $\mathbf{A}_P$ [Eq. (27.15)] is zero. The (first) term in Eq. (27.15) is the vector potential ordinarily used to describe the magnetic effects in the quasistatic limit (electrokinetics).

27.2 CLASSICAL PAULI-FIERZ TRANSFORMATION

In the Pauli-Fierz description [204] of the concept "transverse field tied to a classical particle", a transformation which neglects the higher-order terms in $\mathbf{v}_q$ is used to investigate the physical content and limitation of the concept. In Section 27.3 we shall extend the analysis to the quantum theory.

The potentials in Eqs. (27.14) and (27.15) are derived from the Lorenz gauge, so $\mathbf{A}_P(\mathbf{r}, t)$ has a nonzero divergence. The transverse part of $\mathbf{A}_P(\mathbf{r}, t)$, $\mathbf{A}_P^T(\mathbf{r}, t)$, is gauge invariant, and $\mathbf{A}_P^T(\mathbf{r}, t)$ together with the Coulomb potential describe the same electric field in the Coulomb gauge. The transverse part of $\mathbf{A}_P(\mathbf{r}, t)$ can be obtained by a spatial Fourier transform of $|\mathbf{r} - \mathbf{r}_q(t)|^{-1}$ to the wave-vector ($\mathbf{q}$) domain, and then retaining for each $\mathbf{q}$ only the projection (algebraic operation) of the potential onto the plane perpendicular to $\mathbf{q}$. One thus obtains in the continuum description

$$\mathbf{A}_P^T(\mathbf{r}, t) = \frac{q}{\varepsilon_0 c^2} \int_{-\infty}^{\infty} \frac{1}{q^2} (\mathbf{U} - \boldsymbol{\kappa}\boldsymbol{\kappa}) \cdot \mathbf{v_q} e^{i\mathbf{q}\cdot[\mathbf{r}-\mathbf{r}_q(t)]} \frac{d^3 q}{(2\pi)^3}, \qquad (27.16)$$

and with $\mathbf{v_q} = \mathbf{p}_q/m_q \equiv \mathbf{p}/m$, and [Eq. (4.19)]

$$\mathbf{U} - \boldsymbol{\kappa}\boldsymbol{\kappa} = \sum_i \boldsymbol{\varepsilon}_i \boldsymbol{\varepsilon}_i, \qquad (27.17)$$

[real polarization unit vectors chosen $\boldsymbol{\varepsilon}_i = \boldsymbol{\varepsilon}_i^*$], the transverse vector potential becomes [$\omega = cq$]

$$\mathbf{A}_P^T(\mathbf{r}, t) = \left[\frac{\hbar}{2\varepsilon_0 (2\pi)^3}\right]^{1/2} \sum_{i=1,2} \int_{-\infty}^{\infty} \omega^{-1/2} \left[\beta_i(\mathbf{q})\boldsymbol{\varepsilon}_i e^{i\mathbf{q}\cdot\mathbf{r}} + c.c.\right] d^3 q, \qquad (27.18)$$

where

$$\beta_i(\mathbf{q}) = \frac{q}{m\omega} \left[2\varepsilon_0 \hbar\omega(2\pi)^3\right]^{-1/2} \boldsymbol{\varepsilon}_i \cdot \mathbf{p} e^{-i\mathbf{q}\cdot\mathbf{r}_q(t)}. \qquad (27.19)$$

In the absence of an external (prescribed) perturbation, the classical field plus the particle momentum characterized by $\beta_i(\mathbf{q})$ [$i = 1, 2$] and $\mathbf{p}$, are stationary quantities. The attached field $\mathbf{A}_P^T$ persists forever.

27.3 TRANSVERSE FIELD TIED TO A QUANTUM PARTICLE

27.3.1 Introductory remarks

The extension of the Pauli-Fierz theory [204] to the quantum regime requires a certain shift of paradigm. The classical particle momentum $\mathbf{p}$ is replaced by the kinematic momentum $\boldsymbol{\pi} = \mathbf{p} - q\mathbf{A}(\mathbf{r}_q)$, but since $q\mathbf{A}(\mathbf{r}_q)$ gives a second-order (q^2) contribution in the formalism, this part is neglected in the Pauli-Fierz approximation. Although one attains a longitudinal electric self-field

$$\mathbf{E}_L(\mathbf{r}; \omega) = (i\varepsilon_0\omega)^{-1} \mathbf{j}_L(\mathbf{r}; \omega) \qquad (27.20)$$

in the classical approach, longitudinal fields are eliminated in favour of the particle position coordinate in quantum physics. Since, from Eq. (27.9), the transverse electric field is given by

$$\mathbf{E}_T(\mathbf{r}; \omega) = i\mu_0\omega \int_{-\infty}^{\infty} g(|\mathbf{r} - \mathbf{r}'|; \omega) \mathbf{j}_T(\mathbf{r}'; \omega) d^3 r', \qquad (27.21)$$

one must seek the "transverse field tied to the particle" in the relation between $\mathbf{j}_T(\mathbf{r}; \omega)$ and $\mathbf{j}(\mathbf{r}'; \omega)$, as we have indicated in Chapter 3 { see also my books on the Quantum Theory of Near-Field Electrodynamics [157], and LIGHT – the Physics of the Photon [158] }.

The Pauli-Fierz theory originally was introduced to study the emission of low-frequency radiation during electric collisions, and is essentially limited to long-wavelength modes. Thus, in the particle (or atomic) case one assumes that only q-modes which are essentially constant in space over the domain of a (strongly) localized particle (atom) are considered. For these modes one takes $\mathbf{q} \cdot \mathbf{r}_q \approx 0$ in Eq. (27.19), giving as starting point for the extension to quantum physics

$$\beta_i(\mathbf{q}) = \frac{q}{m\omega} \left[2\varepsilon_0 \hbar\omega(2\pi)^3 \right]^{-1/2} \varepsilon_i \cdot \mathbf{p}, \quad \text{LW} \tag{27.22}$$

and

$$\mathbf{A}_P^T(\mathbf{r}) = \frac{q}{m\varepsilon_0 c^2} \sum_{i=1,2} \int_{\text{LW}} q^{-2}\varepsilon_i\varepsilon_i \cdot \mathbf{p}\, e^{i\mathbf{q}\cdot\mathbf{r}} \frac{d^3q}{(2\pi)^3}, \tag{27.23}$$

where LW means that the integral is over long wavelength (LW) modes, only.

In a prescribed external (ext) scalar potential, $V_{\text{ext}}(\mathbf{r}_q)$, the LW Coulomb-gauge Hamiltonian operator has the form

$$\hat{H} = \frac{1}{2m} \left[\hat{\mathbf{p}} - q\hat{\mathbf{A}}(0) \right]^2 + \varepsilon_{\text{SE}}^{\text{LW}} + qV_{\text{ext}}(\mathbf{r}_q) + \int_{\text{LW}} \sum_{i=1,2} \hbar\omega \left[\hat{a}_i^\dagger(\mathbf{q})\hat{a}_i(\mathbf{q}) + \frac{1}{2} \right] d^3q, \tag{27.24}$$

where

$$\varepsilon_{\text{SE}}^{\text{LW}} = \frac{q^2}{2\varepsilon_0(2\pi)^3} \int_{\text{LV}} \frac{d^3q}{q^2} < \infty \tag{27.25}$$

is the Coulomb self-energy (SE) of the particle. Limiting ourselves to long-wavelength modes, this energy is finite. With all modes included $[\int_{\text{LW}} \rightarrow \int_{-\infty}^{\infty}]$ the Coulomb self-energy is infinite. The operators $\hat{a}_i(\mathbf{q})$ and $\hat{a}_i^\dagger(\mathbf{q})$ characterize the *total* field, as does $\hat{\mathbf{A}}(0)$. The dynamical momentum variable is represented by the operator $-i\hbar\boldsymbol{\nabla}_{\mathbf{r}_q}$ in the quantum theory.

27.3.2 Pauli-Fierz quantum transformation

In a quantum setting $\mathbf{p} \Rightarrow \hat{\mathbf{p}} = i\hbar\boldsymbol{\nabla}_{\mathbf{r}_q}$, and $\beta_i(\mathbf{q}) \Rightarrow \hat{\beta}_i(\mathbf{q})$ therefore is an operator in particle space. The goal of our mission is to seek a transformation $(\hat{T})$ which enables one to subtract the transverse quantum field tied to the particle from the total field. Thus, we must determine $\hat{T}$ in such a manner that

$$\hat{T}\hat{a}_i(\mathbf{q})\hat{T}^\dagger = \hat{a}_i(\mathbf{q}) + \hat{\beta}_i(\mathbf{q}), \tag{27.26}$$

where $\hat{\beta}_i$ is the particle operator originating from Eq. (27.22). The transformation in Eq. (27.26) defines an operator translation, and the $\hat{T}$ giving such a translation is well known, e.g. in quantum optics [63, 64]. Thus,

$$\hat{T} = \exp\left\{ \int_{\text{LW}} \sum_{i=1,2} \left[\hat{\beta}_i^\dagger(\mathbf{q})\hat{a}_i(\mathbf{q}) - \hat{\beta}_i(\mathbf{q})\hat{a}_i^\dagger(\mathbf{q}) \right] d^3q \right\}. \tag{27.27}$$

In preparation for the physical discussion, to be given in Subsections 27.3.3 and 27.3.4, it is useful to introduce the *quantum field operator*

$$\hat{\mathbf{Z}}(\mathbf{r}) \equiv \left[\frac{\hbar}{2\varepsilon_0(2\pi)^3}\right]^{1/2} \sum_{i=1,2} \int_{\mathrm{LW}} \omega^{-1/2}\left[\frac{1}{i\omega}\boldsymbol{\varepsilon}_i(\boldsymbol{\kappa})\hat{a}_i(\mathbf{q})e^{i\mathbf{q}\cdot\mathbf{r}} - \frac{1}{i\omega}\boldsymbol{\varepsilon}_i(\boldsymbol{\kappa})\hat{a}_i^\dagger(\mathbf{q})e^{-i\mathbf{q}\cdot\mathbf{r}}\right]d^3q,$$

$$(27.28)$$

and thus write the unitary transformation in the following form:

$$\hat{T} = \exp\left[\frac{i}{\hbar}\frac{q}{m}\hat{\mathbf{p}}\cdot\hat{\mathbf{Z}}(\mathbf{0})\right],\qquad(27.29)$$

remembering the LW expression for $\beta_i(\mathbf{q})$ [Eq. (27.22)].

In the Coulomb gauge, the total transverse vector potential and the transverse electric field are represented by the operators

$$\hat{\mathbf{A}}_T(\mathbf{r}) = \left[\frac{\hbar}{2\varepsilon_0(2\pi)^3}\right]^{1/2} \sum_{i=1,2} \int_{\mathrm{LW}} \omega^{-1/2}\left[\hat{a}_i(\mathbf{q})\boldsymbol{\varepsilon}_i e^{i\mathbf{q}\cdot\mathbf{r}} + \hat{a}_i^\dagger(\mathbf{q})\boldsymbol{\varepsilon}_i e^{-i\mathbf{q}\cdot\mathbf{r}}\right]d^3q, \quad(27.30)$$

$$\hat{\mathbf{E}}_T(\mathbf{r}) = \left[\frac{\hbar}{2\varepsilon_0(2\pi)^3}\right]^{1/2} \sum_{i=1,2} \int_{\mathrm{LW}} i\omega^{1/2}\left[\hat{a}_i(\mathbf{q})\boldsymbol{\varepsilon}_i e^{i\mathbf{q}\cdot\mathbf{r}} - \hat{a}_i^\dagger(\mathbf{q})\boldsymbol{\varepsilon}_i e^{-i\mathbf{q}\cdot\mathbf{r}}\right]d^3q, \quad(27.31)$$

for real polarization unit vectors. Let us denote the Coulomb gauge picture by superscript (1). Thus,

$$\hat{\mathbf{A}}_T(\mathbf{r}) = \hat{\mathbf{A}}_T^{(1)}(\mathbf{r}),\qquad(27.32)$$

$$\hat{\mathbf{E}}_T(\mathbf{r}) = \hat{\mathbf{E}}_T^{(1)}(\mathbf{r}),\qquad(27.33)$$

The operators representing these quantities in the transformed representation, denoted by superscript (2), are

$$\hat{\mathbf{A}}_T^{(2)}(\mathbf{r}) = \hat{T}\hat{\mathbf{A}}_T^{(1)}\hat{T}^\dagger = \hat{\mathbf{A}}_T(\mathbf{r}) + \hat{\mathbf{A}}_P^T(\mathbf{r}),\qquad(27.34)$$

$$\hat{\mathbf{E}}_T^{(2)}(\mathbf{r}) = \hat{T}\hat{\mathbf{E}}_T^{(1)}\hat{T}^\dagger = \hat{\mathbf{E}}_T(\mathbf{r}),\qquad(27.35)$$

as the reader may convince herself of, utilizing Eqs. (27.19), (27.26) (27.30) and (27.35). The operators $\hat{\mathbf{A}}_T(\mathbf{r})$ and $\hat{\mathbf{E}}_T(\mathbf{r})$ appearing in Eqs. (27.32) and (27.33) are given by Eqs. (27.30) and (27.31). Because $\hat{\mathbf{A}}_P(\mathbf{r})$ depends only on $\hat{\mathbf{p}}$, which commutes with $\hat{T}$, $\hat{\mathbf{A}}_P^T(\mathbf{r})$ is invariant with respect to $\hat{T}$. Hence,

$$\hat{\mathbf{A}}_P^T(\mathbf{r}) = \hat{\mathbf{A}}_P^{T(1)}(\mathbf{r}) = \hat{\mathbf{A}}_P^{T(2)}(\mathbf{r}),\qquad(27.36)$$

represents the transverse field tied to the particle in both pictures. This is not surprising due to the fact that the transverse vector potential is gauge invariant. Neither is it surprising that the transverse electric field operator is the same in the two representations: The deviation of the electric field $\mathbf{E}_T(\mathbf{r},t) = -\partial\Phi_P(\mathbf{r},t)/\partial t$ [Eq. (27.15)] from the Coulomb field is of order $(|\mathbf{v}|/c)^2$ and the calculation of $\beta_i(\mathbf{q})$ takes into account only linear terms, $|\mathbf{v}|/c$.

We now turn our attention towards the transformation of the momentum ($\hat{\mathbf{p}}$) and position ($\hat{\mathbf{r}}$) operators. Since $\hat{T}$ only depends on $\hat{\mathbf{p}}$, $\hat{a}_i(\mathbf{q})$ and $\hat{a}_i^\dagger(\mathbf{q})$; see Eqs. (27.28) and (27.29), it is obvious that $\hat{\mathbf{p}} = -i\hbar\boldsymbol{\nabla}_{\mathbf{r}_q}$ represents the same physical quantity in both representations, i.e.,

$$\hat{\mathbf{p}} = \hat{\mathbf{p}}^{(1)} = \hat{\mathbf{p}}^{(2)}. \tag{27.37}$$

In the Coulomb gauge the operator $\hat{\mathbf{r}}$ is just a multiplication by $\mathbf{r}$, so that

$$\hat{\mathbf{r}}^{(1)} = \mathbf{r}. \tag{27.38}$$

With the help of the commutation relation between the particle position and momentum operators it readily follows that

$$\hat{\mathbf{r}}^{(2)} = \hat{T}\hat{\mathbf{r}}^{(1)}\hat{T}^\dagger = \mathbf{r} + \frac{q}{m}\hat{\mathbf{Z}}(\mathbf{0}). \tag{27.39}$$

The last term in Eq. (27.39) has the same form in both representations since $\hat{\mathbf{Z}}(\mathbf{0})$ commutes with itself. Hence

$$\hat{\boldsymbol{\eta}} \equiv \frac{q}{m}\hat{\mathbf{Z}}(\mathbf{0}) = \hat{\boldsymbol{\eta}}^{(1)} = \hat{\boldsymbol{\eta}}^{(2)}. \tag{27.40}$$

By now, we are prepared to discuss in a qualitative sense the physics appearing in the Pauli-Fierz theory.

27.3.3 New Hamiltonian and physical discussion of vibrational motion

The Hamiltonian operator in the Coulomb gauge $\hat{H}^{(1)} \equiv \hat{H}$, given in explicit form in Eq. (27.24), we now transform to the new representation. By means of Eqs. (27.32) [for $\mathbf{r} = \mathbf{0}$] (27.35), (27.37) and (27.26) [and its Hermitian conjugate] we obtain

$$\hat{H}^{(2)} = \hat{T}\mathbf{H}^{(1)}\hat{T}^\dagger = \frac{1}{2m}\left[\hat{\mathbf{p}} - q\hat{\mathbf{A}}_T(\mathbf{0}) - q\hat{\mathbf{A}}_P^T(\mathbf{0})\right]^2 + \varepsilon_{\mathrm{SE}}^{\mathrm{LW}} + qV_{\mathrm{ext}}(\mathbf{r} + \boldsymbol{\eta})$$

$$+ \int_{\mathrm{LW}} \sum_{i=1,2} \hbar\omega\left\{[\hat{a}_i^\dagger(\mathbf{q}) + \hat{\beta}_i^\dagger(\mathbf{q})][\hat{a}_i(\mathbf{q}) + \hat{\beta}_i(\mathbf{q})] + \frac{1}{2}\right\} d^3q. \tag{27.41}$$

The operator $\hat{\mathbf{A}}_T(\mathbf{0})$ thus describes in the new Picture the difference between the total transverse field and the transverse field tied to the particle, $\hat{\mathbf{A}}_P^T(\mathbf{0})$. In the long-wavelength regime, it thus has turned out to be possible to separate a field part tied to the particle from the total field. This conforms to Pauli's and Fierz's original motivation for the $\hat{T}$-transformation. In the new picture the radiation from the source particle does not involve the tied transverse photons, of course.

To lowest order in q, one may approximate $\hat{H}$ of Eq. (27.24) by its free-field part $\hat{H}_F$ [the last (integral) term of Eq. (27.24)]. The time derivate of $\hat{a}_i(\mathbf{q})$ $[\hat{a}_i^\dagger(\mathbf{q})]$ hence can be approximated by $-i\omega\hat{a}_i(\mathbf{q})$ $[i\omega\hat{a}_i^\dagger(\mathbf{q}]$. Remembering the difference between $\hat{\mathbf{A}}_T(\mathbf{r})$ [Eq. (27.30)] and $\hat{\mathbf{E}}_T(\mathbf{r})$ [Eq. (27.31)], one is led to the following (operator) equation of motion for $\hat{\boldsymbol{\eta}}$ [Eq. (27.40)]:

$$m\frac{d^2\hat{\boldsymbol{\eta}}}{dt^2} = q\hat{\mathbf{E}}_T(\mathbf{0}). \tag{27.42}$$

This equation clearly signals that $\hat{\boldsymbol{\eta}}$ represents the motion the particle would have *if* it were subjected alone to the transverse electric field at the position $\mathbf{r} = \mathbf{0}$. In view of Eq. (27.39), the (operator) $\hat{\mathbf{r}} = \hat{\mathbf{r}}^{2)} - \hat{\boldsymbol{\eta}}$, in a qualitative sense represents the average position about which the particle performs a vibrational motion given by Eq. (27.42).

27.3.4 Tied photons' correction of the bare particle mass

It is obvious that there is no sharp limit separating the low-frequency modes from the rest of the mode spectrum. For calculational purposes let us (somewhat arbitrarily) limit the LW spectrum to modes $|\mathbf{q}| < q_{\mathrm{M(ax)}}$. The integration in Eq. (27.23) can then be carried out without difficulty giving

$$q\hat{\mathbf{A}}_P^T = \frac{\delta m}{m}\hat{\mathbf{p}},\tag{27.43}$$

where the mass δm is

$$\delta m = \frac{q^2 q_M}{3\varepsilon_0 \pi^2 c^2}.\tag{27.44}$$

The first term is Eq. (27.41) now up to second order in q equals

$$\frac{1}{2m}\left[\hat{\mathbf{p}}\left(1 - \frac{\delta m}{m}\right) - q\hat{\mathbf{A}}_T(\mathbf{0})\right]^2 \sim \frac{\hat{\mathbf{p}}^2}{2m}\left(1 - \frac{2\delta m}{m}\right) - \frac{q}{m}\hat{\mathbf{p}}\cdot\hat{\mathbf{A}}_T(\mathbf{0}) + \frac{q^2}{2m}[\hat{\mathbf{A}}_T(\mathbf{0})]^2.\tag{27.45}$$

The last integral in Eq. (27.41) can be calculated using the expression for $\beta_i(\mathbf{q})$ [Eq. (27.22)]. As the reader may prove to herself one obtains

$$\int_{|\mathbf{q}|<q_M}\sum_{i=1,2}\hbar\omega\left\{[\hat{a}_i^\dagger(\mathbf{q}) + \hat{\beta}_i^\dagger(\mathbf{q})][\hat{a}_i(\mathbf{q}) + \hat{\beta}_i(\mathbf{q})] + \frac{1}{2}\right\}d^3q =$$

$$\int_{|q|<q_M}\sum_{i=1,2}\hbar\omega\left[\hat{a}_i^\dagger(\mathbf{q})\hat{a}_i(\mathbf{q}) + \frac{1}{2}\right]d^3q + \frac{q}{m}\hat{\mathbf{p}}\cdot\mathbf{A}_T(\mathbf{0}) + \frac{\hat{\mathbf{p}}^2}{2m}\frac{\delta m}{m},\tag{27.46}$$

When the results given in Eqs. (27.45) and (27.46) are added, one obtains to second order in q the following expression for the transformed Hamiltonian operator

$$\hat{H}^{(2)} = \frac{\hat{\mathbf{p}}^2}{2m}\left(1 - \frac{\delta m}{m}\right) + qV_{\mathrm{ext}}(\mathbf{r} + \boldsymbol{\eta}) + \varepsilon_{\mathrm{SE}}^{\mathrm{LW}} + \frac{q^2}{2m}\left[\hat{\mathbf{A}}_T(\mathbf{0})\right]^2$$

$$+ \int_{\mathrm{LW}}\sum_{i=1,2}\hbar\omega\left[\hat{a}_i^\dagger(\mathbf{q})\hat{a}_i(\mathbf{q}) + \frac{1}{2}\right].\tag{27.47}$$

From the Heisenberg equation of motion, viz.,

$$\dot{\mathbf{r}} = \frac{1}{i\hbar}\left[\mathbf{r}, \hat{H}^{(2)}\right] = \frac{\hat{\mathbf{p}}}{m}\left(1 - \frac{\delta m}{m}\right),\tag{27.48}$$

it thus follows that the tied photon cloud gives the particle an effective mass

$$m^* = m + \delta m. \tag{27.49}$$

The mass correction is a sum of two contributions which together are negative: (i) The positive energy of the transverse field tied to the particle [related to the $\hat{\beta}_i^\dagger \hat{\beta}_i$-term in the last integral of Eq. (27.41), and (ii) the negative interaction energy of the charged particle with its own transverse field; the term $-2\hat{\mathbf{p}} \cdot \hat{\mathbf{A}}_T(\mathbf{0})$ stemming from the term $-(q/m)\hat{\mathbf{p}} \cdot \hat{\mathbf{A}}_T(\mathbf{0}) = -(1/2m)[2\hat{\mathbf{p}} \cdot \hat{\mathbf{A}}_T(\mathbf{0})]$ in Eq. (27.45).

27.4 SPATIAL PHOTON CONFINEMENT VERSUS PHOTON PARTICLE DRESSING

27.4.1 The Power-Zienau-Woolley transformation

In preparation for our introduction of a quantum electrodynamic description with a new Hamiltonian particularly useful in the propagator picture, and, in turn, convenient for obtaining a link between the transverse photon cloud tied (at long wavelength) to a particle and the limitations on the spatial photon confinement in source emission processes, let us return to the standard Lagrangian in the Coulomb gauge. With reference to the Lagrangian approach to electrodynamics treated in Chapter 12, we start from the Coulomb gauge Lagrangian for a system of charged particles, numbered with subscripts $\alpha, \beta, ...$, and with the silent assumption that the charge distribution is well localized in space (around our coordinate origo) at all times. However, it should be remarked that it is possible to extend the description to include "unbound" particles which can be displaced far from the origin.

The standard Coulomb Lagrangian of the field + particle system, viz.,

$$L = \sum_\alpha \frac{1}{2} m_\alpha \dot{\mathbf{r}}_\alpha^2 - V_C + \int \mathcal{L}_C d^3 r, \tag{27.50}$$

with Coulomb energy

$$V_C = \sum_{\alpha > \beta} \frac{q_\alpha q_\beta}{4\pi\varepsilon_0 |\mathbf{r}_\alpha - \mathbf{r}_\beta|} + \sum_d \varepsilon_C^\alpha, \tag{27.51}$$

ε_C^α being the self-energy of particle α [infinite unless only long-wavelength modes are included; Eq. (27.25)], and field plus interaction Lagrangian density [Eq. (12.16)]

$$\mathcal{L}_C = \frac{\varepsilon_0}{2}\left[\dot{\mathbf{A}}_T^2 - c^2(\boldsymbol{\nabla} \times \mathbf{A}_T)^2\right] + \mathbf{j} \cdot \mathbf{A}_T. \tag{27.52}$$

The Power-Zienau-Woolley (PZW) Lagrangian [217, 279]

$$L^{\mathrm{PZW}} = L + \frac{dF}{dt} \tag{27.53}$$

is obtained with the choice

$$F = -\int \mathbf{P}(\mathbf{r}) \cdot \mathbf{A}_T(\mathbf{r}) d^3 r = -\int \mathbf{P}_T(\mathbf{r}) \cdot \mathbf{A}_T(\mathbf{r}) d^3 r, \tag{27.54}$$

$\mathbf{P}(\mathbf{r})$ being the microscopic polarization, and $\mathbf{P}_T(\mathbf{r})$ its divergence-free part. The expression for F makes it obvious to group dF/dt with the interaction (I) Lagrangian. Hence,

$$L^{\text{PZW}} = L_I + \frac{dF}{dt} = \int \left[(\mathbf{j}_T - \dot{\mathbf{P}}_T) \cdot \mathbf{A}_T - \mathbf{P}_T \cdot \dot{\mathbf{A}}_T \right] d^3r, \qquad (27.55)$$

since one may replace $\mathbf{j}$ by $\mathbf{j}_T$ in the integrand without changing the integral. By projecting out the transverse part of Eq. (11.17) it appears that

$$\mathbf{j}_T - \dot{\mathbf{P}}_T = \nabla \times \mathbf{M}, \qquad (27.56)$$

so that

$$L_I^{\text{PZW}} = \int (\nabla \times \mathbf{M}) \cdot \mathbf{A}_T d^3r - \int \mathbf{P}_T \cdot \dot{\mathbf{A}}_T d^3r. \qquad (27.57)$$

Remembering that the integrations are over the entire space, integration by parts of the first term, and use of $\nabla \times \mathbf{A}_T = \mathbf{B}$ and $\dot{\mathbf{A}}_T = -\mathbf{E}_T$, one obtains finally the following form of the PZW interaction Lagrangian:

$$L_I^{\text{PZW}} = \int \mathbf{P}_T \cdot \mathbf{E}_T d^3r + \int \mathbf{M} \cdot \mathbf{B} d^3r. \qquad (27.58)$$

The complete expression for the Power-Zienau-Woolley Lagrangian, i.e.,

$$L^{\text{PZW}} = \sum_\alpha \frac{1}{2} m_\alpha \dot{\mathbf{r}}_\alpha^2 - \sum_{\alpha > \beta} \frac{q_\alpha q_\beta}{4\pi\varepsilon_0 |\mathbf{r}_\alpha - \mathbf{r}_\beta|} - \sum_d \varepsilon_C^\alpha$$

$$+ \frac{\varepsilon_0}{2} \int [\mathbf{E}_T^2 - c^2 \mathbf{B}^2] d^3r + \int \mathbf{P}_T \cdot \mathbf{E}_T d^3r + \int \mathbf{M} \cdot \mathbf{B} d^3r \qquad (27.59)$$

thus has a form where the transverse vector potential has been eliminated in favour of the magnetic field and the transverse part of the electric field.

27.4.2   PZW conjugate momenta and Hamiltonian

As described in [63] it is possible to make the particular integral choices for the polarization field

$$\mathbf{P}(\mathbf{r}) = \sum_\alpha \int_0^1 q_\alpha \mathbf{r}_\alpha \delta(\mathbf{r} - u\mathbf{r}_\alpha) du, \qquad (27.60)$$

and the magnetization field,

$$\mathbf{M}(\mathbf{r}) = \sum_\alpha \int_0^1 q_\alpha \mathbf{r}_\alpha \times \dot{\mathbf{r}}_\alpha \delta(\mathbf{r} - u\mathbf{r}_\alpha) u \, du, \qquad (27.61)$$

quantities which of course only depend on position $(\mathbf{r}_\alpha)$ and velocity $(\dot{\mathbf{r}}_\alpha)$ of the particles. Since the magnetic interaction term thus is

$$\int \mathbf{M} \cdot \mathbf{B} d^3r = \sum_\alpha \int_0^1 q_\alpha (\mathbf{r}_\alpha \times \dot{\mathbf{r}}_\alpha) \cdot \mathbf{B}(\mathbf{r}) \delta(\mathbf{r} - \mathbf{r}_\alpha u) u \, du$$

$$= \sum_\alpha \dot{\mathbf{r}}_\alpha \cdot \left\{ \int_0^1 q_\alpha [\mathbf{B}(\mathbf{r}_\alpha u) \times \mathbf{r}_\alpha] u \, du \right\}, \qquad (27.62)$$

the conjugate particle momentum is in the PZW formulation given by

$$\mathbf{p}_\alpha^{\mathrm{PZW}} \equiv \frac{\partial L^{\mathrm{PSW}}}{\partial \dot{\mathbf{r}}_\alpha} = m_\alpha \dot{\mathbf{r}}_\alpha + \int_0^1 q_\alpha [\mathbf{B}(\mathbf{r}_\alpha u) \times \mathbf{r}_\alpha] u \, du. \tag{27.63}$$

For a well-localized particle system, the contribution from the magnetic effects $(\sim |\mathbf{r}_\alpha|)$ can be quite small. If the magnetic contribution is neglected, $\mathbf{p}_\alpha^{\mathrm{PZW}}$ is just the mechanical momentum $m_\alpha \dot{\mathbf{r}}_\alpha$. The conjugate field momentum

$$\boldsymbol{\pi}^{\mathrm{PZW}} \equiv \frac{\partial \mathcal{L}^{\mathrm{PZW}}}{\partial \dot{\mathbf{A}}_T(\mathbf{r})} = \varepsilon_0 \dot{\mathbf{A}}_T(\mathbf{r}) - \mathbf{P}_T(\mathbf{r}) = -\varepsilon_0 \mathbf{E}_T(\mathbf{r}) - \mathbf{P}_T(\mathbf{r}) \equiv -\mathbf{D}_T(\mathbf{r}), \tag{27.64}$$

where $\mathbf{D}_T(\mathbf{r})$ is the transverse part of the microscopic displacement field. In Chapter 28 we analyze the role of the spatial photon confinement and the transverse photon dressing of plasmariton quasiparticles, a type of collective quanta in a jellium plasma and there the displacement field turns out to play an important part in understanding these phenomena.

The PZW Hamiltonian [extended to the quantum level, $H \Rightarrow \hat{H}$, in the usual manner], defined by

$$H^{\mathrm{PZW}} = \sum_\alpha \dot{\mathbf{r}}_\alpha \cdot \mathbf{p}_\alpha + \int \dot{\mathbf{A}}_T(\mathbf{r}) \cdot \boldsymbol{\pi}^{\mathrm{PZW}}(\mathbf{r}) d^3 r - L^{\mathrm{PZW}}, \tag{27.65}$$

is easily calculated. It consist of three parts

$$H^{\mathrm{PZW}} = H_P^{\mathrm{PZW}} + H_F^{\mathrm{PZW}} + H_I^{\mathrm{PZW}}. \tag{27.66}$$

The particle part is given by

$$H_P^{\mathrm{PZW}} = \sum_\alpha \frac{1}{2m_\alpha}(\mathbf{P}_\alpha^{\mathrm{PZW}})^2 + V_C + \frac{1}{2\varepsilon_0} \int \mathbf{P}_T^2(\mathbf{r}) d^3 r, \tag{27.67}$$

and it differs from the corresponding one in the Coulomb gauge by a transverse self-field contribution to the particle energy:

$$\varepsilon_{\mathrm{SE}}^{\mathrm{PZW}} = \frac{1}{2\varepsilon_0} \int \mathbf{P}_T^2(\mathbf{r}) d^3 r. \tag{27.68}$$

This self-energy is somewhat similar to the self-energy given in Eq. (27.25). For a bound (confined) system of particles $\varepsilon_{\mathrm{SE}}^{\mathrm{PZW}}$ is finite. The field Hamiltonian,

$$H_F^{\mathrm{PZW}}(= H_F) = \frac{\varepsilon_0}{2} \int \left[\left(\frac{\boldsymbol{\pi}^{\mathrm{PZW}}(\mathbf{r})}{\varepsilon_0} \right)^2 + c^2 (\boldsymbol{\nabla} \times \mathbf{A}_T(\mathbf{r}))^2 \right] d^3 r, \tag{27.69}$$

has the *same form* as in the Coulomb gauge. The interesting PZW interaction Hamiltonian is given by

$$H_I^{\mathrm{PZW}} = \frac{1}{\varepsilon_0} \int \boldsymbol{\pi}^{\mathrm{PZW}}(\mathbf{r}) \cdot \mathbf{P}_T(\mathbf{r}) d^3 r + \sum_\alpha \frac{q_\alpha^2}{2m_\alpha}[\mathbf{A}_T(\mathbf{r}_\alpha)]^2 - \sum_\alpha \frac{q_\alpha}{m_\alpha} \mathbf{P}_\alpha^{\mathrm{PZW}} \cdot \mathbf{A}_T(\mathbf{r}_\alpha). \tag{27.70}$$

The PZW interaction Hamiltonian deviates in form from that in the Coulomb gauge by an interaction term between the transverse displacement field [cf. Eq. (27.64)] and the transverse polarization of the charge system. This fact indicates that there might be a quite close relation to the interaction energy of a charged particle with its own (tied) transverse field, discussed in the framework of the Pauli-Fierz theory [Subsection 27.3.3]. We shall enter deeper into this issue in our discussion of the propagator picture of quantum electrodynamics in Section 27.5.

Let us finish our discussion of the PZW theory with a remark on the quantum representation, With each change in the quantum representation is associated a unitary transformation $(\hat{T})$ of the form [?]

$$\hat{T} = \exp\left\{\frac{i}{\hbar}\hat{F}(\{\hat{\mathbf{r}}_\alpha\}; \hat{\mathbf{A}}_T; t)\right\}, \tag{27.71}$$

where $\hat{F}$ is an operator function of the set of generalized particle coordinates, $\{\mathbf{r}_\alpha\}$, and the transverse (gauge independent) vector potential. The PZW $F \Rightarrow \hat{F}$, Eq. (27.54) relates to a particular example for $\hat{T}$.

27.5 SELF-FIELD ELECTRODYNAMICS IN PROPAGATOR PICTURE

27.5.1 Retarded T-photon radiation. Tied T-photon

Let us return to the propagator description given in Eq. (27.9), and let us follow main aspects of my theory presented originally in 1998 [147]. For a detailed account the reader is referred to this paper, and its photon wave mechanical companion [149]. As has been discussed in previous chapters of the book, the longitudinal part of the microscopic electric field $\mathbf{E}_L(\mathbf{r}; \omega) = \mathbf{J}_L(\mathbf{r}; \omega)/(i\varepsilon_0\omega)$ is in a quantum mechanical setting eliminated in favor of the massive particle position coordinates, leaving us with a transverse field

$$\mathbf{E}_T(\mathbf{r}; \omega) = i\mu_0\omega \int_{-\infty}^{\infty} g(|\mathbf{r} - \mathbf{r}'|; \omega)\mathbf{j}_T(\mathbf{r}'; \omega)d^3r'. \tag{27.72}$$

The picture of the emission process offered by Eq. (27.72), I have called the "photon eye view", because a T-photon "sees" the transverse current density domain as source region. The "electron eye view" is obtained by replacing $\mathbf{j}_T(\mathbf{r}; \omega)$ with $\mathbf{j}(\mathbf{r}; \omega)$. A spatially nonlocal relation connects the two; see Chapter 3.

Following the analysis given in a microscopic framework of "Principle Volume and Self-field Dyadics" in my Springer book [157], one may conclude that Eq. (27.72) in every respect is analogous to the following integral equation:

$$\mathbf{E}_T(\mathbf{r}; \omega) = i\mu_0\omega \int^{V} \mathbf{G}_T(\mathbf{r} - \mathbf{r}'; \omega) \cdot \mathbf{j}(\mathbf{r}'; \omega)d^3r' +$$

$$(i\mu_0\omega)^{-1} \lim_{\delta \to 0} \int_{V_\delta}^{V_T} \frac{1}{4\pi R^3}(\mathbf{U} - 3\mathbf{e}_\mathbf{R}\mathbf{e}_\mathbf{R}) \cdot \mathbf{j}_T(\mathbf{r}'; \omega)d^3r' +$$

$$(i\varepsilon_0\omega)^{-1}\mathbf{L} \cdot \mathbf{j}_T(\mathbf{r}; \omega). \tag{27.73}$$

The first integral in Eq. (27.73) describes the T-photon radiation from the current density domain, V. The two last terms account for the tied (non-radiative) T-photon dynamics. Only the sum of these terms has a physical meaning. The balance between the two terms depends on the form chosen for the contraction volume V_δ ($\delta \to 0$) around the R^3-singularity. To each form is connected a self-field dyadic

$$\mathbf{L} = \frac{1}{4\pi} \int^{S_\delta} \frac{\hat{\mathbf{n}}(\mathbf{r}')\hat{\mathbf{R}}}{R^2} dS' \tag{27.74}$$

where the integration runs over the surface (S_δ) of the exclusion volume. On a given point $\mathbf{r}'$ on this surface $\hat{\mathbf{n}}(\mathbf{r}')$ is an inwards directed unit vector, and again $\mathbf{R} = |\mathbf{r} - \mathbf{r}'|$, $R = |\mathbf{R}|$.

In a sense, a spherical contraction volume has a privileged status, and gives the tied T-photon described a particular simple form. Thus, it turns out [157] that the spatially nonlocal part [second integral of Eq. (27.73) is zero] and that the self-field dyadic explicitly is given by

$$\mathbf{L} = \frac{1}{3}\mathbf{U}. \tag{27.75}$$

For what follows this simple local form of the self-field dyadics of the tied T-photon dynamics, represented by the transverse electric self-field (SF), viz.,

$$\mathbf{E}_T^{\mathrm{SF}}(\mathbf{r};\omega) = (3i\varepsilon_0\omega)^{-1}\mathbf{j}_T(\mathbf{r};\omega), \tag{27.76}$$

is particularly useful. In passing note that $\mathbf{E}_T^{\mathrm{SF}}(\mathbf{r};\omega)$ is nonvanishing in the domain occupied by the transverse current density domain. Intuitively, this confirms the idea that the attached transverse photon cloud surrounds the bare particle.

27.5.2 Lagrange formalism based on the propagator picture

Inspired by the Power-Zinau-Woolley transformation, discussed in Subsection 27.4.1, we now introduce a propagator (superscript: G) Lagrangian

$$L^G = L + \frac{dF}{dt}, \tag{27.77}$$

with the choice

$$F = -\frac{1}{3} \int \mathbf{P}_T(\mathbf{r}) \cdot \mathbf{A}_T(\mathbf{r}) d^3r. \tag{27.78}$$

With no loss of essential generality, we limit ourselves to a single, spinless electron bound to a fixed nucleus (located at the origo of the coordinate system). The Coulomb energy thus is given by

$$V_C = -\frac{e^2}{4\pi\varepsilon_0 r_e} + \varepsilon_C, \tag{27.79}$$

where $\mathbf{r}_e$ is the electron position vector, and ε_C [$= \varepsilon_C^{\mathrm{el}} + \varepsilon_C^{\mathrm{nucl}}$] is the sum of the Coulomb self-energies of the electron and the point-like nucleus.

The propagator Lagrangian hence takes the following form:

$$L^G = \frac{1}{2}m\dot{\mathbf{r}}_e^2 - V_C + \int \mathcal{L}^G d^3r,$$

(27.80)

with field + interaction Lagrangian density

$$\mathcal{L}^G = \frac{\varepsilon_0}{2}\left\{\dot{\mathbf{A}}_T^2(\mathbf{r}) - c^2\left[\boldsymbol{\nabla}\times\mathbf{A}_T(\mathbf{r})\right]^2\right\} + \frac{2}{3}\mathbf{j}_T(\mathbf{r})\cdot\mathbf{A}_T(\mathbf{r}) - \frac{1}{3}\mathbf{P}_T(\mathbf{r})\cdot\dot{\mathbf{A}}_T(\mathbf{r}).$$

(27.81)

From Eqs. (27.80) and (27.81) one obtains the conjugate particle momentum [electron charge here $-e$]

$$\mathbf{p}^G \equiv \frac{\partial L^G}{\partial \dot{\mathbf{r}}_e} = m\dot{\mathbf{r}}_e - \frac{2e}{3}\mathbf{A}_T(\mathbf{r}_e),$$

(27.82)

and the canonical field momentum

$$\boldsymbol{\Pi}^G(\mathbf{r}) \equiv \frac{\partial L^G}{\partial \dot{\mathbf{A}}_T} = \varepsilon_0\dot{\mathbf{A}}_T(\mathbf{r}) - \frac{1}{3}\mathbf{P}_T(\mathbf{r}).$$

(27.83)

In the propagator Hamiltonian

$$H^G = \mathbf{p}^G\cdot\dot{\mathbf{r}}_e + \int \boldsymbol{\Pi}^G(\mathbf{r})\cdot\mathbf{A}_T(\mathbf{r})d^3r - L^G$$

(27.84)

Eqs. (27.82) and (27.83) are used to eliminate the quantities $\dot{\mathbf{r}}_e$ and $\dot{\mathbf{A}}_T(\mathbf{r})$ in favor of the new canonical momenta, The resulting Hamiltonian

$$H^G = H_P^G + H_F^G + H_I^G,$$

(27.85)

consists of three parts. The particle part,

$$H_P^G = \frac{1}{2m}(\mathbf{p}^G)^2 + V_C + \frac{1}{2\varepsilon_0}\int\left[\frac{1}{3}\mathbf{P}_T(\mathbf{r})\right]^2 d^3r,$$

(27.86)

has a new transverse self-field contribution

$$\varepsilon_{\mathrm{SE}}^G = \frac{1}{2\varepsilon_0}\int \frac{1}{9}\mathbf{P}_T^2(\mathbf{r})d^3r$$

(27.87)

which is a factor of 9 smaller than that of the Power-Zinau-Woolley representation. The propagator field Hamiltonian

$$H_F^G = \frac{\varepsilon_0}{2}\int\left[\left(\frac{\boldsymbol{\Pi}^G(\mathbf{r})}{\varepsilon_0}\right)^2 + c^2\left(\boldsymbol{\nabla}\times\mathbf{A}_T\right)^2\right]d^3r,$$

(27.88)

has the beforehand expected structure, obtained from H_F^{PZW} by replacing $\boldsymbol{\pi}^{\mathrm{PZW}}(\mathbf{r}) \Rightarrow \boldsymbol{\Pi}^G(\mathbf{r})$. The utmost interesting interaction Hamiltonian is given by

$$H_I^G = \int \frac{1}{3}\mathbf{P}_T(\mathbf{r})\cdot\boldsymbol{\Pi}^G(\mathbf{r})d^3r + \frac{e^2}{2m}\left(\frac{2}{3}\mathbf{A}_T(\mathbf{r}_e)\right)^2 +$$
$$\frac{e}{2m}\left[\mathbf{p}^G\cdot\left(\frac{2}{3}\mathbf{A}_T(\mathbf{r}_e)\right) + \frac{2}{3}\mathbf{A}_T(\mathbf{r}_e)\cdot\mathbf{p}^G\right].$$

(27.89)

27.5.3 Radiated and tied T-photon fields

The propagator field momentum $\mathbf{\Pi}^G(\mathbf{r})$ is related in a simple manner to the retarded part of the transverse electric field, $\mathbf{E}_T(\mathbf{r})$. In order to realize this, we first note that the transverse self-field is given by

$$\mathbf{E}_T^{\mathrm{SF}}(\mathbf{r}) = -\frac{1}{3\varepsilon_0}\mathbf{P}_T(\mathbf{r}), \qquad (27.90)$$

and since $\dot{\mathbf{A}}_T = -\mathbf{E}_T$, the propagator field momentum can be written as

$$\mathbf{\Pi}^G(\mathbf{r}) = -\varepsilon_0\mathbf{E}_T(\mathbf{r}) + \varepsilon_0\mathbf{E}_T^{\mathrm{SF}}(\mathbf{r}), \qquad (27.91)$$

so that

$$-\mathbf{\Pi}^G(\mathbf{r}) = \varepsilon_0\mathbf{E}_T^R(\mathbf{r}), \qquad (27.92)$$

where $\mathbf{E}_T^R(\mathbf{r})$ is the radiated (R) part of the electric field. In the propagator formalism we have thus in the long wavelength region separated the tied field of the photon cloud [Eq. (27.76)] form the total field:

$$\mathbf{E}_T(\mathbf{r}) = \mathbf{E}_T^R(\mathbf{r}) + \mathbf{E}_T^{\mathrm{SF}}(\mathbf{r}). \qquad (27.93)$$

As in the Pauli-Fierz theory, in the propagator formalism the integral term in Eq. (27.89) describes an interaction (interference) between the radiated and tied fields, viz.,

$$\frac{1}{3\varepsilon_0}\int \mathbf{P}_T(\mathbf{r})\cdot\mathbf{\Pi}^G(\mathbf{r})d^3r = \varepsilon_0\int \mathbf{E}_T^{\mathrm{SF}}(\mathbf{r})\cdot\mathbf{E}_T^R(\mathbf{r})d^3r. \qquad (27.94)$$

If one introduces an effective (eff) transverse polarization field defined by

$$\mathbf{P}_T^{\mathrm{eff}}(\mathbf{r}) \equiv \frac{1}{3}\mathbf{P}_T(\mathbf{r}), \qquad (27.95)$$

it appears that

$$\mathbf{D}_T^{\mathrm{eff}}(\mathbf{r}) = -\mathbf{\Pi}^G(\mathbf{r}) = \varepsilon_0\mathbf{E}_T(\mathbf{r}) + \mathbf{P}_T^{\mathrm{eff}}(\mathbf{r}) \qquad (27.96)$$

has the status of an effective transverse microscopic displacement field.

27.5.4 Heisenberg equation for the retarded field in the propagator formalism

Let us upgrade the formalism to the quantum operator level. From the Heisenberg equations of motions for the radiated electric field operator $(\hat{\mathbf{E}}_T^R)$ and the magnetic field operator $(\hat{\mathbf{B}})$ it is possible to establish the following set of inhomogeneous Maxwell operator equations [147]:

$$\boldsymbol{\nabla} \times \hat{\mathbf{E}}_T^R(\mathbf{r},t) + \frac{\partial}{\partial t}\hat{\mathbf{B}}(\mathbf{r},t) = \frac{1}{3\varepsilon_0}\boldsymbol{\nabla} \times \hat{\mathbf{P}}_T(\mathbf{r},t), \qquad (27.97)$$

$$\boldsymbol{\nabla} \times \hat{\mathbf{B}}(\mathbf{r},t) - \frac{1}{c^2}\frac{\partial}{\partial t}\hat{\mathbf{E}}_T^R(\mathbf{r},t) = \frac{2\mu_0}{3}\frac{\partial}{\partial t}\hat{\mathbf{P}}_T(\mathbf{r},t). \qquad (27.98)$$

The Maxwell operator equation in Eq. (27.97) originates in the Heisenberg equation for $\hat{\mathbf{B}}$, i.e.,

$$i\hbar\frac{d}{dt}\hat{\mathbf{B}} = [\hat{\mathbf{B}}, \hat{H}_F^G] + [\hat{\mathbf{B}}, \hat{H}_I^G], \qquad (27.99)$$

as the reader may show to herself. The other Maxwell operator equation [Eq. (27.98)] is derived from an analogous Heisenberg equation for $\mathbf{E}_T^R$.

CHAPTER 28

The Plasmariton Quasiparticle. Surface Plasmaritons

The present chapter describes aspects of ongoing research on jellium quasiparticles, in particular the first- and second-quantized theories of these particles. A door to this new research field was opened a little by ideas first formed in my work on this book. New ideas often come by tortuous paths. In this case notions started intuitively when I worked on the extinction theorem for the complex Klein-Gordon equation. [Sections 16.2–16.4]. In the wake of my early ideas, my coworker Jesper Jung and I have developed a fully "fledged bird" describing the quantum theory of plasmariton quasiparticles [125], and their role in the electrodynamics of surface plasmaritons [128]. The grown-up bird now pass some of the main results back to the reader of this book.

28.1 FIRST-QUANTIZED PLASMARITON THEORY

We have in Section 20.3 seen that the plasmariton field $\boldsymbol{\Phi}(\mathbf{r},t)$, satisfies the Klein-Gordon equation

$$\left(\nabla^2 - \frac{1}{c^2}\frac{\partial^2}{\partial t^2}\right)\boldsymbol{\Phi}(\mathbf{r},t) - Q_T^2\boldsymbol{\Phi}(\mathbf{r},t) = \mathbf{0}, \tag{28.1}$$

where $Q_T \equiv \omega_p/c$ is the plasmariton Compton wave number. The vectorial plasmariton field is subject to the transversality condition

$$\boldsymbol{\nabla}\cdot\boldsymbol{\Phi}(\mathbf{r},t) = 0 \tag{28.2}$$

in every space-time point.

As for photons, the first-quantized theory of the plasmariton is based on the analytical part, $\boldsymbol{\Phi}^{(+)}(\mathbf{r},t)$, of the plasmariton field $\boldsymbol{\Phi}(\mathbf{r},t)$. The analytic part given as

$$\boldsymbol{\Phi}^{(+)}(\mathbf{r},t) = \frac{1}{2\pi}\int_{-\infty}^{\infty}\Theta(\omega)\boldsymbol{\Phi}(\mathbf{r},\omega)e^{-i\omega t}d\omega, \tag{28.3}$$

DOI: 10.1201/9781003029458-28

where $\Theta(\omega)$ is the Heaviside unit step function, we call the plasmariton wave function. This also satisfies the Klein-Gordon equation

$$\left(\nabla^2 - \frac{1}{c^2}\frac{\partial^2}{\partial t^2}\right)\boldsymbol{\Phi}^{(+)}(\mathbf{r},t) - Q_T^2\boldsymbol{\Phi}^{(+)}(\mathbf{r},t) = \mathbf{0}, \tag{28.4}$$

as the reader my realize by inserting Eq. (28.3) on the right-hand side of Eq. (28.4) and utilizing Eq. (28.1). In the $(\mathbf{q},\omega)$-domain, Eq. (28.4) takes the form

$$\left[c^2(q^2 + Q_T^2) - \omega^2\right]\boldsymbol{\Phi}^{(+)}(\mathbf{q},\omega) = \mathbf{0}, \tag{28.5}$$

or equivalent

$$\left[c\sqrt{(q^2 + Q_T^2)} + \omega\right]\left[c\sqrt{(q^2 + Q_T^2)} - \omega\right]\boldsymbol{\Phi}^{(+)}(\mathbf{q},\omega) = \mathbf{0}. \tag{28.6}$$

Since $\boldsymbol{\Phi}^{(+)}(\mathbf{q},\omega) = \Theta(\omega)\boldsymbol{\Phi}(\mathbf{q},\omega) = \mathbf{0}$ for $\omega < 0$, Eq. (28.6) is identical to

$$\left[c\sqrt{(q^2 + Q_T^2)} - \omega\right]\boldsymbol{\Phi}^{(+)}(\mathbf{q},\omega) = \mathbf{0}. \tag{28.7}$$

The relativistic energy-momentum relation for the plasmariton field, viz.,

$$\hbar\omega = \left[(c\hbar q)^2 + (\hbar\omega_p)^2\right]^{1/2}, \tag{28.8}$$

indicates that the plasmariton Hamilton operator in the wave-vector domain is given by

$$\hat{H}(q) = c\hbar\sqrt{q^2 + Q_T^2}. \tag{28.9}$$

In the $(\mathbf{q};t)$-domain, the plasmariton wave function hence satisfies the dynamical (first-order in time) evolution equation

$$\hat{H}(q)\boldsymbol{\Phi}^{(+)}(q;t) = i\hbar\frac{\partial}{\partial t}\boldsymbol{\Phi}^{(+)}(\mathbf{q};t). \tag{28.10}$$

The definition given in Eq. (2.71), extended to the form

$$\sqrt{Q_T^2 - \nabla^2}\mathbf{F}(\mathbf{r}) \equiv \int_{-\infty}^{\infty}\sqrt{Q_T^2 + q^2}\mathbf{F}(\mathbf{q})e^{i\mathbf{q}\cdot\mathbf{r}}\frac{d^3q}{(2\pi)^3}, \tag{28.11}$$

allows us to write the plasmariton evolution equation in the Schrödinger-like space-time form

$$c\hbar\sqrt{Q_T^2 - \nabla^2}\boldsymbol{\Phi}^{(+)}(\mathbf{r},t) = i\hbar\frac{\partial}{\partial t}\boldsymbol{\Phi}^{(+)}(\mathbf{r},t). \tag{28.12}$$

From the Proca equation [218] for a massive transverse photon field, with vector potential $\mathbf{A}_T(\mathbf{r},t)$, one can establish a dynamical equation for the related wave function $\mathbf{A}_T^{(+)}(\mathbf{r},t)$ form-identical to Eq. (28.12); see [156, 158].

It might be noted that the Maxwell-Proca theory can be obtained from the Stueckelberg Lagrangian density [244, 245], which adds a term with $A_\mu A^\mu$ to the standard theory [160, 161].

28.2   PLASMARITON SPIN AND HELICITY

As a bosonic quasiparticle one would expect that the plasmariton is a spin-one particle. In order to prove that this assertion is correct one rewrites the curl of the plasmariton wave function as

$$\nabla \times \mathbf{\Phi}^{(+)}(\mathbf{r}, t) = i^{-1}(\hat{\boldsymbol{\sigma}} \cdot \nabla)\mathbf{\Phi}^{(+)}(\mathbf{r}, t), \tag{28.13}$$

where $\hat{\boldsymbol{\sigma}} = (\hat{\sigma}_1, \hat{\sigma}_2, \hat{\sigma}_3)$ is a dimensionless spin operator, which elements are the matrices [Eq. (2.44)]

$$(\hat{\sigma}_i)_{jk} = i^{-1}\varepsilon_{ijk}, \tag{28.14}$$

ε_{ijk} being the Levi-Civita symbol. I leave it to the reader to show that Eq. (28.13) is correct. The spin operator $\hat{\boldsymbol{\sigma}}$ already has been introduced in Section 2.1 in relation to the first-quantized description of transverse photons. It was mentioned there that $\hat{\boldsymbol{\sigma}}$ is a spin-one vector. To realize that $\hat{\boldsymbol{\sigma}}$ has the form of an angular momentum operator in quantum mechanics let us start from the explicit forms of the Cartesian components of $\boldsymbol{\sigma}$, viz.

$$\hat{\sigma}_x = \begin{pmatrix} 0 & 0 & 0 \\ 0 & 0 & -i \\ 0 & i & 0 \end{pmatrix}, \tag{28.15}$$

$$\hat{\sigma}_y = \begin{pmatrix} 0 & 0 & i \\ 0 & 0 & 0 \\ -i & 0 & 0 \end{pmatrix}, \tag{28.16}$$

$$\hat{\sigma}_x = \begin{pmatrix} 0 & -i & 0 \\ i & 0 & 0 \\ 0 & 0 & 0 \end{pmatrix}. \tag{28.17}$$

Simple matrix manipulations show that the components of σ satisfy the commutator relations

$$[\sigma_i, \sigma_j] = i\varepsilon_{ijk}\sigma_k, \tag{28.18}$$

which in compact notation can be written as

$$\boldsymbol{\sigma} \times \boldsymbol{\sigma} = i\boldsymbol{\sigma}. \tag{28.19}$$

The set of commutator relations in Eq. (28.18) has a form which *defines* an angular momentum operator in quantum mechanics.

Up to an overall phase factor, the normalized eigenvectors of the spin-operator components are

$$|\pm, C_x\rangle = \frac{1}{\sqrt{2}} \begin{pmatrix} 0 \\ \mp i \\ 1 \end{pmatrix}, \ |0; C_x\rangle = \begin{pmatrix} 1 \\ 0 \\ 0 \end{pmatrix}, \tag{28.20}$$

$$|\pm, C_y\rangle = \frac{1}{\sqrt{2}} \begin{pmatrix} \pm i \\ 0 \\ 1 \end{pmatrix}, \quad |0; C_y\rangle = \begin{pmatrix} 0 \\ 1 \\ 0 \end{pmatrix}, \tag{28.21}$$

$$|\pm, C_z\rangle = \frac{1}{\sqrt{2}} \begin{pmatrix} 1 \\ \pm i \\ 0 \end{pmatrix}, \quad |0; C_z\rangle = \begin{pmatrix} 0 \\ 0 \\ 1 \end{pmatrix}. \tag{28.22}$$

The quantity C_i added inside the various kets is meant to remember us that the results belong to the ith component in the Cartesian (C) representation. A discussion of the spherical representation of the components of $\boldsymbol{\sigma}$, and the relation between the two representations is given in [158]. The Cartesian components of the spin operator, $\hat{\mathbf{s}} = \hbar\hat{\boldsymbol{\sigma}}$, all have eigenvalues $\pm\hbar$ and 0, as the reader may convince herself of. In Eqs. (28.20)–(28.22) the upper and lower signs correspond to states of spin $+\hbar$ and $-\hbar$, respectively. The operator

$$\hat{s}^2 = \hat{s}_x^2 + \hat{s}_y^2 + \hat{s}_z^2 = 2\hbar^2 \begin{pmatrix} 1 & 0 & 0 \\ 0 & 1 & 0 \\ 0 & 0 & 1 \end{pmatrix}, \tag{28.23}$$

which characterizes the length of the spin vector, has the triply degenerate eigenvalue $2\hbar^2$. The results above are in agreement with the general results for the eigenvalues (ev) for spin s particle, namely

$$\mathrm{ev}(\hat{s}^2) = s(s+1)\hbar^2, \tag{28.24}$$
$$\mathrm{ev}(\hat{s}_z) = m\hbar, \quad -s \le m \le s, \tag{28.25}$$

as one realizes setting $s = 1$. The transversality of the wave function, $\boldsymbol{\nabla} \cdot \boldsymbol{\Phi}^{(+)}(\mathbf{r}, t) = 0$, means that the longitudinally polarized states $|0; C_i\rangle$, $i = 1 - 3$, have no physical meaning for plasmaritons. So although the plasmariton quasiparticle is a spin-one particle, it is an imperfect one [the $m = 0$ state does not exist].

Before proceeding the general spin discussion, let us take a brief look at the so-called projected (superscript: proj) plasmariton spin operator, $\hat{s}^{\mathrm{proj}}$, which ith component is defined by

$$\hat{s}_i^{\mathrm{proj}} \equiv (\mathbf{U} - \hat{\mathbf{q}}\hat{\mathbf{q}}) \cdot \hat{\mathbf{s}}_i \cdot (\mathbf{U} - \hat{\mathbf{q}}\hat{\mathbf{q}}) \tag{28.26}$$

in the matrix representation of the spin operator belonging to a plasmariton wave vector. A simple calculation shows that

$$\hat{s}_i^{\mathrm{proj}} = \frac{i\hbar}{q^2} q_i \Sigma, \tag{28.27}$$

where

$$\Sigma = \begin{pmatrix} 0 & -q_z & q_y \\ q_z & 0 & -q_x \\ -q_y & q_x & 0 \end{pmatrix} \tag{28.28}$$

is an antisymmetric matrix *independent of the index i*. This independence of i implies that the components of the projected plasmariton spin operator commute mutually, i.e.,

$$\hat{\mathbf{s}}^{\mathrm{proj}} \times \hat{\mathbf{s}}^{\mathrm{proj}} = \mathbf{0}. \tag{28.29}$$

In principle, the components of $\hat{\mathbf{s}}^{\mathrm{proj}}$ thus can be measured simultaneously, although it certainly is not clear how this possibly might be done.

Now, since $\boldsymbol{\nabla} \cdot \boldsymbol{\Phi}^{(+)} = 0$, one gets

$$\nabla^2\boldsymbol{\Phi}^{(+)}(\mathbf{r}, t) = -\boldsymbol{\nabla} \times (\boldsymbol{\nabla} \times \boldsymbol{\Phi}^{(+)}(\mathbf{r}, t)) = (\hat{\boldsymbol{\sigma}} \cdot \boldsymbol{\nabla})^2\boldsymbol{\Phi}^{(+)}(\mathbf{r}, t). \tag{28.30}$$

The last step follows using Eq. (28.13) twice times, By introduction of the momentum operator $\hat{\mathbf{p}} = -i\hbar\boldsymbol{\nabla}$ in Eq. (28.30) one obtains

$$-\hbar^2\nabla^2\boldsymbol{\Phi}^{(+)}(\mathbf{r}, t) = (\hat{\boldsymbol{\sigma}} \cdot \hat{\mathbf{p}})^2\,\boldsymbol{\Phi}^{(+)}(\mathbf{r}, t). \tag{28.31}$$

A combination of Eqs. (28.5) and Eq. (28.31), and use of the relation $c\hbar Q_T = \hbar\omega_p = M_T c^2$, where M_T is the plasmariton rest mass [see Subsection 20.3.1] show that the Klein-Gordon equation for the plasmariton wave function can be recast in the form

$$\left[\left(i\hbar\frac{\partial}{\partial t}\right)^2 + (c\hat{\mathbf{p}} \cdot \hat{\boldsymbol{\sigma}})^2 - (M_T c^2)^2\right]\boldsymbol{\Phi}^{(+)}(\mathbf{r}, t) = \mathbf{0}. \tag{28.32}$$

In particle physics the Klein-Gordon equation for spin-0 bosons is in focus, and this equation's form for spin-1 and higher-order spin is not discussed in textbooks. Since, to the best of our present knowledge, there does not exist any massive fundamental particles with spin-1, the Klein-Gordon equation for spin-1 particles seldom takes a prominent position in relativistic quantum mechanics. In plasmariton wave mechanics the spin-1 Klein-Gordon equation has a fundamental status. The appearance of the spin for the plasmariton is a consequence of the fact that the T-photon (a spin-one particle) is an ingredient of the plasmariton quasiparticle. The Proca equation for massive photons [Subsection 20.3.2] can of course be given the form in Eq. (28.32) replacing M_T by the photon mass, M_{PH}.

The projection of the spin ($\mathbf{s} = \hbar\boldsymbol{\sigma}$) on the quasiparticle momentum direction ($\mathbf{p}/p$) is the plasmariton's helicity (h). The helicity operator

$$\hat{h} \equiv \hat{p}^{-1}(\hat{\mathbf{p}} \cdot \hat{\mathbf{s}}) \tag{28.33}$$

in the $\mathbf{r}$-representation can be introduced in Eq. (28.32) noting that

$$c\hat{\mathbf{p}} \cdot \hat{\boldsymbol{\sigma}} = \frac{c}{\hbar}\hat{p}\hat{h}, \tag{28.34}$$

where $\hat{p} = \hbar[-\nabla^2]^{1/2}$. The definition in Eq. (2.71) implies that the helicity operator in the momentum ($\mathbf{p}$) representation is given by

$$\hat{h} = \frac{\mathbf{p}}{p} \cdot \hat{\mathbf{s}}, \tag{28.35}$$

$\mathbf{p}/p$ being the ordinary unit vector in the momentum direction. In the matrix representation, where the $\hat{\sigma}_i$'s are given by Eqs. (28.15)–(28.17), one has

$$\frac{\hat{h}}{\hbar} = \frac{1}{p}\sum_i p_i\hat{\sigma}_i = \frac{1}{q}\sum_i q_i\hat{\sigma}_i = \frac{1}{q}\boldsymbol{\Sigma}, \tag{28.36}$$

where $\boldsymbol{\Sigma}$ is given by Eq. (28.28). Solutions of the eigenvalue equation for $\boldsymbol{\Sigma}$ leads after elementary calculations to the conclusion that the eigenvalues for the helicity operator are $\hbar$ and $-\hbar$, and that the associated normalized eigenvectors are the helicity unit vectors $\mathbf{e}_+(\mathbf{q}/q)$ and $\mathbf{e}_-(\mathbf{q}/q)$. Hence,

$$\hat{h}\mathbf{e}_\pm(\mathbf{q}/q) = \pm\hbar\mathbf{e}_\pm(\mathbf{q}/q). \tag{28.37}$$

The fact that $\boldsymbol{\Sigma}$ is independent of the index i [see Eq. (28.28)] in a sense underlines the fundamental status of the plasmariton helicity (operator) in comparison to the spin (operator). The fundamental status of $\hat{h}$ flourishes for massless particles, where there is no rest frame for the particle, and then no genuine spin concept since this is the internal angular momentum in the particle's rest frame.

28.3 HAMILTONIAN FORMALISM WITH $-\mathbf{D}_T$ AS CANONICAL FIELD MOMENTUM

28.3.1 Charged jellium oscillator

It appears from the Lagrangian/Hamiltonian approach for the dynamics of transverse photons that the transverse current density

$$\mathbf{J}_T(\mathbf{r}, t) = \frac{\partial}{\partial t}\mathbf{P}_T(\mathbf{r}, t) \tag{28.38}$$

plays a fundamental role in the particle-field interaction. In Eq. (28.38) $\mathbf{P}_T(\mathbf{r}, t)$ is a microscopic transverse polarization field. The relation in Eq. (28.38) is general because it is possible to eliminate the microscopic magnetization field ($\mathcal{M}(\mathbf{r}, t)$) from the well-known equation

$$\mathbf{J}_T(\mathbf{r}, t) = \frac{\partial}{\partial t}\mathcal{P}_T(\mathbf{r}, t) + \boldsymbol{\nabla} \times \mathcal{M}(\mathbf{r}, t) \tag{28.39}$$

by a certain transformation; see e.g. [157]. The quantity $\mathcal{P}_T(\mathbf{r}, t)$ equals the electric dipole moment per unit volume, whereas $\mathbf{P}_T$ includes all multipole moments. In the long-wavelength limit ($q \to 0$) Eq. (28.1) is reduced to a harmonic oscillator equation

$$\left(\frac{\partial^2}{\partial t^2} + \omega_p^2\right)\boldsymbol{\Phi}_T(\mathbf{r}, t) = \mathbf{0}, \tag{28.40}$$

remembering that the plasmariton field is transverse [Eq. (28.2)]. It is obvious from Eq. (28.40), and the elementary (textbook) theory of collective jellium excitations in the limit $q \to 0$, that the time evolution of the microscopic polarization $\mathbf{P} = \mathbf{P}_T$ is equivalent to that of a simple electrically charged harmonic oscillator. The mass (M) and charge (Q) are

$$M = nm, \quad Q = ne, \tag{28.41}$$

where n is the uniform electron density, and m and e are the mass and charge of the electron.

A Lagrangian polarization density

$$\mathcal{L}_P = \frac{M}{2Q^2}\dot{\mathbf{P}}^2 - \frac{1}{2\varepsilon_0}\mathbf{P}^2 \tag{28.42}$$

inserted into the "particle" Lagrangian equation

$$\frac{d}{dt}\left(\frac{\partial \mathcal{L}_P}{\partial \dot{\mathbf{P}}}\right) - \frac{\partial \mathcal{L}_P}{\partial \mathbf{P}} = \mathbf{0} \tag{28.43}$$

gives one the correct vectorial oscillator equation for the polarization field, viz.,

$$\ddot{\mathbf{P}} + \omega_p^2 \mathbf{P} = \mathbf{0}, \tag{28.44}$$

where

$$\omega_p = \left(\frac{Q^2}{M\varepsilon_0}\right)^{1/2} = \left(\frac{ne^2}{m\varepsilon_0}\right)^{1/2}. \tag{28.45}$$

The canonical momentum $(\mathbf{\Pi})$ of the polarization field is given as

$$\mathbf{\Pi} \equiv \frac{\partial \mathcal{L}_P}{\partial \dot{\mathbf{P}}} = \frac{M}{Q^2}\dot{\mathbf{P}}. \tag{28.46}$$

In turn, the particle Hamiltonian density defined by

$$\mathcal{H}_P \equiv \mathbf{\Pi} \cdot \dot{\mathbf{P}} - \mathcal{L}_P, \tag{28.47}$$

takes the explicit form

$$\mathcal{H}_P = \frac{\mathbf{\Pi} \cdot \mathbf{\Pi}}{2MQ^2} + \frac{1}{2\varepsilon_0}\mathbf{P} \cdot \mathbf{P} = \frac{1}{Q^2}\left(\frac{\mathbf{\Pi}^2}{2M} + \frac{1}{2}M\omega_p^2\mathbf{P}^2\right). \tag{28.48}$$

Apart from the factor Q^{-2} the "kinetic" and "potential" energies, have the usual particle form, the resonance frequency being the long-wavelength plasma frequency, ω_p.

In Section 3.5, we used an interaction Lagrangian density of the form

$$\mathcal{L}_I^{\text{OLD}} = \mathbf{J} \cdot \mathbf{A}_T, \tag{28.49}$$

[or equivalently $\mathbf{J}_T \cdot \mathbf{A}_T$]. It follows from Eq. (12.10) that $\mathcal{L}_I^{\text{OLD}}$ may be replaced by a new one

$$\mathcal{L}_I^{\text{NEW}}[\equiv \mathcal{L}_I] = \mathcal{L}_I^{\text{OLD}} + \frac{d}{dt}(-\mathbf{P} \cdot \mathbf{A}_T) \tag{28.50}$$

without changing the physical content of the formalism. Remembering that the position coordinate ($\mathbf{r}$ is a vectorial index in the description, use of $\mathbf{J}_T = \partial\mathbf{P}_T/\partial t \equiv \dot{\mathbf{P}}_T$ [Eq. (28.38)] and $d\mathbf{A}_T/dt = \partial\mathbf{A}_T/\partial t \equiv \dot{\mathbf{A}}_T$, one obtains a new interaction Lagrangian density

$$\mathcal{L}_I = -\mathbf{P}\cdot\dot{\mathbf{A}}_T, \tag{28.51}$$

where $\mathbf{P}[\equiv \mathbf{P}_T]$ is the polarization field belonging to the plasmariton. The total Lagrangian density

$$\mathcal{L} = \mathcal{L}_F + \mathcal{L}_P + \mathcal{L}_I \tag{28.52}$$

with the new interaction Lagrangian density hence takes the explicit form

$$\mathcal{L} = \frac{\varepsilon_0}{2}\left[\dot{\mathbf{A}}_T^2 - c^2(\boldsymbol{\nabla}\times\mathbf{A}_T)^2\right] + \frac{M}{2Q}\dot{\mathbf{P}}^2 - \frac{1}{2\varepsilon_0}\mathbf{P}^2 - \mathbf{P}\cdot\dot{\mathbf{A}}_T. \tag{28.53}$$

The momentum ($\mathbf{M}$ conjugate to $\mathbf{A}_T$ now becomes

$$\mathbf{M} \equiv \frac{\partial\mathcal{L}}{\partial\dot{\mathbf{A}}_T} = \varepsilon_0\dot{\mathbf{A}}_T - \mathbf{P}_T \tag{28.54}$$

$[\mathbf{P} = \mathbf{P}_T]$, and then

$$\mathbf{M} = -(\varepsilon_0\mathbf{E}_T + \mathbf{P}_T) = -\mathbf{D}_T \tag{28.55}$$

where $\mathbf{D}_T$ is the transverse part of the microscopic displacement field. In the plasmariton case $\mathbf{D}_T = \mathbf{D}$. Since $\mathbf{M} = \mathbf{M}_T$, obviously, one may replace $\mathbf{M}$ by $\mathbf{M}_T$ in Eq. (28.55), if wished.

The total Hamiltonian density

$$\mathcal{H} \equiv \mathbf{M}\cdot\dot{\mathbf{A}}_T + \boldsymbol{\Pi}\cdot\dot{\mathbf{P}} - \mathcal{L} \tag{28.56}$$

now can be expressed in terms of the "coordinates" $(\mathbf{P}, \mathbf{A}_T)$ and conjugate "momenta" $(\boldsymbol{\Pi}, \mathbf{M})$ eliminating $\dot{\mathbf{P}}$ and $\dot{\mathbf{A}}_T$ by means of Eqs. (28.46) and (28.54). Thus, one obtains

$$\mathcal{H} = \left\{\frac{1}{2\varepsilon_0}\mathbf{M}^2 + \frac{\varepsilon_0 c^2}{2}(\boldsymbol{\nabla}\times\mathbf{A}_T)^2\right\} + \frac{1}{Q^2}\left[\frac{\boldsymbol{\Pi}^2}{2M} + \frac{1}{2}M\omega_p^2\mathbf{P}^2\right] + \frac{1}{2\varepsilon_0}\mathbf{P}^2 + \frac{1}{\varepsilon_0}\mathbf{M}\cdot\mathbf{P} \tag{28.57}$$

in an approach where the canonical field momentum is the displacement field multiplied by a factor -1.

28.3.2 Structure of the Hamiltonian Density

The question now arises, what is the physical interpretation of the new Hamiltonian density in Eq. (28.57)? The term in the $\{\ldots\}$-parenthesis may be characterized as a quasi-free transverse photon Hamiltonian density, where the word quasi refers to the fact that it is the $\mathbf{D}_T$-field and not the $\mathbf{E}_T$-field which enters the field momentum term, cf. Eq. (28.54).

The expression in the [...]-parenthesis is identical to the Hamiltonian of the charged oscillator discussed in Subsection 28.3.1 [Eq. (28.48)]. However, because the term $\mathbf{P}^2/(2\varepsilon_0)$ only depends on the particle (electron) position coordinates, here the plasmariton polarization density, this term must be grouped with the oscillator part in [...]. Since

$$\frac{1}{Q^2}\left[\frac{\mathbf{\Pi}^2}{2M} + \frac{1}{2}M\omega_p^2\mathbf{P}^2\right] + \frac{1}{2\varepsilon_0}\mathbf{P}^2 = \frac{1}{Q^2}\left[\frac{\mathbf{\Pi}^2}{2M} + \frac{1}{2}M\tilde{\omega}_p^2\mathbf{P}^2\right], \tag{28.58}$$

where

$$\tilde{\omega}_p = \sqrt{2}\omega_p, \tag{28.59}$$

it appears that the polarization field still satisfies a simple harmonic oscillator equation, yet with the resonance frequency ω_p replaced by $\omega_p\sqrt{2}$. The result in Eq. (28.58) follows immediately utilizing Eq. (28.45) written in the form $(2\varepsilon_0)^{-1} = M\omega_p^2/(2Q^2)$.

Without altering *any* observational (physical) results one may see the term $\mathbf{P}^2/(2\varepsilon_0)$ in $\hat{\mathcal{H}}$ [Eq. (28.57)] as an addition to the field part of the Hamiltonian. For this to make sense, and at the same time maintain that retardation (with speed c) cannot enter, this term $\mathbf{P}^2/(2\varepsilon_0)$ must represent a self-field energy density. If one, in line with previous notation, denote the transverse electric self-field by $\mathbf{E}_T^{\text{SF}}$, the connection takes the form $[\mathbf{P} = \mathbf{P}_T]$

$$\frac{\mathbf{P}_T^2}{2\varepsilon_0} = \frac{\varepsilon_0}{2}(\mathbf{E}_T^{\text{SF}})^2 \equiv \mathcal{H}_{\text{TIED}}^{T\text{-photon}}. \tag{28.60}$$

In view of the propagator description used in Section 27.5 to study the quantized light emission from a one-electron atom, the term in Eq. (28.60) has the status of a T-photon field attached (tied) to the plasmariton, as indicated by the notation $\mathcal{H}_{\text{TIED}}^{T\text{-photon}}$. It follows from Eq. (28.60) that $\mathbf{P}_T = \pm\varepsilon_0\mathbf{E}_T^{\text{SF}}$. To understand that only the minus sign makes sense, we write the transverse electric field as a sum of retarded (R) and self-field (SF) parts, i.e.,

$$\mathbf{E}_T = \mathbf{E}_T^R + \mathbf{E}_T^{\text{SF}}, \tag{28.61}$$

and inserts this decomposition into the expression given in Eq. (28.55) for the transverse displacement field. Hence

$$\mathbf{D}_T = \varepsilon_0(\mathbf{E}_T^R + \mathbf{E}_T^{\text{SF}}) + \mathbf{P}_T. \tag{28.62}$$

Referring again to the circumstance that $-\mathbf{D}_T$ plays the role of a *free-field* canonical momentum in $\mathcal{H}$ [Eq. (28.57)], $\mathbf{D}_T$ *must be* Einstein retarded, like $\mathbf{E}_T^R$; see also the analysis in the following section. For consistency, two conclusions may in turn be drawn from Eq. (28.62), viz., (*i*)

$$\mathbf{D}_T = \varepsilon_0\mathbf{E}_T^R, \tag{28.63}$$

and (ii)

$$\mathbf{E}_T^{\mathrm{SF}} = -\frac{1}{\varepsilon_0}\mathbf{P}_T. \tag{28.64}$$

The part

$$\mathcal{H}_I = \frac{1}{\varepsilon_0}\mathbf{M}\cdot\mathbf{P} = \frac{1}{\varepsilon_0}\mathbf{M}_T\cdot\mathbf{P}_T \tag{28.65}$$

in Eq. (28.57), is the interaction Hamiltonian density in our new description. In view of the results in Eqs. (28.63) and (28.64) [and remembering that $\mathbf{D}_T = -\mathbf{M}_T$] the form

$$\mathcal{H}_I = \varepsilon_0\mathbf{E}_T^R\cdot\mathbf{E}_T^{\mathrm{SF}} \tag{28.66}$$

shows that the Hamiltonian density, $\mathcal{H}_I$, appears as an interaction term between the retarded and the tied parts of the transverse electric field of the plasmariton.

In Fig. 28.1 we illustrate the tied T-photon principle in three quite distinct physical systems.

28.4   SECOND-QUANTIZED PLASMARITON THEORY IN THE $\mathbf{D}_T$-FORMULATION

28.4.1   Plasmariton Hamilton operator

In a second-quantized description of plasmaritons the quantities $\mathbf{A}_T$, $\mathbf{M}$ and $\mathbf{P}$ become operators. In a plane-wave expansion over a quantization volume V these operators are given by $[\hat{\mathbf{q}} = \mathbf{q}/q]$

$$\hat{\mathbf{A}}_T(\mathbf{r}) = \sum_{\mathbf{q},s}\left(\frac{\hbar}{2\varepsilon_0 cqV}\right)^{1/2}\boldsymbol{\varepsilon}_s(\hat{\mathbf{q}})\left[\hat{a}_{\mathbf{q}s}e^{i\mathbf{q}\cdot\mathbf{r}} + \hat{a}_{\mathbf{q},s}^{\dagger}e^{-i\mathbf{q}\cdot\mathbf{r}}\right], \tag{28.67}$$

$$\hat{\mathbf{M}}(\mathbf{r}) = -i\varepsilon_0\sum_{\mathbf{q},s}\left(\frac{\hbar cq}{2\varepsilon_0 V}\right)^{1/2}\boldsymbol{\varepsilon}_s(\hat{\mathbf{q}})\left[\hat{a}_{\mathbf{q}s}e^{i\mathbf{q}\cdot\mathbf{r}} - \hat{a}_{\mathbf{q},s}^{\dagger}e^{-i\mathbf{q}\cdot\mathbf{r}}\right], \tag{28.68}$$

$$\hat{\mathbf{P}}(\mathbf{r}) = \frac{P_0}{V^{1/2}}\sum_{\mathbf{q},s}\boldsymbol{\varepsilon}_s(\hat{\mathbf{q}})\left[\hat{b}_{\mathbf{q}s}e^{i\mathbf{q}\cdot\mathbf{r}} + \hat{b}_{\mathbf{q},s}^{\dagger}e^{-i\mathbf{q}\cdot\mathbf{r}}\right]. \tag{28.69}$$

The expansion for the canonical momentum $\mathbf{M} = -\mathbf{D}_T = -\varepsilon_0\mathbf{E}_T^R$, follows directly from the well-known expansion of the free electric field operator, The operators $\hat{a}_{\mathbf{q}s}$ and $\hat{a}_{\mathbf{q}s}^{\dagger}$ associated with the various modes satisfy the usual boson commutation rules

$$[\hat{a}_{\mathbf{q}s}, \hat{a}_{\mathbf{q}'s'}^{\dagger}] = \delta_{\mathbf{q}\mathbf{q}'}\delta_{ss'}, \tag{28.70}$$

$$[\hat{a}_{\mathbf{q}s}, \hat{a}_{\mathbf{q}'s'}] = [\hat{a}_{\mathbf{q}s}^{\dagger}, \hat{a}_{\mathbf{q}'s'}^{\dagger}] = 0. \tag{28.71}$$

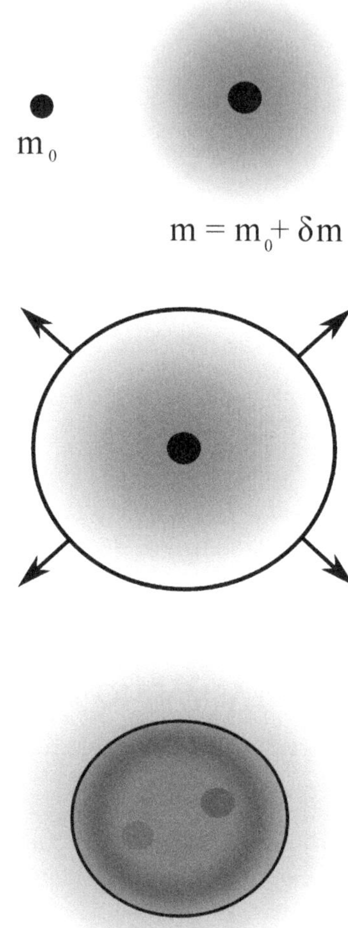

Figure 28.1 Schematic illustration of how the tied T-photon principle can occur in different physical systems. All tied T-photon clouds in violet. a): Nonrelativistic Bethe cloud renormalization of the bare electron mass m_0 to $m = m_0 + \delta m$ [δm finite due to a high-frequency ω-mode cut-off at the relativistic limit $m_0 c^2/\hbar$]. [24] b): Tied T-photon cloud as it appears in my propagator theory describing the quantized light emission from a single-electron atom. The black circle (sphere) gives the trailing edge of the radiated field, propagating outwards (as indicated by the arrows) with the vacuum speed of light. The best possible T-photon localization is determined by the tied T-photon cloud. c): T-photon cloud tied to a localized plasmariton quasiparticle.

In consequence, the field part of the Hamiltonian operator

$$\hat{H}_F = \int_V \mathcal{H}_F d^3 r \tag{28.72}$$

takes the expected "sum of oscillator" form

$$\hat{H}_F = \sum_{\mathbf{q},s} \hbar c q \left(\hat{a}^\dagger_{\mathbf{q}s} \hat{a}_{\mathbf{q}s} + \frac{1}{2} \right). \tag{28.73}$$

The polarization-field mode operators $\hat{b}_{\mathbf{q}s}$ and $\hat{b}^\dagger_{\mathbf{q}s}$ satisfy the boson commutator relations

$$[\hat{b}_{\mathbf{q}s}, \hat{b}^\dagger_{\mathbf{q}'s'}] = \delta_{\mathbf{q}\mathbf{q}'} \delta_{ss'}, \tag{28.74}$$

$$[\hat{b}_{\mathbf{q}s}, \hat{b}_{\mathbf{q}'s'}] = [\hat{b}^\dagger_{\mathbf{q}s}, \hat{b}^\dagger_{\mathbf{q}'s'}] = 0. \tag{28.75}$$

With inclusion of the self-field energy, the oscillator expression in Eq. (28.58) necessarily implies that the Hamilton operator of the polarization field, viz.,

$$\hat{H}_P = \int_V \hat{\mathcal{H}}_P d^3 r, \tag{28.76}$$

attains the following "sum of nondispersive oscillator" form

$$\hat{H}_P = \hbar \tilde{\omega}_p \sum_{\mathbf{q}s} \left(\hat{b}^\dagger_{\mathbf{q}s} \hat{b}_{\mathbf{q}s} + \frac{1}{2} \right). \tag{28.77}$$

The classical connection between the oscillator energy and amplitude (P_0), namely $\hbar \tilde{\omega}_p = M \tilde{\omega}_p^2 P_0^2 / (2Q^2)$ implies that

$$P_0 = (\varepsilon_0 \hbar \tilde{\omega}_p)^{1/2}. \tag{28.78}$$

A somewhat lengthy calculation, which details the reader may find in [125], shows that the interaction Hamiltonian operator

$$\hat{H}_I = \int_V \hat{\mathcal{H}}_I d^3 r = \frac{1}{\varepsilon_0} \int_V \hat{\mathbf{M}} \cdot \hat{\mathbf{P}} d^3 r \tag{28.79}$$

has the plane-wave expansion

$$\hat{H}_I = \frac{i\hbar}{\sqrt{2}} \sum_{\mathbf{q}s} (c q \tilde{\omega}_p)^{1/2} \left[\hat{a}^\dagger_{\mathbf{q}s} \hat{b}_{\mathbf{q}s} + \hat{a}^\dagger_{-\mathbf{q}s} \hat{b}^\dagger_{\mathbf{q}s} - \hat{a}_{-\mathbf{q}s} \hat{b}_{\mathbf{q}s} - \hat{a}_{\mathbf{q}s} \hat{b}^\dagger_{\mathbf{q}s} \right]. \tag{28.80}$$

In passing it should be noticed that all operators between $\hat{a}$'s and $\hat{b}$' commute. By gathering the results in Eqs. (28.73), (28.77) and (28.80) we reach the following Hamilton operator for the plasmariton field in the $\mathbf{D}_T$-formulation:

$$\hat{H} = \sum_{\mathbf{q}s} \left\{ \hbar c q \left(\hat{a}^\dagger_{\mathbf{q}s} \hat{a}_{\mathbf{q}s} + \frac{1}{2} \right) + \hbar \tilde{\omega}_p \left(\hat{b}^\dagger_{\mathbf{q}s} \hat{b}_{\mathbf{q}s} + \frac{1}{2} \right) \right.$$
$$\left. + \frac{i\hbar}{\sqrt{2}} (c q \tilde{\omega}_p)^{1/2} \left[\hat{a}^\dagger_{\mathbf{q}s} \hat{b}_{\mathbf{q}s} + \hat{a}^\dagger_{-\mathbf{q}s} \hat{b}^\dagger_{\mathbf{q}s} - \hat{a}_{-\mathbf{q}s} \hat{b}_{\mathbf{q}s} - \hat{a}_{\mathbf{q}s} \hat{b}^\dagger_{\mathbf{q}s} \right] \right\}. \tag{28.81}$$

28.4.2 Diagonalization of the plasmariton Hamiltonian

In the view of the expansions given for the operators $\hat{\mathbf{A}}_T, \hat{\mathbf{M}}$ and $\hat{\mathbf{P}}$ in Eqs. (28.67)–(28.69) one can introduce a plasmariton fied operator $\hat{\mathbf{W}}(\mathbf{r})$ at each point $\mathbf{r}$ of space as a superposition of plane-wave terms, viz.,

$$\hat{\mathbf{W}}(\mathbf{r}) = \sum_{\mathbf{q}s} W_{qs}\hat{\boldsymbol{\varepsilon}}_s(\hat{\mathbf{q}}) \left[\hat{\alpha}_{\mathbf{q}s}(t)e^{i\mathbf{q}\cdot\mathbf{r}} + \alpha^{\dagger}_{\mathbf{q}s}(t)e^{-i\mathbf{q}\cdot\mathbf{r}} \right], \tag{28.82}$$

where W_{qs} is an unspecified scalar amplitude depending on the magnitude of the plane wave vector and the polarization index. Since the various $\mathbf{q}$-modes are *free* plasmariton modes (eigenmodes), the mode annihilation operator must have a time dependence give by

$$\hat{\alpha}_{\mathbf{q}s}(t) = \hat{\alpha}_{\mathbf{q}s}(0)e^{-i\omega_q t}, \tag{28.83}$$

where ω_q is a yet unknown frequency depending on the magnitude of $\mathbf{q}$. From the Heisenberg equation [Eq. (3.61)] for $\hat{\alpha}_{\mathbf{q}s}(t)$, namely

$$i\hbar\frac{d}{dt}\hat{\alpha}_{\mathbf{q}s}(t) = [\hat{\alpha}_{\mathbf{q}s}(t), \hat{H}], \tag{28.84}$$

and from the time evolution of the annihilation operator [Eq. (28.83)] it appears that $\hat{\alpha}_{\mathbf{q}s}(t)$ satisfies

$$\left[\hat{\alpha}_{\mathbf{q}s}, \hat{H} \right] = \hbar\omega_q\hat{\alpha}_{\mathbf{q}s}. \tag{28.85}$$

The form of the interaction Hamilton in Eq. (28.80) suggests that a linear combination

$$\hat{\alpha}_{\mathbf{q}s} = w\hat{a}_{\mathbf{q}s} + x\hat{b}_{\mathbf{q}s} + y\hat{a}^{\dagger}_{-\mathbf{q}s} + z\hat{b}^{\dagger}_{-\mathbf{q}s}, \tag{28.86}$$

where w, x, y and z are so far unknown constants, might solve Eq. (28.85). In order to obtain a solution these constants are subject to a set of mutually constraints. Thus, first of all

$$[\hat{\alpha}_{\mathbf{q}s}, \hbar^{-1}\hat{H}_F] = cq(w\hat{a}_{\mathbf{q}s} - y\hat{a}^{\dagger}_{-\mathbf{q}s}), \tag{28.87}$$

$$[\hat{\alpha}_{\mathbf{q}s}, \hbar^{-1}\hat{H}_P] = \tilde{\omega}_p(x\hat{b}_{\mathbf{q}s} - z\hat{b}^{\dagger}_{-\mathbf{q}s}), \tag{28.88}$$

and

$$[\hat{\alpha}_{\mathbf{q}s}, \hbar^{-1}\hat{H}_I] = A[w\hat{b}_{\mathbf{q}s} + x(\hat{a}^{\dagger}_{-\mathbf{q}s} - \hat{a}_{\mathbf{q}s}) + y\hat{b}_{\mathbf{q}s}], \tag{28.89}$$

where

$$A = \frac{i}{\sqrt{2}}(cq\tilde{\omega}_p)^{1/2}. \tag{28.90}$$

By inserting Eqs. (28.86)–(28.89) into Eq. (28.85), s solution requires that the individual coefficients to $\hat{a}_{\mathbf{q}s}$, $\hat{b}_{\mathbf{q}s}$, $\hat{a}_{-\mathbf{q}s}^{\dagger}$ and $\hat{b}_{-\mathbf{q}s}^{\dagger}$ all vanish. This implies that the constants satisfy the following set of homogeneous equations:

$$
\begin{pmatrix}
cq - \omega & -A & 0 & 0 \\
A & \tilde{\omega}_p - \omega & A & 0 \\
0 & A & -(cq + \omega) & 0 \\
0 & 0 & 0 & -(\omega_p + \omega)
\end{pmatrix}
\begin{pmatrix} w \\ x \\ y \\ z \end{pmatrix}
=
\begin{pmatrix} 0 \\ 0 \\ 0 \\ 0 \end{pmatrix},
\tag{28.91}
$$

with $\omega_{qs} \equiv \omega$. A solution different from $(w, x, y, z) = (0, 0, 0, 0)$ requires that the determinant of Eq. (28.91) is zero, i.e.

$$
\begin{vmatrix}
cq - \omega & -A & 0 & 0 \\
A & \tilde{\omega}_p - \omega & A & 0 \\
0 & A & -(cq + \omega) & 0 \\
0 & 0 & 0 & -(\omega_p + \omega)
\end{vmatrix}
= 0.
\tag{28.92}
$$

Remembering that only positive values of ω are of interest, Eq. (28.92) leads to the relation

$$
\omega^2 = \omega \tilde{\omega}_p + (cq)^2.
\tag{28.93}
$$

In the "dressed oscillator" picture, Eq. (28.93) is, on implicit form, the dispersion relation for plasmaritons. Since $\omega > 0$,

$$
\omega = \frac{\tilde{\omega}_p}{2} \left\{ 1 + \left[1 + \left(\frac{2cq}{\tilde{\omega}_p} \right)^2 \right]^{1/2} \right\},
\tag{28.94}
$$

giving the expected limiting values $\omega = \tilde{\omega}_p$ $(q \to 0)$, and $\omega = cq$ $(q \to \infty)$.

Utilizing the commutation relations in Eqs. (28.70), (28.71), (28.74) and (28.75), and the fact that all particle and field operators commute mutually, one obtains

$$
[\hat{\alpha}_{\mathbf{q}s}, \hat{\alpha}_{\mathbf{q}s}^{\dagger}] = |w|^2 + |x|^2 - |y|^2 - |z|^2.
\tag{28.95}
$$

Since the w, x, y and z constants satisfy the homogeneous set of equations in Eq. (28.91) one may express x, y and z in terms of w, say. The free parameter w then can be adjusted so that

$$
[\hat{\alpha}_{\mathbf{q}s}, \hat{\alpha}_{\mathbf{q}s}^{\dagger}] = 1.
\tag{28.96}
$$

For the α-oscillators, obtained by diagonalization of the Hamiltonian belonging to the coupled particle and field oscillators, Eq. (28.96) implies that $\hat{\alpha}_{\mathbf{q}s}$ and $\hat{\alpha}_{\mathbf{q}s}^{\dagger}$ play the role of annihilation and creation operators. Hence,

$$
\hat{N} = \sum_{\mathbf{q}s} \hat{\alpha}_{\mathbf{q}s}^{\dagger} \hat{\alpha}_{\mathbf{q}s}
\tag{28.97}
$$

is the global number operator,

$$\hat{\mathbf{P}} = \sum_{\mathbf{q}s} \hbar\mathbf{q}\,\hat{\alpha}^{\dagger}_{\mathbf{q}s}\hat{\alpha}_{\mathbf{q}s} \tag{28.98}$$

is the global momentum operator, and

$$\hat{H} = \sum_{\mathbf{q}s} \hbar\omega_q \left(\hat{\alpha}^{\dagger}_{\mathbf{q}s}\hat{\alpha}_{\mathbf{q}s} + \frac{1}{2} \right) \tag{28.99}$$

is the global Hamilton operator, including for each $(\mathbf{q}, s)$-mode a zero-point energy $\hbar\omega_q/2$.

In the so-called Rotating-Wave Approximation (RWA) one neglects the energy nonconserving terms $\hat{a}^{\dagger}_{-\mathbf{q}s}\hat{b}^{\dagger}_{\mathbf{q}s}$ and $\hat{a}_{\mathbf{q}s}\hat{b}_{\mathbf{q}s}$ in the interaction Hamiltonian [Eq. (28.80)]. These terms correspond to two-particle generation $(\hat{a}^{\dagger}_{-\mathbf{q}s}\hat{b}^{\dagger}_{\mathbf{q}s})$ and absorption $(\hat{a}_{\mathbf{q}s}\hat{b}_{\mathbf{q}s})$ processes. It is known, for instance for atom-field interaction studies, that these processes are less important near resonance [182, 185, 230, 235]. RWA usually simplifies calculations substantially. In the plasmariton case, it appears from Eq. (28.81) that only terms proportional to $\hat{a}_{\mathbf{q}s}$ and $\hat{b}_{\mathbf{q}s}$ are needed in the RWA ansatz for $\hat{\alpha}_{\mathbf{q}s}$. Thus, with

$$\hat{\alpha}^{\mathrm{RWA}}_{\mathbf{q}s} = w\hat{a}_{\mathbf{q}s} + x\hat{b}_{\mathbf{q}s}, \tag{28.100}$$

the 4x4 determinant condition in Eq. (28.92) is replaced by the 2x2-condition

$$\begin{vmatrix} cq - \omega & -A \\ A & \tilde{\omega}_p - \omega \end{vmatrix} = 0, \tag{28.101}$$

giving the following RWA dispersion relation

$$\omega^{\mathrm{RWA}} = \frac{1}{2} \left\{ \tilde{\omega}_p + cq + \left[\tilde{\omega}_p^2 + (cq)^2 \right]^{1/2} \right\}. \tag{28.102}$$

28.5 SURFACE PLASMARITONS WITH ELECTRODYNAMICALLY ACTIVE SELVEDGE

28.5.1 Electrodynamically active surface sheet

In Chapter 19, we studied the classical theory of surface plasmaritons within the framework of the SCIB model for adjacent jellium-vacuum half-spaces. In this specular reflection model the field-induced component of the surface current density perpendicular to the surface is zero, that is [in the notation of Chapter 19]

$$J_z(z \to 0^+; q_{\|}, \omega) = 0. \tag{28.103}$$

In classical electrodynamics all textbooks and most research literature claim that Eq. (28.103) holds universally. The absence of a surface current density in the z-direction is correct of course for $\omega = 0$, but not in general for $\omega \neq 0$. The reader may find a detailed analysis of the dynamical jump (boundary) conditions for the electromagnetic field in [119, 132, 141, 280]. Using for a given $(q_\parallel, \omega)$-component $[\mathbf{q}_\parallel = q_\parallel \hat{x}]$ the notation $\|F\| \equiv F(z = 0^+) - F(z = 0^-)$, the dynamic jump conditions for p-polarized fields, which are the relevant ones for surface plasmaritons, are [141]

$$\|E_x\| = \frac{1}{\varepsilon_0} \frac{q_\parallel}{\omega} J_z^S, \tag{28.104}$$

$$\|D_z\| = \frac{q_\parallel}{\omega} J_x^S, \tag{28.105}$$

$$\|B_y\| = -\mu_0 J_x^S, \tag{28.106}$$

where $\mathbf{J}^S$ is the surface current density. The underlying assumption that a current can flow (exist) in an infinitely thin sheet is of course incorrect physically, but, roughly speaking, the assumption may be a good approximation when the characteristic wavelength of the electromagnetic field is much larger than the physical selvedge [see Subsection 28.5.2]. The jellium-vacuum interface is said to be *passive* if the approximation $\mathbf{J}^S(0; q_\parallel, \omega) = \mathbf{0}$ can be used. The surface plasmariton dispersion relation $[q_\parallel = q_\parallel(\omega)]$ derived in Section 19.2, and given on implicit form in Eq. (19.41), and explicitly in Eq. (19.43), was derived assuming that the interface is passive.

As a first step towards deriving the surface plasmariton dispersion relation (eigenmode condition) for an *electrodynamically active* surface, i.e. a surface for which either $J_x^S(0; q_\parallel, \omega)$ or $J_z^S(0; q_\parallel, \omega)$ [or both] are nonvanishing, one has to use two of the three jump conditions in Eqs. (28.104)–(28.106). Since J_z^S only enters Eq. (28.104), this relation must be one of the two. Since $\|B_y\| = -\mu_0(\omega/q_\parallel)\|D_z\|$, one may use either Eq. (28.105) or Eq. (28.106) as the second relation. Both choices will lead to the same eigenmode condition.

Using the p-polarized ansatz [Vacuum: V, jellium: J]

$$\mathbf{E}^V(z; q_\parallel, \omega) = A_V \left[\hat{\mathbf{x}} + \frac{q_\parallel}{q_\perp^0} \hat{\mathbf{z}}\right] e^{-iq_\perp^0 z}, \tag{28.107}$$

$$\mathbf{E}^J(z; q_\parallel, \omega) = A_J \left[\hat{\mathbf{x}} - \frac{q_\parallel}{q_\perp} \hat{\mathbf{z}}\right] e^{iq_\perp^0 z}, \tag{28.108}$$

where

$$q_\perp^0 = \left[\left(\frac{\omega}{c}\right)^2 - q_\parallel^2\right]^{1/2}, \tag{28.109}$$

$$q_\perp = \left[\left(\frac{\omega}{c}\right)^2 \varepsilon(\omega) - q_\parallel^2\right]^{1/2}, \tag{28.110}$$

together with Eq. (28.104), and say, Eq. (28.105), a first step in the determination of the surface plasmariton dispersion relation [in what follows for a spatially local jellium, i.e., $\varepsilon = \varepsilon(\omega)$] can be taken. In passing, it should be noticed that the ansatz used guarantees that the electric field is divergence-free $[\nabla \cdot \mathbf{E} = 0]$ in both the vacuum and jellium domains. The jump conditions and the ansatz thus lead to the following two equations:

$$A_J - A_V = \frac{1}{\varepsilon_0} \frac{q_\parallel}{\omega} J_z^S, \tag{28.111}$$

$$\frac{\varepsilon(\omega)}{q_\perp} A_J + \frac{1}{q_\perp^0} A_V = -\frac{1}{\varepsilon_0 \omega} J_x^S. \tag{28.112}$$

For a passive surface ($J_x^S = J_z^S = 0$), the eigenmodes are obtained setting the determinant of Eqs. (28.111) and (28.112) to zero. Thus,

$$\begin{bmatrix} 1 & -1 \\ \frac{\varepsilon(\omega)}{q_\perp} & \frac{1}{q_\perp^0} \end{bmatrix} = 0 \Leftrightarrow q_\perp + q_\perp^0 \varepsilon(\omega) = 0, \tag{28.113}$$

gives the implicit dispersion relation in Eq. (19.41), of course.

28.5.2 Microscopic selvedge electrodynamics. Dispersion relation

At the outmost atomic layers of a metal (or semiconductor) surface, below treated in the jellium approximation, the electron density changes from zero (in vacuum) to its homogeneous bulk value. It is the microscopic current density, $\mathbf{J}^{\text{SE}}(z, q_\parallel, \omega)$, in this so-called selvedge (SE) region one relates to the ideal surface current density $\mathbf{J}^S(0; q_\parallel, \omega)$ introduced in Subsection 28.5.1. In the absence of an external driving field the microscopic electric field inside the selvedge profile, $\mathbf{E}^{\text{SE}}(z, q_\parallel, \omega)$, is connected to the selvedge current density by an integral relation [147]

$$\mathbf{E}^{\text{SE}}(z) = i\mu_0 \omega \int_{\text{SE}} \mathbf{G}(z, z') \cdot \mathbf{J}^{\text{SE}}(z') dz'. \tag{28.114}$$

For notational simplicity the reference to $q_\parallel$ and ω has been omitted from the various quantities. The integration in Eq. (28.114) extends over the selvedge region. The propagator, $\mathbf{G}(z, z')$, has two pieces, viz. (i) a part related to the direct propagation between the source (z') and observation (z) planes, and (ii) an indirect part reaching the plane of observation upon reflection from the bulk jellium. As discussed in Section 19.1, the SCIB model is most useful in describing the reflection from the bulk jellium because the associated propagator can be constructed. The relevant Green tensor $\mathbf{G}(z, z') = \mathbf{G}^{<<}(z, z')$ is given in Subsection 19.1.3.

In order to obtain an integral equation for the selvedge field one combines Eq. (28.114) with the constitutive equation

$$\mathbf{J}^{\text{SE}}(z) = \int_{\text{SE}} \boldsymbol{\sigma}^{\text{SE}}(z, z') \cdot \mathbf{E}^{\text{SE}}(z') dz', \tag{28.115}$$

where

$$\boldsymbol{\sigma}^{\text{SE}}(z, z') \equiv \boldsymbol{\sigma}(z, z') - \boldsymbol{\sigma}^{\text{BULK}}(z, z') \tag{28.116}$$

is the selvedge conductivity tensor defined as the difference between the exact linear conductivity tensor of the inhomogeneous jellium $\boldsymbol{\sigma}(z, z')$, and the bulk conductivity tensor, $\boldsymbol{\sigma}^{\mathrm{BULK}}(z, z')$. In terms of the dyadic kernel

$$\mathbf{K}(z, z') = i\mu_0\omega \int_{\mathrm{SE}} \mathbf{G}(z, z'') \cdot \boldsymbol{\sigma}^{\mathrm{SE}}(z'', z')dz'', \qquad (28.117)$$

the homogeneous integral equation for the local field inside the selvedge takes the form

$$\mathbf{E}^{\mathrm{SE}}(z) = \int_{\mathrm{SE}} \mathbf{K}(z, z') \cdot \mathbf{E}^{\mathrm{SE}}(z')dz'. \qquad (28.118)$$

It appears from Eq. (22.36) that the selvedge conductivity tensor has the following sum over tensor-product structure:

$$\boldsymbol{\sigma}^{\mathrm{SE}}(z, z') = \sum_N A_N \mathbf{B}_N(z)\mathbf{C}_N(z') - \sum_n a_n \mathbf{b}_n(z)\mathbf{c}_n(z'), \qquad (28.119)$$

where the first and second sum relates to the tensor product structure of $\boldsymbol{\sigma}(z, z')$ and $\boldsymbol{\sigma}^{\mathrm{BULK}}(z, z')$, respectively. The quantities N and n run over pairs of (many-body) energy eigenstates, see Eq. (22.36). The structure of Eq. (28.119) allows one to convert the integral for the selvedge field [Eq. (28.118)] into a matrix problem for determination of the selvedge eigenmodes, as described in detail in [154].

The sheet surface current density $\mathbf{J}^S$, introduced in Subsection 28.5.1, is to be identified as the integral of the selvedge current density over the selvedge profile, i.e.,

$$\mathbf{J}^S = \int_{\mathrm{SE}} \mathbf{J}^{\mathrm{SE}}(z)dz = \int_{\mathrm{SE}} \boldsymbol{\sigma}^{\mathrm{SE}}(z, z') \cdot \mathbf{E}^{\mathrm{SE}}(z')dz'. \qquad (28.120)$$

For consistency, the electric field used in Subsection 28.5.1 [Eq. (28.107)] on the vacuum side at $z = 0^-$, viz.,

$$\mathbf{E}^V(z = 0^-) = A_V \left(\hat{\mathbf{x}} + \frac{q_\parallel}{q_\perp^0}\hat{\mathbf{z}} \right) \qquad (28.121)$$

necessarily must be connected to the local selvedge field, $\mathbf{E}^{\mathrm{SE}}(z)$, by the following integral relation:

$$\mathbf{E}^V(z = 0^-) = \frac{\int_{\mathrm{SE}} \boldsymbol{\sigma}^{\mathrm{SE}}(z, z') \cdot \mathbf{E}^{\mathrm{SE}}(z')dz'dz}{\int_{\mathrm{SE}} \boldsymbol{\sigma}^{\mathrm{SE}}(z, z')dz'dz}. \qquad (28.122)$$

Thus,

$$\mathbf{J}^S = \mathbf{S} \cdot \mathbf{E}^V(z = 0^-), \qquad (28.123)$$

where

$$\mathbf{S} = \int_{\mathrm{SE}} \boldsymbol{\sigma}^{\mathrm{SE}}(z, z')dz'dz. \qquad (28.124)$$

It is obvious that an *ab initio* microscopic calculation of $\mathbf{S}$ is an extremely challenging task.

In order to determine the surface plasmariton dispersion relation, let us write Eq. (28.123) in the 2x2-matrix form, viz.,

$$\begin{pmatrix} J_x^S \\ J_z^S \end{pmatrix} = \begin{pmatrix} S_{xx} & S_{xz} \\ S_{zx} & S_{zz} \end{pmatrix} \begin{pmatrix} 1 \\ \frac{q_\parallel}{q_\perp^0} \end{pmatrix} A_V . \tag{28.125}$$

By inserting the expressions for J_x^S and J_z^S into Eq. (28.111) and (28.112) one obtains two homogeneous equations among A_J and A_V. To have a nonzero solution for the field amplitudes it is required that the related determinant is zero. This demand gives one the following dispersion relation (on implicit form):

$$q_\perp + \varepsilon(\omega)q_\perp^0 + \frac{q_\perp^0}{\varepsilon_0\omega}\left[\varepsilon(\omega)q_\parallel\left(S_{zx} + \frac{q_\parallel}{q_\perp^0}S_{zz}\right) + q_\perp\left(S_{xx} + \frac{q_\parallel}{q_\perp^0}S_{xz}\right)\right] = 0. \tag{28.126}$$

For a passive surface $[\mathbf{S} = \mathbf{0}]$, the dispersion relation is reduced to the one given in Eq. (28.113). The determination of the dispersion for surface plasmaritons with active surface electrodynamics hence has been reduced to a microscopic calculation of $\mathbf{S}$ from Eq. (28.124).

The short length of the selvedge profile often makes it possible to neglect field retardation effects across the selvedge. In case, the vacuum propagator takes the nonretarded (NR) form

$$\mathbf{G}^{\mathrm{NR}}(Z; q_\parallel, \omega) = -q_0^{-2}\delta(z)\hat{\mathbf{z}}\hat{\mathbf{z}} + \frac{q_\parallel}{2q_0^2}e^{-q_\parallel|Z|}\left[\hat{\mathbf{z}}\hat{\mathbf{z}} - \hat{\mathbf{q}}_\parallel\hat{\mathbf{q}}_\parallel - i(\hat{\mathbf{q}}_\parallel\hat{\mathbf{z}} + \hat{\mathbf{z}}\hat{\mathbf{q}}_\parallel)\mathrm{sgn}(Z)\right]$$

$$= -q_0^{-2}\delta_L(Z; q_\parallel) \tag{28.127}$$

in disk contraction [157]. A further reduction, often used in electrodynamics reflection and transmission studies from selvedges, is obtained keeping only the singular term of Eq. (28.127). In this so-called self-field (SF) approximation where $\mathbf{G}(Z; q_\parallel, \omega)$ is replaced by

$$\mathbf{G}_{\mathrm{SF}}(Z; \omega) = -q_0^{-2}\delta(z)\hat{\mathbf{z}}\hat{\mathbf{z}}, \tag{28.128}$$

a spatially local relation is obtained between the electric selvedge field and the surface current density, namely

$$\mathbf{E}_{\mathrm{SF}}^{\mathrm{SE}}(z) = \frac{1}{i\varepsilon_0\omega}J_z^{\mathrm{SE}}\hat{\mathbf{z}}. \tag{28.129}$$

In the SF approximation, the selvedge field is perpendicular to the plane of the surface, and only the z-component of the surface current density enters the analysis. From Eq. (28.117) it appears that the self-field kernel is given by

$$\mathbf{K}_{\mathrm{SF}}(z, z') = \frac{1}{i\varepsilon_0\omega}\left[\sigma_{zx}^{\mathrm{SE}}(z, z')\hat{\mathbf{z}}\hat{\mathbf{x}} + \sigma_{zz}^{\mathrm{SE}}(z, z')\hat{\mathbf{z}}\hat{\mathbf{z}}\right]. \tag{28.130}$$

The absence of $\sigma_{xx}^{\mathrm{SE}}(z, z')$ and $\sigma_{xz}^{\mathrm{SE}}(z, z')$ from the description implies that $S_{xx} = S_{xz} = 0$ in the dispersion relation in Eq. (28.126). The dispersion relation then can be written as follows:

$$q_\perp + \varepsilon_{\text{eff}}(q_\parallel, \omega) q_\perp^0 = 0, \tag{28.131}$$

where

$$\varepsilon_{\text{eff}}(q_\parallel, \omega) = \varepsilon(\omega) \left[1 + \frac{q_\parallel}{\varepsilon_0 \omega} \left(S_{zx}(q_\parallel, \omega) + \frac{q_\parallel}{q_\perp^0} S_{zz}(q_\parallel, \omega) \right) \right]. \tag{28.132}$$

It appears that the dispersion relation in Eq. (28.131) has the same *form* as that of the passive case [Eq. (28.113)], just with the local dielectric function $\varepsilon(\omega)$ replaced by $\varepsilon_{\text{eff}}(q_\parallel, \omega)$ [Eq. (28.132)]. The spatial dispersion in $\varepsilon_{\text{eff}}(q_\parallel, \omega)$ originates in the selvedge dynamics. For both the passive and active case it was assumed that the spatial dispersion in the bulk jellium is negligible. If one includes (i) spatial dispersion in the bulk, (ii) use passive boundary conditions and (iii) neglects the L-mode ($D \to 0$), one obtains from Eq. (19.40) a surface plasmariton dispersion relation

$$q_\perp + \varepsilon_T(q, \omega) q_\perp^0 = 0. \tag{28.133}$$

A comparison of Eqs. (28.131) and (28.133) shows that the spatial dispersion of the bulk and the active selvedge enter the formalism in a similar fashion, at least formally.

28.5.3   Connection to the plasmariton quasiparticle picture

In order to make contact to the plasmariton quasiparticle picture discussed in Sections 28.1–28.4, one must determine the transverse self-field of the surface plasmariton, and from this the tied T-photon cloud associated with the surface current density can be identified.

It appears from the loop equation for the selvedge field given in Eq. (28.118) that the self-field part of the selvedge field must satisfy the integral equation

$$\mathbf{E}_{\text{SF}}^{\text{SE}}(z) = \int_{\text{SE}} \mathbf{K}_{\text{SF}}(z, z') \cdot \mathbf{E}_{\text{SF}}^{\text{SE}}(z') dz', \tag{28.134}$$

where $\mathbf{K}_{\text{SF}}(z, z')$ is the self-field part of the kernel $\mathbf{K}(z, z')$ determined retaining only the nonretarded part of the propagator, viz., $\mathbf{G}^{\text{NR}}(Z; q_\parallel, \omega) = -q_0^{-2} \delta_L(Z; q_\parallel)$. Hence,

$$\mathbf{K}_{\text{SF}}(z) = \frac{1}{i\varepsilon_0 \omega} \int_{\text{SE}} \delta_L(z - z'') \cdot \sigma^{\text{SE}}(z'', z') dz''. \tag{28.135}$$

Let us now return to the ideal model described in terms of a completely confined surface current density, $\mathbf{J}^S$. This current density can always be written as a sum of longitudinal (L) and transverse (T) parts, and in planes outside $z = 0$ one hae

$$\mathbf{J}_L(z) = \delta_L(z) \cdot \mathbf{J}^S, \quad z \neq 0 \tag{28.136}$$

$$\mathbf{J}_T(z) = \delta_T(z) \cdot \mathbf{J}^S, \quad z \neq 0 \tag{28.137}$$

For $z \neq 0$, $\boldsymbol{\delta}_T(z) = -\boldsymbol{\delta}_L(z)$, implying that

$$\mathbf{J}_T(z) = -\mathbf{J}_L(z), \quad z \neq 0 \tag{28.138}$$

outside the plasmariton surface current density. The transverse electric field generated by $\mathbf{J}_T(z)$ is given by

$$\mathbf{E}_T(z) = i\mu_0\omega \int_{-\infty}^{\infty} g(z - z')\mathbf{J}_T(z')dz'. \tag{28.139}$$

The transverse electric field is obtained from Eq. (28.139) letting $c \rightarrow \infty$ in the Huygens propagator. Hence, one obtains

$$\mathbf{E}_T^{\mathrm{SF}}(z) = \frac{i\mu_0\omega}{2q_{\parallel}} \int_{-\infty}^{\infty} e^{-q_{\parallel}|z-z'|}\mathbf{J}_T(z')dz' \tag{28.140}$$

in planes inside the jellium halfspace.

Let us finish this subsection with a few remarks related to a covariant description of photon wave mechanics established in [155], see also [158]. It appears from the covariant theory that $\mathbf{E}_T^{\mathrm{SF}}(z)$ describes the part of the T-photon emission from $\mathbf{J}_T(z)$ which is coupled back to the $\mathbf{J}_T(z)$-domain. For instance, it was shown that the cycle-average $(< \dots >)$ momentum flow of the T-photon field in the z-direction is given by [162]

$$\hat{\mathbf{z}}\cdot < \mathbf{g}_{TT}(z) >= -\hat{\mathbf{z}}\cdot < \mathbf{g}_{LT}(z) > \tag{28.141}$$

where $\mathbf{g}_{TT}$ is associated with $\mathbf{E}_T \times \mathbf{B}$, and $\mathbf{g}_{LT}$ with $\mathbf{E}_L \times \mathbf{B}$. Eq. (28.141) implies that there is no net momentum flow perpendicular to the plane of the surface current. Furthermore, it turns out from the covariant analysis that

$$< \mathbf{g}_{TT}(z) > \cdot\hat{\mathbf{z}} < 0. \tag{28.142}$$

Roughly speaking this inequality indicates that the T-photons are pull back towards the surface plasmariton current density sheet. In this sense, $\mathbf{E}_T^{\mathrm{SF}}(z)$, given by Eq. (28.140) describes the T-photon cloud which is tied to the surface plasmariton [128]; see also Fig. 28.2.

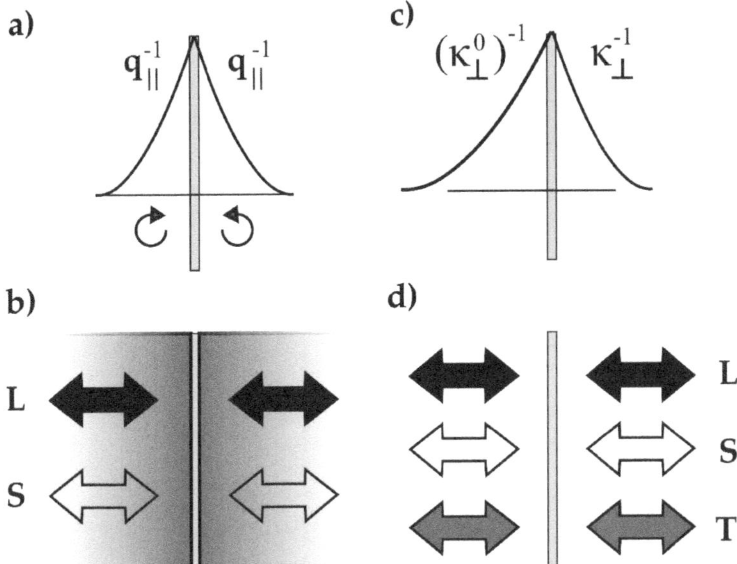

Figure 28.2 Graphical representation of the two types of decay length which relates to a T-photon tied to a surface plasmariton. Fig. 5a: The one-dimensional confined source of the T-photon has an extension charaterized by the decay length $q_{||}^{-1}$ (the same in vacuum and jellium), and the self field is circular polarized in opposite directions. As $q_{||} \to \infty$, one approaches (via the electrostatic region) the geometrical optical limit where essentially complete spatial confinement can be obtained [see Ref. [158] Section 5.2]. Fig 5b: The extension of the T-photon source region (in violet) is in the Coulomb gauge determined by the dynamic Coulomb field extension of the electrons in the current density sheet. In the covariant description this dynamic field is replaced by the L- and S-photon dynamics (black and white double arrows, repsectively). Fig 5c: The range of the tied 1D T-photon domain is characterized by the two decay lengths $(\kappa_\perp^0)^{-1}$ (in vacuum) and $\kappa_\perp^{-1}$ (in the jellium). Fig. 5d: As indicated by the three double arrows a dynamically balanced interaction between the L-, S- and T-photons in the surface plasmariton state prevents the T-photon (red double arrow) from propagating to infinty (be radiated).

VIII

Photons in Curved Structures

CHAPTER 29

The Classical Light Particle: Geodesics and Free Fall

29.1 INHOMOGENEOUS VACUUM. EIKONAL EQUATION

Let us consider a nonconducting isotropic medium, and let us assume that the relative permittivity, ε, and permeability, μ are frequency independent. This assumption may be satisfied approximately in a certain frequency range, but it is in most cases a too dramatic simplification. Nevertheless, for the point of view discussed below the assumption is sufficient. What is important here is that we allow the related refractive index

$$n(\mathbf{r}) = [\varepsilon(\mathbf{r})\mu(\mathbf{r})]^{1/2} \tag{29.1}$$

to be inhomogeneous. By neglecting the frequency (ω) dependence of $n(\mathbf{r})$ there is no longer any physically ACTIVE interaction between field and matter. In the perspective of photon physics, and with an eye to the space-time metric in general relativity, I have in my book "Light – The Physics of the Photon" [158] suggested to use the words INHOMOGENEOUS VACUUM for a solid-state medium described by Eq. (29.1). In an inhomogeneous vacuum the local phase velocity of light is

$$v(\mathbf{r}) = \frac{c}{n(\mathbf{r})}, \tag{29.2}$$

and the basic field equations are in the space-frequency domain given by

$$\nabla \times \mathbf{E}(\mathbf{r};\omega) = i\omega\mathbf{B}(\mathbf{r};\omega), \tag{29.3}$$

$$\nabla \times \mathbf{B}(\mathbf{r};\omega) = \frac{i\omega}{c^2}n^2(\mathbf{r})\mathbf{E}(\mathbf{r};\omega), \tag{29.4}$$

$$\nabla \cdot [n^2(\mathbf{r})\mathbf{E}(\mathbf{r};\omega)] = 0, \tag{29.5}$$

$$\nabla \cdot \mathbf{B}(\mathbf{r};\omega) = 0. \tag{29.6}$$

From a classical light-particle point of view a special type of electromagnetic fields are of central importance, viz., those of the form

$$\mathbf{E}(\mathbf{r};\omega) = \mathbf{E}^0(\mathbf{r})e^{iq_0 S(\mathbf{r})}, \tag{29.7}$$

$$\mathbf{B}(\mathbf{r};\omega) = \mathbf{B}^0(\mathbf{r})e^{iq_0 S(\mathbf{r})}, \tag{29.8}$$

DOI: 10.1201/9781003029458-29

where $q_0 = \omega/c$. The quantity $S(\mathbf{r})$ is the so-called *eikonal* [46]. The term eikonal (from Greek $\sim$ image) was introduced in 1895 by Bruns [51] to describe functions related to the characteristic function of Hamilton [65]. For sufficiently high frequencies the ansatz in Eqs. (29.7) and (29.8), when inserted into the fundamental field equations given in Eqs. (29.3) and (29.4), leads to the following frequency independent relations:

$$\boldsymbol{\nabla} S(\mathbf{r}) \times \mathbf{E}^0(\mathbf{r}) = c\mathbf{B}^0(\mathbf{r}), \tag{29.9}$$

$$\boldsymbol{\nabla} S(\mathbf{r}) \times \mathbf{B}^0(\mathbf{r}) = -\frac{1}{c}n^2(\mathbf{r})\mathbf{E}^0(\mathbf{r}). \tag{29.10}$$

From Eqs. (29.5) and (29.6), the high-frequency approximation shows that $\mathbf{E}^0 \cdot \boldsymbol{\nabla} S(\mathbf{r}) = \mathbf{B}^0(\mathbf{r}) \cdot \boldsymbol{\nabla} S(\mathbf{r}) = 0$, equations which also follow from Eqs. (29.9) and (29.10) by scalar multiplication with $\boldsymbol{\nabla} S(\mathbf{r})$. It appears from a combination of Eqs. (29.9) and (29.10), and with the use of $\mathbf{E}^0(\mathbf{r}) \cdot \boldsymbol{\nabla} S(\mathbf{r}) = 0$, that these equations only have nontrivial solution provided

$$\boldsymbol{\nabla} S(\mathbf{r}) \cdot \boldsymbol{\nabla} S(\mathbf{r}) = n^2(\mathbf{r}). \tag{29.11}$$

The above equation is known as the *eikonal equation* in geometrical optics.

29.2   CLASSICAL LIGHT-PARTICLE STRIP ON A GEODETIC BUNDLE

The surfaces given by

$$S(\mathbf{r}) = C, \tag{29.12}$$

where C is a constant (parameter), are called the *geometrical wave surfaces*. The orthogonal trajectories to these surfaces are called the (geometrical) *light rays*. The local unit vector of a light ray is given by

$$\mathbf{t}(\mathbf{r}) = \frac{\boldsymbol{\nabla} S(\mathbf{r})}{|S(\mathbf{r})|} = \frac{\boldsymbol{\nabla} S(\mathbf{r})}{n(\mathbf{r})}. \tag{29.13}$$

The last expression in Eq. (29.13) follows from the eikonal equation.

Let us consider the cycle-average field momentum at the space-point $\mathbf{r}$, namely,

$$<\mathbf{P}>(\mathbf{r}) = \frac{\varepsilon_0}{2}\,\mathrm{Re}[\mathbf{E}^*(\mathbf{r};\omega) \times \mathbf{B}(\mathbf{r};\omega)] = \frac{\varepsilon_0}{2}\,\mathrm{Re}[(\mathbf{E}^0(\mathbf{r}))^* \times \mathbf{B}^0(\mathbf{r})]. \tag{29.14}$$

A multiplication of Eq. (29.9) from the right by $\mathbf{E}^0(\mathbf{r})$ now gives

$$<\mathbf{P}>(\mathbf{r}) = \frac{\varepsilon_0}{c}\left|\mathbf{E}^0(\mathbf{r})\right|^2 \boldsymbol{\nabla} S(\mathbf{r}) = \frac{\varepsilon_0}{2c}n(\mathbf{r})\left|\mathbf{E}^0(\mathbf{r})\right|^2 \mathbf{t}(\mathbf{r}). \tag{29.15}$$

The cycle-average local field momentum hence is seen to point in the direction of the local unit vector of the light particle. In a particle-like description we now seek to localize the field along the $\mathbf{t}(\mathbf{r})$-direction. Such a localization is accompanied by a weak localization parallel to the relevant wave-front bundle. Hence our classical light particle is a strip.

Partial localization is obtained by making a Fourier integral transformation of the electric field, as this appears in the eikonal appraoch. Thus

$$\mathbf{E}^0(t - c^{-1}S(\mathbf{r})) = \int_{-\infty}^{\infty} \mathbf{E}^0(\mathbf{r};\omega)e^{-i\omega(t-c^{-1}S(\mathbf{r}))}\frac{d\omega}{2\pi}. \tag{29.16}$$

Only if one assumes (somewhat incorrectly) that the eikonal approximation $\mathbf{E}^0(\mathbf{r};\omega) = \mathbf{E}^0(\mathbf{r})$ can be used in Eq. (29.16) one obtains the desired complete localization, along the light ray, i.e.,

$$\mathbf{E}^0(t - c^{-1}S(\mathbf{r})) = \mathbf{E}^0(\mathbf{r})\delta(t - c^{-1}S(\mathbf{r})). \tag{29.17}$$

29.3 GEODETIC LINE IN INHOMOGENEOUS VACUUM

As a step towards general relativity it is rewarding to consider the eikonal equation and the related light rays from a somewhat different point of view. Hence, let

$$ds = (g_{ij}(\mathbf{r})dx^i dx^j)^{1/2}, \quad i,j = 1 - 3 \tag{29.18}$$

be the infinitesimal line element in a three-dimensional Riemannian space, here written in Cartesian coordinates. The quantity $g_{ij}(\mathbf{r}) \equiv g_{ij}(\{x^i\})$ is the ij-component of the metric tensor $\{g_{ij}\}$. In Eq. (29.18) the summation over repeated indices is implicit.

A *geodetic line* between two points in space, say P_1 and P_2, is a line drawn in such a manner that the line integral $\int_{P_1}^{P_2} ds$ is *stationary* [not necessarily a minimum, as often claimed]. The equation for the geodetic line is obtained from the variational principle

$$\delta \int_{P_1}^{P_2} (g_{ij}dx^i dx^j)^{1/2} = 0. \tag{29.19}$$

For an isotropic inhomogeneous vacuum the metric tensor must be diagonal, and the length of the Riemannian line element, $ds = ds(\mathbf{r})$, only can be determined by the local $(\mathbf{r})$ wavelength of light, which is proportional to $n^{-1}(\mathbf{r})$. Thus, the line element must be proportional to the refractive index, $n(\mathbf{r})$, that is

$$ds(\mathbf{r}) = n(\mathbf{r})(dx^2 + dy^2 + dz^2)^{1/2}. \tag{29.20}$$

The related metric tensor, with components

$$g_{ij}(\mathbf{r}) = n^2(\mathbf{r})\delta_{ij}, \tag{29.21}$$

is reduced to the Euclidean metric tensor in vacuum, $g_{ij}(\mathbf{r}) = \delta_{ij}$.

Parametrized by an appropriate σ the variational problem in Eq. (29.19) takes a well-known form, viz.,

$$\delta \int_{P_1}^{P_2} L d\sigma = \int_{P_1}^{P_2} \delta L d\sigma = 0, \tag{29.22}$$

where

$$L = L\left(\{x^i\}, \left\{\frac{dx^i}{d\sigma}\right\}\right) = \left[g_{ij}(\{x^i\})\frac{dx^i}{d\sigma}\frac{dx^j}{d\sigma}\right]^{1/2} \tag{29.23}$$

is the Lagrangian density. In order to satisfy Eq. (29.22), the Euler-Lagrange equation

$$\frac{\partial L}{\partial x^i} - \frac{d}{d\sigma}\left[\frac{\partial L}{\partial(dx^i/d\sigma)}\right] = 0 \tag{29.24}$$

must hold for $i = 1, 2, 3$. By means of the explicit expression for L given in Eq. (29.23), with $g_{ij}(\mathbf{r})$ given by Eq. (29.21), one obtains after some algebra

$$\frac{d}{ds}\left(n\frac{dx^i(s)}{ds}\right) = \frac{\partial n}{\partial x^i}, \quad i = 1, 2, 3 \tag{29.25}$$

This equation is just the ith component of the differential equation for the light ray, $\mathbf{r} = \mathbf{r}(s)$, viz., [46, 158]

$$\frac{d}{ds}\left(n(\mathbf{r})\frac{d\mathbf{r}(s)}{ds}\right) = \boldsymbol{\nabla}n(\mathbf{r}), \tag{29.26}$$

where s is the arc length (measured from a selected point P_0 [$s = 0$]). The tangential unit vector in Eq. (29.13) is $\mathbf{t} = d\mathbf{r}(s)/ds$, giving $n(\mathbf{r})d\mathbf{r}(s)/ds = \boldsymbol{\nabla}S(\mathbf{r})$. Differentiation of this relation with respect to s leads to Eq. (29.25), as the reader may prove to herself, or by consulting [46, 158]. The result in Eq. (29.26) is well known, and is in many textbooks derived without making use of the variational principle.

29.4 TIME-LIKE METRIC GEODESICS IN GENERAL RELATIVITY

In a generalization from flat Minkowski space (also called pseudo-Euclidean space) to curved four-dimensional Riemann space form the metric tensor of vacuum $g_{ij} = \delta_{ij}$ [or $g_{ij} = n^2(\mathbf{r})\delta_{ij}$ of an inhomogeneous vacuum] is replaced by $\{g_{\mu\nu}(x)\}$, $\mu, \nu = 0 - 3$. Setting the speed of light to unity ($c = 1$ units), the infinitesimal proper time is given by

$$d\tau = +(-g_{\mu\nu}(x)dx^\mu dx^\nu)^{1/2}, \tag{29.27}$$

and related to the invariant infinitesimal squared distance (ds^2) by $ds^2 = -d\tau^2$. The set of curvilinear coordinates is $x = \{x^\mu\}$. For time-like separation of neighbouring points, $\{x^\mu\}$ and $\{x^\mu + dx^\mu\}$, $d\tau > 0$.

The world line of a free massive test particle between two time-like separated points (P and Q) is obtained from the variation principle

$$\delta \int_P^Q d\tau = 0. \tag{29.28}$$

As in flat space a Lagrangian for the variational problem can be introduced, and the four Lagrangian equations which are to be satisfied lead one to the following geodesic equation of motion for x^μ [158, 258]:

$$\frac{d^2 x^\mu}{d\tau^2} + \Gamma^\mu_{\alpha\beta} \frac{dx^\alpha}{d\tau} \frac{dx^\beta}{d\tau} = 0, \tag{29.29}$$

where the so-called Christoffel symbol of the second kind is given by [notation $\partial_\alpha \equiv \partial/\partial x^\alpha$]

$$\Gamma^\mu_{\alpha\beta} = \frac{1}{2} g^{\mu\lambda} (\partial_\alpha g_{\lambda\beta} + \partial_\beta g_{\lambda\alpha} - \partial_\lambda g_{\alpha\beta}). \tag{29.30}$$

Together, $\mu = 0 - 3$, the time-like geodesics [109, 255] equations constitute the equation of motion for a test particle freely falling in a gravitational field. Thus, the quantity

$$\{F^\mu_G(x)\} = - \left\{ \Gamma^\mu_{\alpha\beta} \frac{dx^\alpha}{d\tau} \frac{dx^\beta}{d\tau} \right\} \tag{29.31}$$

expresses the gravitational field strength in terms of the Christoffel symbol. In Minkowski space, where $\Gamma^\mu_{\alpha\beta} = 0$, one regains the usual equation of motion for a free particle, viz., $d^2\{x^\mu\}/d\tau^2 = 0$.

29.5 NULL GEODESICS AND MOTION OF A CLASSICAL LIGHT PARTICLE

For a light particle the proper time τ is zero. However, instead of τ it is possible to use another (affine) parameter, say σ, giving the geodesic equation

$$\frac{d^2 x^\mu}{d\sigma^2} + \Gamma^\mu_{\alpha\beta} \frac{dx^\alpha}{d\sigma} \frac{dx^\beta}{d\sigma} = 0. \quad \mu = 0 - 3 \tag{29.32}$$

The equation must be supplemented by the constraint

$$-g_{\mu\nu} \frac{dx^\mu}{d\sigma} \frac{dx^\nu}{d\sigma} = 0 \tag{29.33}$$

since $d\tau = 0$ in Eq. (29.27). In particular Eq. (29.33) tells us that the time (dt) it takes for a light particle to travel a distance $d\mathbf{x}(\equiv d\mathbf{r})$ is determined by the quadratic equation $[c = 1]$

$$g_{00} dt^2 + 2g_{i0} dx^i dt + g_{ij} dx^i dx^j = 0, \tag{29.34}$$

remembering that $g_{\mu\nu} = g_{\nu\mu}$. The solution is

$$dt = \frac{1}{g_{00}} \{ -g_{i0} dx^i + [(g_{i0} g_{j0} - g_{00} g_{ij}) dx^i dx^j]^{1/2} \}, \tag{29.35}$$

utilizing that $(g_{i0} dx^i)^2 = (g_{i0} dx^i)(g_{j0} dx^j) = g_{i0} g_{j0} dx^i dx^j$. The time required for the light particle to travel along an arbitrary path between given points may be calculated by integrating dt along the path.

CHAPTER 30

Electrodynamics in General Relativity and Non-Inertial Frames

30.1 COVARIANT MAXWELL EQUATIONS IN THE FIELD AND POTENTIAL FORMULATIONS

30.1.1 Field description

Let us recall that in the absence of gravitational fields the covariant set of Maxwell equations can be written as

$$\partial_\alpha F^{\alpha\beta}(x) = -\mu_0 J^\beta(x), \tag{30.1}$$

$$\partial_\alpha F_{\beta\gamma}(x) + \partial_\beta F_{\gamma\alpha}(x) + \partial_\gamma F_{\alpha\beta}(x) = 0, \tag{30.2}$$

where $\{F^{\alpha\beta}(x)\}$ is the electromagnetic field tensor; See Section 2.1, and $\{J^\beta(x)\}$ is current density four-vector. In general relativity one *defines* $F^{\alpha\beta}(x)$ and $J^\beta(x)$ by the requirement that they are (i) tensor fields, and (ii) that they reduce to our well-known quantities in locally inertial Minkowskian coordinates. To make Eqs. (30.1) and (30.2) generally covariant one must replace the normal derivatives (∂_α) by covariant derivatives (in the following denoted by ∇_α), i.e.

$$\partial_\alpha \Rightarrow \nabla_\alpha. \tag{30.3}$$

The relation between ∂_α and ∇_α is given by the manner in which it acts on a contravariant (and covariant) vector field. The action on the contravariant vector $\{V^\mu\}$ is given in Eq. (31.81). Thus, the set of Maxwell equations in Minkowskian space is replaced by

$$\nabla_\alpha F^{\alpha\beta}(x) = -\mu_0 J^\beta(x), \tag{30.4}$$

$$\nabla_\alpha F_{\beta\gamma}(x) + \nabla_\beta F_{\gamma\alpha}(x) + \nabla_\gamma F_{\alpha\beta}(x) = 0. \tag{30.5}$$

If $\{\zeta^\alpha(x)\}$ are the coordinates in a local inertial system and $\{x^\alpha\}$ generally relativistic coordinates, and $\{\tilde{F}^{\alpha\beta}\}$ and $\{\tilde{J}^\alpha\}$ are the values (measured) in the Minkowskian

DOI: 10.1201/9781003029458-30

frame

$$F^{\alpha\beta} = \frac{\partial x^\alpha}{\partial \zeta^\mu} \frac{\partial x^\beta}{\partial \zeta^\nu} \tilde{F}^{\mu\nu}, \tag{30.6}$$

$$J^\alpha = \frac{\partial x^\alpha}{\partial \zeta^\mu} \tilde{J}^\mu. \tag{30.7}$$

Indices are to be raised and lowered with the metric tensor $\{g_{\alpha\beta}\}$ of curved space-time [Section 29.4]. Thus, for instance

$$F_{\alpha\beta} = g_{\alpha\mu} g_{\beta\nu} F^{\mu\nu}. \tag{30.8}$$

It is possible to write the covariant Maxwell equations in Eq. (30.4) and (30.5) in terms of normal derivatives. Hence, if $g(x)$ is the invariant determinant of the metric tensor, the set of Maxwell equations takes the form

$$\partial_\alpha(\sqrt{-g(x)} F^{\alpha\beta}(x)) = -\mu_0 \sqrt{-g(x)} J^\beta(x), \tag{30.9}$$

$$\partial_\alpha F_{\beta\gamma}(x) + \partial_\beta F_{\gamma\alpha}(x) + \partial_\gamma F_{\alpha\beta}(x) = 0, \tag{30.10}$$

as shown e.g. in [158, 258]. In our studies of electrodynamics in structures possessing only spatial (3D) curvature, the forms with normal derivatives are particularly useful.

30.1.2 Potential description

In flat space-time the field tensor was *defined* in terms of the generalized curl of the gauge potential $\{A_\alpha(x)\}$; see Eq. (2.8). In this equation one makes the substitution in Eq. (30.3). However, since the gauge potential is antisymmetric, $A_{\alpha\beta} = -A_{\beta\alpha}$, it can be shown that curls with normal and covariant derivatives are identical [158, 258]. Hence,

$$F_{\alpha\beta}(x) = \nabla_\alpha A_\beta(x) - \nabla_\beta A_\alpha(x) = \partial_\alpha A_\beta(x) - \partial_\beta A_\alpha(x). \tag{30.11}$$

By inserting Eq. (30.11) into the inhomogeneous Maxwell equation (here Eq. (30.4)) written as $\nabla^\alpha F_{\alpha\beta} = -\mu_0 J_\beta$) one obtains

$$\nabla^\alpha \nabla_\alpha A_\beta(x) - \nabla^\alpha \nabla_\beta A_\alpha(x) = -\mu_0 J_\beta(x). \tag{30.12}$$

The homogeneous Maxwell equations can be written equivalently, in terms of covariant and normal derivatives since $F_{\alpha\beta}(x)$ is antisymmetric [158, 258], that is

$$\nabla_\alpha F_{\beta\alpha}(x) + \nabla_\beta F_{\gamma\alpha}(x) + \nabla_\gamma F_{\alpha\beta}(x) = \partial_\alpha F_{\beta\gamma}(x) + \partial_\beta F_{\gamma\alpha}(x) + \delta_\gamma F_{\alpha\beta} = 0. \tag{30.13}$$

Various gauges may be used for the four-potential. Among these the Lorenz gauge is of particular interest for what follows. The covariant Lorenz gauge condition takes the form

$$\nabla^\alpha A_\alpha(x) = 0. \tag{30.14}$$

In order to rewrite the inhomogeneous Maxwell Eq. (30.12) in the Lorenz gauge one must realize that covariant derivatives do not commute. The local commutator acting on the four-potential turns out to be given by [158, 258]

$$[\nabla^\alpha, \nabla_\beta] A_\mu(x) = R^{\nu\alpha}_{\mu\beta}(x) A_\nu(x), \tag{30.15}$$

where $R^{\nu\alpha}_{\mu\beta}(x)$ is the Riemann curvature tensor. With the help of Eq. (30.15) the second term on the left side of Eq. (30.12) can be rewritten as

$$\nabla^\alpha \nabla_\beta A_\alpha(x) = \nabla_\beta \nabla^\alpha A_\alpha(x) + R^{\nu\alpha}_{\alpha\beta}(x) A_\nu(x). \tag{30.16}$$

The truncated curvature tensor

$$R^{\nu\alpha}_{\alpha\beta}(x) = R^\nu_\beta(x) \tag{30.17}$$

is called the Ricci tensor. The first term on the right side of Eq. (30.16) vanishes in the covariant Lorenz gauge. With $\nabla^\alpha \nabla_\beta A_\alpha = R^\nu_\beta A_\nu$, and a subsequent label interchange $\alpha \leftrightarrow \beta$, the inhomogeneous Maxwell equation takes the form

$$\nabla^\beta \nabla_\beta A_\alpha(x) - R^\beta_\alpha(x) A_\beta(x) = -\mu_0 J_\alpha(x) \tag{30.18}$$

in the Lorenz gauge.

Inherent in Eq. (30.18) one must have charge conservation. In order to realize what form charge conservation takes in general relativity, let us return to Eq. (30.9) and utilize Eq. (30.11) written in its contravariant form. Thus, from

$$\begin{aligned}
\partial_\alpha(\sqrt{-g}F^{\alpha\beta}) &= (\partial_\alpha\sqrt{-g})F^{\alpha\beta} + \sqrt{-g}\partial_\alpha F^{\alpha\beta} \\
&= (\partial_\alpha\sqrt{-g})(\partial^\alpha A^\beta - \partial^\beta A^\alpha) + \sqrt{-g}(\partial_\alpha\partial^\alpha A^\beta - \partial_\alpha\partial^\beta A^\alpha) = -\mu_0\sqrt{-g}J^\beta,
\end{aligned} \tag{30.19}$$

one obtains in the Lorenz gauge [remembering that normal derivatives commute]

$$(\partial_\alpha\sqrt{-g})(\partial^\alpha A^\beta - \partial^\beta A^\alpha) + \sqrt{-g}\partial_\alpha\partial^\alpha A^\beta = -\mu_0\sqrt{-g}J^\beta, \tag{30.20}$$

Operating on Eq. (30.20) with ∂_β one gets

$$(\partial_\beta\partial_\alpha\sqrt{-g})(\partial^\alpha A^\beta - \partial^\beta A^\alpha) = -\mu_0\partial_\beta(\sqrt{-g}J^\beta). \tag{30.21}$$

By the notational interchange $\alpha \leftrightarrow \beta$ on the *left side* of Eq. (30.21) one obtains

$$(\partial_\alpha\partial_\beta\sqrt{-g})(\partial^\beta A^\alpha - \partial^\alpha A^\beta) = -\mu_0\partial_\beta(\sqrt{-g}J^\beta). \tag{30.22}$$

By addition of Eqs. (30.21) and (30.22) it is realized that [dummy β changed to α on the right side of these equations]

$$\partial_\alpha\left((\sqrt{-g(x)}J^\alpha(x)\right) = 0. \tag{30.23}$$

Sine the covariant divergence of a vector $\{V^\alpha\}$ is given by [158]

$$\nabla_\alpha V^\alpha(x) = \frac{1}{\sqrt{-g(x)}} \partial_\alpha \left(\sqrt{-g(x)} V^\alpha(x) \right) \tag{30.24}$$

it appears that the electric charge conservation in general relativity must have the form

$$\nabla_\alpha \left((\sqrt{-g(x)} J^\alpha(x) \right) = 0. \tag{30.25}$$

This agrees with the fact that [158]

$$J^\alpha(x) = \frac{1}{\sqrt{-g(x)}} \sum_n e_n \int \delta^{(4)}(x - x_n) dx_n^\alpha, \tag{30.26}$$

for a system of point charges e_n located at x_n $[\delta^{(4)}(x - x_n)$ is the four-dimensional delta function].

30.2 SOME KEY RESULTS FOR THE FOUR-POTENTIAL

30.2.1 Plane-mode decomposition of the covariant potential

Before discussing the photon concept in general relativity and in non-inertial frames in Minkoskian space, we need a few of the fundamental results related to the covariant plane-mode expansion of the free-space four-potential $\{A^\mu(x)\}$.

Let us start from the general 4D-Fourier plane-mode expansion of the μth component of the contravarient four-potential, namely,

$$A^\mu(x) = (2\pi)^{-4} \int_{-\infty}^{\infty} A^\mu(q) e^{iq_\nu x^\nu} d^4q \tag{30.27}$$

In free space $A^\mu(q)$ can only be nonvanishing if the squared dispersion relation

$$q_\mu q^\mu = |\mathbf{q}|^2 - (q^0)^2 = 0 \tag{30.28}$$

is satisfied. This fact reduces the 4D integral to the following 3D integral [158]:

$$A^\mu(x) = \left[\frac{\hbar}{2\varepsilon_0 c (2\pi)^3} \right]^{1/2} \int_{-\infty}^{\infty} |\mathbf{q}|^{-1/2} \left(\alpha^\mu(\mathbf{q}) e^{iq_\nu x^\nu} + c.c. \right) d^3q, \tag{30.29}$$

where $\{q_\nu\} = (-|\mathbf{q}|, \mathbf{q})$. Next, we expand $\{\alpha^\mu(\mathbf{q})\}$ after the four ($r = 0 - 3$) linearly independent polarization unit vectors $\{\varepsilon_r^\mu\}$ satisfying the covariant orthonormality and completeness relations

$$\varepsilon_{r\mu}(\mathbf{q})\varepsilon_s^\mu(\mathbf{q}) = \eta_r \delta_{rs}, \quad r, s = 0 - 3 \tag{30.30}$$

$$\sum_r \eta_r \varepsilon_r^\mu(\mathbf{q})\varepsilon_r^\nu(\mathbf{q}) = \eta^{\mu\nu}, \tag{30.31}$$

where the only nonvanishing elements of $\eta^{\mu\nu}$ are $\eta^{00} = -1$, $\eta^{11} = \eta^{22} = \eta^{33} + 1$. Furthermore, $\eta_i = -\eta_0 = 1$ ($i = 1 - 3$). In turn this expansion of $\{\alpha^{\mu}(\mathbf{q})\}$ gives the continuous mode decomposition

$$A^{\mu}(x) = \sum_{r=0}^{3} \int_{-\infty}^{\infty} [\alpha_r(\mathbf{q}) f_r^{\mu}(x) + c.c.] \, d^3q, \tag{30.32}$$

with

$$f_r^{\mu}(x) = \left[\frac{\hbar}{2\varepsilon_0 c|\mathbf{q}|(2\pi)^3}\right]^{1/2} \varepsilon_r^{\mu}(\mathbf{q}) e^{iq_r^{\nu} x_{\nu}}. \tag{30.33}$$

The four-vector mode function

$$\mathbf{f}_r(x) \equiv \{f_r^{\mu}(x)\} = \left[\frac{\hbar}{2\varepsilon_0 c|\mathbf{q}|(2\pi)^3}\right]^{1/2} \{\varepsilon_r^{\mu}(\mathbf{q})\} \, e^{iq_r^{\nu} x_{\nu}} \tag{30.34}$$

is a key quantity in what follows.

30.2.2   Orthonormalization of mode functions

On the basis of the relativistic *definition* of the scalar product, viz., [106,158,186,258]

$$\langle \mathbf{f}_r | \mathbf{f}_s \rangle \equiv \frac{\varepsilon_0}{i\hbar} \int_{-\infty}^{\infty} \left[\left(\frac{\partial}{\partial t} \mathbf{f}_r^{\dagger}\right) \mathbf{f}_s - \mathbf{f}_r^{\dagger} \frac{\partial}{\partial t} \mathbf{f}_s\right] d^3r, \tag{30.35}$$

which is discussed in some detail in [158], one obtains using Eqs. (30.30) and (30.34)

$$\langle \mathbf{f}_r | \mathbf{f}_s \rangle = \eta_r \delta_{rs} \delta(\mathbf{q}_r - \mathbf{q}_s). \tag{30.36}$$

Since

$$\frac{\partial}{\partial t} \mathbf{f}_r(x) = -i\omega_r \mathbf{f}_r(x), \quad \omega_r = c|\mathbf{q}_r|, \tag{30.37}$$

and thus

$$\frac{\partial}{\partial t} \mathbf{f}_r^*(x) = i\omega_r \mathbf{f}_r^*(x), \tag{30.38}$$

it appears that

$$\langle \mathbf{f}_r^* | \mathbf{f}_s^* \rangle = -\eta_r \delta_{rs} \delta(\mathbf{q}_r - \mathbf{q}_s). \tag{30.39}$$

Finally, the reader may show that

$$\langle \mathbf{f}_r | \mathbf{f}_s^* \rangle = 0. \tag{30.40}$$

Eqs. (30.36), (30.39) and (30.40) are crucial for our discussion of the photon particle concept to follow.

30.3 ON THE PHOTON PARTICLE CONCEPT

30.3.1 Bogolubov transformation

Let us now generalize the considerations in Section 30.2 and expand the four-potential $\{A^\mu(x)\}$ after a complete set of positive $[\mathbf{f}_i(x)]$- and negative $[\mathbf{f}_i^*(x)]$-frequency mode solutions to the free-space wave equation $\partial_\nu\partial^\nu\{A^\mu(x)\} = 0$. In discrete mode quantization one has

$$\mathcal{A}(x) \equiv \{A^\mu(x)\} = \sum_i [\alpha_i \mathbf{f}_i(x) + \alpha_i^* \mathbf{f}_i^*(x)]. \tag{30.41}$$

The mode functions satisfy the following generalized orthonormalization condition:

$$\langle \mathbf{f}_i | \mathbf{f}_j \rangle = \eta_i \delta_{ij}, \tag{30.42}$$

$$\left\langle \mathbf{f}_i^* | \mathbf{f}_j^* \right\rangle = -\eta_i \delta_{ij}, \tag{30.43}$$

$$\left\langle \mathbf{f}_i | \mathbf{f}_j^* \right\rangle = 0, \tag{30.44}$$

where $\eta_i = +1$ for $i = 1 - 3$, and $\eta_i = -1$ for $i = 0$. The inner product in Eq. (30.42) is the relativistic scalar product

$$\langle \mathbf{f}_i | \mathbf{f}_j \rangle = \frac{\varepsilon_0}{i\hbar} \int_{-\infty}^{\infty} \left[\left(\frac{\partial}{\partial t} \mathbf{f}_i^\dagger \right) \mathbf{f}_j - \mathbf{f}_i^\dagger \frac{\partial}{\partial t} \mathbf{f}_j \right] d^3 r, \tag{30.45}$$

and from this the inner product $\left\langle \mathbf{f}_i^* | \mathbf{f}_j^* \right\rangle$ and $\left\langle \mathbf{f}_i | \mathbf{f}_j^* \right\rangle$ can be written down. If the plane-mode expansion discussed in Section 30.2 is used (a special choice) the index $i = (r, \mathbf{q})$.

The choice of mode functions is not unique. Consider thus a second complete set of positive $[\mathbf{g}_j(x)]$ - and negative $[\mathbf{g}_j^*(x)]$-frequency modes, satisfying the generalized orthonormalization condition

$$\langle \mathbf{g}_i | \mathbf{g}_j \rangle = \eta_i \delta_{ij}, \tag{30.46}$$

$$\left\langle \mathbf{g}_i^* | \mathbf{g}_j^* \right\rangle = -\eta_i \delta_{ij}, \tag{30.47}$$

$$\left\langle \mathbf{g}_i | \mathbf{g}_j^* \right\rangle = 0. \tag{30.48}$$

The four-potential, $\boldsymbol{A}(x)$, may be expanded in the new set also:

$$\mathcal{A}(x) = \sum_j [\beta_j \mathbf{g}_j(x) + \beta_j^* \mathbf{g}_j^*(x)]. \tag{30.49}$$

Since the two sets $(\mathbf{f}_i, \mathbf{f}_i^*)$ and $(\mathbf{g}_j, \mathbf{g}_j^*)$ are complete, the new mode functions can be expanded in terms of the old, and *vice versa*. Hence,

$$\mathbf{g}_j(x) = \sum_i [\alpha_{ji} \mathbf{f}_i(x) + \beta_{ji} \mathbf{f}_i^*(x)], \tag{30.50}$$

$$\mathbf{f}_i(x) = \sum_j [\sigma_{ij} \mathbf{g}_j(x) + \tau_{ij} \mathbf{g}_j^*(x)]. \tag{30.51}$$

The α_{ji}, β_{ji}, σ_{ij} and τ_{ij} coefficients relate to inner products as follows:

$$\alpha_{ji} = \eta_i \langle \mathbf{f}_i | \mathbf{g}_j \rangle, \tag{30.52}$$

$$\beta_{ji} = -\eta_i \langle \mathbf{f}_i^* | \mathbf{g}_j \rangle, \tag{30.53}$$

$$\sigma_{ij} = \eta_j \langle \mathbf{g}_j | \mathbf{f}_i \rangle, \tag{30.54}$$

$$\tau_{ij} = -\eta_j \langle \mathbf{g}_j^* | \mathbf{f}_i \rangle. \tag{30.55}$$

The results in Eqs. (30.52)–(30.55) are obtained using the generalized orthonormalization conditions in Eqs. (30.42) and (30.44) and Eqs. (30.46)–(30.48). Since the scalar product between the two arbitrary mode function $\mathbf{F}_1$ and $\mathbf{F}_2$ satisfies

$$\langle \mathbf{F}_2 | \mathbf{F}_1 \rangle = \langle \mathbf{F}_1 | \mathbf{F}_2 \rangle^*, \tag{30.56}$$

σ_{ij} and τ_{ij} can be related to α_{ji} and β_{ji}. Thus,

$$\sigma_{ij} = \eta_i \eta_j \alpha_{ji}^*, \tag{30.57}$$

$$\tau_{ij} = -\eta_i \eta_j \beta_{ji}^*, \tag{30.58}$$

and hence

$$\mathbf{f}_i = \sum_j \eta_i \eta_j [\alpha_{ji}^* \mathbf{g}_j - \beta_{ji}^* \mathbf{g}_j^*]. \tag{30.59}$$

The relations in Eqs. (30.50) and (30.59) are known as the Bogolubov transformations [27, 35, 36, 38], and the quantities α_{ij} and β_{ji} are called the Bogolubov coefficients.

By transferring Eq. (30.50) to Dirac notation ($\mathbf{F} \to |\mathbf{F}\rangle$) one obtains with the help of Eqs. (30.52) and (30.53) the expression

$$|\mathbf{g}_i\rangle = \sum_i [\alpha_{ji} |\mathbf{f}_i\rangle + \beta_{ji} |\mathbf{f}_i^*\rangle] = \sum_i \left\{ \eta_i [|\mathbf{f}_i\rangle \langle \mathbf{f}_i| - |\mathbf{f}_i^*\rangle \langle \mathbf{f}_i^*|] |\mathbf{g}_j\rangle \right\}. \tag{30.60}$$

In turn Eq. (30.59) implies that

$$\sum_i \eta_i [|\mathbf{f}_i\rangle \langle \mathbf{f}_i| - |\mathbf{f}_i^*\rangle \langle \mathbf{f}_i^*|] = \mathbf{1}, \tag{30.61}$$

where $\mathbf{1}$ is the identity operator.

30.3.2 Quantization

The classical expression for the four-potential given in Eq. (30.41) is promoted to the quantum level ($\mathcal{A}(x) \Rightarrow \hat{\mathcal{A}}(x)$) by replacing the expansion coefficients by annihilation and creation operators $\hat{a}_i$ and $\hat{a}_i^\dagger$, i.e.,

$$\alpha_i \Rightarrow \hat{a}_i, \quad \alpha_i^* \Rightarrow \hat{a}_i^\dagger, \tag{30.62}$$

and requiring that the operators satisfy the usual Boson commutation relations

$$[\hat{a}_i, \hat{a}_j^\dagger] = \eta_i \delta_{ij}, \tag{30.63}$$

$$[\hat{a}_i, \hat{a}_j] = [\hat{a}_i^\dagger, \hat{a}_j^\dagger] = 0. \tag{30.64}$$

The vacuum state *defined with respect to the set of* $\mathbf{f}_i$-modes, $|0_f\rangle$, is defined by

$$\hat{a}_i\,|0_f\rangle = 0, \quad \forall i. \tag{30.65}$$

For each i-mode one has a number operator

$$\hat{n}_{f,i} = \hat{a}_i^\dagger \hat{a}_i, \tag{30.66}$$

and a single-mode Fock state with $n_{f,i}$ quanta, $|n_{f,i}\rangle$ is obtained by repeated action with $\hat{a}_i^\dagger$ on the ground state in the usual manner. On normalized form

$$|n_{f,i}\rangle = \frac{1}{\sqrt{n_{f,i}!}}\left(\hat{a}_i^\dagger\right)^{n_{f,i}}|0_f\rangle. \tag{30.67}$$

The four-potential operator

$$\hat{\boldsymbol{A}}(x) = \sum_i [\hat{a}_i \mathbf{f}_i(x) + \hat{a}_i^\dagger \mathbf{f}_i^*(x)] \tag{30.68}$$

may be expanded in terms of the $\mathbf{g}_j$-modes also:

$$\hat{\boldsymbol{A}}(x) = \sum_j [\hat{b}_j \mathbf{g}_j(x) + \hat{b}_j^\dagger \mathbf{g}_j^*(x)]. \tag{30.69}$$

Since the basis sets $(\mathbf{f}_i, \mathbf{f}_i^*)$ and $(\mathbf{g}_j, \mathbf{g}_j^*)$ are equivalent to the annihilation $(\hat{b}_j)$ and creation $(\hat{b}_j^\dagger)$ operators necessarily must satisfy the same Boson commutation relations as the $\hat{a}_i$ and $\hat{a}_i^\dagger$-operators, viz.,

$$[\hat{b}_i, \hat{b}_j^\dagger] = \eta_j \delta_{ij}, \tag{30.70}$$

etc. The vacuum state related to the $\mathbf{g}_j$-modes, $|0_g\rangle$, is defined by

$$\hat{b}_j\,|0_g\rangle = 0, \quad \forall j, \tag{30.71}$$

and the mode number operator, $\hat{n}_{g,j} = \hat{b}_j^\dagger \hat{b}_j$ satisfies the eigenvalue equation

$$\hat{n}_{g,j}\,|n_{g,j}\rangle = n_{g,j}\,|n_{g,j}\rangle, \tag{30.72}$$

where $n_{g,j}$ is the number of excitations in the Fock state $|n_{g,j}\rangle$. This state can be obtained from the vacuum state as before. Thus, in normalized form

$$|n_{g,j}\rangle = \frac{1}{\sqrt{n_{g,j}!}}\left(\hat{b}_j^\dagger\right)^{n_{g,j}}|0_g\rangle. \tag{30.73}$$

30.3.3 The non-unique vacuum

Let us now imagine that the electromagnetic field is in the f-vacuum state, $|0_f\rangle$. By this we mean that an observer using the f-modes in her quantization procedure will say that there are no photons in the field. How will another equivalent observer characterize the $|0_f\rangle$ state? In particular, will the g-observer also come to the conclusion that there are no photons in the electromagnetic field? In order to answer this question we take as a starting point the fact that the two-mode decompositions giving the four-potential, $\hat{\mathcal{A}}(x)$, must satisfy the equation

$$\sum_i [\hat{a}_i \mathbf{f}_i + \hat{a}_i^\dagger \mathbf{f}_i^*] = \sum_j [\hat{b}_j \mathbf{g}_j + \hat{b}_j^\dagger \mathbf{g}_j^*]. \tag{30.74}$$

The number of photons in a given electromagnetic field is related to the relevant global number operator. Thus, for the g-observer

$$\hat{N}_g = \sum_j \hat{b}_j^\dagger \hat{b}_j. \tag{30.75}$$

To determine the mean number of g-photons in the f-vacuum state, one must calculate the matrix element

$$N_g = \langle 0_f | \sum_j \hat{b}_j^\dagger \hat{b}_j |0_f\rangle . \tag{30.76}$$

Since we only know how the operators $\hat{a}_i$ and $\hat{a}_i^\dagger$ act on f-Fock states, we must express the $\hat{b}_j$ and $\hat{b}_j^\dagger$ operators in terms of the a-operators. To do this one uses the Bogolubov transformation in Eq. (30.59) to express the left-hand side of Eq. (30.74) in terms of the g-modes. Hence

$$\sum_i [\hat{a}_i \mathbf{f}_i + \hat{a}_i^\dagger \mathbf{f}_i^*] = \sum_{i,j} [\hat{a}_i \eta_i \eta_j (\alpha_{ji}^* \mathbf{g}_j - \beta_{ji}^* \mathbf{g}_j^*) + \hat{a}_i^\dagger \eta_i \eta_j (\alpha_{ji} \mathbf{g}_j^* - \beta_{ji} \mathbf{g}_j)]. \tag{30.77}$$

Since the coefficients of $\mathbf{g}_j$ and $\mathbf{g}_j^*$ in this expression, according to Eq. (30.74) must be equal to $\hat{b}_j$ and $\hat{b}_j^\dagger$, one obtains

$$\hat{b}_j = \sum_i \eta_i \eta_j [\alpha_{ji}^* \hat{a}_i - \beta_{ji} \hat{a}_i^\dagger], \tag{30.78}$$

$$\hat{b}_j^\dagger = \sum_i \eta_i \eta_j [-\beta_{ji}^* \hat{a}_i + \alpha_{ji} \hat{a}_i^\dagger]. \tag{30.79}$$

The mean number of g-photons observed in a single g_i-mode in the f-vacuum hence becomes in a first step

$$\begin{aligned}
\langle 0_f | \hat{n}_{g,j} |0_f\rangle &= \langle 0_f | \hat{b}_j^\dagger \hat{b}_j |0_f\rangle \\
&= \sum_{k,l} \langle 0_f | \eta_k \eta_j (-\beta_{jk}^* \hat{a}_k + \alpha_{jk} \hat{a}_k^\dagger) \eta_l \eta_j (\alpha_{jl}^* \hat{a}_l - \beta_{jl} \hat{a}_l^\dagger) |0_f\rangle \\
&= \sum_{k,l} \eta_k \eta_l \beta_{jk}^* \beta_{jl} \langle 0_f | \hat{a}_k \hat{a}_l^\dagger |0_f\rangle , \tag{30.80}
\end{aligned}$$

as one realize using the results $\hat{a}_i |0_f\rangle = 0$ and $\langle 0_f | \hat{a}_k^\dagger = 0$. Since use of the commutation relation in Eq. (30.63) gives

$$\langle 0_f | \hat{a}_k \hat{a}_l^\dagger |0_f\rangle = \langle 0_f | \hat{a}_l^\dagger \hat{a}_k + \eta_k \delta_{kl} |0_f\rangle = \eta_k \delta_{kl}, \tag{30.81}$$

one obtains in the second step

$$\langle 0_f| \, \hat{n}_{g,j} \, |0_f\rangle = \sum_{k,l} \eta_k^2 \eta_l \beta_{jk}^* \beta_{jl} \delta_{kl}. \tag{30.82}$$

Because $\eta_k^2 = 1$, one finally obtains [with the dummy index replacement $j \to i$ and $k \to j$]

$$\langle 0_f| \, \hat{n}_{g,i} \, |0_f\rangle = \sum_h \eta_j |\beta_{ij}|^2 \tag{30.83}$$

We now can conclude that what look like (is) a vacuum state ($|0_f\rangle$) from the perspective of the f-observer, will in the ith mode contain an average number of photons ($\sum_j \eta_j |\beta_{ij}|^2$) seen in the perspective of the g-observer unless all the β_{ij} coefficients are zero. The reason for this originates in the circumstance that β_{ij} describes an admixture of creation (annihilation) operators from one basis into annihilation (creation) operators in the other basis. One may exemplify the conclusion made above considering the g-vacuum state defined by

$$\hat{b}_i \, |0_g\rangle = 0, \quad \forall i \tag{30.84}$$

and letting the annihilation operator $\hat{a}_i$ operate on this state. An inversion of Eqs. (30.78) and (30.79) leads to the general relations

$$\hat{a}_i = \sum_j [\alpha_{ji}\hat{b}_j + \beta_{ji}^*\hat{b}_j^\dagger], \tag{30.85}$$

$$\hat{a}_i^\dagger = \sum_j [\beta_{ji}\hat{b}_j + \alpha_{ji}^*\hat{b}_j^\dagger]. \tag{30.86}$$

Thus, by means of Eq. (30.85) one obtains

$$\hat{a}_i \, |0_g\rangle = \sum_j \beta_{ji}^*\hat{b}_j^\dagger \, |0_g\rangle \tag{30.87}$$

in view of Eq. (30.84). Acting with the $\hat{a}_i$ annihilation operator on the g-vacuum state, thus results in a linear superposition of one-photon states $|1_{g,j}\rangle = \hat{b}_j^\dagger \, |0_g\rangle$ for an observer using the g-mode basis.

In Minkowski space all inertial observers will agree on what one may call the "natural" vacuum since a boost from one inertial frame to another does not imply an admixture of positive and negative frequencies, i.e., $\beta_{ij} = 0$ for $\forall i, j$. However, even in Minkoski space inertial and accelerated observers will not agree on what is the photon vacuum state, as we shall exemplify in Section 30.4. In general relativity the situation is complicated because inertial observers become free-falling observers, and will not agree on what one should understand as a photon vacuum.

The variance of the $\mathbf{g}_i$-mode number operator, $\hat{n}_{g,i}$ in the f-vacuum can be obtained from the squared standard deviation, $\Delta n_{g,i}$, viz.,

$$(\Delta n_{g,i})^2 = \langle 0_f| \, [\hat{n}_{g,i} - \langle 0_f| \, \hat{n}_{g,i} \, |0_f\rangle]^2 \, |0_f\rangle \,. \tag{30.88}$$

A somewhat lengthy calculation [158] gives the result

$$(\Delta n_{g,i})^2 = \left| \sum_j \eta_j \alpha_{ij} \beta_{ij}^* \right|^2 + \left(\sum_j \eta_j |\alpha_{ij}|^2 \right) \left(\sum_j \eta_j |\beta_{ij}|^2 \right). \qquad (30.89)$$

It appears from Eq. (30.89) that the variance depends on both the α_{ij} and β_{ij} sets of coefficients.

30.4 CLASSICAL UNRUH EFFECT

In recent years much attention has been given to the interaction in Minkowski space of uniformly accelerating systems with quantum fields, particularly with regards to the thermal Unruh effect [27, 251, 252]. In contrast, relatively little attention has been given to the interaction of uniformly accelerating systems with classical fields. For instance, in the context of the electrodynamics of solid-state plasmas (and in a broader context mesoscopic systems) it is of fundamental interest to understand how a uniformly accelerated plasma treated on the basis of microscopic electrodynamics interacts with injected charged particles. The insight obtained from classical studies I believe can be used to gain further insight in quantum field aspects making the bridge from classical to quantum electrodynamics with the help of photon wave mechanics, e.g. Below we shall limit ourselves to a study of the electromagnetic field reflection from a uniformly accelerated perfect metallic mirror in Minkowskian space.

30.4.1 Incident and reflected fields in Rindler coordinates

Let us consider a mirror that moves with uniform proper acceleration $\{a^\mu\}$ such that

$$a_\mu a^\mu = \frac{du_\mu}{d\tau} \frac{du^\mu}{d\tau} = \left(\frac{d^2 \mathbf{x}}{d\tau^2} \right)^2 - \left(c \frac{d^2 t}{d\tau^2} \right)^2 = a^2 \qquad (30.90)$$

where a is a constant, τ is the proper time along the mirror's worldline, $\{x_\mu(\tau)\}$, and $\{u_\mu\} = \{dx_\mu/d\tau\}$ is the four-velocity of the mirror. For simplicity, we only consider the reflection of a monochromatic plane wave at normal incidence on a flat mirror. For propagation along a z-axis, $\mathbf{x} = (0, 0, z)$, Eq. (30.90) reduces to

$$\left(\frac{d^2 z}{d\tau^2} \right)^2 - \left(\frac{d^2 t}{d\tau^2} \right)^2 = a^2. \qquad (30.91)$$

Without loss of essential generality, it is assumed that the position of the mirror surface is $z_0 = c^2/a$ at $t = 0$. Furthermore, it is assumed that the incident (superscript: I) field is linearly polarized, and propagates with wave vector $\mathbf{q} = q\hat{\mathbf{z}}$ along the positive z-axis. With the incident electric field directed along the x-axis, one has in vacuum

$$\hat{\mathbf{x}} \cdot \mathbf{E}^I(z, t) = c\hat{\mathbf{y}} \cdot \mathbf{B}^I(z, t) = E_0 \cos[q(z - ct)]. \qquad (30.92)$$

The reflected field $\mathbf{E}^R(z,t) = (E_x^R, 0, 0)$ can be assumed to be of the form

$$E_x^R = -cB_y^R = -E_0 f(z + ct). \tag{30.93}$$

The fields in Eqs. (30.92) and (30.93) satisfy the free-space Maxwell equations. The general contravariant electromagnetic field tensor thus reduces to

$$\{F^{\mu\nu}\} = \begin{pmatrix} 0 & Ex/c & 0 & 0 \\ -E_x/c & 0 & 0 & -B_y \\ 0 & 0 & 0 & 0 \\ 0 & B_y & 0 & 0 \end{pmatrix}. \tag{30.94}$$

We now introduce Rindler (R) coordinates (z_R, t_R), related to Minkowski coordinates (z, t) by [27, 56, 221]

$$z_R = (z^2 - c^2 t^2)^{1/2}, \quad t_R = \frac{c}{2a} \ln\left(\frac{z + ct}{z - ct}\right). \tag{30.95}$$

As the reader may show, the inverse Rindler transformation is given by

$$z = z_R \cosh\left(\frac{a}{c} t_R\right), \quad t = \frac{1}{c} z_R \sinh\left(\frac{a}{c} t_R\right). \tag{30.96}$$

It appears that the Rindler transform allows one to study the incident and reflected electromagnetic fields in a frame where the mirror is at rest at all times. The Minkowski metric, $ds^2 = dx^2 + dy^2 + dz^2 - c^2 dt^2$, is transformed to the Rindler metric

$$ds^2 = dx^2 + dy^2 + dz_R^2 - \left(\frac{a}{c}\right)^2 z_R^2 dt_R^2. \tag{30.97}$$

The electromagnetic field tensor is transformed by the formula given in (30.6), i.e.,

$$F_R^{\mu\nu} = \frac{\partial x_R^\mu}{\partial x^\alpha} \frac{\partial x_R^\nu}{\partial x^\beta} F^{\alpha\beta}, \tag{30.98}$$

with the index replacement $\mu\nu \Leftrightarrow \alpha\beta$. Hence,

$$F_R^{01} = -F_R^{10} = \frac{\partial t_R}{\partial t} F^{01} + \frac{\partial(ct_R)}{dz} F^{31} = \frac{c}{az_R}\left[E_x \cosh\left(\frac{a}{c} t_R\right) - cB_y \sinh\left(\frac{a}{c} t_R\right)\right], \tag{30.99}$$

$$F_R^{31} = -F_R^{13} = \frac{\partial z_R}{\partial(ct)} F^{01} + \frac{\partial z_R}{dz} F^{31} = \frac{1}{c}\left[-E_x \sinh\left(\frac{a}{c} t_R\right) + cB_y \cosh\left(\frac{a}{c} t_R\right)\right], \tag{30.100}$$

and

$$F_R^{\mu\nu} = 0 \quad \text{otherwise.} \tag{30.101}$$

In order to express $c^{-1}E_{R,x} = F_R^{01}$ and $B_{R,y} = F_R^{31}$ in Rindler coordinates, one has to transform $E_x = cB_y$ to Rindler coordinates; see Eqs. (30.99) and (30.100). It appears from Eqs. (30.92) and (30.93) that one needs to express the quantities $z \mp ct$ in Rindler coordinates. The inverse Rindler transformation given in Eq. (30.96), readily gives

$$z \mp ct = z_R \exp\left(\mp \frac{a}{c} t_R\right). \tag{30.102}$$

The nonvanishing incident and reflected field components in Minkowski metric thus take the forms

$$E_x^I = cB_y^I = E_0 \cos\left[q z_R \exp\left(-\frac{a}{c} t_R\right)\right], \tag{30.103}$$

$$E_x^R = - cB_y^R = -E_0 f\left[z_R \exp\left(\frac{a}{c} t_R\right)\right] \tag{30.104}$$

in Rindler coordinates. By inserting the expressions in Eqs. (30.103) and (30.104) into Eqs. (30.99) and (30.100) one obtains the following results for the incident (I) and reflected (R) fields:

$$E_{R,x}^I = \frac{c^2}{a z_R} E_0 \exp\left(-\frac{a}{c} t_R\right) \cos\left[q z_R \exp\left(-\frac{a}{c} t_R\right)\right], \tag{30.105}$$

$$cB_{R,y}^I = E_0 \exp\left(-\frac{a}{c} t_R\right) \cos\left[q z_R \exp\left(-\frac{a}{c} t_R\right)\right], \tag{30.106}$$

and

$$E_{R,x}^R = - \frac{c^2}{a z_R} E_0 \exp\left(\frac{a}{c} t_R\right) f\left[z_R \exp\left(\frac{a}{c} t_R\right)\right], \tag{30.107}$$

$$cB_{R,y}^R = E_0 \exp\left(\frac{a}{c} t_R\right) f\left[z_R \exp\left(\frac{a}{c} t_R\right)\right]. \tag{30.108}$$

It is a good exercise to check that the Rindler expression for the incident and reflected fields satisfy the generally covariant Maxwell equations given in Eqs. (30.9) and (30.10). Let us consider first the free-space $[J^\beta(x) = 0]$ form Eq. (30.9) takes in our case. Since

$$\sqrt{-g_R} = \frac{a}{c^2} z_R \tag{30.109}$$

for the Rindler metric [Eq. (30.95)], it is easy to show that Eq. (30.9) reduces to

$$\frac{1}{c^2} \frac{\partial}{\partial t_R}(z_R E_{R,x}) + \frac{\partial}{\partial z_R}(z_R B_{R,y}) = 0. \tag{30.110}$$

A straightforward calculation shows that the sets $(E_{R,x}^I, B_{R,y}^I)$ and $(E_{R,x}^R, B_{R,y}^R)$ both satisfy Eq. (30.110).

In order to obtain the Rindler form of the Maxwell equations in Eq. (30.10) for our one-dimensional case we start from the contravariant Rindler field tensor

$$\{F^{\mu\nu}\} = \begin{pmatrix} 0 & E_{R,x}/c & 0 & 0 \\ -E_{R,x}/c & 0 & 0 & -B_{R,y} \\ 0 & 0 & 0 & 0 \\ 0 & B_{R,y} & 0 & 0 \end{pmatrix}, \tag{30.111}$$

and use the diagonal Rindler metric tensor with elements [see Eq. (30.97)]

$$g_{00} = -\left(\frac{a}{c}\right)^2 z_R^2, \quad g_{11} = g_{22} = g_{33} = 1. \tag{30.112}$$

The elements of the covariant Rindler field tensor, obtained from

$$F_{\mu\nu} = g_{\mu\mu}g_{\nu\nu}F^{\mu\nu}, \tag{30.113}$$

cf. Eq. (30.8), hence shows that

$$\{F_{\mu\nu}\} = \begin{pmatrix} 0 & F_{01} & 0 & 0 \\ -F_{01} & 0 & 0 & F_{13} \\ 0 & 0 & 0 & 0 \\ 0 & -F_{13} & 0 & 0 \end{pmatrix}, \tag{30.114}$$

where

$$F_{01} = -\left(\frac{a}{c}\right)^2 z_R^2 \frac{E_{R,x}}{c}, \tag{30.115}$$

$$F_{13} = -B_{R,y}. \tag{30.116}$$

A little algebra reveals that Eq. (30.10) is reduced to

$$\frac{\partial}{\partial t_R} B_{R,y} + \left(\frac{a}{c}\right)^2 \frac{\partial}{\partial z_R}(z_R^2 E_{R,x}) = 0. \tag{30.117}$$

I leave it to the reader to show that the incident and reflected fields [Eqs. (30.103)–(30.106)] both satisfy Eq. (30.117).

30.4.2 Boundary conditions at the mirror surface

For a perfect (infinite conductivity) metal mirror there is no field penetration into the mirror. This implies that the sum of the incident and reflected electric fields vanishes at the mirror surface. This boundary condition allows one to obtain an explicit expression for the f-function [Eq. (30.107)] at the mirror surface. In the Rindler coordinates, the mirror's world line is given by $[z_R(\tau), t_R(\tau)]$, with velocity $[dz_R(\tau)/d\tau, dt_R(\tau)/d\tau]$. It thus appears from the first equality in Eq. (30.94) that the Rindler transformation allows us to study the incident and reflected fields in a uniformly accelerated frame where the mirror is at rest at all times. The position of the mirror, chosen at the initial time ($t = 0$) to be located at $z_0 = c^2/a$ [see the text below Eq. (30.91)], hence is the Rindler coordinate position (z_R) at all times. A schematic illustration of the scattering configuration is shown in Fig. 30.1.

For a perfect metal the boundary condition for the electric field in the Rindler coordination of our two-dimensional Minkowski space thus takes the form

$$(E_{R,x}^I + E_{R,x}^R)|_{z_R=c^2/a} = 0. \tag{30.118}$$

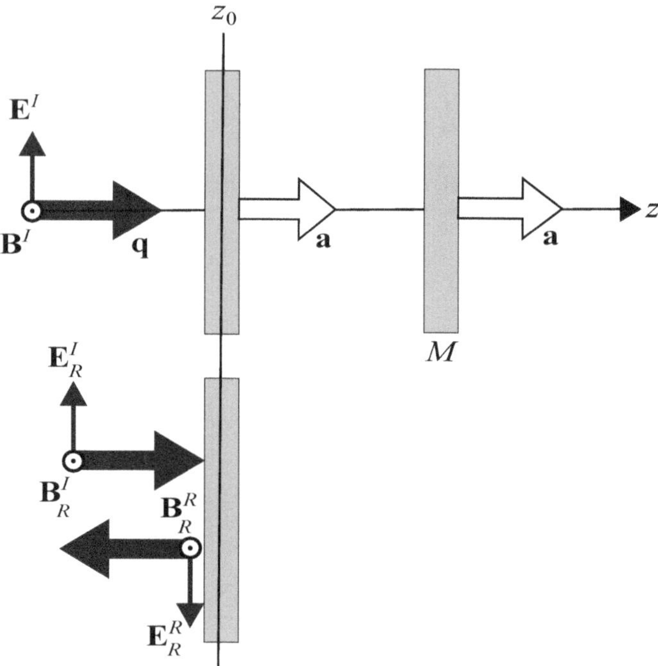

Figure 30.1 Schematic illustration of a classical Unruh effect. Upper part: A linearly polarized monochromatic plane electromagnetic wave [Field: $(\mathbf{E}^I, \mathbf{B}^I)$; wave vector $\mathbf{q}$] is incident (I) at perpendicular incidence on a perfectly conducting flat mirror (M). The mirror moves with uniform acceleration, $\mathbf{a} = a\hat{\mathbf{z}}$, in Minkowskian space, and the position of the mirror surface is $z_0 = c^2/a$ at $t = 0$. Lower part: The reflection of the electromagnetic field $[\mathbf{E}_R^I \to \mathbf{E}_R^R, \mathbf{B}_R^I \to \mathbf{B}_R^R]$ as seen in Rindler (R) coordinates, where the position of the mirror is $Z_R = c^2/a$ at all times.

By inserting the expressions given in Eq. (30.103) and (30.105) into Eq. (30.118), it follows that the f-function at the metal surface is given by

$$f\left[\frac{c^2}{a}\exp\left(\frac{a}{c}t_R\right)\right] = \exp\left(-\frac{2a}{c}t_R\right)\cos\left[q\frac{c^2}{a}\exp\left(-\frac{a}{c}t_R\right)\right]. \qquad (30.119)$$

An educated guess suggest that the replacement

$$\frac{c^2}{a} \Rightarrow \frac{c^4}{a^2 z_R^2} \qquad (30.120)$$

in Eq. (30.119) will give one the f-function in every space point outside the mirror

$$f\left[z_R\exp\left(\frac{a}{c}t_R\right)\right] = \frac{c^4}{a^2 z_R^2}\exp\left(-\frac{2a}{c}t_R\right)\cos\left[\frac{c^4 q}{a^2 z_R}\exp\left(-\frac{a}{c}t_R\right)\right]. \qquad (30.121)$$

A straightforward calculation shows that the electromagnetic field related to Eq. (30.121) indeed satisfies the covariant set of Maxwell equations, given in Eqs. (30.110) and (30.117). Eq. (30.121) of course also reduces to Eq. (30.119) at the metal-vacuum interface.

30.4.3   Reflected electromagnetic field in Minkowski coordinates

The f-function in Eq. (30.121) easily can be transformed to Minkowski coordinates by means of the relation

$$\frac{1}{z_R}\exp\left(-\frac{a}{c}t_R\right) = \frac{1}{z+ct} \tag{30.122}$$

and its square. To derive the result in Eq. (30.122) one uses Eq. (30.95) [for z_R] and Eq. (30.102) [for the minus sign]. Hence, one obtains

$$f(z+ct) = \frac{c^4}{a^2}\frac{1}{(z+ct)^2}\cos\left[\frac{c^4 q}{a^2(z+ct)}\right]. \tag{30.123}$$

With this f-function, the reflected electric and magnetic fields follow directly from Eq. (30.93). Thus,

$$E_x^R = -cB_y^R = -\frac{E_0 c^4}{a^2}\frac{1}{(z+ct)^2}\cos\left[\frac{c^4 q}{a^2(z+ct)}\right]. \tag{30.124}$$

The classical Riemann-Silberstein vectors for the reflected field are given by

$$\mathbf{F}_\pm^R = \sqrt{\varepsilon_0}E_x^R\mathbf{e}_\mp(\hat{\mathbf{z}}), \tag{30.125}$$

with the helicity unit vector choices $\mathbf{e}_\mp(\hat{\mathbf{z}}) = 2^{-1/2}(\hat{\mathbf{x}}\mp\hat{\mathbf{y}})$.

CHAPTER 31

Electrodynamics in Spatially Curved Structures

31.1 INHOMOGENEOUS MAXWELL EQUATION IN THE LORENZ AND COULOMB GAUGES

Let us return to the action of the commutator of covariant derivatives on the four-potential [Eq. (30.15)], here written in the form

$$[\nabla_\alpha, \nabla_\beta] A_\gamma(x) = R_{\gamma\nu\alpha\beta}(x) A^\nu(x). \tag{31.1}$$

In the absence of curvature in time, the covariant derivatives ∇_α and ∇_β commute for α and/or β equal to zero. In relation to Eq.(31.1) this implies that $R_{\gamma\nu 0\beta} = R_{\gamma\nu\alpha 0} = 0$, and in the symmetric Ricci tensor

$$R_{\nu\beta} = g^{\alpha\gamma} R_{\gamma\nu\alpha\beta}, \tag{31.2}$$

and $R_{0\nu} = R_{\nu 0} = 0$, $\nu = 1 - 4$. As it is a sense is obvious from the outset, the Ricci tensor is nonvanishing only in the spatial subdomain of the 4D space-time domain. For what follows it is convenient sometimes to make use of the following compact notation for the symmetric Ricci tensor:

$$\mathbf{R} \equiv \{R_j^i\} = \begin{pmatrix} R_1^1 & R_2^1 & R_3^1 \\ R_1^2 & R_2^2 & R_3^2 \\ R_1^3 & R_2^3 & R_3^3 \end{pmatrix} \tag{31.3}$$

In the Lorenz gauge, Eq. (30.18) hence is reduced to

$$\left(\nabla^j \nabla_j - \frac{1}{c^2} \frac{\partial^2}{\partial t^2}\right) A_i - R_i^j A_j = -\mu_0 J_i \tag{31.4}$$

for $i = 1 - 3$. The quantity

$$\nabla^j \nabla_j \equiv \nabla^2 \tag{31.5}$$

DOI: 10.1201/9781003029458-31

usually is called the Laplace-Beltrami operator. In the space-frequency domain the wave equations for the vector and scalar potentials thus may be written in the compact forms $[q_0 = \omega/c]$

$$[(\nabla^2 + q_0)\mathbf{U} - \mathbf{R}] \cdot \mathbf{A} = -\mu_0 \mathbf{J}, \tag{31.6}$$

$$(\nabla^2 + q_0^2)A_0 = -\mu_0 J_0, \tag{31.7}$$

with gauge condition $\nabla^i A_i + i q_0 A_0 = 0$.

Before making a specific choice of gauge the general inhomogeneous Maxwell equation for A_α is given by

$$\nabla^\beta \nabla_\beta A_\alpha(x) - \nabla_\alpha \nabla^\beta A_\beta(x) - R_\alpha^\beta(x) A_\beta(x) = -\mu_0 J_\alpha(x). \tag{31.8}$$

That this is the correct wave equation follows immediately from Eqs.(30.16)–(30.18). Thus, one just has to add the term $-\nabla_\alpha \nabla^\beta A_\beta(x)$ to the left-hand side of the Lorenz-gauge wave equation [Eq. (30.18)]. The non-covariant Coulomb gauge condition is defined by

$$\nabla^i A_i(x) = 0, \tag{31.9}$$

where the summation runs over $i = 1 - 3$. In this gauge

$$\nabla_\alpha \nabla^\beta A_\beta = \nabla_\alpha(\nabla^0 A_0) = \partial_\alpha(\nabla^0 A_0), \tag{31.10}$$

remembering that the covariant gradient $\{\nabla_\alpha\}$, of a scalar is just the normal gradient $\{\partial_\alpha\}$.

In the presence of only spatial curvature $[\nabla^0 = -\nabla_0 = -c^{-1}\partial/\partial t]$, the inhomogeneous wave equation in Eq. (31.8) reduces to the form

$$\left(\nabla^j \nabla_j - \frac{1}{c^2}\frac{\partial^2}{\partial t^2}\right) A_\alpha + \frac{1}{c}\frac{\partial}{\partial t}(\partial_\alpha A_0) - R_\alpha^\beta A_\beta = -\mu_0 J_\alpha. \tag{31.11}$$

In the space-frequency domain one then obtains the following coupled equations for the vector and scalar potentials in the Coulomb gauge:

$$[(\nabla^2 + q_0^2)\mathbf{U} - \mathbf{R}] \cdot \mathbf{A} - i q_0 \partial A_0 = -\mu_0 \mathbf{J}, \tag{31.12}$$

$$\nabla^2 A_0 = -\mu_0 J_0, \tag{31.13}$$

where $\partial = (\partial_1, \partial_2, \partial_3)$ is the normal spatial gradient operator. In the absence of spatial curvature $(\mathbf{R} = \mathbf{0})$, the set of equations in Eq. (31.12) and (31.13) are reduced to the well-known set [63], since the Laplace-Beltrami operator $(\nabla^j \nabla_j \equiv \nabla^2)$ becomes the usual Laplace operator.

31.2 METRIC TENSOR OF MESOSCOPIC LAYERS AND CURVED SURFACES

The interplay between electrodynamics and geometry is a deep and fundamental one, even if we limit ourselves to spatial curvature phenomena. In a laboratory setting curved space-time geometries are not of immediate importance, but spatially curved structures are. The theoretical framework of microscopic electrodynamics in spatially curved structures stands on two legs: (i) The microscopic Maxwell equations, and (ii) the Schrödinger (or Dirac) equation. The particle-field interaction is described in terms of the Minimal Coupling Principle giving a four-momentum operator

$$\{\pi_\mu\} \equiv \frac{\hbar}{i}\{\nabla_\mu^G\} = \left\{\frac{\hbar}{i}\partial_\mu - qA_\mu(x)\right\} \tag{31.14}$$

of the particle in question. The quantity $\{\nabla_\mu^G\}$ is the so-called gauge-covariant derivative (operator) [158, 258].

31.2.1 Mesoscopic layer. Surface-induced metric tensor

In light of the importance of the electrodynamics of mesoscopic media in recent years, we now consider a curved surface (S), and assume that a mesoscopic medium is enclosed by the parallel surface S and $S + \Delta S$. Parametrized in terms of arbitrary curvilinear coordinates q^i $(i = 1 - 3)$, the position vector in the layer is given by

$$\mathbf{R}(q^1, q^2, q^3) = \mathbf{r}(q^1, q^2) + q^3\mathbf{n}(q^1, q^2), \tag{31.15}$$

where $\mathbf{r}(q^1, q^2)$ is the position vector to the point (q^1, q^2) on the surface S, and $\mathbf{n}(q^1, q^2)$ is a local unit normal vector on S at (q^1, q^2). In the immediate vicinity of S the coordinate direction (q^3) is assumed to be along $\mathbf{n}(q^1, q^2)$. A schematic illustration of the layer coordination is shown in Fig. 31.1.

With metric

$$ds^2 = G_{ij}(q^1, q^2, q^3)dq^i dq^j \tag{31.16}$$

the ijth element of the metric tensor of mesoscopic layer is given by

$$G_{ij}(q^1, q^2, q^3) = \frac{\partial \mathbf{R}}{\partial q^i} \cdot \frac{\partial \mathbf{R}}{\partial q^j}. \tag{31.17}$$

The metric tensor is symmetric, and

$$G_{13} = G_{23} = 0. \tag{31.18}$$

The proofs for Eq. (31.18) run parallel for G_{13} and G_{23}. Thus,

$$G_{13} = \left(\frac{\partial \mathbf{r}}{\partial q^1} + q^3\frac{\partial \mathbf{n}}{\partial q^1}\right) \cdot \mathbf{n} = 0, \tag{31.19}$$

since the infinitesimal vectors $\partial \mathbf{r}/\partial q^1$ and $\partial \mathbf{n}/dq^1$ ly in the surface tangent plane at (q^1, q^2). Furthermore, it is obvious that

$$G_{33}[= \mathbf{n} \cdot \mathbf{n}] = 1. \tag{31.20}$$

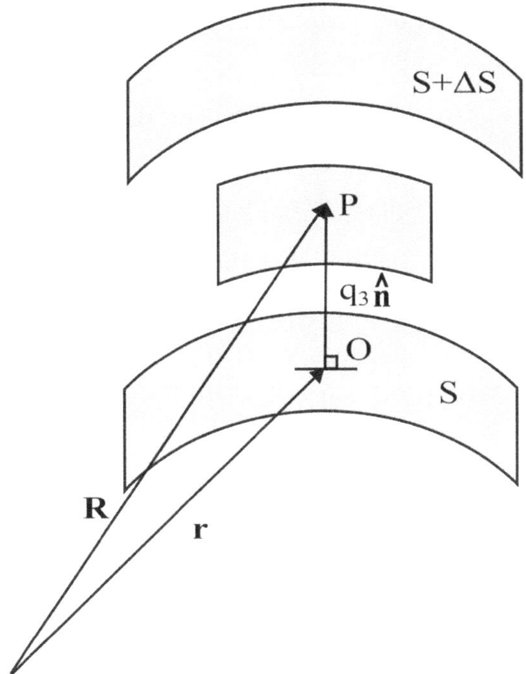

Figure 31.1 Parametrization in arbitrary curve linear coordinates (q_1, q_2, q_3) of a position vector $\mathbf{R} = \mathbf{R}(q_1, q_2, q_3)$, inside a layer of mesoscopic thickness. The layer is confined between the spatially curved surfaces S and $S + \Delta S$. The position vector to a point $\mathcal{O}$ in S is $\mathbf{r}(q_1, q_2)$. The point $\mathcal{P}$ inside the mesoscopic layer has position vector $\mathbf{R}(q_1, q_2, q_3) = \mathbf{r}(q_1, q_2) + q_3\hat{\mathbf{n}}(q_1, q_2)$, where $\hat{\mathbf{n}}(q_1, q_2)$ is the local (at $\mathcal{O}$) normal unit vector. The metric tensor of the mesoscopic layer, $\{G_{ij}(q_1, q_2, q_3)\}$, squeezed on S, namely $\{G_{ij}(q_1, q_2, q_3 \to 0)\}$, is the one related to studies of electromagnetics in spatially curved surfaces, see Section 31.6.

For the subsequent analysis it is useful to change the coordinate labeling from (q^1, q^2, q^3) to (u, v, w) and use the following abbreviations:

$$\frac{\partial \mathbf{r}(u, v)}{\partial u} \equiv \mathbf{r}_u, \quad \frac{\partial \mathbf{r}(u, v)}{\partial v} \equiv \mathbf{r}_v, \tag{31.21}$$

and

$$\frac{\partial \mathbf{n}(u, v)}{\partial u} \equiv \mathbf{n}_u, \quad \frac{\partial \mathbf{n}(u, v)}{\partial v} \equiv \mathbf{n}_v. \tag{31.22}$$

The induced metric in the surface, viz.,

$$\{g_{ij}(u, v)\} = \left\{ \frac{\partial \mathbf{r}(u, v)}{\partial q^i} \cdot \frac{\partial \mathbf{r}(u, v)}{\partial q^j} \right\}, \quad i, j = u \text{ or } v \tag{31.23}$$

has four elements g_{11}, g_{22} and $g_{12} = g_{21}$. In terms of these, and with the abbreviations in Eqs. (31.21) and (31.22), one obtains

$$G_{11} = g_{11} + 2w\mathbf{r}_u \cdot \mathbf{n}_u + w^2\mathbf{n}_u \cdot \mathbf{n}_u, \tag{31.24}$$

$$G_{22} = g_{22} + 2w\mathbf{r}_v \cdot \mathbf{n}_v + w^2\mathbf{n}_v \cdot \mathbf{n}_v, \tag{31.25}$$

$$G_{12}[= G_{21}] = g_{12} + w(\mathbf{r}_u \cdot \mathbf{n}_v + \mathbf{r}_v \cdot \mathbf{n}_u) + w^2\mathbf{n}_u \cdot \mathbf{n}_v. \tag{31.26}$$

It appears from the previous analysis that the metric tensor of our mesoscopic layer, namely,

$$\{G_{ij}\} = \begin{pmatrix} G_{11} & G_{12} & 0 \\ G_{21} & G_{22} & 0 \\ 0 & 0 & 1 \end{pmatrix}, \quad G_{21} = G_{12} \tag{31.27}$$

squeezed on S becomes

$$\{G_{ij}\}_{w\to 0} = \begin{pmatrix} g_{11} & g_{12} & 0 \\ g_{21} & g_{22} & 0 \\ 0 & 0 & 1 \end{pmatrix}, \quad g_{21} = g_{12} \tag{31.28}$$

31.2.2 Extrinsic surface curvature tensor

The $\mathbf{r}_u(u, v)$ and $\mathbf{r}_v(u, v)$ are linearly independent tangent vectors in the surface (S) at (u, v). Since also $\mathbf{n}(u, v)$ and $\mathbf{n}(u, v)$ ly in the tangent plane to S at (u, v), one may express $\mathbf{n}_u$ and $\mathbf{n}_v$ as linear combinations of $\mathbf{r}_u$ and $\mathbf{r}_v$. Thus,

$$\mathbf{n}_u = a\mathbf{r}_u + b\mathbf{r}_v, \tag{31.29}$$

$$\mathbf{n}_v = c\mathbf{r}_u + d\mathbf{r}_v. \tag{31.30}$$

In compact notation [with $a = u$ or v, and $b = u$ or v] and Eqs. (31.29) and (31.30) are written as

$$\mathbf{n}_a = K_a^b \mathbf{r}_b \tag{31.31}$$

with the Einstein summation convention. K_a^b is the (a, b)-element of the so-called *extrinsic surface curvature tensor*

$$\mathbf{K} \equiv \{K_a^b\} = \begin{pmatrix} K_u^u & K_u^v \\ K_v^u & K_v^v \end{pmatrix}. \tag{31.32}$$

In relation to $\mathbf{K}$ one defines the *mean curvature*

$$H \equiv \frac{1}{2}\mathrm{Tr}\mathbf{K} = \frac{1}{2}(K_u^u + K_v^v) \tag{31.33}$$

and the *Gaussian curvature* (K)

$$K \equiv \det \mathbf{K} = K_u^u K_v^v - K_u^v K_v^u. \tag{31.34}$$

In the next subsection we shall realize that the metric tensor $\{G_{ij}\}$ can be expressed in terms of $\{g_{ij}\}$ and $\{K_a^b\}$.

31.2.3 The surface metric tensor $\{G_{uv}(u, v, w)\}$

The essential part of the metric tensor of our mesoscopic layer is contained in the 2x2-matrix

$$\mathbf{G}^{(2)} \equiv \begin{pmatrix} G_{11} & G_{12} \\ G_{21} & G_{22} \end{pmatrix}, \tag{31.35}$$

with the elements given by Eqs. (31.24)–(31.26). In order to introduce the extrinsic surface curvature tensor in these equations one makes use of Eq. (31.31) to eliminate the $\mathbf{n}_u$ and $\mathbf{n}_v$'s in favor of the $\mathbf{r}_u$ and $\mathbf{r}_v$'s. In turn, the scalar products $[\mathbf{r}_u \cdot \mathbf{r}_u, \mathbf{r}_v \cdot \mathbf{r}_v, \mathbf{r}_u \cdot \mathbf{r}_v]$ which result from this elimination, are elements of the $\{g_{uv}\}$ metric tensor [Eq. (31.23)].

In order to carry out the calculation sketched above it is useful to introduce the symmetric matrix

$$\mathbf{h} = \begin{pmatrix} h_{11} & h_{12} \\ h_{21} & h_{22} \end{pmatrix}, \quad h_{21} = h_{12} \tag{31.36}$$

with the elements defined by

$$h_{ab} \equiv -\mathbf{n}_a \cdot \mathbf{r}_b. \tag{31.37}$$

The symmetry of $\mathbf{h}$ can be proved noting that $\mathbf{n} \cdot \mathbf{r}_a = \mathbf{n} \cdot \mathbf{r}_b = 0$. Thus,

$$0 = \frac{\partial}{\partial b}(\mathbf{n} \cdot \mathbf{r}_a) - \frac{\partial}{\partial a}(\mathbf{n} \cdot \mathbf{r}_b) = \frac{\partial \mathbf{n}}{\partial b} \cdot \mathbf{r}_a - \frac{\partial \mathbf{n}}{\partial a} \cdot \mathbf{r}_b = h_{ab} - h_{ba}, \tag{31.38}$$

q.e.d. The matrix part of $\mathbf{G}^{(2)}$ which is proportional to w obviously is given by $-2w\mathbf{h}$. The matrix part proportional to w^2 is determined noting that

$$\mathbf{n}_a \cdot \mathbf{n}_b = \mathbf{n}_a \cdot (K_b^c \mathbf{r}_c) = -h_{ac} K_b^c = -(\mathbf{h} \cdot \mathbf{K})_{ab}, \tag{31.39}$$

and hence given by $-w^2 \mathbf{h} \cdot \mathbf{K}$. Altogether, it appears that

$$\mathbf{G}^{(2)} = \mathbf{g} - 2w\mathbf{h} - w^2 \mathbf{h} \cdot \mathbf{K} \tag{31.40}$$

to order w^2. The $\mathbf{h}$-matrix can be eliminated in favor of $\mathbf{g}$ and $\mathbf{K}$. Since

$$h_{ab} = -\mathbf{n}_a \cdot \mathbf{r}_b = -K_a^c \mathbf{r}_c \cdot \mathbf{r}_b = -K_a^c g_{cb}, \tag{31.41}$$

one obtains

$$\mathbf{h} = -\mathbf{K} \cdot \mathbf{g} = -\mathbf{g} \cdot \mathbf{K}. \tag{31.42}$$

By insertion of Eq. (31.42) into Eq. (31.40), and by utilizing that $\mathbf{h} \cdot \mathbf{K} = \mathbf{K} \cdot \mathbf{h}$, one finally gets the result

$$\mathbf{G}^{(2)} = \mathbf{F} \cdot \mathbf{g}, \tag{31.43}$$

where

$$\mathbf{F} = \mathbf{U} + 2w\mathbf{K} + w^2\mathbf{K}\cdot\mathbf{K}. \tag{31.44}$$

In the following one also needs an expression for the determinant of $\mathbf{G}^{(2)}$, valid to order w^2, and its square root. Now

$$G \equiv \det \mathbf{G}^{(2)} = (\det \mathbf{F})(\det \mathbf{g}) \equiv Fg, \tag{31.45}$$

and a straightforward but somewhat tedious calculation gives

$$F = 1 + 4Hw + (4H^2 + 2K)w^2 + O(w^3). \tag{31.46}$$

The square root of F must to order w^2 have the form

$$f = \sqrt{F} = 1 + \alpha w + \beta w^2, \tag{31.47}$$

giving $\alpha = 2H$ and $\beta = K$. In consequence one obtains

$$\sqrt{G} = f\sqrt{g}, \tag{31.48}$$

where

$$f = 1 + 2Hw + Kw^2 + O(w^3). \tag{31.49}$$

The f-function hence has a linear term (in w) which only depends on the mean curvature [Eq. (31.33)], and a quadratic term proportional to the Gaussian curvature [Eq. (31.34)].

31.2.4 Weingarten equations

By introduction of the following quantities

$$(E, F, G) = (\mathbf{r}_u \cdot \mathbf{r}_u, \mathbf{r}_u \cdot \mathbf{r}_v, \mathbf{r}_v \cdot \mathbf{r}_v), \tag{31.50}$$
$$(L, M, N) = -(\mathbf{r}_u \cdot \mathbf{n}_u, \mathbf{r}_u \cdot \mathbf{n}_v(= \mathbf{r}_v \cdot \mathbf{n}_u), \mathbf{r}_v \cdot \mathbf{n}_v), \tag{31.51}$$

it is easy to prove that the four elements of the extrinsic surface curvature tensor are given by

$$K_u^u = \frac{FM - GL}{EG - F^2}, \quad K_v^v = \frac{FM - EN}{EG - F^2}, \tag{31.52}$$

and

$$K_v^u = \frac{FN - GM}{EG - F^2}, \quad K_u^v = \frac{FL - EM}{EG - F^2}. \tag{31.53}$$

These relations [in connection with Eq. (31.31)] were established in 1861 German mathematician J. Weingarten [262], and are called the Weingarten equations.

31.3 TWO-DIMENSIONAL CURVATURE AND RICCI TENSORS. CURVATURE SCALAR

In this section we describe curvature in two dimensions. On general grounds it is known that the Riemann curvature tensor in two dimensions only has *one* independent component, and altogether four nonvanishing components [258]. The related Ricci tensor is proportional to the 2D metric tensor $\{g_{\mu\nu}\}$ studied in Subsection 31.2.1. The results obtained below are of great importance in establishing the general framework for the microscopic electrodynamics of curved surfaces.

31.3.1 Affine connection

In order to determine the elements of the curvature tensor $\{R_{\mu\nu\alpha\beta}\}$, we start by calculating the affine connection $\{\Gamma^{\mu}_{\alpha\beta}\}$, where [158, 258]

$$\Gamma^{\mu}_{\alpha\beta} = \frac{1}{2}g^{\mu\lambda}(\partial_{\alpha}g_{\lambda\beta} + \partial_{\beta}g_{\lambda\alpha} - \partial_{\lambda}g_{\alpha\beta}). \tag{31.54}$$

The affine connection, which is not a tensor, is sometimes called the Christoffel symbol. The elements of the reciprocal $\{g^{\mu\nu}\}$ metric tensor are given by

$$g^{\mu\lambda}g_{\lambda\nu} = \delta^{\mu}_{\nu}. \tag{31.55}$$

The affine connection has 8 terms of which 6 are different. The elements are

$$\Gamma^{1}_{11} = \frac{1}{2}\left[g^{11}\partial_1 g_{11} + g^{12}(2\partial_1 g_{21} - \partial_2 g_{11})\right], \tag{31.56}$$

$$\Gamma^{1}_{12} = \Gamma^{1}_{21} = \frac{1}{2}\left[g^{11}\partial_2 g_{11} + g^{12}\partial_1 g_{22}\right], \tag{31.57}$$

$$\Gamma^{1}_{22} = \frac{1}{2}\left[g^{11}(2\partial_2 g_{12} - \partial_1 g_{22}) + g^{12}\partial_2 g_{22}\right], \tag{31.58}$$

and the remaining four elements are obtained making the replacement $g^{11} \Rightarrow g^{21}$ and $g^{12} \Rightarrow g^{22}$ in the reciprocal metric tensor appearing in Eqs. (31.56) and (31.58). Thus

$$\Gamma^{2}_{11} = \Gamma^{1}_{11}(g^{11} \Rightarrow g^{21}, g^{12} \Rightarrow g^{22}), \tag{31.59}$$

$$\Gamma^{2}_{12} = \Gamma^{2}_{21} = \Gamma^{1}_{12}(g^{11} \Rightarrow g^{21}, g^{12} \Rightarrow g^{22}), \tag{31.60}$$

$$\Gamma^{2}_{22} = \Gamma^{1}_{22}(g^{11} \Rightarrow g^{21}, g^{12} \Rightarrow g^{22}), \tag{31.61}$$

The affine connection is of importance in itself, e.g. in relation to the geodesic equation; see Eq. (29.29) and [158]. In the subsequent Subsection 31.3.2 it is used for the calculation of the Riemann curvature tensor.

31.3.2 Riemann curvature tensor

The elements of the fully covariant two-dimensional curvature tensor is given by [158, 258]

$$R_{\mu\nu\alpha\beta} = \frac{1}{2}(\partial_{\beta}\partial_{\nu}g_{\mu\alpha} - \partial_{\beta}\partial_{\mu}g_{\nu\alpha} - \partial_{\alpha}\partial_{\nu}g_{\mu\beta} + \partial_{\alpha}\partial_{\mu}g_{\nu\beta}) + g_{\eta\sigma}(\Gamma^{\eta}_{\alpha\mu}\Gamma^{\sigma}_{\nu\beta} - \Gamma^{\eta}_{\beta\mu}\Gamma^{\sigma}_{\nu\alpha}), \tag{31.62}$$

where μ, ν, α and β can take on the values 1 and 2. A little algebra reveals that all terms with an uneven number of 1-indices (and thus also 2-indices) are zero. Of the remaining $2^4/2 = 8$ terms 4 are zero:

$$R_{1111} = R_{2222} = R_{1122} = R_{2211} = 0. \tag{31.63}$$

The last 4 elements are nonvanishing, and related as follows:

$$R_{1212} = -R_{2112} = -R_{1221} = R_{2121} (\neq 0) \tag{31.64}$$

in agreement with what follows from general considerations [258], one is left with only one independent component, say

$$R_{1212} = \frac{1}{2}(\partial_1^2 g_{22} + \partial_2^2 g_{11}) - \partial_1 \partial_2 g_{12} + g_{\eta\sigma}(\Gamma_{11}^\eta \Gamma_{22}^\sigma - \Gamma_{21}^\eta \Gamma_{21}^\sigma), \tag{31.65}$$

with $(\eta, \sigma) = (1, 1), (1, 2), (2, 1)$ and $(2, 2)$.

The reader may convince herself that the surface curvature tensor can be written in the compact form

$$R_{\mu\nu\alpha\beta} = (g_{\mu\alpha} g_{\nu\beta} - g_{\mu\beta} g_{\nu\alpha})\frac{R_{1212}}{g}, \tag{31.66}$$

where $g = g_{11} g_{22} - g_{12} g_{21}$ is the relevant determinant.

31.3.3 Ricci tensor and curvature scalar

We have realized (see Section 31.1) that it is the Ricci tensor which appears in the inhomogeneous Maxwell equations. Its covariant form is given in Eq. (31.2). In the 2D case the Ricci tensor is obtained by contraction of the Riemann tensor in Eq. (31.66):

$$R_{\nu\beta} = g^{\mu\alpha}(g_{\mu\alpha} g_{\nu\beta} - g_{\mu\beta} g_{\nu\alpha})\frac{R_{1212}}{g}. \tag{31.67}$$

By utilizing that

$$g^{\alpha\mu} g_{\mu\beta} = \delta_\beta^\alpha \tag{31.68}$$

one obtains

$$R_{\nu\beta} = \left[\left(\sum_\mu \delta_\alpha^\mu\right) g_{\nu\beta} - \delta_\beta^\alpha g_{\nu\alpha}\right]\frac{R_{1212}}{g}, \tag{31.69}$$

and hence a Ricci tensor

$$R_{\nu\beta} = g_{\nu\beta}\frac{R_{1212}}{g}. \tag{31.70}$$

The curvature scalar (R), defined by contraction of the Ricci tensor,

$$R = g^{\nu\beta} R_{\nu\beta}, \tag{31.71}$$

thus becomes

$$R = \frac{2}{g} R_{1212},\tag{31.72}$$

since $g^{\nu\beta} g_{\nu\beta} = 2$. The covariant surface Ricci tensor hence can be written in the form

$$R_{ab} = g_{ab} \frac{R}{2},\tag{31.73}$$

using the index notation of Subsection 31.2.2. The mixed Ricci tensor

$$R_b^a = g^{ac} g_{cb} \frac{R}{2} = \frac{R}{2} \delta_b^a,\tag{31.74}$$

given in 2x2 matrix form by

$$\{R_b^a\} = \frac{R}{2} \begin{pmatrix} 1 & 0 \\ 0 & 1 \end{pmatrix},\tag{31.75}$$

is the one that may be used in the potential description of surface electrodynamics, cf. the Lorenz and Coulomb forms of the inhomogeneous wave equations; see Eqs. (31.3), (31.6) and (31.12).

31.4 LAPLACE-BELTRAMI OPERATOR

31.4.1 Contracted affine connection, $\Gamma^\mu_{\mu\lambda}$. Covariant divergence

Let us take as a starting point the general expression for the affine connection, viz. [Eq. (29.30)]

$$\Gamma^\mu_{\nu\lambda} = \frac{1}{2} g^{\mu\sigma} (\partial_\nu g_{\sigma\lambda} + \partial_\lambda g_{\nu\sigma} - \partial_\sigma g_{\lambda\nu}).\tag{31.76}$$

By contraction $(\nu = \mu)$ one obtains

$$\Gamma^\mu_{\mu\lambda} = \frac{1}{2} g^{\mu\sigma} \partial_\lambda g_{\mu\sigma}\tag{31.77}$$

since the sum of the first and second term in the parenthesis of Eq. (31.76) upon contraction vanishes. From matrix calculus [258] it appears that

$$\partial_\lambda g = g g^{\mu\sigma} \partial_\lambda g_{\mu\sigma}.\tag{31.78}$$

With the signature choice used in this book the quantity g designates the determinant of the metric tensor $\{g_{\mu\nu}\}$. In the flat space-time limit the signature $(-1, 1, 1, 1)$ gives $\det\{g_{\mu\nu}\} = -1$ so that $-g$ is a positive quantity. By combining Eqs. (31.77) and (31.78) one gets

$$\Gamma^\mu_{\mu\lambda} = \frac{1}{2(-g)} \partial_\lambda(-g) = \partial_\lambda \ln \sqrt{-g},\tag{31.79}$$

and hence the following compact expression for the contracted affine connection:

$$\Gamma^{\mu}_{\mu\lambda} = \frac{1}{\sqrt{-g}}\partial_{\lambda}\sqrt{-g}. \tag{31.80}$$

The covariant derivative of a contravariant vector, $\{V^{\mu}\}$ is given by [158, 258]

$$\nabla_{\nu}V^{\mu} = \partial_{\nu}V^{\mu} + \Gamma^{\mu}_{\nu\lambda}V^{\lambda}. \tag{31.81}$$

Contraction in turn gives one the covariant divergence

$$\nabla_{\mu}V^{\mu} = \partial_{\mu}V^{\mu} + \Gamma^{\mu}_{\mu\lambda}V^{\lambda} = \partial_{\mu}V^{\mu} + \frac{1}{\sqrt{-g}}(\partial_{\lambda}\sqrt{-g})V^{\lambda}. \tag{31.82}$$

The expression for the divergence can obviously be written in the compact form

$$\nabla_{\mu}V^{\mu} = \frac{1}{\sqrt{-g}}\partial_{\mu}(\sqrt{-g}V^{\mu}). \tag{31.83}$$

31.4.2 Spatial Laplace-Beltrami operator

The result in Eq. (31.83) now is specialized to the case, where $\{V^{\mu}\}$ is the contravariant gradient of a scalar, S. Thus,

$$V^{\mu} = \nabla^{\mu}S = \partial^{\mu}S, \tag{31.84}$$

in view of the fact that the contravariant (covariant) gradient of a scalar is just the usual gradient. The four-dimensional Laplace-Beltrami operator hence can be written as follows:

$$\nabla_{\mu}\nabla^{\mu}(\ldots) = \frac{1}{\sqrt{-g}}\partial_{\mu}[\sqrt{-g}\partial^{\mu}(\ldots)]. \tag{31.85}$$

The notation $(\ldots)$ indicates the manner in which the Laplace-Beltrami operator is to act on the given scalar (field).

In the presence of spatial curvature alone the Laplace-Beltrami operator is reduced to

$$\nabla_{j}\nabla^{j}(\ldots) = \frac{1}{\sqrt{G}}\partial_{j}[\sqrt{G}\partial^{j}(\ldots)], \quad j = 1 - 3 \tag{31.86}$$

where $G(>0)$ is the determinant of the three-dimensional determinant of the metric tensor of space, $\{G_{ij}\}$.

31.5 ELECTRODYNAMICS IN MESOSCOPIC LAYERS

In the Lorenz gauge we have seen that the inhomogeneous wave equations for the vector and the scalar potentials in the space-frequency domains have the forms given in Eqs. (31.6) and (31.7). In order to apply these equations to a study of the internal electrodynamics of a mesoscopic layer, one must determine the Laplace-Beltrami operator and the Ricci tensor for such a layer.

The Laplace-Beltrami operator for the mesoscopic layer follows from Eq. (31.35) by utilizing that the square root of the determinant of the metric tensor is given by $\sqrt{G} = f\sqrt{g}$, where $f = 1 + 2Hw + Kw^2 + O(w^3)$; see the analysis in Section 31.2 [resulting in Eqs. (31.48) and (31.49)].

Let us as a starting point rewrite Eq. (31.86) in the form

$$\nabla_j \nabla^j(...) = \partial_j \partial^j(...) + \left[\frac{1}{\sqrt{G}}\partial_j(\sqrt{G})\right]\partial^j(...). \tag{31.87}$$

The deviation between the covariant Laplace-Beltrami operator $[\nabla_j \nabla^j]$ and the usual Laplace operator $[\partial_j \partial^j]$ thus is given by the last term in Eq. (31.87). Since

$$\partial_j \sqrt{G} = \delta_j(f\sqrt{g}) = \begin{cases} f\partial_j\sqrt{g}; & j = 1,2 \\ 2(H + Kw)\sqrt{g}; & j = 3 \end{cases} \tag{31.88}$$

one obtains

$$\nabla_j \nabla^j(...) = \partial_a \partial^a(...) + \partial_w \partial^w(...) + \left[\frac{1}{\sqrt{g}}\partial_a\sqrt{g}\right]\partial^a(...) + \frac{2}{f(w)}(H + Kw)\partial^w(...). \tag{31.89}$$

The index a runs over the two surface coordinates u and w, [with summation over $a = u$ or v kept implicit], and as before w is the coordinate perpendicular to the surface [see Eq. (31.15) with $(q^1, q^2) = (u, v)$ and $q^3 = w$]. If wished, one may write $\partial_w \partial^w = \partial_w^2$.

It can be shown that the Ricci tensor given in Eq. (31.3) for a mesoscopic layer is reduced to the form

$$\{R_j^i(u, v, w)\} = \begin{pmatrix} R_1^1 & R_2^1 & 0 \\ R_1^2 & R_2^2 & 0 \\ 0 & 0 & 0 \end{pmatrix}. \tag{31.90}$$

The symmetric Ricci tensor $[R_1^2 = R_2^1]$ has three independent elements, which all depend on u, v and w. The explicit expressions for these elements are somewhat complicated and not needed in our subsequent discussion of the electrodynamics of curved surfaces [Section 31.6].

Remembering that the vector potential $\mathbf{A}$ is a function of the three variables u, v and w, i.e.

$$\mathbf{A}(u, v, w) = \begin{pmatrix} A_u(u, v, w) \\ A_v(u, v, w) \\ A_w(u, v, w) \end{pmatrix}, \tag{31.91}$$

the Lorenz gauge wave equations for the components of $\mathbf{A}$ become

$$(\nabla^2 + q_0^2)A_u - R_u^u A_u - R_v^u A_v = -\mu_0 J_u, \tag{31.92}$$
$$(\nabla^2 + q_0^2)A_v - R_u^v A_u - R_v^v A_v = -\mu_0 J_v, \tag{31.93}$$
$$(\nabla^2 + q_0^2)A_w = -\mu_0 J_w. \tag{31.94}$$

The presence of the off-diagonal elements in the Ricci tensor implies that the wave equations for A_w and A_v are coupled due to the curvature. Besides this, the current density

$$\mathbf{J}(u,v,w) = \begin{pmatrix} J_u(u,v,w) \\ J_v(u,v,w) \\ J_w(u,v,w) \end{pmatrix}, \tag{31.95}$$

in a selfconsistenr approach in general will give rise to a coupling of A_u, A_v and A_w, as we shall discuss in Chapter 32.

31.6 TRANSITION TO CURVED SURFACES

In the limit $w \to 0$, the mesoscopic layer contracts to a curved surface. While this transition is well-defined from a mathematical point of view, strict physical surfaces do not exist. This non-existence basically has nothing to do with spatial curvature as such, In flat mesoscopic layers the possibilities for replacing such layers with infinitely thin sheets carrying a so-called surface current density distribution have been studied e.g. in relation to electromagnetic diffraction from mesoscopic holes in metallic and semiconducting screens, and in connection to the theory of electromagnetic surface waves (surface plasmaritons) carrying an interface current density. A central issue in such studies concerns a self-consistent modelling of the mesoscopic layer's microscopic current density by a surface current density. Quantum mechanical spatial nonlocality of the microscopic conductivity implies that the surface layer model inherently suffers from a number fundamental shortcomings. In a given application these shortcomings may be unimportant for the overall result.

Let us now consider some of the problems one is facing when attempting to reduce the electromagnetic theory of curved mesoscopic layers to a curved surface theory. To this end we first consider the action of the Laplace-Beltrami operator on a component $A_i(u,v,w) \equiv A_i(u,v,w)$, $I = u, v, w$ of the vector potential in the limit $\omega \to 0$. From Eq. (31.89) one obtains

$$\nabla_j \nabla^j A_i(u,v,w \to 0) = (\partial_u^2 + \partial_v^2 + \partial_w^2) A_i(u,v,w \to 0) +$$
$$\left[\frac{1}{\sqrt{g}} \partial_a \sqrt{g} \right] \partial^a A_i(u,v,w \to 0) + 2H(u,v)\partial^w A_i(u,v,w \to 0),$$
$$\tag{31.96}$$

where, as hitherto, the summation over $a = u$ and v is kept implicit. Utilizing the notation $A_i(u,v) \equiv A_i(u,v,w \to 0)$ one has

$$\nabla_j \nabla^j A_i(u,v,w \to 0) = (\partial_w^2 + \partial_v^2) A_i(u,v) + \left[\frac{1}{\sqrt{g}} \partial_a \sqrt{g} \right] \partial^a A_i(u,v)$$
$$+ f_i^{(1)}(u,v) + f_i^{(2)}(u,v), \tag{31.97}$$

where

$$f_i^{(1)}(u,v) = 2H(u,v) \lim_{w \to 0} \partial^w A_i(u,v,w), \tag{31.98}$$

and

$$f_i^{(2)}(u, v) = \lim_{w \to 0} \partial_w^2 A_i(u, v, w). \tag{31.99}$$

Note that only the local mean curvature $H = H(u, v)$ enters the Laplace-Beltrami operator's action on $A_i(u, v, w \to 0)$.

It appears from Eqs. (31.98) and (31.99) that a knowledge of the first and second derivative of $A_i(u, v, w)$ in the normal (w) direction of the surface is needed. These quantities depend on (u, v) and CANNOT be determined without a knowledge of the microscopic surface current density at a given frequency. A self-consistent calculation of the surface current density requires that the field-coupled Schrödinger (Pauli, Dirac) equation is combined with the wave equation for the vector and scalar potentials.

It was realized in Section 31.3 that the off-diagonal elements of the two-dimensional Ricci tensor vanish $[R_v^u = R_u^v = 0]$. This implies that the two equations [Eqs. (31.92) and (31.93)] for A_u and A_v decouple in the limit $w \to 0$. Furthermore, we have seen that the two on-diagonal elements are equal, and given by $R_u^u = R_v^v = R/2$, where R is the curvature scalar [see Eq. (31.75)]. Expressed in terms of the only independent component of the Riemann curvature tensor here taken as R_{1212}, the surface curvature scalar is $R = (2/g)R_{1212}$.

It must be remembered that by decoupling we refer to curvature decoupling. The wave equations in Eqs. (31.92) and (31.93) are of course in general coupled through an interdependence of the current density components J_u and J_v. The individual components (J_u, J_v, J_w, J_0) of the four-current density depend on the curvature through the determinant $g(x)$ of the metric tensor; see Eq. (30.26).

For curved surfaces Eqs. (31.92) and (31.93), take the common form

$$\left(\nabla_j \nabla^j + q_0^2 - \frac{R}{2} \right) A_a(u, v, w \to 0) = -\mu_0 J_a(u, v, w \to 0), \tag{31.100}$$

where $a = u$ or v. The remaining components of the four-potential satisfy the Riemann surface curvature independent wave equations

$$(\nabla_j \nabla^j + +q_0^2) A_\alpha(u, v, w \to 0) = -\mu_0 J_\alpha(u, v, w \to 0), \tag{31.101}$$

for $\alpha = w$ or 0. In order to enter deeper into the analysis of Eqs. (31.100) and (31.101) one must in a nonrelativistic framework turn the attention towards the Schrödinger equation. From a fundamental point of view massive particles (e.g. electrons) are coupled to the electromagnetic field via the Minimal Coupling Substitution in the free Schrödinger equation.

CHAPTER 32

Field-Coupled Surface Shrödinger Equation. Möbius-Band Electrodynamics

32.1 SCHRÖDINGER EQUATION IN JELLIUM APPROXIMATION. GEOMETRICAL POTENTIAL

In the following it is assumed that the electrons inside the mesoscopic layer, in a one-electron approximation, are subject to forces perpendicular to the surface, S, i.e. in the w-direction. The associated potential is denoted by $V(w)$. This potential is different from zero only in the two thin ($\sim \lambda_F$) transition layers holding the electrons inside the mesoscopic layer. The vector-potential independent Schrödinger equation thus is given by

$$i\hbar\partial_t\psi = -\frac{\hbar^2}{2m}\nabla_i\nabla^i\psi + V(w)\psi. \tag{32.1}$$

To obtain a surface (S) Schrödinger equation the Laplace-Beltrami operator is divided into parts acting along ($\parallel$) and ($\perp$) to S. Thus,

$$\nabla_i\nabla^i = \nabla_\parallel^2 + \nabla_\perp^2, \tag{32.2}$$

where

$$\nabla_\parallel^2(...) = \frac{1}{\sqrt{G}}\partial_a\left[\sqrt{G}\partial^a(...)\right], \tag{32.3}$$

$$\nabla_\perp^2(...) = \frac{1}{\sqrt{G}}\partial_w\left[\sqrt{G}\partial^w(...)\right]. \tag{32.4}$$

DOI: 10.1201/9781003029458-32

If one, in the usual well-known manner, subjects the wave function along S to periodic boundary conditions, the wave function is normalizable with a norm (N) given by

$$N = \int |\psi|^2 \sqrt{G} d^3 x, \tag{32.5}$$

where the integration extends over the mesoscopic layer. The 3D-invariant volume element is $\sqrt{G} d^3 x$. Since $\sqrt{G} = f\sqrt{g}$ [with $f = 1 + 2Hw + Kw^2 + O(w^3)$] for our mesoscopic layer the norm becomes

$$N = \int f|\psi|^2 \sqrt{g} du dv dw = \int |\chi|^2 \sqrt{g} du dv dw, \tag{32.6}$$

where

$$\chi(u, v, w) = f^{1/2}(w)\psi(u, v, w). \tag{32.7}$$

In order to derive the Schrödinger equation in the surface S, one considers the limit $w \to 0$. The action of the "parallel" Laplace-Beltrami operator readily gives

$$\lim_{w \to 0} \nabla_{\parallel}^2 \psi = \lim_{w \to 0} \left[\frac{1}{\sqrt{G}} \partial_a \left(\sqrt{G} \partial^a \psi \right) \right]$$

$$= \lim_{w \to 0} \left\{ \frac{1}{f\sqrt{g}} \partial_a \left[f\sqrt{g} \partial^a \left(f^{-1/2} \chi \right) \right] \right\} = \frac{1}{\sqrt{g}} \partial_a \left(\sqrt{g} \partial^a \chi \right). \tag{32.8}$$

A straightforward calculation leads to the following result for the action of the "perpendicular" Laplace-Beltrami operator:

$$\lim_{w \to 0} \nabla_{\perp}^2 \psi = \lim_{w \to 0} \left\{ \frac{1}{f} \partial_w \left[f \partial^w \left(f^{-1/2} \chi \right) \right] \right\} = \partial_w \partial^w \chi + (H^2 - K)\chi. \tag{32.9}$$

In view of the considerations above we find in the limit $w \to 0$ a Schrödinger equation

$$i\hbar \partial_t \chi = -\frac{\hbar^2}{2m} \left[\frac{1}{\sqrt{g}} \partial_a (\sqrt{g} \partial^\alpha \chi) \right] - \frac{\hbar^2}{2m} \partial_w^2 \chi + V_g \chi + V(w)\chi, \tag{32.10}$$

where

$$V_g = -\frac{\hbar^2}{2m} (H^2 - K) \tag{32.11}$$

is the so-called geometrical (g) potential energy.

It is now possible to make the product ansatz

$$\chi(u, v, w) = \chi_{\parallel}(u, v)\chi_{\perp}(w) \tag{32.12}$$

since $V_g = V_g(u, v)$ and $V = V(w)$. If the electron system does not behave like a jellium along the surface it is in general not possible to separate the wave function into a product of one part which is independent of w and one part which only depends on this coordinate. In the jellium approximation the electrons are subjected only to

V_g in S. The ansatz in Eq. (32.12) separates Eq. (32.10) into a surface Schrödinger equation

$$i\hbar\frac{\partial}{\partial t}\chi_\parallel(u,v) = -\frac{\hbar^2}{2m}\frac{1}{\sqrt{g(u,v)}}\partial_a\left[\sqrt{g(u,v)}\partial^\alpha\chi_\parallel(u,v)\right]$$
$$-\frac{\hbar^2}{2m}\left(H^2(u,v) - K(u,v)\right)\chi_\parallel(u,v), \tag{32.13}$$

and a Schrödinger equation for the dynamics perpendicular to S

$$i\hbar\frac{\partial}{\partial t}\chi_\perp(w) = -\frac{\hbar^2}{2m}\frac{\partial^2}{\partial w^2}\chi_\perp(w) + V(w)\chi_\perp(w). \tag{32.14}$$

From a physical point of view it is not possible to take the limit $V(w \to 0)$. Hence, one may shrink w until quantum size effects appear. A field-induced motion of the electrons in the transverse (w) direction exhibits quantum-well (QW) behaviour. In a microscopic calculation the "spill-out" of the stationary-state electron wave function plays an important role when "$w \to 0$". The spill-out length (λ_F) is of the order $\lambda_F = 2\pi/k_F$, where k_F is the Fermi wave number of the electron. If the thickness of the mesoscopic layer is sufficiently small there will only be one bound state left in the QW, and the electron dynamics becomes essentially two dimensional and can be described by the diamagnetic part of the field-matter interaction. Including the dynamic Pauli spin phenomenon a (small) field-induced electron motion in the w-direction is still present, as one may realize from the spin electrodynamics in a mesoscopic ring [197] generalized to two-dimensional dynamics.

32.2   SURFACE ELECTRON DYNAMICS IN AN ELECTROMAGNETIC FIELD

Let us return to the free Schrödinger equation for a charged particle ($\sim$ electron) in curved space, viz.,

$$i\hbar\partial_t\psi = -\frac{\hbar^2}{2m}\nabla_i\nabla^i\psi, \tag{32.15}$$

whereas before $\nabla_i\nabla^i$ is the Laplace-Beltrami operator. In the Coulomb gauge the coupling to an electromagnetic field changes Eq. (32.14) to

$$i\hbar\partial_t\psi = -\frac{\hbar^2}{2m}D_iD^i\psi + V\psi, \tag{32.16}$$

where

$$D_i = \nabla_i - \frac{ie}{\hbar}A_i \tag{32.17}$$

is the relevant Minimal Coupling Substitution, and V the Coulomb energy of the particle, the charge of which is e. By combining Eqs. (32.16), (32.17) and the Coulomb gauge condition

$$\nabla_iA^i = 0, \tag{32.18}$$

the field-coupled Schrödinger equation takes the following form:

$$i\hbar\partial_t\psi = \left[-\frac{\hbar^2}{2m}\nabla_i\nabla^i + V\right]\psi + \frac{i\hbar e}{m}A_i\nabla_i\psi + \frac{e^2}{2m}A_iA^i\psi. \tag{32.19}$$

We have seen in Section 32.1 that the field-unperturbed Schrödinger equation can be expressed with $\chi(u,v,w) = f^{1/2}(w)\psi(u,v,w)$, Eq. (32.7), as wave function. As shown below, the same is the case for the field-coupled Schrödinger equation. For the nonlinear term one obviously obtains

$$\lim_{w\to 0} A_iA^i\psi = A_iA^i\chi, \tag{32.20}$$

since $f(w\to 0) = 1$. For the linear term one gets

$$\begin{aligned}A^i\nabla_i\psi &= A^a\partial_a(f^{-1/2}\chi) + A^w\partial_w(f^{-1/2}\chi)\\ &= f^{-1/2}A^a\partial_a\chi + A^w(\partial_w f^{-1/2})\chi + f^{-1/2}A^w\partial_w\chi,\end{aligned} \tag{32.21}$$

and then

$$\lim_{w\to w} A^i\nabla_i\psi = A^a\partial_a\chi - HA^w\chi + A^w\partial_w\chi, \tag{32.22}$$

because

$$\lim_{w\to 0}\partial_w f^{-1/2} = -H. \tag{32.23}$$

In the limit $w\to 0$, the field-coupled Schrödinger equation hence takes the following form:

$$\begin{aligned}i\hbar\partial_t\chi = &-\frac{\hbar^2}{2m}\left[\frac{1}{\sqrt{g}}\partial_a(\sqrt{g}\partial^a\chi)\right] - \frac{\hbar^2}{2m}\partial_w^2\chi + V_g\chi + V(w)\chi\\ &+ \frac{i\hbar e}{m}[A^a\partial_a\chi - HA^w\chi + A^w\partial_w\chi] + \frac{e^2}{m}(A_aA^a\chi + A_wA^w\chi).\end{aligned} \tag{32.24}$$

It is obvious that Eq. (32.24) cannot be separated into a surface part and a part describing the dynamics perpendicular to the surface unless certain approximations are made for the prevailing vector potential. A rigorous analysis of the physical criteria for a possible separation of the dynamics requires a self-consistent study of the inhomogeneous Maxwell equations in the Coulomb gauge (see Chapter 31) and the field-coupled Schrödinger equation [Eq. (32.24)]. In the physical limit "$w\to 0$", it is often a good approximation to neglect the variation of the vector potential across the w-profile, and use the mean value, $<...>_w$, in the calculation, i.e.,

$$< \mathbf{A}(u,v,w) >_w \equiv \mathbf{A}(u,v). \tag{32.25}$$

Furthermore, if the mesoscopic layer electrodynamically essentially behaves like a one-level QW, the vector potential perpendicular to the surface may be negligible. Thus,

$$\mathbf{A}(u,v) \approx (A_u(u,w), A_v(u,v), 0). \tag{32.26}$$

The approximations in Eqs. (32.25) and (32.26) enable one to obtain a field-coupled surface Schrödinger equation

$$
i\hbar\partial_t\chi_\|(u,v) = -\frac{\hbar^2}{2m}\left[\frac{1}{\sqrt{g(u,v)}}\partial_a\left(\sqrt{g(u,v)}\partial^a\chi_\|(u,v)\right)\right]
$$
$$
-\frac{\hbar^2}{2m}(H^2(u,v) - K(u,v))\chi_\|(u,v) + \frac{i\hbar e}{2m}A^a(u,v)\partial_a\chi_\|(u,v)
$$
$$
+\frac{e^2}{m}\left(A^a(u,v)A_a(u,v)\chi_\|(u,v)\right). \tag{32.27}
$$

For small field strengths one can neglect the nonlinear term $(e^2/m)A^aA_a\chi_\|$; linear electrodynamics. With $A^w(u,v) = 0$, the Schrödinger equation for $\chi_\perp(w)$ is still given by Eq. (32.14), and thus unaffected by an impressed vector potential.

32.3   METRIC AND EXTRINSIC CURVATURE TENSORS

A general Möbius strip (band) need not be developable, meaning that it cannot be mapped isometrically to a plane rectangular strip. By isometrically we mean that the mapping preserves all intrinsic distances. It may be reasonable to suggest that some Möbius nanostructures to a good approximation are developable. The rectangular strip therefore deforms in such a way that the metrical properties are barely changed. A necessary and sufficient condition for a surface to be developable is that its Gaussian curvature vanishes everywhere.

For the present purpose it is convenient to use an arc-length parametrization of our assumed developable Möbius band. We take $\mathcal{R}(u)$ as a parametrization of the strips centerline curve. Furthermore, we use the local Frenet-Serret basis vectors $\hat{\mathbf{t}}(u)$, $\hat{\mathbf{n}}(u)$, $\hat{\mathbf{b}}(u)$ in our description. These vectors constitute an orthonormal frame at each point $\mathcal{R}(u)$, and $\hat{\mathbf{t}}(u)$ is the unit tangent vector, $\hat{\mathbf{b}}(u)$ the unit binormal, and $\hat{\mathbf{n}}(u)$ the principal normal. For the general reader it may be useful to give the Frenet-Serret formulas [91, 236] satisfied by the frame unit vectors some attention.

The local curvature $\kappa = \kappa(u)$ of the centerline is given by the definition

$$
\frac{d\hat{\mathbf{t}}}{du} = \kappa\hat{\mathbf{n}}. \tag{32.28}
$$

Since $\hat{\mathbf{b}}(u) = \hat{\mathbf{t}}(u) \times \hat{\mathbf{n}}(u)$ by definition one obtains

$$
\frac{d\hat{\mathbf{b}}}{du} = \frac{d\hat{\mathbf{t}}}{du} \times \hat{\mathbf{n}} + \hat{\mathbf{t}} \times \frac{d\hat{\mathbf{n}}}{du} = \hat{\mathbf{t}} \times \frac{d\hat{\mathbf{n}}}{du} \tag{32.29}
$$

by use of Eq. (32.28). It appears from Eq. (32.29) that $d\hat{\mathbf{b}}/du$ is perpendicular to $\hat{\mathbf{t}}$, and of course also $\hat{\mathbf{b}}$. Thus, $d\hat{\mathbf{b}}/du$ must be in the direction of $\hat{\mathbf{n}}$, i.e.

$$
\frac{d\hat{\mathbf{b}}}{du} = -\tau\hat{\mathbf{n}}, \tag{32.30}
$$

where the proportionality factor (times -1) defines the local torsion $\tau = \tau(u)$ of the centerline. The derivative $d\hat{\mathbf{n}}/du$ now can be expressed in terms of κ and τ. Hence,

$$\frac{d\hat{\mathbf{n}}}{du} = \frac{d\hat{\mathbf{b}}}{du} \times \hat{\mathbf{t}} + \hat{\mathbf{b}} \times \frac{d\hat{\mathbf{t}}}{du} = -\tau \hat{\mathbf{n}} \times \hat{\mathbf{t}} + \kappa \hat{\mathbf{b}} \times \hat{\mathbf{n}}, \tag{32.31}$$

and then

$$\frac{d\hat{\mathbf{n}}}{du} = \tau \hat{\mathbf{b}} - \kappa \hat{\mathbf{t}} \tag{32.32}$$

Eqs. (32.28), (32.30) and (32.32) constitute the so-called Frenet-Serret relations.

We can parametrize the Möbius strip by means of the local Frenet-Serret coordinates as follows:

$$\mathbf{r}(u,v) = \mathcal{R}(u) + v \left[\hat{\mathbf{b}}(u) + \frac{\tau(u)}{\kappa(u)} \hat{\mathbf{t}}(u) \right]. \tag{32.33}$$

For a strip of length L and width of $2w$, $u \in [0, L]$ and $v \in [-w, w]$. It appears from Eq. (32.33) that the shape of a developable Möbius strip is completely determined by its centerline, $\mathcal{R} = \mathcal{R}(u)$. In Section 32.4 where the electrodynamics of a quantum wire $(v = 0)$ is discussed, we shall present an adequate parametric form for $\mathcal{R}(u)$, and give a few details related to the establishment of the parametrization in Eq. (32.33).

The metric tensor of the Möbius strip is obtained by applying Eq. (31.23) with $\mathbf{r}(u,v)$ given by Eq. (32.33). To this end, we need the derivatives $\partial \mathbf{r}(u,v)/\partial u$ and $\partial \mathbf{r}(u,v)\partial v$. Since $\partial \mathcal{R}(u)/\partial u = \hat{\mathbf{t}}(u)$, using Eqs. (32.28) and (32.30) one obtains

$$\begin{aligned}\frac{\partial \mathbf{r}(u,v)}{\partial u} &= \hat{\mathbf{t}}(u) + v \left[\frac{d\hat{\mathbf{b}}(u)}{du} + \frac{d}{du} \left(\frac{\tau(u)}{\kappa(u)} \right) \hat{\mathbf{t}}(u) + \frac{\tau(u)}{\kappa(u)} \frac{d\hat{\mathbf{t}}(u)}{du} \right] \\ &= \hat{\mathbf{t}}(u) \left[1 + v \frac{d}{du} \left(\frac{\tau(u)}{\kappa(u)} \right) \right],\end{aligned} \tag{32.34}$$

and

$$\frac{\partial \mathbf{r}(u,v)}{\partial v} = \hat{\mathbf{b}}(u) + \frac{\tau(u)}{\kappa(u)} \hat{\mathbf{t}}(u). \tag{32.35}$$

The orthonormality of the Frenet-Serret basis vectors readily leads to the following metric tensor $[a, b = u$ or $v]$:

$$\{g_{ab}\} = \begin{pmatrix} f^2(u,v) & \eta(u)f(u,v) \\ \eta(u)f(u,v) & 1 + \eta^2(u) \end{pmatrix} \tag{32.36}$$

where

$$\eta(u) = \frac{\tau(u)}{\kappa(u)}, \tag{32.37}$$

and

$$f(u,v) = 1 + v \frac{d\eta(u)}{du}. \tag{32.38}$$

The determinant of the metric tensor is given by

$$g(u, v) \equiv \mathrm{Det}\{g_{ab}\} = \frac{1}{\mathrm{Det}\{g^{ab}\}} = f^2(u, v). \tag{32.39}$$

The elements K_a^b of the extrinsic curvature tensor are determined from the definition $\mathbf{n}_a = K_a^b \mathbf{r}_b$ [Eq. (31.31)]. From

$$\frac{\partial \mathbf{n}}{\partial u} \equiv \mathbf{n}_u = K_u^u \mathbf{r}_u + K_u^v \mathbf{r}_v = K_u^u \frac{\partial \mathbf{r}}{\partial u} + K_u^v \frac{\partial \mathbf{r}}{\partial v} \tag{32.40}$$

one gets

$$\tau \hat{\mathbf{b}} - \kappa \hat{\mathbf{t}} = K_u^u \left(1 + v \frac{d\eta}{du}\right) \hat{\mathbf{t}} + K_u^v (\hat{\mathbf{b}} + \eta \hat{\mathbf{t}}), \tag{32.41}$$

and thus

$$K_u^v = \tau, \tag{32.42}$$

$$K_u^u = -\frac{\kappa(1 + \eta^2)}{1 + v\frac{d\eta}{du}}. \tag{32.43}$$

Since $\mathbf{n}_v \equiv \partial \hat{\mathbf{n}} / \partial v = \mathbf{0}$, one has the condition

$$\mathbf{0} = K_v^u \mathbf{r}_u + K_v^v \mathbf{r}_v = K_v^u \left(1 + v \frac{d\eta}{du}\right) \hat{\mathbf{t}} + K_v^v (\hat{\mathbf{b}} + \eta \hat{\mathbf{t}}). \tag{32.44}$$

In consequence,

$$K_v^u = K_v^v = 0. \tag{32.45}$$

The extrinsic curvature tensor hence attains in explicit form

$$\mathbf{K} = \{K_a^b\} = \begin{pmatrix} -\frac{\kappa(1+\eta^2)}{1+v\frac{d\eta}{du}} & \tau \\ 0 & 0 \end{pmatrix}, \tag{32.46}$$

giving a (signed) mean curvature

$$H \equiv \frac{1}{2}(K_u^u + K_v^v) = -\frac{\kappa}{2} \frac{1 + \eta^2}{1 + v\frac{d\eta}{du}}, \tag{32.47}$$

and a Gaussian curvature

$$K \equiv \det\{K_a^b\} = 0, \tag{32.48}$$

a result which holds for a developable surface in general, and thus for our Möbius strip.

32.4 ELECTRODYNAMICS OF A MESOSCOPIC MÖBIUS QUANTUM WIRE

32.4.1 Time-independent Schrödinger equation

In Section 32.4, we came to the conclusion that a jellium electron confined to a surface under certain approximations satisfies the field-unperturbed Schrödinger equation given in Eq. (32.13). For a Möbius strip where $\sqrt{g(u,v)} = f(u,v)$ [Eq. (32.39)], and $K(u,v) = 0$ [Eq. (32.48)], Eq. (32.13) is reduced to the form [as before with the time dependence of $\chi_\parallel$ kept implicit; $\chi_\parallel(u,v,t) \equiv \chi_\parallel(u,v)$]

$$i\hbar\frac{\partial}{\partial t}\chi_\parallel(u,v) = -\frac{\hbar^2}{2m}\left[\frac{\partial_a f(u,v)}{f(u,v)}\partial^a + \partial_a\partial^a + H^2(u,v)\right]\chi_\parallel(u,v). \tag{32.49}$$

In deriving Eq. (32.49) the relation

$$f^{-1}\partial_a(f\partial^a\chi_\parallel) = \partial_a\partial^a\chi_\parallel + f^{-1}(\partial_a f)\partial^a\chi_\parallel \tag{32.50}$$

was used. Since

$$\partial_u f(u,v) = v\frac{d^2\eta(u)}{du^2}, \tag{32.51}$$

$$\partial_v f(u,v) = \frac{d\eta(u)}{du}, \tag{32.52}$$

the Möbius strip Schrödinger equation can be written as follows:

$$i\hbar\frac{\partial}{\partial t}\chi_\parallel(u,v) = -\frac{\hbar^2}{2m}\left\{\frac{1}{f(u,v)}\left[v\frac{d^2\eta(u)}{du^2}\frac{\partial}{\partial u} + \frac{d\eta(u)}{du}\frac{\partial}{\partial v}\right]\chi_\parallel(u,v)+ \right.$$
$$\left.\left(\frac{\partial^2}{\partial u^2} + \frac{\partial^2}{\partial v^2}\right)\chi_\parallel(u,v) + H^2(u,v)\chi_\parallel(u,v)\right\} \tag{32.53}$$

with

$$H^2(u,v) = \left(\frac{\kappa(u)}{2}\right)^2\left[\frac{1+\eta^2(u)}{1+v\frac{d\eta(u)}{du}}\right]^2. \tag{32.54}$$

In order to study the electrodynamics of the Möbius strip the energy eigenstates $[i\hbar\partial/\partial t \Rightarrow E(= \hbar\omega)]$ related to Eq. (32.53) must be determined. In the general case these states only can be determined by numerical methods, not to be discussed here.

The time-independent wave equation for a quantum wire is obtained in the limit $v \to 0$, remembering that in a strict physical sense a quantum wire with complete electron confinement in the transverse (v) direction is an approximation, c.f. the analogous remarks given for the transition from mesoscopic layer dynamics to surface dynamics. For $v \to 0$, $f(u,v) \to 1$ and $H(u,v) \to -(\kappa(u)/2)(1+\eta^2(u))$, and $\chi_\parallel \to \psi(u)$. The time-independent Schrödinger equation for quantum wire with complete transverse electron confinement hence takes the form

$$\left[\frac{d^2}{du^2} + k^2 + T(u)\right]\psi(u) = 0, \tag{32.55}$$

where

$$T(u) = \left[\frac{\kappa(u)}{2}(1 + \eta^2(u))\right]^2,$$

(32.56)

$$k^2 = \frac{2mE}{\hbar^2},$$

(32.57)

E being the stationary-state energy. The quantity $-[\hbar^2/(2m)]T(u)$ we here call the geometrical potential.

32.4.2 Energy eigensolutions

In order to obtain the solutions to Eq. (32.55) we take as a starting point the solutions to the simpler eigenvalue problem

$$\frac{d^2\phi_n(u)}{du^2} + k_n^2\phi_n(u) = 0,$$

(32.58)

which satisfies the boundary condition for a closed wire of length L, $\phi_n(0) = \phi_n(L)$. The solutions can be taken as left and right running (circulating) waves, i.e., on normalized form

$$\phi_n(u) = L^{-1/2}e^{ik_n u},$$

(32.59)

with

$$k_n = \frac{2\pi n}{L}, \quad n = 0, \pm 1, \pm 2, \dots$$

(32.60)

These solutions form a complete orthonormalized set,

$$\int_0^L \phi_p^*(u)\phi_n(u)du = \delta_{np},$$

(32.61)

where δ_{np} is the Krönecker delta. The wave function $\psi(u)$ in Eq. (32.55) can be expanded after this set. Hence

$$\psi(u) = \sum_n c_n\phi_n(u).$$

(32.62)

By insertion of this expansion into Eq. (32.55), one obtains

$$\sum_n [-k_n^2 + k^2 + T(u)]c_n\phi_n(u) = 0.$$

(32.63)

Next, by (i) multiplication of Eq. (32.63) by $\phi_p^*(u)$, (ii) integration over the interval $[0, L]$, and (iii) use of the orthonormality condition [Eq. (32.61)], one gets

$$\sum_n [(k^2 - k_p^2)\delta_{pn} - T_{pn}]c_n = 0,$$

(32.64)

where

$$T_{pn} = \int_0^L \phi_p^*(u)T(u)\phi_n(u)du. \tag{32.65}$$

Since Eq. (32.64) holds for each p, one ends up with a set of linear homogeneous equations among the c_n-coefficients.

The energy eigenvalues for the Möbius quantum wire thus are determined by the condition that the associated determinant $(\det\{...\})$ vanishes,

$$\mathrm{Det}(k^2 - k_p^2)\delta_{pn} - T_{pn} = 0. \tag{32.66}$$

The condition in Eq. (32.66) gives the spectrum of eigenvalues for $k^2 = 2mE/\hbar^2$, and thereafter the energy eigenfunctions of the given wire, characterized by the curvature $(\kappa(u))$, torsion $(\tau(u))$ and wire length L.

In general, numerical analysis is needed to determine the spectrum of energy eigenvalues. In [123], a numerical calculation was carried out starting from the following parametric Cartesian form [233]:

$$x(u) = A\sin\left(\frac{2\pi u}{L}\right), \tag{32.67}$$

$$y(u) = A\left[1 - \cos\left(\frac{2\pi u}{L}\right)\right]^3, \tag{32.68}$$

$$z(u) = A\sin\left(\frac{2\pi u}{L}\right)\left[1 - \cos\left(\frac{2\pi u}{L}\right)\right]. \tag{32.69}$$

The constant A is related to a given Möbius wire length L via

$$\frac{A}{L} = \left\{\int_0^{2\pi}[(x_N'(t))^2 + (y_N'(t))^2 + (z_N'(t))^2]^{1/2}dt\right\}^{-1}, \tag{32.70}$$

where

$$\begin{pmatrix} x_N(t) \\ y_N(t) \\ z_N(t) \end{pmatrix} = \begin{pmatrix} \sin t \\ (1 - \cos t)^3 \\ \sin t(1 - \cos t) \end{pmatrix}. \tag{32.71}$$

The curvature and torsion are obtained from the following formulas:

$$\kappa(u) = \frac{|\mathbf{r}'(u) \times \mathbf{r}''(u)|}{|\mathbf{r}'(u)|^3}, \tag{32.72}$$

$$\tau(u) = \frac{\mathbf{r}'(u) \times \mathbf{r}''(u) \cdot \mathbf{r}'''(u)}{|\mathbf{r}'(u) \times \mathbf{r}''(u)|^2}, \tag{32.73}$$

where $\mathbf{r}'(u)$, $\mathbf{r}''(u)$ and $\mathbf{r}'''(u)$ are respectively, the first, second and third derivative of the position vector $\mathbf{r}(u) = [x(u), y(u), z(u)]$ given by the parametrization in Eqs. (32.67)–(32.69).

A detailed discussion of the energy eigenvalues is discussed in Ref. [123]. For a 50 nm long GaAs Möbius quantum wire, shown in Fig. 32.1, the eight lowest lying energy eigenfunctions are plotted in Fig. 32.2. For comparison the corresponding untwisted quantum ring eigenfunctions are also shown.

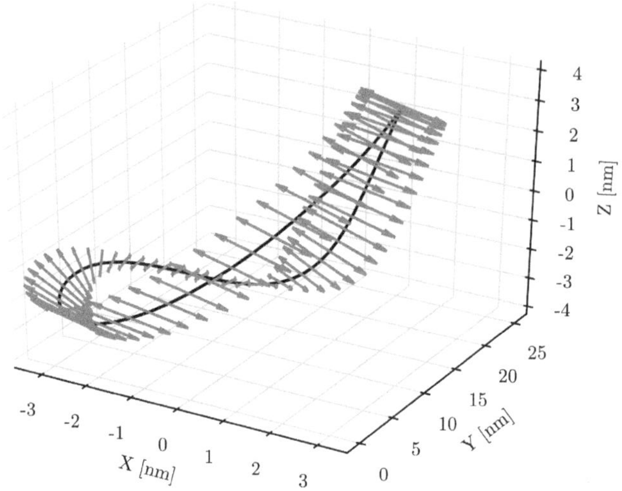

Figure 32.1 Möbius wire of length $L = 50$ nm with parametric form given by Eqs. (32.67)–(32.69). The parameterization of the related finite width Möbius strip is given by $\mathbf{r}(u,v) = \mathbf{r}(u) + v[\hat{\mathbf{b}}(u) + \eta(u)\mathbf{r}'(u)]$. The set of binormal unit vectors ($\hat{\mathbf{b}}(u)$), shown in blue, also illustrates a Möbius band of finite width. The binormal flips around when u passes 0, as required by the Möbius band condition for finite w.

32.4.3 Miroscopic conductivity tensor

In linear electrodynamics the conductivity tensor plays a central role as we have seen in Part V. For a quantum wire the linear constitutive equation takes the form

$$\mathbf{J}(u;\omega) = \int_0^L \boldsymbol{\sigma}(u,u';\omega) \cdot \mathbf{E}(u';\omega)du' \tag{32.74}$$

in the space-frequency domain. The current density at u, $\mathbf{J}(u;\omega)$, is related to the prevailing electric field at u', $\mathbf{E}(u';\omega)$ via the nonlocal conductivity tensor $\boldsymbol{\sigma}(u,u';\omega)$, given by

$$\boldsymbol{\sigma}(u,u';\omega) = \frac{2\hbar}{i} \sum_{i,j} \frac{f_j - f_i}{\varepsilon_j - \varepsilon_i} \frac{1}{\hbar\omega + \varepsilon_j - \varepsilon_i} \mathbf{J}_{i\to j}(u)\mathbf{J}_{j\to i}(u'). \tag{32.75}$$

Irreversible electronic loss mechanisms can at a phenomenological level sometimes be taken into account in the framework of a simple relaxation time (τ) approach making the replacement to $\omega \Rightarrow \omega + i/\tau$ in Eq. (32.75). The transition current density from state i to state j ($i \to j$) is given by ($e < 0$)

$$\mathbf{J}_{i\to j}(u) = \frac{e\hbar}{2mi} \left[\psi_j^*(u)\frac{d}{du}\psi_i(u) - \psi_i(u)\frac{d}{du}\psi_j^*(u) \right] \hat{\mathbf{t}}(u), \tag{32.76}$$

where $\hat{\mathbf{t}}(u) = d\mathbf{r}(u)/du$ is the local tangential unit vector at u. The transition current density $j \to i$ is the complex conjugate of the one given in Eq. (32.76), viz.,

$$\mathbf{J}_{j\to i}(u) = \mathbf{J}_{i\to j}^*(u). \tag{32.77}$$

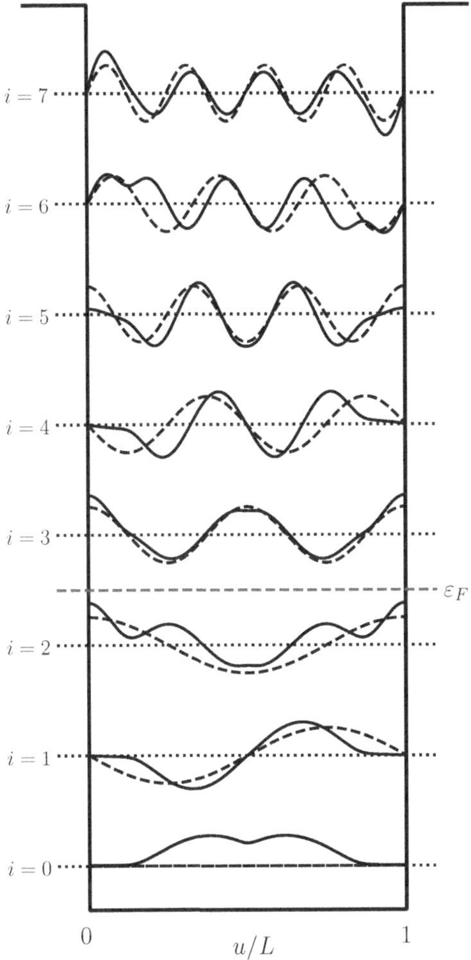

Figure 32.2 The eight lowest lying energy eigenfunctions for a Möbius quantum wire (solid lines). For comparison, the corresponding quantum ring eigenfunctions are also shown (dashed lines). The green dashed line indicates that the Fermi energy, ε_F, is located between the $i = 2$ and $i = 3$ energy eigenstates.

The energy eigenvalues $\varepsilon_i(\varepsilon_j)$ and the associated wave functions $\psi_i(\mathbf{r})$ $[\psi_j(\mathbf{r})]$ enter Eqs. (32.75) and (32.76), respectively, and the Fermi-Dirac factors are $f_i(f_j)$.

It appears from the foregoing analysis that the i and jth energy eigenstates are given by [c.f. Eq. (32.62)]

$$\psi_i(u) = \sum_n c_n^{(i)} \phi_n(u), \tag{32.78}$$

$$\psi_j(u) = \sum_m c_m^{(j)} \phi_m(u), \tag{32.79}$$

where $c_n^{(i)}$ and $c_m^{(j)}$ denote the relevant coefficients obtained from the eigenvalue problem [Eq. (32.64)]. Utilizing Eqs. (32.78) and (32.79) the transition current densities

$(i \to j)$ can be written in the form

$$\mathbf{J}_{i \to j}(u) = \frac{e\hbar}{2m} \hat{\mathbf{t}}(u) \sum_{n,m} (k_n + k_m) c_n^{(i)} [c_m^{(j)}]^* \phi_n(u) [\phi_m(u)]^*, \tag{32.80}$$

where

$$k_n + k_m = \frac{2\pi}{L}(n + m). \tag{32.81}$$

The result in Eqs. (32.80) and (32.81) relates to the running (circulating) eigenstates in Eq. (32.59).

When the quantum wire is mesoscopic in size, the wire behaves like an electric dipole (ED) absorber and radiator, in general. In the ED approximation one needs only the integral ($[0,L]$) of the transition current density, i.e.

$$J_{i \to j}^{\mathrm{ED}} = \frac{e\hbar}{2m} \sum_{n,m} (k_n + k_m) c_n^{(i)} [c_m^{(j)}]^* \frac{1}{L} \int_0^L \exp[i(k_n - k_m)u] \hat{\mathbf{t}}(u) du. \tag{32.82}$$

The integral $L^{-1} \int_0^L (\ldots) du$ is the Fourier transform of $\hat{\mathbf{t}}(u)$ belonging to the wave number difference $k_n - k_m$:

$$\hat{\mathbf{t}}(k_n - k_m) = \frac{1}{L} \int_0^L \hat{\mathbf{t}}(u) \exp[i(k_n - k_m)u] du. \tag{32.83}$$

In the ED limit, and with a relaxation time included, the frequency-dependent wire conductivity thus is given by

$$\boldsymbol{\sigma}^{\mathrm{ED}}(\omega) = \frac{2\hbar}{i} \sum_{i,j} \frac{f_j - f_i}{\varepsilon_j - \varepsilon_i} \frac{1}{\hbar(\omega + \frac{i}{\tau}) + \varepsilon_j - \varepsilon_i} \mathbf{J}_{i \to j}^{\mathrm{ED}} [\mathbf{J}_{i \to j}^{\mathrm{ED}}]^*. \tag{32.84}$$

In order to discuss the symmetry properties of $\boldsymbol{\sigma}^{\mathrm{ED}}(\omega)$, it is convenient to combine the set of basis functions in Eq. (32.59) pairwise $(e^{ik_n u} \pm e^{-ik_n u})$ to obtain a set of real even (E) and odd (O) standing-wave (normalized) basis functions, viz.,

$$\Phi_n^{(E)}(u) = \left(\frac{2}{L}\right)^{1/2} \cos(k_n u), \quad n = 0, 1, 2, .. \tag{32.85}$$

$$\Phi_n^{(O)}(u) = \left(\frac{2}{L}\right)^{1/2} \sin(k_n u). \quad n = 0, 1, 2, .. \tag{32.86}$$

In terms of these basis functions the transition current density takes the form

$$\mathbf{J}_{n \to m} = \frac{e\hbar}{2mi} \left[\Phi_m(u) \frac{d}{du} \Phi_n(u) - \Phi_n(u) \frac{d}{du} \Phi_m(u) \right] \hat{\mathbf{t}}(u), \tag{32.87}$$

where $\Phi_l(u)$ is either $\Phi_l^{(E)}$ or $\Phi_l^{(O)}(u)$, $l = m, n$. It appears from Eqs. (32.67)–(32-69) that the Cartesian components of $\hat{\mathbf{t}}(u)$ have the following symmetries around $u = L/2 : t_x, t_z$ [even], t_y [odd].

In the ED approximation, integrals of the type

$$J_{n\to m}^{\mathrm{ED}} = \int_0^L \mathbf{J}_{n\to m}(u)\,du = \frac{e\hbar}{2mi}\int_0^L \left[\Phi_m(u)\frac{d}{du}\Phi_n(u) - \Phi_n(u)\frac{d}{du}\Phi_m(u)\right]\hat{\mathbf{t}}(u)\,du$$

$$(32.88)$$

appear in the expression for $\boldsymbol{\sigma}^{\mathrm{ED}}(\omega)$. Now, it is seen that the factor $\Phi_m d\Phi_n/du - \Phi_n d\Phi_m/du$ must be (i) even to prevent the x and z components of $\mathbf{J}_{n\to m}^{\mathrm{ED}}$ from vanishing, and (ii) odd to prevent the y component of $\mathbf{J}_{n\to m}^{\mathrm{ED}}$ from becoming zero. In turn, it follows from the dyadic product $\mathbf{J}_{n\to m}^{\mathrm{ED}}\mathbf{J}_{m\to n}^{\mathrm{ED}}$ that the ED conductivity is symmetric $(\sigma_{zx}^{\mathrm{ED}} = \sigma_{zx}^{\mathrm{ED}})$ with the general form

$$\boldsymbol{\sigma}^{\mathrm{ED}}(\omega) = \begin{pmatrix} \sigma_{xx}^{\mathrm{ED}}(\omega) & 0 & \sigma_{xz}^{\mathrm{ED}}(\omega) \\ 0 & \sigma_{yy}^{\mathrm{ED}}(\omega) & 0 \\ \sigma_{xz}^{\mathrm{ED}}(\omega) & 0 & \sigma_{zz}^{\mathrm{ED}}(\omega) \end{pmatrix}.$$

$$(32.89)$$

For comparison, it appears that the ED conductivity tensor of a circular quantum wire (placed in the xz plane) has an antisymmetric form with $\sigma_{yy}^{\mathrm{ED}}(\omega) = 0$ [123, 197], that is

$$\boldsymbol{\sigma}^{\mathrm{ED}}(\omega) = \begin{pmatrix} \sigma_{xx}^{\mathrm{ED}}(\omega) & 0 & \sigma_{xz}^{\mathrm{ED}}(\omega) \\ 0 & 0 & 0 \\ -\sigma_{xz}^{\mathrm{ED}}(\omega) & 0 & \sigma_{zz}^{\mathrm{ED}}(\omega) \end{pmatrix}.$$

$$(32.90)$$

Electrodynamically, a mesoscopic circular wire resembles to some extent a magnetic point dipole for which the related response is known to be antisymmetric [157].

IX

Extension of Quantum Electrodynamics to Electro-Weak Interaction

CHAPTER 33

Parity Conservation and Violation

33.1   SPATIAL INVERSION

The parity operation (P), also called the spatial inversion, is a discrete transformation in which a coordinate vector $\mathbf{r}$ is replaced by $-\mathbf{r}$

$$P: \; \mathbf{r} \to -\mathbf{r}. \tag{33.1}$$

A *polar vector* ($\mathbf{a}$) is one which transforms in the same way as the coordinate vector $\mathbf{r}$ under the parity operation:

$$P: \mathbf{a} \to -\mathbf{a}. \tag{33.2}$$

The vector product of two polar vectors $\mathbf{a}$ and $\mathbf{b}$ defines a *pseudovector* (also called an *axial vector*), transforming under P as follows:

$$P: \mathbf{a} \times \mathbf{b} \to +\mathbf{a} \times \mathbf{b}. \tag{33.3}$$

In forming scalar products we must now distinguish between a *scalar*, $\mathbf{a} \cdot \mathbf{b}$, and a *pseudoscalar*, $\mathbf{a} \cdot \mathbf{b} \times \mathbf{c}$, transforming under P in the following manner:

$$P: \mathbf{a} \cdot \mathbf{b} \to +\mathbf{a} \cdot \mathbf{b}, \tag{33.4}$$

$$P: \mathbf{a} \cdot \mathbf{b} \times \mathbf{c} \to -\mathbf{a} \cdot \mathbf{b} \times \mathbf{c}. \tag{33.5}$$

33.2   PARITY CONSERVATION IN ELECTRODYNAMICS. FORM INVARIANCE

Let us now study how electrodynamics behave under the parity operation, limiting ourselves soon to a single massive particle.

DOI: 10.1201/9781003029458 33

33.2.1 Classical electrodynamics (CED)

The fundamental equations in CED are (i) the Maxwell-Lorentz equations, which we here repeat for convenience:

$$\boldsymbol{\nabla} \cdot \mathbf{E} = \frac{1}{\varepsilon_0}\rho^e, \tag{33.6}$$

$$\boldsymbol{\nabla} \cdot \mathbf{B} = 0, \tag{33.7}$$

$$\boldsymbol{\nabla} \times \mathbf{E} + \frac{\partial \mathbf{B}}{\partial t} = \mathbf{0}, \tag{33.8}$$

$$\boldsymbol{\nabla} \times \mathbf{B} - \frac{1}{c^2}\frac{\partial \mathbf{E}}{\partial t} = \mu_0 \mathbf{J}^e, \tag{33.9}$$

ρ^e and $\mathbf{J}^e$ denoting the electrical (e) charge density and current density (below for a single particle), and (ii) the Newton-Lorentz equation for a particle having mass m, charge e, position $\mathbf{r}(t)$ and velocity $\mathbf{v}(t) = d\mathbf{r}/dt$, which in its relativistic form is

$$\frac{d\mathbf{p}}{dt} = e(\mathbf{E} + \mathbf{v} \times \mathbf{B}), \tag{33.10}$$

where

$$\mathbf{p} = \gamma m \mathbf{v} = \frac{m\mathbf{v}}{\left[1 - \left(\frac{v}{c}\right)^2\right]^{1/2}} \tag{33.11}$$

is the momentum of the particle. It is the force $\mathbf{F} = d\mathbf{p}/dt$ on the charged particle which defines the electric and magnetic fields. The parity operation in Eq. (33.1) shows that the velocity $\mathbf{v}$ is a polar vector: $\mathbf{v} \to -\mathbf{v}$, and thus also the momentum: $\mathbf{p} \to -\mathbf{p}$. From the Newton-Lorentz equation follows that $\mathbf{E}$ and $\mathbf{B}$ are polar and pseudo-vectors, respectively:

$$P : \mathbf{E}(\mathbf{r}, t) \to -\mathbf{E}(-\mathbf{r}, t), \tag{33.12}$$
$$P : \mathbf{B}(\mathbf{r}, t) \to +\mathbf{B}(-\mathbf{r}, t). \tag{33.13}$$

Let us now proceed to the Maxwell-Lorentz equations. Under the parity operation $P : \boldsymbol{\nabla} \to -\boldsymbol{\nabla}$, and therefore it appears that the left-hand sides of the two homogeneous equations transform as a pseudoscalar ($\boldsymbol{\nabla} \cdot \mathbf{b}$) and a pseudovector ($\boldsymbol{\nabla} \times \mathbf{E} + \partial \mathbf{B}/\partial t$). In turn, this implies that the two homogeneous Maxwell equations are *form invariant*. The left-hand sides of the two inhomogeneous field equations clearly transform as a scalar ($\boldsymbol{\nabla} \cdot \mathbf{E}$) and a polar vector ($\boldsymbol{\nabla} \times \mathbf{B} - c^2 \partial \mathbf{E}/\partial t$). Provided these inhomogeneous equations are correct, the charge density must transform as a scalar

$$P : \rho^e \to +\rho^e. \tag{33.14}$$

Perhaps, in itself, this is a natural assumption. The current density transforms as a polar vector, i.e.,

$$P : \mathbf{J}^e \to -\mathbf{J}^e. \tag{33.15}$$

The transformation for $\mathbf{J}^e$ is in agreement with its definition as charge times velocity.

An important lesson to be learned from the considerations above is the following: *The fundamental equations of classical electrodynamics are FORM INVARIANT under the parity transformation.*

33.2.2 Symmetrized electrodynamics

In Part X, we shall study magnetic monopoles. In order to incorporate the magnetic monopole concept in the framework of electrodynamics the form of the set of Maxwell-Lorentz equations must be extended. The new symmetrized set is given in Section 38.1. Specifically, Eqs. (33.7) and (33.8) have to be replaced by

$$\boldsymbol{\nabla} \cdot (c\mathbf{B}) = \frac{1}{\varepsilon_0} \rho^m \tag{33.16}$$

and

$$-\boldsymbol{\nabla} \times \mathbf{E} - \frac{1}{c}\frac{\partial}{\partial t}(c\mathbf{B}) = c\mu_0 \mathbf{J}^m, \tag{33.17}$$

where ρ^m and $\mathbf{J}^m$ are the magnetic (m) charge and current densities. Furthermore, the Lorentz equation of motion given in Eq. (33.10), must be replaced by

$$\frac{d\mathbf{p}}{dt} = e(\mathbf{E} + \mathbf{v} \times \mathbf{B}) + g(c\mathbf{B} - \frac{1}{c}\mathbf{v} \times \mathbf{E}) \tag{33.18}$$

for a particle carrying electric charge e and magnetic charge g.

It appears from Eqs. (33.16) and (33.17) that ρ^m transforms as a *pseudoscalar* (a quantity changing sign under P) i.e.,

$$P : \rho^m \to -\rho^m, \tag{33.19}$$

and $\mathbf{J}^m$ as a pseudovector,

$$P : \mathbf{J}^m \to +\mathbf{J}^m. \tag{33.20}$$

The result in Eq. (33.19) of course is in agreement with the fact that the magnetic monopole part of Eq. (33.18) gives $P : g \to -g$ for a single pole.

33.2.3 Quantum electrodynamics

In QED, let us assume that there exists a unitary parity operator $\hat{P}\hat{P}^\dagger = U$. Its representation on the Fock space of photons we denote by

$$\hat{P} |n\rangle = |n'\rangle. \tag{33.21}$$

The transformation rule for matrix elements, here applied to a transverse vector field operator $\hat{\mathbf{V}}_T$, gives

$$\langle m'| \hat{\mathbf{V}}'_T |n'\rangle = \langle m| \hat{\mathbf{V}}_T |n\rangle = \langle m'| \hat{P}^\dagger \hat{\mathbf{V}}_T \hat{P} |n'\rangle, \tag{33.22}$$

which implies that

$$\hat{\mathbf{V}}'_T = \hat{P}\hat{\mathbf{V}}_T\hat{P}^\dagger. \tag{33.23}$$

For the inversion of the transverse electric and magnetic photon field operators, we thus must have

$$\hat{P}\hat{\mathbf{E}}_T(\mathbf{r},t)\hat{P}^\dagger = -\,\hat{\mathbf{E}}_T(-\mathbf{r},t), \tag{33.24}$$

$$\hat{P}\hat{\mathbf{B}}(\mathbf{r},t)\hat{P}^\dagger = +\,\hat{\mathbf{B}}(-\mathbf{r},t), \tag{33.25}$$

in order to connect correctly to classical electrodynamics. QED too is form-invariant under the parity transformation.

33.3 TIME REVERSAL IN ELECTRODYNAMICS

The effect of time reversal $(t \to -t)$ in electrodynamics can be examined as follows. The particle momentum $\mathbf{p} = \gamma m\mathbf{v}$ is odd under time reversal since $\mathbf{v} = d\mathbf{x}/dt \to d\mathbf{x}/d(-t) = -\mathbf{v}$. Next, $d\mathbf{p}/dt \to d(-\mathbf{p})/d(-t) = d\mathbf{p}/dt$. Denoting the time reversal operation by T, it then appears from the Newton-Lorentz equation [Eq. (33.10)] that

$$T : \mathbf{E}(\mathbf{r},t) \to \mathbf{E}(\mathbf{r},-t), \tag{33.26}$$

$$T : \mathbf{B}(\mathbf{r},t) \to -\mathbf{B}(\mathbf{r},-t). \tag{33.27}$$

Finally, it then follows from the Maxwell-Lorentz Eqs. (33.6) and (33.9) that

$$T : \rho^e(\mathbf{r},t) \to \rho^e(\mathbf{r},-t), \tag{33.28}$$

$$T : \mathbf{J}^e(\mathbf{r},t) \to -\mathbf{J}^e(\mathbf{r},-t). \tag{33.29}$$

In symmetrized electrodynamics it readily appears from Eqs. (33.16) and (33.17) that the magnetic monopole charge and current densities transform under time reversal as follows:

$$T : \rho^m(\mathbf{r},t) \to -\rho^m(\mathbf{r},-t), \tag{33.30}$$

$$T : \mathbf{J}^m(\mathbf{r},t) \to +\mathbf{J}^m(\mathbf{r},-t). \tag{33.31}$$

In QED, time reversal with operator $\hat{T}$, gives

$$\hat{T}\hat{\mathbf{E}}_T(\mathbf{r},t)\hat{T}^\dagger = \hat{\mathbf{E}}_T(\mathbf{r},-t), \tag{33.32}$$

$$\hat{T}\hat{\mathbf{B}}(\mathbf{r},t)\hat{T}^\dagger = -\hat{\mathbf{B}}(\mathbf{r},-t). \tag{33.33}$$

Altogether it appears that CED and QED both are form-invariant under time reversal.

33.4 SPATIAL INVERSION AND TIME REVERSAL IN PHOTON WAVE MECHANICS

In order to examine the discrete P and T symmetries in the first- and second-quantized wave mechanical description of the photon, based on the transverse part of the vector potential, let us begin with a study of the transformation properties of the four-vector potential.

It's gauge invariant part, $\mathbf{A}_T(\mathbf{r}, t)$, relates to the magnetic field via $\boldsymbol{\nabla} \times \mathbf{A}_T = \mathbf{B}$. The P transformation properties of $\boldsymbol{\nabla}$ and $\mathbf{B}$ immediately show that

$$P : \mathbf{A}_T(\mathbf{r}, t) \to -\mathbf{A}_T(-\mathbf{r}, t), \tag{33.34}$$

implying that $\mathbf{A}_T$ is a polar vector, and odd under time reversal, i.e.,

$$T : \mathbf{A}_T(\mathbf{r}, t) \to -\mathbf{A}_T(\mathbf{r}, -t). \tag{33.35}$$

Since

$$\mathbf{E}(\mathbf{r}, t) = -\boldsymbol{\nabla}\phi(\mathbf{r}, t) - \frac{\partial \mathbf{A}_L(\mathbf{r}, t)}{\partial t} - \frac{\partial \mathbf{A}_T(\mathbf{r}, t)}{\partial t}, \tag{33.36}$$

it appears from the transformation properties of $\mathbf{E}(\mathbf{r}, t)$, $\boldsymbol{\nabla}$ and $\partial\partial t$ that the longitudinal part $\mathbf{A}_L(\mathbf{r}, t)$, as expected, transforms in the same way under P and T as $\mathbf{A}_T(\mathbf{r}, t)$, and the scalar potential, $\phi(\mathbf{r}, t)$, transforms as follows:

$$P : \phi(\mathbf{r}, t) \to \phi(-\mathbf{r}, t), \tag{33.37}$$
$$T : \phi(\mathbf{r}, t) \to \phi(\mathbf{r}, -t). \tag{33.38}$$

Hence, $\phi(\mathbf{r}, t)$ is a scalar, even under time reversal.

Photon wave mechanics based on the transverse vector potential is governed by the free dynamical evolution equations [see Eqs. (1.12), e.g.]

$$i\frac{\partial}{\partial t}\mathbf{A}_{T,\pm}^{(+)}(\mathbf{r}, t) = \pm c\boldsymbol{\nabla} \times \mathbf{A}_{T,\pm}^{(+)}(\mathbf{r}, t), \tag{33.39}$$

the subscript $\perp$ denoting the positive $(+)$ and negative $(-)$ helicity species. The superscript $(+)$ indicates that only the analytic part of $\mathbf{A}_{T,\pm}$ enters the evolution equations. It follows from the P and T transformations in Eqs. (33.34) and (33.35) that Eq. (33.39) is not invariant under the parity transformation, nor under time reversal. However, it is obvious that the free evolution equations are form-invariant under the combined PT transformation. This lowering of the symmetry is not surprising due to the fact that Eq. (33.39) was obtained from the wave equation $\partial^\mu \partial_\mu \mathbf{A}_T(x) = 0$ [in a manner described in Section 2.1], which is invariant under both P and T. The extension of Eq. (33.39) to the quantum field level does change the conclusion concerning the discrete symmetry of the dynamical operator evolution equations.

When the interaction between charged massive particles and photons is included Eq. (33.39) is replaced by

$$i\frac{\partial}{\partial t}\mathbf{A}_{T,\pm}^{(+)}(\mathbf{r}, t) \mp c\boldsymbol{\nabla} \times \mathbf{A}_{T,\pm}^{(+)}(\mathbf{r}, t) = -\frac{1}{2\varepsilon_0 c} \int_{-\infty}^{\infty} \mathbf{e}_\pm(\boldsymbol{\kappa})\mathbf{e}_\pm^*(\boldsymbol{\kappa}) \cdot \mathbf{J}_T(\mathbf{q}; t)e^{i\mathbf{q}\cdot\mathbf{r}}\frac{d^3q}{(2\pi)^3 q}, \tag{33.40}$$

and an analogous equation among the $\hat{\mathbf{A}}_{T,\pm}(\mathbf{r},t)$ and $\hat{\mathbf{J}}_{T,\pm}(\mathbf{q},t)$ operators at the QED level. The notation used in the Heisenberg picture in Eq. (33.39) is described in Section 3.6. It follows immediately from P and T transformations of $\mathbf{J}^e(\mathbf{r},t)$, given in Eqs. (33.15) and (33.29) that the inhomogeneous term in Eq. (33.40) is invariant only under the combined PT transformation, the correct result.

The conclusion reached thus is the following: *The fundamental equations of photon wave mechanics in both CED and QED are FORM INVARIANT under the combined parity and time reversal transformations.*

33.5 PARITY VIOLATION IN WEAK INTERACTION

We have realized in this chapter that QED is form invariant under the parity transformation, and we know that QED is also rotationally invariant [see Part II and III]. Since the parity transformation can be made up of a mirror reflection and a rotation, parity violation is equivalent to a breakdown of mirror reflection symmetry.

33.5.1 Parity violation observed in optics

In experiments in bismuth and lead vapour it has been shown that the refractive index of light passing through the vapour is different for left and right circularly polarized light [47,48,208]. Thus, the observations show that the plane of polarization of linearly polarized light is rotated. The effect is very small and for 1 metre of dense vapour the angle of rotation is of the order of 10^{-7} radians. The effect is associated with the circumstance that the mirror symmetry is broken by weak interaction effects in the bismuth and lead atoms.

Optically similar plane of rotation effects are observed in substances consisting of molecules with a given chirality, and in the Faraday effect, where the chirality of say electrons is induced by an externally impressed DC magnetic field. However, one should remember that in the later two examples CED (possible QED) can account for the observed plane of rotation effects. Furthermore, the rotation angles are at least several orders of magnitude larger than in the parity violation phenomenon.

Other optical experiments reveal the violation of mirror symmetry. For instance, it has been shown that a caesium atom placed in counterpropagating right and left circularly polarized laser beams in resonance on the $6S_{1/2}$ and $7S_{1/2}$ transition gives rise to fluorescent light in which the mirror symmetry is broken. As discussed in a beautiful paper by E. P. Wigner [273] the reflection symmetry is restored in a "magic mirror", or a CP mirror, in which matter is replaced by antimatter (electron by positron, e.g.), via charge conjugation (C); see also the articles on symmetry principles in [274]. The related charge conjugation operator is $\hat{C}$. In QED the charge conjugation operator has the following effect on the transverse vector field operator $\hat{\mathbf{V}}_T$ [$\hat{\mathbf{V}}_T$ being $\hat{\mathbf{A}}_T$ $\hat{\mathbf{E}}_T$ or $\hat{\mathbf{B}}$]:

$$\hat{C}\hat{\mathbf{V}}_T(\mathbf{r},t)\hat{C}^\dagger = -\hat{\mathbf{V}}_T(\mathbf{r},t). \tag{33.41}$$

Roughly speaking, the amplitude M of a parity-violation process can be decomposed as

$$M = M_{\mathrm{QED}} + M_{\mathrm{WEAK}}, \tag{33.42}$$

and under mirror reflection, $M \to M'$,

$$M' = M_{\mathrm{QED}} - M_{\mathrm{WEAK}}, \tag{33.43}$$

ignoring a component in the electro-weak interaction which *is* mirror symmetric. The corresponding transition probabilities differ by

$$|M'|^2 - |M|^2 = -2(M_{\mathrm{QED}} M_{\mathrm{WEAK}}^* + c.c.). \tag{33.44}$$

33.5.2 Electric-dipole transition in optics

Let us assume that we have an effective Hamiltonian operator

$$\hat{H} = \hat{H}_{\mathrm{QED}} + \hat{H}_{\mathrm{WEAK}}. \tag{33.45}$$

If the eigenstates of $\hat{H}_{\mathrm{QED}}$ are denoted by $\left|\chi_j^{\mathrm{QED}}\right\rangle$, i.e.,

$$\hat{H}_{\mathrm{QED}} \left|\chi_j^{\mathrm{QED}}\right\rangle = E_j^{\mathrm{QED}} \left|\chi_j^{\mathrm{QED}}\right\rangle, \tag{33.46}$$

the eigenstates of $\hat{H}$ are

$$|\chi_j\rangle = \left|\chi_j^{\mathrm{QED}}\right\rangle + \left|\chi_j^{\mathrm{WEAK}}\right\rangle, \tag{33.47}$$

where $\left|\chi_j^{\mathrm{WEAK}}\right\rangle$ in first-order non-degenerate perturbation theory is given by

$$\left|\chi_j^{\mathrm{WEAK}}\right\rangle = \sum_{i \neq j} \frac{\left|\chi_i^{\mathrm{QED}}\right\rangle \left\langle \chi_i^{\mathrm{QED}}\right| \hat{H}^{\mathrm{WEAK}} \left|\chi_j^{\mathrm{QED}}\right\rangle}{E_j^{\mathrm{QED}} - E_i^{\mathrm{QED}}}. \tag{33.48}$$

Since the states $|\chi_j\rangle$ are not eigenstates of parity, there will be a small nonvanishing electric-dipole (ED) transition moment (W_{ED}),

$$W_{\mathrm{ED}} = \langle \chi_f | \hat{d} | \chi_i \rangle, \tag{33.49}$$

between states of the same parity, $\hat{d}$ being the ED operator component in the relevant direction (say z, e.g.).

CHAPTER 34

Abelian and Non-Abelian Gauge Fields

34.1 COMPLEX SCALAR FIELDS. COVARIANT DERIVATIVE

34.1.1 Free fields. Geometrical interpretation

With the aim of introducing the subject of *non-Abelian gauge fields* to the general reader with a background in electrodynamics (and optics), we begin with a study of complex scalar fields and their coupling to the electromagnetic field. Part of this analysis has already been presented in relation to our description of the extinction theorem for the complex Klein-Gordon field (Section 16.2).

Let us thus consider a complex scalar field, ϕ, with the two real components ϕ_1 and ϕ_2. From

$$\phi = \frac{1}{\sqrt{2}}(\phi_1 + i\phi_2), \tag{34.1}$$

$$\phi^* = \frac{1}{\sqrt{2}}(\phi_1 - i\phi_2), \tag{34.2}$$

we now can construct a real action (Lagrangian density)

$$\mathcal{L} = (\partial_\mu \phi)(\partial^\mu \phi^*) - m^2 \phi\phi^*, \tag{34.3}$$

setting for notational simplicity $c = \hbar = 1$ here, and in the following. If one regards ϕ and ϕ^* as independent fields, the Euler-Lagrange equations give the two Klein-Gordon equations

$$(\partial_\mu \partial^\mu + m^2)\phi = 0, \tag{34.4}$$

$$(\partial_\mu \partial^\mu + m^2)\phi^* = 0. \tag{34.5}$$

If Λ is a real constant (independent of space-time), it is clear that the Lagrangian is invariant under the transformation

$$\phi \to \phi' = e^{-i\Lambda}\phi, \quad \phi^* \to (\phi')^* = e^{i\Lambda}\phi^*. \tag{34.6}$$

DOI: 10.1201/9781003029458-34

This is known as a *gauge transformation of the first kind*. Since the transformation does not involve space-time, one may characterize it as "internal". The infinitesimal forms of the transformations in Eq. (34.6), obtained from $\phi \to \phi + \delta\phi$ and its complex conjugate, give

$$\delta\phi = -i\Lambda\phi, \quad \delta\phi^* = i\Lambda\phi^*. \tag{34.7}$$

The two complex Klein-Gordon Eqs. (34.4) and (34.5) lead to the four-current density given in Eq. (16.31), and the Klein-Gordon particle charge in Eq. (16.34).

Not least in view of the subsequent study of non-Abelian gauge fields, it is useful to express the gauge transformation in geometric form. Thus, in our internal space we introduce the two-dimensional vector

$$\boldsymbol{\phi} = \hat{\mathbf{i}}\phi_1 + \hat{\mathbf{j}}\phi_2, \tag{34.8}$$

where $\hat{\mathbf{i}}$ and $\hat{\mathbf{j}}$ are internal orthonormal basis vectors. In the geometric description the Lagrangian in Eq. (34.3) becomes

$$\mathcal{L} = (\partial_\mu\boldsymbol{\phi}) \cdot (\partial^\mu\boldsymbol{\phi}) - m^2\boldsymbol{\phi} \cdot \boldsymbol{\phi}. \tag{34.9}$$

The gauge transformation in Eq. (34.6), with ϕ and ϕ^* given by Eqs. (34.1) and (34.2), i.e., $\phi_1' + i\phi_2' = \exp(-i\Lambda)(\phi_1 + i\phi_2)$, $\phi_1' - i\phi_2' = \exp(+i\Lambda)(\phi_1 - i\phi_2)$ thus can be written in the form

$$\phi_1' = \phi_1 \cos\Lambda + \phi_2 \sin\Lambda, \quad \phi_2' = -\phi_1 \sin\Lambda + \phi_2 \cos\Lambda, \tag{34.10}$$

as the reader may verify to herself. With an internal 3-axis with unit vector $\hat{\mathbf{k}} = \hat{\mathbf{i}} \times \hat{\mathbf{j}}$, Eq. (34.10) describes a clockwise rotation of $\boldsymbol{\phi}$ [Eq. (34.8)] around $\hat{\mathbf{k}}$ through an angle Λ.

Rotations in two dimensions form the Abelian group $SO(2)$. Since the gauge transformation equivalently is represented as $\exp(i\Lambda)$ it follows from

$$e^{i\Lambda}(e^{i\Lambda})^* = 1, \tag{34.11}$$

that $\exp(i\Lambda)$ is a unitary matrix of dimension 1×1, that is the group $U(1)$. In the study of complex scalar field, space-time independent gauge transformations relate to the group $SO(2) \approx U(1)$.

Since Λ above is a constant; the gauge transformation is the same in all space-time points. We often use the name *global* for this simple gauge transformation. In geometrical terms the rotation in the internal space is the same in all space-time points.

Let us now see what happens if we allow the gauge function to depend on space-time, i.e.,

$$\Lambda = \Lambda(\{x^\mu\}). \tag{34.12}$$

This type of transformation is called a *local* gauge transformation, or a *gauge transformation of the second kind*. The reader may remember that the names "transformations of the first and second kind" also were used when we characterized the photon as a local gauge particle in Section 2.1.

The local gauge transformation does not change the form of the infinitesimal transformation, $\delta\phi = -i\Lambda(\{x^\mu\})\phi$. However, to see what happens to the Lagrangian $\mathcal{L}$, Eq. (34.3), we need the transformation of $\partial_\mu\phi$ [and $\partial^\mu\phi^*$]. Thus, from

$$\partial_\mu\phi \to \partial_\mu\phi - i\partial_\mu(\Lambda\phi) = \partial_\mu\phi - i(\partial_\mu\Lambda)\phi - i\Lambda(\partial_\mu\phi), \tag{34.13}$$

one obtains

$$\delta(\partial_\mu\phi) = -i\Lambda(\partial_\mu\phi) - i(\partial_\mu\Lambda)\phi, \tag{34.14}$$

and similarly

$$\delta(\partial^\mu\phi^*) = i\Lambda(\partial^\mu\phi^*) + i(\partial^\mu\Lambda)\phi^*. \tag{34.15}$$

For a constant Λ, the terms $\partial_\mu\Lambda$ in Eqs. (34.14) and (34.15) vanish, giving the corresponding global transformations.

34.1.2 Fields coupled to a gauge field. Electromagnetic four-potential

The presence of the extra terms $\delta_\mu\Lambda$ and $\delta^\mu\Lambda$ implies that the Lagrangian is no longer invariant. The change in $\mathcal{L}$ [Eq. (34.3)] is

$$\begin{aligned}
\delta\mathcal{L} &= \delta[(\partial_\mu\phi)(\partial^\mu\phi^*)] - m^2\delta(\phi\phi^*) \\
&= [\delta(\partial_\mu\phi)](\partial^\mu\phi^*) + (\partial_\mu\phi)[\delta(\partial^\mu\phi^*)] \\
&= (\partial_\mu\Lambda)[i(\phi^*\partial^\mu\phi - \phi\partial^\mu\phi^*)] = (\partial_\mu\Lambda)j^\mu,
\end{aligned} \tag{34.16}$$

where j^μ is the contravariant current density, see Eq. (16.31). To restore the invariance under gauge transformations of the second kind, we introduce a four-vector $\{A_\mu\}$, later to be identified as the four potential, and add an extra term $\mathcal{L}_1$ to $\mathcal{L}$:

$$\mathcal{L}_1 = -ej^\mu A_\mu. \tag{34.17}$$

Inspired by the analysis in Section 2.1, we *demand* that A_μ transforms as [c.f. Eq. (2.4)]

$$A_\mu \to A_\mu + \frac{1}{e}\partial_\mu\Lambda. \tag{34.18}$$

The quantity e appears at this stage as a coupling constant. The variation in $\mathcal{L}_1$ is

$$\delta\mathcal{L}_1 = -e(\delta j^\mu)A_\mu - j^\mu\partial_\mu\Lambda. \tag{34.19}$$

The addition of $\mathcal{L}_1$ does not make the Lagrangian gauge invariant. Thus, from

$$\delta\mathcal{L} + \delta\mathcal{L}_1 = -e(\delta j^\mu)A_\mu, \tag{34.20}$$

it appears that another extra term

$$\mathcal{L}_2 = e^2 A_\mu A^\mu \phi^* \phi \tag{34.21}$$

will make $\mathcal{L} + \mathcal{L}_1 + \mathcal{L}_2$ gauge invariant. Thus, since $\delta(\phi^*\phi) = 0$, a result already used in Eq. (34.16), we obtain

$$\delta\mathcal{L}_2 = e^2\delta(A_\mu A^\mu)\phi^*\phi = 2e^2 A_\mu \delta A^\mu \phi^*\phi = 2e^2 A_\mu(\partial^\mu\Lambda)\phi^*\phi. \tag{34.22}$$

Because

$$\delta j^\mu = i\delta(\phi^*\partial^\mu\phi - \phi\partial^\mu\phi^*) = 2\phi^*\phi\partial^\mu\Lambda, \tag{34.23}$$

it follows that

$$\delta\mathcal{L} + \delta\mathcal{L}_1 + \delta\mathcal{L}_2 = 0. \tag{34.24}$$

The Lagrangian $\mathcal{L} + \mathcal{L}_1 + \mathcal{L}_2$ therefore is invariant, and has the explicit form

$$\begin{aligned}
\mathcal{L} + \mathcal{L}_1 + \mathcal{L}_2 &= (\partial_\mu\phi)(\partial^\mu\phi^*) - m^2\phi^*\phi - ie(\phi^*\partial^\mu\phi - \phi\partial^\mu\phi^*)A_\mu + e^2 A_\mu A^\mu \phi^*\phi \\
&= (\partial_\mu\phi + ieA_\mu\phi)(\partial^\mu\phi^* - ieA^\mu\phi^*) - m^2\phi^*\phi.
\end{aligned} \tag{34.25}$$

In the last step the relation $(\partial^\mu\phi)A^\mu = (\partial_\mu\phi)A^\mu$ has been used

Comparing Eqs. (34.3) and (34.25), we note that $\partial_\mu\phi$ has been replaced by the *gauge-covariant derivative* $D_\mu\phi$:

$$D_\mu\phi \equiv (\partial_\mu + ieA_\mu)\phi, \tag{34.26}$$

and consequently

$$(D^\mu\phi)^* = (\partial^\mu + ieA^\mu)^*\phi^* = (\partial^\mu - ieA_\mu)\phi^*. \tag{34.27}$$

The term covariant derivative refers to the circumstance that $D_\mu\phi$ transforms covariantly, like ϕ itself under infinitesimal gauge transformation [Eq. (34.7)]. The proof is readily obtained. Thus, utilizing Eqs. (34.14) and (34.18) one obtains

$$\delta(D_\mu\phi) = \delta(\partial_\mu\phi) + ie(\delta A_\mu)\psi + ieA_\mu\delta\psi - -i\Lambda(\partial\mu\phi + ie\Lambda_\mu\phi) = i\Lambda(D_\mu\phi), \tag{34.28}$$

as predicted.

In hindsight, we presume that the gauge field itself contributes to the Lagrangian density. Under the gauge transformation in Eq. (34.18), the quantity

$$\mathcal{L}_3 = -\frac{1}{4}F^{\mu\nu}F_{\mu\nu}, \tag{34.29}$$

with $F_{\mu\nu} = \partial_\mu A_\nu - \partial_\nu A_\mu$, is gauge invariant. The quantity $F_{\mu\nu}$ will be recognized as the electromagnetic field tensor introduced in Section 2.1 [Eq. (2.8)]. With $\varepsilon_0 c^2 \equiv 1$, $\mathcal{L}_3$ is the electromagnetic field Lagrangian density in its standard form [Eq. (12.18)].

The total Lagrangian density, $\mathcal{L} + \mathcal{L}_1 + \mathcal{L}_2 + \mathcal{L}_3$ hence is

$$\mathcal{L}_{\text{tot}} = (D_\mu\phi)(D^\mu\phi)^* - m^2\phi^*\phi - \frac{1}{4}F^{\mu\nu}F_{\mu\nu}. \tag{34.30}$$

The invariance of the complex scalar field ϕ under local rotations in its internal space thus allows one to introduce the electromagnetic field in a manner with interesting and far-reaching perspectives. The quantity ϕ hence describes a field with charge e [whereas ϕ^* describes a field with charge $-e$]. The quantity e also is a *coupling constant* which couples the gauge potential A_μ to the current density j^μ with strength e.

The inhomogeneous Maxwell equations are obtained from the Euler-Lagrange equation

$$\frac{\partial \mathcal{L}_{\text{tot}}}{\partial A_\mu} - \partial_\nu \left[\frac{\partial \mathcal{L}_{\text{tot}}}{\partial(\partial_\nu A_\mu)} \right] = 0. \tag{34.31}$$

As the reader may verify, Eq. (34.31) leads to

$$\partial_\nu F^{\mu\nu} = -e \mathcal{J}^\mu, \tag{34.32}$$

where

$$\mathcal{J}^\mu = i \left[\phi^* D^\mu \phi - \phi (D^\mu \phi)^* \right] \tag{34.33}$$

is the covariant current density replacing j^μ [Eq. (16.45)'s contravariant partner]. Since

$$\partial_\mu \mathcal{J}^\mu = 0, \tag{34.34}$$

the covariant current density is conserved in the presence of an electromagnetic field.

In conclusion we come to a new interpretation of the electromagnetic field: *It is a gauge field which has to be introduced to guarantee that our complex scalar-field formalism is invariant under local U(1) gauge transformations.*

34.2 YANG-MILLS THEORY. ISOSPIN GAUGE FIELDS

34.2.1 Generalization from U(1) to SU(2)

In Subsection 34.1.1, we gave a geometrical interpretation of the scalar field formalism. Thus, in a two-dimensional internal space we introduced a vector $\boldsymbol{\phi} = \hat{\mathbf{i}}\phi_1 + \hat{\mathbf{j}}\phi_2$ [Eq. (34.8)]. Let us now generalize this geometrical picture by considering the 2D-rotations as rotations about a 3-axis, with unit vector $\hat{\mathbf{k}} = \hat{\mathbf{i}} \times \hat{\mathbf{j}}$. The gauge transformation given in geometrical form in Eqs. (34.10) we then generalize to

$$\boldsymbol{\phi} \to \boldsymbol{\phi}' = \boldsymbol{\phi} - \boldsymbol{\Lambda} \times \boldsymbol{\phi}, \tag{34.35}$$

where the vectorial angle $\boldsymbol{\Lambda}$ means that we consider an angle of rotation $|\boldsymbol{\Lambda}|$ around an axis given by the unit vector $\boldsymbol{\Lambda}/|\boldsymbol{\Lambda}|$.

The rotations form a group SO(3) which is non-Abelian; the vector product $\mathbf{a} \times \mathbf{b} = -\mathbf{b} \times \mathbf{a}$ is not commutative. The non-Abelian nature results in complications in the formulation, and has physical consequences. The infinitesimal form of transformation, viz.,

$$\delta\boldsymbol{\phi} = -\boldsymbol{\Lambda} \times \boldsymbol{\phi}, \tag{34.36}$$

holds for both first and second kind of gauge transformations. When Λ depends on $\{x^\mu\}$, the transformation of the derivative $\partial_\mu\phi$ is

$$\partial_\mu\phi \to \partial_\mu\phi' = \partial_\mu\phi - (\partial_\mu\Lambda) \times \phi - \Lambda \times \partial_\mu\phi, \tag{34.37}$$

and thus

$$\delta(\partial_\mu\phi) = -\Lambda \times \partial_\mu\phi - (\partial_\mu\Lambda) \times \phi. \tag{34.38}$$

Only a global transformation, $\partial_\mu\Lambda = 0$, $\partial_\mu\phi$ transforms in the same manner as ϕ. To obtain a gauge invariant Lagrangian density, we must replace the derivative $\partial_\mu\phi$ by a covariant derivative $D_\mu\phi$ satisfying

$$\delta(D_\mu\phi) = -\Lambda \times (D_\mu\phi), \tag{34.39}$$

c.f. Eq. (34.28). Inspired by the scalar case [Eq. (34.26)] we make the following ansatz:

$$D_\mu\phi = \partial_\mu\phi + g\mathbf{W}_\mu \times \phi. \tag{34.40}$$

The quantity g is a coupling constant, analogous to e in the electromagnetic case. $\mathbf{W}_\mu$ is a vectorial gauge potential, where each component ($\mu = 0 - 3$) in itself is a 3D-vector. The analogous gauge potential in the electromagnetic case is A_μ.

In order that $D_\mu\phi$, given by Eq. (34.40), possibly may satisfy Eq. (34.39), $\mathbf{W}_m$ must be gauge dependent, like A_μ. We make the guess for the transformation of $\mathbf{W}_\mu$:

$$\mathbf{W}_\mu \to \mathbf{W}'_\mu = \mathbf{W}_\mu - \Lambda \times \mathbf{W}_\mu + \frac{1}{g}\partial_\mu\Lambda. \tag{34.41}$$

The last term $g^{-1}\partial_\mu\Lambda$ is the analogous of $e^{-1}\partial_\mu\Lambda$ [Eq. (34.18)]. The "extra" term, $-\Lambda \times \mathbf{W}_\mu$ relates to the circumstance that $\mathbf{W}_\mu$ is a 3D-vector which one rotates in the internal space. The infinitesimal form of the transformation in Eq. (34.41) is

$$\delta\mathbf{W}_\mu = -\Lambda \times \mathbf{W}_\mu + \frac{1}{g}\partial_\mu\Lambda. \tag{34.42}$$

Let us now prove that the covariant derivative rule [Eq. (34.39)] is satisfied with the guess in Eq. (34.40) and Eq. (34.41). We have

$$\begin{aligned}
\delta(D_\mu\phi) &= \delta(\partial_\mu\phi) + g(\delta\mathbf{W}_\mu) \times \phi + g\mathbf{W}_\mu \times (\delta\phi) \\
&= -\Lambda\times\partial_\mu\phi - (\partial_\mu\Lambda)\times\phi - g(\Lambda\times\mathbf{W}_\mu)\times\phi + (\partial_\mu\Lambda)\times\phi - g\mathbf{W}_\mu\times(\Lambda\times\phi) \\
&= -\Lambda \times \partial_\mu\phi - g[(\Lambda \times \mathbf{W}_\mu) \times \phi + \mathbf{W}_\mu \times (\Lambda \times \phi)].
\end{aligned} \tag{34.43}$$

After a little algebra on the terms in the square bracket, we finally obtain

$$\delta(D_\mu\phi) = -\Lambda \times (\partial_\mu\phi + g\mathbf{W}_\mu \times \phi) = -\Lambda \times (D_\mu\phi), \tag{34.44}$$

as required.

34.2.2 Field strength $\mathbf{W}_{\mu\nu}$

Above we have seen that $\mathbf{W}_\mu$ for SU(2) is the analogue to A_μ for U(1). What then is the analogue to the field strength $F_{\mu\nu} = \partial_\mu A_\nu - \partial_\nu A_\mu$? Let us call it $\mathbf{W}_{\mu\nu}$. We know that $\mathbf{W}_{\mu\nu}$ must be a vector (like ϕ) in the internal space under SO(3), and therefore it must transform like ϕ itself, giving the infinitesimal transformation

$$\delta(\mathbf{W}_{\mu\nu}) = -\boldsymbol{\Lambda} \times \mathbf{W}_{\mu\nu}. \tag{34.45}$$

Note that $F_{\mu\nu}$ is a scalar under SO(2). From the form $F_{\mu\nu} = \partial_\mu A_\nu - \partial_\nu A_\mu$ one might be tempted to guess (wrongly!) that $\mathbf{W}_{\mu\nu}$ is given by $\mathbf{W}_{\mu\nu} = \partial_\mu \mathbf{W}_\nu - \partial_\nu \mathbf{W}_\mu$. Since, by means of Eq. (34.42)

$$\begin{aligned}
\delta(\partial_\mu \mathbf{W}_\nu - \partial_\nu \mathbf{W}_\mu) &= \partial_\mu(-\boldsymbol{\Lambda} \times \mathbf{W}_\nu + \frac{1}{g}\partial_\nu\boldsymbol{\Lambda}) - \partial_\nu(-\boldsymbol{\Lambda} \times \mathbf{W}_\mu + \frac{1}{g}\partial_\mu\boldsymbol{\Lambda}) \\
&= \partial_\nu(\boldsymbol{\Lambda} \times \mathbf{W}_\mu) - \partial_\mu(\boldsymbol{\Lambda} \times \mathbf{W}_\nu),
\end{aligned} \tag{34.46}$$

and then

$$\delta(\partial_\mu \mathbf{W}_\nu - \partial_\nu \mathbf{W}_\mu) = -\boldsymbol{\Lambda} \times (\partial_\mu \mathbf{W}_\nu - \partial_\nu \mathbf{W}_\mu) - [(\partial_\mu\boldsymbol{\Lambda}) \times \mathbf{W}_\nu - (\delta_\nu\boldsymbol{\Lambda}) \times \mathbf{W}_\mu]. \tag{34.47}$$

So, our guess was wrong because of the presence of the term in the square bracket of Eq. (34.47). The quantity in the square bracket we rewrite using the transformation in Eq. (34.42). Thus, we insert the expressions for $\partial_\mu\boldsymbol{\Lambda}$ and $\partial_\nu\boldsymbol{\Lambda}$ in the bracket. Hence, we get the variations $\delta\mathbf{W}_\mu$ and $\delta\mathbf{W}_\nu$ in play:

$$\begin{aligned}
(\delta_\mu\boldsymbol{\Lambda}) \times \mathbf{W}_\nu - (\partial_\nu\boldsymbol{\Lambda}) \times \mathbf{W}_\mu &= g(\delta\mathbf{W}_\mu \times \mathbf{W}_\nu - \delta\mathbf{W}_\nu \times \mathbf{W}_\mu) \\
&\quad + g[(\boldsymbol{\Lambda} \times \mathbf{W}_\mu) \times \mathbf{W}_\nu - (\boldsymbol{\Lambda} \times \mathbf{W}_\nu) \times \mathbf{W}_\mu] \\
&= g\delta(\mathbf{W}_\mu \times \mathbf{W}_\nu) + g\boldsymbol{\Lambda} \times (\mathbf{W}_\mu \times \mathbf{W}_\nu). \tag{34.48}
\end{aligned}$$

By combination of Eqs. (34.47) and (34.48) we finally obtain

$$\delta(\partial_\mu \mathbf{W}_\nu - \partial_\nu \mathbf{W}_\mu + g\mathbf{W}_\mu \times \mathbf{W}_\nu) = -\boldsymbol{\Lambda} \times (\partial_\mu \mathbf{W}_\nu - \partial_\nu \mathbf{W}_\mu + g\mathbf{W}_\mu \times \mathbf{W}_\nu) \tag{34.49}$$

Thus a field strength defined by

$$\mathbf{W}_{\mu\nu} = \partial_\mu \mathbf{W}_\nu - \partial_\nu \mathbf{W}_\mu + g\mathbf{W}_\mu \times \mathbf{W}_\nu \tag{34.50}$$

transforms as a vector under SO(3) in the internal space. The last term in Eq. (34.50) stems from the fact that the group SU(2) is non-Abelian.

34.2.3 Inhomogenous "Maxwell Equation" for the isospin gauge field

The formalism described in Subsection 34.2.1 is closely related to the Yang-Mills theory first put forward in 1954 in the context of strong interactions [55, 284]. The vector (ϕ) quantity introduced is like an isospin. The internal symmetry space may likewise be referred to as isospin space (or isospace).

Let us return to the isospin gauge field. The field strength $\mathbf{W}_{\mu\nu}$ is a vector, and the product $\mathbf{W}_{\mu\nu} \cdot \mathbf{W}^{\mu\nu}$ is a scalar. In analogy to the electromagnetic case we must assume that the "free" gauge field gives a contribution to the Lagrangian density ($\mathcal{L}$), which is therefore

$$\mathcal{L} = \frac{1}{2}(D_\mu\boldsymbol{\phi}) \cdot (D^\mu\boldsymbol{\phi}) - \frac{m^2}{2}\boldsymbol{\phi} \cdot \boldsymbol{\phi} - \frac{1}{4}\mathbf{W}_{\mu\nu} \cdot \mathbf{W}^{\mu\nu}, \tag{34.51}$$

with $D_\mu\boldsymbol{\phi}$ and $\mathbf{W}_{\mu\nu}$ given by Eqs. (34.40) and (34.50), respectively. The equation of motion for the gauge field is obtained from the Euler-Lagrange equation

$$\partial_\nu \left[\frac{\partial\mathcal{L}}{\partial(\partial_\nu\mathbf{W}_\mu)} \right] - \frac{\partial\mathcal{L}}{\partial\mathbf{W}_\mu} = \mathbf{0}. \tag{34.52}$$

After some algebra one obtains

$$\partial^\nu\mathbf{W}_{\mu\nu} + g\mathbf{W}^\nu \times \mathbf{W}_{\mu\nu} = g[(\partial_\mu\boldsymbol{\phi}) \times \boldsymbol{\phi} + g(\mathbf{W}_\mu \times \boldsymbol{\phi}) \times \boldsymbol{\phi}]. \tag{34.53}$$

This equation can be simplified formally with the help of the expression for the covariant derivative of $\boldsymbol{\phi}$ [Eq. (34.40)], resulting in the following "field" equation:

$$D^\nu\mathbf{W}_{\mu\nu} = g\mathbf{J}_\mu, \tag{34.54}$$

where

$$\mathbf{J}_\mu = (D_\mu\boldsymbol{\phi}) \times \boldsymbol{\phi}. \tag{34.55}$$

Eq. (34.54) is analogous to the inhomogeneous Maxwell equation for the field strength tensor $F_{\mu\nu}$, viz, $\partial^\nu F_{\mu\nu} = -e\mathcal{J}_\mu$ [Eq. (34.32)]. The quantity $\mathbf{J}_\mu$ is the source (matter) term [current density vector for component μ]. In the absence of matter ($\boldsymbol{\phi} \to \mathbf{0}$) one is left with the free non-Abelian gauge field equation.

$$D^\nu\mathbf{W}_{\mu\nu} = \mathbf{0}, \tag{34.56}$$

equivalent to

$$\partial^\nu\mathbf{W}_{\mu\nu} = -g\mathbf{W}^\nu \times \mathbf{W}_{\mu\nu}. \tag{34.57}$$

This last equation indicates that $\mathbf{W}_{\mu\nu}$ acts as a source for itself, because $\mathbf{W}_{\mu\nu}$ carries isospin. For comparison, the electromagnetic field $F_{\mu\nu}$ carries no charge, and thus cannot be its own source. The reader should here recall that $\mathbf{W}_{\mu\nu}$ is its own source as a consequence of the fact that the symmetry group SO(3) is non-Abelian. At the quantum level the field $\mathbf{W}_{\mu\nu}$ may emit (absorb) gauge particles.

The isospin field carries isospin $I = 1$. To understand this one observes that the formula for rotation of an isovector through an angle $\boldsymbol{\Lambda}$ in isospace, given by Eq. (34.35), is the infinitesimal form of

$$\boldsymbol{\phi} \to \boldsymbol{\phi}' = \exp[(i\mathbf{I} \cdot \boldsymbol{\Lambda})\boldsymbol{\phi}], \tag{34.58}$$

where the matrix generators $\mathbf{I}$ are given by (reinserting $\hbar$)

$$I_1 = \hbar \begin{pmatrix} 0 & 0 & 0 \\ 0 & 0 & -i \\ 0 & i & 0 \end{pmatrix}, \quad I_2 = \hbar \begin{pmatrix} 0 & 0 & i \\ 0 & 0 & 0 \\ -i & 0 & 0 \end{pmatrix}, \quad I_3 = \hbar \begin{pmatrix} 0 & -i & 0 \\ i & 0 & 0 \\ 0 & 0 & 0 \end{pmatrix}. \tag{34.59}$$

The matrices satisfy the angular momentum commutator relations

$$[I_i, I_j] = i\hbar\epsilon_{ijk}I_k. \tag{34.60}$$

The components of the isospin operator all have the eigenvalues 0 and $\pm\hbar$. Hence, the isospin is $I = 1$ for the massless gauge field. Note that the spin operator components of the photon in its Cartesian description have the form given in Eq. (34.59). The transversality of the photon field implies that the spin 0 does not appear in the physical description. Also the photon is massless (it moves with the speed of light!).

34.2.4 Massive vector boson field. Proca equation

Numerous experimental results show that the vector boson (W) particles are very massive [55]. Here, simple general arguments relate to the spatial range of the weak force [The long range of the electromagnetic force results form the zero mass of the photon].

However, if one adds a mass (m_W) term $(m_W^2/2)\mathbf{W}_\mu \cdot \mathbf{W}^\nu$ to the Lagrangian in Eq. (34.51) the field equation in Eq. (34.54) is changed to

$$D^\nu \mathbf{W}_{\mu\nu} = g\mathbf{J}_\mu + m_W^2 \mathbf{W}_\mu. \tag{34.61}$$

The mass term is not gauge invariant, as the reader readily may show using Eq. (34.41). The basic question therefore is the following: Can one in some way or another uphold the gauge invariance principle and still have a vector boson mass m_W? the affirmative answer to this question emerges when the concept of *Spontaneous Symmetry Breaking* is brought in play; see Chapter 36. The gauge invariance for the case $m_W = 0$ originates in the non-Abelian character of the theory, and in the expression for the W-boson field in Eq. (34.49). The last term $g\mathbf{W}_\mu \times \mathbf{W}_\nu$ is a consequence of the non-Abelian character of $\mathrm{SU}(2) \sim \mathrm{SO}(3)$.

Let us temporarily close our eyes for the gauge invariance requirement, and consider a massive free vector boson field. Starting from the free-field Lagrangian density

$$\mathcal{L} = -\frac{1}{4}\mathbf{W}_{\mu\nu} \cdot \mathbf{W}^{\mu\nu} + \frac{1}{2}m_W^2 \mathbf{W}_\mu \cdot \mathbf{W}^\nu, \tag{34.62}$$

where

$$\mathbf{W}_{\mu\nu} = \partial_\mu \mathbf{W}_\nu - \partial_\nu \mathbf{W}_\mu, \tag{34.63}$$

and using the Euler-Lagrange equation leads to the following equation of motion for the μth component of the vector field $\{\mathbf{W}^\mu\}$

$$\Box W^\mu - \partial^\mu(\partial_\nu W^\nu) - m_W^2 W^\mu = 0. \tag{34.64}$$

This is the so-called Proca equation. An analogous equation of motion is satisfied for the photon potential A^μ, assuming that the photon has a rest mass (a very radical idea, breaking with relativity). On taking the divergence of Eq. (34.64) it appears that the Lorenz condition

$$\partial_\mu W^\mu = 0 \tag{34.65}$$

must be satisfied. This is in contrast to the massless photon case, where the Lorenz condition $\partial_\mu A^\mu = 0$ is imposed as a subsidiary condition. On account of Eq. (34.65) the Proca equation is reduced to

$$(\Box - m_W^2)W^\mu = 0. \tag{34.66}$$

The W-boson quantum particle description now is established in the usual manner. Hence one expands the W-field operator, $\hat{W}^\mu(x)$, in a complete set of plane waves:

$$\hat{W}^\mu(x) = \hat{W}_+^\mu(x) + \hat{W}_-^\mu(x), \tag{34.67}$$

where

$$\hat{W}_+^\mu(x) = \sum_{\mathbf{k},r} \left(\frac{1}{2V\omega_k}\right)^{1/2} \varepsilon_r^\mu(\mathbf{k})\hat{a}_r(\mathbf{k})e^{ikx}, \tag{34.68}$$

and

$$\hat{W}_-^\mu(x) = \sum_{\mathbf{k},r} \left(\frac{1}{2V\omega_k}\right)^{1/2} \varepsilon_r^\mu(\mathbf{k})\hat{b}_r^\dagger(\mathbf{k})e^{-ikx}, \tag{34.69}$$

and $\omega_k = (|\mathbf{k}|^2 + m_W^2)^{1/2}$. The polarization unit vectors $\{\varepsilon_r^\mu\}$, $r = 1 - 3$, form a complete set of orthonormal unit vectors, i.e.,

$$\varepsilon_{r\mu}(\mathbf{k})\varepsilon_s^\mu(\mathbf{k}) = \eta_r\delta_{rs}, \tag{34.70}$$

with $\eta_1 = \eta_2 = \eta_3 = 1$ and $\eta_0 = -1$. The Lorenz condition is $k_\mu\varepsilon_r^\mu = 0$. The operators $\hat{a}_r$ and $\hat{b}_r$ one interpreters as annihilation and creation operators for positively charged (W^+) and negatively charged (W^-) vector bosons of mass m_W, momentum $\mathbf{k}$ and polarization vector $\{\varepsilon_r^\mu(\mathbf{k})\}$, respectively.

In the frame in which $k = (\omega_k, 0, 0, |\mathbf{k}|)$, a suitable choice of polarization vectors is

$$\varepsilon_1^\mu(\mathbf{k}) = (0, 1, 0, 0), \tag{34.71}$$
$$\varepsilon_2^\mu(\mathbf{k}) = (0, 0, 1, 0), \tag{34.72}$$
$$\varepsilon_3^\mu(\mathbf{k}) = m_W^{-1}(|\mathbf{q}|, 0, 0, \omega_k/c). \tag{34.73}$$

The discussion of the free vector boson given on the basis of the Proca equation is analogous to the one usually presented for a massive photon [158]. Also the W-boson propagator will be identical to that of the massive photon, of course [158].

CHAPTER 35

Hidden Gauge Invariance: U(1) Case

35.1 SCREENING CURRENTS. GENERATION OF "PHOTON MASS"

35.1.1 Introductory remarks

We have seen that the gauge invariance apparently is rigorously linked to the masslessness of the gauge quanta. Let us return to the photon case, and look for way around this apparently solid conclusion.

Using the signature $(+,-,-,-)$ and $c = \hbar = 1$ ($\mu_0 = 1$), the Maxwell wave equation for the vector potential component, A^μ, takes the form

$$\Box A^\mu - \partial^\mu(\partial_\nu A^\nu) = j^\mu, \tag{35.1}$$

where

$$j^\mu = iq[\phi^*\partial^\mu\phi - (\partial^\mu\phi^*)\phi] - 2q^2 A^\mu|\phi|^2 \tag{35.2}$$

is the charge current density for a charge q, cf. the contravariant version of Eq. (16.45) [multiplied by q, and with $D^\mu = \partial^\mu + iqA^\mu$]. In Eq. (35.1) $\Box = \partial^2/\partial t^2 - \nabla^2$. We now ask the following question: Is it possible for the electromagnetic field, A^μ, which satisfies Eq. (35.1) *also* to satisfy a massive vector equation of the type

$$(\Box + M^2)A^\mu = 0, \tag{35.3}$$

where $M(\equiv M_{\rm ph})$ is a "photon mass".

Below we briefly discuss the diamagnetic (screening) response in the normal and superconducting metallic state as a forerunner for the "relativistic superconducting" theory described in Section 35.2.

35.1.2 Diamagnetic response in the normal metallic state

In order to address the question posed in Subsection 35.1.1, let us now discuss the kind of inspiration we may obtain by a closer look at microscopic response theory in the

DOI: 10.1201/9781003029458-35

weak-field (linear) approximation. As a starting point we take the linear constitutive equation

$$\mathbf{J}(\mathbf{r};\omega) = \int_{-\infty}^{\infty} \mathbf{S}(\mathbf{r},\mathbf{r}';\omega) \cdot \mathbf{E}_T(\mathbf{r}';\omega) d^3 r' \tag{35.4}$$

for a many-electron system. The analysis which results in the form given in Eq. (35.4) can be found in my book on the quantum theory of near-field electrodynamics [157]. Here, it is sufficient to note that it is the transverse part ($\mathbf{E}_T$) of the local electric field ($\mathbf{E} = \mathbf{E}_T + \mathbf{E}_L$) which appears in Eq. (35.4). This is so because the longitudinal part ($\mathbf{E}_L$) of the electric field is eliminated in favor of the electron position variables in the standard formulation [63, 157, 158]. The many-body spatially nonlocal conductivity is given by [153, 157]

$$\mathbf{S}(\mathbf{r},\mathbf{r}';\omega) = \frac{\hbar}{i} \sum_{I,J} \frac{P_J - P_I}{E_J - E_I} \frac{1}{\hbar\omega + E_J - E_I} \mathbf{j}_{I \to J}(\mathbf{r}) \mathbf{j}_{J \to I}(\mathbf{r}'), \tag{35.5}$$

reinserting $\hbar$. The notation in Eq. (35.5) is as follows: P_I (P_J) is the probability that the eigenstate $I(J)$ with the many-body energy $E_I(E_J)$ is occupied. The transition current densities $\mathbf{j}_{I \to J}(\mathbf{r})$ and $\mathbf{j}_{J \to I}(\mathbf{r}')$, which are taken at the spatial points $\mathbf{r}$ and $\mathbf{r}'$, relate to the transitions from I to J, and J to I, respectively. It can be shown that terms with $J = I$ do not constitute to $\mathbf{S}(\mathbf{r},\mathbf{r}';\omega)$, [143]. At sufficiently high frequencies ($\hbar\omega >> |E_J - E_I|$ for all relevant pairs of levels) the expression for the microscopic response tensor is reduced to

$$\mathbf{S}_{\mathrm{dia}}(\mathbf{r},\mathbf{r}';\omega) = \frac{1}{i\omega} \sum_{I,J} \frac{P_J - P_I}{E_J - E_I} \mathbf{j}_{I \to J}(\mathbf{r}) \mathbf{j}_{J \to I}(\mathbf{r}'). \tag{35.6}$$

We have given $\mathbf{S}$ a subscript *dia* to indicate that the high-frequency response tensor may be recognized as the diamagnetic part of the total (paramagnetic + diamagnetic) response. A somewhat complicated calculation [157, 158] shows that $\mathbf{S}_{\mathrm{dia}}(\mathbf{r},\mathbf{r}';\omega)$ can be in the spatially local form

$$\mathbf{S}_{\mathrm{dia}}(\mathbf{r},\mathbf{r}';\omega) = \frac{iq^2}{m\omega} N(\mathbf{r})\delta(\mathbf{r} - \mathbf{r}')\mathbf{U}, \tag{35.7}$$

where $N(\mathbf{r})$ is the proper electron density in thermal equilibrium. By inserting Eq. (35.7) into Eq. (35.4) it appears that the diamagnetic current density is given by

$$\mathbf{J}(\mathbf{r};\omega) = \frac{iq^2}{m\omega} N(\mathbf{r})\mathbf{E}_T(\mathbf{r};\omega). \tag{35.8}$$

Let us next (*i*) neglect that spatial fluctuations ($\Delta(\mathbf{r})$) of $N(\mathbf{r}) = N_0 + \Delta(\mathbf{r})$ around the mean value N_0, and (*ii*) introduce the transverse part of the vector potential $\mathbf{A}_T(\mathbf{r};\omega) = \mathbf{E}_T(\mathbf{r};\omega)/(i\omega)$. The diamagnetic current density then achieves a form which is of inspiration for the question posed in Subsection 35.1.1. Thus,

$$\mathbf{J}(\mathbf{r};\omega) = -\frac{q^2 N_0}{m} \mathbf{A}_T(\mathbf{r};\omega). \tag{35.9}$$

The current density in Eq. (35.9) clearly is divergence-free $[\mathbf{J} = \mathbf{J}_T]$. With the current density given by the approximate high-frequency form, the wave equation for the transverse vector potential becomes

$$\left(\nabla^2 + q_0^2 - \frac{\mu_0 N_0 q^2}{m}\right)\mathbf{A}_T(\mathbf{r};\omega) = \mathbf{0}. \tag{35.10}$$

With the notation used in Subsection 35.1.1, the wave equation for $\mathbf{A}_T$ takes the following form in the space-time domain

$$\left(\square + \frac{q^2 N_0}{m}\right)\mathbf{A}_T(\mathbf{r}, t) = \mathbf{0}. \tag{35.11}$$

A comparison to Eq. (35.3) shows that the diamagnetic phenomenon approximately $(N(\mathbf{r}) \approx N_0)$ leads to a massive vector equation for the transverse vector potential.

35.1.3 Notes on the weak-field response of a superconductor

Let us consider a single charged particle (charge q) with the nonrelativistic wave function $\psi(\mathbf{r}, t)$ subjected to a four-potential $(\phi, \mathbf{A})$. The charge probability density,

$$\rho(\mathbf{r}, t) = q|\psi(\mathbf{r}, t)|^2, \tag{35.12}$$

satisfies the equation of charge continuity

$$\nabla \cdot \mathbf{J}(\mathbf{r}, t) + \frac{\partial}{\partial t}\rho(\mathbf{r}, t) = 0. \tag{35.13}$$

The probability current density, $\mathbf{J}(\mathbf{r}, t)$, is in the standard scheme divided into two parts, viz.,

$$\mathbf{J}(\mathbf{r}, t) = \mathbf{J}_0(\mathbf{r}, t) + \mathbf{J}_A(\mathbf{r}, t), \tag{35.14}$$

where

$$\mathbf{J}_0(\mathbf{r}, t) = \frac{q\hbar}{2mi}[\psi^*(\mathbf{r}, t)\nabla\psi(\mathbf{r}, t) - \psi(\mathbf{r}, t)\nabla\psi^*(\mathbf{r}, t)], \tag{35.15}$$

and

$$\mathbf{J}_A(\mathbf{r}, t) = -\frac{q^2}{m}\mathbf{A}(\mathbf{r}, t)|\psi(\mathbf{r}, t)|^2. \tag{35.16}$$

We note that the term $\mathbf{J}_A(\mathbf{r}, t)$, which explicit dependence on $\mathbf{A}(\mathbf{r}, t)$ is linear, is gauge dependent. The sum of the $\mathbf{J}_0$ and $\mathbf{J}_A$ terms *is* gauge invariant, however, and only the total current density relates to physically observable quantities.

Since the transverse part of the vector potential, $\mathbf{A}_T(\mathbf{r}, t)$, is gauge invariant let us divide the current density into the following two parts

$$\mathbf{J}(\mathbf{r}, t) = \mathbf{J}^{\mathrm{par}}(\mathbf{r}, t) + \mathbf{J}^{\mathrm{dia}}(\mathbf{r}, t) \tag{35.17}$$

where

$$\mathbf{J}^{\mathrm{par}}(\mathbf{r}, t) = \frac{q\hbar}{2mi}[\psi^*(\mathbf{r}, t)\boldsymbol{\nabla}\psi(\mathbf{r}, t) - \psi(\mathbf{r}, t)\boldsymbol{\nabla}\psi^*(\mathbf{r}, t)] - \frac{q^2}{m}\mathbf{A}_L(\mathbf{r}, t)|\psi(\mathbf{r}, t)|^2 \quad (35.18)$$

is called the paramagnetic (par) current density, and

$$\mathbf{J}^{\mathrm{dia}}(\mathbf{r}, t) = -\frac{q^2}{m}\mathbf{A}_T(\mathbf{r}, t)|\psi(\mathbf{r}, t)|^2 \quad (35.19)$$

is named the diamagnetic (dia) current density. The para- and diamagnetic current densities are both manifest gauge independent. The Meissner effect in BCS super-conductors is related to $\mathbf{J}^{\mathrm{dia}}$ [11, 67, 138, 140, 200, 220, 231], extended to a sum over all electrons, of course. See also Chapter 25.

In the phenomenological London theory of superconductivity [179, 180] it is suggested that the wave function Ψ_{sup} of the "superfluid" electrons is "rigid" with respect to perturbations due to the presence of a weak magnetic field ($\mathbf{B}$). Since $0 = \boldsymbol{\nabla} \cdot \mathbf{B} = \boldsymbol{\nabla} \cdot \boldsymbol{\nabla} \times \mathbf{A} = \boldsymbol{\nabla} \cdot \boldsymbol{\nabla} \times \mathbf{A}_T$, magnetic perturbations only affect the transverse excitations of the system. The longitudinal part, $\mathbf{A}_L$, of $\mathbf{A}$ couples to longitudinal excitations, and the superconductor wave function of the simple BCS theory is not rigid with respect to longitudinal collective excitations [231]. The rigidity of the superconductor against weak transverse excitations ($\mathbf{A}_L = \mathbf{0}$) implies that the superconducting current density is given by

$$\mathbf{J}(\mathbf{r}, t) = -\frac{q^2}{m}N_{\mathrm{sup}}\mathbf{A}_T(\mathbf{r}, t), \quad (35.20)$$

where N_{sup} is the electron density in the superconductor. With N_{sup} replaced by some appropriate average value we are (again) led to a massive vector equation for $\mathbf{A}_T$, here with photon mass $M = q^2 N_{\mathrm{sup}}/m$.

35.2 ELECTROMAGNETIC SUPERCONDUCTING VACUUM

35.2.1 The Higgs model

In the preceding discussion we have considered the diamagnetic coupling mechanism at high frequencies in the normal metallic state and in superconductors. Roughly speaking, we have noted that the transverse photon appears as massive if the electron density is counted as constant through the system, the mass being associated with the diamagnetic screening current. The "rigidity" of the wave function in both cases is related to non-covariant descriptions.

Inspired by these examples, we now return to the question whether it in principle is possible for the photon to obtain a mass in its propagation through what we are pleased to call vacuum? To a certain extent, the following considerations of this "electromagnetic superconducting (or high-frequency) vacuum" are "only" a pedagogical exercise, by way of preparation for the application to electro-weak interactions. However, the reader may find (as myself) that the considerations below broaden our perspective on fundamental photon physics.

Let us return to the combination of Eqs. (35.1) and (35.2) as this appears in the covariant Lorenz gauge, $\partial_\mu A^\mu = 0$, viz.,

$$\Box A^\mu = iq[\phi^* \partial^\mu \phi - (\partial^\mu \phi^*)\phi] - 2q^2 A^\mu |\phi|^2. \tag{35.21}$$

Now, in a configuration where the complex field

$$\phi = \frac{1}{\sqrt{2}}(\phi_1 + i\phi_2) \tag{35.22}$$

is a space-time constant, with modulus

$$|\phi| \equiv \frac{1}{\sqrt{2}}f, \tag{35.23}$$

the equation for A^μ becomes

$$[\Box + (qf)^2]A^\mu = 0, \tag{35.24}$$

which is the covariant wave equation for a free vector particle of mass $M = qf$.

This fundamental observation was first made in 1964 by Englert and Brout [78], Higgs [112–114] and Guralnik et. al. [107] indicated that what one considers as a "physical vacuum" may carry a complex scalar field ϕ, *provided* this field is constant in space-time. In this vacuum the photons ($T1$, $T2$, L, S) are driven by an "electromagnetic vacuum (vac) current" with density

$$\mathbf{j}^\mu_{\text{vac}} = -q^2 f^2 A^\mu. \tag{35.25}$$

How can this be? The essential idea of Higgs is to suppose that vacuum, $|\tilde{0}\rangle$, is a state in which some quantum field, the Higgs field $\hat{\phi}$, has a non-vanishing expectation value, i.e.,

$$\langle \tilde{0}| \hat{\phi} |\tilde{0}\rangle \neq 0. \tag{35.26}$$

It was suggested by Nambu [191] that the *physical vacuum* of a quantum field theory is in some sense analogous to the *ground state* of an interacting many-body system. The ground state here is a stable (equilibrium) configuration of minimum energy.

35.2.2 Spontaneous symmetry breaking

In our previous discussion of QED we have accepted a standard normal expression such as (simplified notation used)

$$\hat{\phi} = \int \frac{1}{2\omega_q} \left[\hat{a}^\dagger(q)e^{iqx} + \hat{a}(q)e^{-iqx} \right] \frac{d^3q}{(2\pi)^3} \tag{35.27}$$

for all fields, and we use

$$\hat{a}(q) |0\rangle = 0 \tag{35.28}$$

to define the vacuum state, $|0\rangle$. In consequence

$$\langle 0| \hat{\phi} |0\rangle = 0. \tag{35.29}$$

The ground state in QED is very symmetrical, and we are used to quantum field ground states possessing a high degree of symmetry. Nevertheless, it is possible that the *true ground state* is asymmetric or "loop sided". Although this does not seem to be the case in photon physics, it is certainly worthwhile to reflect on this loop-sided model.

There are examples of stable loop-sided ground state configurations in many-body physics even though the basic interactions between the particles are symmetrical. We have exemplified this in our study of spin waves in ferromagnets [Chapter 24]. In such situations, in which the ground state configuration does not display the symmetry of the Hamiltonian, we may say that the symmetry is *spontaneously broken*. However, the high symmetry is still there but it is *hidden*.

In order to illustrate the main point of loop-sided $U(1)$ field theory let us consider the so-called complex ϕ^4-theory. The field-unperturbed Lagrangian density here is given by

$$\mathcal{L} = (\partial_\mu \phi)(\partial^\mu \phi^*) - V(|\phi|^2), \tag{35.30}$$

where $[106, 157, 186, 224]$

$$V(|\phi|^2) = -\frac{1}{2}\mu^2 \phi^* \phi + \frac{1}{2}\left(\frac{\mu}{f}\right)^2 (\phi^* \phi)^2. \tag{35.31}$$

This form of the potential was first discussed in 1961 in the classic paper by Goldstone [100]. The particular constants μ and f multiplying the two terms have been chosen for later convenience [The f-factor is the one appearing in the mass for the free vector particle; see Subsection 35.2.1].

The potential has the global $U(1)$ symmetry

$$\phi \to \phi' = e^{i\alpha}\phi, \tag{35.32}$$

or equivalently

$$\phi_1 \to \phi_1' = \phi_1 \cos\alpha - \phi_2 \sin\alpha, \tag{35.33}$$
$$\phi_2 \to \phi_2' = \phi_1 \sin\alpha + \phi_2 \cos\alpha. \tag{35.34}$$

In the $\phi_1 - \phi_2$ plane the potential is symmetric under arbitrary α-rotations. At the origo, $\phi_1 = \phi_2 = 0$ $V(|\phi|^2)$ has a local maximum. The field configuration of minimum energy is the circular locus

$$\phi_1^2 + \phi_2^2 = f^2, \tag{35.35}$$

remembering $\phi = (\phi_1 + i\phi_2)/\sqrt{2}$ [Eq. (34.1)]. For the chosen $V(|\phi|^2)$ there are therefore infinitely many possibilities for the stable configuration. These form a one-parameter family, say

$$\phi_1 = f\cos\theta, \quad \phi_2 = f\sin\theta \tag{35.36}$$

for arbitrary θ. Once a particular ground state is chosen (θ selected "spontaneously") the $U(1)$ apparently is lost. However, it is clear that the $U(1)$ symmetry is just "hidden".

In the quantum field theory, with potential operator

$$\hat{V}(\hat{\phi}) = -\frac{1}{2}\mu^2\hat{\phi}^\dagger\hat{\phi} + \frac{1}{2}\left(\frac{\mu}{f}\right)^2\left(\hat{\phi}^\dagger\hat{\phi}\right)^2, \tag{35.37}$$

the state $|0\rangle$ [defined in Eq. (35.28)], for which

$$\langle 0|\,\hat{\phi}_1\,|0\rangle = \langle 0|\,\hat{\phi}_2\,|0\rangle = 0, \tag{35.38}$$

is *not the ground state (true vacuum)*. The true vacuum $|\tilde{0}\rangle$ is *any* state lying somewhere on the circle, given in Eq. (35.35). Hence,

$$[\langle\tilde{0}|\,\hat{\phi}_1\,|\tilde{0}\rangle]^2 + [\langle\tilde{0}|\,\hat{\phi}_2\,|\tilde{0}\rangle]^2 = f^2. \tag{35.39}$$

From the one-parameter family in Eq. (35.36) a chosen vacuum state, denoted by $|\tilde{0}\rangle_\theta$, thus satisfies

$$\begin{aligned}
{}_\theta\langle\tilde{0}|\,\hat{\phi}\,|\tilde{0}\rangle_\theta &= {}_\theta\langle\tilde{0}|\,\hat{\phi}_1 + i\hat{\phi}_2\,|\tilde{0}\rangle_\theta \\
&= \frac{f}{\sqrt{2}}(\cos\theta + i\sin\theta)_\theta\langle\tilde{0}|\tilde{0}\rangle_\theta = \frac{1}{\sqrt{2}}fe^{i\theta}
\end{aligned} \tag{35.40}$$

Any phase angle θ will give a ground state. However, *a choice of a definite phase for the actual vacuum breaks the $U(1)$ symmetry spontaneously and renders it hidden.*

35.3 ABELIAN HIGGS MODEL

35.3.1 Massive gauge field, $\hat{A}^\mu$

The potential operator of the ϕ^4-theory given in Eq. (35.37), is locally $U(1)$ phase invariant, since it involves no derivatives of the $\hat{\phi}$-field operator. This implies that the minimal-coupled Lagrangian density operator involving $\hat{\phi}$ and $\hat{A}^\mu$, viz.,

$$\hat{\mathcal{L}} = \left[(\partial^\mu + iq\hat{A}^\mu)\hat{\phi}\right]^\dagger\left[(\partial_\mu + iq\hat{A}_\mu)\hat{\phi}\right] - \frac{1}{4}\hat{F}_{\mu\nu}\hat{F}^{\mu\nu} + \frac{1}{2}\mu^2\hat{\phi}^\dagger\hat{\phi} - \frac{1}{2}\left(\frac{\mu}{f}\right)^2(\hat{\phi}^\dagger\hat{\phi})^2, \tag{35.41}$$

is locally $U(1)$ phase invariant. Note that we have added the Lagrangian operator for the free photon field, $-\frac{1}{4}\hat{F}_{\mu\nu}\hat{F}^{\mu\nu}$. The $\mathcal{L}$ in Eq. (35.41) is the Lagrangian of the *Abelian Higgs Model* [113]; see also [158, 224].

The equation of motion for $\hat{A}^\mu$, which follows from Eq. (35.41) is

$$\Box\hat{A}^\mu - \partial^\mu(\partial_\nu A^\nu) = iq[\hat{\phi}^\dagger\partial^\mu\hat{\phi} - (\partial^\mu\hat{\phi}^\dagger)\hat{\phi}] - 2q^2\hat{A}^\mu\hat{\phi}^\dagger\hat{\phi}, \qquad (35.42)$$

i.e. Eq. (35.1) [with Eq. (35.2) inserted] promoted to a quantum field equation.

Although the Lagrangian in Eq. (35.41) is locally $U(1)$ invariant, that is, it is gauge invariant, the non-zero vacuum expectation value for $\hat{\phi}$ leads to infinitely many possible vacuum states. The selection of *one* breaks the $U(1)$ symmetry "spontaneously". Can we make sure that different choices of the vacuum state lead to the same physics (observations)? In order to investigate this question, let us see what happens for a scalar field operator

$$\hat{\phi}(x) = \frac{f}{\sqrt{2}}\exp\left[i\hat{\alpha}(x)\right], \qquad (35.43)$$

with $\hat{\alpha}^\dagger(x) = \hat{\alpha}(x)$. Inserting Eq. (35.43) into Eq. (35.42), we find

$$\Box\hat{A}^\mu - \partial^\mu(\partial_\nu A^\nu) = -(gf)^2\left[\hat{A}^\mu + q^{-1}\partial^\mu\hat{\alpha}\right]. \qquad (35.44)$$

Let us now consider a gauge transformation of $\hat{A}^\mu$ of the usual form, viz.,

$$\hat{A}^\mu \to \hat{\underset{\sim}{A}}^\mu \equiv \hat{A}^\mu - \partial^\mu\hat{\chi} \qquad (35.45)$$

with the arbitrary gauge function $\hat{\chi}$. If we chose a gauge function

$$\hat{\chi} = -q^{-1}\hat{\alpha}, \qquad (35.46)$$

and remember that the left-hand side of Eq. (35.44) is gauge invariant, the following equation is obtained for $\hat{\underset{\sim}{A}}^\mu$:

$$\Box\hat{\underset{\sim}{A}}^\mu - \partial^\mu(\partial_\nu\hat{\underset{\sim}{A}}^\nu) = -(qf)^2\hat{\underset{\sim}{A}}^\mu, \qquad (35.47)$$

or equivalently $[qf = M]$

$$(\Box + M^2)\hat{\underset{\sim}{A}}^\mu - \partial^\mu(\partial_\nu\hat{\underset{\sim}{A}}^\nu) = 0. \qquad (35.48)$$

The quantity $\hat{\underset{\sim}{A}}^\mu$ hence can be regarded as an acceptable vector potential. By taking the divergence of Eq. (35.48) one obtains, since $M \neq 0$,

$$\partial_\mu\hat{\underset{\sim}{A}}^\mu = 0, \qquad (35.49)$$

and then finally

$$(\Box + M^2)\hat{\underset{\sim}{A}}^\mu = 0. \qquad (35.50)$$

We can now answer the question posed between Eqs. (35.2) and (35.3): Yes, it is possible for the electromagnetic gauge field, A^μ, to satisfy a massive vector equation (of the type given in Eq. (35.3), at least for scalar fields of the form $\hat{\phi} = (f/\sqrt{2})\exp(i\hat{\alpha})$.

Before proceeding one should note that strictly speaking the gauge condition cannot be valid as an operator identity in quantum field theory. Instead one tries to use the condition in Eq. (35.49) to select the physical state $|\psi\rangle$ by requiring these states to be eigenvectors of $\partial_\mu \hat{\underline{A}}^\mu$ with eigenvalue zero [63, 157]:

$$\partial_\mu \hat{\underline{A}}^\mu |\psi\rangle = 0. \tag{35.51}$$

It turns out that Eq. (35.51) is too strong a requirement. Instead it appears that one should take [63, 157]

$$\partial_\mu \hat{\underline{A}}^{(+)\mu} |\psi\rangle = 0. \tag{35.52}$$

where $\hat{\underline{A}}^{(+)\mu}$ is the positive-frequency part of $\hat{\underline{A}}^\mu$. We do not wish to get involved here in the full quantum field problem of how to quantize a field with gauge arbitrariness. The results obtained with "$\partial_\mu \hat{\underline{A}}^\mu |\psi\rangle = 0$" will be correct, see e.g. [63].

To take the last step in the analysis above we ask the following question: What is the gauge specified for our original $\hat{A}^\mu$ [on the right-hand side of Eq. (35.45)]. To answer this question, the gauge transformation in Eq. (35.45) is inserted into Eq. (35.49). Hence, it follows that $\hat{A}^\mu$ is in a gauge specified by

$$\partial_\mu \hat{A}^\mu = -q^{-1}\Box\hat{\alpha}. \tag{35.53}$$

A gauge function χ gives (as usual) an associated local phase change in the scalar field, here

$$\hat{\phi} \to \hat{\underline{\phi}} = e^{iq\hat{\chi}}\hat{\phi} = e^{-i\hat{\alpha}}\hat{\phi} = \frac{f}{\sqrt{2}}. \tag{35.54}$$

The two last members of Eq. (35.54) follow by use of Eq. (35.46) and Eq. (35.43), respectively. We may now conclude that if one starts with a certain non-zero local phase $\hat{\alpha}(x)$ one can always make a particular choice for the vector potential so as to the reduce the phase of $\hat{\phi}$ to zero.

35.3.2 Higgs bosons

The previous considerations, based on the scalar field oprator $\hat{\phi} = (f/\sqrt{2})\exp(i\hat{\alpha})$, Eq. (35.43), have shown that the phase ($\hat{\alpha}$) freedom, corresponding to "motion around the circle with radius $f/\sqrt{2}$", can be transferred to $\hat{A}^\mu$ to render the vector boson field massive ($M = qf$). The different ϕ-states on this circle constitute the set of infinitely many ground states (vacua) with spontaneously broken symmetry. Physically all these states are physically equivalent, and from a given $\hat{\phi}_\alpha = (f/\sqrt{2}\exp(i\hat{\alpha})$, the one with phase $\hat{\alpha} = 0$, $\hat{\phi}_0 = f/\sqrt{2}$, can be "reached" with a specific choice of local gauge function $\hat{\chi} = -q^{-1}\hat{\alpha}$. The associated vector potential $\hat{A}^\mu$ is in a gauge specified by $\partial_\mu \hat{A}^\mu = -q^{-1}\Box\hat{\alpha}$; Eq. (35.53).

In order to generalize the physical approach it is only necessary to relax the "radial" condition $\left|\hat{\phi}\right| = f/\sqrt{2}$, and use a dynamic (real) ansatz

$$\hat{\phi}(x) = \frac{1}{\sqrt{2}}(f + \hat{H}(x)), \tag{35.55}$$

where

$$\langle \tilde{0}| \, \hat{H}(x) \, |\tilde{0}\rangle = 0 \tag{35.56}$$

in the ground state (true vacuum), $|\tilde{0}\rangle$.

It readily follows, that the inclusion of the dynamical $\hat{H}(x)$-field implies that a new term, $-q^2(2f\hat{H} + \hat{H}^2)$, must be added to Eq. (35.47). For what follows, let us make the notational simplification $\hat{A}'^{\mu} \to \hat{A}^{\mu}$. Hence the master equation $\hat{A}^{\mu}(x)$ is

$$(\Box + M^2)\hat{A}^{\mu}(x) - \partial^{\mu}(\partial_{\nu}A^{\nu}(x)) = -2q^2 f\hat{H}(x)\hat{A}^{\mu}(x) - q^2\hat{H}^2(x)\hat{A}^{\mu}(x). \tag{35.57}$$

The neutral scalar field $\hat{H}(x)$ is called the Higgs field, and the quanta of $\hat{H}(x)$ are Higgs bosons. The Higgs particles thus are associated with the radial degree of freedom of $\hat{\phi}$. It is expected that it will make sense to develop a perturbation theory of Eq. (35.57), i.e., look at small oscillations of $\hat{\phi}$ about the *stable* true vacuum located at $\hat{\phi} = f/\sqrt{2}$.

CHAPTER 36

Electro-Weak Theory: An Introduction

36.1 GOLDSTONE BOSON IN $SU(1)$

As a forerunner of our considerations on massive and massless particles in the non-Abelian $SO(3)$ group, let us return to the scalar field in its general form, viz.,

$$\phi(x) = \frac{1}{\sqrt{2}}(f + H(x))e^{i\theta(x)}. \tag{36.1}$$

As we have seen in Section 35.3, a specific choice of the gauge function enables one to make $\phi(x)$ real $[\theta(x) = 0]$. In the quantum theory $[H \to \hat{H}, \theta \to \hat{\theta}]$, where ϕ becomes an operator, one has

$$\langle 0| \hat{H}(x) |0\rangle = \langle 0| \hat{\theta}(x) |0\rangle = 0, \tag{36.2}$$

and $\langle 0| \hat{\phi}(x) |0\rangle = f/\sqrt{2}$.

We have previously realized that the "radial" dynamics is related to the Higgs boson field. The "angular" dynamics is connected to a so-called Goldstone boson field [20, 100, 101, 248]. Is it possible to associate a particle aspect to these fields, and in case, what are the masses of the Goldstone and Higgs particles in the framework of the ϕ^4-theory?

To throw light on this question, we return to the classical Lagrangian in Eq. (35.30) inserting the form of the scalar field given in Eq. (36.1). Since the potential energy in Eq. (35.31) only is a function of $\phi\phi^*$, the phase $\theta(x)$ does not appear in $V(|\phi|^2)$. One readily obtains

$$2\mu^{-2}V = -\left(\frac{f}{2}\right)^2 + H^2 + \frac{H^3}{f} + \frac{H^4}{(2f)^2}. \tag{36.3}$$

the constant term $-(f/2)^2$ plays no physical role and can be omitted. For sufficiently small H the term proportional to H^3 and H^4 may be neglected. In the harmonic approximation

$$V = \frac{1}{2}\mu^2 H^2. \tag{36.4}$$

DOI: 10.1201/9781003029458-36

Since $(\partial_\mu \Theta)(\partial^\mu H) = (\partial^\mu \Theta)(\partial_\mu H)$, it appears that the kinetic energy, given in terms of θ and H, is

$$T \equiv (\partial_\mu \phi)(\partial^\mu \phi^*) = \frac{1}{2}\left[(\partial_\mu H)(\partial^\mu H) + (f + H)^2(\partial_\mu \theta)(\partial^\mu \theta)\right]$$

$$\approx \frac{1}{2}\left[(\partial_\mu H)(\partial^\mu H) + f^2(\partial_\mu \Theta)(\partial^\mu \theta)\right]. \tag{36.5}$$

Since there is no term in θ^2 in the $\mathcal{L} = T - V$, the θ-particle, called the Goldstone particle, is massless. The Higgs particle is massive and its mass is $m_H = \mu$, cf. Eq. (36.4).

36.2 HIGGS AND GOLDSTONE MODES IN $SO(3)$

In the following we consider a specific non-Abelian group, viz. $SO(3)$. As an example let us assume that the isovector scalar field $\boldsymbol{\phi} = (\phi_1, \phi_2, \phi_3)$ relates to a Lagrangian density

$$\mathcal{L} = \frac{1}{2}\left[(\partial_\mu \phi_i)(\partial^\mu \phi_i) + \frac{1}{2}\mu^2 \phi_i \phi_i - \frac{1}{2}\left(\frac{\mu}{f}\right)^2 (\phi_i \phi_i)^2\right] \tag{36.6}$$

for a *real* field. As usual the summation convention over $i = 1 - 3$ applies. The Lagrangian in Eq. (36.6) is a generalization of the $SU(1)$ Lagrangian in Eq. (35.30). The factor $1/2$ is front of the square bracket stems from our replacement of the complex ϕ-field $(SU(1))$ with a real $\boldsymbol{\phi}$-field $(SO(3))$.

The equilibrium positions, $\boldsymbol{\phi}_0$, satisfy

$$|\boldsymbol{\phi}_0| = (\phi_{i0}\phi_{i0})^{1/2} = \frac{f}{\sqrt{2}}, \tag{36.7}$$

as the reader may show. For what follows, we choose a vacuum which points in the 3-direcion in isospin space, i.e.,

$$\boldsymbol{\phi}_0 = \frac{f}{\sqrt{2}}\mathbf{e}_3. \tag{36.8}$$

The gauge in which the vacuum position is the one in Eq. (36.8) is called the *unitary gauge* (or the U-gauge). In this gauge the Lagrangian

$$\mathcal{L} = \frac{1}{2}[(\partial_\mu \phi_i)(\partial^\mu \phi_i) - V] \tag{36.9}$$

has a potential energy given by

$$V = -\frac{\mu^2}{2}\left[\phi_1^2 + \phi_2^2 + \frac{1}{2}(f + H)^2\right] + \frac{1}{2}\left(\frac{\mu}{f}\right)^2\left[\phi_1^2 + \phi_2^2 + \frac{1}{2}(f + H)^2\right]^2, \tag{36.10}$$

since $\phi_3 = (f + H)/\sqrt{2}$. The sum of terms in V/μ^2 which are quadratic in ϕ_1 and ϕ_2 vanishes,

$$-\frac{1}{2}(\phi_1^2 + \phi_2^2) + \frac{1}{2f^2}(\phi_1^2 + \phi_2^2)f^2 = 0. \tag{36.11}$$

Thus, the masses of the two associated Goldstone particles are zero. The mass of the Higgs particle is [as in $U(1)$] $M_H = \mu$.

36.3   WEAK ISOSPIN DOUBLET. $SU(2)_L$ TRANSFORMATIONS

36.3.1   Global phase transformations

With $\hbar = c = 1$, the Dirac equation may be written as

$$(i\gamma^\mu \partial_\mu - m)\Psi(x) = 0, \tag{36.12}$$

where γ^μ ($\mu = 0 - 3$) are the well-known gamma coefficients [158, 224]. These are 4x4 matrices. The associated free-fermion Lagrangian density is given by

$$\mathcal{L}_0 = \bar{\Psi}(x)(i\gamma^\mu \partial_\mu - m)\Psi(x), \tag{36.13}$$

where $\bar{\Psi}(x) = \Psi^\dagger \gamma^0$. In the following we shall assume that all leptons (e, μ, τ) are massless, and for a given species, the right (R)- and left (L)-handed helicity components of

$$\Psi = \Psi_R + \Psi_L \tag{36.14}$$

are [224]

$$\Psi_R = \frac{1}{2}(1 + \gamma_5)\Psi, \tag{36.15}$$

$$\Psi_L = \frac{1}{2}(1 - \gamma_5)\Psi, \tag{36.16}$$

where $\gamma_5(= \gamma^5) = i\gamma^0\gamma^1\gamma^2\gamma^3$. Let us simplify our notation a bit, and denote the wave functions of the right- and left-handed components by e_R and e_L, and the associated electron neutrino wave function by ν_R and ν_L. Using the "A-slash" notation $\slashed{A} \equiv \gamma^\mu A_\mu$, the free Lagrangian density for the electron plus the electron neutrino becomes

$$\mathcal{L}_0 = i\left[\bar{e}_R \slashed{\partial} e_R + \bar{e}_L \slashed{\partial} e_L + \bar{\nu}_L \slashed{\partial} \nu_L\right]. \tag{36.17}$$

The electron has L and R components, but the electron neutrino only exists in a L helicity state. Equivalent expressions for the free Lagrangian hold for the sets of μ and τ species. If needed, a Lagrangian with summation over the e, μ and τ species readily can be written down.

The theory of the electro-weak interaction begins with the postulate that the left-handed fermions of a given flavour pair such as the electron and the electron neutrino are *physically equivalent* from the view point of weak interactions. This physical symmetry is represented by combining the fields ν_L and e_L into a so-called *weak isospin doublet* [here ν_L and e_L]

$$L \equiv \begin{pmatrix} \nu_L \\ e_L \end{pmatrix}, \quad \left[\bar{L} = (\bar{\nu}_L, \bar{e}_L)\right]. \tag{36.18}$$

The remaining particle

$$R \equiv e_R, \quad \left[\bar{R} = \bar{e}_R\right] \tag{36.19}$$

is supposed to be a *weak isoscalar*.

The free Lagrangian in Eq. (36.17) thus may be written in the form

$$\mathcal{L}_0 = i\left[\bar{L}\displaystyle{\not}\partial L + \bar{R}\displaystyle{\not}\partial R\right].\tag{36.20}$$

The physical symmetry of the doublet is represented by unitary transformations (U), viz.,

$$L \to L' = UL,\tag{36.21}$$

where U is an element of the group matrices $SU(2)_L$. The subscript L is a reminder that the matrices act only on the left-handed fermion components. The elements of $SU(2)_L$ may be represented by the conventional representation for two-components spinors which describe spin-1/2 particles in the weakly relativistic Pauli theory. Hence,

$$U(\boldsymbol{\alpha}) = \exp\left(\frac{i}{2}\alpha_i \tau_i\right),\tag{36.22}$$

where $\boldsymbol{\alpha} = (\alpha_1, \alpha_2, \alpha_3)$ are any three real numbers, and

$$\tau_1 = \begin{pmatrix} 0 & 1 \\ 1 & 0 \end{pmatrix},\ \tau_2 = \begin{pmatrix} 0 & -i \\ i & 0 \end{pmatrix},\ \tau_3 = \begin{pmatrix} 1 & 0 \\ 0 & -1 \end{pmatrix}\tag{36.23}$$

are the three Hermitian Pauli matrices, which satisfy the commutation relations $[\tau_i, \tau_j] = 2i\varepsilon_{ijk}\tau_k$. The unitary 2×2 $U(\boldsymbol{\alpha})$ matrices have $\mathrm{Det}U(\boldsymbol{\alpha}) = +1$. The global U-transformations implies that

$$i\bar{L}'\displaystyle{\not}\partial L' = i\bar{L}U^{-1}\displaystyle{\not}\partial U\bar{L} = i\bar{L}\displaystyle{\not}\partial L,\tag{36.24}$$

so that the term $i\bar{L}\displaystyle{\not}\partial L$ in $\mathcal{L}_0$ [Eq. (36.20)] is an invariant. The right-handed lepton (here electron) field is defined to be a *weak isoscalar*, i.e. invariant under any $SU(2)$ transformation:

$$R \to R' = R.\tag{36.25}$$

In conclusion, it follows that the $SU(2)_L$ transformation leaves the free lepton Lagrangian density, $\mathcal{L}_0$ [Eq. (36.20)], invariant.

36.3.2 Charged and neutral leptonic currents

In order to make a preparation for the introduction and analysis of leptonic currents, let us suppose that a Lagrangian $(\mathcal{L})$ involves a set of fields $\{\Psi_r\}$, which could be bosons or fermions, and suppose that $\mathcal{L}$ is invariant under the small transformation (constant small parameter: ε)

$$\delta\Psi_r = -i\varepsilon T_{rs}\Psi_s\tag{36.26}$$

for some set of numerical coefficients. The invariance of $\mathcal{L}$, i.e. $\delta\mathcal{L} = 0$, implies that

$$\frac{\partial \mathcal{L}}{\partial \Psi_r}\delta\Psi_r + \frac{\partial \mathcal{L}}{\partial(\partial^\mu \Psi_r)}\partial^\mu(\delta\Psi_r) = 0.\tag{36.27}$$

By combining Eq. (36.27) with the Lagrange equation,

$$\frac{\partial \mathcal{L}}{\partial \Psi_r} = \partial^\mu \left(\frac{\partial \mathcal{L}}{\partial (\partial^\mu \Psi_r)} \right), \tag{36.28}$$

multiplied by $\delta \Psi_r$, one obtains

$$\partial^\mu \left(\frac{\partial \mathcal{L}}{\partial (\partial^\mu \Psi_r)} \delta \Psi_r \right) = \partial^\mu \left[-i\varepsilon \frac{\partial \mathcal{L}}{\partial (\partial^\mu \Psi_r)} T_{rs} \Psi_s \right] = 0. \tag{36.29}$$

The result in Eq. (36.29) is a current conservation law of the covariant form $\partial^\mu J_\mu = 0$, disregarding (dividing by) the irrelevant small parameter. Hence, the conserved four-current is

$$\mathbf{j}_\mu = -i \frac{\partial \mathcal{L}}{\partial (\partial^\mu \Psi_r)} T_{rs} \Psi_s. \tag{36.30}$$

Let us apply this result to the weak isospin doublet, L, which free Lagrangian is

$$\mathcal{L} = i\bar{L}\slashed{\partial} L = i\bar{L}\gamma_\mu \partial^\mu L. \tag{36.31}$$

In this case the small symmetry transformation is given by the three parameters, $\boldsymbol{\alpha} = (\alpha_1, \alpha_2, \alpha_3)$, matrix as

$$\delta L = \frac{i}{2} \alpha_j \tau_j L, \tag{36.32}$$

cf. Eqs. (36.21) and (36.22). Since $\partial \mathcal{L}/\partial(\partial^\mu L) = i\bar{L}\gamma_\mu$, we obtain the following three conserved currents (denoted below by capital J's):

$$J_i^\mu(x) = \frac{1}{2}\bar{L}(x)\gamma^\mu \tau_i L(x), \quad i = 1, 2, 3. \tag{36.33}$$

The three conserved currents in Eq. (36.33) are called *weak isospin currents*. The corresponding conserved quantities

$$I_i^W = \int J_i^0(x)d^3x = \frac{1}{2}\int \bar{L}(x)\gamma^0 \tau_i L(x)d^3x = \frac{1}{2}\int L^\dagger \tau_i L d^3x \tag{36.34}$$

are called *weak isospin charges*. The last expression in Eq. (36.34) is obtained noting that $\bar{L}\gamma^0 = L^\dagger\gamma^0\gamma^0 = L^\dagger$, since $(\gamma^0)^2 = 1$.

By inserting the weak isospin doublet [Eq. (36.18)] and the Pauli matrix for τ_1 [Eq. (36.23)] into Eq. (36.33) one obtains

$$J_1^\mu = \frac{1}{2}(\bar{\nu}_L, \bar{e}_L)\gamma^\mu \begin{pmatrix} 0 & 1 \\ 1 & 0 \end{pmatrix} \begin{pmatrix} \nu_L \\ e_L \end{pmatrix} = \frac{1}{2}(\bar{\nu}_L\gamma^\mu e_L + \bar{e}_L\gamma^\mu \nu_L). \tag{36.35}$$

For J_2^μ one gets

$$J_2^\mu = \frac{i}{2}(\bar{e}_L\gamma^\mu \nu_L - \bar{\nu}_L\gamma^\mu e_L). \tag{36.36}$$

Furthermore, $J^\mu \equiv 2(J_1^\mu - iJ_2^\mu) = 2\bar{e}_L\gamma^\mu\nu_L$ and $(J^\mu)^\dagger = 2\bar{\nu}_L\gamma^\mu e_L$. The currents J_1^μ, and J_2^μ, as well as J^μ and $(J^\mu)^\dagger$ are called *charged* weak isospin currents. These currents couple electrically neutral leptons (here ν_L) with electrically charged leptons (here e_L).

Most remarkable, is the conservation of a third current

$$J_3^\mu = \frac{1}{2}\bar{L}\gamma^\mu\tau_3 L = \frac{1}{2}(\bar{\nu}_L, \bar{e}_L)\gamma^\mu \begin{pmatrix} 1 & 0 \\ 0 & -1 \end{pmatrix} \begin{pmatrix} \nu_L \\ e_L \end{pmatrix} = \frac{1}{2}(\bar{\nu}_L\gamma^\mu\nu_L - \bar{e}_L\gamma^\mu e_L). \quad (36.37)$$

This, so-called *neutal* weak isospin current, couples either electrically neutral or electrically charged leptons [here the types ν_L and e_L]. The fact that J_3^μ contains a contribution $\bar{\tilde{e}}_L\gamma^\mu e_L$, which is a part of the electromagnetic current s^μ, given below, is a first indication that electromagnetic and weak processes will be interconnected.

36.3.3 Weak hypercharge and hypercharge currents

In order to determine the expression for the electromagnetic current (density) one makes the minimal coupling substitution $\partial_\mu \to D_\mu = \partial_\mu - iA_\mu$ [$q = -e$, $\hbar = c = 1$] in the free fermion Lagrangian $\mathcal{L}_0$ [Eq. (36.13)]. This gives one an interaction Lagrangian $e\bar{\Psi}\gamma^\mu\Psi A_\mu = -s^\mu A_\mu$, where

$$s^\mu = -e\bar{\Psi}\gamma^\mu\Psi, \quad (36.38)$$

is identified as the electromagnetic current (density). Composed into L and R components

$$s^\mu = -e(\bar{e}_L + \bar{e}_R)\gamma^\mu(e_L + e_R) = -e(\bar{e}_L\gamma^\mu e_L + \bar{e}_R\gamma^\mu e_R), \quad (36.39)$$

since $\bar{e}_L\gamma^\mu e_R = \bar{e}_R\gamma^\mu e_L = 0$. To prove that the cross-coupling between the L and R states vanishes, let us first determine the relation between $\bar{e}_L$ and $\bar{e}$. Since $\gamma^\mu\gamma^0 = -\gamma^0\gamma^\mu$, one obtains

$$\bar{e}_L \equiv e_L^\dagger\gamma^0 = e^\dagger\frac{1}{2}(1 - \gamma^5)\gamma^0 = e^\dagger\gamma^0\frac{1}{2}(1 + \gamma^5) = \bar{e}\frac{1}{2}(1 + \gamma^5), \quad (36.40)$$

and then

$$\bar{e}_L\gamma^\mu e_R = \frac{1}{4}\bar{e}(1 + \gamma^5)\gamma^\mu(1 + \gamma^5)e = \frac{1}{4}\bar{e}(1 + \gamma^5)(1 - \gamma^5)\gamma^\mu e = 0, \quad (36.41)$$

remembering that $(\gamma^5)^2 = 1$. I leave it to the reader to show that $\bar{e}_R\gamma^\mu e_L = 0$.

The weak hypercharge current, J_Y^μ, is defined by

$$J_Y^\mu(x) = e^{-1}s^\mu(x) - J_3^\mu(x), \quad (36.42)$$

also can be given the form

$$J_Y^\mu(x) = -\bar{e}_L\gamma^\mu e_L - \bar{e}_R\gamma^\mu e_R - \frac{1}{2}(\bar{\nu}\gamma^\mu\nu_L - \bar{e}_L\gamma^\mu e_L)$$

$$= \frac{1}{2}(\bar{\nu}_l, \bar{e}_L)\gamma^\mu \begin{pmatrix} \nu_L \\ e_L \end{pmatrix} - \bar{e}_R\gamma^\mu e_R = -\frac{1}{2}\bar{L}\gamma^\mu L - \bar{R}\gamma^\mu R. \quad (36.43)$$

The corresponding weak hypercharge

$$Y = \int J_Y^0(x)d^3x \qquad (36.44)$$

is related to the electric charge Q (in units of e) by

$$Y = \frac{Q}{e} - I_3^W . \qquad (36.45)$$

The conservation of the electric charge and the weak isospin charge I_3^W implies conservation of the weak hypercharge Y. It appears from Eqs. (36.34) and (36.37) that the weak isospin charges for the pure electron neutrino and electron states have the values $+\frac{1}{2}$ and $-\frac{1}{2}$, respectively. Since the electron neutrino is uncharged ($Q = 0$) the hypercharge for the left-handed neutrino is $Y = -\frac{1}{2}$. The left-handed electron state also has $Y = -\frac{1}{2}$ $[-1 - (-\frac{1}{2})$ from Eq. (36.45)]. The remaining right-handed electron is an isosinglet and thus associated with the value $I_3^W = 0$, giving $Y = -1$ for e_R.

36.4 ELECTRO-WEAK INTERACTION: HIGGS-FIELD CURRENTS

36.4.1 $SU(2)_L \otimes U(1)$ as a local symmetry

Let us return to the global phase transformation for the weak isospin doublet, namely

$$\begin{pmatrix} \nu_L \\ e_L \end{pmatrix} \rightarrow \begin{pmatrix} \nu_L \\ e_L \end{pmatrix}' = \exp\left(\frac{i}{2}\boldsymbol{\alpha} \cdot \boldsymbol{\tau}\right) \begin{pmatrix} \nu_L \\ e_L \end{pmatrix} \qquad (36.46)$$

cf. the discussion in Subsection 36.3.1. We know from the Yang-Mills theory described in Section 34.2, how to make the global $SU(2)_L$ invariance into a local one. In order to unify the electromagnetic and weak interaction it is necessary to include an additional $U(1)$ gauge group, as first suggested by Glashow [96]. When the group structure is enlarged to $SU(2)_L \otimes U(1)$ the unification falls marvelously into place. The unified standard theory of weak and electromagnetic interactions was constructed independently by Weinberg [257] and Salam [227]; see also [228, 259, 260]

Making the global $SU(2)_L$ invariance into a local phase invariance entails the introduction of three gauge fields transforming as a $t = 1$ triplet under the group. Two of the gauge fields are associated with transitions between the doublet members. Since these (ν_L and e_L) differ by one unit of charge, these two gauge fields will have charges ± 1 (in units of e). The third gauge boson of the triplet will be electrically neutral, and since it is massive it cannot be identified with the photon. By enlarging the $SU(2)_L$ group to $SU(2)_L \otimes U(1)$ it turns out that one can accommodate a massive neutral boson *and* a massless neutral boson (the photon) at the same time. The right-handed fermion (here electron) particle which is supposed to be a weak isoscalar, and assumed to be a *singlet,* under the weak isospin group must also be included in the formalism. For clarity, the weak isospin t (with "z-component" t_3), the hypercharge

(Y) and the electrical charge Q (in units of e) are shown for the ν_L, e_L and e_R fermions in the table below.

	t	t_3	Y	Q
ν_L	1/2	1/2	-1/2	0
e_L	1/2	-1/2	-1/2	-1
e_R	0	0	-1	-1

36.4.2 Covariant derivatives. Weak hypercharge and isospinor

We know that the introduction of appropriate covariant derivatives allows one to change the global $SU(2)_L$ and $U(1)$ invariances into local ones; see Chapt's 2 and 34. Extending the prescription to quantum fields, i.e.,

$$\partial^\mu \to D^\mu \Rightarrow \hat{D}^\mu, \tag{36.47}$$

the covariant derivative for the $SU(2)_L \otimes U(1)$ symmetry takes the general form

$$\hat{D}^\mu = \partial^\mu + ig\mathbf{t} \cdot \hat{\mathbf{W}}^\mu + ig'y\hat{B}^\mu \tag{36.48}$$

in a representation specified by $\mathbf{t} = (t_1, t_2, t_3)$ and y. The three $\hat{\mathbf{W}}^\mu$'s are a $SU(2)_L$ gauge field, and $\hat{B}^\mu$ is a $U(1)$ gauge field, and the constants g and g' are coupling constants, analogous to the coupling constant q (charge) in electrodynamics.

For the left-handed (LH) neutrino-electron doublet $\mathbf{t} = \boldsymbol{\tau}/2$ $[\boldsymbol{\tau} = (\tau_1, \tau_2, \tau_3)$, the Pauli vector] and $y = Y$ $[Y$, the hypercharge; here $Y = -1/2]$, giving

$$\hat{D}^\mu = \partial^\mu + i\frac{g}{2}\boldsymbol{\tau} \cdot \hat{\mathbf{W}}^\mu - i\frac{g'}{2}\hat{B}^\mu. \quad \text{LH} \tag{36.49}$$

For the right-handed (RH) electron lepton $t = 0$ [isoscalar] and $y = Y = -1$, so that

$$\hat{D}^\mu = \partial^\mu - iy'\hat{B}^\mu. \quad \text{RH} \tag{36.50}$$

The next step is the introduction of the covariant current associated with the complex (Higgs field) isospinor $\hat{\phi}$. Let us remember that in the electromagnetic case the current

$$\hat{j}^\mu = iq[\hat{\phi}^\dagger(\partial^\mu\hat{\phi}) - (\partial^\mu\hat{\phi})^\dagger\hat{\phi}] \tag{36.51}$$

was made gauge invariant by the replacement $\partial^\mu \to \hat{D}^\mu = \partial^\mu + iq\hat{A}^\mu$. Corresponding to $U(1)$ and $SU(2)_L$, the weak hypercharge currents

$$\hat{j}^\mu_y(\hat{\phi}) = ig'y[\hat{\phi}^\dagger(\partial^\mu\hat{\phi}) - (\partial^\mu\hat{\phi})^\dagger\hat{\phi}], \tag{36.52}$$

and

$$\hat{j}^\mu_i(\hat{\phi}) = i\frac{g}{2}[\hat{\phi}^\dagger\tau_i(\partial^\mu\hat{\phi}) - (\partial^\mu\hat{\phi})^\dagger\tau_i\hat{\phi}], \tag{36.53}$$

are made gauge invariant by the replacement

$$\partial^\mu \to \hat{D}^\mu = \partial^\mu + i\frac{g}{2}\boldsymbol{\tau}\cdot\hat{\mathbf{W}}^\mu + ig'y\hat{B}^\mu. \tag{36.54}$$

Note that Eq. (36.54) is slightly different from Eq. (36.49) since we have left unspecified the hypercharge, y, of the Higgs field. In Section 36.5 we shall argue that the Higgs field is a complex scalar doublet ($t = \frac{1}{2}$), and fix the value of its hypercharge.

In the present elementary description we shall leave out all contributions from leptons to the weak isospin current. From the analysis in Section 34.2 we know that the W's themselves must contribute a self-interaction term $[\hat{j}_i^\mu(W)]$ to the current.

By insertion of the replacement given in Eq. (36.54) into Eq. (36.52), the gauge invariant Higgs hypercharge current is obtained. In explict form the result reads as follows:

$$\hat{j}_y^\mu(\hat{\phi}) = ig'y[\hat{\phi}^\dagger(\partial^\mu\hat{\phi}) - (\partial^\mu\hat{\phi})^\dagger\hat{\phi}] - gg'y(\hat{\phi}^\dagger\boldsymbol{\tau}\hat{\phi})\cdot\hat{\mathbf{W}}^\mu - 2(g')^2y^2\hat{\phi}^\dagger\hat{B}^\mu\hat{\phi}, \tag{36.55}$$

remembering that $\hat{\mathbf{W}}^\mu$, $\hat{B}^\mu$ and $\boldsymbol{\tau}$ are hermitian quantities.

The current appropriate to the weak Higgs isospinor is obtianed inserting Eq. (36.54) in Eq. (36.53). Hence, one obtains

$$\hat{j}_i^\mu(\hat{\phi}) = i\frac{g}{2}[\hat{\phi}^\dagger\tau_i(\partial^\mu\hat{\phi}) - (\partial^\mu\hat{\phi}^\dagger)\tau_i\hat{\phi}] - \frac{g^2}{2}\hat{\phi}^\dagger\hat{\phi}\hat{W}_i^\mu - gg'y(\hat{\phi}^\dagger\tau_i\hat{\phi})\hat{B}^\mu. \tag{36.56}$$

In order to obtain the term proportional to g^2, we have made use of the anticommutator relation specific to the Pauli 2x2 matrices, viz.,

$$\{\tau_i, \tau_j\} = 2\delta_{ij}, \tag{36.57}$$

giving

$$i\frac{g}{2}\left[\hat{\phi}^\dagger\tau_i\left(\frac{ig}{2}\boldsymbol{\tau}\cdot\hat{\mathbf{W}}^\mu\right)\hat{\phi} - \left(\frac{ig}{2}\boldsymbol{\tau}\cdot\hat{\mathbf{W}}^\mu\hat{\phi}\right)^\dagger\hat{\phi}\right] = -\frac{g^2}{4}\hat{\phi}^\dagger(\tau_i\boldsymbol{\tau} + \boldsymbol{\tau}\tau_i)\cdot\hat{\mathbf{W}}^\mu\hat{\phi}$$

$$= -\frac{g^2}{4}\hat{\phi}^\dagger(2\tau_i^2)\hat{W}_i^\mu\hat{\phi} = -\frac{g^2}{2}\hat{\phi}^\dagger\hat{\phi}\hat{W}_i^\mu. \tag{36.58}$$

36.5 ELECTRO-WEAK INTERACTION: VACUUM SCREENING CURRENTS AND MASSES

36.5.1 Higgs doublet. Higgs vacuum in U gauge

In Chapter 35, we explained how the complex Higgs scalar field in the $U(1)$ case could give a gauge boson a mass. We illustrated the Higgs mechanism for the photon, although we do not want a "real" photon to have a non-zero mass in free space. In order to generalize the study to the $SU(2)_L \otimes U(1)$ case we introduce a complex scalar doublet following Weinberg and Salam [227, 228, 257, 259, 260]. Thus our Higgs field may be written in the form

$$\hat{\phi} = \frac{1}{\sqrt{2}}\begin{pmatrix}\hat{\phi}_1(x) + i\hat{\phi}_2(x) \\ \hat{\phi}_3(x) + i\hat{\phi}_4(x)\end{pmatrix}, \tag{36.59}$$

with $t = 1/2$ and $t_3 = +1/2$ and $-1/2$. The hypercharge assignment for the Higgs will be determined below. In Eq. (36.59) four real scalar fields, $\hat{\phi}_j$ $(j = 1 - 4)$ appear.

Let us consider the doublet form

$$\hat{\phi}(x) = \exp\left[\frac{i}{2}\boldsymbol{\tau}\cdot\hat{\boldsymbol{\alpha}}(x)\right]\begin{pmatrix} 0 \\ \hat{P}(x) \end{pmatrix} \tag{36.60}$$

in which three *real* phase fields $\hat{\boldsymbol{\alpha}}(x) = (\hat{\alpha}_1(x), \hat{\alpha}_2(x), \hat{\alpha}_3(x))$ appear, together with the *real* field $\hat{P}(x)$. Note that $\hat{\boldsymbol{\alpha}}$ and $\hat{P}$ in general depend on x, the space-time point. The quantity $\hat{\mathbf{U}}(x) = \exp\left[\frac{i}{2}\boldsymbol{\tau}\cdot\hat{\boldsymbol{\alpha}}(x)\right]$ is a unitary 2x2 matrix operator. The arbitrary isospinor in Eq. (36.59) thus can be regarded as arising from a certain isospin rotation, parametrized by $\hat{\boldsymbol{\alpha}}(x)$, and performed on an isospinor $(0, \hat{P}(x))$ with only a $t_3 = -1/2$ component. The dependence of $\hat{\boldsymbol{\alpha}}$ and x corresponds to carrying out *independent* isospin rotations at each space-time point. In a sense, the form of Eq. (36.60) represents a generalization of the $U(1)$ form $\hat{\phi}(x) = \left|\hat{\phi}(x)\right|\exp[i\alpha(x)]$.

In the $U(1)$ case the vacuum was given by $\hat{\phi}(x) = (f/\sqrt{2})\exp[i\hat{\alpha}(x)]$, Eq. (35.43). In the $SU(2)_L \otimes U(1)$ case the vacuum value of $\hat{\phi}(x)$ is given by

$$\hat{\phi}(x) = \begin{pmatrix} 0 \\ f/\sqrt{2} \end{pmatrix}\exp\left[\frac{i}{2}\boldsymbol{\tau}\cdot\hat{\boldsymbol{\alpha}}(x)\right], \tag{36.61}$$

and we may simplify the following calculations by the choice

$$\hat{\phi}(x) = \begin{pmatrix} 0 \\ f/\sqrt{2} \end{pmatrix}. \tag{36.62}$$

The gauge in which only a $t_3 = -1/2$ component is present is called a unitary gauge or just the U gauge. This field operator choice gives the mean value

$$\langle\tilde{0}|\,\hat{\phi}\,|\tilde{0}\rangle = \begin{pmatrix} 0 \\ f/\sqrt{2} \end{pmatrix} \tag{36.63}$$

in the associated Higgs ground state, $|\tilde{0}\rangle$.

36.5.2 Vacuum screening currents. Mass matrix

In the vacuum of particle physics we do not allow charged fields a non-zero value. For the Higgs-field choice $\hat{\phi}(x) = (0, \hat{P}(x))$, which has $t_3[= I_3^W] = -1/2$, Eq. (36.45) gives a hypercharge $Y = -t_3 = 1/2$, since $Q = 0$ for the Higgs particle. Thus, the designation $y = 1/2$ is used in the weak currents $j_Y^\mu(\hat{\phi})$ and $j_i^\mu(\hat{\phi})$, given in Eqs. (36.55) and (36.56).

To obtain the vacuum screening currents Eq. (36.62) is inserted into Eqs. (36.55) and (36.56) using the designation $y = Y = 1/2$. Since $\hat{\phi}^\dagger\hat{\phi} = f^2/2$, and

$$\hat{\phi}^\dagger\tau_i\hat{\phi} = -\frac{f^2}{2}\delta_{i3}, \tag{36.64}$$

one gets

$$\hat{j}_Y^{\mu}(\text{VAC}) = \frac{f^2}{4}[gg'\hat{W}_3^{\mu} - (g')^2\hat{B}^{\mu}] \tag{36.65}$$

and

$$\hat{j}_i^{\mu}(\text{VAC}) = \frac{f^2}{4}[-g^2\hat{W}_i^{\mu} + gg'\delta_{i3}\hat{B}^{\mu}]. \tag{36.66}$$

With the "mass" definitions $M = gf/2$ and $M' = g'f/2$, Eqs. (36.65) and (36.66) become

$$\hat{j}_Y^{\mu}(\text{VAC}) = MM'\hat{W}_3^{\mu} - (M')^2\hat{B}^{\mu}, \tag{36.67}$$

$$\hat{j}_i^{\mu}(\text{VAC}) = -M^2\hat{W}_i^{\mu} + MM'\delta_{i3}\hat{B}^{\mu}. \tag{36.68}$$

For $i = 1, 2$, the weak isospin vacuum current has the form

$$\hat{j}_i^{\mu}(\text{VAC}) = -M_W^2\hat{W}_i^{\mu}, \ i = 1, 2 \tag{36.69}$$

with the mass identification

$$M_W(= M) = \frac{gf}{2}. \tag{36.70}$$

This mass thus is that of the $W^{\pm}$-bosons $[\hat{W}^{\pm} = 2^{-1/2}(\hat{W}_1^{\mu} \pm i\hat{W}_2^{\mu})]$.

Due to the presence of the Higgs field, the associated operator equation of motion takes the form

$$(\Box + M_W^2)\hat{W}_i^{\mu} - \partial^{\mu}\partial_{\nu}\hat{W}_i^{\nu} = \hat{j}_i^{\mu}(\hat{W}), \ i = 1, 2 \tag{36.71}$$

where we have reintroduced the self-interaction current $\hat{j}_i^{\mu}(W)$. The equation of motions for $\hat{W}_3^{\mu}$ and $\hat{B}^{\mu}$ are

$$\Box\hat{W}_3^{\mu} - \partial^{\mu}\partial_{\nu}\hat{W}_3^{\nu} = -M^2\hat{W}_3^{\mu} + MM'\hat{B}^{\mu} + \hat{j}_3^{\mu}(\hat{W}), \tag{36.72}$$

and

$$\Box\hat{B}^{\mu} - \partial^{\mu}\partial_{\nu}\hat{B}^{\nu} = -(M')^2\hat{B}^{\mu} + MM'\hat{W}_3^{\mu}. \tag{36.73}$$

Note that the equation for $\hat{B}^{\mu}$ (in vacuum) has no self-field coupling because $\hat{B}^{\mu}$ is an Abelian gauge field. The equations of motions for $\hat{W}_3^{\mu}$ and $\hat{B}^{\mu}$ are *coupled* through the squared mass matrix

$$\mathcal{M}^2 \equiv \begin{pmatrix} -M^2 & MM' \\ MM' & -(M')^2 \end{pmatrix}. \tag{36.74}$$

This means the fields $\hat{W}_3^{\mu}$ and $\hat{B}^{\mu}$ do not have a definite mass, and thus these are *not physical (measurable) fields*.

36.5.3 Massive neutral vector boson (Z^0) and massless photon

The physical fields are certain linear combinations of $\hat{W}_3^\mu$ and $\hat{B}^\mu$, and to obtain these one diagonalizes the $\mathcal{M}^2$-matrix. This gives the mass eigenvalues

$$M_0 = 0 \tag{36.75}$$

$$M_Z = \frac{f}{2}[g^2 + (g')^2]^{1/2}. \tag{36.76}$$

The associaated normalized eigenvectors are

$$\hat{A}^\mu = [g^2 + (g')^2]^{-1/2}(g'\hat{W}_3^\mu + g\hat{B}^\mu), \tag{36.77}$$

and

$$\hat{Z}^\mu = [g^2 + (g')^2]^{-1/2}(g\hat{W}_3^\mu - g'\hat{B}^\mu), \tag{36.78}$$

for M^0 and M_Z, respectively. The equation of motions for $\hat{A}^\mu$ and $\hat{Z}^\mu$ are uncoupled and given by

$$(\delta_\nu^\mu\Box - \partial^\mu\partial_\nu)\,\hat{A}^\nu = \frac{g'}{[g^2 + (g')^2]^{1/2}}\hat{j}_3^\mu(\hat{W}), \tag{36.79}$$

and

$$\left[\delta_\nu^\mu(\Box + M_Z^2) - \partial^\mu\partial_\nu\right]\hat{Z}^\nu = \frac{g}{[g^2 + (g')^2]^{1/2}}\hat{j}_3^\mu(\hat{W}), \tag{36.80}$$

The massless ($M_0 = 0$) particle is the photon coupled via the self-interaction term, $\hat{j}_3^\mu(\hat{W})$, to the charged W's. We elaborate on the identification of $\hat{A}^\mu$ as the photon field below. The neutral vector boson of mass M_Z [Eq. (36.76)] is usually denoted by Z^0.

36.6 WEAK MIXING ANGLE. CONNECTION BETWEEN WEAK AND ELECTROMAGNETIC COUPLING STRENGTHS

It is customary to introduce the weak mixing angle (also called the Glashow-Weinberg angle) θ_W, defined in terms of the fundamental coupling constants g and g', as follows:

$$\tan\theta_W = \frac{g'}{g}, \tag{36.81}$$

$$\cos\theta_W = \frac{g}{[g^2 + (g')^2]^{1/2}}, \tag{36.82}$$

so that

$$\sin\theta_W = \frac{g'}{[g^2 + (g')^2]^{1/2}}. \tag{36.83}$$

The mixing angle specifies the mixture of the $\hat{Z}_\mu$ and $\hat{A}_\mu$ fields in $\hat{W}_{3\mu}$ and $\hat{B}_\mu$; see Eqs. (36.77) and (36.78). Thus, from

$$\hat{A}_\mu = \hat{W}_{3\mu} \sin\theta_W + \hat{B}_\mu \cos\theta_W, \tag{36.84}$$

$$\hat{Z}_\mu = \hat{W}_{3\mu} \cos\theta_W - \hat{B}_\mu \sin\theta_W, \tag{36.85}$$

one obtains the inverse relations

$$\hat{W}_{3\mu} = \hat{A}_\mu \sin\theta_W + \hat{Z}_\mu \cos\theta_W, \tag{36.86}$$

$$\hat{B}_\mu = \hat{A}_\mu \cos\theta_W - \hat{Z}_\mu \sin\theta_W. \tag{36.87}$$

We have realized in Subsection 36.5.3 that the $\hat{A}_\mu$ field is massless. We now demand that the gauge field $\hat{A}_\mu$ *is* the electromagnetic field, and hence coupled to electric charges in the usual way of QED, i.e. through the usual interaction term $-\hat{s}^\mu \hat{A}_\mu$. As we shall realize now, this demand leads to a connection between the coupling constants g (g') and e. In order to determine this connection, it is sufficient to consider the part ($\hat{\mathcal{L}}_I(3)$) of interaction Lagrangian which involves $\hat{W}_{3\mu}$ and $\hat{B}_\mu$, viz.,

$$\hat{\mathcal{L}}_I(3) = -g\hat{J}_3^\mu \hat{W}_{3\mu} - g'\hat{J}_Y^\mu \hat{B}_\mu. \tag{36.88}$$

The electric current is brought in play using Eq. (36.42) to eliminate the hypercharge current in favor of $\hat{s}^\mu$ and $\hat{J}_3^\mu$. Hence,

$$\hat{\mathcal{L}}_I(3) = -\hat{J}_3^\mu (g\hat{W}_{3\mu} - g'\hat{B}_\mu) - \frac{g'}{e}\hat{s}^\mu \hat{B}_\mu. \tag{36.89}$$

Next we introduce the physical fields $\hat{A}_\mu$ and $\hat{Z}_\mu$ [using Eqs. (36.86) and (36.87)] in the interaction Lagrangian $\hat{\mathcal{L}}_I(3) \equiv \hat{\mathcal{L}}_I(x;3)$, which thus may be written in the form

$$\hat{\mathcal{L}}_I(x;3) = \hat{J}_3^\mu(x)\hat{A}_\mu(x)(g'\cos\theta_W - g\sin\theta_W) - \hat{s}^\mu(x)\hat{A}_\mu(x)\frac{g'}{e}\cos\theta_W$$
$$+ \left[\frac{g'}{e}\sin\theta_W \hat{s}^\mu(x) - (g\cos\theta_W + g'\sin\theta_W)\hat{J}_3^\mu(x) \right] \hat{Z}_\mu(x). \tag{36.90}$$

The term proportional to $\hat{J}_3^\mu \hat{A}_\mu$ vanishes since Eqs. (36.82) and (36.83) give $g'\cos\theta_W = g\sin\theta_W$. From QED we know that the coefficient to $-\hat{s}^\mu \hat{A}_\mu$ must be $+1$. Hence, we require

$$g'\cos\theta_W = g\sin\theta_W = e. \tag{36.91}$$

From this relation between the coupling coefficients g, g' and e, we obtain the following result for the $\hat{\mathcal{L}}_I(x;3)$ part of the interaction Lagrangian:

$$\hat{\mathcal{L}}_I(x;3) = -\hat{s}^\mu(x)\hat{A}_\mu(x) - \frac{g}{\cos\theta_W}\left[\hat{J}_3^\mu(x) - \frac{1}{e}\sin^2\theta_W \hat{s}^\mu(x) \right]\hat{Z}_\mu(x). \tag{36.92}$$

36.7 TOTAL INTERACTION LAGRANGIAN DENSITY

In the description in 36.6 the contributions to the interaction Lagrangian from the vector boson coupling to the $\hat{J}_1^\mu$ and $\hat{J}_2^\mu$ parts of the lepton current were omitted. Thus, the total interaction Lagrangian, $\hat{\mathcal{L}}_I(x)$, is

$$\hat{\mathcal{L}}_I(x) = -g\hat{J}_i^\mu(x)\hat{W}_{i\mu}(x) - g'J_Y^\mu(x)\hat{B}_\mu(x), \tag{36.93}$$

with summation over $i = 1 - 3$, as before.

Let us now introduce (see Subsection 36.3.2) the leptonic currents

$$\hat{J}^\mu(x) = 2[\hat{J}_1^\mu(x) - i\hat{J}_2^\mu(x)], \tag{36.94}$$

$$(\hat{J}^\mu(x))^\dagger = 2[\hat{J}_1^\mu(x) + i\hat{J}_2^\mu(x)]. \tag{36.95}$$

These leptonic currents appear in the intermediate vector boson (IVB) theory [186], a forerunner of the modern electro-weak theory. We also introduce the non-Hermitian gauge field and its adjoint

$$\hat{W}_\mu(x) = \frac{1}{\sqrt{2}}[\hat{W}_{1\mu} - i\hat{W}_{2\mu}], \tag{36.96}$$

$$\hat{W}_\mu^\dagger(x) = \frac{1}{\sqrt{2}}[\hat{W}_{1\mu} + i\hat{W}_{2\mu}]. \tag{36.97}$$

In terms of $\hat{J}^\mu(x)$, $(\hat{J}^\mu(x))^\dagger$, $\hat{W}_\mu(x)$ and $\hat{W}_\mu^\dagger(x)$, the sum of the first two terms $(i = 1, 2)$ of the interaction Lagrangian becomes

$$-g\sum_{i=1}^{2} \hat{J}_i^\mu(x)\hat{W}_{i\mu}(x) = -\frac{g}{2\sqrt{2}}\left[\hat{J}^\mu(x)\hat{W}_\mu^\dagger(x) + (\hat{J}^\mu(x))^\dagger\hat{W}_\mu(x)\right]. \tag{36.98}$$

By now we have obtained the final expression for the interaction Lagrangian density, namely,

$$\hat{\mathcal{L}}_I(x) = -\hat{s}^\mu(x)\hat{A}_\mu(x) - \frac{g}{2\sqrt{2}}\left[\hat{J}^\mu(x)\hat{W}_\mu^\dagger(x) + (\hat{J}^\mu(x))^\dagger\hat{W}_\mu(x)\right]$$

$$-\frac{g}{\cos\theta_W}\left[\hat{J}_3^\mu(x) - \frac{1}{e}\sin^2\theta_W\,\hat{s}^\mu(x)\right]\hat{Z}_\mu(x). \tag{36.99}$$

This $SU(2) \otimes U(1)$ gauge-invariant interaction was introduced by Glashow in 1961 [96]. The quanta of the gauge field $W(x)$ are the well-known positively charged (W^+) and negatively charged (W^-) vector bosons, briefly discussed in Subsection 34.2.3.

X

Magnetic Monopoles

CHAPTER 37

On the Dirac Monopole and Its Attached String

37.1 QUANTIZED SINGULARITIES IN THE ELECTROMAGNETIC FIELD

In 1931 Dirac introduced the hypothesis of a new particle, the magnetic monopole, in an article with the same title as the one of our Section 37.1 [71]. In 1948 Dirac published a comprehensive quantum theory of electrically charged particles and magnetic monopoles interaction via the electromagnetic field [72]. Although Dirac himself in 1948 characterized his 1931 idea of magnetic poles as "a very primitive (incomplete) theory", the original work is monumental because he came to the conclusion that if the new particle was to fit into conventional quantum mechanics it must satisfy a quantization condition [setting $4\pi\varepsilon_0 = 1$]

$$eg = \frac{n}{2}\hbar c, \tag{37.1}$$

where g is the charge of the magnetic monopole, e is the electric charge of any other particle and n is some integer. Thus the mere existence of just one magnetic monopole would require all electric charges to be quantized [in unit of $\hbar c/(2g)$].

37.1.1 Non-integrable phases for wave functions

Let us now briefly consider the line of reasoning put forward by Dirac in 1931. The core of his argument is an analysis of the phase for a wave function $\psi = \psi(\mathbf{r}, t)$, expressed in the form

$$\psi(\mathbf{r}, t) = A(\mathbf{r}, t)e^{i\gamma(\mathbf{r},t)}, \tag{37.2}$$

where A and γ are real functions of $(\mathbf{r}, t)$, denoting the amplitude and the phase of the wave function. We assume that ψ shall be normalized. The addition of an arbitrary constant C to the phase, $\gamma(\mathbf{r}, t) \rightarrow \gamma(\mathbf{r}, t) + C$ has no physical meaning, only the difference between the values of $\gamma(\mathbf{r}, t)$ at two different points is of importance. Dirac went a step further and assumed that this difference is not definite unless the two points are neighbouring. For two distant points there will then be a definite phase

DOI: 10.1201/9781003029458-37

difference only relative to some curve joining them. Thus, *the total change in phase obtained going round a closed curve need not vanish.*

Next, Dirac considers two different wave functions ψ_n and ψ_m. The overlap integral

$$I = \int \psi_m^* \psi_n d^3 r \tag{37.3}$$

is a generally complex number, which absolute square has a physical meaning. In order that the integral shall have definite modules the integrand, $\psi_m^* \psi_n$ must have a *definite phase difference between any two points*, whether neighbouring or not. Thus *the change in phase in $\psi_m^* \psi_n$ round a closed curve must vanish.* In turn, this requires that the change in ψ_n round a closed curve shall be equal and opposite to that of ψ_m^*, and hence the same as that in ψ_m. Conclusion: THE CHANGE IN PHASE ROUND ANY CLOSED CURVE MUST BE THE SAME FOR ALL WAVE FUNCTIONS. It follows from this conclusion that the change in phase must be something determined by the dynamical system (field) with which the particles interact. For a charged single particle, the non-integrability of the phase must be connected with the electromagnetic field in which the particle moves.

37.1.2 Phase with definite derivatives

Dirac now expresses ψ as a product

$$\psi(\mathbf{r}, t) = \psi_1(\mathbf{r}, t) e^{i\beta(\mathbf{r}, t)}, \tag{37.4}$$

where $\psi_1(\mathbf{r}, t)$ is a wave function with a *definite phase* at each point. The uncertainty in the phase he thus puts in the factor $\exp[i\beta(\mathbf{r}, t)]$. The definite phase difference between "neighbouring" points is mathematically expressed by the requirement that $\beta(\mathbf{r}, t)$ must have definite derivatives

$$\boldsymbol{\kappa} = \boldsymbol{\nabla}\beta, \quad \kappa_0 = \frac{\partial\beta}{\partial t}, \tag{37.5}$$

at each point $(\mathbf{r}, t)$. The change in phase a round a closed curve in space is, according to Stokes' theorem, given by

$$\oint_C \boldsymbol{\kappa} \cdot d\mathbf{s} = \int_S (\boldsymbol{\nabla} \times \boldsymbol{\kappa}) \cdot d\mathbf{S}, \tag{37.6}$$

where the integration of $\boldsymbol{\nabla} \times \boldsymbol{\kappa}$ is over a (two-dimensional) surface (S) whose boundary is the closed curve C.

From Eqs. (37.5) one obtains

$$\frac{\hbar}{i}\boldsymbol{\nabla}\psi = e^{i\beta}\left(\frac{\hbar}{i}\boldsymbol{\nabla} + \hbar\boldsymbol{\kappa}\right)\psi_1, \tag{37.7}$$

and

$$i\hbar\frac{\partial\psi}{\partial t} = e^{i\beta}\left(i\hbar\frac{\partial}{\partial t} - \hbar\kappa_0\right)\psi_1. \tag{37.8}$$

These equations show that if ψ satisfies any wave equation, involving the momentum and energy operators $\hat{\mathbf{p}} = -i\hbar\boldsymbol{\nabla}$ and $\hat{E} = i\hbar\partial/\partial t$, ψ_1 will satisfy the corresponding wave equation with the replacements $\hat{\mathbf{p}} \to \hat{\mathbf{p}} + \hbar\boldsymbol{\kappa}$ and $\hat{E} = \hat{E} - \hbar\kappa_0$.

Let us assume that ψ satisfies the wave equation for a free particle, then ψ_1 will satisfy the usual wave equation for a particle with charge e in an electromagnetic field whose four-potential is

$$\mathbf{A} = -\frac{\hbar}{e}\boldsymbol{\kappa}, \quad A_0 = \frac{\hbar}{e}\kappa_0. \tag{37.9}$$

This result follows from the Minimal Coupling Substitution $\{\hat{p}_\mu\} \Rightarrow \{\hat{p}_\mu - eA_\mu\}$ to be obeyed in the replacement $\psi \to \psi_1$.

Dirac points out that the connection between non-integrability of the phase and the electromagnetic field is not new, being essentially just Weyl's Principle of Gauge Invariance [267]. A more general kind of non-integrability was also contained in the works of Iwanenko and Fock [86, 117]

37.1.3 Nodal singularities

The amazing achievement of Dirac springs from a seemingly simple fact, namely, that the phase is always undetermined to the extent of an *arbitrary integral multiple of* 2π. At first sight, one would say that such a factor cannot have any observable physical consequences. However, Dirac pointed out, careful as he was, that this 2π-issue requires a reconsideration of the connection between the κ's and the four potential. Dirac's analysis leads to the magnetic magnetic monopole concept and the quantization condition given in Eq. (37.1).

We have realized in Section 37.1 that the change of phase a round a closed curve is the same for all wave functions. However, now we realize that this change in phase may be different for different wave functions by arbitrary multiples of 2π. This fact shows that it might not be possible to uphold the electromagnetic field interpretation related to the accepted form of the Maxwell equations.

To examine this question, Dirac considers a closed loop so small that the total phase shift is much less than 2π for different wave functions. In this case the previous analysis holds *provided the wave function does not vanish*. For a complex wave function to vanish two conditions have to be satisfied. For a wave function in three spatial dimensions, this in general implies that the points in which this happens will lie along a line, called a Dirac *nodal line*. For a wave function having a nodal line passing through our small closed curve, all we shall be able to say is that the change in phase compared to a loop with no nodal lines will be close to $2\pi n$, where n is some integer, positive or negative.

A large closed curve now can be treated by dividing it up into a network of small closed curves lying in a surface whose boundary is the large closed curve. Summation over all nodal lines, n, that pass through the given surface (S) will therefore result in an extra phase shift $2\pi \sum n$. The total change in phase Φ, a round the large closed curve will therefore be

$$\Phi = 2\pi \sum n - \frac{e}{\hbar} \int_S (\boldsymbol{\nabla} \times \mathbf{A}) \cdot d\mathbf{S}, \tag{37.10}$$

where $2\pi \sum n$ is the part which may be different for different wave functions.

As we know, the sign of the integer n will be associated with the direction encircling the nodal line. The expression in Eq. (37.10) can be extended to a closed surface $S = S_1 + S_2$, where S_1 and S_2 are two open surfaces which are bounded by the large closed curve. Now, we can say that $\sum n$ will vanish unless some nodal lines have end points inside S. Since this result applies to any closed surface, Dirac concluded *the endpoints of nodal lines must be the same for all wave functions, and that these endpoints are points of singularity in the electromagnetic field.* A nodal line ending at a singular point nowadays is called a DIRAC STRING. A Dirac string stretches off to infinity. IS THERE A COSMOLOGICAL PROBLEM HERE WHICH REQUIRES ATTENTION?

37.1.4 Magnetic monopoles

Let us consider $(\hbar/e)\Phi$ in Eq. (37.10) as a flux of a certain unknown magnetic field $\mathcal{B}$ though a closed surface S. Since

$$\frac{\hbar}{e}\Phi = \int_S \mathcal{B} \cdot d\mathbf{S} = \int_V \nabla \cdot \mathcal{B} d\mathbf{S} \neq 0, \tag{37.11}$$

where V is the volume enclosed by S. Since $\nabla \cdot \mathbf{B} = 0$ in the classical Maxwell equations, Eq. (37.11) can only be satisfied if one or several magnetic monopoles are present in V. Let us assume that we have only a single magnetic monopole of magnetic charge g in V. The magnetic field from the monopole located at $\mathbf{r} = \mathbf{0}$ is given by

$$\mathbf{B} = \frac{g}{4\pi\varepsilon_0 c}\frac{\mathbf{r}}{r^2}, \tag{37.12}$$

and the total magnetic flux originating in the $2\pi n$ closed-loop phase shift from the monopole therefore satisfies the relation [MKSA units]

$$\frac{g}{\varepsilon_0 c} = 2\pi n\frac{\hbar}{e}. \tag{37.13}$$

Setting $4\pi\varepsilon_0 = 1$, one obtains the Dirac quantization condition postulated in Eq. (37.1).

37.1.5 String not penetrated by a magnetic pole

Dirac's discussion depends on the introduction of a singular vector potential $\mathbf{A}$ to represent the field of a fixed monopole. Thus, $\mathbf{A}$ corresponds not to an isolated magnetic monopole but rather to a magnetic flux line (Dirac string) extending from, say, the origo of our coordinate system to infinity. Essentially, the string may be any curved line, and in Section 38.4 we analyze the situation for a magnetic dipole flux line extending from zero (the monopole position) to infinity along the negative z-axis.

Let us assume that the magnetic flux line is not penetrated by a magnetic monopole charge [we may let the monopole recede towards plus infinity]. A linear magnetic dipole line extending from $-\infty$ to ∞ is just an Aharanov-Bohm flux line; see Sections 11.2 and 11.3.

37.2   DOUBLE-POTENTIAL FORMALISM

In his quantum electrodynamic theory published in 1948 [72], Dirac sets up a general electromagnetic theory for magnetic monopoles with their attached strings. To get a theory which can be transferred to quantum mechanics, and then to quantum electrodynamics, electromagnetic potentials are required. The usual way of introducing them consists of putting the electromagnetic field tensor elements in the form $F_{\mu\nu}(x) = \partial_\mu A_\nu(x) - \partial_\nu A_\mu(x)$, but this is no longer possible when there are magnetic monopoles present. The field produced by a magnetic monopole can be described in this way only if A_μ is allowed to be singular along an arbitrary line [the Dirac string]. In an attempt to uphold as much as possible of the standard minimal coupling form, $\hat{\mathbf{p}} - e\hat{\mathbf{A}}$, Dirac replaced the expression for $F_{\mu\nu}(x)$, given above, by an equation of the form

$$F_{\mu\nu} = \partial_\mu A_\nu(x) - \partial_\nu A_\mu(x) + 4\pi \sum_g (G^\dagger)_{\mu\nu}, \tag{37.14}$$

where each $(G^\dagger)_{\mu\nu}$ is a field quantity which vanishes everywhere on one of the two-dimensional sheets traced out by a given string in space-time. The summation in Eq. (37.14) is over all the sheets, one of which is associated with a single magnetic monopole. The tensor $\{G_{\mu\nu}\}$ is associated with the field tensor and its dual tensor $\{(F^\dagger)_{\mu\nu}\}$ via [72]

$$(F^\dagger)_{\mu\nu} G^{\mu\nu} = F_{\mu\nu}(G^\dagger)^{\mu\nu}. \tag{37.15}$$

The singularity in A_μ is clearly an unphysical feature, and to avoid this Cabibbo and Ferrari [53] developed the double-potential formalism, which we shall describe in some detail in Section 39.4. The field tensor is now given by

$$F_{\mu\nu} = \partial_\mu A_\nu - \partial_\nu A_\mu + \varepsilon^{\mu\nu\rho\sigma} \partial_\rho B_\sigma, \tag{37.16}$$

where A_μ and B_μ are the two potentials of Cabibbo and Ferrari, and $\varepsilon^{\mu\nu\rho\sigma}$ is the completely antisymmetric Levi-Civita tensor.

37.3   MINIMAL COUPLING SUBSTITUTION RESTORED

Cabibbo and Ferrari [53] did not address the question whether a kind of minimal coupling substitution can be established in the framework of the double-potential description. Perhaps, the reason is due to the fact their approach, after having introduced the A_μ and B_μ potentials, represents a generalization of Mandelstams's treatment of quantum electrodynamics without potentials to the case in which both electric and magnetic monopoles are present.

Recently [159], it has been shown by the author of this book that a specific space-time Fourier integral transformation of $\mathbf{A}_T^e$ and $\mathbf{A}_T^m$, the transverse parts of the electric (e) and magnetic (m) vector potentials, allows one to express the transverse parts of the electric and magnetic fields in the mixed potential form [see Section 39.4]

$$c^{-1}\mathbf{E}_T(\mathbf{r}, t) = \boldsymbol{\nabla} \times [\boldsymbol{\mathcal{A}}_T^e(\mathbf{r}, t) - \mathbf{A}_T^m(\mathbf{r}, t)], \tag{37.17}$$

and

$$\mathbf{B}_T(\mathbf{r}, t) = \boldsymbol{\nabla} \times [\mathbf{A}_T^e(\mathbf{r}, t) + \boldsymbol{\mathcal{A}}_T^m(\mathbf{r}, t)], \tag{37.18}$$

where $\boldsymbol{\mathcal{A}}_T^e(\mathbf{r}, t)$ and $\boldsymbol{\mathcal{A}}_T^m(\mathbf{r}, t)$ are the nonlocally transformed potentials. As shown in Section 39.5, Minimal Coupling Principle now is upheld in the forms

$$\hat{\mathbf{p}}^e \to \hat{\mathbf{p}}^e - e[\mathbf{A}_T^e + \boldsymbol{\mathcal{A}}_T^m], \tag{37.19}$$

and

$$\hat{\mathbf{p}}^m \to \hat{\mathbf{p}}^m - g[\mathbf{A}_T^m - \boldsymbol{\mathcal{A}}_T^e], \tag{37.20}$$

finally.

37.4 FLUX QUANTIZATION IN A SUPERCONDUCTING RING

Let us suppose that Cooper-paired electrons, with pair charge $q = 2e$, have a pair concentration $n = \psi^* \psi =$constant, where $\psi = \psi(\mathbf{r}, t)$ is the particle probability density for the boson gas with a large number of bosons in the same ring orbital. By writing the wave function (and its complex conjugate) in the form

$$\psi = n^{1/2} e^{i\theta(\mathbf{r})}, \quad \psi^* = n^{1/2} e^{-i\theta(\mathbf{r})}, \tag{37.21}$$

the reader may show that the electric current density is given by [m is the electron mass]

$$\mathbf{j} = \frac{nq}{m}(\hbar \boldsymbol{\nabla}\theta(\mathbf{r}) - q\mathbf{A}). \tag{37.22}$$

By taking the curl on both sides of Eq. (37.22) one obtains the London equation [179–181]

$$\boldsymbol{\nabla} \times \mathbf{j} = -\frac{nq^2}{m}\mathbf{B}. \tag{37.23}$$

Roughly speaking the Meissner effect [187] tells us that $\mathbf{B}$ and $\mathbf{j}$ are zero in the interior of the ring. Thus,

$$\hbar\boldsymbol{\nabla}\theta = q\mathbf{A}. \tag{37.24}$$

Going once around inside the ring, one obtains a phase change

$$\oint_C \boldsymbol{\nabla}\theta \cdot d\mathbf{l} = \theta_2 - \theta_1 = 2\pi n, \tag{37.25}$$

where n is an integer. To obtain the last expression in Eq. (37.25) we have used that ψ must be single-valued. By Stokes' theorem

$$\oint_C \mathbf{A} \cdot d\mathbf{l} \left[= \frac{\hbar}{q}\int_C \boldsymbol{\nabla}\theta \cdot d\mathbf{l} = \frac{\hbar}{q}2\pi n \right] = \int_S \mathbf{B} \cdot d\mathbf{S} \equiv \Phi, \tag{37.26}$$

where Φ is the magnetic flux through an open surface bounded by C. The flux is composed of a flux Φ_{ext} from external sources and the flux Φ_{surface} which flows in the inner ring surface: $\Phi = \Phi_{\text{ext}} + \Phi_{\text{surface}}$. The magnetic flux thus is quantized, i.e.,

$$\Phi = 2\pi n \frac{\hbar}{q}. \tag{37.27}$$

Suppose now that a magnetic monopole of charge g is situated in the center of the ring. This gives a magnetic flux through the circular ring $g/(cr^2)2\pi r^2 = 2\pi g/c$ $[4\pi\varepsilon_0 = 1]$. This flux must equal the one in Eq. (37.27). Hence we obtain, remembering that $q = 2e$,

$$eg = \frac{n}{2}\hbar c. \tag{37.28}$$

This is the famous Dirac quantization result [Eq. (37.1)], here for a Boson wave function with a concentration of Cooper-paired electrons given by $n = \psi\psi^* =$const.

CHAPTER 38

Magnetic Monopoles: Non-Retarded Electrodynamics

38.1 SYMMETRIZED ELECTRODYNAMICS

The introduction of the concept of *duality* in electrodynamics is inspired by the symmetry between the electric and magnetic (times c) fields of the source-free Maxwell equations, viz.,

$$\boldsymbol{\nabla} \cdot \mathbf{E} = \boldsymbol{\nabla} \cdot (c\mathbf{B}) = 0, \tag{38.1}$$

$$\boldsymbol{\nabla} \times \mathbf{E} = -\frac{1}{c}\frac{\partial}{\partial t}(c\mathbf{B}), \quad \boldsymbol{\nabla} \times (c\mathbf{B}) = \frac{1}{c}\frac{\partial}{\partial t}\mathbf{E}. \tag{38.2}$$

Thus, this set is form-invariant under the transformation $\mathbf{E} \to c\mathbf{B}$ and $c\mathbf{B} \to -\mathbf{E}$, obviously.

The symmetry can be maintained in the presence of four-currents provided one is willing to introduce both electric and magnetic monopoles. The Maxwell-Lorentz equations hence are extended to the following symmetrized form:

$$\boldsymbol{\nabla} \cdot \mathbf{E} = \frac{1}{\varepsilon_0}\rho^e, \tag{38.3}$$

$$\boldsymbol{\nabla} \cdot (c\mathbf{B}) = \frac{1}{\varepsilon_0}\rho^m, \tag{38.4}$$

$$-\boldsymbol{\nabla} \times \mathbf{E} = \frac{1}{c}\frac{\partial}{\partial t}(c\mathbf{B}) + c\mu_0\mathbf{J}^m \tag{38.5}$$

$$\boldsymbol{\nabla} \times (c\mathbf{B}) = \frac{1}{c}\frac{\partial}{\partial t}\mathbf{E} + c\mu_0\mathbf{J}^e, \tag{38.6}$$

where ρ^e and ρ^m are the microscopic electric (e) and magnetic (m) charge densities, and $\mathbf{J}^e$ and $\mathbf{J}^m$ are the related current densities. The prefactors to ρ^m and $\mathbf{J}^m$ here have been chosen such that the electric and magnetic charges (and thus also the associated current densities) have the same dimensionality. As a consequence of Eqs.

DOI: 10.1201/9781003029458-38

(38.4) and (38.5), and Eqs. (38.3) and (38.6), the electric and magnetic charges satisfy separate equations of continuity, i.e.,

$$\nabla \cdot \mathbf{J}^e + \frac{1}{c}\frac{\partial}{\partial t}(c\rho^e) = 0, \tag{38.7}$$

$$\nabla \cdot \mathbf{J}^m + \frac{1}{c}\frac{\partial}{\partial t}(c\rho^m) = 0. \tag{38.8}$$

The symmetrized Maxwell-Lorentz equations are form-invariant under the following quite general duality transformation through the arbitrary pseudorotation angle Θ:

$$\mathbf{E}_{\mathrm{NEW}} = \mathbf{E}\cos\Theta + c\mathbf{B}\sin\Theta, \tag{38.9}$$

$$c\mathbf{B}_{\mathrm{NEW}} = c\mathbf{B}\cos\Theta - \mathbf{E}\sin\Theta, \tag{38.10}$$

$$\rho^e_{\mathrm{NEW}} = \rho^e\cos\Theta + \rho^m\sin\Theta, \tag{38.11}$$

$$\rho^m_{\mathrm{NEW}} = \rho^m\cos\Theta - \rho^e\sin\Theta, \tag{38.12}$$

$$\mathbf{J}^e_{\mathrm{NEW}} = \mathbf{J}^e\cos\Theta + \mathbf{J}^m\sin\Theta, \tag{38.13}$$

$$\mathbf{J}^m_{\mathrm{NEW}} = \mathbf{J}^m\cos\Theta - \mathbf{J}^e\sin\Theta. \tag{38.14}$$

I urge the reader to show that the NEW set given by Eqs. (38.9)–(38.14) satisfies the same form of Maxwell-Lorentz equations as the old set [Eqs. (38.3)–(38.6)].

Although no magnetic counterpart to electric charge is known experimentally, it is certainly from a theoretical point of view inspiring to retain the *qualitative* idea of reciprocity between electric charge and magnetic charge, as I shall advocate in what follows.

It appears from Eqs. (38.11) and (38.12) that if one *believes* in a fundamental status of symmetrized electrodynamics it is a matter of convention to speak of a particle possessing an electric charge, but not a magnetic charge (or *vice versa*). Thus, the choice $\Theta = \pi/2$ gives $\rho^e_{\mathrm{NEW}} = \rho^m$ and $\rho^m_{\mathrm{NEW}} = -\rho^e$. If one assumes that *all* particles have the same ratio (K) of magnetic and electric charge, so that

$$\frac{\rho^m}{\rho^e} = K, \tag{38.15}$$

the choice $\tan\Theta = K$ in the charge duality transformations in Eqs (38.12) and (38.14) leads to $\rho^m_{\mathrm{NEW}} = 0$ and $\mathbf{J}^m_{\mathrm{NEW}} = \mathbf{0}$. The condition in Eq. (38.15) reduces the set of symmetrized (extended) Maxwell-Lorentz equations to the form usually accepted on current experimental ground.

38.2 MONOPOLES AND DYONS. CHARGE QUANTIZATION

38.2.1 Nonrelativistic classical considerations

Let us consider the nonrelativistic behaviour of a particle with mass m, carrying an electric charge e_1 and a magnetic charge g_1, which moves with velocity $\mathbf{v}$ in the field of a stationary body that possesses the charges e_2 and g_2. Historically, it was

Schwinger who proposed to associte both electric and magnetic charge in a single particle, referred to as a dyon [234]. The equation of motion is

$$m\frac{d\mathbf{v}}{dt} = e_1(\mathbf{E} + \mathbf{v} \times \mathbf{B}) + g_1(c\mathbf{B} - \frac{1}{c}\mathbf{v} \times \mathbf{E}), \tag{38.16}$$

where the last term on the right-hand side is the Lorentz force acting on the magnetic monopole part of the dyon. The form of this force is obtained from the usual Lorentz force [first term on the right side of Eq. (38.16)] making the replacements $e_1 \to g_1$, $\mathbf{E} \to c\mathbf{B}$ and $\mathbf{B} \to -\mathbf{E}/c$.

Now, we come to an important point seen in the perspective of photon physics (see Chapter 39), in that we assume that the dyon-dyon interaction is non-retarded. This means that

$$\mathbf{E} = \frac{e_2}{4\pi\varepsilon_0}\frac{\mathbf{r}}{r^3}, \tag{38.17}$$

$$c\mathbf{B} = \frac{g_2}{4\pi\varepsilon_0}\frac{\mathbf{r}}{r^3}, \tag{38.18}$$

where the stationary body is placed at $\mathbf{r} = \mathbf{0}$. By inserting the expressions given in Eqs. (38.17) and (38.18) into Eq. (38.16), one obtains

$$m\frac{d\mathbf{v}}{dt} = \frac{1}{4\pi\varepsilon_0}\left[(e_1e_2 + g_1g_2)\frac{\mathbf{r}}{r^3} + (e_1g_2 - e_2g_1)\frac{1}{c}\mathbf{v} \times \frac{\mathbf{r}}{r^3}\right]. \tag{38.19}$$

Written in this form, it is made explicit that the equation of motion involves two invariant quantities namely,

$$e_1e_2 + g_1g_2 = K_1, \tag{38.20}$$

$$e_1g_2 - e_2g_1 = K_2. \tag{38.21}$$

Under the invariance transformation through the arbitrary pseudorotation angle Θ, it appears from Eqs. (38.11) and (38.12), with $\rho^e = (e_1 \equiv e^1 \text{ and } e_2 \equiv e^2)$ and $\rho^m = (g_1 \equiv g^1 \text{ and } g_2 \equiv g^2)$ that

$$e^1_{\text{NEW}}e^2_{\text{NEW}} + g^1_{\text{NEW}}g^2_{\text{NEW}} = e^1e^2 + g^1g^2, \tag{38.22}$$

$$e^1_{\text{NEW}}g^2_{\text{NEW}} - e^2_{\text{NEW}}g^1_{\text{NEW}} = e^1g^2 - e^2g^1. \tag{38.23}$$

In the special case where a single stationary magnetic monopole ($e_2 = 0$, $g_2 \equiv g$) interacts with an electric monopole ($e_1 \equiv e$, $g_1 = 0$), one has $K_1 = 0$ and

$$eg = K \tag{38.24}$$

setting $K_2 = K$. The classical result given in Eq. (38.24), upon an adequate quantization of the angular momemtum, leads to the simplest form of charge quantization pointed out by Dirac long ago [71].

The equation of motion for the angular momentum is

$$\mathbf{r} \times m\frac{d\mathbf{v}}{dt} = \frac{1}{4\pi\varepsilon_0 c}(e_1g_2 - e_2g_1)\frac{d}{dt}\left(\frac{\mathbf{r}}{r}\right). \tag{38.25}$$

In order to obtain the result in Eq. (38.25), we have used that

$$\frac{1}{r^3}\mathbf{r} \times (\mathbf{v} \times \mathbf{r}) = \frac{\mathbf{v}}{r} - \frac{1}{r^3}\mathbf{r}(\mathbf{r} \cdot \mathbf{v}) = \frac{\mathbf{v}}{r} - \frac{1}{r^2}\mathbf{r}\frac{dr}{dt} = \frac{d}{dt}\left(\frac{\mathbf{r}}{r}\right) \qquad (38.26)$$

From Eq. (38.25) follows the conservation of an interesting angular momentum ($\mathbf{J}$) concept for the dyon-dyon system, viz.,

$$\mathbf{J} = \mathbf{r} \times m\mathbf{v} - \frac{1}{4\pi\varepsilon_0 c}(e_1 g_2 - e_2 g_1)\frac{\mathbf{r}}{r}. \qquad (38.27)$$

Note that the last term is nonvanishing also for $\mathbf{v} = 0$. Hence, in a certain sense this term may be considered as a kind of spin (internal angular momentum of the dyon-dyon pair).

38.2.2 Quantization of angular momentum

The component of the angular momentum along the dual-charged particle's connection line is given by

$$\frac{1}{r}\mathbf{r} \cdot \mathbf{J} = \frac{1}{4\pi\varepsilon_0 c}(e_2 g_1 - e_1 g_2). \qquad (38.28)$$

The quantization of this component of the angular momentum then gives the following charge quantization

$$\frac{1}{4\pi\varepsilon_0}\frac{e_2 g_1 - e_1 g_2}{\hbar c} = \frac{n}{2}, \quad n = 0, \pm 1, \pm 2, \dots . \qquad (38.29)$$

In the quantization we have included both integer and integer $+ \, 1/2$ values. We shall reflect on this point in Chapter 39. For the electric-magnetic dipole case we obtain the celebrated Dirac result (setting $4\pi\varepsilon_0 = 1$)

$$\frac{eg}{\hbar c} = \frac{n}{2}. \qquad (38.30)$$

The existence of just one magnetic monopole thus would "explain" the quantization of electric charge (and *vice versa*).

It was suggested by Goldhaber [99] and Schwinger [234] that it could be useful to regard the quantity on the left side of Eq. (38.29) as the radial component of a spin angular momentum vector $\mathbf{S}$, that is

$$\frac{1}{4\pi\varepsilon_0}\frac{e_2 g_1 - e_1 g_2}{\hbar c} = \frac{1}{r}\mathbf{S} \cdot \mathbf{r}. \qquad (38.31)$$

If the canonical momentum ($\mathbf{p}$) is introduced by the definition

$$m\mathbf{v} = \mathbf{p} + \frac{1}{r^2}\mathbf{S} \times \mathbf{r}, \qquad (38.32)$$

one obtains the familiar form for the composition of the total angular momentum ($\mathbf{J}$), namely

$$\mathbf{J} = \mathbf{r} \times \mathbf{p} + \mathbf{S} = \mathbf{L} + \mathbf{S}, \qquad (38.33)$$

where $\mathbf{L}$ is the orbital angular momentum. For the quantum case $\mathbf{p} \Rightarrow \hat{\mathbf{p}} = -i\hbar\boldsymbol{\nabla}$.

The conservation of the total angular momentum implies that a Hamiltonian formulation of the usual type with particle kinetic energy (T) (here the total energy, H), $H = T = (m/2)\mathbf{v}^2$. From the expression $\mathbf{r} \times \mathbf{p} = \mathbf{L}$, which relates the component of $\mathbf{p}$ which is perpendicular to $\mathbf{L}$, one gets

$$\mathbf{p} = \frac{\mathbf{r}}{r}p_r + \frac{\mathbf{L} \times \mathbf{r}}{r^2}, \tag{38.34}$$

where p_r is the radial component of the canonical momentum. By inserting this expression for $\mathbf{p}$ into Eq. (38.32) one obtains

$$m\mathbf{v} = \frac{\mathbf{r}}{r}p_r + \frac{1}{r^2}(\mathbf{L} + \mathbf{S}) \times \mathbf{r}, \tag{38.35}$$

and, then remembering that $\mathbf{L} + \mathbf{S} = \mathbf{J}$, one infers the Hamiltonian

$$H = \frac{1}{2}m\mathbf{v}^2 = \frac{1}{2m}\left[p_r^2 + \frac{1}{r^4}(\mathbf{J} \times \mathbf{r}) \cdot (\mathbf{J} \times \mathbf{r})\right] = \frac{1}{2m}\left\{p_r^2 + \frac{1}{r^2}\left[\mathbf{J}^2 - \frac{1}{r^2}(\mathbf{J} \cdot \mathbf{r})^2\right]\right\}. \tag{38.36}$$

In the quantum setting [with $\nu = n/2$]

$$\mathbf{J} \cdot \frac{\mathbf{r}}{r} = \mathbf{S}\frac{\mathbf{r}}{r} = \nu\hbar, \tag{38.37}$$

and a corresponding total angular momentum spectrum

$$\mathbf{J}^2 = j(j + 1)\hbar^2, \quad j = |\nu|, |\nu| + 1, \dots . \tag{38.38}$$

In the quantum spin formalism, $\mathbf{S}$ is taken as an angular momentum with independent degrees of freedom. The associated operator, $\mathbf{S}$, thus obeys the usual commutation relations $\hat{\mathbf{S}} \times \hat{\mathbf{S}} = i\hbar\hat{\mathbf{S}}$.

The ratio between the reduced Compton wavelength, $\lambda_C = \hbar/(m_0 c)$ and the classical electron radius, $r_0 = e^2/(4\pi\varepsilon_0 m_0 c^2)$, is given by

$$\frac{\lambda_C}{r_0} = \frac{2\varepsilon_0 ch}{e^2} \equiv \frac{1}{\alpha} \approx 137, \tag{38.39}$$

where $\alpha \approx 1/137$ is the fine structure constant. From the observed unit of electric charge via $e^2/(2\varepsilon_0 ch) = \alpha \approx 1/137$, we deduce a unit of magnetic charge [choosing $n = 1$, and reinserting the $(4\pi\varepsilon_0)^{-1}$ factor in Eq. (38.29)] from

$$\frac{g^2}{2\varepsilon_0 ch} = \frac{1}{4\alpha} \approx \frac{137}{4}. \tag{38.40}$$

Hence, the forces between magnetic charges are super strong in comparison to the forces between electrons [$10^4 - 10^5$ times as strong].

38.3   THE DIRAC MONOPOLE AND ITS STRING

A quantum theory for magnetic monopoles must relate to the vector potential, or in a covariant formulation to the four-potential. In fact, it was a quantum theory with vector potential which long ago led Dirac to the conclusion that the presence of just one magnetic monopole would lead to a quantization of electric charge in which only integral multiples of a fundamental unit could occur [72]. Dirac's idea was to seek to fit the quantum physics of magnetic monopoles into the already well-established scheme based on potentials.

Although the Dirac theory can be criticized on various grounds, personally I have never doubted that his theory got hold of an "explanation" for the mysterious charge quantization principle. Prior to Dirac's work charge quantization, despite the word quantization, was not linked in any way to quantization in quantum electrodynamics. Nowadays the observed electric charge quantization most often is ascribed to gauge symmetry enlarged to a certain non-Abelian group [see Chapter 40].

Let us consider a magnetic monopole of strength g placed at the origin of our coordinate system. The non-retarded magnetic field from the pole is radial and given by

$$\mathbf{B} = \frac{g}{4\pi\varepsilon_0 c}\frac{\mathbf{r}}{r^3} = -\frac{g}{4\pi\varepsilon_0 c}\boldsymbol{\nabla}\left(\frac{1}{r}\right) \tag{38.41}$$

at an observation point located at $\mathbf{r}$. Since $\nabla^2(r^{-1}) = -4\pi\delta(\mathbf{r})$, we have

$$\boldsymbol{\nabla}\cdot\mathbf{B} = \frac{g}{\varepsilon_0 c}\delta(\mathbf{r}). \tag{38.42}$$

The total flux through a sphere surrounding the origin is

$$\Phi = 4\pi r^2 \mathbf{B}\cdot\frac{\mathbf{r}}{r} = \frac{g}{\varepsilon_0 c}. \tag{38.43}$$

In order to relate to quantum physics, consider a free particle with electric charge e in the near field of this monopole. The wave function of this particle in the absence of the monopole in usual notation is

$$\Psi_0 = |\Psi_0|\exp\left[\frac{i}{\hbar}(\mathbf{p}\cdot\mathbf{r} - Et)\right]. \tag{38.44}$$

In the presence of the monopole field, the minimal substitution $\mathbf{p} \to \mathbf{p} - e\mathbf{A}$ replaces Ψ_0 by

$$\Psi = \Psi_0\exp\left(-\frac{ie}{\hbar}\mathbf{A}\cdot\mathbf{r}\right). \tag{38.45}$$

Hence, the wave function phase α is changed as follows:

$$\alpha \to \alpha - \frac{e}{\hbar}\mathbf{A}\cdot\mathbf{r}. \tag{38.46}$$

Described in usual spherical coordinates (r, θ, ϕ), the total phase shift around a closed path with fixed r and θ, and with ϕ ranging from 0 to 2π (around the z-axis) is given by

$$\Delta\alpha = \frac{e}{\hbar} \oint \mathbf{A} \cdot d\mathbf{l} = \frac{e}{\hbar} \int \boldsymbol{\nabla} \times \mathbf{A} \cdot d\mathbf{S} = \frac{e}{\hbar} \int \mathbf{B} \cdot d\mathbf{S} = \frac{e}{\hbar}\Phi(r, \theta). \qquad (38.47)$$

Here $\Phi(r, \theta)$ is the flux through a cap (area element $d\mathbf{S}$) limited by the closed path. Keeping r fixed, the flux through the cap changes as θ is varied. For $\theta \to 0$ the loops shrinks to a point, and the flux thus approaches

$$\Phi(r, 0) = 0. \qquad (38.48)$$

For $\theta \to \pi$, Eq. (38.43) tells us that

$$\Phi(r, \pi) = \frac{g}{\varepsilon_0 c}. \qquad (38.49)$$

However, as $\theta \to \pi$ the loop has again shrunk to a point, and the line integral of $\mathbf{A}$ becomes zero, giving (wrongly) $\Phi(r, \pi) = 0$. The calculation in Eq. (38.47) therefore must be wrong. The only step in which Eq. (38.47) can be wrong is the first (from the line to the surface integral). We can therefore conclude that the vector potential ($\mathbf{A}$) must be *singular at* $\theta = \pi$. Since this argument holds for spheres of all possible radii, it follows that $\mathbf{A}$ is singular along the entire negative z-axis. This is known as the *Dirac string*. By suitable choice of coordinates the string may be chosen to be along any direction, and clearly not necessarily straight, since for each r one can make a different choice for the orientation of the polar axis. However, the string must be continuous.

In order to fit into standard quantum theory, Dirac held the point of view (not easily defended) that the wave function must vanish on the string (here the negative z-axis). In turn, the phaseshift therefore is indeterminate, and thus there is no necessity that $\Delta\alpha \to 0$ for $\Theta \to \pi$. However, in order for Ψ to be single-valued one must have $\Delta\alpha = 2\pi n$, where n is an integer. From Eqs. (38.47) and (38.49) one then obtains

$$\frac{eg}{\hbar c} = 2\pi\varepsilon_0 n. \qquad (38.50)$$

This is the Dirac quantization condition, identical to the condition given in Eq. (38.30) [there with $4\pi\varepsilon_0 = 1$].

38.4 UNION OF MONOPOLE VECTOR POTENTIAL SECTIONS

Let us now derive an expression for the singular vector potential describing the magnetic monopole and its attached string. To begin with let us recall the standard expression for the vector potential of a magnetic dipole, viz. [157],

$$\mathbf{A}(\mathbf{r}; \omega) = \frac{\mu_0}{4\pi} \frac{e^{iq_0 R}}{R} \left(iq_0 - \frac{1}{R} \right) \frac{\mathbf{R}}{R} \times \mathbf{m}(\omega), \qquad (38.51)$$

where $\mathbf{R} = \mathbf{r} - \mathbf{r}_0$ for a dipole of moment $\mathbf{m}$ placed at the space point $\mathbf{r}_0$. For an infinitesimally long dipole of vectorial length $d\mathbf{l}$ and magnetic moment

$$d\mathbf{m} = cg d\mathbf{l}, \tag{38.52}$$

(c comes from our choice of units; see Section 38.1) the associated vector potential takes the following form in the magnetostatic limit ($q_0 = \omega/c \to 0$):

$$d\mathbf{A} = -\frac{g}{4\pi\varepsilon_0 c}\frac{1}{R^3}\mathbf{R} \times d\mathbf{l}. \tag{38.53}$$

For a fundamental monopole g is frequency independent. A Dirac magnetic monopole with it attached string L will have a *singular* vector potential derived from

$$\mathbf{A}(\mathbf{r}) = \frac{g}{4\pi\varepsilon_0 c}\int_L \frac{d\mathbf{l}' \times (\mathbf{r} - \mathbf{r}_0)}{|\mathbf{r} - \mathbf{r}_0|^3}. \tag{38.54}$$

Suppose for definiteness that the monopole is located at the origin ($\mathbf{r}_0 = \mathbf{0}$) and the string along the negative part of the z-axis (extending to minus infinity). If matter is *assumed* to be magnetically neutral it should not be necessary to refer the compensating charge $-g$ if the latter is sufficiently remote from the origin. Note that a Dirac string of finite length will carry also a charge $-g$. On integral form the vector potential hence is given by

$$\mathbf{A}(\mathbf{r}) = -\frac{g}{4\pi\varepsilon_0 c}\int_{-\infty}^{0}\frac{(\mathbf{r} - z_0\hat{\mathbf{z}}) \times \hat{\mathbf{z}}\, dz_0}{|\mathbf{r} - z_0\hat{\mathbf{z}}|^3} = \frac{g}{4\pi\varepsilon_0 c}(x\hat{\mathbf{y}} - y\hat{\mathbf{x}})\int_{-\infty}^{0}\frac{dz_0}{[x^2 + y^2 + (z - z_0)^2]^{3/2}}. \tag{38.55}$$

The integral in Eq. (38.55) can be carried out completely, using the formula

$$\int \frac{d\alpha}{[x^2 + y^2 + \alpha^2]^{3/2}} = \frac{\alpha}{x^2 + y^2}[x^2 + y^2 + \alpha^2]^{-1/2}. \tag{38.56}$$

Thus, one obtains

$$\mathbf{A}(\mathbf{r}) = \frac{g}{4\pi\varepsilon_0 c}\frac{1 - \frac{z}{r}}{x^2 + y^2}(x\hat{\mathbf{y}} - y\hat{\mathbf{x}}) \tag{38.57}$$

and finally

$$\mathbf{A}(\mathbf{r}) = \frac{g}{4\pi\varepsilon_0 c}\frac{x\hat{\mathbf{y}} - y\hat{\mathbf{x}}}{r(r + z)}. \tag{38.58}$$

The vector potential is manifest singular along the negative part of the z-axis, where $r = |z|$. The reader may check that the associated magnetic field is that of a magnetic monopole outside the negative part of the z-axis, i.e.,

$$\mathbf{B}(\mathbf{r}) = \frac{g}{4\pi\varepsilon_0 c}\frac{\mathbf{r}}{r^3}, \quad \mathbf{r} \neq z\hat{\mathbf{z}} \ (z < 0). \tag{38.59}$$

In spherical coordinates (with polar axis along z)

$$A_r = A_\theta = 0, \ A_\phi = \frac{g}{4\pi\varepsilon_0 c} \frac{1}{r} \frac{1 - \cos\theta}{\sin\theta}. \tag{38.60}$$

If the Dirac string were chosen to be along the positive part of the z axis, the reader may verify that

$$A_r = A_\theta = 0, \ A_\phi = -\frac{g}{4\pi\varepsilon_0 c} \frac{1}{r} \frac{1 + \cos\theta}{\sin\theta}. \tag{38.61}$$

We know that the only physical singularity in $\mathbf{A}$ is at the origin. So, how do we get rid of the Dirac string problem? It is obvious that if we divide the sphere surrounding the monopole into two overlapping regions R_a and R_b, viz.,

$$R_a: \ r > 0, \ 0 \le \theta \le \frac{\pi}{2} + \delta, \ o \le \phi \le 2\pi, \tag{38.62}$$

$$R_b: \ r > 0, \ \frac{\pi}{2} - \delta \le \theta \le \pi, \ o \le \phi \le 2\pi, \tag{38.63}$$

where δ is an angle such that $0 < \delta < \pi/2$, the space outside of the monopole then may be considered as the union of the overlapping R_a and R_b regions. In both regions A_r and A_θ are zero, and the nonvanishing component of the vector potential ($\mathbf{A}_\phi$) are in the two sections given by

$$A_\phi^a = \frac{g}{4\pi\varepsilon_0 c} \frac{1}{r} \frac{1 - \cos\theta}{\sin\theta}, \ (r, \theta, \phi) \in R_a, \tag{38.64}$$

$$A_\phi^b = -\frac{g}{4\pi\varepsilon_0 c} \frac{1}{r} \frac{1 + \cos\theta}{\sin\theta}, \ (r, \theta, \phi) \in R_b, \tag{38.65}$$

To work in calculations with two sectional expressions for the vector potential which both are finite (nonsingular) in their own domains, let us look at the difference between the potentials in the region of overlap. This is given by

$$A_\phi^b - A_\phi^a = -\frac{2g}{4\pi\varepsilon_0 c} \frac{1}{r \sin\theta}. \tag{38.66}$$

In order to understand this result, we introduce the quantity

$$S = \exp\left(\frac{2ige}{4\pi\varepsilon_0 c\hbar}\phi\right). \tag{38.67}$$

Remembering that $\boldsymbol{\nabla}\phi = \hat{\boldsymbol{\phi}}(r\sin\theta)^{-1}\partial/\partial\phi$, it is realized that

$$A_\phi^b - A_\phi^a = \frac{i\hbar}{e} S\hat{\boldsymbol{\phi}} \cdot \boldsymbol{\nabla}_\phi S^{-1}, \tag{38.68}$$

or equivalently

$$(A_\phi^b - A_\phi^a)\hat{\boldsymbol{\phi}} = \frac{i\hbar}{e} S\boldsymbol{\nabla}_\phi S^{-1}. \tag{38.69}$$

Since

$$\boldsymbol{\nabla}_\phi \times (S\boldsymbol{\nabla}_\phi S^{-1}) = \mathbf{0}, \tag{38.70}$$

it follows that the two sectional expressions for A_ϕ are related by a gauge transformation. This fact guarantees the physical equivalence of A_ϕ^a and A_ϕ^b. Furthermore, the requirement that the gauge transform function S be single-valued under an integer number (n) of 2π-rotations, i.e.,

$$S(\phi + 2\pi n) = S(\phi), \tag{38.71}$$

clearly leads to the Dirac quantization condition in Eq. (38.50). To check that Eqs. (38.64) and (38.65) together do represent a magnetic monopole, we calculate the magnetic flux through a sphere surrounding the origin, viz.,

$$\Phi = \int_{R_a} \boldsymbol{\nabla} \times \mathbf{A} \cdot d\mathbf{S} + \int_{R_b} \boldsymbol{\nabla} \times \mathbf{A} \cdot d\mathbf{S}. \tag{38.72}$$

For definiteness, let us assume that R_a and R_b do not overlap, but have a common boundary at the equator $(\theta = \pi/2)$. By means of Stokes' theorem, and remembering that the equator bounds R_a and R_b with opposite orientations, one obtains

$$\Phi = \oint_{\theta=\pi/2} \mathbf{A}^a \cdot d\mathbf{l}^a - \oint_{\theta=\pi/2} \mathbf{A}^b \cdot d\mathbf{l}^b = \frac{i}{e} \oint S\frac{d}{d\phi}S^{-1} d\phi = \frac{i}{e} \oint \frac{d}{d\phi}(\ln S^{-1}) d\phi = \frac{g}{\varepsilon_0 c}, \tag{38.73}$$

in agreement with the result in Eq. (38.43).

38.5 COVARIANT SYMMETRIZED ELECTRODYNAMICS

In Eqs. (38.3)–(38.6) the set of symmetrized Maxwell-Lorentz equations was expressed in three-vector notation. In order to give this set in covariant notation, we return to the covariant antisymmetric field tensor $\{F_{\mu\nu}\}$, given in Eq. (2.8). Written out in explicit form

$$\{F_{\mu\nu}\} = \frac{1}{c}\begin{pmatrix} 0 & -E_1 & -E_2 & -E_3 \\ E_1 & 0 & cB_3 & -cB_2 \\ E_2 & -cB_3 & 0 & cB_1 \\ E_3 & cB_2 & -cB_1 & 0 \end{pmatrix} \tag{38.74}$$

Its antisymmetric dual partner,

$$\{G_{\mu\nu}\} = \{F_{\mu\nu}(\mathbf{E} \to c\mathbf{B}, \mathbf{B} \to -\frac{1}{c}\mathbf{E}), \tag{38.75}$$

thus has the explicit form

$$\{G_{\nu\mu}\} = \frac{1}{c}\begin{pmatrix} 0 & -cB_1 & -cB_2 & -cB_3 \\ cB_1 & 0 & -E_3 & E_2 \\ cB_2 & E_3 & 0 & -E_1 \\ cB_3 & -E_2 & E_1 & 0 \end{pmatrix} \tag{38.76}$$

with the electric and magnetic four-current densities given by $\{J_\nu^e\} = (-c\rho^e, \mathbf{J}^e)$ and $\{J_\nu^m\} = (-c\rho^m, \mathbf{J}^m)$, and with $\{\partial^\mu\} = (-c^{-1}\partial/\partial t, \mathbf{\nabla})$, it appears that the symmetrized Maxwell-Lorentz equations are given by $[\nu = 0 - 3]$

$$\partial^\mu F_{\mu\nu} = -\mu_0 J_\nu^e, \tag{38.77}$$

$$\partial^\mu G_{\mu\nu} = -\mu_0 J_\nu^m. \tag{38.78}$$

Above the matrix tensor signature $(-1, 1, 1, 1)$ has been employed.

For studies of the electrodynamics of an electrically charged particle in the field of a (stationary) magnetic monopole on the basis of the two-sectional approach discussed in 38.4, it is important to know how the two sections of the moving charged particle scalar wave function denoted by Ψ^a and Ψ^b are related. Since we know that the vector potentials in the two sections are connected by a gauge transformation, given in covariant notation by Eq. (38.69), and remembering that $A_r^a = A_\theta^a = A_r^b = A_\theta^b = 0$, it follows that the gauge function $\chi(x)$ introduced in Eq. (2.1) in polar coordinates depends solely on the azimuth angle ϕ, i.e. $\chi(x) = \chi(\phi)$. After a moment's reflection, one guess that

$$S \equiv \exp\left(\frac{2ige}{4\pi\varepsilon_0 c\hbar}\phi\right) = \exp\left(i\frac{e}{\hbar}\chi\right), \tag{38.79}$$

so that

$$\chi(\phi) = \frac{2g}{4\pi\varepsilon_0 c}\phi. \tag{38.80}$$

Let us check if this is correct by calculating $\partial_\mu\chi(\phi)$. Obviously, only the component $\partial_\phi\chi$ is different from zero. One obtains

$$\mathbf{\nabla}\chi(\phi) = \hat{\boldsymbol{\phi}}(\hat{\boldsymbol{\phi}}\cdot\mathbf{\nabla}_\phi)\chi(\phi) = \frac{\hat{\boldsymbol{\phi}}}{r\sin\theta}\frac{d\chi(\phi)}{d\phi} = \frac{2g}{4\pi\varepsilon_0 c}\frac{\hat{\boldsymbol{\phi}}}{r\sin\theta}, \tag{38.81}$$

showing that [cf. Eq. (38.66)]

$$\mathbf{A}^a = \mathbf{A}^b + \mathbf{\nabla}\chi(\phi), \tag{38.82}$$

and

$$\Psi^a = \exp\left(i\frac{e}{\hbar}\chi\right)\Psi^b. \tag{38.83}$$

Eqs. (38.82) and (38.83) is the expected result. The Dirac equation for the electrically charged particle in the magnetic monopole field consequently divide into section parts, i.e.,

$$i\hbar\frac{\partial}{\partial t}\Psi^{a,b}(\mathbf{r}, t) = \hat{H}^{a,b}\Psi^{a,b}(\mathbf{r}, t), \tag{38.84}$$

where in minimal coupling prescription

$$\hat{H}^{a,b} = c\boldsymbol{\alpha}\cdot\left(\frac{\hbar}{i}\mathbf{\nabla} - e\mathbf{A}^{a,b}\right) + \beta mc^2, \tag{38.85}$$

the particle rest mass being m. The "section" approach has been used, e.g., by Yamagishi [283] in a reexamination of earlier studies of fermion-monopole systems.

CHAPTER 39

Photon Wave Mechanical Monopole Theory

39.1 LONGITUDINAL AND TRANSVERSE DYNAMICS IN SYMMETRIZED ELECTRODYNAMICS

For what follows it is convenient to divide the extended Maxwell-Lorentz equations into sets describing respectively the longitudinal (L) and transverse (T) dynamics. We start from the following unique (up to a physically unimportant space-independent constant) separation of the fields and particle current densities: $\mathbf{E} = \mathbf{E}_L + \mathbf{E}_T$, $\mathbf{B} = \mathbf{B}_L + \mathbf{B}_T$, $\mathbf{J}^e = \mathbf{J}^e_L + \mathbf{J}^e_T$ and $\mathbf{J}^m = \mathbf{J}^m_L + \mathbf{J}^m_T$.

39.1.1 Longitudinal parts of the field equations. Dynamical particle position varialbles

It appears from the symmetrized set of Maxwell-Lorentz equations [given in Eqs. (38.3)–(38.6)] that the curl-free L-parts satisfy the following set of field equations:

$$\boldsymbol{\nabla} \cdot \mathbf{E}_L = \frac{1}{\varepsilon_0} \rho^e, \tag{39.1}$$

$$\boldsymbol{\nabla} \cdot (c\mathbf{B}_L) = \frac{1}{\varepsilon_0} \rho^m, \tag{39.2}$$

$$\frac{\partial}{\partial t}(c\mathbf{B}_L) = -\frac{1}{\varepsilon_0} \mathbf{J}^m_L, \tag{39.3}$$

$$\frac{\partial}{\partial t}\mathbf{E}_L = -\frac{1}{\varepsilon_0} \mathbf{J}^e_L. \tag{39.4}$$

Note that there is no direct cross coupling between the electric and magnetic monopole L-dynamics. The equations of continuity for the electric and magnetic charges [Eqs. (38.7) and (38.8)] are obtained by combining, respectively, Eqs. (39.1) and (39.4), and Eqs. (39.2) and (39.3).

In the framework of classical symmetrized electrodynamics the electric and magnetic monopoles are assumed to be point particles. If one denotes the various (particle label: α) electric and magnetic monopole charges by e_α and g_α, and the position

DOI: 10.1201/9781003029458-39

vectors of these by $\mathbf{r}_\alpha^e(t)$ and $\mathbf{r}_\alpha^m(t)$, the four-current densities become

$$\{J^{e,\mu}\} = [c\rho^e(\mathbf{r},t), \mathbf{J}^e(\mathbf{r},t)] = \sum_\alpha e_\alpha \left[c, \frac{d}{dt}\mathbf{r}_\alpha^e(t)\right]\delta(\mathbf{r}-\mathbf{r}_\alpha^e(t)), \tag{39.5}$$

$$\{J^{m,\mu}\} = [c\rho^m(\mathbf{r},t), \mathbf{J}^m(\mathbf{r},t)] = \sum_\alpha g_\alpha \left[c, \frac{d}{dt}\mathbf{r}_\alpha^m(t)\right]\delta(\mathbf{r}-\mathbf{r}_\alpha^m(t)), \tag{39.6}$$

where δ is the Dirac delta function.

From Eqs. (39.1) and (39.2), combined with the $\mu = 0$ components of Eqs. (39.5) and (39.6), one may express the longitudinal fields in terms of the particle-position variables. Thus

$$\mathbf{E}_L(\mathbf{r},t) = \frac{1}{4\pi\varepsilon_0}\sum_\alpha e_\alpha \frac{\mathbf{r}-\mathbf{r}_\alpha^e(t)}{|\mathbf{r}-\mathbf{r}_\alpha^e(t)|^3}, \tag{39.7}$$

$$c\mathbf{B}_L(\mathbf{r},t) = \frac{1}{4\pi\varepsilon_0}\sum_\alpha g_\alpha \frac{\mathbf{r}-\mathbf{r}_\alpha^m(t)}{|\mathbf{r}-\mathbf{r}_\alpha^m(t)|^3}. \tag{39.8}$$

From these well-known results [Eq. (39.7) is a textbook result, and Eq. (39.8) follows from analogous calculations] it appears that $\mathbf{E}_L$ and $c\mathbf{B}_L$ are just the instantaneous electric and magnetic Coulomb fields of the charge density distributions. In a quantum mechanical context the elimination of $\mathbf{E}_L$ and $c\mathbf{B}_L$ in favour of the particle-position coordinates implies that in the (extended) Dirac (or Schrödinger) equation the particles do not "see" their own longitudinal fields.

39.1.2 Transverse parts of the field equations. Wave equations

Although the division of the vector fields into T and L parts is not relativistically invariant this decomposition most often is convenient in photon wave mechanics. The transverse parts of the symmetrized Maxwell-Lorentz equations, obtained from Eqs. (38.5) and (38.6), obviously are

$$-\boldsymbol{\nabla}\times\mathbf{E}_T = \frac{1}{c}\frac{\partial}{\partial t}(c\mathbf{B}_T) + c\mu_0\mathbf{J}_T^m, \tag{39.9}$$

$$\boldsymbol{\nabla}\times(c\mathbf{B}_T) = \frac{1}{c}\frac{\partial}{\partial t}\mathbf{E}_T + c\mu_0\mathbf{J}_T^e. \tag{39.10}$$

By a combination of the set of transverse field equations, one may obtain the following inhomogenous wave equations for $\mathbf{E}_T$ and $c\mathbf{B}_T$:

$$\Box\mathbf{E}_T = \mu_0\left[\frac{\partial}{\partial t}\mathbf{J}_T^e + c\boldsymbol{\nabla}\times\mathbf{J}_T^m\right], \tag{39.11}$$

$$\Box(c\mathbf{B}_T) = \mu_0\left[\frac{\partial}{\partial t}\mathbf{J}_T^m - c\boldsymbol{\nabla}\times\mathbf{J}_T^e\right], \tag{39.12}$$

where $\Box = \nabla^2 - c^{-2}\partial^2/\partial t^2$. Even in the presence of electric and magnetic monopoles there exists of course only *one* electromagnetic field, and each of the transverse fields, $\mathbf{E}_T$ and $c\mathbf{B}_T$, are driven by the dynamics of both types of monopoles.

39.2   PROPAGATOR APPROACH

In a scattering theory formulation, one may use the retarded outgoing scalar Huygens propagator $g(R,\tau) = \delta\left(\frac{R}{c} - \tau\right)/(4\pi R)$ to write the solutions to Eqs. (39.11) and (39.12) in the form

$$\mathbf{E}_T(\mathbf{r},t) = \mathbf{E}_T^0(\mathbf{r},t) - \mu_0 \int_{-\infty}^{\infty} g(R,\tau) \left[\frac{\partial}{\partial t'}\mathbf{J}_T^e(\mathbf{r}',t') + c\boldsymbol{\nabla}' \times \mathbf{J}^m(\mathbf{r}',t')\right] d^3r'dt',$$

(39.13)

and

$$c\mathbf{B}_T(\mathbf{r},t) = c\mathbf{B}_T^0(\mathbf{r},t) - \mu_0 \int_{-\infty}^{\infty} g(R,\tau) \left[\frac{\partial}{\partial t'}\mathbf{J}_T^m(\mathbf{r}',t') - c\boldsymbol{\nabla}' \times \mathbf{J}^e(\mathbf{r}',t')\right] d^3r'dt',$$

(39.14)

where $\mathbf{E}_T^0$ and $\mathbf{B}_T^0$ are the electric and magnetic components of the incoming field, and as ususal $\mathbf{R} = \mathbf{r} - \mathbf{r}'$ ($R = |\mathbf{R}|$) and $\tau = t - t'$. In writing down the solutions we used that $\boldsymbol{\nabla} \times \mathbf{J}_T^{e,m} = \boldsymbol{\nabla} \times \mathbf{J}^{e,m}$.

For what follows it is useful to rewrite Eqs. (39.13) and (39.14) in terms of the dyadic transverse (subscript T) electric field propagator, $\mathbf{G}_T(\mathbf{R},\tau)$, and the dyadic magnetic (subscript M) field propagator, $\mathbf{G}_M(\mathbf{R},\tau)$. We reach our goal by making use of the following two relations:

$$\int_{-\infty}^{\infty} g(R,\tau)\frac{\partial}{\partial t'}\mathbf{J}_T(\mathbf{r}',t')d^3r' = \int_{-\infty}^{\infty} \mathbf{G}_T(\mathbf{R},\tau)\cdot\frac{\partial}{\partial t'}\mathbf{J}(\mathbf{r}',t')d^3r' + \mathbf{W}_T^{\text{SF}}(\mathbf{r},t), \quad (39.15)$$

$$\int_{-\infty}^{\infty} g(R,\tau)\boldsymbol{\nabla}' \times \mathbf{J}(\mathbf{r}',t)d^3r' = -\frac{1}{c}\int_{-\infty}^{\infty} \mathbf{G}_M(\mathbf{R},\tau)\cdot\frac{\partial}{\partial t'}\mathbf{J}(\mathbf{r}',t')d^3r', \quad (39.16)$$

where $\mathbf{J} = \mathbf{J}^e$ or $\mathbf{J}^m$, and $\mathbf{W}_T^{\text{SF}}(\mathbf{r},t)$ is the transverse self-field (superscript SF) contribution to the scattered (electric or magnetic) field. The interested reader may find the quite elaborate derivation of the results in Eqs. (39.15) and (39.16) in Chapter 14 of my book "Quantum Theory of Near-Field Electrodynamics" [157]. By utilizing the two relations above, Eqs. (39.13) and (39.14) can be written in forms involving in the integrals only time derivatives of the electric and magnetic current densities. Hence,

$$\mathbf{E}_T(\mathbf{r},t) = \mathbf{E}_T^0(\mathbf{r},t) + \mathbf{E}_T^{\text{SF}}(\mathbf{r},t) \tag{39.17}$$
$$- \mu_0 \int_{-\infty}^{\infty} \left[\mathbf{G}_T(\mathbf{R},\tau)\cdot\frac{\partial}{\partial t'}\mathbf{J}^e(\mathbf{r}',t') - \mathbf{G}_M(\mathbf{R},\tau)\cdot\frac{\partial}{\partial t'}\mathbf{J}^m(\mathbf{r}',t')\right] d^3r'dt'$$

and

$$c\mathbf{B}_T(\mathbf{r},t) = c\mathbf{B}_T^0(\mathbf{r},t) + c\mathbf{B}_T^{\text{SF}}(\mathbf{r},t) \tag{39.18}$$
$$- \mu_0 \int_{-\infty}^{\infty} \left[\mathbf{G}_T(\mathbf{R},\tau)\cdot\frac{\partial}{\partial t'}\mathbf{J}^m(\mathbf{r}',t') + \mathbf{G}_M(\mathbf{R},\tau)\cdot\frac{\partial}{\partial t'}\mathbf{J}^e(\mathbf{r}',t')\right] d^3r'dt'.$$

We have already encountered the explicit expression for the transverse dyadic electric field propagator in Section 3.3 [Eq. (3.31)], and in Section 14.1 [Eq. (14.4)]. The expresssion for the magnetic field operator may be found in my Physics Reports review "On the Theory of Spatial Localization of Photons" [153]. In spherical coordinates centred on the singular point $\mathbf{R} = \mathbf{0}$, the expressions for $\mathbf{G}_T(\mathbf{R}, \tau)$ and $\mathbf{G}_M(\mathbf{R}, \tau)$ are particularly simple, viz.,

$$\mathbf{G}_T(\mathbf{R}, \tau) = \frac{1}{4\pi R}\delta\left(\frac{R}{c} - \tau\right)(\mathbf{U} - \mathbf{e_R}\mathbf{e_R}) - \frac{c^2\tau}{4\pi R^3}\Theta(\tau)\Theta\left(\frac{R}{c} - \tau\right)(\mathbf{U} - 3\mathbf{e_R}\mathbf{e_R}),$$

$$(39.19)$$

[Eq. (3.31)], and

$$\mathbf{G}_M(\mathbf{R}, \tau) = \left[\frac{1}{4\pi R}\delta\left(\frac{R}{c} - \tau\right) - \frac{c}{4\pi R^2}\Theta\left(\tau - \frac{R}{c}\right)\right]\mathbf{U} \times \mathbf{e_R}, \qquad (39.20)$$

where Θ is the Heaviside unit step function and $\mathbf{e_R}$ $[\equiv \hat{\mathbf{R}}] = \mathbf{R}/R$ a unit vector in the radial direction. The $\mathbf{G}_M$ tensor only behaves as a genuine magnetic field *propagator* in the absence of persistent electric and magnetic currents, as discussed in some detail in [153]. Such currents are of no interest here, and we thus assume they are absent. In Fig. 39.1 a schematic illustration of the Dirac-String, the union of sections and the propagator approaches to magnetic monopoles electrodynamics are presented.

The transverse self-field, depends on the contraction geometry used in the propagator formalism around the singular point [157]. In the spherical contraction scheme one has [157]

$$\mathbf{E}_T^{\mathrm{SF}}(\mathbf{r}, t) = -\frac{\mathbf{P}_T^e(\mathbf{r}, t)}{3\varepsilon_0}, \qquad (39.21)$$

$$c\mathbf{B}_T^{\mathrm{SF}}(\mathbf{r}, t) = -\frac{\mathbf{P}_T^m(\mathbf{r}, t)}{3\varepsilon_0}, \qquad (39.22)$$

where $\mathbf{P}_T$ (with superscript e or m) is the transverse part of the generalized polarizability, $\mathbf{P}$, related to the current density via the definition [159]

$$\mathbf{J}(\mathbf{r}, t) \equiv \frac{\partial \mathbf{P}(\mathbf{r}, t)}{\partial t}. \qquad (39.23)$$

It is well known that the self-fields are nonvanishing in the near-field zones (also called the rim zones [150]) of the electric and magnetic current density distributions.

39.3 PHOTON WAVE MECHANICS

39.3.1 Energy wave function approach

The first-quantized theory of photons can be formulated in a number of physically equivalent ways, each having its own advantages [26, 153, 156–158, 198, 242]. In Chapter 2, the theory of photon wave mechanics was based on the transverse part of the vector potential, and in Chapter 3 this first-quantized approach was extended to the

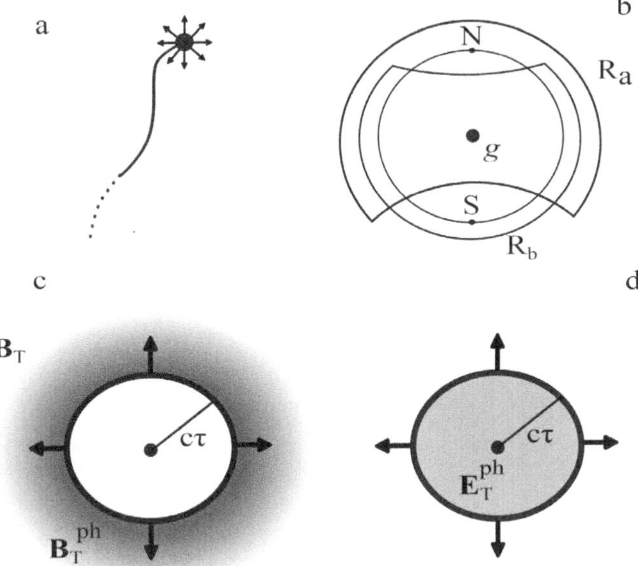

Figure 39.1 Schematic illustration showing important approaches to theoretical studies of magnetic monopole electrodynamics. a: Dirac monopole with its attached string. b: Sphere surrounding a magnetic monopole (g) divided into overlapping regions R_a and R_b. The north (N) and South (S) poles are left our from R_b and R_a, respectively, since the vector potential is manifest singular along the positive respective the negative part of the z-axis. c: In the propagator approach the magnetic part $(\mathbf{B}_T^{Ph})$ of the photon field is nonvanishing only in front of the light cone, and the electric part $(\mathbf{E}_T^{Ph})$ is nonzero only behind the light cone. The outwards propagating field energy hence is confined to the light cone as indicated by the thick-line circles of radii $c\tau$.

quantum field level. In Section 4.3, we touched upon an alternative approach to QED based on the so called photon energy wave function's extension to the operator level. In the globally coherent field quantum state, and with the neglect of vacuum fluctuations, one reaches the evolution equation for a single transverse photon in the classical photon energy wave function formalism. Below this classical theory will be used to study the photon scattering from a distribution of electric and magnetic monopoles. A substantial part of the following description is based on my article entitled "Electrodynamics with magnetic monopoles: Photon wave mechanical theory" [159].

Let us begin by recalling the introduction of the photon energy wave function. For a detailed account the reader may be referred to the review articles in [26, 153], and my books [157, 158].

Starting from the two complex transverse Riemann-Silberstein vectors

$$\mathbf{F}_\pm(\mathbf{r}, t) = \sqrt{\frac{\varepsilon_0}{2}} \left[\mathbf{E}_T(\mathbf{r}, t) \pm ic\mathbf{B}_T(\mathbf{r}, t) \right], \tag{39.24}$$

which relates to electromagnetic fields composed of positive ($\mathbf{F}_+$) and negative ($\mathbf{F}_-$) helicity species, one projects out their analytical (i.e. positive-frequency) parts, i.e.,

$$\mathbf{F}_\pm^{(+)}(\mathbf{r}, t) = \int_0^\infty \mathbf{F}_\pm(\mathbf{r}; \omega) e^{-i\omega t} \frac{d\omega}{2\pi}, \qquad (39.25)$$

where the $\mathbf{F}_\pm(\mathbf{r}; \omega)$'s are the Fourier integral transforms of the $\mathbf{F}_\pm(\mathbf{r}, t)$'s. In Eq. (39.25) the analytical part of $\mathbf{F}_\pm(\mathbf{r}, t)$ has been indicated by a suoerscript $(+)$. Subsequently, we characterize in spinorial notation the state of the photon by the six-component object

$$\boldsymbol{\Psi}(\mathbf{r}, t) = \begin{pmatrix} \mathbf{F}_+^{(+)}(\mathbf{r}, t) \\ \mathbf{F}_-^{(+)}(\mathbf{r}, t) \end{pmatrix}. \qquad (39.26)$$

The name Riemann-Silberstein vector has its roots in the Riemann's and Silberstein's cast of the Maxwell equations into complex form; see Bateman [19] (and also [3, 207, 239, 256]).

The reader may wonder why we relate the present photon formalism to the analytical signal. The reason is simply that this underlines the close relation between the wave mechanical theories of massive and massless elementary particles: Positive-frequency parts describe *particles* and negative-frequency parts relate to *antiparticles*. To the best of our knowledge the photon and its antiparticle are physically identical. From the classical electrodynamical theory this identity stems from the fact that informations carried by the positive and negative frequencies in the Maxwell-Lorentz equations are identical. Of course, if the photon and the antiphoton are physically identical, there is no absolute need for connecting a photon wave mechanical formalism alone to the positive-frequency part of the electromagnetic spectrum.

Based on the scattering theory approach discussed in Section 39.2, a prescribed incident free-photon wave function

$$\boldsymbol{\Phi}^0(\mathbf{r}, t) = \begin{pmatrix} \mathbf{F}_+^{0(+)}(\mathbf{r}, t) \\ \mathbf{F}_-^{0(+)}(\mathbf{r}, t) \end{pmatrix} = \sqrt{\frac{\varepsilon_0}{2}} \begin{pmatrix} \mathbf{E}_T^{0(+)}(\mathbf{r}, t) + ic\mathbf{B}_T^{0(+)}(\mathbf{r}, t) \\ \mathbf{E}_T^{0(+)}(\mathbf{r}, t) - ic\mathbf{B}_T^{0(+)}(\mathbf{r}, t) \end{pmatrix}, \qquad (39.27)$$

thus gives rise to a scattered (superscript s) photon state

$$\boldsymbol{\Psi}^s(\mathbf{r}, t) = \boldsymbol{\Psi}(\mathbf{r}, t) - \boldsymbol{\Phi}^0(\mathbf{r}, t) = \begin{pmatrix} \mathbf{F}_+^{s(+)}(\mathbf{r}, t) \\ \mathbf{F}_-^{s(+)}(\mathbf{r}, t) \end{pmatrix}. \qquad (39.28)$$

The explicit expressions for the analytical Riemann-Silberstein vectors entering $\boldsymbol{\Psi}^s(\mathbf{r}, t)$ are obtained from the analytical parts of the integrals in Eq. (39.13) and (39.14). Hence

$$\mathbf{F}_\pm^{s(+)} = -\mu_0 \sqrt{\frac{\varepsilon_0}{2}} \int_{-\infty}^\infty g(R, \tau) \left(\frac{\partial}{\partial t'} \mp i\boldsymbol{\nabla}' \times \right) \boldsymbol{\mathcal{J}}_\pm^{(+)}(\mathbf{r}', t') d^3 r' dt', \qquad (39.29)$$

where

$$\boldsymbol{\mathcal{J}}_\pm^{(+)}(\mathbf{r}, t) = \int_0^\infty [\mathbf{J}_T^e(\mathbf{r}; \omega) \pm i\mathbf{J}_T^m(\mathbf{r}; \omega)] e^{-i\omega t} \frac{d\omega}{2\pi}, \qquad (39.30)$$

$\mathbf{J}_T^e(\mathbf{r};\omega)$ and $\mathbf{J}_T^m(\mathbf{r};\omega)$ being the Fourier integral transforms of $\mathbf{J}_T^e(\mathbf{r},t)$ and $\mathbf{J}_T^m(\mathbf{r},t)$. With the help of Eq. (39.16), it appears that Eq. (39.29) can be written in the following compact form:

$$\mathbf{F}_{\pm}^{s(+)} = -\mu_0\sqrt{\frac{\varepsilon_0}{2}} \int_{-\infty}^{\infty} \boldsymbol{\mathcal{G}}_{\pm}(\mathbf{R},\tau) \cdot \frac{\partial}{\partial t'} \boldsymbol{\mathcal{J}}_{\pm}^{(+)}(\mathbf{r}',t')d^3r'dt', \tag{39.31}$$

with generalized propagator

$$\boldsymbol{\mathcal{G}}(\mathbf{R},\tau) = g(R,\tau) \pm i\mathbf{G}_M(\mathbf{R},\tau). \tag{39.32}$$

Upon completion of the scattering process at time T [we may assume that the induced current densities $\mathbf{J}_T^e(\mathbf{r},t)$ and $\mathbf{J}_T^m(\mathbf{r},t)$ both vanish outside the finite-time interval $(0,T)$], the scattered photon state takes some (asymptotic) form, say

$$\boldsymbol{\Phi}^s(\mathbf{r},t) = \boldsymbol{\Psi}^s(\mathbf{r},t(> T)). \tag{39.33}$$

The final photon state

$$\boldsymbol{\Phi}(\mathbf{r},t) = \boldsymbol{\Phi}^0(\mathbf{r},t) + \boldsymbol{\Phi}^s(\mathbf{r},t), \tag{39.34}$$

as well as the incident photon state are free-photon states, which implies that they have constant energies, say E and E_0. In the photon energy wave function formalism these energies are

$$E_0 = \int_{-\infty}^{\infty} \left[\boldsymbol{\Phi}^0(\mathbf{r},t)\right]^{\dagger} \cdot \boldsymbol{\Phi}^0(\mathbf{r},t)d^3r, \tag{39.35}$$

and

$$E = \int_{-\infty}^{\infty} \boldsymbol{\Phi}^{\dagger}(\mathbf{r},t) \cdot \boldsymbol{\Phi}(\mathbf{r},t)d^3r, \tag{39.36}$$

where (as before in the book) $\dagger$ stands for Hermitian conjugation. In the first-quantized description the free-photon energy equals the energy in the classical electromagnetic field [149].

39.3.2 Dynamical equations for the helicity eigenvectors

Combining the transverse parts of the symmetrized Maxwell-Lorentz equations, given in Eqs. (39.9) and (39.10), it appears that the transverse Riemann-Silberstein vectors satisfying the dynamical equations

$$i\hbar\frac{\partial}{\partial t}\mathbf{F}_{\pm}(\mathbf{r},t) = \pm c\hbar\boldsymbol{\nabla} \times \mathbf{F}_{\pm}(\mathbf{r},t) - \frac{i\hbar}{\sqrt{2\varepsilon_0}} \left[\mathbf{J}_T^e(\mathbf{r},t) \pm i\mathbf{J}_T^m(\mathbf{r},t)\right]. \tag{39.37}$$

In the framework of the first-quantized theory Planck's constant (divided by 2π) is just a multiplicative constant.

The vectorial character of the $\mathbf{F}_{\pm}$'s incorporates the spin ($\mathbf{s}$) of the photon (and antiphoton). A spin one can be represented by a vector in a three-dimensional complex vector space. Now, it is convenient to introduce a dimensionless spin-one operator of the photon, $\hat{\boldsymbol{\Sigma}} = \hat{\mathbf{s}}/\hbar$. In a (3×3)-matrix representation the Cartesian component (k) of $\boldsymbol{\Sigma}$ is given by [26, 158]

$$\left(\hat{\Sigma}_k\right)_{ij} = i^{-1}\varepsilon_{ijk}, \tag{39.38}$$

where ε_{ijk} is the completely antisymmetric Levi-Civita tensor. With the help of $\hat{\boldsymbol{\Sigma}}$, the analytical part of Eq. (39.37) can be written in the following form [26, 158]:

$$i\hbar\frac{\partial}{\partial t}\mathbf{F}_{\pm}^{(+)}(\mathbf{r}, t) = \pm c\left(\hat{\boldsymbol{\Sigma}} \cdot \hat{\mathbf{p}}\right)\mathbf{F}_{\pm}^{(+)}(\mathbf{r}, t) - \frac{i\hbar}{\sqrt{2\varepsilon_0}}\mathcal{J}_{\pm}^{(+)}(\mathbf{r}, t), \tag{39.39}$$

where $\mathcal{J}_{\pm}^{(+)}(\mathbf{r}, t)$ is given in Eq. (39.30), and $\hat{\mathbf{p}} = (\hbar/i)\boldsymbol{\nabla}$ is the momentum operator in the $\mathbf{r}$-representation. The scalar product

$$\hat{\boldsymbol{\Sigma}} \cdot \hat{\mathbf{p}} = \hat{p}\hat{h}, \tag{39.40}$$

relates to the photon helicity operator $\hat{h}$ [the plane-wave components of $\mathbf{F}_{+}^{(+)}$ and $\mathbf{F}_{-}^{(+)}$ are eigenvectors of $\hat{h}$ with eigenvalues $+1$ and -1. In $\mathbf{r}$-space

$$\hat{p} = (\hat{\mathbf{p}} \cdot \hat{\mathbf{p}})^{1/2} = \frac{\hbar}{i}\sqrt{\nabla^2} \tag{39.41}$$

in a symbolic notation dating back to the Landua-Peirls quantum theory of the photon [156]. The operator is defined via its action in $\mathbf{p}$-space. It can be shown that the two helicity species are eigenvectors of the helicity operator, that is

$$\hat{h}\mathbf{F}_{\pm}^{(+)}(\mathbf{r}, t) = \pm\mathbf{F}_{\pm}^{(+)}(\mathbf{r}, t). \tag{39.42}$$

39.3.3 Local energy conservation in the photon field

Taking as a starting point the expression for the photon energy density, namely

$$\boldsymbol{\Psi}^{\dagger} \cdot \boldsymbol{\Psi} = \sum_{s=+,-}\left(\mathbf{F}_s^{(+)}\right)^* \cdot \mathbf{F}_s^{(+)}, \tag{39.43}$$

a continuity equation for the photon energy can be derived. The summation in Eq. (39.43) is over the two helicity species, and the Riemann-Silberstein vectors for the two species satisfy the dynamical equations given in Eq. (39.37). The time derivative of the photon energy density is given by

$$\frac{\partial}{\partial t}\left(\boldsymbol{\Psi}^{\dagger} \cdot \boldsymbol{\Psi}\right) = \sum_{s=+,-}\left\{\left[\frac{\partial}{\partial t}\left(\mathbf{F}_s^{(+)}\right)^*\right] \cdot \mathbf{F}_s^{(+)} + \left(\mathbf{F}_s^{(+)}\right)^* \cdot \frac{\partial}{\partial t}\mathbf{F}_s^{(+)}\right\}, \tag{39.44}$$

and if one here eliminates $\partial(\mathbf{F}_s^{(+)})^*/\partial t$ and $\partial\mathbf{F}_s^{(+)}/\partial t$ by means of the analytical parts of the dynamical equations in (39.37) and their complex conjugates, the following result is obtained:

$$\frac{\partial}{\partial t}\left[\left(\mathbf{F}_{\pm}^{(+)}\right)^* \cdot \mathbf{F}_{\pm}^{(+)}\right] = \pm ic\boldsymbol{\nabla} \cdot \left[\left(\mathbf{F}_{\pm}^{(+)}\right)^* \times \mathbf{F}_{\pm}^{(+)}\right]$$
$$- \frac{1}{\sqrt{2\varepsilon_0}}\left[\left(\mathbf{F}_{\pm}^{(+)}\right)^* \cdot \mathcal{J}_{\pm}^{(+)} + \mathbf{F}_{\pm}^{(+)} \cdot \left(\mathcal{J}_{\pm}^{(+)}\right)^*\right], \tag{39.45}$$

with

$$\boldsymbol{\mathcal{J}}_{\pm}^{(+)} = \mathbf{J}_T^{e(+)} \pm i\mathbf{J}_T^{M(+)}. \tag{39.46}$$

By combining Eqs. (39.43) and (39.45) one gets

$$\frac{\partial}{\partial t}\left(\boldsymbol{\Psi}^{\dagger}\cdot\boldsymbol{\Psi}\right) = ic\boldsymbol{\nabla}\cdot\left[\left(\mathbf{F}_{+}^{(+)}\right)^{*}\times\mathbf{F}_{+}^{(+)} - \left(\mathbf{F}_{-}^{(+)}\right)^{*}\times\mathbf{F}_{-}^{(+)}\right]$$
$$- \frac{1}{\sqrt{2\varepsilon_0}}\sum_{+,-}\left[\left(\mathbf{F}_{s}^{(+)}\right)^{*}\cdot\boldsymbol{\mathcal{J}}_{s}^{(+)} + \mathbf{F}_{s}^{(+)}\cdot\left(\boldsymbol{\mathcal{J}}_{s}^{(+)}\right)^{*}\right]. \tag{39.47}$$

Since, from Eq. (39.42)

$$\hat{h}\boldsymbol{\Psi} = \hat{h}\begin{pmatrix}\mathbf{F}_{+}^{(+)}\\\mathbf{F}_{-}^{(+)}\end{pmatrix} = \begin{pmatrix}\mathbf{F}_{+}^{(+)}\\-\mathbf{F}_{-}^{(+)}\end{pmatrix}, \tag{39.48}$$

we obtain

$$\boldsymbol{\Psi}^{\dagger}\times(\hat{h}\boldsymbol{\Psi}) = \left(\mathbf{F}_{+}^{(+)}\right)^{*}\times\mathbf{F}_{+}^{(+)} - \left(\mathbf{F}_{-}^{(+)}\right)^{*}\times\mathbf{F}_{-}^{(+)}. \tag{39.49}$$

By introduction of the six-component current density

$$\boldsymbol{\mathcal{J}} = \begin{pmatrix}\boldsymbol{\mathcal{J}}_{+}^{(+)}\\\boldsymbol{\mathcal{J}}_{-}^{(+)}\end{pmatrix}, \tag{39.50}$$

it appears that

$$\boldsymbol{\Psi}^{\dagger}\cdot\boldsymbol{\mathcal{J}} + \boldsymbol{\mathcal{J}}^{\dagger}\cdot\boldsymbol{\Psi} = \sum_{s=+,-}\left[\left(\mathbf{F}_{s}^{(+)}\right)^{*}\cdot\boldsymbol{\mathcal{J}}_{s}^{(+)} + \mathbf{F}_{s}^{(+)}\cdot\left(\boldsymbol{\mathcal{J}}_{s}^{(+)}\right)^{*}\right]. \tag{39.51}$$

The expressions in Eqs. (39.49) and (39.51) allow one to rewrite Eq. (39.47) in compact spinorial notation. Thus,

$$\frac{\partial}{\partial t}\left(\boldsymbol{\Psi}^{\dagger}\cdot\boldsymbol{\Psi}\right) = -\boldsymbol{\nabla}\cdot\left[\frac{c}{i}\boldsymbol{\Psi}^{\dagger}\times(\hat{h}\boldsymbol{\Psi})\right] - \frac{1}{\sqrt{2\varepsilon_0}}\left(\boldsymbol{\Psi}^{\dagger}\cdot\boldsymbol{\mathcal{J}} + \boldsymbol{\mathcal{J}}^{\dagger}\cdot\boldsymbol{\Psi}\right). \tag{39.52}$$

Eq. (39.52) is the sought for continuity equation for the photon energy density, $\boldsymbol{\Psi}^{\dagger}\cdot\boldsymbol{\Psi}$. The quantity

$$\boldsymbol{\mathcal{S}}_{\boldsymbol{\Psi}} = \frac{c}{i}\boldsymbol{\Psi}^{\dagger}\times\left(\hat{h}\boldsymbol{\Psi}\right), \tag{39.53}$$

is identified as the energy current density related to the photon energy wave function, and the term $\left(\boldsymbol{\Psi}^{\dagger}\cdot\boldsymbol{\mathcal{J}} + \boldsymbol{\mathcal{J}}^{\dagger}\cdot\boldsymbol{\Psi}\right)/\sqrt{2\varepsilon_0}$ represents the density of work done by the photon field on the electric and magnetic monopoles per unit time.

39.4 DOUBLE-POTENTIAL FORMALISM

We have seen in Section 38.3 that in order to seek to uphold the minimal coupling substitution $\boldsymbol{\nabla} = \boldsymbol{\nabla} - (ie/\hbar)\mathbf{A}$, it is necessary to suppose that the magnetic monopole is attached to (placed at the end of) a so-called Dirac string. On the string, stretching off to infinity, the vector potential is singular. Furthermore the string has to be unobservable (a singular vector potential being unacceptable) and Dirac therefore suggested that the particle wave function should vanish on the string. This let, as we have realized, to the charge quantization condition, given in Eq. (38.50).

39.4.1 Electric and magnetic four-potential

It was shown by Cabibbo and Ferrari [53] that the pathological string description can be avoided by introducing a second four-potential. In the double-potential formalism the electric and magnetic fields are given by the dual forms

$$\mathbf{E} = -\frac{\partial}{\partial t}\mathbf{A}_e - \boldsymbol{\nabla}\phi^e - c\boldsymbol{\nabla} \times \mathbf{A}^m, \tag{39.54}$$

$$\mathbf{B} = -\frac{1}{c}\frac{\partial}{\partial t}\mathbf{A}^m - \frac{1}{c}\boldsymbol{\nabla}\phi^m + \boldsymbol{\nabla} \times \mathbf{A}^e, \tag{39.55}$$

where

$$\{A^{e,\mu}\} = \left(\frac{\phi^e}{c}, \mathbf{A}^e\right), \tag{39.56}$$

and

$$\{A^{m,\mu}\} = \left(\frac{\phi^m}{c}, \mathbf{A}^m\right), \tag{39.57}$$

are the electric (superscript e) and magnetic (superscript m) four-potentials. In the absence of magnetic monopoles, $\{A^{m,\mu}\} = 0$, Eqs. (39.54) and (39.55) express $\mathbf{E}$ and $\mathbf{B}$ in terms of the electric vector and scalar potentials in the well-known manner. The addition of the sum $-c^{-1}(\partial \mathbf{A}^m/\partial t + \boldsymbol{\nabla}\phi^m)$ in Eq. (39.55) to $\boldsymbol{\nabla} \times \mathbf{A}^e$ implies that the problem with $\mathbf{B} = \boldsymbol{\nabla} \times \mathbf{A}^e$ can be avoided. On top of this, the Cabibbo-Ferrari double-potential approach adds to the symmetrized Maxwell equations, with $\mathbf{E}$ and $\mathbf{B}$, a corresponding elegant symmetry in the potential formalism. Alone for this reason the Cabibbo-Ferrari description appears attractive.

In the covariant form of symmetrized electrodynamics [Section 38.5] the double-potential field tensor now has elements [with $c = 1$]

$$F_{\mu\nu} = \partial_\mu A_\nu^e - \partial_\nu A_\mu^e + \varepsilon^{\mu\nu\rho\sigma}\partial_\rho A_\sigma^m, \tag{39.58}$$

where $\varepsilon^{\mu\nu\rho\sigma}$ is the completely antisymmetric Levi-Civita tensor in four indices. In Eq. (39.58) the Cabibbo-Ferrari form of $F_{\mu\nu}$ [with signature $(-,+,+,+)$] has been used. The reader may convince herself that Eq. (39.58) gives the forms of Eqs. (39.54) and (39.55) for $\mathbf{E}$ and $\mathbf{B}$ [remember Eq. (38.74)].

The potentials $\{A_\mu^e\}$ and $\{A_\mu^m\}$ are determined by $\{F_{\mu\nu}\}$ up to a group of gauge transformations, given below. For notational simplicity, let us take $A_\mu^e \equiv A_\mu$ and $A_\mu^m \equiv B_\mu$. The gauge transformations contain the individual transformations

$$A_\mu \rightarrow A_\mu + \partial_\mu\Lambda, \tag{39.59}$$
$$B_\mu \rightarrow B_\mu + \partial_\mu\Gamma, \tag{39.60}$$

as well as the mixing transformations

$$A_\mu \rightarrow A_\mu + A_\mu', \tag{39.61}$$
$$B_\mu \rightarrow B_\mu + B_\mu'. \tag{39.62}$$

The new gauge functions A'_μ and B'_μ obviously shall satisfy the zero-field condition

$$\partial_\mu A'_\nu - \partial_\nu A'_\mu + \varepsilon^{\mu\nu\rho\sigma}\partial_\rho B'_\sigma = 0. \tag{39.63}$$

Note that one obtains Eqs. (39.59) and (39.60) setting $\{B'_\mu\} = 0$ and $\{A'_\mu\} = 0$, respectively, in Eq. (39.63).

A division of Eqs. (39.54) and (39.55) into their longitudinal and transverse parts gives respectively,

$$\mathbf{E}_L = -\frac{\partial}{\partial t}\mathbf{A}^e_L - \boldsymbol{\nabla}\phi^e, \tag{39.64}$$

$$c\mathbf{B}_L = -\frac{\partial}{\partial t}\mathbf{A}^m_L - \boldsymbol{\nabla}\phi^m, \tag{39.65}$$

and

$$\mathbf{E}_T = -\frac{\partial}{\partial t}\mathbf{A}^e_T - c\boldsymbol{\nabla}\times\mathbf{A}^m_T, \tag{39.66}$$

$$c\mathbf{B}_T = -\frac{\partial}{\partial t}\mathbf{A}^m_T + c\boldsymbol{\nabla}\times\mathbf{A}^e_T. \tag{39.67}$$

As expected, there is no mixing of the e and m potentials in relation to the calculation of the longitudinal electric and magnetic fields, see Eqs. (39.64) and (39.65). Individual gauge transformations on the electric and magnetic four-potentials do not change the transverse parts, $\mathbf{A}^e_T$ and $\mathbf{A}^m_T$, of the potentials. Within the Lorenz gauge, $\partial^\mu A^e_\mu = \partial^\mu A^m_\mu = 0$, certain gauge transformations are still allowed. Hence, gauge transformations of the kind given in Eqs. (39.59) and (39.60) are still allowed, provided $\Box\Lambda = \Box\Gamma = 0$ (as we already know). For the mixed transformations satisfying Eq. (39.63), the zero-field condition must be restricted by $\partial^\mu A'_\mu = \partial^\mu B'_\mu = 0$ for transformations within the Lorenz gauge.

It appears, from Eqs. (39.54) and (39.55) that the complex field vectors $\mathbf{E} + ic\mathbf{B}$ are functions only of the complex four-potentials

$$\{A^{e,\mu} \pm iA^{m,\mu}\} = \left[\frac{1}{c}(\phi^e \pm i\phi^m), \mathbf{A}^e \pm i\mathbf{A}^m\right]. \tag{39.68}$$

In particular the Riemann-Silberstein vectors are given by

$$\sqrt{\frac{2}{\varepsilon_0}}\mathbf{F}_\pm = -\frac{\partial}{\partial t}\left(\mathbf{A}^e_T \pm i\mathbf{A}^m_T\right) \pm ic\boldsymbol{\nabla}\times\left(\mathbf{A}^e_T \pm i\mathbf{A}^m_T\right) \tag{39.69}$$

in symmetrized electrodynamics.

39.4.2 Nonlocal transformation of transverse vector potentials

Notwithstanding the fact that the double-potential formalism allows one to avoid the pathological string concept of Dirac, it is not obvious whether a kind of minimal coupling substitution, can be established in the framework of the double-potential description. Cabibbo and Ferrari do not address this point since in their treatment of quantum electrodynamics no use is made of potentials. Of course, their theory does not contain pathological elements, like the string singularities. The Cabibbo-Ferrari theory is an extension of Mandelstam's treatment to the case in which both electric and magnetic monopoles are present [53].

Below we shall demonstrate how a certain nonlocal space-time transformation of the potentials allows one to express the transverse parts of the electric and magnetic fields as curls of the original and the transformed transverse vector potentials. This nonlocal approach appears to be first established by the present book author [159]. The usefulness of the transformation in a broader context is briefly described in Sections 39.5-39.7.

In addition to the gauge-invariant transverse vector potentials $\mathbf{A}_T^e$ and $\mathbf{A}_T^m$ we now introduce two new transverse vector potentials denoted by $\mathcal{A}_T^e$ and $\mathcal{A}_T^m$ (calligraphic $\mathbf{A}$'s). Let the old and new vector potentials be given by the following Fourier integral transformations of their space and time dependencies:

$$\mathbf{A}_T^\alpha(\mathbf{r}, t) = (2\pi)^{-4} \int_{-\infty}^{\infty} \mathbf{A}_T^\alpha(\mathbf{q}, \omega) e^{i(\mathbf{q} \cdot \mathbf{r} - \omega t)} d^3 q d\omega, \qquad (39.70)$$

$$\mathcal{A}_T^\alpha(\mathbf{r}, t) = (2\pi)^{-4} \int_{-\infty}^{\infty} \mathcal{A}_T^\alpha(\mathbf{q}, \omega) e^{i(\mathbf{q} \cdot \mathbf{r} - \omega t)} d^3 q d\omega, \qquad (39.71)$$

with $\alpha = e$ or m. The vector potentials $\mathbf{A}_T^\alpha(\mathbf{r}, t)$ and $\mathcal{A}_T^\alpha(\mathbf{r}, t)$ we now connect via a relation between the Fourier amplitudes, viz.,

$$\mathbf{A}_T^\alpha(\mathbf{q}, \omega) = \frac{cq}{\omega} \hat{\mathbf{q}} \times \mathcal{A}_T^\alpha(\mathbf{q}, \omega), \qquad (39.72)$$

where $\hat{\mathbf{q}} = \mathbf{q}/q$. Since we have demanded that the new potentials shall be transverse [implying that $\hat{\mathbf{q}} \cdot \mathcal{A}_T^\alpha(\mathbf{q}, \omega) = 0$], we obtain from Eq. (29.72) the inverse connection

$$\mathcal{A}_T^\alpha(\mathbf{q}, \omega) = -\frac{\omega}{cq} \hat{\mathbf{q}} \times \mathbf{A}_T^\alpha(\mathbf{q}, \omega). \qquad (39.73)$$

By making use of the fact that multiplication by $-i\omega$ and $i\mathbf{q}$ in the four-dimensional Fourier space corresponds to the operations $\partial/\partial t$ and ∇ in the space-time domain, it is obvious that the old $[\mathbf{A}_T^\alpha(\mathbf{r}, t)]$ and new $[\mathcal{A}_T^\alpha(\mathbf{r}, t)]$ transverse vector potentials are related as follows:

$$-\frac{1}{c}\frac{\partial}{\partial t} \mathbf{A}_T^\alpha(\mathbf{r}, t) = \nabla \times \mathcal{A}_T^\alpha(\mathbf{r}, t). \qquad (39.74)$$

By utilizing Eq. (39.74) in Eqs. (39.66) and (39.67) the transverse part of the electric and magnetic fields can be expressed in the mixed potential terms

$$c^{-1}\mathbf{E}_T(\mathbf{r}, t) = \nabla \times \left[\mathcal{A}_T^e(\mathbf{r}, t) - \mathbf{A}_T^m(\mathbf{r}, t) \right], \qquad (39.75)$$

$$\mathbf{B}_T(\mathbf{r}, t) = \nabla \times \left[\mathbf{A}_T^e(\mathbf{r}, t) + \mathcal{A}_T^m(\mathbf{r}, t) \right]. \qquad (39.76)$$

The space-time nonlocal transformations of the potentials thus have enabled us to express $\mathbf{E}_T/c$ and $\mathbf{B}_T$ as *curls* of simple combinations of the old and the new transverse electric and magnetic vector potentials. The expressions for $\mathbf{B}_T$ signals that one might be able to uphold the "usual" form of the Minimal Coupling Principle without use of the Dirac string concept.

39.5 MOMENTUM OF PARTICLE (E+M) - PHOTON SYSTEM

The introduction of the new transverse vector potentials, $\mathcal{A}_T^e$ and $\mathcal{A}_T^m$, enables one to obtain a unified view of the momentum of a system of electric (E) and magnetic (M) monopoles coupled to a photon field.

39.5.1 Total field momentum

Within the framework of the extended microscopic first-quantized Maxwell-Lorentz theory the total field momentum is given by

$$\mathbf{P}(t) = \varepsilon_0 \int_{-\infty}^{\infty} \mathbf{E}(\mathbf{r}, t) \times \mathbf{B}(\mathbf{r}, t) d^3 r. \tag{39.77}$$

The total field momentum is time dependent in general, as indicated. For a closed system of photons plus electric and magnetic monopoles the total particle-field momentum is conserved, of course. In the microscopic Maxwell-Lorentz approach one just adds to Eq. (39.77) the kinetic momenta of the electric and magnetic monopole particles to obtain the conserved total momentum.

To analyse the structure of $\mathbf{P}(t)$ in Eq. (39.77) we divide the $\mathbf{E}$ and $\mathbf{B}$ fields into their longitudinal and transverse parts. Hence,

$$\mathbf{P}(t) = \mathbf{P}_{LL}(t) + \mathbf{P}_{LT}(t) + \mathbf{P}_{TL}(t) + \mathbf{P}_{TT}(t), \tag{39.78}$$

where

$$\mathbf{P}_{IJ}(t) = \varepsilon_0 \int_{-\infty}^{\infty} \mathbf{E}_I(\mathbf{r}, t) \times \mathbf{B}_J(\mathbf{r}, t) d^3 r. \tag{39.79}$$

The (IJ) subscripts refer to each of the four combinations (LL), (LT), (TL) and (TT) in Eq. (39.78). In the absence of magnetic monopoles $\mathbf{P}(t)$ has the two well-known parts [63, 157], that is

$$\mathbf{P}(t \mid \rho^m = 0) = \mathbf{P}_{LT}(t \mid \rho^m = 0) + \mathbf{P}_{TT}(t \mid \rho^m = 0), \tag{39.80}$$

remembering that the magnetic field has no longitudinal part when $\rho^m = 0$; see Eq. (39.2).

39.5.2 The LL and TT parts of the momentum

For what follows it is convenient to transfer the integral given in $\mathbf{P}_{IJ}(t)$ in Eq. (39.79) from direct space to an integral over wave-vector ($\mathbf{q}$) space. Using the Parseval-Planchevel identity one obtains

$$\mathbf{P}_{IJ}(t) = \frac{\varepsilon_0}{(2\pi)^3} \int_{-\infty}^{\infty} \mathbf{E}_I^*(\mathbf{q}; t) \times \mathbf{B}_J(\mathbf{q}; t) d^3 q \tag{39.81}$$

Since $\mathbf{E}_L(\mathbf{q}; t)$ and $\mathbf{B}_L(\mathbf{q}; t)$ both are along the $\hat{\mathbf{q}}$ direction, the LL part of the momentum vanishes (at all times):

$$\mathbf{P}_{LL}(t) = \mathbf{0}. \tag{39.82}$$

Although there is no field momentum associated with pure longitudinal fields, the classical analysis of non-retarded electrodynamics [Eq. (38.17) and (38.18)] shows that an (E,M) pair possesses a nonvanishing LL angular momentum.

For the massless photon in the state $\boldsymbol{\Psi}(\mathbf{r}, t)$, given in six-component form in Eq. (39.26), the relation between the momentum density, $\boldsymbol{\mathcal{P}}_{TT}(\mathbf{r}, t)$, and the energy current density, $\mathcal{S}_{\boldsymbol{\Psi}}(\mathbf{r}, t)$, necessarily is given by

$$\mathcal{S}_{\boldsymbol{\Psi}}(\mathbf{r}, t) = c^2 \boldsymbol{\mathcal{P}}_{TT}(\mathbf{r}, t). \tag{39.83}$$

The momentum of the photon field, namely

$$\mathbf{P}_{TT}(t) = \int_{-\infty}^{\infty} \boldsymbol{\mathcal{P}}_{TT}(\mathbf{r}, t) d^3 r, \tag{39.84}$$

thus can be related to the photon energy wave function using the expression for $\mathcal{S}_{\boldsymbol{\Psi}}(\mathbf{r}, t)$ given in Eq. (39.53). Hence

$$\mathbf{P}_{TT}(t) = \frac{1}{ic} \int_{-\infty}^{\infty} \boldsymbol{\Psi}^{\dagger}(\mathbf{r}, t) \times \left[\hat{h} \boldsymbol{\Psi}(\mathbf{r}, t) \right] d^3 r. \tag{39.85}$$

The photon-field momentum $\mathbf{P}_{TT}(t)$ is time dependent as long as the photon is coupled to the monopoles. The momenta of the incoming $[\boldsymbol{\Phi}^0(\mathbf{r}, t)]$ and outgoing $[\boldsymbol{\Phi}(\mathbf{r}, t)]$ free-photon states are both time independent. This is self-evident on physical grounds, but can of course be shown explicitly using the dynamical equations for the free Riemann-Silberstein vectors [Eqs. (39,37) with $\mathbf{J}_T^e = \mathbf{J}_T^m = 0$]. For instance, it appears from the Parseval-Planchevel relation that the outgoing (superscript out) photon field momentum, $\mathbf{P}_{TT}^{\text{out}}$, can be expressed in the form

$$\mathbf{P}_{TT}^{\text{out}} = \frac{1}{ic} \int_{-\infty}^{\infty} \boldsymbol{\Phi}^{\dagger}(\mathbf{q}; t) \times \left[\hat{h} \boldsymbol{\Phi}(\mathbf{q}; t) \right] \frac{d^3 q}{(2\pi)^3}. \tag{39.86}$$

The reader may show to herself that the dynamical free-space equations for the transverse Riemann-Silberstien vectors lead to the solution

$$\boldsymbol{\Phi}(\mathbf{q}; t) = \boldsymbol{\Phi}(\mathbf{q}; t = 0) \exp(-icqt). \tag{39.87}$$

The time dependence of $\boldsymbol{\Phi}(\mathbf{q}; t)$ immediately shows that the integrand of Eq. (39.86) is time independent. Consequently $\mathbf{P}_{TT}^{\text{out}}$ is constant in time. The calculation leading to Eq. (39.87) may also be found in [26, 158]. The momentum of the incoming (superscript in) photon field, $\mathbf{P}_{TT}^{\text{in}}$, is of course also time independent.

39.5.3 Electromagnetic parts of the canonical particle momenta. Minimal Coupling Principle without singularities

The most interesting parts of the field momentum are $\mathbf{P}_{LT}(t)$ and $\mathbf{P}_{TL}(t)$ as we shall realize now. For simplicity, we shall confine the analysis to the case where just one electric monopole (charge e) and one magnetic monopole (charge g) is present. The instantaneous positions of the monopoles are given by the position vectors $\mathbf{r}^e(t)$ and $\mathbf{r}^m(t)$.

Let us first consider the field momentum contribution

$$\mathbf{P}_{LT}(t) = \varepsilon_0 \int_{-\infty}^{\infty} \mathbf{E}_L^*(\mathbf{q};t) \times \mathbf{B}_T(\mathbf{q};t) \frac{d^3q}{(2\pi)^3}. \tag{39.88}$$

From the Maxwell-Lorentz equation in Eq. (39.1) one obtains in the wave-vector domain $i\mathbf{q} \cdot \mathbf{E}_L(\mathbf{q};t) = \rho^e(\mathbf{q};t)/\varepsilon_0$, and thus

$$\mathbf{E}_L(\mathbf{q};t) = \hat{\mathbf{q}}\hat{\mathbf{q}} \cdot \mathbf{E}_L(\mathbf{q};t) = \frac{\hat{\mathbf{q}}}{i\varepsilon_0 q}\rho^e(\mathbf{q};t), \tag{39.89}$$

where

$$\rho^e(\mathbf{q};t) = e \int_{-\infty}^{\infty} \delta(\mathbf{r} - \mathbf{r}_e(t))e^{-i\mathbf{q}\cdot\mathbf{r}}d^3r = e\exp[-i\mathbf{q} \cdot \mathbf{r}^e(t)]. \tag{39.90}$$

The usefulness of the nonlocal transformation of the transverse vector potentials now shows up. The $\mathbf{q}$-space form of Eq. (39.76) is given by

$$\mathbf{B}_T(\mathbf{q};t) = iq\hat{\mathbf{q}} \times \left[\mathbf{A}_T^e(\mathbf{q};t) + \boldsymbol{\mathcal{A}}_T^m(\mathbf{q};t)\right], \tag{39.91}$$

and a combination of Eqs. (39.88)–(39.91) then leads to

$$\begin{aligned}
\mathbf{P}_{LT}(t) &= -e\int_{-\infty}^{\infty} \hat{\mathbf{q}} \times \left\{\hat{\mathbf{q}} \times \left[\mathbf{A}_T^e(\mathbf{q};t) + \boldsymbol{\mathcal{A}}_T^m(\mathbf{q};t)\right]e^{i\mathbf{q}\cdot\mathbf{r}^e(t)}\right\}\frac{d^3q}{(2\pi)^3} \\
&= e\int_{-\infty}^{\infty} \left[\mathbf{A}_T^e(\mathbf{q};t) + \boldsymbol{\mathcal{A}}_T^m(\mathbf{q};t)\right]e^{i\mathbf{q}\cdot\mathbf{r}^e(t)}\frac{d^3q}{(2\pi)^3}
\end{aligned} \tag{39.92}$$

The last expression in Eq. (39.92) is just the Fourier integral representative of

$$\mathbf{P}_{LT}(t) = e\left[\mathbf{A}_T^e(\mathbf{r}^e(t),t) + \boldsymbol{\mathcal{A}}_T^m(\mathbf{r}^e(t),t)\right]. \tag{39.93}$$

This expression shows that $\mathbf{P}_{LT}(t)$ solely is a function of the position coordinate of the electric monopole, which in itself is a function of time. Furthermore, also an explicit time dependence of the transverse vector potentials at the particle position enters $\mathbf{P}_{LT}(t)$. Consequently, one may characterize $\mathbf{P}_{LT}(t)$ as an *electromagnetic momentum* associated with the electric monopole particle. If $\boldsymbol{\pi}^e$ denotes the *mechanical momentum* of the particle, the quantity

$$\mathbf{p}^e = \boldsymbol{\pi}^e + e\left[\mathbf{A}_T^e(\mathbf{r}^e(t),t) + \boldsymbol{\mathcal{A}}_T^m(\mathbf{r}^e(t),t)\right] \tag{39.94}$$

represents the total momentum of the electric monopole particle, also called the *canonical momentum*. In a quantum physical setting $\mathbf{p}^e$ translate into a momentum operator, $(\hbar/i)\boldsymbol{\nabla}^e$, for the e-monopole.

In retrospect, it appears that the double-potential formalism of Cabibbo-Ferrari [53] in combination with the nonlocal transformation introduced by the present author [159] have enabled one to uphold the Minimal Coupling Principle,

$$\mathbf{p}^e \to \mathbf{p}^e - e[\mathbf{A}_T^e + \boldsymbol{\mathcal{A}}_T^m], \tag{39.95}$$

for the e-particle. The Hamiltonian formalism thus can be upheld without introduction of the Dirac string concept.

By now it is obvious that

$$\mathbf{P}_{TL}(t) = \varepsilon_0 \int_{-\infty}^{\infty} \mathbf{E}_T(\mathbf{r}, t) \times \mathbf{B}_L(\mathbf{r}, t) d^3 r \qquad (39.96)$$

may be identified as the electromagnetic momentum associated with the magnetic monopole. If one makes use of the expression given for $\mathbf{E}_T$ in Eq. (39.75), and notes that $\mathbf{E}_T \times \mathbf{B}_L = -\mathbf{B}_L \times \mathbf{E}_T$, a moment of reflection shows that $\mathbf{P}_{LT}(t)$ can be written in the following form:

$$\mathbf{P}_{LT}(t) = g\left[\mathbf{A}_T^m(\mathbf{r}^m(t), t) - \mathcal{A}_T^e(\mathbf{r}^m(t), t)\right]. \qquad (39.97)$$

The quantity

$$\mathbf{p}^m = \boldsymbol{\pi}^m + g\left[\mathbf{A}_T^m(\mathbf{r}^m(t), t) - \mathcal{A}_T^e(\mathbf{r}^m(t), t)\right], \qquad (39.98)$$

where $\boldsymbol{\pi}^m$ is the mechanical momentum of the g-particle, hence is the total (canonical) momentum of the magnetic monopole.

Since the terms $\mathcal{A}_T^m(\mathbf{r}^e(t), t)$ and $\mathcal{A}_T^e(\mathbf{r}^m(t), t)$ are proportional to g and e, respectively, the new contributions to $\mathbf{p}^e$ and $\mathbf{p}^m$ both are proportional to the product of eg. A graphical representation of the physics of Eqs. (39.94) and (39.98) is shown in Fig. 39.2. The analysis in Subsection 39.5.3 easily can be generalized to assemblies of electric and magnetic monopoles, see [159].

39.6   ANGULAR MOMENTUM OF PARTICLE ($E + M$) - PHOTON SYSTEM

39.6.1   Total field angular momentum

We have realized in Section 38.2 that the angular momentum of a pair of dyons (or a pair of electric and magnetic monopoles) has a contribution which one may characterize as a kind of spin (internal angular momentum). The result obtained was based on the assumption that the dyon-dyon (monopole-monopole) interaction is non-retarded. In the following we shall see how the inclusion of field retardation extends the previous theory. Furthermore, our starting point will be different from that presented in Section 38.2.

It appears from the (symmetrized) set of Maxwell-Lorentz equations that the angular momentum. of the global particle-field system can be written as a sum of the particle angular momentum and a field part given by [63, 157]

$$\mathbf{I}(t|\mathbf{r}_0) = \varepsilon_0 \int_{-\infty}^{\infty} (\mathbf{r} - \mathbf{r}_0) \times [\mathbf{E}(\mathbf{r}, t) \times \mathbf{B}(\mathbf{r}, t)] d^3 r \qquad (39.99)$$

in a classical approach. The angular momentum $\mathbf{I}(t|\mathbf{r}_0)$ given in Eq. (39.99) is just the integrated moment of the momentum density, $\varepsilon_0 \mathbf{E} \times \mathbf{B}$, with respect to a fixed reference point $\mathbf{r}_0$. In analogy with the structural division made for the field momentum in Subsection 39.5.1, $\mathbf{I}(t|\mathbf{r}_0)$ is decomposed as follows:

$$\mathbf{I}(t|\mathbf{r}_0) = \sum_{I,J} \mathbf{I}_{IJ}(t|\mathbf{r}_0), \qquad (39.100)$$

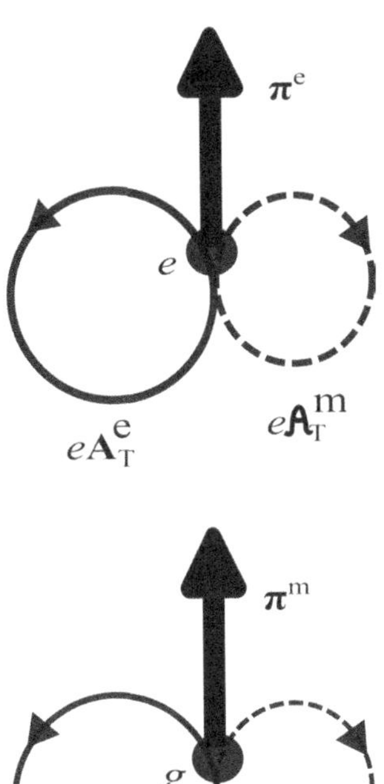

Figure 39.2 Schematic illustrations highlighting the three parts of the canonical momenta $(\mathbf{p}^{e},\mathbf{p}^{m})$ of a pair of interacting electric (e) and magnetic (g) monopoles. Mechanical momenta: π^{e}, π^{m}. Electromagnetic self-field momenta: $e\mathbf{A}_{T}^{e}, g\mathbf{A}_{T}^{m}$. The self-field contributions originate in the transverse (T) photon clouds attached to the bare monopoles. Coupling momenta $e\boldsymbol{\mathcal{A}}_{T}^{m}, -g\boldsymbol{\mathcal{A}}_{T}^{e}$. The photon-mediated coupling is proportional to the charge product e.g. The old $[\mathbf{A}_{T}^{e}, \mathbf{A}_{T}^{m}]$ and new $[\boldsymbol{\mathcal{A}}_{T}^{e}, \boldsymbol{\mathcal{A}}_{T}^{m}]$ transverse vector potentials are related via the space-time nonlocal equation $-c^{-1}\partial \mathbf{A}_{T}^{\alpha}/\partial t = \nabla \times \boldsymbol{\mathcal{A}}_{T}^{\alpha}$ [see. Eq. (39.74)]; $\alpha = e$ or g.

where

$$\mathbf{I}_{IJ}(t|\mathbf{r}_0) = \varepsilon_0 \int_{-\infty}^{\infty} (\mathbf{r} - \mathbf{r}_0) \times [\mathbf{E}_I(\mathbf{r}, t) \times \mathbf{B}_J(\mathbf{r}, t)]\, d^3r = I_{IJ}(t|\mathbf{0}) - \mathbf{r}_0 \times \mathbf{P}_{IJ}(t).$$

$$(39.101)$$

The last expression in Eq. (39.101) shows that it is sufficient to analyze the various contributions to the field angular moment about the origo of our coordinate system, viz.,

$$\mathbf{I}_{IJ}(t|\mathbf{0}) \, [\equiv \mathbf{I}_{IJ}(t)] = \varepsilon_0 \int_{-\infty}^{\infty} \mathbf{r} \times [\mathbf{E}_I(\mathbf{r}, t) \times \mathbf{B}_J(\mathbf{r}, t)]\, d^3r. \qquad (39.102)$$

As before, the (IJ) subscript runs over the four combinations (LL), (LT), (TL) and (TT). In the absence of magnetic monopoles

$$\mathbf{I}_{LL} = \mathbf{I}_{TL} = \mathbf{0}. \qquad (39.103)$$

39.6.2 Dynamic Saha-Wilson part

In the remaining part of 39.6 we limit our analysis to the case of a single pair of monopoles, with charges e and g. Generalizations to dyon pairs and an assembly of monopoles (dyons) are straightforward. Since, $\mathbf{P}_{LL}(t) = \mathbf{0}$, it appears from Eq. (39.102) that the $\mathbf{I}_{LL}$ part is independent of choice of reference point ($\mathbf{r}_0$). Hence,

$$\mathbf{I}_{LL}(t) = \varepsilon_0 \int_{-\infty}^{\infty} \mathbf{r} \times [\mathbf{E}_L(\mathbf{r}, t) \times \mathbf{B}(\mathbf{r}, t)]\, d^3 r. \tag{39.104}$$

Let

$$\mathbf{R}_{ge}(t) = \mathbf{r}^e(t) - \mathbf{r}^m(t) \tag{39.105}$$

be the instantaneous vectorial distance between the electric and magnetic monopoles, these having position vectors $\mathbf{r}^e(t)$ and $\mathbf{r}^m(t)$, respectively.

The calculation of $\mathbf{I}_{LL}(t)$ is carried out using the expression for $\mathbf{E}_L(\mathbf{r}, t)$ and $\mathbf{B}_L(\mathbf{r}, t)$ given in Eqs. (39.7) and (39.8), with the summations over α not in force. Since the integration over space is to be carried out at a fixed time, one may without loss of generality take $\mathbf{r}^m(t) = \mathbf{0}$, so that $\mathbf{R}_{ge}(t) = \mathbf{r}^e(t)$ here. Thus, one obtains

$$\mathbf{I}_{LL} = \frac{eg}{(4\pi)^2 \varepsilon_0 c} \int_{-\infty}^{\infty} \mathbf{r} \times \left[\frac{(\mathbf{r} - \mathbf{R}_{ge}(t)) \times \mathbf{r}}{|\mathbf{r} - \mathbf{R}_{ge}(t)|^3 r^3} \right] d^3 r$$

$$= \frac{eg}{(4\pi)^2 \varepsilon_0 c} \int_{-\infty}^{\infty} \mathbf{r} \times \left[\frac{\mathbf{r} - \mathbf{R}_{ge}(t)}{|\mathbf{r} - \mathbf{R}_{ge}(t)|^3} \right] \cdot \left[\frac{\mathbf{U} - \hat{\mathbf{r}}\hat{\mathbf{r}}}{r} \right] d^3 r. \tag{39.106}$$

By carrying out the integration in Eq. (39.106) [if needed the reader may find some details of the integration in [159]], one finally obtains

$$\mathbf{I}_{LL}(t) = \frac{eg}{4\pi\varepsilon_0 c} \hat{\mathbf{R}}_{eg}(t), \tag{39.107}$$

where $\hat{\mathbf{R}}_{eg}(t) = \mathbf{R}_{eg}(t)/R_{eg}(t)$ is a unit vector pointing from the electric monopole towards the magnetic monopole, i.e. $\hat{\mathbf{R}}_{eg}(t) = -\hat{\mathbf{R}}_{ge}(t)$. The result in Eq. (39.107) is a generalization of the results obtained originally by Saha [225, 226] and Wilson [275] for a static system (time independent $\mathbf{R}_{eg}$). The result in Eq. (39.107) is in agreement with Eq. (38.27) for $e_2 = 0$, $g_2 \equiv g$, $e_1 \equiv e$, $g_1 = 0$, remembering $\mathbf{r}/r = -\hat{\mathbf{R}}_{eg}(t)$.

39.6.3 Electromagnetic parts of the canonical particle angular momentum

The angular momentum

$$\mathbf{I}_{LT}(t) = \varepsilon_0 \int_{-\infty}^{\infty} \mathbf{r} \times [\mathbf{E}_L(\mathbf{r}, t) \times \mathbf{B}_T(\mathbf{r}, t)]\, d^3 r =$$

$$\varepsilon_0 \int_{-\infty}^{\infty} \mathbf{r} \times \{\mathbf{E}_L(\mathbf{r}, t) \times [\boldsymbol{\nabla} \times (\mathbf{A}_T^e(\mathbf{r}, t) + \boldsymbol{\mathcal{A}}_T^m(\mathbf{r}, t))]\}\, d^3 r \tag{39.108}$$

may be calculated in a manner completely identical to the one used in the absence of magnetic monopoles [$\boldsymbol{\mathcal{A}}_T^m(\mathbf{r}, t) = \mathbf{0}$]. Details of the calculation can be found in [157, 158], e.g. The final result for a single electric monopole is

$$\mathbf{I}_{LT}(t) = e\mathbf{r}^e(t) \times [\mathbf{A}_T^e(\mathbf{r}^e(t), t) + \boldsymbol{\mathcal{A}}_T^m(\mathbf{r}^e(t), t)]. \tag{39.109}$$

As expected, on the knowledge of the result for the electromagnetic momentum $\mathbf{P}_{LT}(t)$ which is given in Eq. (39.93), $\mathbf{I}_{LT}(t)$ becomes a function of the instantaneous position coordinate $(\mathbf{r}^e(t))$ of the electric monopole. Apart from an explicit appearance of $\mathbf{r}^e(t)$, the position coordinate enters via the transverse vector potential, $\mathbf{A}_T^e + \boldsymbol{\mathcal{A}}_T^m$, prevailing at a given time at the position. The sum of the mechanical and electromagnetic angular momenta equals the angular momentum of the canonical particle momentum that is

$$\mathbf{r}^e(t) \times \boldsymbol{\pi}^e(t) + \mathbf{I}_{LT}(t) = \mathbf{r}^e(t) \times \mathbf{p}^e(t), \tag{39.110}$$

with $\mathbf{p}^e(t)$ given by Eq. (39.94).

The field angular momentum, $\mathbf{I}_{TL}$ is given by

$$\mathbf{I}_{TL}(t) = \varepsilon_0 \int_{-\infty}^{\infty} \mathbf{r} \times [\mathbf{E}_T(\mathbf{r}, t) \times \mathbf{B}_L(\mathbf{r}, t)]\, d^3r =$$

$$\varepsilon_0 \int_{-\infty}^{\infty} \mathbf{r} \times \{c\mathbf{B}_L(\mathbf{r}, t) \times [\boldsymbol{\nabla} \times (\mathbf{A}_T^m(\mathbf{r}, t) - \boldsymbol{\mathcal{A}}_T^e(\mathbf{r}, t))]\}\, d^3r \tag{39.111}$$

The last expression in Eq. (39.111) is obtained using Eq. (39.75) and $\mathbf{E}_T \times \mathbf{B}_L = -\mathbf{B}_L \times \mathbf{E}_T$. A comparison of Eqs. (39.108) and (39.111), and with the result given in Eq. (39.109), we obviously obtain

$$\mathbf{I}_{TL}(t) = g\mathbf{r}^m(t) \times [\mathbf{A}_T^m(\mathbf{r}^m(t), t) - \boldsymbol{\mathcal{A}}_T^e(\mathbf{r}^m(t), t)]. \tag{39.112}$$

Grouped together with the mechanical momentum of the magnetic monopole, one gets a canonical particle momentum

$$\mathbf{r}^m(t) \times \boldsymbol{\pi}^m(t) + \mathbf{I}_{TL}(t) = \mathbf{r}^m(t) \times \mathbf{p}^m(t), \tag{39.113}$$

where $\mathbf{p}^m(t)$ is the canonical momentum of the magnetic monopole; Eq. (39.98).

39.6.4   Photon-field angular momentum

The angular momentum of the photon field is given by,

$$\mathbf{I}_{TT}(t) = \varepsilon_0 \int_{-\infty}^{\infty} \mathbf{r} \times [\mathbf{E}_T(\mathbf{r}, t) \times \mathbf{B}_T(\mathbf{r}, t)]\, d^3r. \tag{39.114}$$

A glance of Eqs. (39.83)–(39.85) allows one to conclude that $\mathbf{I}_{TT}(t)$ can be expressed in terms of the energy current density of photon wave function. Hence.

$$\mathbf{I}_{TT}(t) = \frac{1}{ic} \int_{-\infty}^{\infty} \mathbf{r} \times \left\{ \boldsymbol{\Psi}^\dagger(\mathbf{r}, t) \times [\hat{h}\boldsymbol{\Psi}(\mathbf{r}, t)] \right\} d^3r. \tag{39.115}$$

The angular momentum of the outgoing free-photon state $\boldsymbol{\Phi}(\mathbf{r}, t)$, is time independent due to angular momentum conservation. An explicit calculation based on Eqs. (39.86) and (39.87) shows that $d\mathbf{I}_{TT}/dt = \mathbf{0}$ for the free photon.

39.7   RETARDED NEAR FIELD OF A MAGNETIC MONOPOLES

In the near field of a dynamic magnetic monopole the magnetic field has both longitudinal and transverse parts, see Eqs. (39.8) and (39.18). From Eq. (39.3) one obtains for $\mathbf{B}_L(\mathbf{r}, t)$ the expression

$$c\mathbf{B}_L(\mathbf{r}, t) = -\frac{1}{\varepsilon_0} \int_{-\infty}^{t} \mathbf{J}_L^m(\mathbf{r}, t')dt' = -\frac{1}{\varepsilon_0}\mathbf{P}_L^m(\mathbf{r}, t), \qquad (39.116)$$

remembering the microscopic definition $\mathbf{J} = \partial\mathbf{P}/\partial t \Rightarrow \mathbf{J}_L^m = \partial\mathbf{P}_L^m/\partial t$, and assuming that there is no permanent source polarization $[\mathbf{P}_L^m(\mathbf{r}, -\infty) = \mathbf{0}]$. With the help of the longitudinal delta function in spherical contraction, Eq. (39.116) for the longitudinal magnetic field can be written in the integral form

$$c\mathbf{B}_L(\mathbf{r}, t) = -\frac{1}{4\pi\varepsilon_0} \int_{-\infty}^{\infty} \frac{1}{R^3}(\mathbf{U} - 3\mathbf{e_R}\mathbf{e_R}) \cdot \mathbf{P}^m(\mathbf{r}', t)d^3r' \qquad (39.117)$$

for $R = |\mathbf{r} - \mathbf{r}'| \neq 0$.

The transverse part of the magnetic near field, denoted by $\mathbf{B}_T^{\mathrm{NF}}(\mathbf{r}, t)$, is calculated from Eq. (39.18). The self-field contribution, $c\mathbf{B}_T^{\mathrm{SF}} = -\mathbf{P}_T^m/(3\varepsilon_0)$, given in Eq. (39.22), relates to the spatial confinement problem for a transverse photon emitted from a magnetic current density distribution $\mathbf{J}_T^m(\mathbf{r}, t)$, a phenomenon we leave out in what follows. Since the magnetic propagator $\mathbf{G}_M(\mathbf{R}, \tau)$ has no near-field part [see Eq. (39.20)], one is left with the following integral expression for the transverse part of the magnetic field:

$$c\mathbf{B}_T^{\mathrm{NF}}(\mathbf{r}, t) = -\mu_0 \int_{-\infty}^{\infty} \mathbf{G}_T^{\mathrm{NF}}(\mathbf{R}, \tau) \cdot \frac{\partial}{\partial t'}\mathbf{J}^m(\mathbf{r}', t')dt'd^3r', \qquad (39.118)$$

where

$$\mathbf{G}_T^{\mathrm{NF}}(\mathbf{R}, \tau) = -\frac{c^2\tau}{4\pi R^3}\Theta(\tau)\Theta\left(\frac{R}{c} - \tau\right)(\mathbf{U} - 3\mathbf{e_R}\mathbf{e_R}). \qquad (39.119)$$

As shown in [159], the expression for $c\mathbf{B}_T^{\mathrm{NF}}(\mathbf{r}, t)$ can be written in the form

$$c\mathbf{B}_T^{\mathrm{NF}}(\mathbf{r}, t) = \frac{1}{\varepsilon_0} \int_{-\infty}^{\infty} \frac{1}{4\pi R^3}(\mathbf{U} - 3\mathbf{e_R}\mathbf{e_R})$$
$$\cdot \left[\mathbf{P}^m(\mathbf{r}', t) - \mathbf{P}^m\left(\mathbf{r}', t - \frac{R}{c}\right) - \frac{R}{c}\mathbf{J}^m\left(\mathbf{r}', t - \frac{R}{c}\right)\right]d^3r'. \qquad (39.120)$$

By addition of the results in Eqs. (39.117) and (39.120) it appears that the total magnetic near field of the magnetic monopole is given by

$$c[\mathbf{B}_L(\mathbf{r}, t) + \mathbf{B}_T^{\mathrm{NF}}(\mathbf{r}, t)] = -\frac{1}{4\pi\varepsilon_0} \int_{-\infty}^{\infty} \frac{1}{R^3}(\mathbf{U} - 3\mathbf{e_R}\mathbf{e_R})$$
$$\cdot \left[\mathbf{P}^m\left(\mathbf{r}', t - \frac{R}{c}\right) + \frac{R}{c}\mathbf{J}^m\left(\mathbf{r}', t - \frac{R}{c}\right)\right]d^3r'. \qquad (39.121)$$

It is obvious from Eq. (39.121) that the total magnetic near field satisfies the Einstein causality criterion, viz.,

$$\mathbf{B}_L(\mathbf{r},t) + \mathbf{B}_T^{\mathrm{NF}}(\mathbf{r},t) = [\mathbf{B}_L + \mathbf{B}_T^{\mathrm{NF}}]\left(\mathbf{r}, t - \frac{R}{c}\right). \tag{39.122}$$

In a quantum physical setting a magnetic monopole is never completely at rest, and therefore it most often will be insufficient to treat the electric-magnetic monopole near-field interaction on the basis of non-retarded electrodynamics.

CHAPTER 40

Magnetic Monopoles in Non-Abelian Gauge Symmetry

40.1 ISOVECTOR MAXWELL EQUATIONS

In the following we set

$$\hbar = c = 1, \tag{40.1}$$

as it is common in magnetic monopole research, e.g. In the non-Abelian extension of the $U(1)$ description of electrodynamics we make the translations

$$g \Rightarrow e, \ \mathbf{W}_\mu \Rightarrow \mathbf{A}_\mu, \ \mathbf{W}_{\mu\nu} \Rightarrow \mathbf{F}_{\mu\nu} \tag{40.2}$$

in the general Yang-Mills theory described in Section 34.2. The gauge-covariant derivative

$$D_\mu = \partial_\mu + e\mathbf{A}_\mu \times, \tag{40.3}$$

which here depends on a four-potential, $\{\mathbf{A}_\mu\}$, where each component ($\mu = 0 - 3$) $\mathbf{A}_\mu$ is a three-vector, acts on the real triplet Higgs boson field

$$\boldsymbol{\phi} = (\phi_1, \phi_2, \phi_3), \tag{40.4}$$

as follows:

$$D_\mu\boldsymbol{\phi} = \partial_\mu\boldsymbol{\phi} + e\mathbf{A}_\mu \times \boldsymbol{\phi}, \tag{40.5}$$

cf. Eq. (34.40).

The field strength

$$\mathbf{F}_{\mu\nu} = \partial_\mu\mathbf{A}_\nu - \partial_\nu\mathbf{A}_\mu + e\mathbf{A}_\mu \times \mathbf{A}_\nu, \tag{40.6}$$

484

DOI: 10.1201/9781003029458-40

is expressed in terms of the four-potential in the Yang-Mills form [Eq. (34.50)]. Each $(\mu\nu)$-component is a three-vector quantity, and $\{\mathbf{F}_{\mu\nu}\}$ is antisymmetric. In order to make at a later stage the connection to $U(1)$-electrodynamics, we introduce "electric" and "magnetic" isovectors

$$\mathbf{E}_{\text{iso}} \equiv (\mathbf{E}_1, \mathbf{E}_2, \mathbf{E}_3), \tag{40.7}$$

and

$$\mathbf{B}_{\text{iso}} \equiv (\mathbf{B}_1, \mathbf{B}_2, \mathbf{B}_3). \tag{40.8}$$

In terms of these isovectors vectorial with three-vector components, the field strength is (remembering $c = 1$)

$$\mathbf{F}_{\mu\nu} = \begin{pmatrix} 0 & -\mathbf{E}_1 & -\mathbf{E}_2 & -\mathbf{E}_3 \\ \mathbf{E}_1 & 0 & \mathbf{B}_3 & -\mathbf{B}_2 \\ \mathbf{E}_2 & -\mathbf{B}_3 & 0 & \mathbf{B}_1 \\ \mathbf{E}_3 & \mathbf{B}_2 & -\mathbf{B}_1 & 0 \end{pmatrix} \tag{40.9}$$

In covariant notation the inhomogenous field equation has the form

$$D^\nu \mathbf{F}_{\mu\nu} = e\mathbf{J}_\mu, \tag{40.10}$$

where

$$\mathbf{J}_\mu = (D_\mu\phi) \times \phi, \tag{40.11}$$

as the reader immediately can see making the translations $\mathbf{W}_{\mu\nu} \to \mathbf{F}_{\mu\nu}$ and $g \Rightarrow e$ in Eq. (34.54). By means of Eqs. (40.3) and (40.6) it can be proved that the field strength satisfies a homogeneous relation

$$D_\lambda \mathbf{F}_{\mu\nu} + D_\mu \mathbf{F}_{\nu\lambda} + D_\nu \mathbf{F}_{\lambda\mu} = \mathbf{0}. \tag{40.12}$$

The attentive reader may observe that the field equations in Eqs. (40.10) and (40.12) have formal forms identical to the inhomogeneous and homogeneous covariant Maxwell equations, given in Eqs. (30.1) and (30.2), e.g.

Let us introduce the covariant three-vector ($\mu = 1, 2, 3$) gauge derivative operator

$$\mathbf{D} \equiv (D_1, D_2, D_3), \tag{40.13}$$

and the three-vector *isocurrent density*

$$\mathbf{J}_{\text{iso}} \equiv (\mathbf{J}_1, \mathbf{J}_2, \mathbf{J}_3). \tag{40.14}$$

With metric signature $(-1, 1, 1, 1)$, $\{D_\mu\} = (D_0, \mathbf{D})$ and $\{D^\mu\} = (D^0 = -D_0, \mathbf{D})$, and furthermore $\{\mathbf{J}_\mu\} = (\mathbf{J}_0, \mathbf{J}_{\text{iso}})$ and $\{\mathbf{J}^\mu\} = (\mathbf{J}^0 = -\mathbf{J}_0, \mathbf{J}_{\text{iso}})$. In the three-vector notation the field isovectors $(\mathbf{E}_{\text{iso}}, \mathbf{B}_{\text{iso}})$ satisfy the following set of equations

$$\mathbf{D} \cdot \mathbf{B}_{\text{iso}} = \mathbf{0}, \tag{40.15}$$

$$\mathbf{D} \times \mathbf{E}_{\text{iso}} + D_0 \mathbf{B}_{\text{iso}} = \mathbf{0} \tag{40.16}$$

$$\mathbf{D} \cdot \mathbf{E}_{\text{iso}} = e\mathbf{J}^0, \tag{40.17}$$

$$\mathbf{D} \times \mathbf{B}_{\text{iso}} - D_0 \mathbf{E}_{\text{iso}} = e\mathbf{J}_{\text{iso}}. \tag{40.18}$$

The set appears in a formal form which is identical to the microscopic Maxwell equations in $U(1)$ electrodynamcis. The two homogenous $U(2)$ Maxwell equations [Eqs. (40.15) and (40.16)] are derived from the homogeous equation for the field strength [Eq. (40.12)], $\mathbf{F}_{\mu\nu}$ having the form given in Eq. (40.9). Thus, with no indices equal to zero, and setting $\lambda = 1$ (or if wished $\lambda = 2$ or 3) one obtains $0 = D_1\mathbf{F}_{23} + D_2\mathbf{F}_{31} + D_3\mathbf{F}_{12} = D_1\mathbf{B}_1 + D_2\mathbf{B}_2 + D_3\mathbf{B}_3 = \mathbf{D}\cdot\mathbf{B}_{\text{iso}}$, which is Eq. (40.15). With one of the indices equal to zero, say $\lambda = 0$, and the other indices equal to $(\mu,\nu) = (1,2)$ one obtains $\mathbf{0} = D_0\mathbf{F}_{12} + D_1\mathbf{F}_{20} + D_2\mathbf{F}_{01} = D_0\mathbf{B}_3 + (\mathbf{D}\times\mathbf{E})_3$, i.e. , the third component of Eq. (40.16). With $\lambda = 0$ and $(\mu,\nu) = (1,3)$ and $(2,3)$ the remaining two components of Eq. (40.16) are obtained. The two inhomogenous Eqs. (40.17) and (40.18) are obtained from Eq. (40.10). Hence, as the reader may show, setting $\mu = 0$, Eq. (40.17) follows, and by taking in turn $\mu = 1, 2$ and 3, the three components of Eq. (40.18) are obtained.

40.2   SYMMETRIZED ISOVECTOR MAXWELL EQUATIONS

Inspired by symmetrized $U(1)$ electrodynamics we introduce a dual-field strength

$$\{\mathbf{G}_{\mu\nu}\} \equiv \{\mathbf{F}_{\mu\nu}(\mathbf{E}_{\text{iso}} \to \mathbf{B}_{\text{iso}}, \mathbf{B}_{\text{iso}} \to -\mathbf{E}_{\text{iso}})\} \tag{40.19}$$

explicitly given by the antisymmetric form

$$\{\mathbf{G}_{\mu\nu}\} = \begin{pmatrix} 0 & -\mathbf{B}_1 & -\mathbf{B}_2 & -\mathbf{B}_3 \\ \mathbf{B}_1 & 0 & -\mathbf{E}_3 & \mathbf{E}_2 \\ \mathbf{B}_2 & \mathbf{E}_3 & 0 & -\mathbf{E}_1 \\ \mathbf{B}_3 & -\mathbf{E}_2 & \mathbf{E}_1 & 0 \end{pmatrix} \tag{40.20}$$

In covariant notation the symmetrized field equations now are

$$D^\nu\mathbf{F}_{\mu\nu} = e\mathbf{J}^e_\mu, \tag{40.21}$$

$$D^\nu\mathbf{G}_{\mu\nu} = g\mathbf{J}^m_\mu, \tag{40.22}$$

where $\mathbf{J}^m_\mu$ is the magnetic (superscript m) monopole four-current density. Eq. (40.21) is identical to Eq. (40.10) just with superscript e added to the electric four-current density [given in Eq. (40.11) for the triplet Higgs field].

In the three-vector notation, the homogenous isovector Maxwell equations now are replaced by inhomogeneous equations, derived from Eq. (40.22) in the same manner as Eqs. (40.17) and (40.18) were derived from Eq. (40.10) [or Eq. (40.21)]. Remembering that $\{\mathbf{J}^m_\mu\} = (\mathbf{J}^m_0, \mathbf{J}^m_{\text{iso}})$ and thus $\{\mathbf{J}^{m,\mu}\} = (\mathbf{J}^{0,m} = -\mathbf{J}^m_0, \mathbf{J}^m_{\text{iso}})$ [with signature $(-1,+1,+1,+1)$], in three-vector notation the field isovectors satisfy the following set of symmetrized Maxwell equations

$$\mathbf{D}\cdot\mathbf{E}_{\text{iso}} = e\mathbf{J}^{e,0}, \tag{40.23}$$

$$\mathbf{D}\cdot\mathbf{B}_{\text{iso}} = g\mathbf{J}^{m,0}, \tag{40.24}$$

$$\mathbf{D}\times\mathbf{B}_{\text{iso}} = D_0\mathbf{E}_{\text{iso}} + e\mathbf{J}^e_{\text{iso}}, \tag{40.25}$$

$$-\mathbf{D}\times\mathbf{E}_{\text{iso}} = D_0\mathbf{B}_{\text{iso}} + g\mathbf{J}^m_{\text{iso}}. \tag{40.26}$$

The reader may note that the set of isovector field equations above is form-identical to the symmetrized set of $U(1)$ Maxwell equations [Eqs. (38.3)–(38.6)].

40.3   THE HEDGEHOG MONOPOLE

40.3.1   The 't Hooft-Polyakov approach

In the framework of Maxwell electrodynamics, with Abelian group $U(1)$, magnetic charges may be added to the theory. The addition gives one a more symmetric theory between electricity and magnetism, but although symmetrized $U(1)$ electrodynamics appears attractive from an aesthetic point of view this does not amount to a requirement that magnetic monopoles exist. When the gauge symmetry is enlarged to a certain non-Abelian group, and spontaneous symmetry breaking is introduced 't Hooft [247] and Polyakkov [214,215] showed that the field equations yield a solution which corresponds to a magnetic charge. Since the matter (Higgs) and gauge fields carry electric charge *only*, the origin of the magnetic charge is *topological*. The 1974 discovery by t'Hooft and Polyakov was based on the $SO(3)$ symmetry group, i.e., a model of the type described by Georgi and Glashow [93] two years earlier.

Let us consider a theory with an $O(3)$ symmetry group, containing an isovector Higgs field $\{\phi^a\}$ with group index a and a gauge field $\{F^a_{\mu\nu}\}$. The Higgs field in this approach has *three real components* ($a = 1-3$) and the theory is to some extent similar to the $SU(2)_L \otimes U(1)$ electro-weak theory with its two complex Higgs components.

The Lagrangian density used is

$$\mathcal{L} = \frac{1}{2}\left[(D_\mu\phi^a)(D^\mu\phi^a) + \frac{1}{2}\mu^2(\phi^a\phi^a) - \frac{1}{2}\left(\frac{\mu}{f}\right)^2(\phi^a\phi^a)^2\right] - \frac{1}{4}F^a_{\mu\nu}F^{a\mu\nu}. \qquad (40.27)$$

This form is just the covariant extension of Eq. (36.6), with (subscript) group index i replaced by (superscript) group index a, for convenience. As usual the summation over the repeated group index is kept implicit. In a notation where Eq. (40.27) is written in terms of the Levi-Civita symbol ε^{abc}, the isopin gauge field's a-component is

$$F^a_{\mu\nu} = \partial_\mu A^a_\nu - \partial_\nu A^a_\mu + e\varepsilon^{abc}A^b_\mu A^c_\nu, \qquad (40.28)$$

and the covariant derivative of the Higgs component ϕ^a [Eq. (40.5)] reads

$$D_\mu\phi^a = \partial_\mu\phi^a + e\varepsilon^{abc}A^b_\mu\phi^c. \qquad (40.29)$$

The equation of motion for ϕ^a is obtained from the covariant Euler-Lagrange equation

$$D_\mu\left(\frac{\partial\mathcal{L}}{\partial(D_\mu\phi^a)}\right) = \frac{\partial\mathcal{L}}{\partial\phi^a}, \qquad (40.30)$$

giving by a straightforward calculation

$$D_\mu(D^\mu\phi^a) = \left[\mu^2 - 2\left(\frac{\mu}{f}\right)^2(\phi^b\phi^b)\right]\phi^a, \qquad (40.31)$$

where μ is the mass of the Higgs particle, and with the summation convention over repeated group index b in force.

For our route to the topological magnetic monopole we shall be interested only in time-independent solutions to the field equations, thus setting the gauge potential

$$A_0^a = 0. \tag{40.32}$$

Furthermore, only the behaviour of the fields far from the origin ($r = 0$) is needed in the following. The suggestion made by t'Hooft was to take for $r \to \infty$ a scalar field of the form

$$\phi^a = \frac{f}{\sqrt{2}} \frac{r^a}{r}, \quad r \to \infty. \tag{40.33}$$

In a manner of speaking $\{\phi^a\}$ is "radial". Thus in the x direction in space the field has only an isospin "1" component, in the y direction only a "2" component, and in the z direction only a "3" component. Polyakov named this form a hedgehog. Since

$$\phi^a \phi^a = \frac{f^2}{2} \frac{r^a r^a}{r^2} = \frac{f^2}{2}, \tag{40.34}$$

the magnitude of the field tends towards to the vacuum value $|\phi| = f/\sqrt{2}$; cf. the analysis in Section 36.2.

We now require that the asymptotic form given in Eq. (40.33) satisfies the equation of motion for ϕ^a. By inserting Eq. (40.34), with the index replacement $a \to b$, into the right-hand side of Eq. (40.31), it appears that this side becomes zero, so that one must demand that $D_\mu(D^\mu \phi^a) = 0$. For a time-independent solution we thus look for a gauge potential giving

$$D_i \phi^a = 0, \quad i = 1 - 3. \tag{40.35}$$

By combination of Eqs. (40.29) and (40.35) it is seen that the gauge potential $\{A_i^b\}$ must satisfy the relation

$$e\varepsilon^{abc} A_i^b r^c = \frac{r^a r^i}{r^2} - \delta^{ia}. \tag{40.36}$$

This suggests that A_i^b must be proportional to $e^{-1} r^{-2} r^b$ and contain an appropriate Levi-Civity symbol. The guess

$$A_i^a = -\varepsilon_{iab} \frac{r^b}{er^2} \tag{40.37}$$

for the gauge potential A_i^a solves Eq. (40.36), as the reader may prove. Hint: Define

$$K_i^a = -\varepsilon^{abc} \varepsilon_{ibm} r^c r^m. \tag{40.38}$$

(i) Calculate K_1^1 and K_2^1. (ii) Show $K_2^1 = K_1^2$. (iii) Infer the general term K_i^a from (i) and (ii).

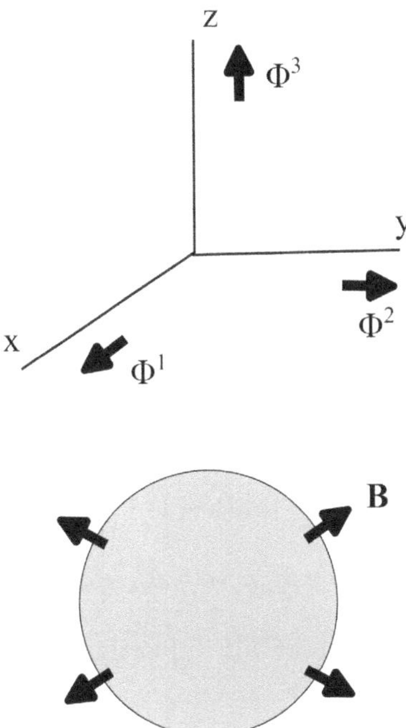

Figure 40.1 Schematic illustration of the asymptotic behaviour of the Higgs scalar field, $\phi^a = (f/\sqrt{2})r^a/r$ and the radial magnetic field $\mathbf{B} = \mathbf{r}/(er^3)$, originating solely in the Higgs field. Along the Cartesian axes in direct space the scalar field has only isospin components ϕ^1, ϕ^2 and ϕ^3, as indicated. The hedgehog monopole is sketched as a grey-toned domain. The monopole has a magnetic charge $g = e^{-1}$. From an electrodynamic point view the Higgs scalar field may be considered as an example of a DYNAMIC STRUCTURAL VACUUM.

Let us repeat our conclusion for the $SO(3)$ theory obtained from the Lagrangian density given in Eq. (40.27): Asymptotically ($r \to \infty$), we have obtained a static solution in which the scalar field is

$$\phi^a = \frac{f}{\sqrt{2}}\frac{r^a}{r}, \quad a = 1-3, \tag{40.39}$$

and the gauge potentials have the non-trivial form

$$A_i^a = -\,\varepsilon_{iab}\frac{r^b}{er^2}, \quad i = 1-3 \tag{40.40}$$

$$A_0^a = 0. \tag{40.41}$$

Hence, asymptotically $|\phi|$ takes on its vacuum value $(\phi^a\phi^a)^{1/2} = f/\sqrt{2}$, and is co-variantly constant, $D_\mu\phi^a = 0$. The asymptotic form of the gauge and scalar fields is indicated in Fig. 40.1.

40.3.2　Generalized electromagnetic field and potential

In order to speak of a magnetic field in the framework of the non-Abelian theory described in Subsection 40.3.1, one must generalize the *definition* of the electromagnetic field $\{F_{\mu\nu}\}$. Following the suggestion by 't Hooft [247], we put

$$F_{\mu\nu} \equiv \frac{1}{|\phi|}\phi^a F_{\mu\nu}^a - \frac{1}{e|\phi|^3}\varepsilon_{abc}\phi^a(D_\mu\phi^b)(D_\nu\phi^c). \tag{40.42}$$

The definition in Eq. (40.42) is (i) gauge invariant, and (ii) yields the usual definition in the gauge where the Higgs fields has only one component everywhere, say along the z direction, provided the definition of the electromagnetic potential $\{A_\mu\}$ is generalized to

$$A_\mu \equiv \frac{1}{|\phi|}\phi^a A_\mu^a. \tag{40.43}$$

Thus, with a gauge rotation to $\phi = |\phi|(0,0,1)$ everywhere, one obtains

$$F_{\mu\nu} = F_{\mu\nu}^3 = \partial_\mu A_\nu^3 - \partial_\nu A_\mu^3, \tag{40.44}$$

which is the usual $U(1)$ definition of $F_{\mu\nu}$, identifying $A_\nu^3 = A_\mu$ as the Maxwell four-potential.

40.3.3 Time-independent vector potential and magnetic field in the near-field zone

Let us consider the gauge invariant definition of the four-potential, given in Eq. (40.43), in the framework of the time-independent $[A_0^\mu = 0]$ hedgehog model. Of course, $A_0 = 0$, and for $i = 1-3$

$$A_i = \frac{1}{|\phi|}\phi^a A_i^a = \frac{r_a}{r}(-\varepsilon_{iab})\frac{r_b}{er^2}. \tag{40.45}$$

Here, and in the following we have changed superscript a $[r^a]$ to subscript a $[r_a]$ for calculational clarity. In Eq. (40.45) nonvanishing contributions to the summation over a and b only comes from (i) $a \neq b$, and (ii) $a \neq i$, $b \neq i$. This fact immediately gives

$$A_i = -\frac{r_a r_b}{er^3}(\varepsilon_{iab} + \varepsilon_{iba}) = 0, \tag{40.46}$$

since $\varepsilon_{iba} = -\varepsilon_{iab}$.

Hence, we have come to the conclusion that all the electromagnetic field is contributed by the Higgs field. This was to be expected since a gauge rotation to

$$\frac{\{\phi^a\}}{|\phi|} = (0,0,1) \tag{40.47}$$

must give one the usual Maxwell field tensor

$$F_{\mu\nu} = \partial_\mu A_\nu^3 - \partial_\nu A_\mu^3, \tag{40.48}$$

and therefore

$$F_{ij} = \partial_i A_j^3 - d_j A_i^3 = 0, \tag{40.49}$$

in agreement with the fact that the presence of only electric charges cannot provide one with a non-retarded magnetic field from a stationary point particle (placed at $\mathbf{r} = \mathbf{0}$); see Eqs. (38.17) and (38.18).

With the metric signature $(+, -, -, -)$ [$\{\partial^\mu\} = (\partial/\partial x_0, -\boldsymbol{\nabla})$, $\{\partial_\mu\} = (\partial/\partial x_0, \boldsymbol{\nabla})$], the part of the time-independent antisymmetric ($F_{\nu\mu} = -F_{\mu\nu}$) field tensor, which relates to the magnetic field, has the form

$$\{F_{ij}\} = \begin{pmatrix} 0 & F_{12} & F_{13} \\ -F_{12} & 0 & F_{23} \\ -F_{13} & -F_{23} & 0 \end{pmatrix} = \begin{pmatrix} 0 & -B_3 & B_2 \\ B_3 & 0 & -B_1 \\ -B_2 & B_1 & 0 \end{pmatrix}. \tag{40.50}$$

We here use the metric signature $(+, -, -, -)$ because it is most often used in hedgehog analyses.

The field tensor of the hedgehog solution is calculated from

$$F_{ij} = \frac{1}{|\phi|} \phi^a F_{ij}^a, \tag{40.51}$$

since Eq. (40.42) one has $D_\mu \phi^a = 0$ and $F_{0\mu} = F_{\mu 0} = 0$. Although it is a straightforward matter to obtain the various components of the field tensor from the expression

$$F_{ij} = \frac{1}{|\phi|} \phi^a \left[\partial_i A_j^a - \partial_j A_i^a + e\varepsilon^{abc} A_i^b A_j^c \right], \tag{40.52}$$

it is instructive to look at the explicit results obtained for the three parts of Eq. (40.52):

$$F_{ij} = \alpha_{ij} + \beta_{ij} + \gamma_{ij}, \tag{40.53}$$

where

$$\alpha_{ij} \equiv \frac{1}{|\phi|} \phi^a \partial_i A_j^a, \tag{40.54}$$

$$\beta_{ij} \equiv -\frac{1}{|\phi|} \phi^a \partial_j A_i^a, \tag{40.55}$$

$$\gamma_{ij} \equiv \frac{e}{|\phi|} \phi^a \varepsilon^{abc} A_i^b A_i^c. \tag{40.56}$$

We start by a calculation of the three terms' contribution to F_{12}. Thus,

$$\alpha_{12} = \frac{1}{|\phi|} \phi^a \partial_1 A_2^a = \frac{r_a}{r} \partial_1 \left(-\varepsilon_{2ab} \frac{r_b}{er^2} \right) = \frac{r_1}{r} \partial_1 \left(-\varepsilon_{213} \frac{r_3}{er^2} \right) + \frac{r_3}{r} \partial_1 \left(-\varepsilon_{231} \frac{r_1}{er^2} \right), \tag{40.57}$$

since only $(a, b) = (1, 3)$ and $(3, 1)$ contribute. Remembering that $-\varepsilon_{213} = \varepsilon_{231} = 1$, one obtains upon simple differentiations

$$\alpha_{12} = -\frac{r_3}{er^3}. \tag{40.58}$$

We urge the reader to show that a calculation similar to the one for α_{12} gives

$$\beta_{12} = -\frac{r_3}{er^3}. \tag{40.59}$$

The two terms of the linear (in A_i^a) part hence have the same magnitude. The non-linear term is calculated as follows:

$$
\begin{aligned}
\gamma_{12} &= \frac{r_a}{r}\left(e\varepsilon^{abc}A_1^b A_2^c\right) = \frac{e}{r}\left[r_1\varepsilon_{1bc}A_1^b A_2^c + r_2\varepsilon_{2bc}A_1^a A_2^c + r_3\varepsilon_{3bc}A_1^b A_2^c\right] \\
&= \frac{e}{r}\left\{r_1\left[\varepsilon_{123}A_1^2 A_2^3 + \varepsilon_{132}A_1^3 A_2^2\right] + r_2\left[\varepsilon_{213}A_1^1 A_2^3 + \varepsilon_{231}A_1^3 A_2^1\right]\right. \\
&\qquad \left. + r_3\left[\varepsilon_{312}A_1^1 A_2^2 + \varepsilon_{321}A_1^2 A_2^1\right]\right\} \\
&= \frac{e}{r}\left[r_1 A_1^2 A_2^3 + r_2 A_1^3 A_2^1 - r_3 A_1^2 A_2^1\right].
\end{aligned}
\tag{40.60}
$$

The last expression is obtained using that (i) $A_i^i = 0$ $\forall i$; See Eq. (40.37), and (ii) $\varepsilon_{123} = \varepsilon_{231} = -\varepsilon_{321} = +1$. Inserting the explicit expressions for A_1^2, A_2^3, A_1^3, A_2^1, taken from Eq. (40.37) into Eq. (40.60) one obtains

$$\gamma_{12} = \frac{r_3}{er^5}\left(r_1^2 + r_2^2 + r_3^2\right) = \frac{r_3}{er^2}. \tag{40.61}$$

By addition of Eqs. (40.58), (40.59) and (40.61) we finally get

$$F_{12} = -\frac{r_3}{er^3} = -\varepsilon_{123}\frac{r_3}{er^3}. \tag{40.62}$$

On second thoughts it becomes clear that the hedgehog field tensor has the following components

$$F_{ij} = -\varepsilon_{ijk}\frac{r_k}{er^3}. \tag{40.63}$$

A comparison to Eq. (40.50) then shows that the kth element of the magnetic field is given by

$$B_k = \frac{r_k}{er^3}. \tag{40.64}$$

One thus may conclude that the hedgehog model leads to a radial magnetic monopole field

$$\mathbf{B} = \frac{1}{e}\frac{\mathbf{r}}{r^3} = g\frac{\mathbf{r}}{r^3}, \tag{40.65}$$

remembering Eq. (38.18), with $g_2 = g$ and $4\pi\varepsilon_0 c = 1$. The hedgehog model hence leads to the quantization condition

$$eg = 1, \tag{40.66}$$

which is twice the Dirac unit, given in Eq. (38.50); [with $4\pi\varepsilon_0 c = 1$, $\hbar = 1$, and for $n = 1$].

40.4 ISOVECTOR DYNAMICS IN RIEMANN-SILBERSTEIN APPROACH

40.4.1 Complex field isovectors

Inspired by the Abelian $U(1)$ photon wave mechanical theory, as formulated in terms of the Riemann-Silberstein (RS) vectors belonging to the positive and negative photon helicities, we now extend this formulation to the non-Ablian $SU(2)$ gauge description. Our starting point is the isovector Maxwell formalism described in Section 40.1 and 40.2.

Let us thus introduce two complex field isovectors by the following definitions:

$$\mathbf{F}_{\pm}^{\text{iso}} \equiv \mathbf{E}^{\text{iso}} \pm i\mathbf{B}^{\text{iso}}, \tag{40.67}$$

remembering we took $c = 1$. Formally, the $\mathbf{F}_{\pm}^{\text{iso}}$'s are analogous in form to the well known RS vectors used in the classical photon energy wave mechanics [63, 158], see also Section 4.3. In the $U(1)$ theory the Riemann-Silberstein vectors may be expressed in a simple manner in terms of the four-potential, $\{A_\mu\}$. In the $SU(2)$ theory the relation to the vectorial four-potential, $\{\mathbf{A}_\mu\}$ is more complicated. Hence,

$$\mathbf{F}_{\pm}^{\text{iso}} = (\mathbf{F}_{10} \pm i\mathbf{F}_{23}, \mathbf{F}_{20} \pm i\mathbf{F}_{31}, \mathbf{F}_{30} \pm i\mathbf{F}_{12}), \tag{40.68}$$

where $[i = 1 - 3]$

$$\mathbf{F}_{i0} = \partial_i \mathbf{A}_0 - \partial_0 \mathbf{A}_i + e\mathbf{A}_i \times \mathbf{A}_0, \tag{40.69}$$

$$\mathbf{F}_{ij} = \partial_i \mathbf{A}_j - \partial_j \mathbf{A}_i + e\mathbf{A}_i \times \mathbf{A}_j, \tag{40.70}$$

with $(i, j) = (2, 3), (3, 1), (1, 2)$.

40.4.2 Dynamical equation for RS isovector

In the absence of magnetic monopoles the electric and magnetic isovectors satisfy the set of isovector Maxwell equations, given in Eqs. (40.15)–(40.18). By adding Eqs. (40.16) and (40.18) [multiplied by, respectively $+i$ and $-i$] it appears that the RS isovector satisfies the inhomogenous dynamical equations

$$iD_0\mathbf{F}_{\pm}^{\text{iso}} = \pm\mathbf{D} \times \mathbf{F}_{\pm}^{\text{iso}} - ie\mathbf{J}_{\text{iso}}. \tag{40.71}$$

From Eqs. (40.15) and (40.17) it is realized that the divergence of the RS isovectors is given by the relation

$$\mathbf{D} \cdot \mathbf{F}_{\pm}^{\text{iso}} = e\mathbf{J}^0. \tag{40.72}$$

The reader should remember that the gauge covariant derivative, D_μ, also depends on the vectorial four-potential [see Eq. (40.3)].

In the symmetrized isovector Maxwell equations, given by the set in Eqs. (40.23)–(40.26), the reader may show that the RS isovectors satisfy the following extended dynamical equations:

$$iD_0\mathbf{F}_{\pm}^{\text{iso}} = \pm\mathbf{D} \times \mathbf{F}_{\pm}^{\text{iso}} - i(e\mathbf{J}_{\text{iso}}^e \pm ig\mathbf{J}_{\text{iso}}^m). \tag{40.73}$$

Furthermore,

$$\mathbf{D} \cdot \mathbf{F}_{\pm}^{\text{iso}} = e\mathbf{J}^{e,0} \pm ig\mathbf{J}^{m,0}. \tag{40.74}$$

40.5   STRUCTURAL VACUUM IN $SO(3)$

40.5.1   Isoscalar feild tensor

It is instructive to reformulate the RS formalism in such a manner that the generalized electromagnetic four-potential $\{A_\mu\}$, Eq. (40.43), appears in explicit form. For this purpose let us rewrite the definition of the generalized electromagnetic field tensor $\{F_{\mu\nu}\}$, which elements are given by

$$F_{\mu\nu} = \hat{\phi} \cdot \left[\mathbf{F}_{\mu\nu} - \frac{1}{e}(D_\mu\hat{\phi}) \times (D_\nu\hat{\phi}) \right], \tag{40.75}$$

where

$$\hat{\phi} = \frac{\phi}{|\phi|} \tag{40.76}$$

is the unit vector of the triplet Higgs boson field. The form given in Eq. (40.75) is just a vectorial rewriting of the 't Hooft expression, given in Eq. (40.42). Let us start by inserting the explicit expression for the covariant derivative, Eq. (40.3), into the $(D_\mu\hat{\phi}) \times (D_\nu\hat{\phi})$ product. Thus,

$$\begin{aligned}
(D_\mu\hat{\phi}) \times (D_\nu\hat{\phi}) &= (\partial_\mu\hat{\phi} + e\mathbf{A}_\mu \times \hat{\phi}) \times (\partial_\nu\hat{\phi} + e\mathbf{A}_\nu \times \hat{\phi}) \\
&= (\partial_\mu\hat{\phi}) \times (\partial_\nu\hat{\phi}) + e\left[(\partial_\mu\hat{\phi}) \times (\mathbf{A}_\nu \times \hat{\phi}) - (\partial_\nu\hat{\phi}) \times (\mathbf{A}_\mu \times \hat{\phi}) \right] \\
&\quad + e^2 (\mathbf{A}_\mu \times \hat{\phi}) \times (\mathbf{A}_\nu \times \hat{\phi}),
\end{aligned} \tag{40.77}$$

and then

$$\frac{1}{e}\hat{\phi} \cdot (D_\mu\hat{\phi}) \times (D_\nu\hat{\phi}) = \frac{1}{e}\hat{\phi} \cdot (\partial_\mu\hat{\phi}) \times (\partial_\nu\hat{\phi}) + L_{\mu\nu} + N_{\mu\nu}, \tag{40.78}$$

where

$$L_{\mu\nu} = \hat{\phi} \cdot \left[(\partial_\mu\hat{\phi}) \times (\mathbf{A}_\nu \times \hat{\phi}) - (\partial_\nu\hat{\phi}) \times (\mathbf{A}_\mu \times \hat{\phi}) \right] \tag{40.79}$$

is the part which is linear (L) in the gauge potential, and

$$N_{\mu\nu} = e\hat{\phi} \cdot (\mathbf{A}_\mu \times \hat{\phi}) \times (\mathbf{A}_\nu \times \hat{\phi}) \tag{40.80}$$

the part which is nonlinear (N) in $\{A_\mu\}$. Elementary manipulations show that expression for $N_{\mu\nu}$ can be simplified to

$$N_{\mu\nu} = e\hat{\phi} \cdot \mathbf{A}_\mu \times \mathbf{A}_\nu. \tag{40.81}$$

The linear term can be simplified utilizing that

$$\hat{\phi} \cdot (\partial_\mu\hat{\phi}) = 0, \tag{40.82}$$

a result which follows from the fact that $\partial_\mu(\hat{\phi} \cdot \hat{\phi}) = 0$. Since

$$\hat{\phi} \cdot (\partial_\mu\hat{\phi}) \times (\mathbf{A}_\nu \times \hat{\phi}) = -\mathbf{A}_\nu \cdot (\partial_\mu\hat{\phi}), \tag{40.83}$$

one obtains

$$L_{\mu\nu} = \mathbf{A}_\mu \cdot (\partial_\nu \hat{\boldsymbol{\phi}}) - \mathbf{A}_\nu \cdot (\partial_\mu \hat{\boldsymbol{\phi}}). \tag{40.84}$$

Altogether,

$$\frac{1}{e}\hat{\boldsymbol{\phi}} \cdot (D_\mu \hat{\boldsymbol{\phi}}) \times (D_\nu \hat{\boldsymbol{\phi}}) = \frac{1}{e}\hat{\boldsymbol{\phi}} \cdot (\partial_\mu \hat{\boldsymbol{\phi}}) \times (\partial_\nu \hat{\boldsymbol{\phi}}) + \mathbf{A}_\mu \cdot (\partial_\nu \hat{\boldsymbol{\phi}}) - \mathbf{A}_\nu \cdot (\partial_\mu \hat{\boldsymbol{\phi}})$$
$$+ e\hat{\boldsymbol{\phi}} \cdot \mathbf{A}_\mu \times \mathbf{A}_\nu. \tag{40.85}$$

The first scalar product in Eq. (40.75) is rewritten as follows:

$$\hat{\boldsymbol{\phi}} \cdot \mathbf{F}_{\mu\nu} = \hat{\boldsymbol{\phi}} \cdot [\partial_\mu \mathbf{A}_\nu - \partial_\nu \mathbf{A}_\mu + e\mathbf{A}_\mu \times \mathbf{A}_\nu]$$
$$= \partial_\mu(\hat{\boldsymbol{\phi}} \cdot \mathbf{A}_\nu) - \partial_\nu(\hat{\boldsymbol{\phi}} \cdot \mathbf{A}_\mu) - (\partial_\mu \hat{\boldsymbol{\phi}}) \cdot \mathbf{A}_\nu + (\partial_\nu \hat{\boldsymbol{\phi}}) \cdot \mathbf{A}_\mu + e\hat{\boldsymbol{\phi}} \cdot \mathbf{A}_\mu \times \mathbf{A}_\nu. \tag{40.86}$$

Remembering the definition of the generalized four-potential [Eq. (40.43)], obviously

$$\partial_\mu(\hat{\boldsymbol{\phi}} \cdot \mathbf{A}_\nu) - \partial_\nu(\hat{\boldsymbol{\phi}} \cdot \mathbf{A}_\mu) = \partial_\mu A_\nu - \partial_\nu A_\mu. \tag{40.87}$$

By subtraction of Eq (40.85) from Eq. (40.86) [with Eq. (40.87) inserted], one finally obtains for $F_{\mu\nu}$ [Eq. (40.75)] the alternative form

$$F_{\mu\nu} = \partial_\mu A_\nu - \partial_\nu A_\mu - \frac{1}{e}\hat{\boldsymbol{\phi}} \cdot (\partial_\mu \hat{\boldsymbol{\phi}}) \times (\partial_\nu \hat{\boldsymbol{\phi}})$$
$$= \partial_\mu A_\nu - \partial_\nu A_\mu - \frac{1}{e|\phi|^3}\varepsilon_{abc}\phi^a(\partial_\mu \phi^b)(\partial_\nu \phi^c). \tag{40.88}$$

Besides the promised "classical" field tensor form $\partial_\mu A_\nu - \partial_\nu A_\mu$, which reduces $F_{\mu\nu}$ to the usual definition of the electromagnetic field when $\hat{\boldsymbol{\phi}}$ becomes fixed in isospace, the expression in Eq. (40.88) has a part which only depends on the usual (non-covariant) derivatives, ∂_μ (∂_ν).

It is worthwhile to remember that the asymptotic time-independent hedgehog gauge potential [Eq. (40.37)] together with the asymptotic scalar field [Eq. (40.39)] gives $\{A_\mu\} = 0$, so that all the generalized electromagnetic field stems from the Higgs field, generally defined here by

$$F_{\mu\nu}^{\text{Higgs}} \equiv -\frac{1}{e}\hat{\boldsymbol{\phi}} \cdot (\partial_\mu \hat{\boldsymbol{\phi}}) \times (\partial_\nu \hat{\boldsymbol{\phi}}), \tag{40.89}$$

giving the hedgehog field tensor $\{F_{ij}\}$, Eq. (40.63), and then the magnetic monopole field $\{B_k\}$, Eq. (40.64).

40.5.2 The non-covariant vacuum

In Chapters 3 and 4, we discussed the concept *Structural Vacuum* in $U(1)$ electrodynamics. I here propose that the concept is generalized to $SO(3)$ by the definition

$$\partial_\mu A_\nu - \partial_\nu A_\mu = 0, \tag{40.90}$$

where $\{A_\mu\} = \hat{\phi} \cdot \mathbf{A}_\mu$. The $SO(3)$ structural vacuum (SV) thus has a non-covariant field structure given by

$$F_{\mu\nu}^{\mathrm{SV}}[= F_{\mu\nu}^{\mathrm{Higgs}}] = -\frac{1}{e}\hat{\phi} \cdot (\partial_\mu \hat{\phi}) \times (\partial_\nu \hat{\phi}). \tag{40.91}$$

In a globally constant (fixed) scalar field $\{\phi^a\}$, say $(0,0,1)$ the $U(1)$ [subgroup of $SO(3)$] definition of structural vacuum is recovered. In particular, the Aharonov-Bohm phenomenon belongs to time-independent $U(1)$ structural vacuum, yet in a restricted spatial domain. In $SO(3)$ the hedgehog is a special example of structural vacuum in the asymptotic $(r \to \infty)$ domain.

It is important to stress that we have *defined* the Structural Vacuum in $SO(3)$ as a non-covariant concept, because of our division of the gauge invariant scalar electromagnetic field tensor $F_{\mu\nu}$ [Eq. (40.75) with its two covariant parts!] into two non-covariant parts [Eq. (40.88)].

40.6   THE MAGNETIC MONOPLE DRIVEN ISOSCALAR MAXWELL EQUATIONS

40.6.1   The $\nabla \cdot \mathbf{B} = 4\pi\rho^m$ equation

Remembering that we have take $4\pi\varepsilon_0 c = 1$, the relation given in the Subsection title above is just the symmetrized Maxwell equation, given in Eq. (38.4).

Let us take as a starting point the scalar field tensor given by the first member of Eq. (40.88), and let us calculate the quantity $\partial_1 F_{23} + \partial_2 F_{31} + \partial_3 F_{12}$. Since

$$\partial_1(\partial_2 A_3 - \partial_3 A_2) + \partial_2(\partial_3 A_1 - \partial_1 A_3) + \partial_3(\partial_1 A_2 - \partial_2 A_1) = 0, \tag{40.92}$$

the four-potential does not contribute to the sum $\partial_1 F_{23} + \partial_2 F_{23} + \partial_3 F_{12}$. Thus,

$$e(\partial_1 F_{23} + \partial_2 F_{31} + \partial_3 F_{12}) = \partial_1 \left[\hat{\phi} \cdot (\partial_2\hat{\phi}) \times (\partial_3\hat{\phi})\right] + \partial_2 \left[\hat{\phi} \cdot (\partial_3\hat{\phi}) \times (\partial_1\hat{\phi})\right]$$
$$+ \partial_3 \left[\hat{\phi} \cdot (\partial_1\hat{\phi}) \times (\partial_2\hat{\phi})\right] = R_1 + R_2, \tag{40.93}$$

where

$$R_1 = \hat{\phi} \cdot \left\{\partial_1 \left[(\partial_2\hat{\phi}) \times (\partial_3\hat{\phi})\right] + \partial_2 \left[(\partial_3\hat{\phi}) \times (\partial_1\hat{\phi})\right] + \partial_3 \left[(\partial_1\hat{\phi}) \times (\partial_2\hat{\phi})\right]\right\}, \tag{40.94}$$

and

$$R_2 = (\partial_1\hat{\phi}) \cdot (\partial_2\hat{\phi}) \times (\partial_3\hat{\phi}) + (\partial_2\hat{\phi}) \cdot (\partial_3\hat{\phi}) \times (\partial_1\hat{\phi}) + (\partial_3\hat{\phi}) \cdot (\partial_1\hat{\phi}) \times (\partial_2\hat{\phi}). \tag{40.95}$$

Elementary differentiations show that the sum in the curly bracket, $\{\dots\}$, of Eq. (40.94) is zero, implying that

$$R_1 = 0. \tag{40.96}$$

The R_2 term [Eq. (40.95)] obviously can be simplified to

$$R_2 = 3(\partial_1\hat{\phi}) \cdot (\partial_2\hat{\phi}) \times (\partial_3\hat{\phi}). \tag{40.97}$$

Since [with $c = 1$] according to Eq. (38.74) one has

$$\partial_1 F_{23} + \partial_2 F_{31} + \partial_3 F_{12} = \partial_1 B_1 + \partial_2 B_2 + \partial_3 B_3 = \boldsymbol{\nabla} \cdot \mathbf{B}, \tag{40.98}$$

one obtains [remembering that $4\pi\varepsilon_0 = 1$] from Eqs. (40.93), and (40.96)–(40.98) the following result

$$\boldsymbol{\nabla} \cdot \mathbf{B} = 4\pi\rho^m, \tag{40.99}$$

i.e. the equation given in the title of Subsection 40.6.1. This is the extended Maxwell equation, given in Eq. (38.4), with *a magnetic charge density given by the triplet Higgs field* as follows:

$$\rho^m = \frac{3}{4\pi e}(\partial_1 \hat{\boldsymbol{\phi}}) \cdot (\partial_2 \hat{\boldsymbol{\phi}}) \times (\partial_3 \hat{\boldsymbol{\phi}}). \tag{40.100}$$

40.6.2   The $\boldsymbol{\nabla} \times \mathbf{E} + \frac{\partial \mathbf{B}}{\partial t} = -4\pi \mathbf{J}^m$ equation

In analogy with our considerations in Subsection 40.6.1, we now calculate the quantity $\partial_0 F_{12} + \partial_1 F_{20} + \partial_2 F_{01}$, with $F_{\mu\nu}$ taken from Eq. (40.88). Again, it turns out, as the reader may show, that the four potential does not contribute to the above quantity. Hence,

$$e(\partial_0 F_{12} + \partial_1 F_{20} + \partial_2 F_{01}) = \partial_0[\hat{\boldsymbol{\phi}} \cdot (\partial_1 \hat{\boldsymbol{\phi}}) \times (\partial_2 \hat{\boldsymbol{\phi}})] + \partial_1[\hat{\boldsymbol{\phi}} \cdot (\partial_2 \hat{\boldsymbol{\phi}}) \times (\partial_0 \hat{\boldsymbol{\phi}})]$$
$$+ \partial_2[\hat{\boldsymbol{\phi}} \cdot (\partial_0 \hat{\boldsymbol{\phi}}) \times (\partial_1 \hat{\boldsymbol{\phi}})]. \tag{40.101}$$

Following analogous steps to those taken in Subsection 40.6.1, we obtain

$$e(\partial_0 F_{12} + \partial_1 F_{20} + \partial_2 F_{01}) = 3(\partial_0 \hat{\boldsymbol{\phi}}) \cdot (\partial_1 \hat{\boldsymbol{\phi}}) \times (\partial_2 \hat{\boldsymbol{\phi}}). \tag{40.102}$$

According to Eq. (38.74), and with the following usual unit choice: $c = 1$, $4\pi\varepsilon_0 = 1$, $\mu_0\varepsilon_0 = c^{-2} = 1 \ [\Rightarrow \mu_0 = \varepsilon_0^{-1}]$, it appears after a few elementary calculations that

$$\left(\boldsymbol{\nabla} \times \mathbf{E} + \frac{\partial \mathbf{B}}{\partial t}\right)_3 = -4\pi J_3^m, \tag{40.103}$$

where

$$J_3^m = -\frac{3}{4\pi e}(\partial_0 \hat{\boldsymbol{\phi}}) \cdot (\partial_1 \hat{\boldsymbol{\phi}}) \times (\partial_2 \hat{\boldsymbol{\phi}}) \tag{40.104}$$

is the 3rd component of a *magnetic current density associated with the triplet Higgs field*.

A moment of contemplating gives one the following extended Maxwell equation, given in Eq. (38.5), viz.,

$$\boldsymbol{\nabla} \times \mathbf{E} + \frac{\partial \mathbf{B}}{\partial t} = -4\pi \mathbf{J}^m, \tag{40.105}$$

where

$$\mathbf{J}^m = -\frac{3}{4\pi e}(\partial_0\hat{\phi}) \cdot \left[(\partial_2\hat{\phi}) \times (\partial_3\hat{\phi}), (\partial_3\hat{\phi}) \times (\partial_1\hat{\phi}), (\partial_1\hat{\phi}) \times (\partial_2\hat{\phi})\right]. \qquad (40.106)$$

The neutral triplet Higgs boson field cannot give rise to electric charge (ρ^e) and current densities ($\mathbf{J}^e$), implying that $\partial^\mu F_{\mu\nu} = 0$ [see Eq. (38.97)], that is in three-vector notation

$$\boldsymbol{\nabla} \cdot \mathbf{E} = 0, \quad \boldsymbol{\nabla} \times \mathbf{B} = \frac{\partial \mathbf{E}}{\partial t}. \qquad (40.107)$$

Our covariant definition of the generalized electromagnetic field tensor $\{F_{\mu\nu}\}$ given in Eq. (40.75) [or in alternative form in Eq. (40.88)] thus has led to a set of generalized Maxwell equations [Eqs. (40.99), (40.105) and (40.107)] coupled to the Higgs field. Without the Higgs field this set is reduced to the free Maxwell equations in $U(1)$; Eqs. (38.1) and (38.2).

40.6.3   Higgs field current conservation

We have seen that the magnetic current depends on the Higgs field only. Moreover, it follows by combining Eqs. (40.99) and (40.105), that this current is conserved, i.e.,

$$\partial_\mu J^{m,\mu} \left[= \frac{\partial \rho^m}{\partial t} + \boldsymbol{\nabla} \cdot \mathbf{J}^m\right] = 0. \qquad (40.108)$$

In the case of the time-independent hedgehog model, where $\hat{\phi} = (x,y,z)/r$, it is easy to show from Eq. (40.100) that the magnetic charge density is identically zero, i.e.

$$\rho^m = 0. \qquad (40.109)$$

This result is in agreement with the fact that the radial magnetic field [Eq. (40.64)] is divergence-free. The time-independent magnetic field is also rotational-free, see Eq. (40.107). For the non-vanishing hedgehog solution we may conclude that

$$\boldsymbol{\nabla} \cdot \mathbf{B} = \boldsymbol{\nabla} \times \mathbf{B} = 0, \qquad (40.110)$$

a result which is quite analogous to a well-known circumstance in near-field electrodynamics [157]. Here $\boldsymbol{\nabla} \times \mathbf{E} = \boldsymbol{\nabla} \cdot \mathbf{E} = 0$ in the near-field zone of an electrical point-dipole, which is electrically neutral [globally on a microscopic scale]. In the electric dipole's near-field zone, the field (photon) retardation is negligible, thus making the electric dipole and the hedgehog cases formally equivalent, cf. Eqs. (38.17) and (38.18) [or Eqs. (39.7) and (39.8)].

Bibliography

[1] V. M. Agranovich and D. L. Mills, editors. *Surface Polaritons. Electromagnetic Waves at Surfaces and Interfaces*. North-Holland, Amsterdam, 1982.

[2] Y. Aharonov and A. Anandan. *Phys. Rev. Lett.*, 58:1593, 1987.

[3] Y Aharonov and D. Bohm. *Phys. Rev.*, 115:485, 1959.

[4] I. J. R. Aitchison and A. J. G. Hey. *Gauge Theories in Particle Physics*. Adam Hilger, Bristol, 2nd edition, 1989.

[5] A. I. Akhiezer, V. G. Bar'yakhtar, and M. I. Kaganov. *Soviet Physics-Uspekhi*, 3:567, 1961.

[6] A. I. Akhiezer, V. G. Bar'yakhtar, and S. V. Peletminskii. *Spin Waves*. North-Holland, Amsterdam, 1968.

[7] G. B. Arfken and H. J. Weber. *Mathematical Methods for Physicists*. Harcourt Science and Technology, San Diego, 5th edition, 2001.

[8] J. A. Armstrong, N. Bloembergen, J. Ducuing, and P. S. Pershan. *Phys. Rev.*, 127:1918, 1962.

[9] G. Bacciagaluppi and A. Valentini. *Quantum Theory at the Crossroads*. Cambridge Univ., Cambridge, 2009.

[10] A. Bagchi. *Phys. Rev. B*, 15:3060, 1977.

[11] J. Bardeen, L. N. Cooper, and J. R. Schrieffer. *Phys. Rev.*, 108:1175, 1957.

[12] S. M. Barnett and P. M. Radmore. *Methods in Theoretical Quantum Optics*. Clarendon, Oxford, 1997.

[13] A. O. Barut and B. Blaive. *Phys. Rev. A*, 45:2810, 1992.

[14] A. O. Barut and J. P. Dowling. *Phys. Rev. A*, 36:649, 1987.

[15] A. O. Barut, J. P. Dowling, and J. F. van Heule. *Phys. Rev. A*, 38:4405, 1988.

[16] A. O. Barut and J. F. Van Huele. *Phys. Rev. A*, 32:3187, 1985.

[17] A. O. Barut and J. Kraus. *Found. Phys.*, 13:189, 1983.

[18] J. Barwick. *Phys. Rev. A*, 17:1912, 1978.

[19] H. Bateman. *The Mathematical Analysis of Electrical and Optical Wave Motion on the Basis of Maxwell Equations.* Cambridge Univ., Cambridge, 1915. Reprinted: 1955 (Dover, New York).

[20] J. Bernstein. *Rev. Mod. Phys.*, 46:7, 1974.

[21] M. V. Berry. *Proc. Roy. Soc. London A*, 392:45, 1984.

[22] M. V. Berry. *Annals of the New York Academy of Sciences*, 755:303, 1995.

[23] M. V. Berry and R. Lim. *J. Phys. A*, 23:L655, 1995.

[24] H. A. Bethe. *Phys. Rev.*, 72:339, 1947.

[25] I. Bialyniciki-Birula. *Acta Phys. Polon.*, A86:97, 1994.

[26] I. Bialyniciki-Birula. *Photon Wave Function.* In E. Wolf, editor, *Progress in Optics, vol. 36, p. 245.* North-Holland, Amsterdam, 1996.

[27] N. D. Birrell and P. C. W. Davies. *Quantum Fields in Curved Space.* Cambridge Univ., New York, 1982.

[28] B. Blaive, A. O. Barut, and R. Boudet. *J. Phys. B: At. Mol. Opt. Phys.*, 24:3121, 1991.

[29] N. Bloembergen. *Proc. IEEE*, 51:124, 1963.

[30] N. Bloembergen. *Nonlinear Optics.* Benjamin, London, 1965.

[31] N. Bloembergen and P. S. Pershan. *Phys. Rev.*, 128:606, 1962.

[32] N. Bloembergen and Y. R. Shen. *Phys. Rev.*, 133:A37, 1964.

[33] A. D. Boardman, editor. *Electromagnetic Surface Modes.* Wiley, Chichester, 1982.

[34] N. N. Bogoliubov. *Nuovo Cimento*, 7:6 and 794, 1958.

[35] N. N. Bogolubov. *Zh. Eksp. Teor. Fiz.*, 34:58, 1958. [Sov. Phys. JETP, 7:51, 1958].

[36] N. N. Bogolubov and D. V. Shirkov. *Introduction to the Theory of Quantized Fields.* Interscience, New York, 1959.

[37] A. Bohm, A. Mostafazadeh, H. Koizumi, Q. Niu, and J. Zwanziger. *The Geometric Phase in Quantum Systems.* Springer, Berlin, 2003.

[38] D. Bohm. *Phys. Rev.*, 85:166, 1952.

[39] D. Bohm and B. J. Hiley. *The undivided universe.* Routledge (Taylor and Francis), New York, 1993.

[40] D. Bohm and D. Pines. *Phys. Rev.*, 92:609, 1953.

[41] N. Bohr. *Atomic Physics and Human Knowledge.* Wiley, New York, 1958.

[42] N. Bohr and L. Rosenfeld. *Mat-fys. Medd Dan. Vid. Selsk.*, 12:8, 1933. English translation by A. Petersen in: R. S. Cohen and J. J. Stackel, editors, *Selected Papers of Leon Rosenfeld.* Reidel, Dordrecht, 1979.

[43] N. Bohr and L. Rosenfeld. *Phys. Rev.*, 78:794, 1950.

[44] M. Born, W. Heisenberg, and P. Jordan. *Z. Phys.*, 35:557, 1925.

[45] M. Born and P. Jordan. *Z. Phys.*, 34:858, 1925.

[46] M. Born and E. Wolf. *Principles of Optics.* Cambridge Univ., Cambridge, 7th (expanded) edition, 1999.

[47] M. A. Bouchiat. In: G. Grynberg and R. Stora, editors, *New trends in atomic physics,* vol. 2. North-Holland, Amsterdam, 1984.

[48] M. A. Bouchiat and L. Pottier. *Science*, 234:1203, 1986.

[49] R. G. Boyd and J. Callaway. *Phys. Rev.*, 138:A1621, 1965.

[50] R. W. Boyd. *Nonlinear Optics.* Academic, San Diego, 1992.

[51] H. Bruns. *Abh. Kgl. Sachs. Ges. Wiss., math-phys. Kl.*, 21:323, 1895.

[52] P. N. Butcher and D. Cotter. *The Elements of Nonlinear Optics.* Cambridge, New York, 1990.

[53] N. Cabibbo and E. Ferrari. *Nuovo Cimento*, XXIII:1147, 1962.

[54] J. Callaway. *Quantum Theory of the Solid State, Part A.* Academic, New York, 1974.

[55] D. Carlsmith. *Particle Physics.* Pearson, Boston, 2013.

[56] S. M. Carroll. *Spacetime and Geometry. An Introduction to General Relativity.* Addison-Wesley, Boston, 2004.

[57] H. B. G. Casimir. *Proc. Kon. Ned. Akad.*, 51:793, 1948.

[58] H. B. G. Casimir. *Physica*, XIX:846, 1953.

[59] R. G. Chambers. *Phys. Rev. Lett.*, 5:3, 1960.

[60] S. H. Charap and E. L. Boyd. *Phys. Rev.*, 133:A811, 1964.

[61] R. Claus, L. Merten, and J. Brandmüller. *Light Scattering by Phonon-Polaritons.* Springer, Berlin, 1975.

[62] C. Cohen-Tannoudji, B. Diu, and F. Laloe. *Quantum Mechanics*, volumes I and II. Wiley, London, 1977.

[63] C. Cohen-Tannoudji, J. Dupont-Roc, and G. Grynberg. *Photons and Atoms. Introduction to Quantum Electrodynamics*. Wiley, New York, 1989.

[64] C. Cohen-Tannoudji, J. Dupont-Roc, and G. Grynberg. *Atom-Photon Interactions. Basic Processes and Applications*. Wiley, New York, 1992.

[65] A. W. Conway and J .L. Synge, editors. *The Mathematical Papers of Sir W. R. Hamilton, vol. 1 (Geometrical Optics)*. Cambridge Univ., Cambridge, 1931.

[66] M. D. Crisp and E. T. Jaynes. *Phys. Rev.*, 179:1253, 1969.

[67] P. G. de Gennes. *Superconductivity of Metals and Alloyes*. Benjamin, New York, 1966.

[68] L. deBroglie. *Journ. de Physique et le Radium, (6)*, 8:225, 1927.

[69] P. A. M. Dirac. *Proc. Roy. Soc. A*, 114:243, 1927.

[70] P. A. M. Dirac. *Proc. Roy. Soc. A*, 123:714, 1929.

[71] P. A. M. Dirac. *Proc. R. Soc. Lond.*, 133:60, 1931.

[72] P. A. M. Dirac. *Phys. Rev.*, 74:817, 1948.

[73] F. J. Dyson. *Phys. Rev.*, 102:1217, 1956.

[74] F. J. Dyson. *Phys. Rev.*, 102:1230, 1956.

[75] A. Einstein. *Ann. Phys.*, 17:132, 1905.

[76] A. Einstein. *Ann. Phys.*, 20:199, 1906.

[77] A. Einstein. *Phys. Z.*, 18:121, 1917.

[78] F. Englert and R. Brout. *Phys. Rev. Lett.*, 13:321, 1964.

[79] P. P. Ewald. *Dissertation, Univ of Munich*, 1912.

[80] P. P. Ewald. *Ann. Phys.*, 49:1, 1915.

[81] U. Fano. *Phys. Rev.*, 103:1202, 1956.

[82] P. J. Feibelman. *Phys. Rev. B*, 12:1319, 1975.

[83] P. J. Feibelman. *Prog. Surf. Sci.*, 12:287, 1982.

[84] E. Fermi. *Rev. Mod. Phys.*, 4:87, 1932.

[85] A. L. Fetter and J. D. Walecka. *Quantum Theory of Many-Particle Systems*. McGraw-Hill, New York, 1971.

[86] V. Fock. *Z. Phys.*, 57:261, 1929.

[87] L. L. Foldy and S. A. Wouthuysen. *Phys. Rev.*, 78:29, 1950.

[88] F. Forstman and R. R. Gerhardts. *Metal Optics Near the Plasma Frequency.* Springer, Berlin, 1986.

[89] P. A. Franken, A. E. Hill, C. W. Peters, and G. Weinreich. *Phys. Rev. Lett.*, 7:118, 1961.

[90] P. A. Franken and J. F. Word. *Rev. Mod. Phys.*, 35:23, 1963.

[91] F. Frenet. *Journal de Mathematiques Pures et Appliquees*, 17:437, 1852.

[92] F. Garcia-Moliner and F. Flores. *Introduction to the Theory of Solid Surfaces.* Cambridge, London, 1979.

[93] M. Georgi and S. L. Glashow. *Phys. Rev. Lett.*, 28:1494, 1972.

[94] R. R. Gerhardts. *Physica Scripta*, 28:235, 1983.

[95] D. M. Gingrich. *Practical Quantum Electrodynamics.* CRC (Taylor and Francis), London, 2006.

[96] S. L. Glashow. *Nucl. Phys.*, 22:579, 1961.

[97] R. J. Glauber. *Phys. Rev.*, 130:2529, 1963.

[98] R. J. Glauber. *Phys. Rev.*, 131:2766, 1963.

[99] A. S. Goldhaber. *Phys. Rev.*, 140:B1407, 1965.

[100] J. Goldstone. *Nuovo Cimento*, 19:154, 1961.

[101] J. Goldstone, A. Salam, and S. Weinberg. *Phys. Rev.*, 127:965, 1962.

[102] W. Gordon. *Z. Phys.*, 50:630, 1928.

[103] M. S. Green. *J. Chem. Phys.*, 20:1281, 1952.

[104] M. S. Green. *J. Chem. Phys.*, 22:398, 1954.

[105] W. Greiner. *Relativistic Quantum Mechanics. Wave Equations.* Springer, Berlin, 1990.

[106] F. Gross. *Relativistic Quantum Mechanics and Field Theory.* Wiley, New York, 1993.

[107] G. S. Guralnik, C. R. Hagen, and T. W. B. Kible. *Phys. Rev. Lett.*, 13:585, 1964.

[108] H. Haken. *Light,* Vol. 1 of *Waves, Photons, Atoms.* North-Holland, Amsterdam, 1981.

[109] S. W. Hawking and G. F. R. Ellis. *The Large Scale Structure of Space-Time.* Cambridge Univ., Cambridge, 1973.

[110] C. Herring. In G. Rado and H. Suhl, editors, *Direct Exchange between well Separated Atoms.* Magnetism, vol. 2B, p. 1, New York, 1966.

[111] C. Herring and M. Flicker. *Phys. Rev.*, 134:A362, 1964.

[112] P. W. Higgs. *Phys. Lett.*, 12:132, 1964.

[113] P. W. Higgs. *Phys. Rev. Lett.*, 13:508, 1964.

[114] P. W. Higgs. *Phys. Rev.*, 145:1156, 1966.

[115] T. Holstein and H. Primakoff. *Phys. Rev.*, 58:1098, 1940.

[116] J. J. Hopfield. *Phys. Rev.*, 112:1555, 1958.

[117] D. Iwanenko and V. Fock. *"C. R."*, 188:1470, 1929.

[118] J. D. Jackson. *Classical Electrodynamics.* Wiley, New York, 1999.

[119] J. Jung and O. Keller. *Phys. Rev. A*, 90:043830, 2014.

[120] J. Jung and O. Keller. *Phys. Rev. A*, 92:012122, 2015.

[121] J. Jung and O. Keller. *J. Opt. Soc. Am. B*, 34:726, 2017.

[122] J. Jung and O. Keller. *Phys. Rev. A*, 98:053825, 2018.

[123] J. Jung and O. Keller. *J. Opt. Soc. Am. B*, 37:3005, 2020.

[124] J. Jung and O. Keller. *J. Phys. B: At. Mol. Phys.*, 54:045401, 2021.

[125] J. Jung and O. Keller. *Phys. Rev. A*, 104:013714, 2021.

[126] J. Jung and O. Keller. *Phys. Rev. A*, 103:063501, 2021.

[127] J. Jung and O. Keller. *Phys. Rev. A*, 104:053508, 2021.

[128] J. Jung and O. Keller. *Phys. Rev. A*, 106:033503, 2022.

[129] J. Jung and O. Keller. *Phys. Rev. A*, 108:023507, 2023.

[130] J. Jung and O. Keller. *To be published,* 2024.

[131] F. Keffer. *Spin Waves.* In: S. Flügge editor, *Encyclopedia,* vol 18/2, p. 1. Springer, Berlin, 1966.

[132] L. V. Keldysh, D. A. Kirzhnitz, and A. A. Maradudin, editors. *The Dielectric Function of Condensed Systems.* North-Holland, Amsterdam, 1989.

[133] O. Keller. *Optica Acta,* 33:673, 1986.

[134] O. Keller. *Phys. Rev. B*, 33:990, 1986.

[135] O. Keller. *Phys. Rev. B*, 34:3883, 1986.

[136] O. Keller. *Phys. Rev. B*, 38:8041, 1988.

[137] O. Keller. *Phys. Rev. B*, 37:10588, 1988.

[138] O. Keller. *Phys. Rev. B*, 42:6049, 1990.

[139] O. Keller. *On the Nonlocal Response Theory of Optical Rectification and Second-Harmonic Generation in Centrosymmetric Superconductors.* In O. Keller, editor, *Nonlinear Optics in Solids,* p. 78. Springer, Berlin, 1990.

[140] O. Keller. *Phys. Rev. B*, 43:10293, 1991.

[141] O. Keller. *J. Opt. Soc. Am. B.*, 12:987, 1995.

[142] O. Keller. *J. Opt. Soc. Am. B*, 12:997, 1995.

[143] O. Keller. *J. Nonl. Phys. Mater.*, 5:109, 1996.

[144] O. Keller. *Local Fields in the Electrodynamics of Mesoscopic Media. Phys. Rep.* 268(2-3):85, 1996.

[145] O. Keller. *Photon Drag in Non-Simply Connected Mesoscopic Media and Quantum Confinement of Light.* In O. Keller, editor, *Notions and Perspectives of Nonlinear Optics,* World Scentific, London, 1996.

[146] O. Keller. *Aspects of Local-Field Electrodynamics in Condensed Matter.* In T. Hakioglu and A. S. Shumovsky, editors, *Quantum Optics and the Spectroscopy of Solids,* p. 1, Kluwer, Dordrecht, 1997.

[147] O. Keller. *Phys. Rev. A*, 58:3407, 1998.

[148] O. Keller. *Electromagnetic Propagators in Mirco- and Mesoscopic Optics.* In D. A. Jelski and T. F. George, editors, *Computational Studies of New Materials,* p. 375, World Scientific, London, 1999.

[149] O. Keller. *Phys. Rev. A*, 62:022111, 2000.

[150] O. Keller. *Optical works of L. V. Lorenz,* In E. Wolf, editor, *Progress in Optics,* vol. 43, p. 195. Elsevier, Amsterdam, 2002.

[151] O. Keller. *J. Microscopy,* 209:272, 2003.

[152] O. Keller. *Phys. Rev. E*, 72:026612, 2005.

[153] O. Keller. *On the Theory of Spatial Localization of Photons. Phys. Rep.* 411(1-3):1, 2005.

[154] O. Keller. *J. Opt. A: Pure Appl. Opt.*, 8:S174, 2006.

[155] O. Keller. *Phys. Rev. A*, 76:062110, 2007.

[156] O. Keller. *Histocical Papers on the Particle Concept of Light,* In E. Wolf, editor, *Progress in Optics*, vol. 50, p. 51. Elsevier, Amsterdam, 2007.

[157] O. Keller. *Quantum Theory of Near-Field Electrodynamics*. Springer, Berlin, 2011.

[158] O. Keller. *Light: The Physics of the Photon*. CRC (Taylor and Francis), London, 2014.

[159] O. Keller. *Phys. Rev. A*, 98:052112, 2018.

[160] O. Keller and L. M. Hively. *J. Phys. Commun.*, 3:115002, 2019.

[161] O. Keller and L. M. Hively. *Physics Essayes*, 32:3, 2019.

[162] O. Keller and D. S. Olesen. *Phys. Rev. A*, 86:053818, 2012.

[163] O. Keller and J. H. Pedersen. In: A. A. Ferwerda, *Scattering and Diffraction*, 18:1029, 1988.

[164] O. Keller and K. Pedersen. *J. Phys. C: Solid State Phys.*, 19:3631, 1986.

[165] O. Keller and P. Sonderkaer. *Proc. SPIE*, 954:344, 1988.

[166] O. Keller and P. Sonderkaer. *J. Phys. D: Appl. Phys.*, 22:343, 1989.

[167] O. Keller and G. Wang. *Phys. Rev. B*, 56:12327, 1997.

[168] C. Kittel. *Quantum Theory of Solids,* 2nd revised edition. Wiley, New York, 1987.

[169] C. Kittel. *Introduction to Solid State Physics,* 7th edition. Wiley, Chicester, 1996.

[170] O. Klein. *Z. Phys.*, 53:157, 1929.

[171] K. L. Kliewer. *Surf. Sci.*, 101:57, 1980.

[172] R. Kubo. *J. Phys. Soc. Japan*, 12:570, 1957.

[173] R. Kubo. *Lectures in Theoretical Physics*, Vol. 1. Wiley-Interscience, New York, 1959.

[174] W. E. Lamb. *Rep. Progr. Phys.*, 14:19, 1951.

[175] W. E. Lamb and R. C. Retherford. *Phys. Rev.*, 72:241, 1947.

[176] W. E. Lamb and R. C. Retherford. *Phys. Rev.*, 79:549, 1950.

[177] L. D. Landau and R. Peierls. *Zeitschr. Phys.*, 69:56, 1931.

[178] J. Lindhard. *K. Dan. Vidensk. Selsk. Mat. Fys. Medd.*, 28:8, 1954.

[179] F. London. *Superfluids* Vol. 1. Wiley, New York, 1950.

[180] F. London and H. London. *Proc. Roy. Soc. (London)*, A149:71, 1935.

[181] F. London and H. London. *Physica*, 2:341, 1935.

[182] R. Loudon. *The Quantum Theory of Light.* Oxford Univ., Oxford, 2nd edition, 1983.

[183] W. H. Louisell. *Quantum Statistical Properties of Radiation.* Wiley, New York, 1990.

[184] G. D. Mahan. *Many-Particle Physics.* Plenum, New York, 2nd edition, 1990.

[185] L. Mandel and E. Wolf. *Optical Coherence and Quantum Optics.* Cambridge Univ., Cambridge, 1995.

[186] F. Mandl and G Shaw. *Quantum Field Theory.* Wiley, Chichester, Revised edition, 1993.

[187] W. Meissner and R. Ochsenfeld. *Naturwiss*, 21:787, 1933.

[188] P. Meystre and W. Sargent. *Quantum Optics.* Springer, Berlin, 1990.

[189] P. M. Morse. *Thermal Physics.* Benjamin, New York, 1964.

[190] P. M. Morse and H. Feshbach. *Methods of Theoretical Physics.* Parts I and II, McGraw-Hill, New York, 1953.

[191] Y. Nambu. *Phys. Rev. Lett.*, 4:380, 1960.

[192] E. Nöther. *Nachr. Ges. Wiss. Göttingen*, 171, 1918.

[193] M Nieto-Vesperinas. *Scattering and Diffraction in Physical Optics.* Wiley, New York, 1991.

[194] H. M. Nussenzveig. *Causality and Dispersion Relations.* Academic, New York, 1972.

[195] T. Ohmura. *Prog. Theor. Phys.*, 16:684, 1956.

[196] D. S. Olesen and O. Keller. *Science China (Physics, Mechanics and Astronomy)*, 55:1383, 2012.

[197] D. S. Olesen and O. Keller. *J. Phys. Commun.*, 4:105012, 2020.

[198] J. R. Oppenheimer. *Phys. Rev.*, 38:725, 1931.

[199] C. W. Oseen. *Ann. Phys.*, 48:1, 1915.

[200] R. D. Parks. , editor. *Superconductivity,* Vol. 1 and 2. Dekker, New York, 1969.

[201] R. Passante and E. A. Power. *Phys. Rev. A*, 35:188, 1987.

[202] D. N. Pattanayak and W. Wolf. *Opt. Commun.*, 6:217, 1972.

[203] C. K. N. Pattel and R. E. Slusher. *Phys. Rev. Lett.*, 22:282, 1969.

[204] W. Pauli and M. Fierz. *Nuovo Cimento*, 15:167, 1938.

[205] J. Perina. *Quantum Statistics of Linear and Nonlinear Optical Phenomena.* Reidel, Dordrecht, 1984.

[206] P. S. Pershan. *Non-Linear Optics* In E. Wolf editor, *Progess in Optics*, vol 5, p. 83. North-Holland, Amsterdam, 1966.

[207] M Peshkin and A. Tonomura. *The Aharonov-Bohm Effect.* Springer, Berlin, 1989.

[208] E. R. Pike and S. Sarkar. *The Quantum Theory of Radiation.* Clarendon, Oxford, 1995.

[209] D. Pines and D. Bohm. *Phys. Rev.*, 85:338, 1952.

[210] D. Pines and P. Nozieres. *The Theory of Quantum Liquids: Normal Fermi Liquids*, Vol. I. Addison-Wesley, Redwood City, California, 1989.

[211] M. Planck. *Verh. d. deutsch Phys. Ges.*, 2:202 and 2:237, 1900.

[212] M. Planck. *Ann. Phys.*, 4:553, 1901.

[213] P. M. Platzman and P. A. Wolff. *Waves and Interactions in Solid State Plasmas.* Academic, New York, 1973.

[214] A. M. Polyakov. *Zh. Eksp. Teor. Fiz.*, 68:1975, 1975.

[215] A. M. Polyakov. *Sov. Phys. JETP*, 41:988, 1976.

[216] E. A. Power. *Introductory Quantum Electrodynamics.* Longmans, Green and Co., London, 1964.

[217] E. A. Power and S. Zienau. *Philos. Trans. London, Ser. A*, 251:427, 1959.

[218] A. Proca and G. A. Proca. *Aleandre Proca 1897-1955: Oeuve Scientifique Publiee.* S.I.A.G., Rome, 1988.

[219] T. Radozycki. *J. Phys. A*, 23:4925, 1990.

[220] G. Rickayzen. *Theory of Superconductivity.* Wiley (Interscience), New York, 1965.

[221] W. Rindler. *Am. J. Phys.*, 34:1174, 1966.

[222] R. H. Ritchie. *Prog. Theor. Phys.*, 29:607, 1963.

[223] F. Rohrlich. *Classical Charged Particles.* World Scientific, New Jersey, 3th (expanded) edition, 2007.

[224] L. H. Ryder. *Quantum Field Theory,* second edition. Cambridge Univ., Cambridge, 1996.

[225] M. N. Saha. *Indian J. Phys.*, 10:145, 1936.

[226] M. N. Saha. *Phys. Rev.*, 75:1968, 1949.

[227] A. Salam. In N. Svartholm, editor, *Elementary Particle Physics,* p. 367. Almqvist and Wiksells, Stockholm, 1968.

[228] A. Salam. *Rev. Mod. Phys.*, 52:525, 1980.

[229] L. I. Schiff. *Quantum Mechanics.* McGraw-Hill, New York, 1968.

[230] W. P. Schleich. *Quantum Optics in Phase Space.* Wiley-VCH, Berlin, 2001.

[231] J. R. Schrieffer. *Theory of Superconductivity.* Benjamin/Cummings, London, 1964. Revised printing, 1983.

[232] M. Schubert and B. Wilhelmi. *Nonlinear Optics and Quantum Electronics.* Wiley, New York, 1986.

[233] G. Schwarz. *Pac. J. Math.*, 143:195, 1990.

[234] J. Schwinger. *Science,* 165:757, 1969.

[235] M. O. Scully and M. Suhail Zubairy. *Quantum Optics.* Cambridge Univ., Cambridge, 1997.

[236] J. A. Serret. *Journal de Mathematiques Pures et Appliquees,* 16:193, 1851.

[237] Y. R. Shen. *The Principles of Nonlinear Optics.* Wiley, New York, 1984.

[238] R. Silberglitt and A. B. Harris. *Phys. Rev.*, 174:640, 1968.

[239] L. Silberstein. *Ann. Phys.*, 22:579, 1907.

[240] L. Silberstein. *Ann. Phys.*, 24:783, 1907.

[241] L. Silberstein. *The Theory of Relativity.* MacMillan, London, 2nd edition, 1924.

[242] J. F. Sipe. *Phys. Rev. A,* 52:1875, 1995.

[243] A. Sommerfeld. *Electrodynamics.* Academic, New York, 1952.

[244] E. C. G. Stueckelberg. *Helv. Phys. Acta.*, 11:225, 1938.

[245] E. C. G. Stueckelberg. *Helv. Phys. Acta.*, 11:299, 1938.

[246] J. L. Synge and W. Conway, editors. *Hamilton's Mathematical Papers* Vol. 1. Cambridge Univ., Cambridge, 1931.

[247] G. t' Hooft. *Nucl. Phys.*, B79:276, 1974.

[248] J. C. Taylor. *Gauge Theories of Weak Interactions.* Cambridge Univ., Cambridge, 1976.

[249] D. ter Haar and H. Vergerland. *Ark. Fys. Semin. Trondheim*, 11, 1977.

[250] J. S. Toll. *Phys. Rev.*, 104:1760, 1956.

[251] W. G. Unruh. *Phys. Rev. D*, 10:3194, 1974.

[252] W. G. Unruh. *Phys. Rev. D*, 14:870, 1976.

[253] J. Valatin. *Nuovo Cimento*, 7:843, 1958.

[254] J. H. van Vleck. *The Theory of Electric and Magnetic Susceptibilities.* Oxford Univ., London, 1932.

[255] R. M. Wald. *General Relativity.* Chicago Univ., London, 1984.

[256] H. Weber. *Die Partiellen Differential-Gleichungen der Mathematischen Physik nach Riemann's Vorlesungen.* F. Vieweg and Sohn, Braunschweig, 1901.

[257] S. Weinberg. *Phys. Rev. Lett.*, 19:1264, 1967.

[258] S. Weinberg. *Gravitation and cosmology: Principles and applications of the general theory of relativity.* Wiley, New York, 1972.

[259] S. Weinberg. *Phys. Rev. Lett.*, 36:294, 1976.

[260] S. Weinberg. *Rev. Mod. Phys.*, 52:513, 1980.

[261] S. Weinberg. *The Quantum Theory of Fields.* Vol. 1 *Foundations.* Cambridge Univ., New York, 1995.

[262] J. Weingarten. *Journal fur die Reine und Angewandte Mathematik*, 59:382, 1861.

[263] M. Weissbluth. *Photon-Atom Interactions.* Academic, San Diego, 1989.

[264] V. Weisskopf and E. P. Wigner. *Z. Phys.*, 63:54, 1930.

[265] V. Weisskopf and E. P. Wigner. *Z. Phys.*, 65:18, 1931.

[266] H. Weyl. *Ann. Phys. (Leipzig)*, 365:481, 1919.

[267] H. Weyl. *Z. Phys.*, 56:330, 1929.

[268] J. A. Wheeler and W. H. Zurek, editors. *Quantum Theory and Measurements.* Princeton Univ., New Jersey, 1983.

[269] E. T. Whittaker. *A History of the Theories of Aether and Electricity,* vol.1: *The Classical Theories.* Nelson, London, 1910.

[270] E. P. Wigner. *Gruppentheorie und ihre Anwendung auf die Quantummechanic der Atomspektren.* Braunschweig (English translation Academic Press, New York, 1959), 1931.

[271] E. P. Wigner. *Ann. Math.,* 40:149, 1939.

[272] E. P. Wigner. *In Theoretical Physics.* International Atomic Energy Agency, Vienna, 1963.

[273] E. P. Wigner. *Scientific American,* 213:28, 1965.

[274] E. P. Wigner. *The Collected Works of Eugene Paul Wigner, Part B* [edited by J. Mehra]. Springer, Berlin, 1995.

[275] H. A. Wilson. *Phys. Rev.,* 75:309, 1949.

[276] E. Wolf. *A Generalized Extinction Theorem and its Role in Scattering Theory.* In E. Wolf, editor, *Coherence and Quantum Optics,* p. 339. Plenum, New York, 1973.

[277] D. A. Woodside. *J. Math. Phys.,* 41:4622, 1999.

[278] D. A. Woodside. *Am. J. Phys.,* 77:438, 2009.

[279] R. G. Woolley. *Mol. Phys.,* 22:1013, 1971.

[280] F. Wooten. *Optical Properties of Solids.* Academic, New York, 1972.

[281] M. Wortis. *Phys. Rev.,* 132:85, 1963.

[282] M. Wortis. *Phys. Rev.,* 138:A1126, 1965.

[283] H. Yamagishi. *Phys. Rev. D,* 27:2383, 1983.

[284] C. N. Yang and R. L. Mills. *Phys. Rev.,* 96:191, 1954.

[285] F. J. Yndurain. *Relativistic Quantum Mechanics and Introduction to Field Theory.* Springer, Berlin, 1996.

Index

Energy density, 98

Energy eigen solutions, 392–393, *394*

Energy flux density, 97

Energy wave functions, 17

Englert, F., 422

Ensemble average, 161–162
    integral equation for, 164–165
    time development of, 163–165

Entangled gauge photon states, 52

Equation of motion, 34, 38, 39, 42, 74

Euler-Lagrange equations, 224, 228, 230, 408

Evolution operator, 126–127
    integral equation, 163–164
    interaction representation, 163–164

Ewald, P. P., 176

Ewald-Oseen extinction, 187
    in electrodynamics, 176–179
    theorem, 148

Exchange parameters, 267

Excited spin states, 270–271

Exponential photon localization, 264–265

Exponential photon source confinement, 261–266

Extended Lagrangians, 140–141

Extended transverse dielectric function, 235

Extented field Lagrangian density, 140

Extinction theorem
    complex Klein-Gordon field, 180–184

Extrinsic surface curvature tensor, 374

F

Fermi-Dirac distribution, 126, 286, 306

Fermi-Dirac occupation factor, 247

Fermi factors, *256*

Fermi Lagrangian density, 139, 140

Fermion operator, 159, 160, 283

Fermi velocity, 226

Fermi wave number, 234, 386

Ferrari, E., 449, 472

Ferromagnets
    spin waves in, 267–281

Feynman diagrams, 304

Feynman propagator, 30, 32, 44, **46,** *150,* 151

Feynman T-photon green tensor, 28–31

Field coordinate, 26–28

Field-coupled Dirac equation
    form-invariance of, 115–116

Field-coupled surface Shrödinger equation, 384–397

Field extinction, 53–56

Field-field propagator, **46**

Field-induced current density, 204–205

Field-matter interaction, 5–6, 23, 38

Field momentum
    correct identification of, 97–98

First-order Born approximation, 252–254

First-quantized plasmariton theory, 324–325

Fixed time loop, 116–117

Flicker, M., 267

Flux quantization, superconducting ring, 450–451

Foldy-Wouthuysen transformation, 158

Forced interaction, 294

Form-invariance, 115–116, 401–404

Fourier integral transform, 100, 170

Four-momentum field conservation
    on covariant integral form, 112–114

Franken, P. A., 282

Free Klein-Gordon equation, 180–183

Free space, fluctuations, 53–66

Frenet-Serret basis vectors, 388, 389

Fundamental photon-electron quantum interaction processes, 198–199

G

Gauge functions, 8

Gauge invariance, 4, 80–81

Gauge photon dynamics, 158

Gauge photons, 7–24
    longitudinal and scalar, 18–20
    quantum theory of, 25–37

Gauge-photon vacuum, 49–52

Gauge-photon wave mechanics, 97–101

Gauge potential formalism, 4

M

Magnetic current density, 497

Magnetic dipole-dipole (MD-MD) interaction, 278

Magnetic dipole moment densities, 117

Magnetic flux tubes, 123–126

Magnetic moment operator, 267

Magnetic monopoles, 452–462
    in non-Abelian gauge symmetry, 484–498
    retarded near field of, 482–483

Magnetic (dipole) string, 117–123
    rectilinear magnetic string, 121–123
    vector potential, curvature and torsion, 117–120

Magnetization current density, 124

Magnetostatic Aharonov-Bohm vector potential, 97–101

Magnon operators, 280

Magnon-photon interaction, 261, 276–277
    local-field effects in, 277–279

Manifest covariant formalism, 139

Manifest gauge invariant tensor, 108

Many-electron theory, 201–206

Massive boson fields, 215

Mathematical digression, 169
    delta function, analytical part, 169

Matter-coupled electromagnetic fields, 110
    extension, 110

Maxwell-Lorentz equations, 402, 452, 453, 461, 463, 469, 477

Maxwell-Lorentz theory, 208

Maxwell operator equations, 38–46

Maxwell stress tensor, 97

Mean field operator, 56–58

Mean fields, 53–66

Mechanical momentum, 477

Meissner effect, 204, 289–293

Mesoscopic layers, 371–376
    electrodynamics in, 380–382
    surface-induced metric tensor, 372–374

Mesoscopic Möbius quantum wire electrodynamics of, 391–397

Metric tensor, 371–376
    extrinsic surface curvature tensor, 374
    surface-induced metric tensor, 372–374
    surface metric tensor, 375–376

Microcausality, 45

Microscopic Ewald-Oseen extinction theorem, 176

Microscopic extinction theorems, 176–188

Microscopic "inhomogeneous" response, 193–194

Microscopic magnetization current density, 118

Minimal coupling principle, 18

Minimal coupling substitution, 449–450

Minkowski coordinates, 369

Minkowski space, 352, 353

Miroscopic conductivity tensor, 394–397

Mixed photon-electron propagators, **46**

Möbius-band electrodynamics, 384–397

Molecular field theory, 268

Momentum density, 98

Momentum of particle, 475
    electromagnetic parts, canonical particle momenta, 476–478
    LL and TT, 475 476
    minimal coupling principle, 476–478
    total field momentum, 475

Monopoles, 453–456
    vector potential, 458–461

N

Nambu, Y., 422

Near-field photon, 20–23

Neumann function, 250

Non-Abelian gauge fields, 408–417

Non-Abelian gauge symmetry, 484–498

Nonlinear dynamical Meissner effect, 289–293

Nonlinear electromagnetic rectification, 282–298